PLEASE STAMP DATE DUE, BOTH BELOW AND ON CARD

DATE DUE	DATE DUE	DATE DUE	DATE DUE

GL-15

Springer Series in Optical Sciences

Volume 179

Founded by
H.K.V. Lotsch

For further volumes:
www.springer.com/series/624

Springer Series in Optical Sciences

The Springer Series in Optical Sciences, under the leadership of Editor-in-Chief William T. Rhodes, Georgia Institute of Technology, USA, provides an expanding selection of research monographs in all major areas of optics: lasers and quantum optics, ultrafast phenomena, optical spectroscopy techniques, optoelectronics, quantum information, information optics, applied laser technology, industrial applications, and other topics of contemporary interest.
With this broad coverage of topics, the series is of use to all research scientists and engineers who need up-to-date reference books.
The editors encourage prospective authors to correspond with them in advance of submitting a manuscript. Submission of manuscripts should be made to the Editor-in-Chief or one of the Editors. See also www.springer.com/series/624.

Editor-in-Chief
William T. Rhodes
School of Electrical and Computer Engineering
Georgia Institute of Technology
Atlanta, GA 30332-0250, USA
bill.rhodes@ece.gatech.edu

Editorial Board
Ali Adibi
School of Electrical and Computer Engineering
Georgia Institute of Technology
Atlanta, GA 30332-0250, USA
adibi@ee.gatech.edu

Toshimitsu Asakura
Faculty of Engineering
Hokkai-Gakuen University
1-1, Minami-26, Nishi 11, Chuo-ku
Sapporo, Hokkaido 064-0926, Japan
asakura@eli.hokkai-s-u.ac.jp

Theodor W. Hänsch
Max-Planck-Institut für Quantenoptik
Hans-Kopfermann-Straße 1
85748 Garching, Germany
t.w.haensch@physik.uni-muenchen.de

Takeshi Kamiya
National Institution for Academic Degrees
Ministry of Education, Culture, Sports
Science and Technology
3-29-1 Otsuka, Bunkyo-ku
Tokyo 112-0012, Japan
kamiyatk@niad.ac.jp

Ferenc Krausz
Ludwig-Maximilians-Universität München
Lehrstuhl für Experimentelle Physik
Am Coulombwall 1
85748 Garching, Germany

Max-Planck-Institut für Quantenoptik
Hans-Kopfermann-Straße 1
85748 Garching, Germany
ferenc.krausz@mpq.mpg.de

Bo A.J. Monemar
Department of Physics and Measurement Technology,
Materials Science Division
Linköping University
58183 Linköping, Sweden
bom@ifm.liu.se

Herbert Venghaus
Fraunhofer Institut für Nachrichtentechnik
Heinrich-Hertz-Institut
Einsteinufer 37
10587 Berlin, Germany
venghaus@hhi.de

Horst Weber
Optisches Institut
Technische Universität Berlin
Straße des 17. Juni 135
10623 Berlin, Germany
weber@physik.tu-berlin.de

Harald Weinfurter
Sektion Physik
Ludwig-Maximilians-Universität München
Schellingstraße 4/III
80799 München, Germany
harald.weinfurter@physik.uni-muenchen.de

Gianluca Gagliardi · Hans-Peter Loock

Editors

Cavity-Enhanced Spectroscopy and Sensing

Springer

Editors

Gianluca Gagliardi
CNR-Istituto Nazionale di Ottica (INO)
Pozzuoli, Italy

Hans-Peter Loock
Dept. of Chemistry
Queen's University
Kingston, Ontario, Canada

ISSN 0342-4111   ISSN 1556-1534 (electronic)
Springer Series in Optical Sciences
ISBN 978-3-642-40002-5   ISBN 978-3-642-40003-2 (eBook)
DOI 10.1007/978-3-642-40003-2
Springer Heidelberg New York Dordrecht London

Printed on acid-free paper

Springer is part of Springer Science+Business Media (www.springer.com)

In memory of my father, Angelo.
(May 2013) Gianluca Gagliardi

Preface

It is with deep satisfaction and pride that we present this book on cavity enhanced spectroscopy and sensing. The reader will find a wide variety of articles on techniques and applications related to the use optical cavities in this book. It would, of course, be impossible to provide a complete survey of the applications of optical cavities. Instead, in the selection of the chapters we aimed for a large breadth of topics that, we hope, is representative of the field in 2013.

Only a few years ago the term "cavity enhanced spectroscopy" would have implied the use of free-space optical cavities that are used primarily for gas detection and spectroscopy through a limited variety of methods—such as cavity ring-down spectroscopy and cavity-enhanced absorption spectroscopy. In parallel other research groups developed novel optical resonators made from fiber optic cables, pure dielectric materials or using silicon nanowires with intended applications as e.g. optical switches or refractive index sensors. In recent years, there appears to be a convergence of these two fields. The selection of chapters in this book shows that cavity-based optical measurements (such as time measurements) are increasingly used with e.g. fiber optic cavities, or microresonators, whereas interrogation methods that are derived from telecommunication technologies, and optical metrology find uses in gas phase spectroscopy. Newly developed techniques involving, for example, optical frequency comb synthesizers, tunable mid-infrared lasers, nonlinear light generators, fiber lasers, supercontinuum sources and active frequency locking of a laser to an optical cavity further accelerate the progress in the field. An excellent historic overview is provided by Romanini et al. as part of the first chapter of this book.

The following chapters introduce an astonishing variety of techniques and methods that all rest on the most prominent properties of optical cavities: first, that they are capable of storing light at high powers for a considerable amount of time and, second, that they act as narrow band-pass (or band rejection) filters. The second property may be considered a nuisance, but it is important for those interested in using cavities as optical switches, or for strain, vibration, and refractive index measurements.

Most contributions in this book use high finesse cavities made from two or more mirrors, but you will also find cavities made from fiber optic loops (Vallance, Wang) fiber optic strands and gratings (Avino & Gagliardi) as well as microtoroids (Wu & Vollmer) and microspheres (Barnes). The choice of light sources and spectral range is just as varied: pulsed lasers have been extended into the 7000–9000 cm^{-1} region using stimulated Raman shifting or difference frequency generation (Kine & Miller, and Cancio et al.), quantum cascade lasers permit ring-down studies in the 600–2250 cm^{-1} region (Welzel, Engeln & Röpcke, and Cancio et al.), and broadband light sources such as supercontinuum sources, lamps and LEDs are reviewed by Ruth, Dixneuf, and Raghunandan. Optical frequency combs are an emerging source of considerable importance to cavity enhanced spectroscopy, because of their unique coexistence of high temporal coherence and stability with an extraordinary spectral coverage. This opens the possibility to simultaneously couple into many different cavity modes with amazingly high spectral resolution over a wide wavelength range (Masłowski). In addition frequency combs can also be used to stabilize "conventional" lasers and thereby increase the sensitivity of a sensing experiment (Avino). On the other hand simple diode lasers can also be locked effectively to the optical resonators either by optical feedback (Morville et al., Welzel et al.) or using electronic feedback circuitry (Barnes, and Avino).

Of particular interest is noise-immune cavity-enhanced optical heterodyne molecular spectroscopy (NICE-OHMS) which is a cavity enhanced absorption method that is record-breaking with regards to sensitivity, detection limit and dynamic range. Chapters by Axner, and by Siller & McCall, present the technical details and the different applications of the method, whereas the introductory chapter by Romanini et al. gives a historic and scientific overview in the context to all cavity-based methods.

The choice of applications for cavity enhanced detection schemes has also grown to include practically all absorption, attenuation and scattering phenomena from optical absorption of trace gases, atmospheric sensing, transient absorption in reaction mixtures, breath analysis, liquid absorption and detection, etc. to even include mechanical deformation, vibration, strain, pressure and temperature. Given the diversity of the research field one may wonder whether those involved in cavity enhanced spectroscopy and sensing still find common ground. Yet, in our experience, the borders across disciplines are surprisingly permeable and each research group occupies typically not just one niche of the field but uses a variety of methods—each adapted to a particular application.

Again, we hope you will find this book instructive, and enjoy learning from some of the world experts about the state-of-the-art in cavity enhanced spectroscopy and sensing.

Napoli, Italy   Gianluca Gagliardi
Kingston, Canada   Hans-Peter Loock

Contents

Contributors

Saverio Avino Istituto Nazionale di Ottica, Consiglio Nazionale delle Ricerche, Pozzuoli (Napoli), Italy

Ove Axner Department of Physics, Umeå University, Umeå, Sweden

Jack A. Barnes Dept. of Chemistry, Queen's University, Kingston, ON, Canada

S. Bartalini Istituto Nazionale di Ottica, Consiglio Nazionale delle Ricerche, Sesto Fiorentino, Italy

P. Cancio Istituto Nazionale di Ottica, Consiglio Nazionale delle Ricerche, Sesto Fiorentino, Italy

K.C. Cossel JILA, National Institute of Standards and Technology and University of Colorado, Department of Physics, University of Colorado, Boulder, CO, USA

Paolo De Natale Istituto Nazionale di Ottica, Consiglio Nazionale delle Ricerche, Sesto Fiorentino, Italy

S. Dixneuf Physics Department & Environmental Research Institute, University College Cork, Cork, Ireland

Patrick Ehlers Department of Physics, Umeå University, Umeå, Sweden

R. Engeln Eindhoven University of Technology, Eindhoven, The Netherlands

Aleksandra Foltynowicz Department of Physics, Umeå University, Umeå, Sweden

Gianluca Gagliardi Istituto Nazionale di Ottica, Consiglio Nazionale delle Ricerche, Pozzuoli (Napoli), Italy

I. Galli Istituto Nazionale di Ottica, Consiglio Nazionale delle Ricerche, Sesto Fiorentino, Italy

Antonio Giorgini Istituto Nazionale di Ottica, Consiglio Nazionale delle Ricerche, Pozzuoli (Napoli), Italy

G. Giusfredi Istituto Nazionale di Ottica, Consiglio Nazionale delle Ricerche, Sesto Fiorentino, Italy

Erik Kerstel LIPhy UMR 5588, Univ. Grenoble 1/CNRS, Grenoble, France; Laboratoire Interdisciplinaire de Physique, UMR 5588, Université J. Fourier (Grenoble I), CNRS, Saint-Martin d'Hères, France

Neal Kline Dept. of Chemistry, The Ohio State University, Columbus, OH, USA

Hans-Peter Loock Dept. of Chemistry, Queen's University, Kingston, ON, Canada

P. Masłowski Institute of Physics, Faculty of Physics, Astronomy and Informatics, Nicolaus Copernicus University, Torun, Poland

D. Mazzotti Istituto Nazionale di Ottica, Consiglio Nazionale delle Ricerche, Sesto Fiorentino, Italy

Benjamin J. McCall Department of Chemistry, University of Illinois, Urbana, IL, USA

Guillaume Méjean LIPhy UMR 5588, Univ. Grenoble 1/CNRS, Grenoble, France

Terry A. Miller Dept. of Chemistry, The Ohio State University, Columbus, OH, USA

Jérôme Morville Institut Lumière Matière, UMR 5306, Université Lyon 1, CNRS, Université de Lyon, Villeurbanne cedex, France

R. Raghunandan Physics Department & Environmental Research Institute, University College Cork, Cork, Ireland

Daniele Romanini LIPhy UMR 5588, Univ. Grenoble 1/CNRS, Grenoble, France; Laboratoire Interdisciplinaire de Physique, UMR 5588, Université J. Fourier (Grenoble I), CNRS, Saint-Martin d'Hères, France

J. Röpcke INP Greifswald e.V., Greifswald, Germany

Cathy M. Rushworth Department of Chemistry, Chemistry Research Laboratory, University of Oxford, Oxford, UK

A.A. Ruth Physics Department & Environmental Research Institute, University College Cork, Cork, Ireland

Isak Silander Department of Physics, Umeå University, Umeå, Sweden

Brian M. Siller Tiger Optics, Warrington, PA, USA

Claire Vallance Department of Chemistry, Chemistry Research Laboratory, University of Oxford, Oxford, UK

Irène Ventrillard LIPhy UMR 5588, Univ. Grenoble 1/CNRS, Grenoble, France

Frank Vollmer Laboratory of Nanophotonics & Biosensing, Max Planck Institute for the Science of Light, Erlangen, Germany

Chuji Wang Department of Physics and Astronomy, Mississippi State University, Mississippi State, MS, USA

Junyang Wang Department of Physics, Umeå University, Umeå, Sweden

S. Welzel Eindhoven University of Technology, Eindhoven, The Netherlands; Dutch Institute for Fundamental Energy Research (DIFFER), Nieuwegein, The Netherlands

Yuqiang Wu Laboratory of Nanophotonics & Biosensing, Max Planck Institute for the Science of Light, Erlangen, Germany

J. Ye JILA, National Institute of Standards and Technology and University of Colorado, Department of Physics, University of Colorado, Boulder, CO, USA

Chapter 1
Introduction to Cavity Enhanced Absorption Spectroscopy

Daniele Romanini, Irène Ventrillard, Guillaume Méjean, Jérôme Morville, and Erik Kerstel

Abstract In this introductory chapter we will begin with an historical outline of the development of cavity enhanced absorption methods, with just enough attention to the applications that either motivated them or became conceivable after their development. Given the number of publications in this domain, we will consider only the first demonstrations, and those works leading to substantial improvement or innovation in the state of the art.

Subsequently, rather than reviewing in detail all principal applications, we will provide a review of the many reviews that have already appeared, even quite recently, dealing preferentially with a specific cavity enhanced implementation or a specific domain of application.

Finally, we will provide wide but mostly intuitive foundations for approaching to cavity enhanced methods, by considering first the physics behind the (static) response of a cavity in the spectral domain, followed by a discussion of the physics of the (transient) coupling of different types of lasers to a cavity, going from the ideal tunable monochromatic wave to the realistic noisy continuous wave laser, to the pulsed nanosecond laser, and finally the broadband femtosecond laser combs. We will try to situate the most widespread cavity enhanced schemes along these detailed discussions.

1.1 Introduction

Cavity Enhanced Absorption Spectroscopy (CEAS) is used in this book mostly in a broad sense, as including all absorption spectroscopic methods exploiting the principal property of high finesse optical cavities, which is the increase of the interaction time of light with matter. The narrow technical meaning of CEAS will be also used

D. Romanini (✉) · I. Ventrillard · G. Méjean · E. Kerstel
LIPhy UMR 5588, Univ. Grenoble 1/CNRS, 38041 Grenoble, France
e-mail: danielromanini@gmail.com

J. Morville
Institut Lumière Matière, UMR5306 Université Lyon 1-CNRS, Université de Lyon, 69622 Villeurbanne cedex, France

G. Gagliardi, H.-P. Loock (eds.), *Cavity-Enhanced Spectroscopy and Sensing*, Springer Series in Optical Sciences 179, DOI 10.1007/978-3-642-40003-2_1, © Springer-Verlag Berlin Heidelberg 2014

when talking about specific detection schemes in which the intracavity absorption is deduced from the intensity of light transmitted by the cavity (usually time averaged), as opposed to ring-down or phase-shift methods where the time dependence of transmitted light is used.

From a fundamental point of view, an optical cavity is first of all a resonator. Resonators have always been associated with sensitive measurements in physics, and are widely present in high-tech instrumentation and in everyday appliances. A common example is the radio tuner, in which an electronic resonator enhances a weak signal from a radio antenna at a specific frequency channel. The sensitivity and selectivity of the oscillation amplitude of a resonator to external perturbations at a specific frequency makes it analogous to a lock-in detection system. The response is notably characterized by an integration time given by the duration of the resonator oscillation after an impulsive excitation. Another property of resonators is to support stable resonance frequencies well determined by some physical parameters. Most modern watches exploit a narrow acoustic resonance of a quartz crystal as a stable reference that can be excited with little power (by an electronic oscillator with poor frequency stability) and enables the precise measurement of time. Temperature control of such a crystal may be used to stabilize its lattice spacing and stiffness and avoid drifts of the resonance frequency. Before electronics and quartz acoustic resonators, spring resonators were widely used for watches, and even earlier a pendulum oscillator, more adequate for clocks than wrist watches. In physics laboratories, electronic oscillations of isolated and cold atoms are used to measure time, and frequency, ever more precisely. Finally, a femtosecond light pulse oscillating inside a laser resonator generates an optical frequency comb, which is today the highest precision ruler available for comparing and linking widely separated optical frequencies.

Before the advent of optical cavities, absorption spectroscopy was hampered by the limited interaction length of the light beam through the sample, by amplitude fluctuations of the source, and by frequency fluctuations of the laser source in the case of the highest spectral resolution measurements (as needed for gas phase samples). Multiple-pass cells provided the most intuitive and appealing solution for increasing the interaction length, but their implementation turns out to be technically complex when hundreds of passes are desired. In particular, their challenging optical alignment requires highly stable mechanics. Large mirrors, and by consequence a large sample volume, are required in order to reduce the detrimental effects of optical interference between neighboring reflections. All the while, source amplitude and frequency noise persist and require an independent solution, further complicating the experimental setup. Optical cavities can provide a simultaneous solution for these different problems, as one can already understand from their general properties outlined above. The enhancement of the light-matter interaction time in a high finesse optical cavity may correspond to an effective interaction length of thousands of passes, and it turns out that increasing this interaction length by a still higher cavity finesse does not add to technical complexity (except for the fabrication of better mirrors). In addition, measurement of the cavity excitation decay (ring-down signal) gives access to the intracavity sample absorption regardless of amplitude source fluctuations. Finally, cavity resonances may be put at work to provide a highly precise frequency scale, significantly improving on the intrinsic frequency stability of a laser source. An apparently smaller advantage, but fundamental for some applications, is that the sample size in a high finesse cavity is ultimately limited by the size of the $TEM_{0,0}$ mode, much smaller than the volume required by a multipass system.

1.1.1 A Short History of Cavity Enhanced Methods

Since Joseph von Fraunhofer demonstrated his first spectroscope in 1814, spectroscopy has been used in a bewildering range of studies in physics and chemistry. Given the broad scope of applications, the technical developments have naturally been pursued with widely different motivations, which is also true for CEAS. In this section we will attempt to sketch a wide panorama of developments that each time resulted in different methods of performing CEAS measurements (in its broad sense).

It is difficult to date the first documented theoretical or experimental demonstration concerning the use of optical cavities to enhance absorption measurements. As underlined above, resonators are present in everyday life and in physics laboratories, mostly due to their properties of frequency selectivity and stability. It was not coincidental that these properties of optical cavities were the first to be studied theoretically and to be exploited in the laboratory. Around 1961–1962, investigation of optical resonators focused on their spectral properties, and in particular on the existence of narrow transmission resonances associated with specific transverse field patterns [1, 2]. As noted by Jackson [3], in the earliest applications optical cavities made of flat mirrors (the original Fabry-Perot etalon) were indeed exploited for their spectral resolving power to deliver high resolution atomic spectra. But already in 1958, Connes had understood the advantages of using spherical concave mirrors for such applications and exposed the confocal cavity configuration [4] and its special properties. Jackson was probably the first to exploit a confocal cavity a couple of years later [3] to obtain cavity-filtered, and also cavity-enhanced, high-resolution spectra of electronic transitions in a beam of Barium atoms, using a spectral lamp as a source and a grating spectrometer for rough spectral selection.

In 1962 the first published issue of Applied Optics presented as a second contribution (p. 17) a paper titled "Atomes à l'intérieur d'un Interféromètre Perot-Fabry" [5] in which Albert Kastler clearly stated the potential interest of using an optical cavity for atomic spectroscopy. Among other relevant statements, the abstract reads: "... the absorption of atoms inside the interferometer is studied. It is shown that this device is equivalent to a long absorption path in an ordinary light beam." (the rest of the paper is in French).

While more theoretical [6–9] and experimental [10, 11] insight was pursued, several years passed before new applications were stimulated by the manufacturing of dielectric mirrors with increasing reflectivity and lower losses, which also demanded special efforts to be characterized, and were going to provide optical cavities with

astonishing finesse several years later [12]. In 1977 Damaschini [13] proposed using a resonator with an intracavity Brewster plate for input and output coupling, and
demonstrated reflectivity measurements with few parts per thousands accuracy by
measuring the transmitted intensity. It was 1980 when Herbelin et al. [14] reported
cavity photon lifetime measurements by the same phase-shift method previously applied to determine excited-state lifetimes in atoms and molecules by laser induced
fluorescence. From the photon lifetime the mirror reflectivity could be accurately
determined at the 100 ppm level. The idea is to modulate the amplitude of the excitation field and measure (by lock-in electronics) the phase shift of the induced modulation appearing at the cavity output, which is inversely proportional to the cavity
photon lifetime. Herbelin et al. discussed the potential of measuring absorption also
from gas or solid samples inserted in the cavity. The major limitation of the phase-
shift measurement was the large fluctuations of the cavity injection level attributed
to acoustic fluctuations of the cavity (but really due to laser phase noise), requiring
long signal averaging. This scheme was later exploited again for CEAS [15, 16].
However, without locking of the laser frequency to the cavity resonance, it appears
that excessive noise condemns this simple and appealing technique to a lower performance compared to other CEAS schemes.

Four years after Herberlin's paper, Anderson et al. [17] tackled the same problem by a direct measurement of the cavity photon lifetime with a sensitivity down
to the 5 ppm level. In their words: "*Our technique relies on the fact that, with no
light incident on the cavity, its output is determined only by its transient response
which is characterized by an exponential decay of the intensity with a time constant
which in turn is determined only by the round trip losses of the cavity, the round
trip path length of the cavity, and the speed of light*". This is perhaps the first cavity
ring-down application, accompanied by a clear statement of its specific advantages
(insensitivity to source intensity fluctuations), which was theoretically investigated
ten years earlier by Kastler [18]. In order to obtain the transient response, the accidental resonant cavity injection by a CW (continuous wave) laser was used, with
a threshold detector on the cavity output signal triggering interruption of the injection by a fast electro-optic switch. As Anderson et al. observed (note 3 of [17]), the
measurement of the decay time has its roots in electronics, in passive RCL circuit
theory and microwave networks. They also discussed the theory of the cavity decay
with a rigorous approach based on the light field rather than its intensity (the fast
lane too often chosen by authors dealing with CEAS, see discussion in Sect. 1.2.1
around Eq. (1.12)). Using a bit of math, they pointed out a general property of
resonators: The decay time reflects intrinsic resonator losses only if the excitation
field is switched off fast enough. Other interesting points are acknowledged: When
several transverse cavity modes are excited at a time, not only the beating of their
frequencies may appear on the decay signal if the mode orthogonality is broken (as
already shown by Goldsborough [10]), in addition the decay may become a sum of
exponentials with slightly different time constants, due to spatial inhomogeneity of
the mirrors surface. Finally, Anderson et al. considered that cavity length fluctuations could impact the measured decay time. But they did not give clear reasons as
to why this effect remains negligible in practice. Luckily enough, photons trapped

in a cavity mode follow the changes of its resonance frequency all along the field decay, thanks to the Doppler frequency shift at reflection from a moving mirror [19].

The figurative term "cavity ring-down" was introduced only around 1985 by Crawford [20] to denote the exponential decay of the intracavity light intensity after impulsive excitation (or abrupt interruption of a continuous excitation). Like Anderson, Crawford et al. were developing accurate reflectance measurements, and they started using a pulsed dye laser to obtain cavity ring-down measurements over the broad laser tuning range. They discussed [20, 21] other advantages of using nanosecond laser pulses, notably that no frequency match is necessary since the pulse linewidth is broader than the cavity mode separation. It was 1988 when O'Keefe and Deacon [22] obtained the spectrum of a weak magnetic-dipole transition in atmospheric oxygen, using the experimental scheme of Crawford, in what is now commonly considered to be the first demonstration of Cavity Ring-Down Spectroscopy (CRDS). Incidentally, they adopted the same term of "cavity ring-down" used by Crawford.

The wide spectral coverage of pulsed lasers, extended to the IR and the UV regions of the spectrum by nonlinear techniques like frequency doubling, Raman shifting, and optical parametric conversion, combined perfectly with the indifference of CRDS to their large intensity fluctuations. This encouraged several groups to implement CRDS for various applications, like the spectroscopy of weak molecular overtone bands in the gas phase [23], of molecular clusters in supersonic jets (in the IR) [24, 25], or the measurement of trace concentrations of radicals in chemical reactors [26] or flames (in the UV) [27]. In order to avoid spectral filtering effects by cavity resonances when using spectrally narrow lasers, Meijer et al. [27] proposed exploiting multi-transverse-modes excitation. They experimentally illustrated this idea by comparing the results from different cavity lengths for which transverse modes are either grouped in widely spaced degenerate groups (confocal cavity), or dispersed to produce a quasi-continuum cavity transmission, especially when using a non optimized beam alignment, far from mode-matched.

Another common situation, when absorption lines are spectrally narrower than the laser pulse, was considered in detail [28] and recognized to generate multiexponential decays due to different losses across the injected spectrum. However a simple approximation was used to show that the integrated absorption of spectral lines may still be correctly recovered in the limit of small absorptions, as was soon experimentally confirmed [29]. These results (and implicitly those in [27]) were contested on the basis of experiments in which no filtering effect was observed [24], but no mode matching, nor a confocal cavity configuration, had been considered. In contrast to these claims, efforts continued to be made towards understanding and avoiding artifacts appearing with multi-transverse-modes cavity excitation (mode beatings, multi-exponential decay) [30–32]. Eventually van Zee et al. [33] achieved shot-noise-limited performance by single cavity mode excitation with a frequency stabilized pulsed OPO laser, with excellent spectral resolution, but at the price of increased experimental complexity.

Meanwhile, multiple-quantum-wells (MQW) diode lasers had started shining CW narrow-line radiation, stabilized and tunable thanks to optical feedback from an external or internal grating, respectively as External Cavity Diode Lasers (ECDL) or as Distributed Feed-Back (DFB) diode lasers. These affordable devices promoted the spreading of a CW-CRDS technique introduced by Romanini et al. in 1997 using a CW single-frequency dye laser [34] and not long thereafter using an ECDL [35, 36]. In this simple scheme the cavity length is modulated to induce the recurrent transient excitation of a single cavity mode passing through resonance with the laser line. When the incoming wave resonates and builds up enough intra-cavity field so that cavity transmission reaches a given threshold, like in Anderson's scheme, a fast acousto-optic switch interrupts the laser beam to produce a ring-down event. As new ideas are never really new, this scheme had been exploited ten years earlier [37] for measuring the absorption of thin films deposited directly on a mirror surface, but the fact of using it for high resolution gas phase spectroscopy did demand some innovation, in particular an electronic control of the cavity length to maintain the frequency dither of a cavity resonance centered around the laser frequency while this is tuned. This allows the optimal and exclusive excitation of $TEM_{0,0}$ cavity modes after the laser beam is at least partially mode matched to the cavity, delivering reproducible ring-down spectra over broad spectral ranges enabling the determination of broad-band absorption features [36], opening the way to a number of important applications. Another advantage is that the acquisition rate is only limited by the duration of the cavity ring-down [38], not by the (low) repetition rate of a pulsed laser.

In order to improve frequency precision in CW-CRDS, and following their developments with single-mode pulsed CRDS, Hodges et al. started in 2006 a frequency-stabilized CRDS (FS-CRDS) program based on precise cavity length control [39, 40] dedicated to the investigation of molecular line shapes of weak and isolated transitions, at the frontier with metrology and fundamental physics. On the other hand Giusfredi et al. [41] demonstrated kHz accuracy in CW-CRDS using frequency comb referencing. More exploratory works based on CW-CRDS, but which are of wider interest, include the studies by Huang and Lehmann about excess noise in CRDS from transverse modes coupled by mirror surface scattering [42], or from an insufficient extinction by the optical switch [43, 44], or still from mirror birefringence and polarization effects [45], and the exploration of coupled cavities resulting in the fast control of a cavity finesse by Courtois and Hodges [46].

It appeared rapidly that CW-CRDS is more adapted than its pulsed sibling for high resolution and high precision spectroscopy. For example, it readily delivered sub-Doppler absorption spectra in a supersonic slit-jet [35, 47, 48] with a noise level $(2 \times 10^{-10}/\text{cm})$, better than the best pulsed-CRDS results [23, 33] even though still far away from the shot noise limit. The main drawback of CW lasers often is their relatively limited tuning range. However high-end ECDLs allow mode-hop-free tuning of 4–5 % of their central frequency, and a series of telecom-packaged DFB diode lasers is a convenient solution to cover a large fraction of the near-IR (from about 1.2 to 1.8 μm), since these fibered devices are easily interchangeable [49]. Other implementations were proposed, e.g. to simplify the setup by avoiding the optical switch [50] (by means of a fast resonance shift, as earlier proposed in another context [51]), to obtain CRDS spectra by fast laser sweeps [52] down to ms time scales [53], to improve the data acquisition rate by frequency locking [38, 54], to increase the signal-to-noise level by heterodyne detection [55, 56] or by tight cavity locking and fast averaging [54] (with improved frequency precision). Finally, brute-force averaging of several CW-CRDS spectra over long time periods, was shown to yield white noise statistics all the way down to a record detection level of 5×10^{-13}/cm (with an extreme cavity finesse), with a potential dynamic range approaching 10^6 [57].

An interesting property of CW-CRDS is that high intracavity power can be exploited to observe non-linear effects such as two-photon transitions or the saturation of molecular transitions (e.g. sub-Doppler Lamb dips) [58]. In 2010 Giusfredi et al. [41] exploited saturation in the mid-IR to obtain the empty cavity (baseline) losses and the sample absorption by fitting a molecular transition saturation model to a single (averaged) ring-down profile, assuming saturation follows adiabatically the decaying intracavity power. Using a high-power frequency-stabilized difference-frequency laser, they managed to use this principle to subtract large (4 %) cavity loss fluctuations affecting their measurements, all the while correcting for the strong saturation and obtain the unsaturated spectral profile of very intense fundamental vibrational transitions of CO_2. It should be underlined that this is an unusual situation, since ring-down measurements have often been shown to be very stable and reproducible [36, 57, 59]. Soon after, the same group applied their treatment of saturation to the first CW-CRDS measurement of a radiocarbon labelled molecule, $^{14}CO_2$, detecting standard atmospheric levels around 1 ppt, with a 43 ppq detection limit [60]. For this result they averaged successive frequency scans to reach a noise level of $\sim 2 \times 10^{-11}$/cm in 1 hour (as deduced from Fig. 2 of [60], corresponding to 10^{-9}/cm/$\sqrt{Hz}$). The appealing perspective is that CRDS will replace large and expensive high-energy accelerator mass-spectrometers for radiocarbon dating. It is also interesting to note that CW-CRDS is already at the core of high-accuracy commercial trace gas and isotope ratio analyzers.

The small size and low power requirement of diode lasers spawned a blooming of compact spectrometers for trace analysis *in situ*, exploiting multiple-pass cells or high finesse cavities. While the ECDL mechanism is rather sensitive to the environment, DFB diode lasers are robust and reliable, and their smaller tuning range is sufficient to monitor one or two molecules at a time. In addition, their size and cost make it easy and affordable to install more than one laser in the same compact setup, sharing the same absorption cell, and using, for example, a modulation frequency multiplexing scheme to observe two spectra simultaneously (as in [61], albeit in a multiple-pass arrangement) or using simple time-division multiplexing [62]. DFB diode lasers initially covered the near IR but have been extended up to 3.4 µm using GaSb MQW structures. Lead-salt diode lasers were available earlier to access stronger transitions even at longer wavelengths, but they were (and are) affected by several problems, like cryogenic operation, not quite single-frequency emission, and poor beam quality. Quantum Cascade Lasers (QCL) operating in the mid IR were introduced in 1994 and evolved over several years from multi-line cryogenic devices to room temperature, reliable, high-power, single-frequency emission lasers (with DFB structure).

All these lasers, perhaps with the exception of lead-salt diodes, were coupled to high finesse cavities, in particular using the "basic" CEAS (in its narrow sense) scheme in which a CW laser is scanned across the cavity modes and the cavity output intensity is recorded without going through any ring-down time determination. The first implementations [63, 64], which also received the name ICOS (Integrated Cavity Output Spectroscopy), were user-friendly but not generous in terms of performance. In particular, long signal averaging was needed to eliminate the cavity mode structure, even when coupling the laser beam to a dense bath of high order transverse cavity modes as previously done with pulsed lasers. However, Paul et al. [65] eventually realized that with large mirrors and strongly off-axis cavity excitation it is possible to obtain a structureless continuous cavity transmission. Papers describing this off-axis ICOS implementation were not keen on discussing the technical details leading to the excellent results reported, but a few years and publications later it appeared that specific cavity configurations with respect to mirror separation, curvature, and (stress induced) astigmatism were exploited [66, 67]. At least initially, only a few researchers appeared to master this seemingly simple technique, while others were not able to go beyond typical ICOS performance levels. Still, this technique was perhaps the first form of CEAS (in the narrow sense) to find its way into commercial trace gas and isotope ratio analyzers.

A similar story may be told with respect to the use of Optical Feedback (OF) to obtain optimal coupling of laser radiation into cavity resonances: A simple V-shape cavity geometry produces frequency selective OF from the resonant field inside the cavity. In a first implementation, long-pulse operation of a DFB diode laser was used to obtain ring-down events at the end each pulse, with poor frequency stability and no easy control to force the same cavity mode to ring-down at the end of each pulse [68, 69]. The fact that one or another cavity mode becomes excited at the end of the laser pulse depends on the OF phase, which in turn is sensitive to sub-wavelengths changes of the laser-cavity distance. Without a control over this phase, this OF-CRDS scheme delivers modest performance but is tolerant to misalignment and vibrations. Eventually, Morville et al. [70] introduced an OF-CEAS scheme based on a linear laser frequency sweep, which allowed to easily implement an active phase control and obtain the injection of all $TEM_{0,0}$ modes in the laser scan, reproducibly and with low noise. An important benefit is that an absorption spectrum is sampled at nearly perfectly equidistant steps in frequency space. OF-CEAS can thus deliver high quality spectra over relatively fast diode laser scans, several times per second, with a small sample volume (like most CEAS techniques, and unlike multipass cells and off-axis ICOS), which also enables sub-second sample exchange times. And this is possible with an optical layout almost as simple as that of a basic absorption spectroscopy setup. Still, this technique demands a good understanding of several physical details, and thus, so far, it is only exploited by few research groups for application to trace gas monitoring in different environments [71, 72], including breath analysis faster than a respiratory cycle [73], isotope ratio analyses [74], or aerosol measurements [75], where particle size distribution could be related to signal fluctuations. At the same time, the OF-CEAS scheme is still being further developed and tested [76–79], and may be exploited for applications other than spectroscopy, notably for the measurement of the Kerr effect in gases at the shot-noise limited level of 10^{-13} radians [80]. Recently, OF-CEAS also started to appear in commercial trace gas analyzers.

Nonetheless, except for fast measurements, there appears to be no major performance benefit when using OF-CEAS over some competing CEAS techniques, but this may be expected to change with application of the technique in the mid IR. OF-CEAS has been shown to be compatible with QCLs [81, 82] and to attain high S/N even with room temperature photodetectors. This contrasts with other CEAS schemes or multiple-pass arrangements that require cooled detectors to compensate for the low signals levels and the decreasing sensitivity of detectors with wavelength.

To conclude with CEAS based on a single-frequency laser, the NICE-OHMS technique (Noise-Immune Cavity-Enhanced Optical Heterodyne Molecular Spectroscopy) introduced by Ye et al. [83] deserves a special mention, with its shot noise limited detection placing it several orders of magnitude above other CEAS implementations. The principle is to use high frequency phase modulation of the laser wave to obtain spectrally symmetric sidebands, with subsequent locking of the carrier and the sidebands to successive cavity modes and demodulation of the cavity output as in heterodyne spectroscopy: a small and narrow intracavity absorption line will then produce a displacement (by the associated dispersion) and/or attenuation, either of the carrier or of one sideband, sensitively detected as an unbalance and change in the heterodyne signal. Like other frequency modulation techniques, this scheme requires good knowledge of the modulation parameters in order to provide quantitative absorption measurements, with some modelling. In return it provided spectacular results in detecting the narrow sub-Doppler saturation dips of overtone transitions [83], down to a 10^{-14}/cm detection limit for 1 s averaging, performance never reproduced since. NICE-OHMS requires sophisticated RF electronics and includes several frequency locking loops. In addition, when considering application to Doppler limited spectroscopy and to trace analysis, it is eventually limited by optical interference fringes [84, 85], like other absorption techniques, as we discuss in Sect. 1.3. On the other hand, given the use of frequency locking, NICE OHMS is also well suited to the mid-IR spectral range where Taubman et al. [85] obtained similar performance with a QCL as demonstrated by OF-CEAS [81]. NICE-OHMS also allows highly sensitive and selective spectroscopy of ions in a plasma when coupled with velocity modulation [86]. Recent developments exploiting fibered optics yield simplification, improved robustness and compactness, while they promise improved performance [87, 88].

The requirement in some applications to monitor a large spectral range was a good motivation to try and couple broadband light sources into a high finesse cavity, with spectral resolution provided by a grating spectrometer (coupled with a CCD[1] detector array) or a Fourier-Transform (FT) spectrometer. This idea is as old as 1987, when Dasgupta et al. [89] tested a Fabry-Perot cavity to enhance absorption through liquid samples in a commercial spectrophotometer, but it was pursued with

---

[1]Charge Coupled Device, a linear or rectangular array of small detectors capable of converting photons into electrons which are accumulated into charge wells before readout

more determination later on, using pulsed (nanosecond) dye lasers by Scherer et al. [90] and Ball et al. [91], high intensity arc lamps by Fiedler et al. [92], LEDs by Ball et al. [93], as well as super-continuum fiber sources by Johnston et al. [94]. All these light sources have rather different characteristics when considering their spatial and temporal coherence.

While these initial demonstrations spawned several applications and variations on this theme, notably for sensitive absorption spectroscopy in liquids [95] or in thin films [96], broadband CEAS is clearly suitable and limited to obtain low resolution absorption profiles covering a broad spectral range. Pushing the spectral resolution means reducing the photon flux per spectral element to such low levels that one needs to integrate the signal over excessively long times, even when using a sensitive CCD detector. Besides, CCD detectors possess a limited number of pixels, thus increasing the resolution implies reducing the spectral window available during one acquisition. On the other hand, FT acquisition of broad band spectra does not suffer from this limitation, but is inherently less sensitive [97, 98] and requires that the spectrum and intensity of the source remain steady during the acquisition of an interferogram, which may take from seconds up to minutes at higher resolutions [93]. Supercontinuum sources with single transverse mode output, are interesting for attaining higher resolution as they allow for an optimized cavity injection by cavity mode matching. In addition, broad-band total-internal-reflection "mirrors" [94, 99, 100] have been introduced by Lehmann [101] to keep up with such broad sources, since normal dielectric mirrors cover 'only' 5 to 20 % of their central wavelength. We should mention that total internal reflection was also exploited to realize completely solid cavities to investigate absorption of the evanescent wave by samples deposited at the surface [102]. It may appear surprising that broad band pulsed dye lasers, with their high spectral density and well collimated beam were not developed further after the first encouraging demonstrations [90, 91]. This was probably due to their unstable spectral profile together with a poor and fluctuating beam profile [90].

Mode-locked femtosecond lasers are another class of broadband sources. A femtosecond pulse generated and maintained by nonlinear effects in a laser cavity produces a highly stable series of replicas at the cavity output. The periodicity of this pulse train corresponds, by the properties of the FT, to a spectrum containing frequencies that are harmonics (multiples) of the repetition rate. This comb-like emission can be matched to the comb of cavity transmission resonances for optimal broad-band cavity injection. In addition, mode-locked lasers have a Gaussian beam, thus transverse mode matching is possible, altogether allowing to push spectral resolution much harder than with the broadband sources previously considered. Even more than nanosecond pulsed lasers, femtosecond lasers can take advantage of nonlinear effects to access almost any spectral region from the mid-IR to the deep UV. Finally, the development of stabilized laser combs makes available a high accuracy frequency scale for spectroscopic measurements.

Matching a frequency comb to a cavity corresponds in the time domain to matching the cavity length to the pulse repetition rate, such that each pulse coincides with a previous one after an integer number N of cavity round trips. This results in the buildup of an intracavity pulse, in a manner analogous to the buildup considered earlier in CW-CRDS, but now occurring simultaneously for all those comb teeth that coincide (one every N) with a cavity mode. For this collective behavior to result in coherent buildup, in the time domain the carrier wave inside the laser pulse envelope must be in phase with the carrier of the previous pulse. In practice however, a delay between carrier and envelope accumulates over time from a small mismatch in group and phase velocity (due to dispersion), and it is not possible to overlap carrier and envelope perfectly from a pulse to the next, resulting in a frustrated buildup. In the frequency domain this corresponds to a non-zero frequency offset f_0 of the laser comb, such that its frequencies are written as $f_n = n f_{\text{rep}} + f_0$ (see also Fig. 1.6). In order to obtain optimal pulse buildup, the offsets of the laser and cavity combs should be matched.[2] Cavity injection by a laser comb was exploited to obtain selection of one in every N modes and increase the laser comb spacing and repetition rate in 1999 [103] but without having a spectroscopic application in mind. The same year the corresponding time-domain concept of pulse stacking was applied using a free electron laser in the mid-IR [104] to increase cavity throughput and obtain CRDS spectra by the phase-shift method, but only at a single frequency at a time (using a monochromator).

A truly broadband demonstration of comb-based CEAS, using a mode-locked fs Ti:Sapphire laser, had to wait until 2002 when Gherman et al. [105] generalized the CW-CRDS injection scheme by using periodic passages through a global comb resonance (transient injection) to obtain CEAS spectra from a spectrograph coupled with a CCD array detector. The same group illustrated the interest of this "Mode-Locked CEAS" (ML-CEAS) for accessing the blue and near-UV range [106, 107] thanks to the efficient frequency doubling of short pulses. These pioneering works were followed in 2006 by a multiplexed CRDS demonstration by Thorpe et al. [108] in the same spectral region, using a broader Ti:Sapphire comb and a dispersion-compensated high-finesse cavity. Indeed, dispersion inside an optical cavity limits the spectral window over which comb teeth can simultaneously match and inject the non-uniformly spaced cavity modes. Using special mirrors can mitigate this problem and indeed allowed Thorpe et al. to obtain cavity injection of a large spectral window. Using a rotating mirror as in [90] and a fast optical switch they obtained spectrally dispersed ring-down profiles on a CCD matrix following a broad-band transient cavity injection. In practice, multiplexed ring-down was not pursued further given the difficulty of obtaining many spectral elements (only 340 pixels in the cited demonstration) and good signal levels from a single passage through resonance. In addition, the particular advantage of a ring-down scheme, namely its ability to eliminate localized amplitude fluctuations, is superfluous given the smooth and stable spectral envelope of laser combs. In fact, a couple of years later the same group implemented a two-dimensional spectral dispersion system integrating a Vir-

---

[2] A shift applies also to the cavity resonances due to dispersion by the mirrors and in the intracavity medium, see Eq. (1.27).

tually Imaged Phased Array[3] (VIPA) to obtain about 1k spectral elements using a direct CEAS scheme [109] and again transient cavity injection.

More recently, as commercial mode-locked laser systems become more rugged and compact, it has become possible to conceive comb-CEAS systems for field measurements. In 2012 Grilli et al. completed a transportable near-UV ML-CEAS spectrometer equipped with two high finesse cavities for the measurement of atmospheric radicals in two spectral windows, attaining the shot-noise limit from 10 ms up to 10 minutes integration times [98]. This led to ppqv (10^{-15} volume mixing ratio) detection limits [110] for strongly absorbing molecules like IO. This instrument has been deployed in Antarctica, where it performed atmospheric measurements under rough conditions during one month [111].

An interesting alternative detection scheme exploits "Vernier" spectral selection between cavity and laser combs adjusted to have slightly detuned mode spacings: In combination with a CCD matrix detector, and using high precision frequency comb locking, individual comb modes could be resolved by Gohle et al. [112], while soon thereafter Thorpe et al. [113] reported using Vernier cavity filtering with a single photodetector.

Another important innovation is dual-comb multi-heterodyne detection, proposed by Schiller in 2002 [114], and later demonstrated in the mid-IR by Keilmann et al. [115, 116]. Two laser combs of slightly different repetition rate are superposed and beat on a single fast photodetector to produce a compressed RF replica of the product of the comb spectral envelopes [117–119], since each pair of teeth (one from each comb) beat at a slightly shifted RF frequency with respect to neighboring pairs. For example Coddington et al. [117] applied this concept to ultrastable erbium-fiber near-IR combs, covering a spectrum of 15.5 THz with 155k spectral elements defined to 1 Hz accuracy. A tunable filter was used to enable piece-wise recording of 1 THz acquisition windows, thus avoiding problems with photodetector saturation and dynamic range limitations. This was followed in 2009 by the first demonstration of dual-comb CEAS by Bernhardt et al. [120]. Recently, Chandler et al. [121] demonstrated the heterodyne concept coupled with pulsed-CRDS, using two high finesse cavities injected by a single ns laser pulse. This promising approach has the advantage of a simple experimental setup with no frequency locking, allows single shot acquisition with time resolution below the ring-down time, and spectral resolution of the cavity modes.

In 2010, comb-CEAS was coupled by Kassi et al. [122] to a commercial FT spectrometer. Using a commercial Ti:Sapphire comb they achieved a fairly high resolution, but over a limited spectral range and with an acquisition time of several minutes, and far from shot noise performance. A broadly tunable mid-IR fs commercial OPO system was also shown to work well with FT detection [123]. Foltynowicz et al. [124] have largely improved on these results by using comb frequency locking and a home-made FT spectrometer to cover a larger spectral interval with a much higher spectral resolution, few seconds acquisition time, and a shot-noise limited detection.

---

[3] See footnote 10.

1.1.2 A Review of Reviews on CEAS Developments and Applications

By now, cavity-ring-down and, more in general, cavity-enhanced spectroscopy have been the subject of a fair number of reviews, notably those of Scherer et al. in 1997 [125], Wheeler et al. in 1998 [126], Berden et al. in 2000 [127] and again in 2002 [128], Paldus et al. [129], Mazurenka et al. [130] and Vallance [131] in 2005. Books on the subject have also been edited, in 1999 by Busch and Busch [132], in 2003 by Van Zee and Looney [133], and by Berden and Engeln in 2009 [134].

The pulsed-CRDS technique as demonstrated by O'Keefe and Deacon [22] was followed by a number of cavity based methods, some of which have already seen their own specialized reviews. In 2003, Ball and Jones [135] reviewed the combination of CRDS and broad-band light sources. In 2007, Wang [136] reviewed the use of CRDS with plasma sources of atoms and elemental isotopes. In 2008, Thorpe and Ye [113] reviewed the combination of CEAS and frequency combs generated by mode-locked femtosecond pulsed lasers, while Foltynowicz et al. [137] described the state of affairs concerning the highly sensitive but complex NICE-OHMS technique. In 2010 Alder et al. [138] published a rather complete review on frequency comb cavity enhanced developments and applications, followed one year later by a shorter presentation of different comb CEAS schemes by Foltynowicz et al. [139]. In their 2010 review of the application of Quantum Cascade Lasers in chemical physics, Curl et al. [140] devote one section to cavity enhanced measurements using QCLs. The same year Waechter et al. [141] reviewed the use of waveguide (fibre) CRDS for chemical sensing, while Schnippering et al. [142] have treated, in 2011, recent advances in evanescent wave cavity-enhanced spectroscopy. The same year Orr et al. [143] wrote about developments and applications of specific schemes for fast-sweep CRDS, and in 2012 Long et al. [144] reviewed all frequency-stabilized CRDS works. Standing alone, is the 2009 tutorial by Lehmann and Huang [145] covering fitting algorithms and all what concerns optimal ring-down signal processing.

Apart from the above reviews that deal primarily with the technical aspects of one or more implementations of cavity-based spectroscopic measurements, a fair number of reviews deal with a restricted field of application. Brown in 2003 [146], Sigrist et al. in 2008 [147], Fiddler et al. in 2009 [148] and Cui et al. in 2012 [149] all looked at the application of cavity-based techniques in trace gas detection, and atmospheric monitoring in particular, while Atkinson in 2003 [150] dealt specifically with applications of CRDS in environmental chemistry. The aforementioned book edited by Berden and Engeln [134] contains a number of chapters dedicated to applications of cavity-based techniques. Applications to the analysis of stable isotope ratios of the light elements were reviewed by Kerstel in 2004 [151] and Kerstel and Gianfrani in 2008 [152]. In 2006, Loock [153] resumed the applications of fiber-based CRDS to microdevices for analysis of liquids. Cheskis and Goldman [154] reviewed in 2009 cavity ring-down spectroscopy of flames, whereas in the same year Wang and Sahay [155] reviewed the use of cavity enhanced techniques for exhaled breath analysis. The use of CRDS to measure the optical properties of aerosols has been the subject of a critical discussion by Miles et al. in 2011 [156]. In their recent review on explosives detection Caygill et al. [157] briefly discuss the role of CRDS.

As the above mentioned reviews already demonstrate, cavity-based techniques have been applied to measure atomic and molecular absorption in the gas, liquid, and solid phases. Furthermore, these techniques make it possible to measure the total extinction due to the combined absorption and scattering by small particles (aerosols) inside the optical cavity. An empty cavity may also be configured to measure mirror reflectivity, and other parameters such as mirrors or sample linear or circular birefringence by detecting the change of polarization state of light transmitted by the cavity. Finally, fiber-based cavities can be used as sensitive miniature strain, acceleration and temperature sensors [158].

Of the many different possible applications, the measurement of the absolute absorption of infrared light by small molecules of environmental and health interests is certainly the most widespread, as it enables trace gas detection in, for example, atmospheric sciences, industrial monitoring, and biomedical research.

We will not attempt here to review all recent results which escaped the reviews cited above, also considering that these should be mostly covered in the following chapters of this book, treating many, if not all, domains of cavity-enhanced application.

It is worthwhile, however, to mention the still fairly recent commercial developments that have led to an explosion of the use of CEAS-equipped instrumentation in measuring (trace) gas concentrations and stable isotope ratios in applications ranging from (Ant)arctic ice-cores and climate research to ecosystem monitoring, carbon sequestration, and land-atmosphere surface exchange of greenhouse gases, and even qualification of ultrapure process gases. There are currently at least four commercial players that offer instruments based on CEAS techniques for the detection of small molecules of environmental and industrial interest, including H_2O, CO_2, N_2O, CO, CH_4, HF, HCl, C_2H_2, H_2S, NH_3, SO_2, OCS, and more, plus H_2O, CO_2, and CH_4 isotopes. Los Gatos Inc. of Mountain View (CA) builds instruments based on off-axis ICOS, while Tiger Optics of Warrington (PA) and Picarro of Santa Clara (CA) both rely on an implementation of CRDS. The French company AP2E of Aix-en-Provence uses the OF-CEAS technique in its trace gas analyzers.

This new generation of diagnostic tools for trace gas and isotope analyses provides high time resolution, high accuracy and precision, and can overcome limitations of more conventional instrumentation (such as gas chromatography or mass spectrometry) relating to robustness, portability, and cost of acquisition and ownership. Therewith they enable new laboratory and field applications of environmental and geochemical interest that in the past were difficult or impossible to carry out. Progress in this field is still rapid, resulting in improved instrument capabilities and an ever-widening range of applications. As a critical note, we mention that the affordability of these instruments and their acceptance by a fast growing class of users that is not necessarily aware of their working details may lead to concerns about calibration, validation, and data integrity.

1.2 The High Finesse Optical Resonator

CEAS and CRDS techniques are based upon the use of optical cavities in order to enhance light interaction with matter present inside the cavity, or even placed on the mirrors. An optical cavity is also referred to as an optical resonator, since the electromagnetic field inside the cavity becomes excited (and increases in amplitude) by incident light at some specific (resonant) frequencies. For simplicity we will initially consider cavity resonances associated with the cavity fundamental transverse mode, which are also referred to as longitudinal modes. We will see that these are uniformly separated in frequency space by the cavity Free Spectral Range which is the reciprocal of the round trip time: $\mathrm{FSR} = c/(2n_\mathrm{r}L_\mathrm{c}) = t_r^{-1}$, where $2L_\mathrm{c}$ is the cavity round trip length, c is the speed of light,[4] and n_r the index of refraction of the intracavity medium (a gas sample, usually). Another important cavity parameter, or actually a common figure of merit, is the finesse $\mathcal{F}$, defined by the ratio of the FSR over the full width at half maximum (FWHM) of the cavity resonances $\Delta\nu_\mathrm{c}$, for which a simple expression (for a symmetric 2-mirrors cavity) will be derived below. Sometimes the quality factor Q (also defined below) is used, given by the resonance frequency over its linewidth, which is equivalent (apart from a 2π factor) to consider the field energy stored at resonance over the energy fraction lost per optical cycle.

In the following, we will first derive the cavity transmission as a function of frequency yielding the longitudinal modes, then we will include the absorption by an intra-cavity sample. We will show that the effective absorption length is enhanced by a factor close to $\mathcal{F}$. Afterwards, we will give a brief introduction to transverse modes, allowing for a more realistic picture of the ways a cavity may be excited by an incident field. Indeed a cavity admits different families of resonances with specific transverse $TEM_{m,n}$ field distributions, where TEM stands for Transverse Electro-Magnetic and m, n are positive integers, and $TEM_{0,0}$ is the fundamental transverse mode [159].

1.2.1 Intensity Transmitted by a Cavity: A Simplified One-Dimensional Model

As previously mentioned, we consider a symmetric linear cavity made of two mirrors facing each other at a distance L_c along an optical axis z. Let us suppose an incoming monochromatic field with a Gaussian profile $E_0(x, y, z)$ in the transverse directions x, y:

$$E_\mathrm{in}(x, y, z, t) = E_0(x, y, z)\, e^{i(\omega t - kz)}.$$

We suppose that this Gaussian beam is mode-matched to the cavity, i.e. it matches in size and waist position the Gaussian profile of the $TEM_{0,0}$ mode. Transverse cavity

---

[4] ...or the light group velocity at the given optical frequency, if intracavity dispersion effects are considered.

modes $TEM_{m,n}$ are eigenfunctions of the cavity propagation operator [159], which means they are unchanged after propagating one round trip inside the cavity, except for a small intensity loss (and a phase shift dependent on the optical frequency and on m, n, which we will neglect for now). Thanks to this property, we can write the field transmitted by an empty cavity E_{out} by considering all cavity round trips ($z > L_c$, if we take the origin $z = 0$ at cavity input):

$$E_{out} = E_0(x, y, z) \sum_{p=0}^{\infty} \mathbf{t}^2 \mathbf{r}^{2p} e^{i\omega(t - 2pL_c/c - L_c/c) - ikz}$$

$$= \frac{\mathbf{t}^2 e^{-i\omega L_c/c}}{1 - \mathbf{r}^2 e^{-i\omega t_r}} E_{in}(x, y, z, t), \tag{1.1}$$

where $\mathbf{t}$ and $\mathbf{r}$ are, respectively, the field transmission and the reflection coefficient of the two (equal) cavity mirrors, whose square modulus yield the corresponding coefficients for the light intensity $\mathcal{T}$ and $\mathcal{R}$. The case of an asymmetric cavity with different mirrors (with $\mathbf{r}_1, \mathbf{r}_2$ and $\mathbf{t}_1, \mathbf{t}_2$) is easily seen to give the same expressions below if we define $\mathcal{R} = \sqrt{\mathbf{r}_1 \mathbf{r}_2}$ and $\mathcal{T} = \sqrt{\mathbf{t}_1 \mathbf{t}_2}$. The term $i\omega(t - 2pL_c/c - L_c/c)$ is the phase change after one passage L_c/c plus p round trips. Taking the square absolute value leads directly to the transmitted intensity, the well-known Airy formula, expressing the comb of resonances uniformly spaced by the FSR:

$$I_{out}(\omega) = \frac{\mathcal{T}^2}{|1 - \mathcal{R} e^{-i\omega t_r}|^2} I_{in}$$

$$= \frac{\mathcal{T}^2}{(1 - \mathcal{R})^2} \frac{I_{in}}{1 + (\frac{2\sqrt{\mathcal{R}}}{1-\mathcal{R}})^2 \sin^2(\omega t_r/2)}. \tag{1.2}$$

It should be noted that this single frequency development can be extended easily to any time dependent input field (still mode matched, for simplicity), by considering its Fourier expansion, which is a sum (or integral) of monochromatic waves with complex amplitude coefficients, and taking the cavity response to each of these. The superposition of all the cavity responses gives the complete time dependent cavity output, according to the superposition principle of linear response systems [30]. While the present description is a static one where the monochromatic wave is infinite in space and time, a more intuitive representation of the intracavity field building up from a wave, which is switched on at a given instant in time, will be presented in Sect. 1.4.1.

Around a resonant frequency, the transmitted intensity given by Eq. (1.2) is well approximated by a Lorentzian function. The FWHM of this resonance, and the cavity finesse, are:

$$\Delta \nu_c = \frac{1}{\pi t_r} \frac{1 - \mathcal{R}}{\sqrt{\mathcal{R}}}, \tag{1.3}$$

$$\mathcal{F} = \frac{FSR}{\Delta \nu_c} = \pi \frac{\sqrt{\mathcal{R}}}{1 - \mathcal{R}}. \tag{1.4}$$

The quality factor defined earlier turns out to be expressed in the following form:

$$Q = \frac{v}{\text{FSR}} \mathcal{F}, \tag{1.5}$$

where v is the optical frequency of excitation.

As expected, the higher the reflectivity of the mirrors, the narrower the cavity modes and the higher the cavity finesse. Furthermore, we can derive the relation between Δv_c and the photon lifetime in the cavity, the ring-down time τ_{RD}. This is mathematically obtained by considering the cavity time response to an impulsive excitation which is written as the convolution of the pulse time profile and the FT of Eq. (1.1) taken for $r \sim 1$. Taking a delta function for the pulse, we obtain a decaying exponential for the field. Switching to the intensity ($I = |E|^2$), the decay becomes twice as fast and we find that the ring-down time is inversely proportional to the cavity mode width (which is related to the time-frequency uncertainty relation of FT theory):

$$\tau_{RD} = \frac{1}{2\pi \Delta v_c} = \frac{L_c}{c} \frac{\sqrt{\mathcal{R}}}{1 - \mathcal{R}}. \tag{1.6}$$

We see that the ring-down time is proportional to cavity finesse and to cavity length. It should be noted that a direct derivation of the ring-down time, by considering the cavity losses per round trip, yields the same result except for the term $\sqrt{\mathcal{R}}$, which is missing. With respect to high finesse cavities, these formulas are practically equivalent ($\mathcal{R} \simeq 1$). However, the first expression has the advantage of correctly going to zero in the limit $\mathcal{R} = 0$.

If we want now to include intracavity absorption in the previous treatment, we may consider that for each passage in the cavity the field is attenuated by the Lambert-Beer absorption factor $\exp(-\alpha L_c/2)$, where α is the absorption coefficient and the factor $1/2$ comes from considering the field rather than the intensity (we neglect the imaginary part corresponding to dispersion and related to the absorption through the Kramers-Kronig relations [160]). At each passage in the cavity, light goes once through the sample and is reflected once by a mirror, we then see that the intra-cavity absorption factor can be associated to the mirror reflection coefficient. Additionally, to be exact we have to consider that the incoming light, which is directly transmitted by both mirrors at the first passage through the cavity, experiences intra-cavity absorption on the cavity length, which gives a global multiplicative factor of $\exp(-\alpha L_c/2)$ on the cavity transmitted field. Then, we do not need to repeat the previous derivation but just apply these substitutions:

$$\mathbf{r} \to \mathbf{r} e^{-\alpha L_c/2} \quad \text{or} \quad \mathcal{R} \to \mathcal{R} e^{-\alpha L_c}$$

$$\text{and} \quad \mathbf{t}^2 \to \mathbf{t}^2 e^{-\alpha L_c/2} \quad \text{or} \quad \mathcal{T}^2 \to \mathcal{T}^2 e^{-\alpha L_c}. \tag{1.7}$$

At resonance, when the optical frequency equals ω_q for the qth longitudinal mode, the phase $\omega_q t_r$ is a multiple[5] of 2π. From Eq. (1.2), we can thus write the

---

[5] We neglect here phase factors associated to the complex coefficients $\mathbf{r}$ and $\mathbf{t}$, which would introduce a tiny change of the effective cavity length. Likewise, we neglect the index of refraction n_r of

cavity transmission at resonance ω_q:

$$I_{out}(\omega_q) = I_{in}(\omega_q)\frac{T^2 e^{-\alpha L_c}}{(1 - \mathcal{R}e^{-\alpha L_c})^2} \tag{1.8}$$

$$\simeq I_{in}(\omega_q) \times 1, \tag{1.9}$$

where the last passage is true for vanishing intracavity sample absorption and if mirror losses are negligible so that $T + \mathcal{R} \simeq 1$. This is a well-known property of optical cavities made of two identical mirrors with equal T and equal $\mathcal{R}$: Cavity transmission at resonance may approach 100 % if total cavity losses are much smaller than mirror transmission.

It is interesting to compare this result to the simplified picture that is all too often encountered in the literature. This considers just the sum of the decreasing intensities over cavity round trips, instead of the corresponding amplitudes as was done here, yielding:

$$I_{out}(\omega_q) = I_{in}(\omega_q)T^2 \sum_p \mathcal{R}^{2p} \tag{1.10}$$

$$= I_{in}(\omega_q)\frac{T^2}{1 - \mathcal{R}^2} \tag{1.11}$$

$$\simeq I_{in}(\omega_q) \times \frac{T}{2}. \tag{1.12}$$

This result is clearly inconsistent with Eq. (1.9), which underlines the fact that field interference cannot be neglected, even when considering cavity response at resonance when the round trip phase is a multiple of 2π and field amplitudes simply add in phase. Mathematically it is evident that considering a round trip sum of intensities corresponds to dropping all crossed field terms:

$$\left|\sum_p \mathbf{r}^{2p}\right|^2 = \sum_{p,m} \mathbf{r}^{2p}\bar{\mathbf{r}}^{2m} \tag{1.13}$$

$$= \sum_p \mathcal{R}^{2p} + \left(\sum_{p \neq m} \mathcal{R}^p \mathcal{R}^m\right), \tag{1.14}$$

which is why a smaller cavity transmission is obtained.

The $T/2$ loss of incident light resulting from this simplistic estimation is actually realized when using a broadband source or a swept laser (as we will see), which may justify that the crossed field terms average to zero as they correspond to the interference of incoherent fields. A simple picture allowing to better understand this situation, is that of an isolated and spectrally broad pulse, shorter than the cavity

---

the sample, which makes the cavity length equal to $n_r L_c$. These effects have no importance here, but they have an impact on cavity dispersion (spectral dependence of FSR, considered later).

round trip time. This pulse suffers an initial attenuation $\mathcal{T}$ when transmitted inside the cavity through the input mirror. Then, if we neglect losses, this fraction $\mathcal{T}$ of the pulse energy remains trapped in the cavity and will completely leak out of it in time, half in the forward and half in the backward direction, which accounts for the factor $1/2$. In the case of an incoherent LED or lamp, the source emission may be considered equivalent to a continuous sequence of short and broad pulses with random relative phases. The duration of these idealized pulses should be taken to correspond to the very short coherence time of the source, which may be proportional to the inverse of the emitted spectral width. While these hand-waving arguments seem reasonable, it is difficult to formalize them in a rigorous mathematical development in the time domain, while it is not particularly hard to obtain a rigorous analytical result by considering the frequency domain cavity response, as presented further down.

Let us now discuss the enhancement of intracavity absorption. From the cavity transmission at resonance Eq. (1.8), one obtains to first order in α:

$$\frac{I_{\text{out}}(\omega_q)}{I_{\text{in}}} \sim \frac{\mathcal{T}^2}{(1-\mathcal{R})^2}\left(1 - \frac{2\alpha L_c}{1-\mathcal{R}}\right) \tag{1.15}$$

$$= \frac{\mathcal{T}^2}{(1-\mathcal{R})^2}\left(1 - \alpha L_{\text{eff}}^{\text{res}}\right), \tag{1.16}$$

which holds if $\alpha L_c \ll 1 - \mathcal{R}$. For a high finesse cavity, with typically $1 - \mathcal{R} \sim 10^{-4}$, this would imply an absorption coefficient below 10^{-7} cm^{-1} (for $L_c \sim 1$ m). We see that for $\alpha \to 0$, the transmission approaches $\mathcal{T}^2/(1-\mathcal{R})^2$, which is close to unity for "low loss" mirrors (with $1 - \mathcal{R} \simeq \mathcal{T}$). In practice, cavity transmission will be smaller than this if the laser linewidth is broader than the cavity resonance, and additionally if mode-matching is not perfect, as the incoming beam will only partially overlap with the $TEM_{0,0}$ mode being considered.

In Eq. (1.16), we introduced the effective absorption length where the cavity enhancement effect appears explicitly:

$$L_{\text{eff}}^{\text{res}} = \frac{2}{1-\mathcal{R}} L_c \simeq \frac{2\mathcal{F}}{\pi} L_c. \tag{1.17}$$

As expected using a resonant cavity, the absorption path length is enhanced by a large factor close to $\mathcal{F}$, as compared to a single pass over the sample length L_c. For a cavity finesse of 10^4, easily available in the visible and near-IR regions, and a cavity length of 1 m, the effective absorption length is already close to 10 km.

We should stress that this $2\mathcal{F}/\pi$ enhancement factor is only realized with a sufficiently monochromatic source probing the maximum of the cavity mode, which requires that a laser source is actively locked to a cavity resonance. In fact, the lock should be so "tight" that the laser frequency noise is reduced down to an emission linewidth well below the cavity mode width (more about this in Sect. 1.4.1).

In the opposite regime, with a broadband source such as a LED or a lamp, a lower enhancement factor is found [161]. To see this, let us consider Eq. (1.2) giving the cavity transmission spectrum when multiplied by the source spectral power density. If we integrate this over a spectral window we obtain the total power transmitted in that window. Since a broad-band source has a smooth spectrum, we may consider it as constant if we take a sufficiently small window, for instance one cavity FSR. We are then left with the transmission function averaged over one FSR [161], yielding:

$$\mathcal{A} = \frac{1}{\text{FSR}} \int_0^{\text{FSR}} I_{\text{out}}(2\pi\nu)\,d\nu = \frac{\mathcal{T}^2}{1-\mathcal{R}^2} I_{\text{in}}, \tag{1.18}$$

where I_{out} has been substituted according to the Airy formula Eq. (1.2). Taking into account the absorption via the substitutions of Eqs. (1.7), we obtain:

$$\mathcal{A} = \frac{\mathcal{T}^2 \exp(-\alpha L_c)}{1 - \mathcal{R}^2 \exp(-2\alpha L_c)} I_{\text{in}} \tag{1.19}$$

$$\simeq \frac{\mathcal{T}^2}{(1-\mathcal{R}^2)} \left(1 - \alpha \frac{L_c}{1-\mathcal{R}}\right) I_{\text{in}}$$

$$\simeq \frac{\mathcal{T}}{2} \left(1 - \alpha L_{\text{eff}}^{\text{BB}}\right) I_{\text{in}}, \tag{1.20}$$

where again we assume small absorption losses ($\alpha L_c \ll 1 - \mathcal{R}$) and high reflectivity mirrors ($\mathcal{R} \lesssim 1$). This confirms the previous non-rigorous result Eq. (1.12) that in broadband CEAS (BB-CEAS) the transmitted intensity is reduced by a factor $\mathcal{T}/2$ as compared to resonant CEAS. Additionally, we learn that the effective broadband absorption length $L_{\text{eff}}^{\text{BB}}$ is reduced by a factor 2:

$$L_{\text{eff}}^{\text{BB}} = \frac{1}{1-\mathcal{R}} L_c \simeq \frac{\mathcal{F}}{\pi} L_c = \frac{L_{\text{eff}}^{\text{res}}}{2}. \tag{1.21}$$

This difference in cavity enhancement comes from the fact that BB-CEAS is sensitive to the resonance profile surface, while resonant CEAS depends only on the peak transmission. When intra-cavity losses increase, the peak amplitude decreases faster than the mode profile area, because at the same time the width of the resonance increases (half as fast), as is easily verified.

The previous path-length enhancement factors are valid for cavity enhanced schemes, where the cavity output intensity is measured while the cavity is coupled with the light source. It is interesting to consider the ring-down case, where the cavity is isolated from the source during the measurement of the light decay time. It is then possible to define an effective path length as:

$$L_{\text{eff}}^{\text{RD}} = c\,\tau_{\text{RD}} = \frac{\mathcal{F}}{\pi} L_c, \tag{1.22}$$

equal to the path length for BB-CEAS. This is consistent with the fact that the decaying ring-down field is not composed only of photons sitting at the center of the resonance, which would require a longer observation time. It rather has the same spectral profile as the cavity mode, and the same as in BB-CEAS at cavity output. One can also consider the first order expansion of τ_{RD} for small absorptions $\delta\alpha$, which is done by substituting Eqs. (1.7) inside Eq. (1.6). This yields $\delta\tau_{RD}/\tau_{RD} = L_{eff}^{RD}\delta\alpha$, as suggested for example by Thorpe et al. [113]. They also explain that in resonant cavity enhanced the backward wave is cancelled by interference with the directly reflected fraction of the input wave, compared with the ringdown case where light escapes from the cavity in both directions, accounting for a reduced sample inter-action and two time smaller enhancement effect. This explanation does not seem to hold if we consider that in the resonant case reflection is null only in case the cav-ity has no losses (only mirror transmission) or its mirrors are chosen to satisfy an impedance matching condition (the input mirror transmission is equal to the sum of all other cavity losses and transmissions), however the enhancement factors derived above do not depend on this detail but only on the cavity finesse, thus the ratio of the enhancement factors for resonant cavity enhanced and CRDS is always 2.

The take-home message is that the enhancement factor in CEAS does not de-pend only on the cavity but also on the light source and the injection scheme. We have studied here static configurations in the two extreme limits where analytical results are straightforward, and in general we may write the enhancement factor as $\beta\mathcal{F}/\pi$ [113, 138], where β is a constant ranging between 1 for BB-CEAS and 2 for monochromatic resonant CEAS. We should also stress that the whole depen-dence on intracavity absorption is modified, since the absorption law for BB-CEAS Eq. (1.19) and resonant CEAS Eq. (1.8) are definitely not the same (see below and Fig. 1.1). In practice, intermediate configurations may exist but are not common. An interesting case is CEAS with a laser frequency locked to resonance, but not tightly enough to reduce its linewidth below that of the resonance (e.g. [124]). The average intensity transmitted will then be sensitive to the distribution of the laser spectrum across the resonance profile, which changes both in width and peak transmission as a function of cavity loss. This case will thus produce an intermediate β value that will vary when adjusting the locking loop. Even worse, the absorption law in this case will also be intermediate between that for the BB-CEAS and resonant CEAS, and will depend on the laser lineshape. Contrary to what claimed in [113, 138], β does not really depend on the laser tuning speed when transient injection schemes are concerned, except in a limit of very slow tuning, which is never realized in practice. Let us discuss this case in some detail.

If we take a transient injection with an ideal monochromatic wave tuning across resonance sufficiently slow that a complete cavity buildup is achieved, the transmit-ted peak will indeed present a sensitivity to cavity losses corresponding to $\beta = 2$. As the tuning speed will increase, the transmission profile will become distorted as discussed in Sect. 1.4.2, the peak intensity will decrease and the enhancement factor might start to decrease. However, there are two points to be considered. First of all, this is an extreme situation demanding a prohibitively slow tuning when we consider the response time and the mode width of a typical high finesse cavity.[6] Second, in swept CEAS schemes one does not consider the peak transmission, but some kind

---

[6]Inferior to 10 kHz/10 µs for a typical high finesse cavity, or about 0.1 s for tuning over one cavity FSR.

Fig. 1.1 Cavity transmission for resonant and integrated CEAS, as well as direct absorption over the effective pathlength $L_{\text{eff}} = (2\mathcal{F}/\pi)L_c = 10$ km, all as functions of the absorption coefficient α. The CEAS signals are normalized to the respective zero absorption values

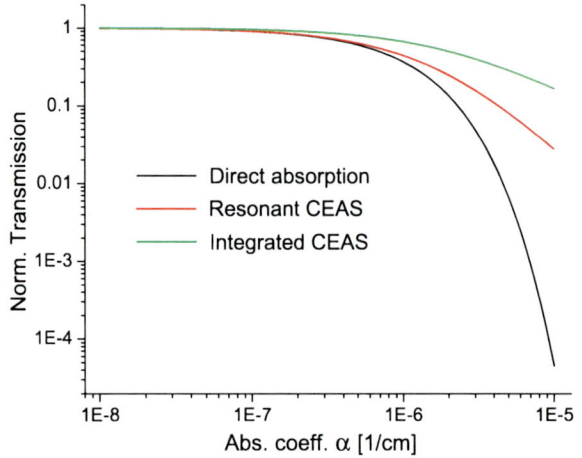

Norm. Transmission

Direct absorption
Resonant CEAS
Integrated CEAS

Abs. coeff. α [1/cm]

of time running average, which corresponds to an integration over a spectral region much broader than cavity resonances. Such "low pass" signal filtering also becomes necessary when using lasers having a non negligible linewidth, producing a transient noisy peak profile at cavity output (more of this in Sect. 1.4.1). In the end, as the cavity output signal is smoothed or averaged over a time window, which corresponds to a tuned frequency range wider than the cavity resonance, we are really in a situation with $\beta = 1$. As before, it is difficult to understand this by considering the signals in the time domain, since we have to convolute the time-domain cavity response function (the linear response Green's function) with the frequency swept source field (including noise), then integrate the result in time. As before, we can find a simpler description in the frequency domain, which will yield the same conclusions as in the time domain, but more easily and directly. We will use again the fact that the average cavity output spectrum is the product of the averaged source spectrum times the cavity transmission spectrum. The source spectrum to be considered is the FT of the laser field over an observation time window that includes the passage through resonance. A monochromatic sine wave frequency-swept at a constant rate over a spectral interval can be readily shown (e.g. by numerical FT) to have a flat power spectral density distribution over that interval, independently of the sweep rate. This clearly shows that swept-CEAS with integrated cavity output is indeed equivalent to BB-CEAS in terms of enhancement factor.

We conclude this section by an interesting observation concerning the attenuation of transmitted intensity by a cavity compared with direct absorption through an hypothetic cell with the same effective absorption length, given by the Lambert-Beer law $T_{\text{LB}} = \exp(-\alpha L_{\text{eff}})$. In order to make this comparison independent of the cavity transmission level, which is usually lower than unity, we consider the cavity transmission relative to its value at zero absorption $\alpha = 0$. We consider the two extreme cases of resonant CEAS, ruled by Eq. (1.8), and BB-CEAS, ruled by Eq. (1.19). For the typical parameters used above, in particular $L_{\text{eff}} \simeq 10$ km for the resonant CEAS, we obtain the curves of Fig. 1.1 which show that BB-CEAS has the smallest loss of signal, thus potentially the largest dynamic range, while resonant CEAS falls below it and direct absorption displays the worst case behavior by far. In practice, one should also consider that the situation is reversed when considering the available signal level: Direct absorption may in principle benefit of the full source light intensity in the limit of small sample absorption, resonant CEAS usually induces some source intensity loss, while we saw that BB-CEAS is characterized by low cavity throughput.

1.2.2 The Real World of Transverse Modes

Until now the concept of longitudinal cavity mode has been developed in a frame equivalent to a basic plane wave model, and the resonances have been found to correspond to a round trip self-reproducing condition (2π phase change). Considering the finite size of the mirrors and their curvature, the modes of a real cavity can certainly not be described by plane waves. An ensemble of functions which can be made to satisfy the round trip self-reproducing condition (by a choice of parameters) are obtained by solving Maxwell's equation for the propagation of the electromagnetic field under the paraxial approximation (small angles and small deviations of the field relative to the resonator optical axis), and by making them satisfy the cavity boundary conditions. These functions are written as the products of Hermite polynomials and a Gaussian distribution function [7, 162].

In Fig. 1.2 the profiles of several Transverse Electro-Magnetic modes are represented. The lowest-order, or fundamental mode, has a pure Gaussian-shaped transverse profile:

$$TEM_{0,0}(x, y, z, t) \propto \exp\left(-\frac{r^2}{w^2(z)} - \frac{ikr^2}{2R_f(z)} - ikz + i\omega t + i\eta(z)\right), \quad (1.23)$$

where for clarity we have written r^2 in place of $x^2 + y^2$, and $k = n_r\omega/c$ where n_r is the refraction index and ω the optical angular frequency. We see that along the propagation axis z we have a plane wave $\exp(-ikz + i\omega t)$ (with a slowly changing phase $\eta(z)$), but this is multiplied by a Gaussian profile with radius ($1/e$ half-width) given by the function $w(z) = w_0\sqrt{1 + z^2/z_0^2}$ which has waist size w_0 at $z = 0$, and diverges appreciably over the Rayleigh length $z_0 = \pi n_r w_0^2/\lambda$. At large distances along the propagation axis $z \gg z_0$, the beam radius increases as w_0z/z_0, thus with a divergence angle $\theta \simeq \lambda/(n_r\pi w_0)$. We see that the divergence of the Gaussian beam is inversely proportional to its waist size. This relation of the divergence to the spot size corresponds to an uncertainty principle (position vs kinetic momentum, akin Heisenberg principle in quantum mechanics), and it should be noted that Gaussian beams possess the smallest uncertainty product and thus are said to be "diffraction limited" [159]. The other Gaussian beam parameters have similar slow dependence on z. They are the wavefront curvature $R_f(z) = z + z_0^2/z$ and the Gouy phase $\eta(z) = \arctan(z/z_0)$ (which is not zero even on the optical axis and differen-

Fig. 1.2 Examples of $TEM_{m,n}$ transverse mode field distributions: (**a**) $TEM_{0,0}$, (**b**) $TEM_{2,0}$, (**c**) $TEM_{0,4}$, (**d**) $TEM_{2,4}$. The quantum numbers m and n correspond to the number of field nodes in the horizontal and vertical direction, respectively

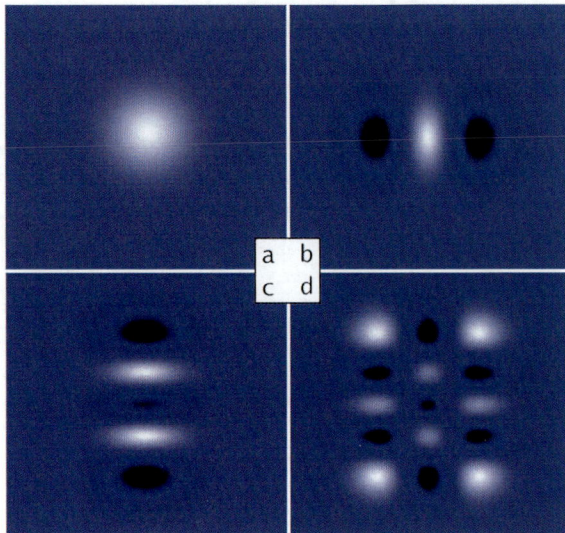

tiates the Gaussian wave from a plane wave). The higher order *TEM* modes have the same general dependence (thus, the same beam radius and wavefront curvature as a function of z) except for an additional factor $H_m(x/w_0)H_n(y/w_0)$ where the Hermite polynomials introduce the transverse node structure (H_m possesses m zeroes), plus a Gouy phase equal to $(1+m+n)\eta(z)$. We see that this last modification implies a shift in the wavefronts that depends on the transverse order $m+n$.

"Mode-matching" a Gaussian shaped laser beam to a cavity consists in matching the waist size and position of the laser beam to the waist size and position of the cavity $TEM_{0,0}$ mode (and, obviously the beam should be aligned on the cavity axis). If the laser beam is not Gaussian, spatial filtering (see Sect. 1.4.2) may be used with some power loss to render it sufficiently close to Gaussian. In several implementations of CEAS, a good mode matching is needed, since excitation of higher transverse modes represent a source of noise or artifacts. Some implementations, however, such as off-axis CEAS (see Sects. 1.4.1 and 1.4.2), do exploit multi-transverse-mode excitation.

The cavity boundary conditions impose first of all that the wavefront matches the mirror surfaces, thus the z_0 parameter and the $z = 0$, waist position are fixed by the cavity geometry. For the symmetric cavity that we considered until now, $z = 0$ is clearly at the cavity center, and the mirror curvature R and cavity length L_c impose that $z_0 = \sqrt{(2R - L_c)L_c}/2$. The resonance condition is found (as before) when the round trip phase change is a multiple of 2π, but now the whole on-axis phase term $kz + (m + n + 1)\eta(z)$ is to be considered in place of kz alone. The resonant angular frequencies of the $TEM_{m,n}$ transverse modes can thus be expressed as [159]:

$$\omega_{m,n,q} = \text{FSR}\left(2\pi q + 4(m + n + 1)\,\text{atan}\sqrt{\frac{L_c}{2R - L_c}}\right), \qquad (1.24)$$

where q is the longitudinal mode number. For a family of longitudinal modes of a given transverse order $m + n$, the resonances constitute a frequency comb with a specific offset but whose periodicity is always given by the same cavity FSR. We may also assume that the finesse of all transverse modes is the same up to rather large transverse orders, if the reflectivity $\mathcal{R}$ is spatially uniform. For sufficiently large transverse orders, depending on mirror transverse dimensions, diffraction losses at the mirror edges will induce a degradation of the finesse.

From Eq. (1.24) it is clear that all frequency combs with the same transverse order coincide. This is a consequence of considering a cavity with cylindrical symmetry. An interesting situation arises when the transverse spacing to the longitudinal spacing is rational: $2 \operatorname{atan}\sqrt{L_c/(2R - L_c)} = \pi M/N$. The cavity spectrum will then consist of a series of resonances spaced by FSR$/N$. In practice this is obtained by adjusting the cavity length for a given set of mirrors. A simple example is the well known confocal cavity where $L_c = R$ and the transverse mode separation is exactly half the FSR ($M/N = 1/2$). The confocal cavity is also known for the fact that a beam injected off axis will follow a trajectory which closes onto itself after 2 round trips in the shape of a bow-tie, which can be understood easily by geometric ray propagation. When considering the more realistic Gaussian beam propagation, the realization of such a closed trajectory is possible via a superposition of transverse modes which are degenerate in frequency. This condition insures that the relative phases of the modes whose superposition reproduces the localized bow-tie trajectory remain constant as a function of time, so that the trajectory itself is stationary (like the modes composing it). This is related to the fact that by analogy with quantum mechanical systems, the cavity modes would also be energy eigenstates, and a stationary state must also be an energy eigenstate or a superposition of energy eigenstates with the same energy.

The general situation of an intracavity trajectory closing after N round trips has been investigated by Herriot and Kogelnik [163] and became the basis of a widespread type of multipass absorption cell. This configuration was later proposed as a basis for the off-axis CEAS scheme [65], with large cavity mirrors allowing the use of very high order transverse modes, which may be excited by an incident beam parallel but distant from the cavity axis. The interest of this configuration is that for a well chosen cavity length the cavity mode spacing may be divided by a large N since the large mirror size supports high order Herriot trajectories. It is surprising that even though N groups of degenerate modes are excited in succession as the laser frequency tunes across one cavity FSR, the resulting folded trajectory appears to be always the same for each group (a generalized bow-tie with N branches) except for fine details relating to the phases of the field along the branches of the trajectory [164]. Going back to the confocal case it is easy to understand how the bow-tie trajectory composed of the even modes which are degenerate with the 00 mode (02, 11, 20, 04, 13, 22, ..., $m + n =$ even) differs from the bow tie formed using $m + n =$ odd modes (obtained for a comb of frequencies shifted by FSR$/2$): The first will be an even function for an inversion with respect to the cavity axis, while the second will be an odd function.

Going back to off-axis CEAS, from the literature it turns out that a preferred configuration is a cavity with astigmatic mirrors with no cylindrical symmetry and which does not admit Herriot-like trajectories but rather a generalization thereof, i.e. closed trajectories whose points of reflection on the mirrors lie on a Lissajous curve. In this case, cavity modes with the same $m + n$ are split into several smaller degenerate subgroups, each describing always the same Lissajous trajectory [165], as before with differences in field phase distributions. At this stage it is preferable to postpone this discussion and conclude on off-axis CEAS at the end of Sect. 1.4.1.

1.3 Detection Limit, Noise, Fringes, and More

In absorption spectroscopy the measured transmission of a cell or an optical cavity can be related directly to the absorption per unit length of the sample, or absorption coefficient α, whose dependence on frequency gives the absorption spectrum. The well-known Lambert Beer law applies to direct absorption through a (multipass) cell, while more complex relations (see Sect. 1.2.1) exist for the case of a sample inside an optical cavity. The absorption coefficient is in turn given by the number of absorbers per unit volume times the absorption cross section, which is composed of the sum of absorption lines from all transitions between a lower to an upper quantized energy level of the absorbing molecule(s) in the sample. Each absorption line has a spectral profile with a shape (and a width) which depends on physical processes like Doppler frequency shifts from disordered molecular motion or time limitation of the transitions by perturbing binary collisions, which depend on the pressure and temperature of the sample. Using CEAS allows either measuring weak molecular transitions when using samples of known (high) concentration, or analyzing (unknown) samples for quantification of weak concentrations of one or more species.

The concentration detection limit of a given CEAS technique is then dependent on the molecular transition being probed, and it will be maximized either by using strong fundamental vibrational transitions in the mid infrared, or even stronger electronic transitions in the visible or near UV. In order to compare different techniques in a way that does not depend on the choice of spectral region (which depends on the application of interest and on the availability of laser sources), it is better to consider the detection limit in terms of the absorption coefficient, that is, the smallest detectable variation of α. To define this, it is customary to consider the rms (root mean square) noise level on the acquired spectra in the absence of molecular absorption or in the limit of very small absorption, i.e. the noise-equivalent absorption coefficient. This can be obtained by fitting a spectrum baseline section with a straight line and taking the rms of the residuals. In case the baseline is not flat because some residual molecular absorption is present, a fit of the structured absorption can also be used. An alternative that does not depend on L_c, is to indicate the smallest fractional intensity change detectable, or minimum detectable absorption, which is equal to the noise equivalent absorption times the sample physical length L_c.

This rms noise equivalent detection limit can present a white noise character, or it can contain oscillations produced by optical interference fringes. These are most often the real limitation to the smallest detectable absorption. If light from the laser beam is scattered or partially reflected by an optical surface then further scattered by other surfaces in the setup to reach the detector, it will interfere with the main signal beam, and produce an oscillation when the source frequency is tuned, with period inversely proportional to the difference in pathlength between the signal and the parasite. What should be realized is that the amplitude of an optical fringe is given be the product of the fields, thus a weak parasite may be strongly amplified by beating with a strong signal, which is the base of heterodyne detection (this comes directly from the detected intensity being the squared modulus of the field). Typical fringes may be in the range of 10^{-5} of the signal, and are thus caused by a parasite having a power level of only 10^{-10} relative to the main signal. To reduce fringes by a factor 10 it is necessary to reduce parasites reaching the detector by a factor 100, which gives a feeling for the difficulty in eliminating this effect.

An illustrative case is the NICE-OHMS technique, which achieved a performance limited by the shot noise level on the detected light signal, giving a rms noise on a measured spectrum corresponding to 10^{-14}/cm absorption with 1 s averaging [83]. That was however the noise level observed over a tiny tuning range, a few MHz being enough to obtain the sub-Doppler saturation Lamb dip of a molecular transition, which was the scope of the demonstration. In contrast, NICE-OHMS applications to Doppler broadened molecular spectra could not get really close to that performance [88]. Indeed the typical period of interference fringes depends on the size of the experimental settings, thus typically these are on the order of $1/1$ m, or 0.01 cm^{-1}. This is just below the Doppler width but largely above the width of saturation dips, which explains why fringes where not such a nuisance in the first experiment.

When comparing different techniques it is therefore important to be careful. Notably, it is necessary to consider the noise level that is finally obtained on the baseline of an acquired spectrum for a given measurement time, independently of intermediate performance indicators. In this respect, the detection limit is usually reported in terms of the noise equivalent absorption for 1 Hz bandpass, i.e. for 1 s averaging per data point (or better per spectral element, another relevant concept discussed below). However, some techniques are able to provide N spectral data points in 1 s by parallel detection or just a fast frequency tuning. In this case, it is also usual to normalize the detection limit by $\sqrt{N}$, since the acquisition of N spectrally distinct points could be considered as equivalent to N independent values to be averaged for a single spectral element. This equivalence holds only in the limit of data affected by white noise, but not in the presence of fringes or other noise sources, usually having a stronger contribution at longer timescales ("$1/f$" type of noise, associated to setup drifts).

A couple of examples will clarify this point. An acquisition of 100 spectral datapoints in 1 s may yield a spectrum with a relatively large noise level σ_α where fringes are not visible. Averaging this spectrum 100 times would decrease the white noise component by a factor 10 and could reveal the fringes, which normally are stationary over relatively long time scales. These fringes would then dominate the noise level of the spectrum, ending up with a total noise level larger than $\sigma_\alpha/\sqrt{N}$. Thus, while it is reasonable to take into account the number of datapoints provided by a setup by consolidating it with the detection limit, the resulting value $\sigma_\alpha/\sqrt{N}$ should not be considered as a true detection limit, but rather as a "figure of merit", useful for comparing with other setups. A similar situation occurs even in the absence of fringes, when averaging a signal does not improve its fluctuations as $1/\sqrt{N}$, due to changing parameters (optical alignment, but also sample pressure and temperature, and more). In that case, spectral baseline drift in shape or level may occur without having any visible fringes.

In ring-down measurements it is often remarked that the signal quality of a single event allows an exponential fit of good quality (with flat residuals displaying uniform noise level). This yields a ring-down time estimate τ with a statistical error bar provided by the (nonlinear) fit starting from the signal noise level (the rms of the fit residuals, in practice). That single-shot noise level σ_τ is correctly estimated if the time data points are uncorrelated [145], which is not the case, for instance, if the analog to digital conversion rate is faster than the photodetector amplifier bandpass (the reverse situation is also not desirable since the undersampled signal will appear more noisy than it is). The single-shot noise is usually found to be smaller than the shot to shot fluctuations, the rms noise from an ensemble of ring-down measurements. It may be useful to provide here a relation for estimating the shot noise level for a single ring-down signal which is obtained in the case of nonlinear fitting with the proper weighting (Eq. (17) in [23], see also [145]):

$$\frac{\sigma_\tau^{SN}}{\tau} = \frac{1}{\sqrt{I_0 \tau}}, \tag{1.25}$$

where $I_0 = \eta P/e$ is the flux of photoelectrons generated in the detector at the beginning of the digitized ring-down signal, dependent on the optical power P [W] (at the beginning of ring-down), the detector sensitivity η [A/W], and the electron charge e [C].

To complicate things, such an ensemble of measurements can be obtained by scanning the laser over a baseline section or just by keeping the laser frequency fixed on a single spectral element. In the first case the effect of fringes may become visible as they give a clear pattern on the data, and the rms of this ensemble would allow a realistic estimate of the total noise affecting the spectral profiles which are delivered by the setup. The second case, if the ensemble is obtained in a short time, delivers rms fluctuations closer to the ideal single-shot value, if no other noise sources affect the signal besides fringes. This is certainly an important test, but clearly does not give a fair assessment of global system performance.

The Allan variance (AV) concept originally developed to characterize the performance of clock in terms of frequency stability [166], has been applied to characterize the performance of absorption spectroscopy [59, 167, 168] but it has the disadvantage of being a one-variable evaluation test, which can be directly applied to a single spectral point at a time, but not to a complete spectral section. It delivers information about the time variability of fringes or of other baseline drift mechanisms, which affect the single spectral point. However, as above for the rms fluctuations of an ensemble of measurements at the same spectral point, the AV of a spectral point is not representative of the spectral noise level delivered by a setup if this includes fringes or other frequency-dependent artifacts. Indeed on sufficiently short time scales a fringe does not move and does not produce an effect at the single point level, while it certainly affects a spectral profile as a whole, as we discussed above. At the end, it is more appropriate to consider the rms value of a baseline section of a spectrum, or the rms of the residuals after fitting a section of a spectrum using an appropriate model of the molecular absorption being present in it [98]. This rms value, if plotted against the averaging time T, yields a curve resembling an AV plot, decreasing as $1/\sqrt{T}$ as long as white noise is dominant, but no longer decreasing as soon as fringes or other spectral artifacts produce deviations from the fit model.

An AV plot (as a function of averaging time) [167] is however useful to quantitatively estimate the precision in the determination of some spectral fit parameters, for instance line intensities that are proportional to the concentrations of the absorbing molecules: If these are kept fixed by providing a stationary sample, the Allan variance will tell over what time interval the sample concentration can be averaged before drifts become dominant, and the AV value for a given averaging time will provide a true estimate of the rms fluctuations of concentration averaged over this time interval. Also, the time after which the AV starts increasing will indicate the stability time of the setup. Calibration procedures will allow to obtain the measurements of highest accuracy if they are realized within this time frame. For example, an unknown sample measurement and a calibrated sample measurement (of the same duration) should fit inside this stability interval in order that the second can be really useful to calibrate the value of the first within an error given by the minimum of the AV [168]. Making a calibration over a longer time span would allow for large instrumental drifts and reduce the effectiveness of the calibration, the AV corresponding to the longer interval will still give an estimate of the (larger) calibration error.

Spectral resolution and "spectral element" are other concepts that play an important role in spectroscopy or even in trace analysis, as long as this depends on fitting of spectral profiles. A spectral element does not necessarily coincide with a single datapoint in an acquired spectrum, as sometimes is assumed. This will be the case only if datapoint separation is larger than the spectral resolution. This is usually represented by an apparatus spectral response function, which may for example correspond to a laser line width, or to a spectrograph resolution in case of a broad band source. More simply one just considers the FWHM of the apparatus function as being the spectral resolution, even though it is clear that its shape is also relevant. A spectrum where datapoints are closer than this FWHM will be oversampled and close-by datapoints will be correlated. This is analogous to sampling a time signal at time intervals shorter than the time response of the signal chain generating it. In some cases datapoints may be at a separation larger than the spectral width that they correspond to. This is the case of CEAS schemes where measurements are obtained only for the $TEM_{0,0}$ cavity modes (and these are spectrally stable in time), where the spectral resolution may be considered as corresponding to the mode width, while their separation is $\mathcal{F}$ times larger. Another example is a 1 MHz linewidth laser which is tuned in jumps of 100 MHz to obtain an absorption spectrum.

1.4 Coupling of Light into a High Finesse Optical Cavity

We will now begin the description of different schemes for coupling light from
various sources, not only lasers, into a high finesse cavity. All these schemes have to
take into account, more or less explicitly, the resonant properties of optical cavities
that we have introduced above. The logical order we have chosen follows loosely
the historical developments outlined earlier.

1.4.1 Single Mode Injection: From the Ideal Monochromatic Laser to the Realistic Noisy Laser

In Sect. 1.2.1 we considered the properties of optical cavities, avoiding as much as
possible any reference to a specific light source. Here we consider how cavity response combines in the spectral or time domain with different types of light sources.

Some of the most refined CEAS schemes are based on frequency locking of a CW
narrow-line laser to a cavity, and are well described by the monochromatic stationary excitation model considered in Sect. 1.2.1, especially if the locking bandwidth
is sufficient to squeeze the laser spectrum inside the cavity resonance linewidth.
These schemes deliver a precise frequency axis, optimized cavity transmission,
and usually high acquisition rates, at the price of increased experimental complexity [38, 55, 83, 169].

Several methods exist for locking a "single frequency" laser to an optical cavity,
and the performance of the locking (such as the fact that the laser linewidth is reduced with respect to that of the free running laser), is generally increases with the
bandwidth of the lock loop. This is a combination of the bandwidth of the system
used to obtain an "error signal" (a signal proportional to the frequency difference
between the laser emission and the cavity resonance) and of the bandpass of the
control system acting on the laser, or on the cavity resonance frequency. There are
not many general reviews dealing with different techniques to generate the error signal [170–172] and efficient methods to control laser frequency or cavity resonance
frequency, thus we just cite the one by Hall [173].

Simpler CEAS schemes demand that the source optical frequency is swept
through the cavity resonances, thus the cavity temporal response has to be taken
into account. For the time being, we will continue considering mode matched cavity
injection and neglecting the evolution of the field in the transverse direction as only
the fundamental $TEM_{0,0}$ cavity mode is excited.

The temporal response of a cavity mode corresponds to the process of building
up the intracavity field by the input field, or else to the field decay when the input
is interrupted. The cavity response time, i.e. the characteristic time of such transient
effects, is the ring-down time. As we have already seen, the field amplitude inside
the cavity is governed by the interference (sum) of fields injected over multiple
round trips. The field transmitted by the input mirror constitutes a source term for
the field circulating intracavity, while the loss term is given at least by the leakage

Fig. 1.3 Transient response
of a cavity mode excited by a
monochromatic laser wave.
The ideal laser is switched on
at the origin of time and the
output of a lossless ideal
cavity is calculated as a
function of time for different
mismatches between the laser
frequency and the resonance.
After a transient behavior,
revealing damped oscillations
with a period proportional to
inverse of the frequencies
mismatch, a stationary state
corresponding to the cavity
mode amplitude is attained

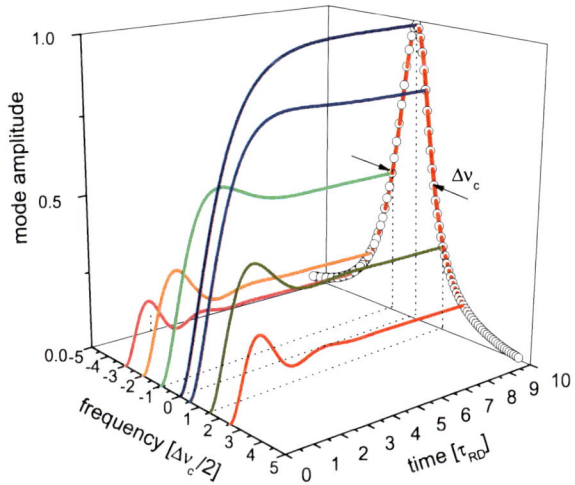

Fig. 1.3 Transient response of a cavity mode excited by a monochromatic laser wave. The ideal laser is switched on at the origin of time and the output of a lossless ideal cavity is calculated as a function of time for different mismatches between the laser frequency and the resonance. After a transient behavior, revealing damped oscillations with a period proportional to inverse of the frequencies mismatch, a stationary state corresponding to the cavity mode amplitude is attained

outside the cavity through both mirrors, if we neglect the absorption and scattering losses, or sample absorption.

At exact resonance, the input field interferes constructively with a previously injected field persisting in the cavity. Starting with an empty cavity, as the incoming field is switched on, the source term dominates over the (initially null) leaking field and the intracavity field grows. After enough round trips the buildup will be large enough to make the loss term equal to the source term. The stationary condition corresponding to maximum transmission of a cavity mode is then obtained.

For an optical frequency slightly detuned above or below the cavity resonance, the input field transmitted by the entrance mirror has a small phase difference with the field injected one round trip earlier. This shift cumulates with the number of considered round trips. If there were no field attenuation, fields consecutively injected over few round trips would interfere constructively while interference would become destructive with fields in the past whose cumulated phase shift is close to π. For longer times, constructive interferences would reappear, and so forth. But due to losses the intracavity field still reaches a stationary state given the exponentially smaller amplitudes of field components injected further in the past. As long as the phase shift is small compared to π the field buildup is still relevant. For larger detuning the oscillation becomes faster and the steady state is established at a lower buildup level. This behavior is illustrated by simulations in Fig. 1.3. For higher reflectivity mirrors, the reduced field damping makes interference effects last longer with tolerance for a smaller mismatch between an optical frequency and the exact resonance resulting in a reduced cavity mode linewidth.

These oscillations before reaching the steady state admit a perhaps surprising physical interpretation. They can be considered as a beating note of the incoming field frequency with the cavity resonance frequency. When the field is switched on abruptly, its frequency is not at once well defined, therefore initially the cavity begins to be excited even at resonance. Before the field frequency becomes defined better with respect to its separation from the cavity resonance, a time equal to the reciprocal of the frequency difference must elapse, and up to this time the field builds up also at resonance. At that point a beating starts appearing between a transient excitation of the resonance, which dies off exponentially, and the buildup at the incoming frequency which continues up to steady state. The same kind of argument allows to understand that the cavity will produce a ring-down decay signal centered at the resonance frequency even though the excitation occurs off resonance [17]. As it can be confirmed by a rigorous calculation based on Laplace transform theory, a condition to observe a free ring-down decay is that the interruption of the incoming field is faster than the ring-down time itself [17]. Then, the cavity response time dominates the field extinction time and also determines the resolution to which the spectrum of the decaying field can be defined, given by the inverse of this time and coinciding with the cavity resonance profile.

We can now consider several input waves of equal intensity and with frequencies uniformly distributed around the resonance, which are switched on together and produce each a buildup as in Fig. 1.3. This corresponds to the case of a spectrally broad source with respect to the cavity resonance. During the first round trip, all input waves are multiplied by t and the intracavity field is weak but uniform: No interference effect has yet occurred. As time passes, interference occurs over more and more round trips. The intracavity spectral components closer to resonance continue to build up for longer times, while the further away the earlier the buildup terminates (with oscillations around a lower and lower asymptotic field level). As a result, interference makes the intracavity field to develop a spectrum narrower and narrower in time until full buildup and a stationary state is attained. In this "spectral evolution" it can be shown that the product of an intracavity field linewidth and the corresponding buildup time (the time from the beginning of the broad band injection, during which the cavity analyzes the input field) conforms to the FT limit ($\geq 1/\pi$).

This picture can however be misleading. It should be realized that the spectral envelope one can deduce from Fig. 1.3 for a given buildup time cannot be defined over time intervals shorter than that very buildup time. Due to the uncertainty relation there is no such thing as an "instantaneous" intracavity spectral envelope. If one tries to analyze the spectral envelope of the cavity output field at a buildup time t_b, one will need an observation time of at least t_b before being able to attain a spectral resolution sufficient to resolve that profile.[7]

Considerations of this type allow an intuitive physical understanding of cavity excitation as well as the realization that even under pulsed excitation, with a broad

---

[7] Any spectrograph will present an effective observation time inversely proportional to its resolution. An interesting case is the grating spectrograph, where the measurement time corresponds to the difference in delay of light paths reaching the observation plane after being diffracted at the opposite edges of the grating.

input spectrum and a short excitation time, the intracavity field (and cavity transmission) has a spectrum composed by narrow resonances whose width can be defined only over the long cavity response time. As an illustration, consider a pulse injected in the cavity and shorter than its round-trip time. The cavity output will appear composed of replicas of this pulse spaced by the round trip time ($2L_c$ divided by the group velocity, which is the propagation speed of the pulse inside the cavity, accounting for dispersion effects). The whole train of decaying pulses will present a comb-like spectrum made of the excited cavity resonances lying under the initial pulse spectral envelope. But if one analyzes each replica individually (by using a fast optical switch), only the envelope will be there, and no cavity modes. Likewise, molecules inside the cavity need to see the whole time series of ringing-down pulses in order to produce an observable response which takes into account the interference effects which create the cavity resonances and the almost null average field value in the frequency intervals between them. As we will see later for the case of frequency combs, the periodicity in frequency of the cavity resonances is not accidental but corresponds by FT to the time periodicity of the pulse train. We can also understand now that the cavity mode spacing in the presence of intracavity dispersion is given by the inverse of the cavity round-trip time calculated using the group velocity.[8]

It is now time to consider a monochromatic laser wave continuously tuned across a fixed cavity resonance, as required by several CEAS techniques. We would expect to recover the spectral cavity mode profile after converting the tuning time to frequency by using the scanning speed W in units Hz/s. However for this to be true, W should be slow enough that for each frequency there is time to achieve the stationary regime of the intracavity field. This adiabatic regime is realized when the duration of the passage through the resonance (width $\Delta\nu_c$) is much longer than τ_{RD}, the cavity response time: $W \ll \Delta\nu_c/\tau_{RD}$. Above this tuning speed as the input frequency approaches the resonance the field starts building up, but without attaining the steady state it starts decreasing by destructive interference after the tuning frequency has gone across the resonance. At the same time a beating appears between the input field and the field partially built-up at the resonance, which exponentially decays. In addition, the phase shift considered above between successively injected fields is now increasing in time quadratically as the frequency mismatch increases linearly, thus the beating has blue-chirped oscillations (see Fig. 1.4). This ringing effect is readily observable with a narrow line laser and moderate cavity finesse [174] and has been the subject of several studies [175–178]. Finally, in the limit of large W the regime of impulsive excitation is attained, with a short buildup time resulting in a small output amplitude, and a time response approaching a single exponential decay with superimposed fast and negligibly small oscillations. This excitation regime has enabled fast CRDS measurement with CW lasers using a simplified experimental setup [50, 51, 53, 179–181], notably avoiding the need for a fast optical switch.

The widespread CW-CRDS scheme exploits the larger cavity buildup generally obtained with a smaller tuning speed and requires a fast optical switch (typically an

---

[8]See also footnotes 4 and 5.

Fig. 1.4 Injection efficiency of a lossless cavity as a function of the relative laser-cavity frequency scanning speed W expressed in natural cavity units $\Delta v_c / \tau_{RD}$. Different sections of this curve are associated with a specific injection regime and transmission profile

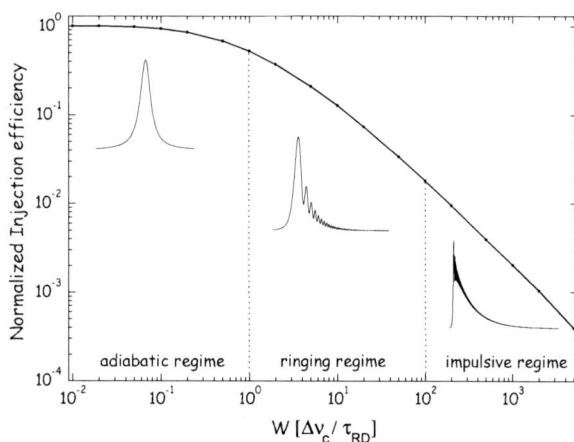

acoustooptic deflector, but see [44]) placed at the cavity input to interrupt the injection when sufficient signal is detected at cavity output during a passage through resonance. In practice, rather than tuning the laser and obtain one resonance per cavity FSR, it is usually more convenient to modulate the cavity mode position across the fixed laser frequency. This is obtained by placing one of the cavity mirrors on a piezoelectric mount to modulate the cavity length. This also enables a finer tuning if the laser frequency can be continuously adjusted. However spectral resolution is no more limited by the cavity mode width (and jitter from vibrations and drift) but by the laser linewidth and jitter, and additionally by the scanning cavity mode frequency. Concerning the temporal profile at cavity output during a passage through resonance, it can be shown that this situation is strictly equivalent to the previous one. The pertinent parameter is the relative laser-cavity frequency change. This equivalence [178] rests on the correspondence $v/L_c = W/v$ where v is the mirror speed and v is the optical frequency at resonance. The growing phase shift now originates from the Doppler shift of the wave at each reflection off the moving mirror. The only difference lies in the spectral analysis of a ring-down event obtained after switching off the laser beam. In case of a laser frequency sweep, the optical frequency of the ring-down signal is determined by the frequency of the resonance, which is stationary even if it may drift over time scales usually longer than the ring-down event. In case of a cavity length sweep, the frequency of the ring-down signal is determined by a resonance frequency which matches the laser frequency when injection occurs, but which continues its scan after laser interruption and over the duration of the ring-down. For a typical finesse used in CRDS and scanning speeds satisfying a compromise between signal level and acquisition rate, one estimates a frequency shift of several % of the cavity FSR during the ring-down. Such a limitation to the spectral resolution, however, can be mitigated using a tracking system that modulates the cavity resonance over a small range around the laser frequency. This also allows increasing the frequency of passages through resonance using a lower frequency scanning speed, while reducing spurious injection events due to transverse modes (which would produce different ring-down times).

Previous discussions, and in particular the observation of the ringing effect, are based on the explicit assumption of a perfect monochromatic source. In practice this situation is approached if the laser coherence time τ_L is longer than the time of passage through the cavity resonance, which in addition should be shorter than the cavity response time. In the frequency domain, this corresponds to a laser linewidth $\Delta\nu_{las}$ narrow compared to the inverse of the time of passage. Unfortunately, especially when considering tunable semiconductor lasers, the situation is reversed: DFB diode lasers and ECDLs have short term linewidths (μs to ms timescales) in the MHz or 0.1 MHz range, respectively, while CRDS often features ring-down times in excess of 10 μs, corresponding to mode linewidths of a few tens of kHz, and below.

The spectral line profile of a CW laser is not a simple concept. It could be considered as a description of the fluctuations of the laser frequency, a sort of distribution of the instantaneous laser frequency values weighted by their temporal occurrence. It reflects the fluctuations of physical parameters determining the laser frequency, such as the laser cavity length, the laser pumping process (the injection current for a diode laser), the dye jet thickness for a dye laser, *etc*. Besides technical fluctuations, the very process of laser emission produces the fundamental Schawlow-Townes linewidth [182], due to spontaneous emission by the laser gain: Spontaneous photons with a random phase add over time to the coherent photon flux circulating inside the laser cavity and produce small phase jumps (random in time and magnitude), and therewith an average diffusion of the phase of the laser field. In this limit it can be shown that the laser lineshape is a Lorentzian of width $\Delta\nu_{las}$ proportional to the ratio of the spontaneous emission rate over the intracavity laser photon flux. The inverse of this is the laser coherence time τ_L corresponding to the time needed in average for a phase diffusion of one radian. The presence of technical noise changes the laser line profile and renders it time dependent, in particular with a distribution of emitted frequencies becoming broader over longer observation times.

In order to get a sound picture of what the laser spectral profile represents and how it relates to the phase noise and its spectral distribution, it would be necessary to refer to the exact definition of power spectral density of a field as the FT of its time autocorrelation function, which is not an easy subject. However, it was recently remarked that this admits a simplified understanding, with a computational scheme allowing to easily go from the phase/frequency noise spectrum to the emission linewidth [183], whose validity was confirmed experimentally [184].

Here, it will be sufficient to retain that a CW laser can be described realistically as a "locally monochromatic" carrier wave affected by a random-walk of the phase (a diffusion process resulting in an average phase drift proportional to $\sqrt{t}$) induced by spontaneous emission, plus a (smoother) technical drift of the carrier frequency itself (e.g. as the laser cavity drifts). This frequency change integrated over time also results in a phase drift, but linear or even quadratic in time. The characteristic time over which phase drift remains small (relative to 1) is defined as the laser coherence time τ_L: For longer times the accumulated phase change becomes distributed uniformly over 2π, and the autocorrelation function drops to zero.

When a CW laser is tuned across a cavity resonance, let us name τ_p the characteristic time given by the cavity linewidth divided by the relative laser-cavity tuning speed, $\tau_p = \Delta \nu_c / W$. This represents the minimum time of passage through resonance that would be realized for a monochromatic wave, while a noisy laser wave with a broader frequency distribution will usually take a longer time. Let us start with τ_p larger than the cavity response time τ_{RD}. If this is also larger than the laser coherence time τ_L, the wave admitted in the cavity will experience phase diffusion during the buildup. When this diffusion attains π the incoming wave will be interfering destructively with the field previously constructed inside the cavity, resulting in the intracavity field annihilation before starting the buildup of a new field with a new relative phase. This process has been reproduced experimentally by inducing abrupt π phase jumps in a long coherence time laser [185]. Such sudden large phase changes were actually observed to induce a cavity field decay sensibly faster than the ring-down time.

Now, a τ_L larger than τ_{RD}, but still smaller than τ_p, gives a regime where the cavity is able to attain a full buildup after which the resonance conditions continues for some time. The intracavity field will then follow adiabatically the slow laser phase/frequency drifts. In this regime the cavity output may still present amplitude fluctuations as the wave central frequency (the carrier) fluctuates relative to the cavity mode linewidth during τ_p. Also, in this regime the cavity output noise spectrum will be limited by the cavity response bandpass given by $1/2\pi \tau_{RD}$, and the output will drop to zero only when the carrier will go completely out of resonance.

On the other hand, as the laser tuning speed is increased until τ_p becomes comparable, then smaller than τ_L, the cavity output will become closer and closer to the monochromatic limit, irrespective of τ_{RD}, and eventually appear unaffected by laser phase or frequency noise [178]. The ringing effect will be observable in this case if the monochromatic limit is achieved for not too high a tuning speed. Otherwise, an impulsive excitation will be obtained with a ringing too small and too fast to be observable.

Let us now turn to the subject of cavity injection efficiency. At a first sight, cavity output intensity would appear to be determined by considering that a part of the laser spectrum has been sampled by the cavity mode, without accounting for the frequency/amplitude noise projection described above. For a laser linewidth $\Delta \nu_{las}$ larger than the cavity resonance linewidth $\Delta \nu_c$, the maximum power fraction transmitted by a cavity is then taken to be $\Delta \nu_c / \Delta \nu_{las}$. This matches observations only if the time of passage is larger than τ_{RD}, and additionally, only considering time-averaged values over an ensemble of transmission profiles in order to wash out the large amplitude fluctuations that will be observed. In fact, because of the spiky character of transmission in the presence of fast laser phase fluctuations (if τ_L is smaller than other timescales) the effective maximum cavity transmission will be several times higher than the average value set by $\Delta \nu_c / \Delta \nu_{las}$, as seen in Fig. 1.5. For frequency tuning speeds such that the time of passage drops below τ_{RD}, this behavior

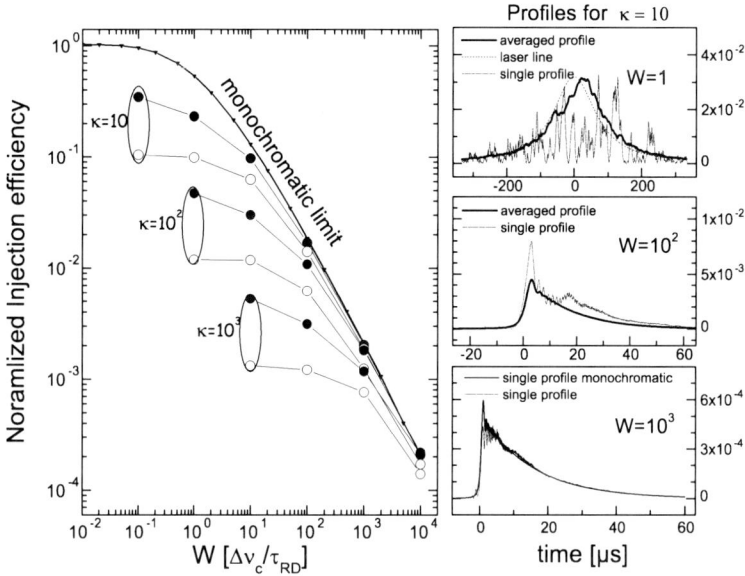

Fig. 1.5 *Left panel*: Injection efficiency for different laser linewidths. Data are extracted from simulated profiles as shown in the *right panel*. *Open circle* corresponds to maxima of averaged profiles, *full circle* corresponds to the averaged value of single profile maxima. *Right panel*: Simulated profiles obtained by a passage through resonance when the laser linewidth is ten times the cavity mode linewidth, $\kappa = \Delta\nu_{las}/\Delta\nu_c = 10$, for different values of the scanning speed (in cavity units). For $W = 1$ and 10^2 one-shot and averaged profiles (over 200 samples) are shown. For $W = 10^3$ a direct comparison of the one-shot profile with the monochromatic limit becomes possible. All simulation come from the model detailed in [178]

persists but the average transmission becomes inversely proportional to W due to a frustrated buildup, then the fluctuations disappear when the monochromatic limit is attained (for τ_L longer than the time of passage).

After these rather qualitative remarks, we conclude this section with a short review of more technical studies which have been devoted to the cavity excitation with CW lasers and especially to the modelling of the intracavity optical field. This was a general need in contexts as diverse as high resolution Fabry-Perot analyzers [174, 175] (in order to interpret asymmetric response profiles), measurement of laser cavity finesse [51, 176] (for values of $\mathcal{F}$ too high for the usual Airy-peak width measurement method and too low for easy ring-down time determination, especially with short cavities), gravitational wave detection based on optical resonators [177, 186] (as the Pound-Drever-Hall error signal used to correct resonator length variations is distorted at high correction speed), in cavity-QED [185] (in which the atom's trajectory could be better controlled by using fast changes of the photon field), and of course in the frame of CEAS developments [50, 178] to improve the understanding of laser-cavity coupling.

In these studies, three different approaches were exploited. One is based on the transfer function formalism in the time domain, where cavity output results as the convolution of the incident field with the cavity impulsive response. Another uses the field expressed as a sum of contributions from each round trip where a fast recursive algorithm can be used to describe the field evolution. The last approach is a continuum version of the previous one, as it uses differential equations for the intracavity field evolution. All these studies agree perfectly with experience in the case of the monochromatic wave approximation, but only three of them address the question of the laser linewidth. Li and co-workers were the first to attempt taking the laser linewidth into account [175] but naively they used the inverse FT of the laser line to describe its field. This actually leads to a temporally limited optical field which is inconsistent with a CW laser. An alternative treatment was proposed by Hahn and co-worker [50] who used a weighted sum over all the monochromatic temporal profiles calculated for frequencies composing the laser lineshape. This clearly does not reproduce the cavity output fluctuations but reproduces quite correctly the coupling efficiency as a function of the scanning speed and for different laser linewidths. Finally, Morville et al. [178] developed a model with the laser linewidth as a phase diffusion process easily included in a field expression as a series as in Eq. (1.1). This enables to reproduce experimental profiles for all configurations, that is to say, for all laser linewidths and scanning speeds, as long as the correct noise model is used in the calculation. For an analytical treatment see also [43].

1.4.2 Multi-mode Injection (Longitudinal and Transverse): Incoherent Pulsed Lasers

Until now, the simple and widespread CEAS scheme also named ICOS [63, 64] where the cavity transmission is monitored (or integrated by using a slow-response photodetector) as the laser is tuned, has not been considered since it is usually implemented without mode matching, given that multi-transverse-mode excitation delivers a smoother cavity output. In that case however, one would expect to observe a periodic transmission pattern, since the same sequence of transverse modes is excited as the laser tunes over each cavity FSR: In principle these transverse modes should be excited always with the same efficiency given by the transverse superposition integral of the incoming beam with each transverse mode profile. In practice, laser phase/frequency noise may induce strong fluctuations well above this periodic structure. Such fluctuations are often reported as a problem in CEAS, and are usually attributed to cavity acoustic fluctuations, which usually play a minor role and only at timescales much longer than the typical duration of a passage through resonance. From the above discussion it is evident that in order to reduce such noise effects it should be enough to use a sufficiently fast laser scan, as illustrated by Bakowski et al. [179]. These authors obtained in CW-CRDS a detection limit close to what other groups demonstrated with similar cavity characteristics but with a much slower acquisition time. On the other hand, excessively fast frequency tuning may induce a distortion of the absorption line profiles when using CEAS, due to the limited cavity response time.

Another CEAS scheme, namely off-axis ICOS, positively exploits transverse mode excitation by off-axis alignment of the laser beam [65–67]. As mentioned earlier in Sect. 1.2.1, a cavity with large astigmatic mirrors is required, with its length adjusted in order to obtain specific trajectories whose impact points on the mirrors form a Lissajous-like figure, rather than just a circle or an ellipse as in the basic Herriot cell. Here we can add to that understanding the fact that a given laser injected into such a cavity is characterized by a finite linewidth and an associated coherence length, as we learned from the previous section. The laser beam launched into an intracavity trajectory folding onto itself after N round trips, with N possibly as large as 100, will have lost its coherence and will possess a random phase with respect to new incoming radiation, resulting in zero average interference. This means that as the laser frequency is tuned the cavity transmission will basically present no resonances. An equivalent point of view is that the cavity FSR is divided by N equally spaced groups of degenerate resonances, and FSR/N may be smaller than the laser linewidth so that the cavity mode structure appears to the laser like an effective continuum transmission. However in practice this would demand very large mirrors and an excessive precision in their surface quality and the adjustment of their distance to satisfy the reentrant condition well enough, especially in the limit of high finesse. Using an astigmatic cavity configuration allows for larger N with still reasonably small mirrors, since the N distinct impact spots are now distributed over a Lissajous folded curve with a large perimeter extending uniformly over the entire mirror surface, rather than being constrained over the shorter perimeter of a circle running close to the mirror edges. As we mentioned in Sect. 1.2.1, the exact degeneracy in N groups of modes is broken with an astigmatic cavity: But this is not a penalty because of sub-groups of still degenerate modes which are strongly clustered to form a regular structure of mode groups uniformly filling the cavity FSR period [164, 165]. In addition, the fact that as the laser scans its beam keeps describing the same folded trajectory, implies that the mirror inhomogeneity does not make a difference. Indeed the mirror spatial regions used for the multiple reflections are constant as the laser frequency scans, contrary to what happens when a non reentrant cavity configuration is used. To conclude about the off-axis ICOS technique, what we just exposed appear to be the principal reasons underlying the excellent performance demonstrated with off-axis ICOS and also justify the need for a large transverse cavity size (using mirrors of up to 5 cm diameter), which was experimentally illustrated by Engel et al. [67]. On the other hand, this large transverse cavity size is a drawback because of the large sample volume, and the need for a large detector surface for collecting the spatially extended cavity output. In the mid-IR such a detector has to be cryogenically cooled in order to obtain a reasonably low noise figure and a high bandpass.

We can now turn to another important class of developments where the laser source usually imposed multimode cavity injection, both transverse and longitudinal. Pulsed (ns) dye lasers allow very wide spectral coverage with a resolution (down to and below 0.05 cm^{-1}) sufficient to study the rotational structure of most light molecules. They are capable of fast continuous tuning by the synchronized tilt of a grating and of an etalon present in the laser cavity. By using different dyes and nonlinear frequency doubling or tripling techniques, or Raman shifting, the whole visible range is accessible and even the near UV or the near IR are at reach. Pulsed ns Ti:Sapphire, Optical Parametric Oscillators (OPO) and other solid state laser systems extend the accessible spectral range and possess similar properties compared to dye lasers, thus should also be included in this discussion.

The high peak power available with ns lasers is often used to produce, exploit, or investigate nonlinear effects in optics or molecular spectroscopy, on the other hand these lasers suffer of a poor shot-to-shot stability not only in intensity but also in the beam profile which is far from Gaussian. In particular ns lasers cannot be used for the simplest absorption spectroscopy measurement with a sensitivity better than 10^{-3} (in terms of the detectable intensity loss). However, CRDS provides high detection sensitivity without penalty from amplitude fluctuations, at least in principle, and it fully profits of the high peak power available: As a laser pulse is transmitted through the input cavity mirror it is strongly attenuated but still enables linear absorption measurements by detecting the cavity transient response, i.e. a ring-down decay profile. The variations of the ring-down time due to sample absorption are usually detectable at the few % level even in the presence of source intensity and spatial fluctuations (and these can be strongly reduced using spatial filtering as discussed below). A 1 % variation in the ring-down of a high finesse cavity then corresponds to a very small sample absorption. Since the visible spectral range and the near UV are populated by weak vibrational overtone transitions of molecules of atmospheric or planetological interest, pulsed CRDS is a working tool highly appreciated by spectroscopists.

When considering how CRDS is implemented using a pulsed laser, we easily realize that another advantage is simplicity: It is enough to align two high reflectivity concave mirrors along the laser beam, at a separation L_c chosen well inside the stability region of the cavity [159]. Then a photodiode with high gain (typically 10^5–10^6 A/W, 1 MHz bandpass) is used to detect the cavity output signal.

We have already discussed how a cavity responds to an external excitation in the limit of a short light pulse. In the frequency domain we have to consider that the spectral width of ns lasers is typically larger than the cavity FSR (assuming $L_c \sim 1$ m) which implies that at any laser frequency at least one longitudinal cavity mode can always be excited. In particular a pulse shorter than the cavity round trip time will possess a spectrum with no structure finer than the cavity FSR. However, more rigorously we should consider the laser coherence time, which is shorter or equal to the pulse duration. Equality of pulse length and coherence time occurs for a FT-limited laser spectrum, but this is rarely realized and in general CRDS presents no real difficulty with laser pulses which are longer than the cavity round trip time [27]. Indeed, even when a pulsed laser is built to be FT limited, we still have to consider that transverse cavity modes may transmit at different frequencies. Thus in order to really observe clear effects of cavity resonances one has to use good mode matching in other to excite only the $TEM_{0,0}$ (longitudinal) cavity modes. It is then possible to observe no cavity transmission when the pulse frequency falls in between two modes [187]. A corollary of this is that if a molecular absorption line falls in between two longitudinal modes, mode matched cavity injection would make this line to be missing in the CRDS spectrum.

Pulsed CRDS easily provides measurement of the ring-down time at the 1 % level after reasonable signal averaging using relatively quick and dirty cavity injection. When using cavities of high finesse, easily exceeding 10^4 in the visible range, with ring-down times $\tau_{RD} \sim 10$ µs when using $L_c = 1$ m ($\tau_{RD} \simeq L_c/[c(1 - \mathcal{R})] = \mathcal{F}/\pi \cdot L_c/c$), the detection limit should then be on the order of 3×10^{-8}/cm, in terms of the smallest detectable, or noise-equivalent, absorbance for each spectral data point. This is easily seen by considering the basic equation stating that for an absorption coefficient α by the intracavity sample the ring-down time is given by $1/\tau_{RD} = c(1 - \mathcal{R})/L_c + c\alpha$, from which we deduce that the "noise equivalent" absorption is $\alpha_{ne} = (\delta\tau/\tau_{RD})/(c\tau_{RD})$.

In order to optimize the sensitivity in CRDS and go beyond such basic 'warranted' detection limit, special care must be taken with respect to mode matching, which requires spatial filtering. Without such extra efforts, pulsed CRDS is usually limited by wide fluctuations in the transverse cavity excitation patterns (fluctuation of the ensemble of excited transverse modes) obtained at each laser pulse, with non-exponential cavity output signals as a result of the beating between transverse modes and their different exponential decays. Transverse cavity modes are orthogonal so that the intensity produced by their superposition should be equal to the sum of the intensities of each mode which are all exponentially decaying after pulsed cavity excitation. However spatially non-uniform detector response (some photomultipliers are particularly bad in this respect), and mirror surface non-uniform loss or transmission, break the orthogonality and reveals the beatings [10] as it has been discussed by several authors [30–32, 42] and even exploited to monitor the transverse cavity excitation and optimize the mode matching [188–190].

Spatial filtering is a corollary of mode matching when using pulsed ns lasers. This may be obtained by using a pinhole, with typical diameter of 50 to 100 µm, through which the laser beam is focussed by choosing a lens producing a spot larger than the pinhole, so that at the focal point the beam may be considered as a plane wave which is limited by the pinhole edges and produces a diffraction Airy pattern in the far field. Focussing on the pinhole is critical: if the focal spot is too large much pulse energy will be lost, and if it is too small the beam will go through the pinhole without attenuation but without spatial filtering either. An adjustable diaphragm can be used in the far field to eliminate the rings of the Airy pattern leaving the central lobe which is close to a Gaussian profile. This cleaned quasi-Gaussian beam may then be imaged to match the cavity $TEM_{0,0}$ mode by another properly chosen lens (at appropriate distances from the pinhole and the cavity), then by looking at the cavity output spot shape and size it is possible to optimize the beam and cavity alignment. Another spatial filtering aid is using Brillouin scattering in a liquid [191], which has the advantage of eliminating the amplified spontaneous emission (ASE), spectrally broad incoherent radiation accounting for a sizable fraction of the laser power. ASE is source of artifacts in CRDS when absolute absorption measurements are desired, since it produces a multispectral ring-down component insensitive to spectrally narrow absorption lines.

1.4.3 Multi-mode Injection (Longitudinal): Coherent Pulsed Lasers, or Frequency Combs

As we mentioned in the historical outline (Sect. 1.1.1), broad-band (BB) implementations of CEAS using LEDs, lamps, or supercontinuum sources, coupled with multispectral detection at cavity output, have been successfully applied to the detection of trace species over a broad spectral range. The developments and multiple applications of BB-CEAS will be described in a dedicated chapter of this book. The principal limitation of spatially and/or spectrally incoherent sources is the inefficient cavity injection, linearly degrading with the cavity mirror transmission coefficient. The resulting power spectral density available at cavity output makes it hard to obtain very high spectral resolution over reasonable acquisition times and using reasonably high finesse cavities. Mode-Locked (ML) lasers are an expensive solution to this problem. Below we will find it convenient to first focus on the different ways of coupling a ML laser to a high finesse cavity, then to separately present different ways of measuring the spectrum transmitted by the cavity, even though the two aspects are not decoupled.

ML lasers are known to generate very short pulses, from picoseconds down to femtoseconds, with a very stable repetition rate. A few types are now available to cover the near infrared and red spectral domains directly, and nonlinear frequency conversion techniques are very efficient due to the high peak power of these sources and allow accessing a much broader spectrum. To fix ideas, we will consider a Ti:Sapphire ML laser with pulse duration $\tau = 30$ fs, a repetition rate $f_{rep} = 80$ MHz, and a spectrum centered around 800 nm. The short pulse duration corresponds, if one considers the FT uncertainty principle, to a broad laser emission spectrum spanning about 30 nm (the FWHM $\Delta\nu$ is then ~ 15 THz, with $\Delta\nu \times \tau = 0.44$ for a FT-limited Gaussian pulse), as broad as the spectrum emitted by a LED. However, again due to FT properties, it is easy to understand that the periodicity in time of laser pulses must also give a periodic signature in the frequency domain. Indeed the broad ML laser spectrum consists of a comb of narrow peaks centered at uniformly spaced frequencies, a periodic structure composed by some 180k modes in our example, whose frequency interval is equal to the laser repetition rate. The uniformity of the mode spacing of an Optical Comb (OC) may be quite remarkable, better than a few parts in 10^{17} [192, 193]. However, as mentioned in Sect. 1.1.1, the frequencies of the OC teeth are not an integer multiple of the repetition rate. This is a consequence of the difference between the phase velocity and the group velocity of the pulse propagating inside the laser cavity, which results in a relative delay between the carrier wave and the pulse envelope which accumulates from pulse to pulse. Consequently, the OC has a frequency offset f_0 as in Fig. 1.6, and the frequency of mode N is given:

$$\nu_N = f_0 + N f_{rep}. \tag{1.26}$$

In a free running OC, laser phase noise results in a short term (ms) linewidth of comb modes ranging from below 100 kHz for fiber oscillators up to around

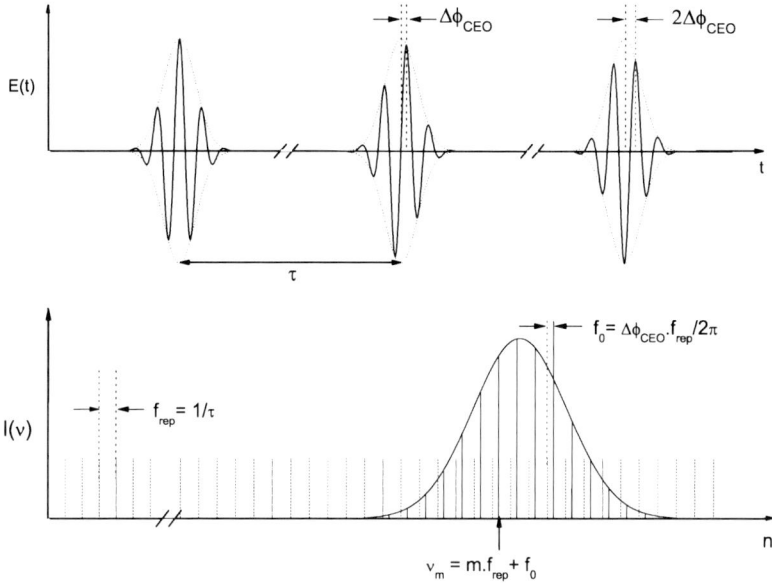

Fig. 1.6 Time-frequency correspondence of pulses delivered by a mode-locked laser (picture was inspired from Ref. [194], but not copied directly)

1 MHz for Ti:Sapphire oscillators. However, stabilization techniques have pushed the linewidth and/or the definition of modes frequencies to subhertz levels [195–197]. The consequences are remarkable in the metrology domain with respect to atomic clocks and optical frequency standards and measurements, and made the object of a Nobel prize in 2006 [198, 199].

Since an OC is clearly an easy match for a comb of cavity resonances, different comb-CEAS coupling schemes have been introduced and developed. We will generically refer to these as OC-CEAS techniques. The basic coupling scheme is obtained when the cavity FSR is exactly equal to an integer multiple of the laser repetition rate. For instance, if the high-finesse cavity length is half the laser cavity length, a choice usually leading to a reasonable cavity size, then one every two laser modes is transmitted by the cavity while the rest will be reflected (a 50 % waste of laser power). In the time domain, each laser pulse entering the cavity (by partial transmission through the input mirror) will add up to the next laser pulse after two cavity round trips, during which it will be partially transmitted by both cavity mirrors and produce two replicas at the output (of slightly decreasing amplitude). The cavity will then multiply the laser repetition rate by a factor 2. This property has been exploited to increase the laser mode spacing by a large factor [103] to allow an easier discrimination of successive frequency channels. If the cavity length is tuned across such a comb matching position, the laser spectrum envelope transmitted as a whole as an intense light burst, which can be periodically reproduced by cavity length modulation. Alternatively, the cavity length may be servo-locked to this position, which is more challenging but yields a continuous transmission with a high signal level. The application of the simpler configuration to absorption spectroscopy was demonstrated in 2002 by Gherman et al. [105]. The second configuration had to wait 2011 to be implemented by a group well established in frequency metrology [124].

On the other hand, when the cavity length L_c is not at such an exact "global resonance" but is at δL from this point, the cavity transmitted spectrum presents a beating pattern which correspond to modes going periodically in and out of resonance [105]. A partial match of the two combs then occurs only by groups of modes spaced by the beating period $\delta \nu_b = c/2\delta L$. This "Vernier effect" can be exploited for a combined cavity injection and dispersion scheme. Gohle et al. in 2007 [112] actually succeeded resolving single comb modes by further combination of Vernier with multiplexed dispersion on a CCD array, while Thorpe et al. [113] reported low resolution Vernier CEAS spectra taken with a single element detector.

Up to this point we did not consider cavity dispersion effects nor the frequency offsets of both cavity and laser combs. Indeed, contrary to the laser OC, the free spectral range (FSR) of the cavity is not perfectly constant. The first reason is a phase shift θ at the reflection on the mirrors (the field coefficient of reflection $\mathbf{r}$ is complex and $\theta = \arg(\mathbf{r})$) and this phase shift depends on wavelength. This is due to the fact that the coating of high reflective mirrors is a stack of alternate $\lambda/4$ dielectric layers of alternate high/low refraction index which constitute a complex interferometric system with a penetration depth of the incoming field which is frequency dependent. Even though it is possible to engineer non periodic dielectric stacks to reduce this effect, this comes at the expense of the reflectivity value, the mirror losses, and also the flatness of the mirror spectral response, and a compromise has to be accepted. A second reason is that, as soon as the intracavity sample presents absorption lines, there is a dispersion associated to these because of the principle of causality (from which one derives the Kramers-Kronig relations between absorption and dispersion for a linear response system). This dispersion effect becomes particularly strong in the proximity of absorption lines, but may present a broad band dependence induced by intense absorption bands lying outside the spectral window of interest.

Both effects lead to a cavity FSR which is frequency dependent, and to an offset of the comb of cavity resonances (this is a local offset appearing when trying to approximate the resonances with a uniform comb, thus, it is also frequency dependent). To show this we consider the resonance condition for a cavity of length L_c:

$$k_m L_c - \theta = m\pi \quad \Rightarrow \quad \frac{2\pi n_r \nu_m}{c} \times L_c = m\pi + \theta \quad \Rightarrow$$

$$\nu_m = m \frac{c}{2n_r L_c} + \frac{\theta c}{2\pi n_r L_c} \quad \Rightarrow \quad \nu_m = m\mathrm{FSR}(\nu_m) + f_{0,\mathrm{cav}}(\nu_m). \quad (1.27)$$

Since FSR and $f_{0,\mathrm{cav}}$ are weakly frequently dependent through the refraction index $n_r(\nu_m)$ and $\theta(\nu_m)$, the value of ν_m is the solution of this nonlinear equation (readily obtained by numerical iteration). For normal high reflectivity mirrors and an empty cavity, one finds that change in the cavity mode spacings add up to a mismatch with

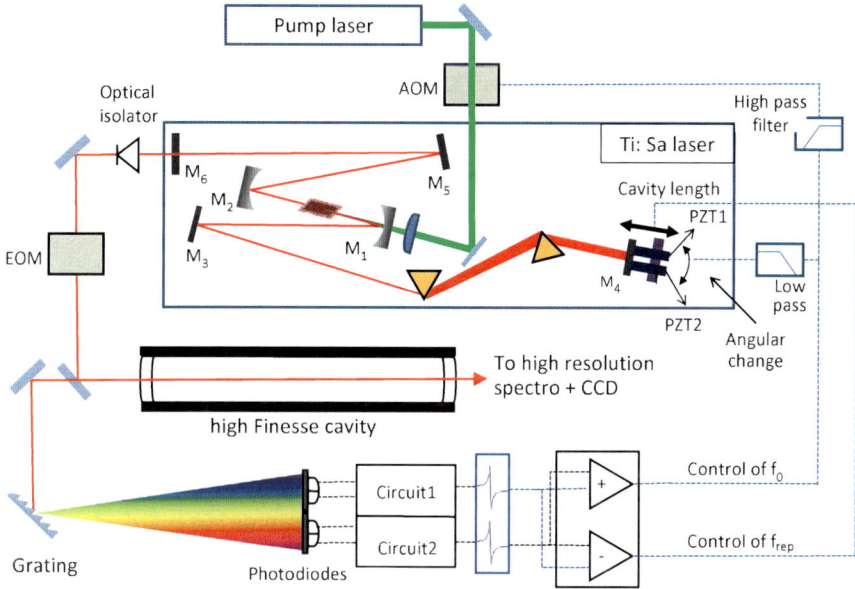

Fig. 1.7 Principle scheme of tight lock of a laser comb to a high-finesse cavity. An optical isolator protects the laser from any feedback. The electro-optic phase modulator EOM is used to obtain the locking error signal from two Pound-Drever-Hall arrangements working in parallel at different wavelengths, selected by a grating (each error signal is still produced by thousands comb teeth). f_{rep} is here controlled by a change in the cavity length by a piezoelectric transducer and f_0 is controlled at low frequency by a piezo tilt of the mirror M_4 which acts on the group velocity dispersion. This is doubled by a high frequency control of f_0 by a change of laser pump intensity using an acoustooptic modulator

the comb teeth in the MHz range for a 1 % frequency variation in the flat central region of the mirrors stopband [200]. We will see that this results in a spectral filtering effect limiting the spectral region over which an OC can be simultaneously transmitted by the cavity (in particular this occurs when using comb to cavity frequency locking) [113].

After these general remarks we would like to review in more detail the different strategies to achieve a coupling between laser and cavity combs, which may be classified as follows:

• Static continuous cavity injection by comb frequency locking;
• Transient injection at the passage through a comb quasi-resonance;
• Vernier effect in the presence of a slight combs spacing mismatch.

Maximizing cavity injection over the broadest possible spectrum requires to lock the two OC parameters, the repetition rate f_{rep} and the frequency offset f_0, for which two error signals are needed. An appealing scheme, represented in Fig. 1.7, is to duplicate and run simultaneously the same standard laser-cavity frequency locking method at two different wavelengths inside the OC spectrum [196, 201, 202]. The frequency locking method of choice is typically the Pound-Drever-Hall (PDH) scheme [171], which can be easily coupled with a grating which spatially separates two spectral windows at the two ends of the OC spectrum. An advantage of the PDH scheme is its wide bandpass which is not limited by the cavity response time as is the case with any scheme which would exploit the cavity transmitted signal. In addition, a high S/N error signal is produced by two frequency modulation sidebands heterodyning with a strong carrier field leaking out of the cavity resonance used for the locking (once the locking kicks in).

On the other hand, a general limiting factor of frequency locking schemes is the bandpass of the servo actuators available to control the OC parameters. The use of piezoelectric actuators to adjust the laser cavity length, for example, may limit the control bandpass on $f_{\rm rep}$ to 10 kHz or less (see however [203]). Some fast controls do exist however (such as using an acoustooptic modulator to introduce a frequency shift equivalent to f_0) and can be used to achieve a reduction of the laser phase noise up to sufficiently high frequencies to obtain a linewidth narrowing of the comb teeth at all timescales. Indeed, a locking acting only on a low frequency region of the laser noise spectrum will allow to have zero drift of the relative comb-cavity frequency differences with very good statistics (typically resulting in an Allan variance dropping steadily with averaging time), but will not reduce the laser linewidth. The CEAS signal will then suffer from the conversion of laser phase noise to amplitude noise which we discussed in Sect. 1.4.1, even though this will be low-pass filtered by the cavity and can be efficiently removed in some detection schemes [124].

Another difficulty with tight frequency locking is associated with cavity dispersion and the frequency walk-off of the cavity modes with respect to the OC teeth. Thus, high transmission can only be maintained over a limited spectral range which tends to become narrower as the cavity finesse is increased or the locking is made tighter which is in a way equivalent to make the OC "finesse" to also increase (this may be defined as the OC $f_{\rm rep}$ divided by the OC modes linewidth). As we mentioned earlier, the use of mirrors with compensation of the group velocity dispersion only mitigates the problem and imposes a compromise on the attainable finesse.

In addition, at the edges of the transmitted OC spectral window, distortion of molecular absorption lines by local dispersion effects is expected and was observed [124]. Even though this can be quite well modelled, it adds complexity and uncertainty to the exploitation of the recorded spectra. This effect is easily understood by considering that at the edges of the transmitted portion of the OC spectrum, the cavity modes are systematically offset by a fraction of their linewidth then cavity transmission of the OC modes does not occur at the peak of the cavity modes. Then, a small shift of the cavity modes by the increase and decrease of the sample refraction index on opposite sides of an absorption line will result in an increase or decrease of transmission thus giving the observed asymmetric profiles [124].

Another effect which does not appear to have been considered until now, concerns the cavity enhancement factor that we discussed in Sect. 1.2.1. As we explained there, if the linewidth of the OC is comparable to that of the cavity resonances, the enhancement factor β lies in between the monochromatic case $2\mathcal{F}/\pi$ and the broad-band case $\mathcal{F}/\pi$, which poses a problem of calibration of the intensity scale of the measured molecular absorption spectra. It appears plausible that static OC-CEAS schemes should be affected by this artifact unless a high-bandpass tight lock guarantees that the OC teeth linewidth is well below that of cavity modes (resulting in a well-defined and favorable $2\mathcal{F}/\pi$ enhancement factor).

Notwithstanding these difficulties, Foltynowicz et al. recently achieved shot-noise limited performance by using static OC-CEAS with a fast scanning FT inter-ferometer coupled with a balanced-detector noise cancelling system, detecting 10k spectral elements in a 6 s measurement time [124], thus covering a 6 THz window around 1.55 μm with 380 MHz resolution.

The second strategy involves transient OC-CEAS coupling, or ML-CEAS as ini-tially named [105], where ML refers to the laser source and not to the way the OC teeth are coupled to the cavity, since no actual frequency locking is involved. In ML-CEAS, the cavity length is modulated around a global comb resonance in a way that each cavity mode is rapidly swept over the corresponding OC tooth. It is also possible, and almost equivalent (see Sect. 1.4.1), to let the cavity length fixed and modulate the OC. One advantage of this scheme is that the whole laser spectrum is transmitted by the cavity with no penalty from cavity dispersion: no broadband spectral filtering, and no local spectral distortion. This also allows using standard high reflectors which may be optimized to attain the highest finesse without con-straints. Even though a broad OC spectrum can be coupled through the cavity over a short time scale (down to a few μs if necessary), cavity injection results in average quite less efficient than in the locked scheme, depending on the fraction of the cavity FSR over which the comb modulation occurs. For a modulation as large as one cav-ity FSR, the injection efficiency is as low as for a broad-band spectrally incoherent source, but still remains the advantage of spatial coherence of an OC, which allows mode matched injection of the cavity $TEM_{0,0}$ modes. By the same token, the cavity enhancement factor is well defined but assumes the lower value of $\mathcal{F}/\pi$ valid for broad-band or swept sources (see Sect. 1.4.1) [98].

This coupling scheme is vastly simpler to implement and more robust than a tight locking technique. In particular, only one comb parameter needs to be actively driven by a fixed modulation (constant amplitude and frequency) plus a bias which is controlled in order to maintain the global resonance of the combs at the center of the modulation range.[9] The width of the modulation usually has to be increased in the presence of stronger acoustic perturbations, and the modulation frequency has to be high enough so that the passages through resonance belong to the monochromatic regime discussed in Sect. 1.4.1. We recall that in this regime all phase/frequency laser noise is not converted into amplitude noise at cavity output, contrary to what occurs at slower tuning speed.

Above, we referred to this scheme by using the term "quasi-resonance" because the active control of a single comb parameter does not allow to achieve optimal comb matching and a true simultaneous resonance of all comb teeth. The method

---

[9]A drift of the other comb parameter (typically f_0) will only affect the match of the combs and the width of the transmitted peak at the passage through resonance, with no impact on the measurement as long as all the comb modes go through resonance with the cavity, even if at different times.

works well even when using a secondary resonance, i.e. if the cavity comb modulation is centered around one of the peaks occurring close to the best comb match resonance (where the only residual mismatch is due to non optimal f_0 plus cavity dispersion). Indeed injection of the whole OC works well as long as the mismatch between the combs spacings is sufficiently small that there is a single beating period in frequency space between the two combs. As long as there is only one group of modes being transmitted at a time, the cavity modulation just induces a sweep of this beating across the OC spectrum. Still in other words, the final result of modulating around such a quasi-resonance is the same as using a more perfect comb match since for each modulation sweep all cavity modes will go through resonance once with their corresponding OC tooth. The only disadvantage is that a larger modulation range is required and the resulting average transmission decreases accordingly.

This ML-CEAS scheme has been implemented by sweeping the cavity length thanks to a piezoelectric transducer [105] providing 6 GHz spectral resolution over a 2 THz spectral window (330 spectral elements) in 40 ms acquisition time, using a moderate cavity finesse. It was then successfully tested for a few applications at different wavelengths with higher cavity finesse, in particular to access blue wavelengths and detect ions and radicals in a plasma discharge [106] or measure weak overtone vibrational transitions [107]. These applications fully profited of the efficient and direct frequency doubling of fs pulses.

Perhaps surprisingly, this simple scheme is capable of shot noise limited performance [98] during routine operation, with low signal levels (10 nW) available at the output of a very high finesse cavity ($\mathcal{F} = 32$k), however sufficient to fill the CCD array detector during the shortest available exposure and readout times (10 ms). This kind of detector is optimal to accumulate in parallel spectral channels the CEAS spectrum dispersed by a grating spectrograph, and is one of the key factors of this performance (given the small readout noise and high quantum efficiency). The other factor is the fast cavity modulation which eliminates spurious laser frequency/phase technical noise. The number of simultaneous spectral elements available is limited by the size of a linear CCD array detector, however two-dimensional dispersion schemes (like VIPA[10]) coupled with a CCD matrix should allow to acquire many more spectral elements with similar performance.

While the previous schemes require a resolving spectrograph at cavity output and a CCD detector, the third OC-CEAS configuration is able to economize these components by exploiting the properties of combs to achieve high spectral resolution with a number of spectral elements limited by the cavity finesse and ultimately by the number of comb teeth. A cavity displacement δL with respect to the optimal comb resonance position may be introduced on purpose, to obtain a cavity FSR slightly different of the laser repetition rate f_{rep}. A comb beating with period $\Delta \nu_b = c/2\delta L$ will be obtained in the spectral domain at cavity output, with each beat containing a group of modes in resonance with OC teeth and producing a

---

[10]Virtually Imaged Phased Array: Basically, a tilted glass etalon which strongly disperses the frequencies of a light beam in the plane of the tilt, usually then coupled with a grating dispersing in the orthogonal direction [204].

transmission peak in frequency. The number of modes being in resonance is smaller for faster beating and larger mismatch, but will also depend on the cavity and OC finesses. If the cavity modes are swept by acting on the cavity length, the position of the beatings will be swept and cover a complete beating period when the modes run over one cavity or laser FSR, whichever is smaller. If the comb mismatch is important, it is possible to selectively transmit a single OC tooth per beating period, but it will be necessary to introduce again a dispersive system and multiplexed (CCD) detection and combine it with the scan of the cavity length to retrieve successive absorption spectra, which will demand to be interlaced to obtain a complete mode-resolved spectrum. For instance, Gohle et al. [112] used a rotating mirror to impress successive spectra over successive lines of a CCD matrix as the cavity length was continuously tuned. The single cavity modes appeared as distinct dots on the CCD image. They demonstrated a resolution of 1 GHz over a bandwidth of 4 THz (4k spectral elements) with an acquisition time of only 10 ms. For this result they used a frequency stabilized OC which should allow in principle to achieve a frequency accuracy well below 1 GHz for each data point.

In the opposite limit of a small mismatch, only one group of modes may be transmitted at a time by the cavity as soon as $\Delta \nu_b > \Delta \nu_{\text{laser}}$ is satisfied, and spectral analysis can be obtained just using one photodiode [113]. In that case, many consecutive teeth are transmitted simultaneously by the cavity. Still Thorpe et al. reported a resolution of 5 GHz over a bandwidth of 1.6 THz (320 spectral elements) with a sweep time of 220 ms. This Vernier effect could be pushed to achieve higher resolving power over a broader laser spectrum by increasing the cavity finesse and the relative cavity-laser stability thanks to a tight locking which induces a comparable or higher OC finesse.

After considering injection schemes, we are now going to outline different detection schemes that have been coupled with OC-CEAS. For the first and second coupling schemes, cavity transmission should be analyzed with a high resolving power spectrometer. Different implementations have been tested, each presenting advantages and drawbacks. The easiest appears to be the diffractive spectrometer equipped with a linear CCD array, possessing a few thousands pixels at most. The number of spectral elements will then be limited, especially if low spectral distortion is desired, since then the spectrograph apparatus function should be wider than the pixel size. A respectable resolution for gas phase spectroscopy is then only possible over a relatively narrow spectral window [105, 110, 205].

To increase the number of spectral elements, spectral dispersion can be performed in two spatial dimensions, over a CCD matrix detector, by using a VIPA in combination with a grating [204, 206]. It is thus possible to increase the number of spectral elements and to analyze a broadband spectrum at high resolution [109]. However, using this technique beyond the near-infrared may be a challenge because of the lower sensitivity (and much higher price) of matrix detectors at longer wavelengths. Another difficulty appears to be that the VIPA is very sensitive to alignment drifts, which demands frequent acquisition of calibrations spectra [138].

An alternative to VIPA is to use a FT Spectrometer as we mentioned earlier. In that case, the detector is a simple photodiode [122, 124], but a mechanically sophisticated Michelson setup is needed to generate the interferogram. Kassi et al.

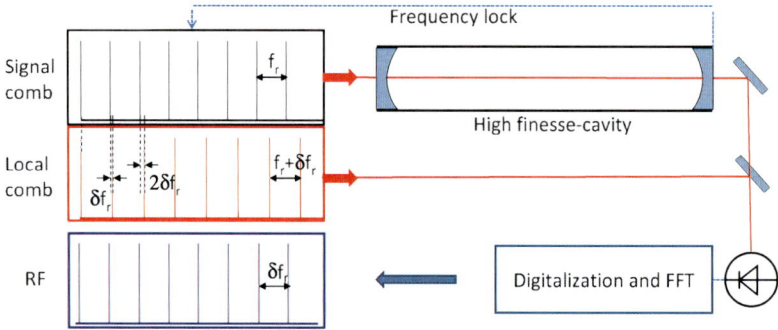

Fig. 1.8 Principle scheme of multiheterodyne detection. The absorption spectrum is shifted and compressed to the radiofrequency domain

in 2010 used commercial FT and OC systems to achieve 3 GHz resolution over 2.4 THz (800 spectral elements), with an acquisition time of minutes. A nice feature of a Michelson interferometer is that two complementary signals can be obtained whose combination eliminates common mode amplitude fluctuations. This is the feature exploited by Foltynowicz et al. [124] to obtain photon shot-noise limited FT cavity-enhanced spectra from a home-made fast FT system, as mentioned above. An important detail is that to obtain undistorted FT spectra, it is necessary that the complete spectrum to be analyzed is available and reproducible during each acquisition step of the interferometer, which either demands that the OC is steadily locked to the cavity or that it is modulated across a global comb resonance at a sufficiently high repetition rate [98, 122], since for the rest, the stability of the spectral envelope of an OC laser is usually excellent. This detection scheme was also readily exploited in the mid-infrared by the use of a commercial FT spectrometer and femtosecond OPO [123].

Spectral analysis is also possible using multi-heterodyne detection (Fig. 1.8), which delivers an FT interferogram without using moving parts and over short measurement times [114–116]. Two frequency combs with slightly different repetition rates are employed, which corresponds to have a different teeth spacing for the two combs. Thus the gap between closest teeth from both combs may be arranged to be continuously increasing going from one to the other end of the optical spectrum covered by the two combs. When superposing the two laser beams on a fast detector, each pair of teeth gives a beating note at a specific frequency in the radio-frequency domain. The product spectrum of the two optical combs is then directly mapped to a compressed RF spectrum. The principle has been demonstrated with direct absorption spectroscopy by Coddington et al. and Giaccari et al. in 2008 [117, 118] and with a high finesse cavity one year later by Bernhardt et al. [120]. In this demonstration, a first comb is locked to a cavity and the transmitted beam is then mixed with the second comb. The acquisition time for a single frequency scan depends on the mismatch between repetition rates of the two laser combs. Bernhardt et al. achieved a resolution of 4.5 GHz over 7.5 THz around 1 μm, corresponding to 1500 spectral elements in a recording time of 550 μs.

1.5 Conclusion

In this long introductory chapter we first presented a historical overview of the development of cavity enhanced techniques for absorption spectroscopy in molecular physics, in trace detection and isotope ratio analysis. This overview shows that most forms of CEAS have reached today a good maturity and are fruitful in many domains of application. However, the field is still rich of possibilities to be developed and exploited. In particular, we think that modelocked laser sources present a most promising playing ground.

We then introduced the basic properties of high finesse optical cavities, which are of interest to different forms of CEAS, and which we therefore treated with special attention to some specific details. We also indulged in a detailed consideration of the interaction of different types of laser sources with an optical cavity in view of optimizing injection for CEAS measurements. This has led us to discuss some implementations in more details than others, certainly biased by our own experience.

We leave to the following chapters of this book the task of discussing many more necessary technical details of different aspects of CEAS, from a specialist point of view and with more attention to the applications. We hope this chapter will serve as a sufficiently good starting point for the interested reader who wished to pursue this fascinating subject matter.

Acknowledgements   We would like to thank Kevin Lehmann and Marco Prevedelli for their critical reading of the manuscript and the useful discussions concerning several subtle issues.

References

1. G.D. Boyd, J.P. Gordon, Bell Syst. Tech. J. **40**(2), 489 (1961). http://www3.alcatel-lucent.com/bstj/vol40-1961/articles/bstj40-2-489.pdf
2. G.D. Boyd, H. Kogelnik, Bell Syst. Tech. J. **41**(4), 1348 (1962). http://www3.alcatel-lucent.com/bstj/vol41-1962/articles/bstj41-4-1347.pdf
3. D.A. Jackson, Proc. R. Soc. A, Math. Phys. Eng. Sci. **263**(1314), 289 (1961). doi:10.1098/rspa.1961.0161. http://rspa.royalsocietypublishing.org/cgi/doi/10.1098/rspa.1961.0161
4. P. Connes, J. Phys. Radium **19**(3), 262 (1958). doi:10.1051/jphysrad:01958001903026200. http://www.edpsciences.org/10.1051/jphysrad:01958001903026200
5. A. Kastler, Appl. Opt. **1**(1), 17 (1962). doi:10.1364/AO.1.000017
6. A.G. Fox, T. Li, Proc. IEEE **51**(1), 80 (1963). doi:10.1109/PROC.1963.1663. http://ieeexplore.ieee.org/lpdocs/epic03/wrapper.htm?arnumber=1443593
7. H. Kogelnik, T. Li, Proc. IEEE **54**(10), 1312 (1966). doi:10.1109/PROC.1966.5119. http://ieeexplore.ieee.org/lpdocs/epic03/wrapper.htm?arnumber=1447049
8. H. Kogelnik, T. Li, Appl. Opt. **5**(10), 1550 (1966). doi:10.1364/AO.5.001550. http://www.opticsinfobase.org/abstract.cfm?URI=ao-5-10-1550

9. J.A. Arnaud, H. Kogelnik, Appl. Opt. **8**(8), 1687 (1969). doi:10.1364/AO.8.001687. http://www.opticsinfobase.org/abstract.cfm?URI=ao-8-8-1687

10. J.P. Goldsborough, Appl. Opt. **3**(2), 267 (1964). doi:10.1364/AO.3.000267

11. R.L. Fork, D.R. Herriott, H. Kogelnik, Appl. Opt. **3**(12), 1471 (1964). doi:10.1364/AO.3.001471. http://www.opticsinfobase.org/abstract.cfm?URI=ao-3-12-1471

12. G. Rempe, R.J. Thompson, H.J. Kimble, R. Lalezari, Opt. Lett. **17**(5), 363 (1992). http://www.ncbi.nlm.nih.gov/pubmed/19784329

13. R. Damaschini, Opt. Commun. **20**(3), 441 (1977). doi:10.1016/0030-4018(77)90225-5. http://linkinghub.elsevier.com/retrieve/pii/0030401877902255

14. J.M. Herbelin, J.a. McKay, M.a. Kwok, R.H. Ueunten, D.S. Urevig, D.J. Spencer, D.J. Benard, Appl. Opt. **19**(1), 144 (1980). http://www.ncbi.nlm.nih.gov/pubmed/20216808

15. R. Engeln, G. von Helden, G. Berden, G. Meijer, Chem. Phys. Lett. **262**(1–2), 105 (1996). doi:10.1016/0009-2614(96)01048-2. http://linkinghub.elsevier.com/retrieve/pii/0009261496010482

16. J.H. van Helden, D.C. Schram, R. Engeln, Chem. Phys. Lett. **400**(4–6), 320 (2004). doi:10.1016/j.cplett.2004.10.081. http://linkinghub.elsevier.com/retrieve/pii/S0009261404016768

17. D.Z. Anderson, J.C. Frisch, C.S. Masser, Appl. Opt. **23**(8), 1238 (1984). http://www.ncbi.nlm.nih.gov/pubmed/18204709

18. A. Kastler, Nouv. Rev. Opt. **5**(3), 133 (1974). http://iopscience.iop.org/0335-7368/5/3/301

19. J.Y. Lee, J.W. Hahn, Appl. Phys. B, Lasers Opt. **79**(3), 371 (2004). doi:10.1007/s00340-004-1550-2. http://www.springerlink.com/index/10.1007/s00340-004-1550-2

20. T.M. Crawford, in *Southwest Conf. on Optics '85*, ed. by S.C. Stotlar. Proceedings of SPIE, vol. 0540 (1985), pp. 295–302. doi:10.1117/12.976129. http://proceedings.spiedigitallibrary.org/proceeding.aspx?doi=10.1117/12.976129

21. S.N. Jabr, T.M. Crawford, J. Opt. Soc. Am. A **1**(12), 1329 (1984). http://adsabs.harvard.edu/abs/1984JOSAA...1.1329J

22. A. O'Keefe, D.A.G. Deacon, Rev. Sci. Instrum. **59**(12), 2544 (1988). doi:10.1063/1.1139895. http://link.aip.org/link/RSINAK/v59/i12/p2544/s1&Agg=doi

23. D. Romanini, K.K. Lehmann, J. Chem. Phys. **99**(9), 6287 (1993). doi:10.1063/1.465866. http://link.aip.org/link/JCPSA6/v99/i9/p6287/s1&Agg=doi

24. J.J. Scherer, D. Voelkel, D.J. Rakestraw, J.B. Paul, C.P. Collier, R.J. Saykally, A. O'Keefe, Chem. Phys. Lett. **245**(2–3), 273 (1995). doi:10.1016/0009-2614(95)00969-B. http://linkinghub.elsevier.com/retrieve/pii/000926149500969B

25. J.J. Scherer, J.B. Paul, C.P. Collier, R.J. Saykally, J. Chem. Phys. **102**(13), 5190 (1995). doi:10.1063/1.469244. http://link.aip.org/link/JCPSA6/v102/i13/p5190/s1&Agg=doi

26. T. Yu, M.C. Lin, J. Am. Chem. Soc. **115**(10), 4371 (1993). doi:10.1021/ja00063a069 http://pubs.acs.org/doi/abs/10.1021/ja00063a069

27. G. Meijer, M.G.H. Boogaarts, R.T. Jongma, D.H. Parker, A.M. Wodtke, Chem. Phys. Lett. **217**(1–2), 112 (1994). doi:10.1016/0009-2614(93)E1361-J. http://linkinghub.elsevier.com/retrieve/pii/0009261493E1361J

28. P. Zalicki, R.N. Zare, J. Chem. Phys. **102**(7), 2708 (1995). doi:10.1063/1.468647. http://link.aip.org/link/JCPSA6/v102/i7/p2708/s1&Agg=doi

29. J.T. Hodges, J.P. Looney, R.D. van Zee, Appl. Opt. **35**(21), 4112 (1996). doi:10.1364/AO.35.004112. http://www.opticsinfobase.org/abstract.cfm?URI=ao-35-21-4112, http://www.opticsinfobase.org/abstract.cfm?&id=46719

30. K.K. Lehmann, D. Romanini, J. Chem. Phys. **105**(23), 10263 (1996). doi:10.1063/1.472955. http://link.aip.org/link/JCPSA6/v105/i23/p10263/s1&Agg=doi

31. J. Martin, B.A. Paldus, P. Zalicki, E.H. Wahl, T.G. Owano, J.S. Harris, C.H. Kruger, R.N. Zare, Chem. Phys. Lett. **258**(1–2), 63 (1996). doi:10.1016/0009-2614(96)00609-4. http://linkinghub.elsevier.com/retrieve/pii/0009261496006094

32. J.T. Hodges, J.P. Looney, R.D. van Zee, J. Chem. Phys. **105**(23), 10278 (1996). doi:10.1063/1.472956. http://link.aip.org/link/JCPSA6/v105/i23/p10278/s1&Agg=doi

33. R.D. van Zee, J.T. Hodges, J.P. Looney, Appl. Opt. **38**(18), 3951 (1999). http://www.ncbi.
    nlm.nih.gov/pubmed/18320004
34. D. Romanini, A.A. Kachanov, N. Sadeghi, F. Stoeckel, Chem. Phys. Lett. **264**(3–4), 316
    (1997). doi:10.1016/S0009-2614(96)01351-6. http://linkinghub.elsevier.com/retrieve/pii/
    S0009261496013516
35. D. Romanini, A.A. Kachanov, F. Stoeckel, Chem. Phys. Lett. **270**(5–6), 538 (1997). doi:10.
    1016/S0009-2614(97)00406-5. http://linkinghub.elsevier.com/retrieve/pii/S000926149700
    4065
36. D. Romanini, A.A. Kachanov, F. Stoeckel, Chem. Phys. Lett. **270**(5–6), 546 (1997).
    doi:10.1016/S0009-2614(97)00407-7. http://linkinghub.elsevier.com/retrieve/pii/S0009261
    497004077
37. R.M. Curran, T.M. Crook, D.J. Zook, MRS Proc. **105**, 175 (1987). doi:10.1557/
    PROC-105-175
38. B.A. Paldus, C.C. Harb, T.G. Spence, B. Willke, J. Xie, J.S. Harris, R.N. Zare, J. Appl.
    Phys. **83**(8), 3991 (1998). doi:10.1063/1.367155. http://link.aip.org/link/JAPIAU/v83/i8/
    p3991/s1&Agg=doi
39. J.T. Hodges, D. Lisak, Appl. Phys. B, Lasers Opt. **85**(2–3), 375 (2006). doi:10.1007/
    s00340-006-2411-y. http://www.springerlink.com/index/10.1007/s00340-006-2411-y
40. D. Lisak, J.T. Hodges, R. Ciurylo, Phys. Rev. A **73**(1), 1 (2006). doi:10.1103/PhysRevA.
    73.012507. http://link.aps.org/doi/10.1103/PhysRevA.73.012507
41. G. Giusfredi, S. Bartalini, S. Borri, P. Cancio, I. Galli, D. Mazzotti, P. De Natale, Phys.
    Rev. Lett. **104**(11), 1 (2010). doi:10.1103/PhysRevLett.104.110801. http://link.aps.org/
    doi/10.1103/PhysRevLett.104.110801
42. H. Huang, K.K. Lehmann, Opt. Express **15**(14), 8745 (2007). doi:10.1364/OE.15.008745.
    http://www.ncbi.nlm.nih.gov/pubmed/19547210
43. H. Huang, K.K. Lehmann, Appl. Phys. B, Lasers Opt. **94**(2), 355 (2009). doi:10.1007/
    s00340-008-3293-y. http://www.springerlink.com/index/10.1007/s00340-008-3293-y
44. H. Huang, K.K. Lehmann, Chem. Phys. Lett. **463**(1–3), 246 (2008). doi:10.1016/j.cplett.
    2008.08.030. http://linkinghub.elsevier.com/retrieve/pii/S0009261408011081
45. H. Huang, K.K. Lehmann, Appl. Opt. **47**(21), 3817 (2008). doi:10.1364/AO.47.003817.
    http://www.ncbi.nlm.nih.gov/pubmed/18641751
46. J. Courtois, J.T. Hodges, Opt. Lett. **37**(16), 3354 (2012). doi:10.1364/OL.37.003354.
    http://www.opticsinfobase.org/abstract.cfm?URI=ol-37-16-3354
47. M. Hippler, M. Quack, Chem. Phys. Lett. **314**(3–4), 273 (1999). doi:10.1016/S0009-
    2614(99)01071-4. http://linkinghub.elsevier.com/retrieve/pii/S0009261499010714
48. L. Biennier, D. Romanini, A.A. Kachanov, A. Campargue, B. Bussery-Honvault, R.
    Bacis, J. Chem. Phys. **112**(14), 6309 (2000). doi:10.1063/1.481192. http://link.aip.org/link/
    JCPSA6/v112/i14/p6309/s1&Agg=doi
49. P. Macko, D. Romanini, S.N. Mikhailenko, O.V. Naumenko, S. Kassi, A. Jenou-
    vrier, V.G. Tyuterev, A. Campargue, J. Mol. Spectrosc. **227**(1), 90 (2004). doi:10.1016/
    j.jms.2004.05.020. http://linkinghub.elsevier.com/retrieve/pii/S002228520400178X
50. J.W. Hahn, Y.S. Yoo, J.Y. Lee, J.W. Kim, H.W. Lee, Appl. Opt. **38**(9), 1859 (1999).
    http://www.ncbi.nlm.nih.gov/pubmed/18305817
51. K. An, C. Yang, R.R. Dasari, M.S. Feld, Opt. Lett. **20**(9), 1068 (1995). doi:10.1364/OL.
    20.001068. http://www.ncbi.nlm.nih.gov/pubmed/19859426, http://www.opticsinfobase.org/
    abstract.cfm?URI=ol-20-9-1068
52. Y. He, B.J. Orr, Chem. Phys. Lett. **319**(1–2), 131 (2000). doi:10.1016/S0009-
    2614(00)00107-X. http://linkinghub.elsevier.com/retrieve/pii/S000926140000107X
53. I. Debecker, A.K. Mohamed, D. Romanini, Opt. Express **13**(8), 523 (2005). http://www.
    opticsinfobase.org/abstract.cfm?URI=OPEX-13-8-2906
54. A. Cygan, D. Lisak, S. Wójtewicz, J. Domyslawska, J.T. Hodges, R. Trawinski, R. Ciurylo,
    Phys. Rev. A **85**(2), 1 (2012). doi:10.1103/PhysRevA.85.022508

55. M.D. Levenson, B.A. Paldus, T.G. Spence, C.C. Harb, J.S.J. Harris, R.N. Zare, Chem. Phys. Lett. **290**(4–6), 335 (1998). doi:10.1016/S0009-2614(98)00500-4. http://linkinghub.elsevier.com/retrieve/pii/S0009261498005004

56. Y. He, B.J. Orr, Chem. Phys. Lett. **335**(3–4), 215 (2001). doi:10.1016/S0009-2614(01)00031-8. http://linkinghub.elsevier.com/retrieve/pii/S0009261401000318

57. S. Kassi, A. Campargue, J. Chem. Phys. **137**(23), 234201 (2012). doi:10.1063/1.4769974. http://www.ncbi.nlm.nih.gov/pubmed/23267478

58. D. Romanini, P. Dupré, R. Jost, Vib. Spectrosc. **19**(1), 93 (1999). doi:10.1016/S0924-2031(99)00018-1. http://linkinghub.elsevier.com/retrieve/pii/S0924203199000181

59. H. Huang, K.K. Lehmann, Appl. Opt. **49**(8), 1378 (2010). doi:10.1364/AO.49.001378. http://www.opticsinfobase.org/abstract.cfm?URI=ao-49-8-1378

60. I. Galli, S. Bartalini, S. Borri, P. Cancio, D. Mazzotti, P. De Natale, G. Giusfredi, Phys. Rev. Lett. **107**(27), 1 (2011). doi:10.1103/PhysRevLett.107.270802. http://link.aps.org/doi/10.1103/PhysRevLett.107.270802

61. L. Gianfrani, G. Gagliardi, M. van Burgel, E.R.T. Kerstel, Opt. Express **11**(13), 1566 (2003)

62. G. Totschnig, D.S. Baer, J. Wang, Appl. Opt. **39**(12), 2009 (2000). http://www.opticsinfobase.org/abstract.cfm?id=60847

63. R. Engeln, G. Berden, R. Peeters, G. Meijer, Rev. Sci. Instrum. **69**(11), 3763 (1998). doi:10.1063/1.1149176. http://link.aip.org/link/RSINAK/v69/i11/p3763/s1&Agg=doi

64. A. O'Keefe, J.J. Scherer, J.B. Paul, Chem. Phys. Lett. **307**(5–6), 343 (1999). doi:10.1016/S0009-2614(99)00547-3. http://linkinghub.elsevier.com/retrieve/pii/S0009261499005473

65. J.B. Paul, L. Lapson, J.G. Anderson, Appl. Opt. **40**(27), 4904 (2001). http://www.ncbi.nlm.nih.gov/pubmed/18360533

66. D.S. Baer, J.B. Paul, M. Gupta, A. O'Keefe, Appl. Phys. B, Lasers Opt. **75**(2–3), 261 (2002). doi:10.1007/s00340-002-0971-z. http://www.springerlink.com/openurl.asp?genre=article&id=doi:10.1007/s00340-002-0971-z

67. G.S. Engel, W.S. Drisdell, F.N. Keutsch, E.J. Moyer, J.G. Anderson, Appl. Opt. **45**(36), 9221 (2006). http://www.ncbi.nlm.nih.gov/pubmed/17151763newsensitivitylimitsforabsorptionmeasurementsinpassiveopticalcavities.pdf

68. J. Morville, M. Chenevier, A.A. Kachanov, D. Romanini, in *Proceedings of SPIE*, vol. 4485, ed. by A.M. Larar, M.G. Mlynczak (2002), pp. 236–243. doi:10.1117/12.454256

69. J. Morville, D. Romanini, A.A. Kachanov, M. Chenevier, Appl. Phys. B, Lasers Opt. **78**(3–4), 465 (2004). doi:10.1007/s00340-003-1363-8. http://www.springerlink.com/openurl.asp?genre=article&id=doi:10.1007/s00340-003-1363-8

70. J. Morville, S. Kassi, M. Chenevier, D. Romanini, Appl. Phys. B, Lasers Opt. **80**(8), 1027 (2005). doi:10.1007/s00340-005-1828-z. http://www.springerlink.com/index/10.1007/s00340-005-1828-z

71. D. Romanini, M. Chenevier, S. Kassi, M. Schmidt, C. Valant, M. Ramonet, J. Lopez, H.J. Jost, Appl. Phys. B, Lasers Opt. **83**(4), 659 (2006). doi:10.1007/s00340-006-2177-2. http://www.springerlink.com/index/10.1007/s00340-006-2177-2

72. S. Kassi, M. Chenevier, L. Gianfrani, A. Salhi, Y. Rouillard, A. Ouvrard, D. Romanini, Opt. Express **14**(23), 11442 (2006). doi:10.1364/OE.14.011442

73. I. Ventrillard, T. Gonthiez, C. Clerici, D. Romanini, J. Biomed. Opt. **14**(6), 64026 (2009). doi:10.1117/1.3269677. http://www.ncbi.nlm.nih.gov/pubmed/20059264

74. E.R.T. Kerstel, R.Q. Iannone, M. Chenevier, S. Kassi, H.J. Jost, D. Romanini, Appl. Phys. B, Lasers Opt. **85**(2–3), 397 (2006). doi:10.1007/s00340-006-2356-1. http://www.springerlink.com/index/10.1007/s00340-006-2356-1

75. T.J.A. Butler, D. Mellon, J. Kim, J. Litman, A.J. Orr-Ewing, J. Phys. Chem. A **113**(16), 3963 (2009). doi:10.1021/jp810310b

76. V. Motto-Ros, J. Morville, P. Rairoux, Appl. Phys. B, Lasers Opt. **87**(3), 531 (2007). doi:10.1007/s00340-007-2618-6. http://www.springerlink.com/index/10.1007/s00340-007-2618-6

77. V. Motto-Ros, M. Durand, J. Morville, Appl. Phys. B, Lasers Opt. **91**(1), 203 (2008). doi:10.1007/s00340-008-2950-5. http://www.springerlink.com/index/10.1007/s00340-008-2950-5
78. D.J. Hamilton, M.G.D. Nix, S.G. Baran, G. Hancock, A.J. Orr-Ewing, Appl. Phys. B, Lasers Opt. **100**(2), 233 (2009). doi:10.1007/s00340-009-3811-6. http://www.springerlink.com/index/10.1007/s00340-009-3811-6
79. M. Hippler, C. Mohr, K.A. Keen, E.D. McNaghten, J. Chem. Phys. **133**(4), 44308 (2010). doi:10.1063/1.3461061. http://www.ncbi.nlm.nih.gov/pubmed/20687651
80. M. Durand, J. Morville, D. Romanini, Phys. Rev. A **82**(3), 031803(R) (2010). doi:10.1103/PhysRevA.82.031803. http://link.aps.org/doi/10.1103/PhysRevA.82.031803
81. G. Maisons, P. Gorrotxategi Carbajo, M. Carras, D. Romanini, Opt. Lett. **35**(21), 3607 (2010). doi:10.1364/OL.35.003607
82. D.J. Hamilton, A.J. Orr-Ewing, Appl. Phys. B, Lasers Opt. **102**(4), 879 (2010). doi:10.1007/s00340-010-4259-4. http://www.springerlink.com/index/10.1007/s00340-010-4259-4
83. J. Ye, L.S. Ma, J.L. Hall, J. Opt. Soc. Am. B **15**(1), 6 (1998). http://www.opticsinfobase.org/abstract.cfm?id=35318
84. N.J. van Leeuwen, A.C. Wilson, J. Opt. Soc. Am. B **21**(10), 1713 (2004). doi:10.1364/JOSAB.21.001713. http://www.opticsinfobase.org/abstract.cfm?URI=JOSAB-21-10-1713
85. M.S. Taubman, T.L. Myers, B.D. Cannon, R.M. Williams, Spectrochim. Acta, Part A, Mol. Biomol. Spectrosc. **60**(14), 3457 (2004). doi:10.1016/j.saa.2003.12.057. http://www.ncbi.nlm.nih.gov/pubmed/15561632
86. B.M. Siller, M.W. Porambo, A.A. Mills, B.J. McCall, Opt. Express **19**(24), 24822 (2011). http://www.ncbi.nlm.nih.gov/pubmed/22109511
87. F.M. Schmidt, A. Foltynowicz, W. Ma, T. Lock, O. Axner, Opt. Express **15**(17), 10822 (2007). doi:10.1364/OE.15.010822
88. P. Ehlers, I. Silander, J. Wang, O. Axner, J. Opt. Soc. Am. B **29**(6), 1305 (2012). doi:10.1364/JOSAB.29.001305. http://www.opticsinfobase.org/abstract.cfm?URI=josab-29-6-1305
89. P.K. Dasgupta, J.S. Rhee, Anal. Chem. **59**, 783 (1987). http://onlinelibrary.wiley.com/doi/10.1002/cbdv.200490137/abstract, http://pubs.acs.org/doi/abs/10.1021/ac00132a022
90. J.J. Scherer, J.B. Paul, H. Jiao, A. O'Keefe, Appl. Opt. **40**(36), 6725 (2001). http://www.ncbi.nlm.nih.gov/pubmed/18364983
91. S.M. Ball, I.M. Povey, E.G. Norton, R.L. Jones, Chem. Phys. Lett. **342**(1–2), 113 (2001). doi:10.1016/S0009-2614(01)00573-5. http://linkinghub.elsevier.com/retrieve/pii/S0009261401005735
92. S.E. Fiedler, A. Hese, A.A. Ruth, Chem. Phys. Lett. **371**(3–4), 284 (2003). doi:10.1016/S0009-2614(03)00263-X. http://linkinghub.elsevier.com/retrieve/pii/S000926140300263X
93. S.M. Ball, J.M. Langridge, R.L. Jones, Chem. Phys. Lett. **398**(1–3), 68 (2004). doi:10.1016/j.cplett.2004.08.144. http://linkinghub.elsevier.com/retrieve/pii/S0009261404014009
94. P.S. Johnston, K.K. Lehmann, Opt. Express **16**(19), 15013 (2008). http://www.ncbi.nlm.nih.gov/pubmed/18795038
95. S.E. Fiedler, A. Hese, A.A. Ruth, Rev. Sci. Instrum. **76**(2), 23107 (2005). doi:10.1063/1.1841872. http://link.aip.org/link/RSINAK/v76/i2/p023107/s1&Agg=doi
96. G.A. Marcus, H.A. Schwettman, Appl. Opt. **41**(24), 5167 (2002). doi:10.1364/AO.41.005167. http://www.opticsinfobase.org/abstract.cfm?URI=ao-41-24-5167
97. N.R. Newbury, I. Coddington, W.C. Swann, Opt. Express **18**(8), 7929 (2010). http://www.ncbi.nlm.nih.gov/pubmed/20588636
98. R. Grilli, G. Méjean, C. Abd Alrahman, I. Ventrillard, S. Kassi, D. Romanini, Phys. Rev. A **85**(5), 1 (2012). doi:10.1103/PhysRevA.85.051804. http://link.aps.org/doi/10.1103/PhysRevA.85.051804
99. H. Moosmüller, Appl. Opt. **37**(34), 8140 (1998). doi:10.1364/AO.37.008140. http://www.opticsinfobase.org/abstract.cfm?URI=ao-37-34-8140

100. G. Engel, W.B. Yan, J. Dudek, K.K. Lehmann, P. Rabinowitz, in *Laser Spectroscopy XIV International Conference*, ed. by R. Blatt, J. Eschner, D. Leibfried, F. Schmidt-Kaler (World Scientific, Singapore, 1999), pp. 314–315

101. K.K. Lehmann, High-finesse optical resonator for cavity ring-down spectroscopy based upon Brewster's angle prism retroreflectors (1999). http://www.boliven.com/patent/US5973864

102. A.C.R. Pipino, J.W. Hudgens, R.E. Huie, Rev. Sci. Instrum. **68**, 2978 (1997). doi:10.1063/1.1148230

103. T. Udem, J. Reichert, R. Holzwarth, T.W. Hansch, Phys. Rev. Lett. **82**(18), 3568 (1999). doi:10.1103/PhysRevLett.82.3568. http://link.aps.org/doi/10.1103/PhysRevLett.82.3568

104. E.R. Crosson, P. Haar, G.A. Marcus, H.A. Schwettman, B.A. Paldus, T.G. Spence, R.N. Zare, Rev. Sci. Instrum. **70**(1), 4 (1999). doi:10.1063/1.1149533. http://link.aip.org/link/RSINAK/v70/i1/p4/s1&Agg=doi

105. T. Gherman, D. Romanini, Opt. Express **10**(19), 1033 (2002). doi:10.1364/OE.10.001033

106. T. Gherman, E. Eslami, D. Romanini, S. Kassi, J.C. Vial, N. Sadeghi, J. Phys. D, Appl. Phys. **37**(17), 2408 (2004). doi:10.1088/0022-3727/37/17/011. http://stacks.iop.org/0022-3727/37/i=17/a=011?key=crossref.762430055a776caaec8cfaa62362d3df

107. T. Gherman, S. Kassi, A. Campargue, D. Romanini, Chem. Phys. Lett. **383**(3–4), 353 (2004). doi:10.1016/j.cplett.2003.10.148. http://linkinghub.elsevier.com/retrieve/pii/S0009261403019766

108. M.J. Thorpe, K.D. Moll, R. Jason Jones, B. Safdi, J. Ye, Science **311**(5767), 1595 (2006). doi:10.1126/science.1123921. http://www.ncbi.nlm.nih.gov/pubmed/16543457

109. M.J. Thorpe, D. Balslev-Clausen, M.S. Kirchner, J. Ye, Opt. Express **16**(4), 2387 (2008). doi:10.1364/OE.16.002387

110. G. Méjean, R. Grilli, C. Abd Alrahman, I. Ventrillard, S. Kassi, D. Romanini, Appl. Phys. Lett. **100**(25), 251110 (2012). doi:10.1063/1.4726190. http://link.aip.org/link/APPLAB/v100/i25/p251110/s1&Agg=doi

111. R. Grilli, M. Legrand, A. Kukui, G. Méjean, S. Preunkert, D. Romanini, Geophys. Res. Lett. **40** (2013). doi:10.1002/grl.50154

112. C. Gohle, B. Stein, A. Schliesser, T. Udem, T.W. Hansch, Phys. Rev. Lett. **99**(26), 1 (2007). doi:10.1103/PhysRevLett.99.263902. http://link.aps.org/doi/10.1103/PhysRevLett.99.263902

113. M.J. Thorpe, J. Ye, Appl. Phys. B, Lasers Opt. **91**(3–4), 397 (2008). doi:10.1007/s00340-008-3019-1. http://www.springerlink.com/index/10.1007/s00340-008-3019-1

114. S. Schiller, Opt. Lett. **27**(9), 766 (2002). doi:10.1364/OL.27.000766

115. F. Keilmann, C. Gohle, R. Holzwarth, Opt. Lett. **29**(13), 1542 (2004). http://www.ncbi.nlm.nih.gov/pubmed/15259740

116. A. Schliesser, M. Brehm, F. Keilmann, D.W. van der Weide, Opt. Express **13**(22), 9029 (2005). doi:10.1364/OPEX.13.009029

117. I. Coddington, W.C. Swann, N.R. Newbury, Phys. Rev. Lett. **100**(1), 11 (2008). doi:10.1103/PhysRevLett.100.013902. http://link.aps.org/doi/10.1103/PhysRevLett.100.013902

118. P. Giaccari, J.D. Deschênes, P. Saucier, J. Genest, P. Tremblay, Opt. Express **16**(6), 4347 (2008). http://www.ncbi.nlm.nih.gov/pubmed/18542532

119. J.D. Deschênes, P. Giaccari, J. Genest, Opt. Express **18**(22), 23358 (2010). doi:10.1364/OE.18.023358

120. B. Bernhardt, A. Ozawa, P. Jacquet, M. Jacquey, Y. Kobayashi, T. Udem, R. Holzwarth, G. Guelachvili, T.W. Hansch, N. Picqué, Nat. Photonics **4**(1), 55 (2009). doi:10.1038/nphoton.2009.217. http://www.nature.com/nphoton/journal/v4/n1/abs/nphoton.2009.217.html

121. D.W. Chandler, K.E. Strecker, J. Chem. Phys. **136**(15), 154201 (2012). doi:10.1063/1.3700473. http://www.ncbi.nlm.nih.gov/pubmed/22519318

122. S. Kassi, K. Didriche, C. Lauzin, X. de Ghellinck d'Elseghem Vaernewijck, A. Rizopoulos, M. Herman, Spectrochim. Acta, Part A, Mol. Biomol. Spectrosc. **75**(1), 142 (2010). doi:10.1016/j.saa.2009.09.058. http://www.ncbi.nlm.nih.gov/pubmed/19880347

123. X. de Ghellinck d'Elseghem Vaernewijck, K. Didriche, C. Lauzin, A. Rizopoulos, M. Herman, S. Kassi, Mol. Phys. **109**(17–18), 2173 (2011). doi:10.1080/00268976.2011.602990. http://www.tandfonline.com/doi/abs/10.1080/00268976.2011.602990

124. A. Foltynowicz, T. Ban, P. Maslowski, F. Adler, J. Ye, Phys. Rev. Lett. **107**(23), 1 (2011). doi:10.1103/PhysRevLett.107.233002

125. J.J. Scherer, J.B. Paul, A. O'Keefe, R.J. Saykally, Chem. Rev. **97**, 25 (1997). http://pubs.acs.org/doi/abs/10.1021/cr930048d

126. M.D. Wheeler, S.M. Newman, A.J. Orr-Ewing, M.N.R. Ashfold, J. Chem. Soc. Faraday Trans. **94**(3), 337 (1998). doi:10.1039/a707686j. http://xlink.rsc.org/?DOI=a707686j

127. G. Berden, P. Peeters, G. Meijer, Int. Rev. Phys. Chem. **19**(4), 565 (2000). http://www.tandfonline.com/doi/abs/10.1080/014423500750040627#.UjxKi38vQll

128. G. Berden, G. Meijer, W. Ubachs, in *Experimental Methods in the Physical Sciences*, vol. 40 (Elsevier, Amsterdam, 2003), pp. 47–82. doi:10.1016/S1079-4042(03)80018-8

129. B.A. Paldus, A.A. Kachanov, Can. J. Phys. **83**(10), 975 (2005). doi:10.1139/p05-054. http://www.nrcresearchpress.com/doi/abs/10.1139/p05-054

130. M.I. Mazurenka, A.J. Orr-Ewing, R. Peverall, G.A.D. Ritchie, Annu. Rep. Prog. Chem., Sect. C, Phys. Chem. **101**, 100 (2005). doi:10.1039/b408909j. http://xlink.rsc.org/?DOI=b408909j

131. C. Vallance, New J. Chem. **29**(7), 867 (2005). doi:10.1039/b504628a. http://xlink.rsc.org/?DOI=b504628a

132. K.W. Busch, M.A. Busch, *Cavity-Ringdown Spectroscopy* (American Chemical Society, Washington, 1999), pp. i–vii. doi:10.1021/bk-1999-0720.fw001. http://pubs.acs.org/doi/abs/10.1021/bk-1999-0720.fw001

133. R.D. van Zee, J.P. Looney (eds.), *Experimental Methods in the Physical Sciences*, vol. 40 (Academic Press, New York, 2003), pp. 1–323. doi:10.1016/S1079-4042(03)80015-2. http://www.sciencedirect.com/science/article/pii/S1079404203800152, http://www.sciencedirect.com/science/bookseries/10794042/40

134. G. Berden, R. Engeln, *Cavity Ring-Down Spectroscopy: Techniques and Applications* (Wiley-Blackwell, West Sussex, 2009)

135. S.M. Ball, R.L. Jones, Chem. Rev. **103**(12), 5239 (2003). doi:10.1021/cr020523k. http://www.ncbi.nlm.nih.gov/pubmed/14664650

136. C. Wang, J. Anal. At. Spectrom. **22**(11), 1347 (2007). doi:10.1039/B701223C

137. A. Foltynowicz, F.M. Schmidt, W. Ma, O. Axner, Appl. Phys. B, Lasers Opt. **92**(3), 313 (2008). doi:10.1007/s00340-008-3126-z. <GotoISI>://000258703600003

138. F. Adler, M.J. Thorpe, K.C. Cossel, J. Ye, Annu. Rev. Anal. Chem. **3**, 175 (2010). doi:10.1146/annurev-anchem-060908-155248. http://www.ncbi.nlm.nih.gov/pubmed/20636039

139. A. Foltynowicz, P. Maslowski, T. Ban, F. Adler, K.C. Cossel, T.C. Briles, J. Ye, Faraday Discuss. **150**, 23 (2011). doi:10.1039/c1fd00005e. http://xlink.rsc.org/?DOI=c1fd00005e

140. R.F. Curl, F. Capasso, C. Gmachl, A.A. Kosterev, J.B. McManus, R. Lewicki, M. Pusharsky, G. Wysocki, F.K. Tittel, Chem. Phys. Lett. **487**(1–3), 1 (2010). doi:10.1016/j.cplett.2009.12.073. <GotoISI>://WOS:000274432400001

141. H. Waechter, J. Litman, A.H. Cheung, J.A. Barnes, H.P. Loock, Sensors **10**(3), 1716 (2010). doi:10.3390/s100301716. <GotoISI>://WOS:000277158300016

142. M. Schnippering, S.R.T. Neil, S.R. Mackenzie, P.R. Unwin, Chem. Soc. Rev. **40**(1), 207 (2011). doi:10.1039/c0cs00017e. <GotoISI>://WOS:000285390900016ISI>://000285390900016

143. B.J. Orr, Y. He, Chem. Phys. Lett. **512**(1–3), 1 (2011). doi:10.1016/j.cplett.2011.05.052. http://linkinghub.elsevier.com/retrieve/pii/S0009261411006592

144. D.A. Long, A. Cygan, R.D. van Zee, M. Okumura, C.E. Miller, D. Lisak, J.T. Hodges, Chem. Phys. Lett. **536**, 1 (2012). doi:10.1016/j.cplett.2012.03.035. http://linkinghub.elsevier.com/retrieve/pii/S0009261412003466

145. K.K. Lehmann, H. Huang, in *Frontiers of Molecular Spectroscopy*, ed. by J. Laane (Elsevier, Amsterdam, 2009), pp. 623–658

146. S.S. Brown, Chem. Rev. **103**(12), 5219 (2003). doi:10.1021/cr020645c. http://www.ncbi. nlm.nih.gov/pubmed/14664649
147. M.W. Sigrist, R. Bartlome, D. Marinov, J.M. Rey, D.E. Vogler, H. Wachter, Appl. Phys. B, Lasers Opt. **90**(2), 289 (2008). doi:10.1007/s00340-007-2875-4. <GotoISI>://WOS: 000252990900019
148. M.N. Fiddler, I. Begashaw, M.A. Mickens, M.S. Collingwood, Z. Assefa, S. Bililign, Sensors **9**(12), 10447 (2009). doi:10.3390/s91210447. http://www.mdpi.com/1424-8220/9/12/10447, <GotoISI>://WOS:000273048800053
149. X. Cui, C. Lengignon, W. Tao, W. Zhao, G. Wysocki, E. Fertein, C. Coeur, A. Cassez, L. Croize, W. Chen, Y. Wang, W. Zhang, X. Gao, W. Liu, Y. Zhang, F. Dong, J. Quant. Spectrosc. Radiat. Transf. **113**(11), 1300 (2012). http://www.sciencedirect.com/science/ article/pii/S0022407311003943
150. D.B. Atkinson, Analyst **128**(2), 117 (2003). doi:10.1039/b206699h. http://xlink.rsc.org/ ?DOI=b206699h
151. E.R.T. Kerstel, in *Handbook of Stable Isotope Analytical Techniques*, vol. 1, ed. by P.A. De Groot (Elsevier, Amsterdam, 2004), pp. 759–787
152. E.R.T. Kerstel, L. Gianfrani, Appl. Phys. B, Lasers Opt. **92**(3), 439 (2008). doi:10.1007/ s00340-008-3128-x. <GotoISI>://000258703600017
153. H.P. Loock, TrAC, Trends Anal. Chem. **25**(7), 655 (2006). doi:10.1016/j.trac.2006.05.003. http://linkinghub.elsevier.com/retrieve/pii/S0165993606001130
154. S. Cheskis, A. Goldman, Prog. Energy Combust. Sci. **35**(4), 365 (2009). doi:10.1016/ j.pecs.2009.02.001. <GotoISI>://WOS:000267195600002
155. C.J. Wang, P. Sahay, Sensors **9**(10), 8230 (2009). doi:10.3390/s91008230. <GotoISI>:// WOS:000271265800034
156. R.E.H. Miles, S. Rudić, A.J. Orr-Ewing, J.P. Reid, Aerosol Sci. Technol. **45**(11), 1360 (2011). doi:10.1080/02786826.2011.596170. http://dx.doi.org/10.1080/02786826.2011.596170
157. J.S. Caygill, F. Davis, S.P.J. Higson, Talanta **88**, 14 (2012). doi:10.1016/j.talanta. 2011.11.043. <GotoISI>://WOS:000301159400002
158. G. Gagliardi, M. Salza, S. Avino, P. Ferraro, P. De Natale, Science (N.Y.) **330**(6007), 1081 (2010). doi:10.1126/science.1195818. http://www.ncbi.nlm.nih.gov/pubmed/21030606
159. A. Yariv, *Quantum Electronics*, 3rd edn. (Wiley, New York, 1989)
160. K.K. Lehmann, in *Cavity-Ringdown Spectroscopy—An Ultratrace-Absorption Measurement Technique*, ed. by K.W. Busch, M.A. Busch (American Chemical Society, Washington, 1999), pp. 106–124. doi:10.1021/bk-1999-0720.ch008. http://pubs.acs.org/doi/abs/ 10.1021/bk-1999-0720.ch008
161. M. Triki, P. Cermak, G. Méjean, D. Romanini, Appl. Phys. B, Lasers Opt. **91**(1), 195 (2008). doi:10.1007/s00340-008-2958-x. http://www.springerlink.com/index/10.1007/s00340-008-2958-x
162. W.T. Silfast, *Laser Fundamentals*, 1st edn. (Cambridge University Press, New York, 1996)
163. D.R. Herriott, H. Kogelnik, R. Kompfner, Appl. Opt. **3**(4), 523 (1964). doi:10.1364/ AO.3.000523. http://www.opticsinfobase.org/abstract.cfm?URI=ao-3-4-523
164. D. Romanini, Modelling the excitation field of an optical resonator. Appl. Phys. B (2013). doi:10.1007/s00340-013-5632-x
165. J. Courtois, A.K. Mohamed, D. Romanini, The degenerate astigmatic cavity. Phys. Rev. A (2013 to appear)
166. W. Riley, *Handbook of Frequency Stability Analysis*, NIST special publication 1065 (1999)
167. P. Werle, R. Miicke, F. Slemr, Appl. Phys., B Photophys. Laser Chem. **57**(2), 131 (1993). doi:10.1007/BF00425997. http://link.springer.com/10.1007/BF00425997
168. P. Werle, Appl. Phys. B **102**(2), 313 (2010). doi:10.1007/s00340-010-4165-9. http://www. springerlink.com/index/10.1007/s00340-010-4165-9
169. L.S. Ma, J.L. Hall, IEEE J. Quantum Electron. **26**(11), 2006 (1990). doi:10.1109/3.62120. http://ieeexplore.ieee.org/lpdocs/epic03/wrapper.htm?arnumber=62120
170. T.W. Hansch, B. Couillaud, Opt. Commun. **35**(3), 441 (1980). doi:10.1016/0030-4018(80)90069-3

171. R.W.P. Drever, J.L. Hall, F.V. Kowalski, J. Hough, G.M. Ford, A.J. Munley, H. Ward, Appl. Phys. B, Lasers Opt. **31**(2), 97 (1983). doi:10.1007/BF00702605. http://www.springerlink. com/index/10.1007/BF00702605
172. D.A. Shaddock, M.B. Gray, D.E. McClelland, Opt. Lett. **24**(21), 1499 (1999). doi:10.1364/OL.24.001499
173. J.L. Hall, M.S. Taubman, J. Ye, Laser stabilization, in *Handbook of Optics, vol. II: Design, Fabrication, and Testing; Sources and Detectors; Radiometry and Photometry*, 3rd edn. (McGraw-Hill, New York, 2010)
174. Z. Li, R.G.T. Bennett, G.E. Stedman, Opt. Commun. **86**(1), 51 (1991). doi:10.1016/0030-4018(91)90242-6. http://linkinghub.elsevier.com/retrieve/pii/0030401891902426
175. Z. Li, G.E. Stedman, H.R. Bilger, Opt. Commun. **100**(1–4), 240 (1993). doi:10.1016/0030-4018(93)90586-T. http://linkinghub.elsevier.com/retrieve/pii/003040189390586T
176. J. Poirson, F. Bretenaker, M. Vallet, A. Le Floch, J. Opt. Soc. Am. B **14**(11), 2811 (1997). doi:10.1364/JOSAB.14.002811. http://www.opticsinfobase.org/abstract.cfm?URI=josab-14-11-2811
177. M.J. Lawrence, B. Willke, M.E. Husman, E.K. Gustafson, R.L. Byer, J. Opt. Soc. Am. B **16**(4), 523 (1999). doi:10.1364/JOSAB.16.000523. http://www.opticsinfobase.org/abstract.cfm?URI=josab-16-4-523
178. J. Morville, D. Romanini, M. Chenevier, A.A. Kachanov, Appl. Opt. **41**(33), 6980 (2002). doi:10.1364/AO.41.006980. http://www.opticsinfobase.org/abstract.cfm?URI=ao-41-33-6980
179. B. Bakowski, L. Corner, G. Hancock, R. Kotchie, R. Peverall, G.A.D. Ritchie, Appl. Phys. B, Lasers Opt. **75**(6–7), 745 (2002). doi:10.1007/s00340-002-1026-1. http://www.springerlink.com/openurl.asp?genre=article&id=doi:10.1007/s00340-002-1026-1
180. Y. He, B.J. Orr, Appl. Phys. B, Lasers Opt. **79**(8), 941 (2004). doi:10.1007/s00340-004-1691-3. http://www.springerlink.com/index/10.1007/s00340-004-1691-3
181. J. Courtois, A.K. Mohamed, D. Romanini, Opt. Express **18**(5), 4845 (2010). http://www.opticsinfobase.org/abstract.cfm?URI=oe-18-5-4845
182. A. Schawlow, C. Townes, Phys. Rev. **112**(6), 1940 (1958). doi:10.1103/PhysRev.112.1940. http://link.aps.org/doi/10.1103/PhysRev.112.1940
183. G. Di Domenico, S. Schilt, P. Thomann, Appl. Opt. **49**(25), 4801 (2010)
184. N. Bucalovic, V. Dolgovskiy, C. Schori, P. Thomann, G. Di Domenico, S. Schilt, Appl. Opt. **51**(20), 4582 (2012). doi:10.1364/AO.51.004582
185. H. Rohde, J. Eschner, F. Schmidt-Kaler, R. Blatt, J. Opt. Soc. Am. B **19**(6), 1425 (2002). doi:10.1364/JOSAB.19.001425. http://www.opticsinfobase.org/abstract.cfm?URI=josab-19-6-1425
186. D. Redding, M. Regehr, L. Sievers, Appl. Opt. **41**(15), 2894 (2002). doi:10.1364/AO.41.002894. http://www.ncbi.nlm.nih.gov/pubmed/12027177
187. J.T. Hodges, J. Looney, R.D. van Zee, Quantitative absorption measurements using cavity-ringdown spectroscopy with pulsed lasers, in *Cavity-Ringdown Spectroscopy. An Ultratrace-Absorption Measurement Technique*, ed. by K.W. Busch, M.A. Busch (American Chemical Society, Washington, 1999). http://pubs.acs.org/isbn/9780841236004
188. D.Z. Anderson, Appl. Opt. **23**(17), 2944 (1984). http://www.ncbi.nlm.nih.gov/pubmed/18213100
189. G. Mueller, Q.Z. Shu, R. Adhikari, D.B. Tanner, D. Reitze, D. Sigg, N. Mavalvala, J. Camp, Opt. Lett. **25**(4), 266 (2000). http://www.ncbi.nlm.nih.gov/pubmed/18059850
190. D.H. Lee, Y. Yoon, E.B. Kim, J.Y. Lee, Y.S. Yoo, J.W. Hahn, Appl. Phys. B, Lasers Opt. **74**(4–5), 435 (2002). doi:10.1007/s003400200802. http://www.springerlink.com/openurl.asp?genre=article&id=doi:10.1007/s003400200802
191. D. Romanini, K.K. Lehmann, J. Chem. Phys. **99**, 6287 (1993). doi:10.1063/1.465866
192. T. Udem, J. Reichert, R. Holzwarth, T.W. Hansch, Opt. Lett. **24**(13), 881 (1999). http://www.ncbi.nlm.nih.gov/pubmed/18073883
193. S.A. Diddams, L.W. Hollberg, L.S. Ma, L. Robertsson, Opt. Lett. **27**(1), 58 (2002). doi:10.1364/OL.27.000058. http://www.opticsinfobase.org/abstract.cfm?URI=ol-27-1-58

194. D.J. Jones, S.A. Diddams, J.K. Ranka, A. Stentz, R.S. Windeler, J.L. Hall, S.T. Cundiff, Science **288**(5466), 635 (2000). doi:10.1126/science.288.5466.635. http://www.sciencemag.org/cgi/doi/10.1126/science.288.5466.635

195. A. Bartels, C.W. Oates, L.W. Hollberg, S.A. Diddams, Opt. Lett. **29**(10), 1081 (2004). doi:10.1364/OL.29.001081. http://www.opticsinfobase.org/abstract.cfm?URI=OL-29-10-1081

196. R. Jason Jones, I. Thomann, J. Ye, Phys. Rev. A **69**(5), 2 (2004). doi:10.1103/PhysRevA.69.051803. http://link.aps.org/doi/10.1103/PhysRevA.69.051803

197. W. Zhang, M. Lours, M. Fischer, R. Holzwarth, G. Santarelli, Y. Le Coq, IEEE Trans. Ultrason. Ferroelectr. Freq. Control **59**(3), 432 (2012). doi:10.1109/TUFFC.2012.2212. http://www.ncbi.nlm.nih.gov/pubmed/22481776

198. J.L. Hall, Rev. Mod. Phys. **78**(4), 1279 (2006). doi:10.1103/RevModPhys.78.1279. http://link.aps.org/doi/10.1103/RevModPhys.78.1279

199. T.W. Hansch, Rev. Mod. Phys. **78**(4), 1297 (2006). doi:10.1103/RevModPhys.78.1297. http://link.aps.org/doi/10.1103/RevModPhys.78.1297

200. T. Gherman, ML-CEAS a new high sensitivity absorption spectroscopy technique using ultra-short laser pulses. Ph.D. thesis, University J. Fourier Grenoble, 2004

201. R. Jason Jones, J.C. Diels, Phys. Rev. Lett. **86**(15), 3288 (2001). doi:10.1103/PhysRevLett.86.3288. http://link.aps.org/doi/10.1103/PhysRevLett.86.3288

202. J.C. Diels, R. Jason Jones, L. Arissian, in *Femtosecond Optical Frequency Comb: Principle, Operation, and Applications*, ed. by J. Ye, S.T. Cundiff (Kluwer Academic/Springer, Norwell, 2005), Chap. 12. http://link.springer.com/chapter/10.1007/0-387-23791-7_12

203. T.C. Briles, D.C. Yost, A. Cingöz, J. Ye, T.R. Schibli, Opt. Express **18**(10), 9739 (2010). doi:10.1364/OE.18.009739. http://www.opticsinfobase.org/abstract.cfm?URI=oe-18-10-9739

204. S. Xiao, A.M. Weiner, Opt. Express **12**(13), 2895 (2004). http://www.ncbi.nlm.nih.gov/pubmed/19483805

205. R. Grilli, G. Méjean, S. Kassi, I. Ventrillard, C. Abd Alrahman, E. Fasci, D. Romanini, Appl. Phys. B, Lasers Opt. **107**(1), 205 (2011). doi:10.1007/s00340-011-4812-9. http://www.springerlink.com/index/10.1007/s00340-011-4812-9

206. S. Xiao, A.M. Weiner, C. Lin, IEEE J. Quantum Electron. **40**(4), 420 (2004). doi:10.1109/JQE.2004.825210. http://ieeexplore.ieee.org/xpls/abs_all.jsp?arnumber=1278611, http://ieeexplore.ieee.org/lpdocs/epic03/wrapper.htm?arnumber=1278611

Chapter 2
Detection and Characterization of Reactive Chemical Intermediates Using Cavity Ringdown Spectroscopy

Neal Kline and Terry A. Miller

Abstract Cavity ringdown spectroscopy is a powerful technique for detecting reactive chemical intermediates in a variety of circumstances. The characterization of the ethyl peroxy radical in a variety of ways using different ringdown techniques is used as an example to illustrate the diverse capabilities. Several results are discussed including the room temperature, moderate resolution $\tilde{A}-\tilde{X}$ spectrum and the jet-cooled, rotationally resolved $\tilde{A}-\tilde{X}$ spectrum of ethyl peroxy. The concept of dual wavelength cavity ringdown spectroscopy is explored and its utility is demonstrated by a measurement of the $\tilde{A}-\tilde{X}$ absorption cross section of ethyl peroxy. The self-reaction kinetics of ethyl peroxy are also studied by means of cavity ringdown spectroscopy with a continuous source. The capability of CRDS to measure dynamical effects is illustrated by work on a closely related radical, hydroxy ethyl peroxy radical.

2.1 Introduction

Since its advent there have been many efforts to use cavity ringdown spectroscopy (CRDS) as a method to detect trace species in the gas phase. For example, Wang and coworkers [1] developed a CRDS spectrometer and conducted preliminary clinical trials to analyze acetone concentrations in breath samples to find a correlation with traditional diabetes biomarkers, since acetone has been identified as a biomarker for diabetes [1]. Snels et al. [2] implemented a cw-CRDS spectrometer for the trace detection of explosive materials in the 1560–1680 nm region. A flash heater was used inside their ringdown cavity to evaporate the solid explosive compounds. The spectra that were obtained showed distinct features belonging to TNT, 2,4-DNT, and 2,6-DNT allowing unambiguous identification while achieving a detection limit of 10–100 ng. The Ravishankara [3, 4] group and the Brown [5] group have explored

N. Kline (✉) · T.A. Miller
Dept. of Chemistry, The Ohio State University, 120 W. 18th Ave., Columbus, OH 43210, USA
e-mail: kline.253@osu.edu

T.A. Miller
e-mail: tamiller@chemistry.ohio-state.edu

G. Gagliardi, H.-P. Loock (eds.), *Cavity-Enhanced Spectroscopy and Sensing*,
Springer Series in Optical Sciences 179, DOI 10.1007/978-3-642-40003-2_2,
© Springer-Verlag Berlin Heidelberg 2014

Table 2.1 Peroxy radicals that have been detected using CRDS

Peroxy radical	Reference
Methyl	[47, 51, 53, 69–71]
Ethyl	[25, 33, 52, 53, 58, 69, 72]
1-Propyl, 2-Propyl	[73–75]
1-Butyl, 2-Butyl, Isobutyl, *t*-butyl	[67]
1-Pentyl, Neopentyl, 2-Methylbutyl, 3-Methylbutyl, 3-Pentyl, 3-Methyl-2-butyl, *t*-Pentyl	[68]
Cyclopentyl, Cyclohexyl	[76]
Cyclopentadienyl	[77]
Phenyl	[78]
Allyl	[79]
Propargyl	[80]
β-Hydroxyethyl	[65, 66]
2,1-Hydroxypropyl	[81]
Acetyl	[82]

the relevance of CRDS to study trace gases in the atmosphere. They demonstrated that CRDS techniques possessed the sensitivity and selectivity to measure trace gas concentrations below 1 ppb in ambient air gas mixtures by monitoring NO_X species. Anderson [6–8] has developed a CRDS spectrometer capable of being flown on NASA's WB-57 high altitude research aircraft. This spectrometer has been used to detect water isotopologues and determine methane isotopologue ratios in the ambient atmosphere. Lehmann [9] has also demonstrated the capability of using CRDS to detect single cell biological agents. An optical fiber resonator was incorporated using the optical scatter of the evanescent field surrounding the fiber to detect the cells. A detectable scattering cross section of $\sim 10 \ \mu m^2$ was observed and opens the possibility for cavity ringdown biosensing studies.

This chapter focuses on the use of CRDS to detect and characterize another class of trace species in the gas phase. These species are chemical intermediates of significance in atmospheric, combustion, and astrophysical processes [10–12]. The purpose of this chapter is to give an overview of how various experimental techniques and apparatuses, all involving CRDS, have been employed in our laboratory to study the spectroscopy of these reactive species, measure their concentration, and ultimately study their chemical reactions. Since a class of reactive intermediates that has been the subject of much work in our lab is the peroxy radical family, we will use the ethyl peroxy radical to illustrate the experimental application of various forms of CRDS to the study of reactive intermediates and show the kind of information that can be obtained from their spectra. Table 2.1 provides a list of other peroxy radicals that have been detected using CRDS techniques.

2.2 The Chemistry and Spectroscopy of Peroxy Radicals

The immense scientific interest in peroxy radicals stems from their importance in combustion processes and atmospheric oxidation cycles. In low temperature combustion processes (<700 K), the mechanism of peroxy radical formation begins with the abstraction of a hydrogen atom from the hydrocarbon fuel (RH) by a hydroxyl radical ($\cdot$OH) to form an alkyl radical (R$\cdot$). The alkyl radical undergoes a 3-body reaction with molecular oxygen to form an alkyl peroxy radical [13, 14].

$$RH + \cdot OH \longrightarrow R\cdot + H_2O \tag{2.1}$$

$$R\cdot + O_2 + M \longrightarrow RO_2\cdot + M \tag{2.2}$$

The production of peroxy radicals sustains the forward rate of combustion as it is involved in a chain-branching sequence of reactions that will eventually lead to the production of $\cdot$OH radicals. Peroxy radical chemistry is also important in the atmospheric oxidation of volatile organic compounds (VOC's). The formation of peroxy radicals in the atmosphere proceeds again through the reaction of alkyl radicals with molecular oxygen. The alkyl radicals are formed by the abstraction of hydrogen from a hydrocarbon by $\cdot$OH or Cl$\cdot$ which are also present in the atmosphere [15]. The alkyl radical then reacts rapidly with molecular oxygen and a third body, M-typically N_2, which stabilizes the peroxy radical product by collision. The presence of peroxy radicals in the atmosphere has several consequences, particularly in the chemistry of the troposphere. For example, O_3 can be produced in the troposphere, where it is considered a pollutant, via peroxy radicals' reaction with NO:

$$NO\cdot + RO_2\cdot \longrightarrow NO_2\cdot + RO\cdot \tag{2.3}$$

$$NO_2 \xrightarrow{\leq 430 \text{ nm}} NO\cdot + O \tag{2.4}$$

$$O(^3P) + O_2 + M \longrightarrow O_3 + M \tag{2.5}$$

Peroxy radicals' reaction with NO produces excess amounts of NO_2 in the troposphere where it can be photolyzed by sunlight to generate oxygen atoms. The oxygen atoms then react with molecular oxygen to produce O_3 [16, 17]. A second reaction of peroxy radicals with NO in the troposphere can produce excess amounts of hydroxyl radical:

$$NO\cdot + RO_2\cdot \longrightarrow NO_2\cdot + RO\cdot \tag{2.6}$$

$$RO\cdot + O_2 \longrightarrow R'C = O + HO_2\cdot \tag{2.7}$$

$$HO_2\cdot + NO\cdot \longrightarrow NO_2\cdot + \cdot OH \tag{2.8}$$

$$\cdot OH + O_3 \longrightarrow HO_2\cdot + O_2 \tag{2.9}$$

The $\cdot$OH can then migrate from the troposphere to the stratosphere, where O_3 is considered a filter against UV radiation, and reduce the amount of O_3 present.

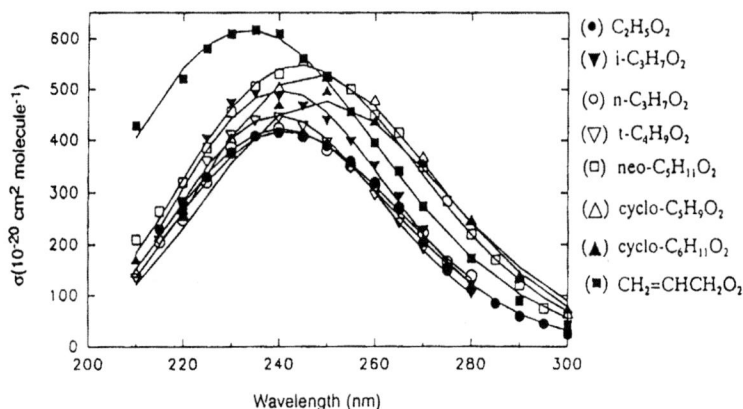

Fig. 2.1 $\widetilde{B}-\widetilde{X}$ absorption spectra of several peroxy radicals. Each spectrum consists of a very broad, featureless absorption band centered near 240 nm and roughly 40 nm wide. Reprinted with permission from [16]. Copyright 1997, John Wiley and Sons

For many years the best way to spectroscopically observe peroxy radicals was to probe their $\widetilde{B}-\widetilde{X}$ absorption located in the UV. Many papers were published in the 1980's and 90's citing this transition as a way to identify peroxy radicals and to monitor their self-reaction kinetics using time-resolved UV absorption spectroscopy [18–22]. Monitoring the $\widetilde{B}-\widetilde{X}$ transition of peroxy radicals also proved to be experimentally facile as it has a relatively large absorption cross section on the order of 10^{-18} cm²/molecule. However, there is a considerable disadvantage to using this transition to distinguish among different peroxy radicals. The $\widetilde{B}-\widetilde{X}$ transition involves promotion of an electron from the second highest occupied molecular orbital (HOMO-1) to the singly occupied molecular orbital (SOMO). The HOMO-1 is bonding and the SOMO is antibonding with respect to the O–O bond. Hence, the $\widetilde{B}-\widetilde{X}$ transition involves the promotion of an electron from a bonding orbital to an antibonding orbital so that when a peroxy radical undergoes a transition to the $\widetilde{B}$ state, the O–O bond dissociates [23]. This gives rise to a repulsive $\widetilde{B}$ state potential energy surface and the resulting spectrum for all peroxy radicals is a broad featureless absorption near 240 nm (Fig. 2.1) which is roughly 40 nm wide. This lack of any observable structure makes the UV transition a poor diagnostic with which to identify different peroxy radicals.

Alternatively, the $\widetilde{A}-\widetilde{X}$ transition of peroxy radicals is located in the NIR and is a HOMO to SOMO transition. The HOMO of peroxy radicals is nonbonding along the O–O bond making the $\widetilde{A}-\widetilde{X}$ transition a promotion of an electron from a nonbonding to antibonding orbital [23] Therefore, the $\widetilde{A}$ state is a bound state and upon excitation to the $\widetilde{A}$ state there is only a lengthening of the O–O bond instead of complete dissociation. These qualities indicate that the $\widetilde{A}-\widetilde{X}$ spectra of peroxy radicals will be well structured and characteristic of a given RO_2 radical. This allows the differentiation among different peroxy radicals and makes the $\widetilde{A}-\widetilde{X}$ transition an excellent diagnostic with which to observe peroxy radicals.

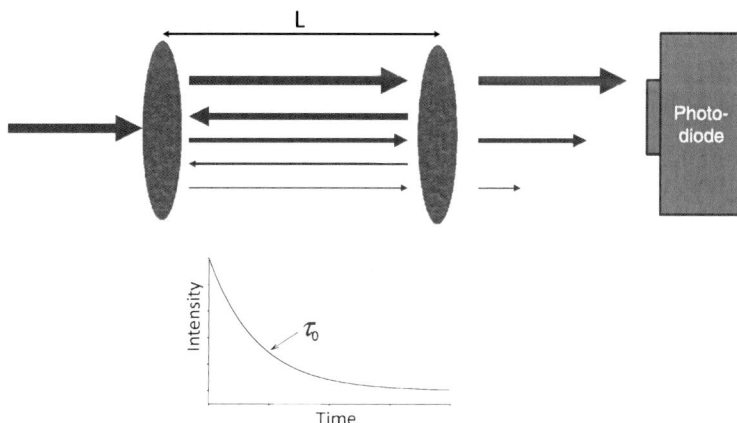

Fig. 2.2 Schematic diagram of the CRDS experiment. A pulse of light is trapped in a cavity formed between two highly reflective mirrors with leaked light being detected by a photodiode. Exponential decay ringdown curve that is obtained from the photodiode is shown below the cavity

However, observing the $\widetilde{A}-\widetilde{X}$ transition may be experimentally difficult as it is based on the highly forbidden $a^1\Delta_g-X^3\Sigma_g^{\,-}$ transition of the O_2 chromophore [24]. Consequently the cross section for this transition is $\approx 10^4$ times smaller than the $\widetilde{B}-\widetilde{X}$ transition. Unfortunately traditional absorption experiments do not have the sensitivity to detect this weak of a transition. With the innovation of CRDS, observation of the $\widetilde{A}-\widetilde{X}$ transition of peroxy radicals has become experimentally feasible because of the long effective path lengths achieved in a typical CRDS experiment compared to a traditional absorption experiment. These long effective path lengths in a short physical dimension provide significantly more sensitivity for reactive intermediates in a chemical reaction whose spatial extent is limited.

2.3 CRDS Spectrometers and Results

Our laboratory has three operating CRDS spectrometers specifically designed for the study of reactive chemical intermediates. These apparatuses differ in a number of ways including frequency range and resolution. They also differ in the nature of the sample investigated and the manner in which reactive species are produced. In this section we first discuss common features of the CRDS technique and then describe the various spectrometers in some detail.

Generally speaking CRDS is an attractive spectroscopic technique because of the long path lengths that can be achieved with a basic ringdown apparatus as shown schematically in Fig. 2.2. In a typical experiment a pulse of light is injected into an optical cavity formed by two highly reflective mirrors. The pulse of light makes on the order of 10^5 passes before decaying to $1/e$ of its initial intensity because a small amount is lost at the mirrors on each pass. This multipass set up generally gives effective path lengths on the order of hundreds of kilometers for a physical dimension measured in centimeters. The amount of time it takes for the light pulse to decay to $1/e$ of its initial intensity is a called a "ringdown" time and can be calculated using the following equation:

$$\tau_0 = \frac{L/c}{(1-R)} \tag{2.10}$$

where τ_0 is the ringdown time of the empty cavity, c is the speed of light, R is the mirror reflectivity, and L is the cavity length. A ringdown time can also be calculated when there is an absorbing species present in the cell:

$$\tau = \frac{L/c}{1 - R + N\sigma(\nu)l} \tag{2.11}$$

with N being the number density of the absorbing species, $\sigma(\nu)$ being the absorption cross section of the species, at frequency ν, and l being the absorber path length. The quantity $N\sigma(\nu_0)l$ can be defined as the absorbance, A, of the species at frequency ν_0, which can be calculated from the measured ringdown times by combining Eqs. (2.10) and (2.11):

$$A = \frac{L}{c\tau} - \frac{L}{c\tau_0} \tag{2.12}$$

To illustrate the capabilities of CRDS, the characterization of a simple reactive intermediate, the ethyl peroxy radical, utilizing the different ringdown techniques and apparatuses will be presented. The different aspects of ethyl peroxy that are probed include: vibrational assignment of the room temperature $\tilde{A}-\tilde{X}$ spectrum of ethyl peroxy, a rotational assignment of the same spectrum obtained under jet-cooled conditions with a high resolution laser source, determination of the absorption cross section of the $\tilde{A}-\tilde{X}$ transition, and measurement of the self-reaction rate of the ethyl peroxy radical. The following sections describe the different ringdown techniques and apparatuses and discuss results obtained.

2.3.1 Room Temperature, Moderate Resolution CRDS of Ethyl Peroxy

As mentioned in the previous section, for analytical purposes the $\tilde{A}-\tilde{X}$ electronic transition of peroxy radicals is the preferred way both to monitor and to characterize peroxy radicals. The simplest and most straight-forward sample to prepare of peroxy radicals is a room temperature sample. Even for a simple peroxy radical like ethyl peroxy, rotational congestion at room temperature is sufficiently high that only rotational contours can be observed, so a relatively simple moderate resolution laser source suffices. Below we describe such a CRDS apparatus and then discuss its application to the ethyl peroxy radical to illustrate the kind of results obtainable using this spectrometer.

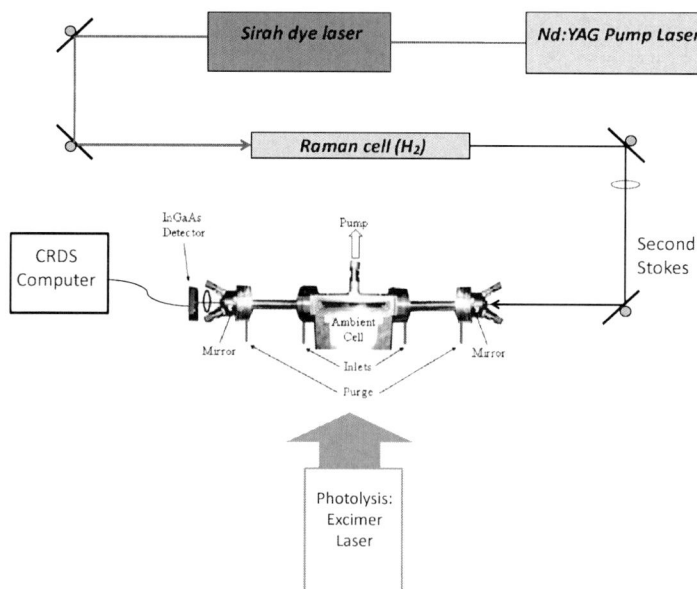

Fig. 2.3 Setup of the room temperature, moderate resolution CRDS experiment. The NIR radiation is generated via isolation of second Stokes stimulated Raman shifting of a visible dye laser output. Radicals are produced inside the ringdown cavity via photolysis using 193 nm light from an excimer laser

2.3.1.1 Experimental Approach and Apparatus

The room temperature, moderate resolution $\widetilde{A}-\widetilde{X}$ electronic transition of ethyl peroxy was first observed [25] in a CRDS apparatus like that depicted in Fig. 2.3. A PrecisionScan Sirah Dye Laser is pumped with the second harmonic of a Nd:YAG laser and tuned over the range 588–645 nm using three different dyes: DCM, Rhodamine B, and Rhodamine 101. The radiation necessary to probe ethyl peroxy is generated using stimulated Raman shifting (SRS) by focusing the output of the dye laser into a 70 cm long, single pass Raman cell filled with approximately 325 psi of molecular hydrogen. Once the radiation exits the Raman cell, the desired NIR second Stokes component (7000–9000 cm^{-1}) is isolated using 1000 nm long pass filters and coupled into the ringdown cell. The cell is 54 cm in length and is terminated by two 6 m radius-of-curvature, plano-concave mirrors ($R \geq 99.995$ %) which are housed in custom made flanges that allow for accurate alignment using finely threaded screws for adjustment of the mirrors. To cover the entire spectral region of interest two sets of mirrors were used which had sufficient overlap to ensure complete wavelength coverage. To prevent damage to the mirrors from the chemistry a constant purge of nitrogen was used to shield them. Once the NIR light exits the cell it is focused onto an amplified photodiode with the detector output recorded by a 12 bit 20 MHz digitizing card. During a typical scan approximately 20 consecutive laser shots are averaged at each dye laser frequency and, using the nonlinear Levenberg-Marquardt algorithm, the digitized signal of the decay of radiation is fit to a single exponential to acquire the ringdown time, initial amplitude, and baseline. The calculated ring-down time is converted to cavity absorption per pass and saved as a point in the spectrum. At each dye laser frequency, a ringdown time is acquired with the photol-ysis excimer laser on and off. This will generate two traces: an on trace containing photolysis products and an off trace that contains background structures (i.e. precur-sor bands, background water lines, etc.). These two traces are subtracted (on-off) to produce a trace dependent only upon the photolysis products.

In addition to the mirror purge inlets, the ringdown cell was constructed with inlets for precursor gases, a Baratron pressure gauge, and an exhaust port leading to a mechanical pump. The cell is also fashioned with two rectangular apertures (2 × 16 cm) in the center of the cell separated by 2.3 cm for UV grade quartz windows which allow for photolysis of the precursor gases. An excimer laser is used to initiate the chemistry necessary to produce ethyl peroxy. The excimer is operated with a gas mixture of ArF to generate 193 nm light. Once the radiation leaves the excimer the beam is shaped by a cylindrical and spherical lens to a rectangular shape (13.0 cm along the cavity axis by 0.5 cm height) and sent through the photolysis windows in the central part of the cell. For this particular set of experiments, the excimer laser is fired 100 μs before each shot of NIR light entered the cell, which allows enough time for the ethyl peroxy radicals to form, but not enough time for them to react, be pumped out, or diffuse to the walls of the cell.

There were two chemical approaches used to produce room-temperature ethyl peroxy radicals in the CRDS apparatus as illustrated by the following reactions,

$$(COCl)_2 \xrightarrow{193\ nm} 2Cl\cdot + 2CO \tag{2.13}$$

$$CH_3CH_3 + Cl\cdot \longrightarrow CH_3\dot{C}H_2 + HCl \tag{2.14}$$

$$(CH_3CH_2)_2C = O \xrightarrow{193\ nm} 2CH_3\dot{C}H_2 + CO \tag{2.15}$$

$$CH_3\dot{C}H_2 + O_2 + N_2 \longrightarrow CH_3CH_2OO\cdot \tag{2.16}$$

In the first method chlorine atoms abstract a hydrogen from ethane to produce ethyl radicals. The chlorine atoms are generated by the 193 nm photolysis of ox-alyl chloride, which has been proven to be an efficient source of chlorine atoms due to its large absorption cross section ($\sigma_{193} = 3.83 \times 18^{-18}$ cm^2/molecule) [26, 27]. In addition oxalyl chloride is a very clean source as the only photolysis products are two chlorine atoms and carbon monoxide. Typical pressures of reactants used for the hydrogen abstraction experiments included 0.5–1.0 torr $(COCl)_2$, 4.0 torr C_2H_6, 80.0 torr O_2, and 150 torr N_2. In the second production method, 3-pentanone ($\sigma_{193} = 2.8 \times 10^{-18}$ cm^2/molecule) [28] is directly photolyzed by the 193 nm light to yield ethyl radicals and carbon monoxide. Pressures for photolysis of 3-pentanone included 2.0–3.0 torr 3-pentanone, 50.0 torr O_2, and 125.0 torr N_2. In both methods the ethyl radicals subsequently react with oxygen and a third body (in our case N_2) to yield the ethyl peroxy radicals. The use of two different production methods is advantageous in that it allows independent verification of the chemistry that sup-ports assignments of the spectral carrier to ethyl peroxy. Presence of ethyl peroxy is

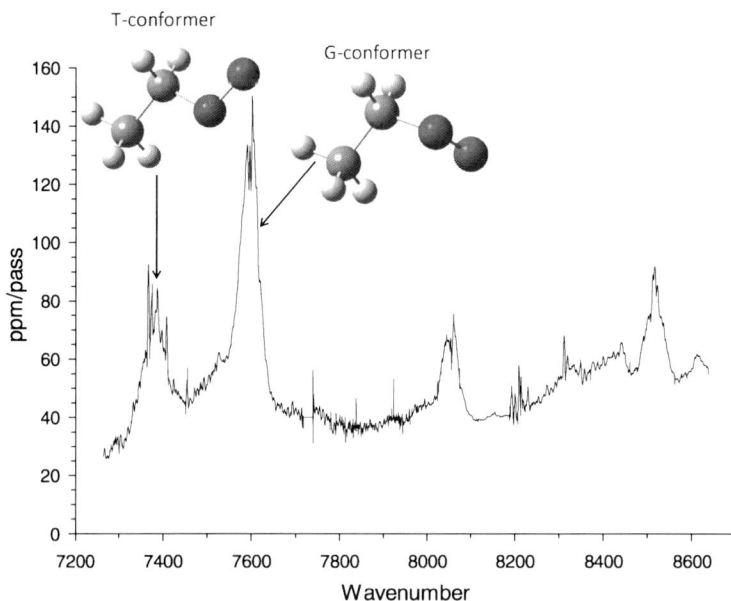

Fig. 2.4 Portion of the room temperature, moderate resolution spectrum of ethyl peroxy obtained from the photolysis of 3-pentanone. Both G and T conformers are present in the spectrum and their origins are indicated by the *arrows* in the figure

also confirmed by performing oxygen-independent experiments and by varying the delay time between firing the excimer and NIR probe beam. The delay can be varied from 100 μs to 1 ms with the ethyl peroxy signal progressively decreasing with increasing delay time indicating the carrier is a transient species and not a long-lived minor by-product which could accumulate in the cell.

2.3.1.2 Observations and Results

There is only one isomeric form of ethyl peroxy but *ab initio* calculations predict [29] two stable conformeric forms. The two conformers are shown in Fig. 2.4 and differ by the orientation of their ∠OOCC dihedral angles. The two conformers are designated gauche (G) and trans (T) and correspond to ±60° and 180° orientations of the ∠OOCC dihedral angle, respectively. According to the calculation of Rienstra-Kiracoffe et al. [29], the G and T conformers are separated by less than kT at room temperature; however, only one conformer had ever been experimentally observed. Clearly the $\tilde{B}-\tilde{X}$ spectra depicted in Fig. 2.1 lacks the resolution to distinguish among conformers. Hunziker and Wendt [30] were the first to observe the $\tilde{A}-\tilde{X}$ spectrum of ethyl peroxy and reported only a single intense origin band at 7593 cm^{-1} and weaker O–O stretch band at 8511 cm^{-1}. Blanksby et al. [31] obtained the photoelectron spectrum of ethyl peroxy and reported only an $\tilde{A}-\tilde{X}$ origin at 7565 ± 30 cm^{-1}.

In Fig. 2.4 the room temperature, moderate resolution $\tilde{A}-\tilde{X}$ spectrum of ethyl peroxy is presented with 0.5 cm^{-1} laser resolution obtained from the photolysis of 3-pentanone. To assign the vibrational structure present in the spectrum Rupper et al. [25] performed high level *ab initio* calculations. On the basis of the calculations two broad bands present at 7362 cm^{-1} and 7592 cm^{-1} are assigned as the origins of the T and G conformer, respectively. One may notice that there is a significant difference in the intensities of the origin bands of the two conformers despite the small predicted difference in their zero point corrected energies, (≤ 100 cm^{-1}). This difference in intensity can be explained by taking into account a degeneracy factor of 2 and 1 for the G and T conformers. The equilibrium population at room temperature is therefore decreased by a factor of ≈ 3 for the higher energy T conformer compared to the G conformer. The authors also assigned fundamental bands pertaining to the $\angle$COO bend (ν_{12}) for both the T and G conformers appearing at 7874 cm^{-1} for the T conformer and 8049 cm^{-1} for the G conformer. In addition to the $\angle$COO bend fundamentals, the authors were able to assign a fundamental transition to ν_{13} of the G conformer, which is the $\angle$CCO bend vibration. Moving further to the blue, the authors observed fundamental transitions belonging to the O–O stretch (ν_9) of both conformers. The bands at 8307 cm^{-1} and 8507 cm^{-1} are assigned as the O–O stretches of the T and G conformer, respectively.

Overall the ethyl peroxy spectrum shows three distinct sections that are generally characteristic of peroxy radicals. There is an origin band region; 400–500 cm^{-1} to the blue is the $\angle$COO bend vibrational region; about 900 cm^{-1} to the blue of the origin is a region that contains vibrations from the O–O stretch of peroxy radicals. Observing transitions pertaining to the $\angle$COO bending and O–O stretching of peroxy radicals is not surprising as the $\tilde{A}-\tilde{X}$ transition is fundamentally the promotion of an electron to the π^* singly occupied orbital localized on the terminal oxygen atom of the peroxy moiety. Therefore, one should expect to see a variation in geometry around the terminal O atom and hence also have fundamental transitions in the $\tilde{A}$ state involving motions of the terminal O atom, which is clearly demonstrated in the ethyl peroxy spectrum.

The authors also noted that there appears to be a large difference in rotational structure of the origin bands for the two conformers. This difference in rotational contours can be mainly attributed to the fact that the two conformers possess different symmetries. The T conformer has C_s symmetry, with the $\tilde{A}^2A'-\tilde{X}^2A''$ transition possessing a pure c-type transition moment; while the G conformer has C_1 symmetry, which means the $\tilde{A}^2A-\tilde{X}^2A$ transition is a mixture of a-, b-, and c-type transition moments. To illustrate the impact of the different symmetries and the presence of the different transition moments, simulations of the rotational envelopes can be performed using rotational constants taken from *ab initio* calculations. Figure 2.5 shows simulations obtained from our SpecView [32] program compared to the experimental scans of the origin region. Figure 2.5 includes simulations done with rotational constants directly taken from *ab initio* calculations and simulations where the constants have been slightly adjusted to better reproduce the experimental data. Since the rotational contours of the two conformers are so distinct, the correlation between the experimental data and the simulations clearly supports the correctness of the conformer assignments for the origin bands based on *ab initio* calculations.

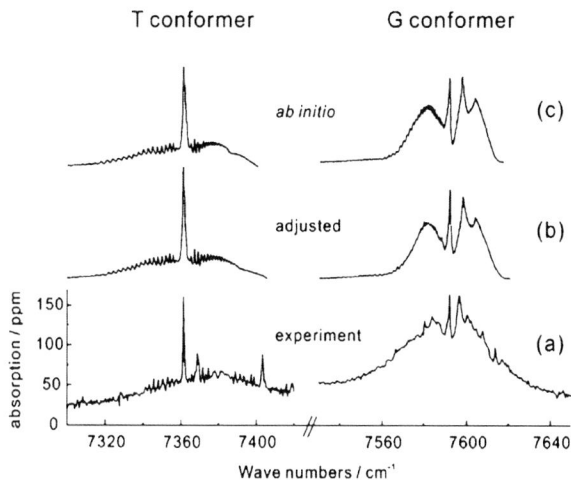

Fig. 2.5 Experimental spectra and simulations [32] of the origin regions of the G and T conformers. In the experimental T conformer spectrum the second and third strongest peaks are a result of sequence transitions Reprinted with permission from [16]. Copyright 2007 American Chemical Society

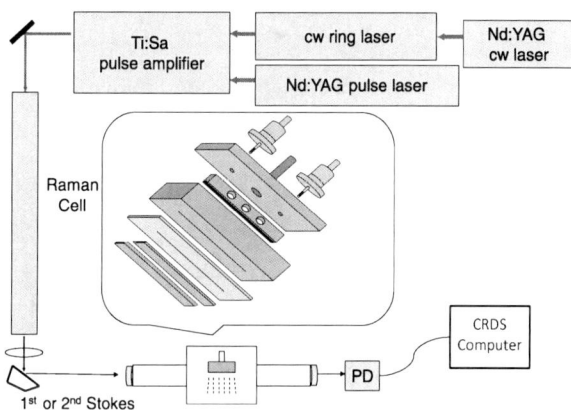

Fig. 2.6 Illustration of high resolution, jet-cooled CRDS apparatus with enlarged diagram of the pulsed discharge apparatus

2.3.2 Jet-Cooled, High Resolution CRDS of Ethyl Peroxy

While obtaining room temperature, low-resolution data on the $\tilde{A}-\tilde{X}$ transition is a useful diagnostic to distinguish between peroxy radicals, one can only obtain rotational contours using this approach. To confirm conformer assignments, determine precisely rotational constants, spin-rotation constants, and other parameters that characterize molecular structure a higher resolution laser source and a method to reduce rotational congestion is needed, which can be achieved by jet-cooling the radicals.

2.3.2.1 Experimental Approach and Apparatus

To obtain the high resolution, rotationally resolved spectrum of ethyl peroxy [33] a separate experimental system was built and is illustrated in Fig. 2.6. The nearly

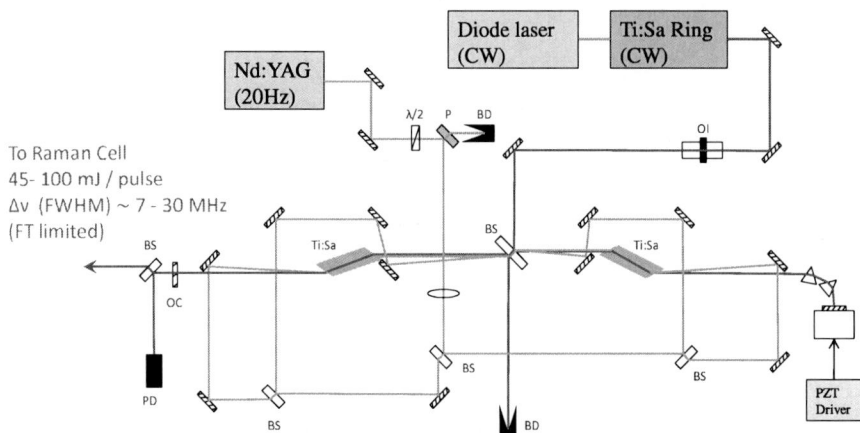

Fig. 2.7 Diagram of the Ti:Sa amplifier. Abbreviations: BD—beam dump, BS—beam splitter, OC—output coupler, λ/2—half wave plate. Reprinted with permission from [37]. Copyright 2007, American Institute of Physics

Fourier transform limited, high resolution laser source is based [34–37] on the amplification of cw ring laser radiation in a custom built Titanium:Sapphire (Ti:Sa) pulse amplifier (Fig. 2.7). The Ti:Sa amplifier is seeded by a tunable single mode cw-Ti:Sa ring laser (pumped by a cw solid state frequency doubled $Nd:YVO_4$ diode laser) with a bandwidth of ≤ 1 MHz. The pulse amplification is achieved by quadruple pumping of two Ti:Sa rods in the amplifier with a frequency doubled Nd:YAG laser operating at 20 Hz. In order to obtain the pulse amplification and narrow linewidth necessary for the experiment, the cavity of the Ti:Sa amplifier needs to be actively locked to match the single-mode cw-laser source. To establish this active control, the ramp-lock-and-fire (RLF) [38–40] technique (which is specific to locking a pulse amplifier to a cw-laser source) has been employed in our set up. Cavity mode matching is achieved through controlling the cavity length by dithering a small flat end mirror mounted on a piezo-electric actuator. In order to actively change the cavity length when conducting an experiment, there needs to be an active feeback loop from the Ti:Sa amplifier to the actuator driver. This feedback loop is established through the use of the amplified output of a photodetector which collects a small amount of the transmitted beam (~ 1 %) from a beamsplitter and relays the information to an analog-to-digital converter (ADC), which is part of a larger digital signal processing (DSP) module which has been programmed for this apparatus. On this DSP module there is also a digital-to-analog converter (DAC) which feeds information directly to the actuator driver that controls the piezo. These components make up the active feedback loop necessary for the Ti:Sa pulse amplifier. The radiation exiting the Ti:Sa amplifier has a linewidth measured from ≈ 8 MHz to ≈ 30 MHz (FWHM), while delivering 40–100 mJ of power and is tuned through the range 800–900 nm. The bandwidth of the radiation leaving the amplifier results mainly from power broadening and depends primarily on the energy of the pump laser. Full details of the Ti:Sa amplifier are given elsewhere [37].

To generate the NIR radiation necessary for the experiment we again employ SRS with a Raman cell filled with 150 psi H_2. The radiation exiting the Raman cell experiences pressure broadening and has a measured linewidth of $\sim$250 MHz (FWHM). Radiation from the SRS is directed into two CaF_2 Pellin-Broca prisms to disperse the radiation and isolate the desired first Stokes component in the 7000–9000 cm^{-1} range. In addition to the SRS, difference frequency mixing (DFM) can also be used to generate radiation in this region. To use DFM, the radiation exiting the pulse amplifier is directed into a BBO crystal where it is mixed with the second harmonic of a seeded Nd-YAG laser. This will produce radiation in the 7000–9000 cm^{-1} region with a bandwidth of 60 MHz, significantly less than the SRS radiation. A small amount of Doppler broadening ($\sim$50 MHz) from the slit jet expansion is present and modestly increases the observed molecular linewidth.

Radiation generated by DFM or SRS is directed into the ringdown cell with the cavity being formed by two highly reflective mirrors ($R \geq 99.999\%$) that are purged continuously with N_2. The cavity length of 67 cm corresponds to a 224 MHz longitudinal mode spacing, which is larger than the bandwidth of the incoming radiation (particularly when difference frequency radiation is used). The fact that the cavity mode spacing is larger than the bandwidth of the radiation, in addition to the varying beam shape of the SRS beam, can make it very difficult to couple the radiation to a specific TEM mode of the cavity. However, these problems can be circumvented by avoiding the use of mode matching optics and making use of both longitudinal and transverse modes to form a pattern dense compared to the radiation bandwidth [41]. The NIR radiation exiting the cell is captured with a photodiode which transmits the data to a PC running a Labview software program for data processing. During a typical experiment, 4 laser shots are averaged per data point to obtain a ringdown curve, with a frequency step size of 50 MHz between data points. A background trace is simultaneously obtained with the electric discharge off and the spectrum of the photolysis products is obtained by subtracting off from on.

To produce the radicals of interest, and to achieve the low rotational temperatures necessary to minimize spectral congestion, $\approx$1 % ethyl iodide precursor molecules are entrained in a mixture of a few percent of molecular oxygen in helium:neon (75 %:25 %) which is supersonically expanded in a pulsed slit-jet discharge [42, 43] (Fig. 2.6). The precursor gas mixture is prepared by bubbling $\sim$500 torr of an O_2 and He:Ne gas mixture through a sample bomb containing C_2H_5I at −45 °C. This slit-jet discharge is an adaptation of the design used by the Nesbitt [44, 45] and Saykally [46] groups. The discharge apparatus consists of a slit body and solenoid mount that are fashioned from aluminum; opening the slit (1 mm × 5 cm) is accomplished by using two General Valve solenoids with a commercial multi-channel driving circuit to actuate a poppet. The radicals are produced in the plasma that is created between the two electrode plates. The plasma is generated by applying a voltage of $\approx$500 V between the two electrodes separated by 1.0 mm yielding a current of 300–400 mA during the opening time of the slit. When the gas mixture flows through the discharge electrons attach to ethyl iodide which dissociates to ethyl radicals and I^-. The ethyl radicals react with the oxygen to give ethyl peroxy. The

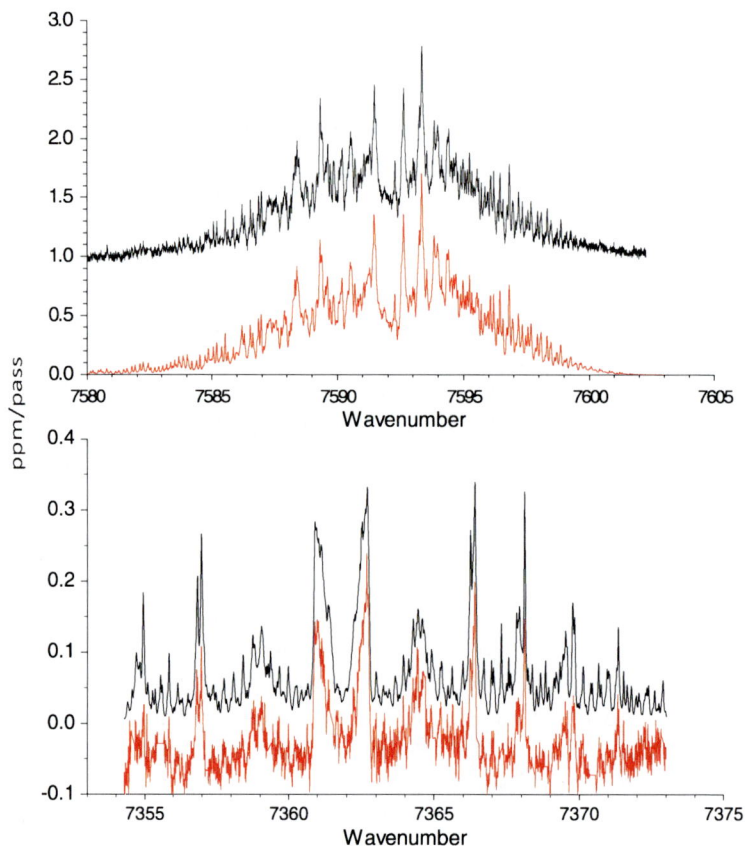

Fig. 2.8 The *top panel* shows the experimental spectrum (*lower-red*) and simulated spectrum (*top-black*) of the G conformer of ethyl peroxy. The *bottom panel* shows the experimental spectrum (*lower-red*) and simulated spectrum (*top-black*) of the T conformer

radicals are cooled via the supersonic expansion to 15–25 K and probed approximately 10 mm downstream from the jet throat, where radical densities have been estimated as $\sim 5 \times 10^{12}$ molecules/cm^3.

2.3.2.2 Observations and Results

The jet cooled, high resolution spectra for the origin transitions of both the G and T conformers of ethyl peroxy are presented in Fig. 2.8 along with simulations based on the spectral analysis. The objective of obtaining high resolution spectra of molecules is to determine molecular parameters such as rotational constants. Traditionally, the optimum values for molecular parameters are obtained by assigning individual transitions in the high resolution spectra and using a least squares fitting (LSF) procedure with our group's SpecView [32] software package. This approach is completely appropriate if one has a well resolved spectrum [47] for which assignments are possible for individual lines. However, the spectra of the two conformers of ethyl peroxy are only partially rotationally resolved, with many features in the spectrum corresponding to multiple (2–5) overlapping transitions. This makes the use of the LSF procedure of fitting the data inappropriate as unique transition assignments are nearly impossible.

To simulate such spectral data and obtain the desired molecular parameters, the evolutionary algorithm (EA) approach is valuable [48]. The EA is an iterative, semi-automated fitting procedure which can be used to produce fits to fully or partially resolved rovibronic spectra, from which one can extract molecular information [48–50]. The EA takes into account line position, shape and intensity (LSF takes into account only line position directly) in order to generate the fits. The molecular parameters for ethyl peroxy are the three rotational constants and four spin rotation constants in each the ground and excited states, the band origin, the rotational temperature, and the two angles (θ and ϕ) which describe the orientation of the transition dipole moment in the inertial axis system. For the EA to work, the user needs to give an initial estimate for the parameters, which can be obtained from *ab initio* calculations, and to set upper and lower bounds on each of the parameters, typically $\pm 10\%$. For the $\widetilde{A}$–$\widetilde{X}$ spectrum of ethyl peroxy, the above mentioned 18 parameters need to be determined. Simulations of the experimental spectrum can be generated based upon multiple, randomly selected sets of parameters satisfying the above criteria. A fitness function evaluates the quality of the match between experiment and the simulation with each parameter set. The EA automatically adjusts (evolves) the parameters to improve the match.

Upon examination of the data in Fig. 2.8, the spectrum of the G conformer (top panel) appears to have a fair amount of congestion present; which is largely attributed to the population of many rotational and spin-rotational levels. The G conformer spectrum has been simulated at a rotational temperature of 16.2 K with a Voigt profile with a Gaussian component of 250 MHz and a Lorentzian component of 1400 MHz. The T conformer was found to have a rotational temperature of 24.4 K, with line shapes consisting of Voigt profiles with a Gaussian component of 300 MHz and Lorentzian component of 1250 MHz. The Lorentzian component of the linewidth is thought to be a result of lifetime broadening in the $\widetilde{A}$ state and is much larger then the Gaussian component of the linewidths determined. From visually comparing the T and G conformer spectra, it is obvious that the T conformer has a simpler spectrum than the G conformer. This can be explained by the symmetries of the molecules once again. As mentioned in Sect. 2.3.1.2, the T conformer possesses C_s symmetry; hence its electronic transition dipole moment lies along the c-axis and two of the spin rotation tensor components are equal to zero. The G conformer has C_1 symmetry with electronic transition dipole components along the a, b, and c axes; moreover all components of the spin-rotation tensor are non-zero resulting in more allowed transitions and a more complicated spectrum.

A full analysis of the high resolution data of the G and T conformers (along with their deuterated analogues) has been presented by Just et al. [33]. A Hamiltonian was defined for each vibronic level, which depended upon the rotational constants and the different components of the spin-rotation tensor. Values for the molecular parameters determined from the spectrum were used to benchmark various *ab initio* calculations.

2.3.3 Dual Wavelength Spectroscopy

The two previous sections detailed experiments in which ethyl peroxy radicals were investigated using a single radiation source. However, some properties of peroxy radicals are studied more effectively by using two radiation sources simultaneously in a "dual wavelength" (2λ) experiment. The advantage of using 2λ spectroscopy is that it allows two different experiments to be conducted at the same time which would otherwise have to performed separately. Performing experiments simultaneously can eliminate the need to make assumptions about experimental conditions, laser power, and other variables that can change throughout the course of a day. The 2λ experiment can also eliminate the need to change radiation sources throughout the course of an experiment as radiation sources with different spectral coverage and different resolutions can be used concurrently. Finally the 2λ experiment can record the spectra of two molecular species simultaneously. These advantages make 2λ spectroscopy very appealing for use in cavity ringdown studies.

2.3.3.1 Dual Wavelength Apparatus

For 2λ spectroscopy to be performed, a cavity ringdown apparatus that can accommodate two different radiation sources probing the same sample needs to be developed. A schematic of such an apparatus is shown in Fig. 2.9. The apparatus has two separate arms, "A" and "B" through which radiation can pass. They are each 70 cm long and intersect at a 9° angle. The radiation coming into arm "A" is the radiation used to monitor peroxy radicals and is produced the same way as described in Sect. 2.3.1.1. A Sirah dye laser operating with DCM in methanol is pumped with the second harmonic of a Nd:YAG laser. The output of the dye laser near 630 nm is focused into a Raman cell filled with H_2 and the resulting second stokes radiation in the NIR is isolated via optical filters. The NIR radiation is coupled into arm "A" of the 2λ system with the optical cavity being formed by two highly reflective mirrors centered at 1.3 μm with radiation exiting the cavity being detected by a fast photodiode. The input to arm "B" can be varied, with the specific type of radiation that is required being dictated by the experiment being performed.

The reaction cell used to produce the peroxy radicals is placed at the intersection of the two arms and has dimensions ($l \times w \times h$) 19.0 cm $\times$ 2.5 cm $\times$ 1.25 cm. There are two rectangular openings for the UV grade photolysis windows measuring 13.5 cm $\times$ 1.25 cm. The radicals are produced by the 193 nm photolysis of an appropriate gas mixture with the photolysis beam being masked to ensure homogeneous radical production in the intersection region (Reaction Region in Fig. 2.9) of arms "A" and "B" with the masked photolysis beam measuring 9.0 cm $\times$ 0.9 cm.

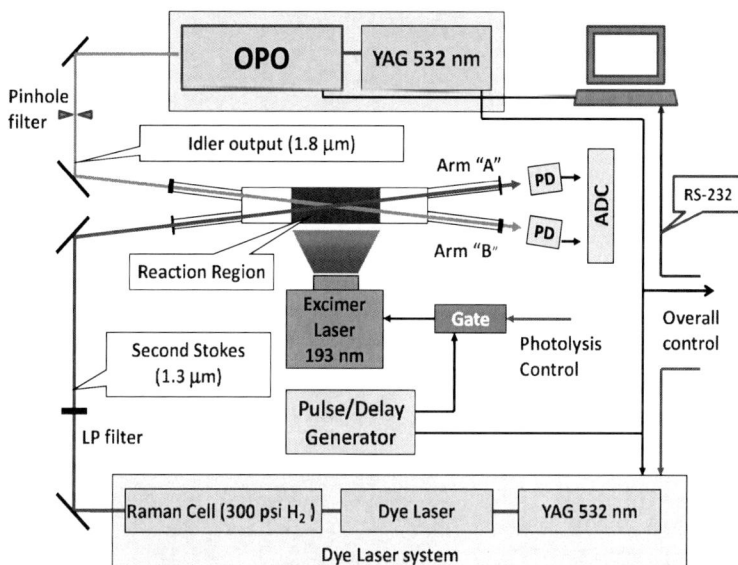

Fig. 2.9 Schematic of dual wavelength CRDS apparatus. Reprinted with permission from [58]. Copyright 2010 American Chemical Society

The use of two arms presents an experimental criteria that needs to be satisfied that was not considered before: arms "A" and "B" need to be "optically equivalent" i.e. in the overlap region between the two arms the concentrations of the gas species integrated over the column length of paths "A" and "B" need to be equal. The equivalence of the two optical paths can be demonstrated by the observation of the 6^1_0 band of methyl peroxy [51]. Two identical mirror sets centered at 1.2 μm were put on both arms "A" and "B" and the Raman shifted radiation from the dye laser was split into two beams of equivalent power and coupled into the arms of the apparatus using identical optics. Methyl peroxy radicals were generated by photolyzing a gas mixture of acetone, oxygen, and nitrogen at 193 nm. The spectra recorded are shown in Fig. 2.10. The red and black curves correspond to absorption spectra recorded from arms "A" and "B", respectively; the green trace at the bottom is the result of the subtraction of the red and black traces. The higher frequency noise and longer term undulations present in the subtracted trace are comparable to the noise in individual traces.

2.3.3.2 Applications of Dual Wavelength Spectroscopy

A successful application of 2λ cavity ringdown spectroscopy is the determination of the $\tilde{A}-\tilde{X}$ absorption cross section of ethyl peroxy. Quantitatively, the absorption

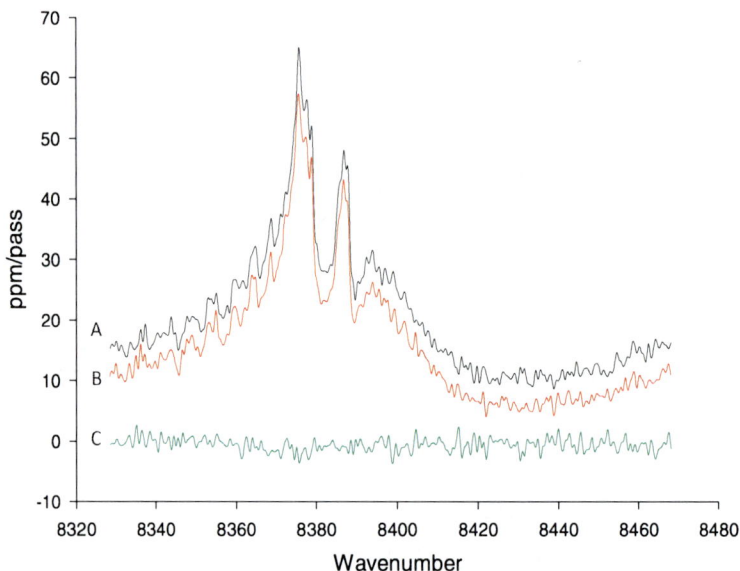

Fig. 2.10 Scan showing the 6^1_0 band of methyl peroxy obtained for each arm of the 2λ system. Both traces appear to be qualitatively identical with the B trace (*red*) offset by -4 ppm/pass for clarity. A subtraction of the two upper traces yields the lower trace C (*green*) with subtraction residuals within experimental error. Reproduced with permission from [58]. Copyright 2010 American Chemical Society

cross section, $\sigma(\nu)$, is defined as

$$S(\nu) = \sigma(\nu) \int_0^l N(x)dx \qquad (2.17)$$

where $S(\nu) = \Delta I(\nu)/I_0(\nu)$ is the fractional absorption of the radiation at frequency ν. The column length integral on the right hand side of the equation accounts for the variation of concentration, $N(x)$, of the absorbing molecules along an optical path of length l. In the case of transient species such as radicals, the integral of $N(x)$ is not known *a priori* and needs to be measured independently. Once it is obtained the absorption cross section can be easily calculated from the measured absorption.

Previous studies have been performed to determine the $\tilde{A}$–$\tilde{X}$ absorption cross section of ethyl peroxy and employed different ways to independently determine the integral $N(x)$. Methods to evaluate the column length integral have included using the time decay of the radicals and the reported self-reaction kinetic rate constants and estimation of radical number density by using the photolysis cross section of oxalyl chloride [25, 52, 53]. An alternative method for determining the column length is the "reporter molecule" approach [54–56]. In the reporter molecule method, the reactive intermediate to be studied is formed in stoichiometric equivalence with a non-reactive molecule which has been well characterized. Ethyl peroxy radicals can be produced by utilizing the production method described in Sect. 2.3.1.1 in which chlorine atoms are used to abstract hydrogen atoms from ethane to generate ethyl radicals which subsequently react with O_2 to give ethyl peroxy. By using this production method, HCl is produced in a 1:1 ratio with ethyl peroxy and its well-known absorption can be used to determine its and the equivalent ethyl peroxy radical's column length.

The 2λ apparatus used for this particular experiment is shown in Fig. 2.9. The G conformer origin of the ethyl peroxy radical is probed using NIR radiation in arm "A" and the rotationally resolved lines P(1), P(5), and P(6) belonging to the vibrational overtone of $H^{37}Cl$ are used as the reporter transitions monitored in arm "B". $H^{37}Cl$ is used as the reporter as opposed to $H^{35}Cl$ because the interference from overtone and combination bands of ethane in the P-branch of the $H^{37}Cl$ isotopologue are minimal when compared with other regions of the HCl spectrum. The monitoring of the P(1), P(5), and P(6) of $H^{37}Cl$ requires radiation in the 1.8 μm region. This radiation is produced using the idler output of an optical parametric oscillator (OPO), operating in single longitudinal mode, [57] pumped with an injection-seeded Nd:YAG laser operating at 10 Hz. The frequency, linewidth, and presence of broadband interference in the OPO output was periodically examined by analyzing the OPO signal bandwidth with a wavemeter. The HWHM of the radiation from the OPO is measured to be ≈0.01 cm^{-1}, with the HWHM of the reporter lines being 0.02–0.034 cm^{-1}. Once the idler beam exits the OPO it is propagated to far field and recollimated by a 1:1 Kepler telescope. The 2 mm paraxial beam exiting the telescope is directed into arm "B" by a $f = 750$ mm lens. The cavity is formed by two highly reflective mirrors centered at 1.8 μm and radiation exiting the cavity is focused onto a fast photodiode for signal processing. The firing of the excimer and probe lasers is initiated by a single pulse from the computer controlling the experiment program via a homemade LabView program. Typically the probe lasers were fired 10 μs (controlled by a delay generator) after the excimer to allow the complete conversion of ethyl radicals to ethyl peroxy. The photodiodes collecting the electrical signals are connected to an ADC card acquiring the data at 10 MS/s rate per channel.

Pressures of precursor gases used in this experiment are 0.8 torr $(COCl)_2$, 0.3 torr ethane, 60.0 torr O_2, and 240 torr N_2. Spectra are recorded by measuring the ringdown time in both cavities. A 40 μs portion of the ringdown curve in arm "A" (peroxy radical) and a 35 μs portion in arm "B" ($H^{37}Cl$) are used to generate absorption spectra. To scan the origin band of the G conformer, the dye laser is scanned over the range 7460–7670 cm^{-1} with a scan step size of 0.2 cm^{-1} with 30 laser shots averaged per frequency point. The $H^{37}Cl$ is recorded by repeatedly scanning the OPO over the same 0.5 cm^{-1}, with a 0.01 cm^{-1} step size, centered at the absorption frequency of the given reporter transition. To minimize effects of backlash, the scanning of the OPO was done unidirectionally; hence, at the end of the 0.5 cm^{-1} scan the OPO frequency was quickly moved to the initial position and the scan over the same range was repeated. This $H^{37}Cl$ scan cycle was repeated until the scan of the peroxy radical was completed. For signal processing, a number of ringdown decay curves are co-added and the cumulative curve is fit to an exponential decay. To eliminate any background structures, on and off traces were simultaneously taken and then subtracted to give the final spectrum.

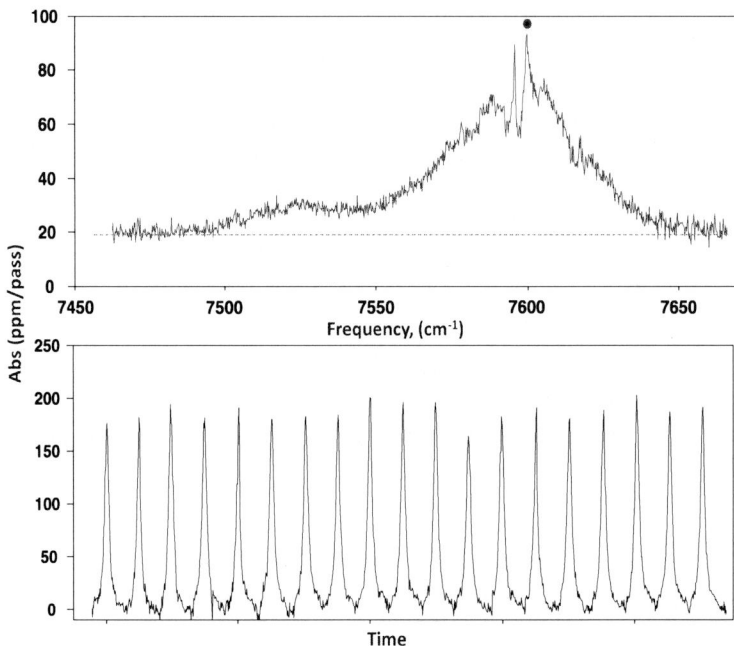

Fig. 2.11 Spectra obtained of P(1) line of $H^{37}Cl$ (*lower*) and origin of the G conformer (*upper*) of ethyl peroxy. The *black dot* in the latter trace marks the peak absorption for ethyl peroxy. Reprinted with permission from [58]. Copyright 2010 American Chemical Society

The simultaneously recorded spectra of the G conformer origin and P(1) reporter line of $H^{37}Cl$ are shown in Fig. 2.11. The top panel shows the ethyl peroxy spectrum while the bottom panel shows repeated 0.5 cm^{-1} scans of the reporter transition taken over the same time span. Small deviations in the peak absorption intensity of the P(1) line, attributed to mechanical backlash of the OPO drive, are compensated by averaging over all the traces. When evaluating the absorption cross section two forms of $\sigma(\nu)$ are generally used, the integral absorption cross-section, σ_I, and peak absorption cross-section, σ_p:

$$\sigma_I = \int \sigma(\nu)d\nu \tag{2.18}$$

$$\sigma_p = \sigma(\nu_{max}, T, P) \tag{2.19}$$

where ν_{max} is the frequency at which the absorption spectrum reaches its peak. To evaluate σ_I, an integration is required over the entire spectral range including regions of the spectrum of low signal/noise (S/N) or regions where there may be interferences from other spectral carriers. Since the σ_p measurement is referenced to a small spectral region where S/N is much higher, a measurement of σ_p is generally more precise than one of σ_I.

Table 2.2 Values of σ_p for ethyl peroxy obtained using various methods. Errors indicate 1σ uncertainties. σ_p values obtained using self-decay rate were done experimentally measuring time decay of ethyl peroxy then using previously reported values of the kinetic rate constant, k_{obs}. Values for k_{obs} were taken from [59, 60] Experimental decay measurements were also done by Atkinson and Spillman [53]; σ_p values were derived from their decay measurements using k_{obs} from Refs. [59, 60]

Method of determination	Reference	σ_p^{EP}, 10^{-21} cm^2
2λ System	[58]	5.29(20)
Self-Decay Rate (Using k_{obs} from [59])	[58]	4.7(14)
Self-Decay Rate (Using k_{obs} from [60])	[53]	3.50(33)
Photolysis Beam Absorption	[25]	3.30(83)
Photolysis Beam Absorption	[58]	6.15(60)

The value reported for the absorption cross section of ethyl peroxy is $\sigma_p = 5.29(20) \times 10^{-21}$ cm^2/molecule [58]. Table 2.2 shows the value for σ_p obtained using the 2λ method compared to alternative ways of determining σ_p and to values obtained in other experiments utilizing the self decay rate of ethyl peroxy or the photolysis beam absorption method. The determination of σ_p from the kinetic decay rate required Melnik et al. [58] to experimentally determine the time decay of ethyl peroxy and then to use previously published values of the effective self-reaction rate constant, k_{obs}, to obtain a value for σ_p. Values for k_{obs} were taken from Lightfoot et al., [59] and Fenter et al. [60]. Experimental decay measurements were also done by Atkinson and Spillman [53] and σ_p values were derived from their decay measurements using k_{obs} from references [59, 60] using the following equation

$$\frac{1}{S(\nu_G, t_d)} = \frac{1}{S(\nu_G, 0)} + \frac{2k_{obs}}{\sigma_p l} t_d \tag{2.20}$$

where t_d is a given delay time after the photolysis laser pulse, $S^{EP}(\nu_G, 0)$ is the fractional absorption of ethyl peroxy at $t = 0$, $S^{EP}(\nu_G, t_d)$ is fractional absorption of ethyl peroxy at some delay time t_d, and l is absorption path length of ethyl peroxy. The evaluation of σ_p was also repeated by Melnik et al. [58] and compared to a previous value reported in [25]. The values of σ_p presented in Table 2.2 qualitatively correlate well with one another. However, the determination of σ_p with the 2λ-CRDS system likely gives an error substantially smaller than any of the other experiments given the scatter and hence uncertainty in the reported values for k_{obs} and the photolysis cross-section.

2.3.4 Self-Reaction Kinetics

While definitive spectroscopic identification and characterization of individual peroxy radicals is desirable, it is also important to accurately model peroxy radical reactions and photodissociation pathways that occur in atmospheric and combustion processes. In order to achieve this goal, many studies have been done to accurately determine the $\tilde{A}-\tilde{X}$ absorption cross section for peroxy radicals, as discussed in the previous section, and to measure kinetic reaction rates of peroxy radicals with themselves as well as other peroxy radicals and reactive intermediates. Kinetic reaction rates are desired by workers in both the atmospheric and combustion communities in order to completely characterize hydrocarbon oxidation cycles in atmospheric and combustion chemistry. In addition, the self-reaction rate of a molecule can be used to calculate the absorption cross section. Thus, it is beneficial to get the most accurate and precise determination possible for this value. As described below the 2λ CRDS apparatus can be used to measure independently the self-reaction rate of ethyl peroxy.

2.3.4.1 Experimental Approach and Apparatus

When the ethyl peroxy radicals are generated in the controlled laboratory environment the following elementary reactions occur [20],

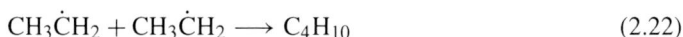

$$CH_3\dot{C}H_2 + O_2 + N_2 \longrightarrow CH_3CH_2OO\cdot \tag{2.21}$$

$$CH_3\dot{C}H_2 + CH_3\dot{C}H_2 \longrightarrow C_4H_{10} \tag{2.22}$$

$$CH_3CH_2OO\cdot + CH_3CH_2OO\cdot \longrightarrow CH_3CH_2O\cdot + O_2 \tag{2.23}$$

$$CH_3CH_2OO\cdot + CH_3CH_2OO\cdot \longrightarrow 2CH_3CHO + C_2H_5OH + O_2 \tag{2.24}$$

$$CH_3CH_2O\cdot + O_2 \longrightarrow CH_3CHO + HOO\cdot \tag{2.25}$$

$$CH_3CH_2OO\cdot + HOO\cdot \longrightarrow C_2H_5OOH + O_2 \tag{2.26}$$

$$HOO\cdot + HOO\cdot \longrightarrow H_2O_2 + O_2 \tag{2.27}$$

The self-reaction rate of ethyl peroxy, k, receives contributions from both Eqs. (2.23) and (2.24). But in the laboratory environment, the *total destruction* of ethyl peroxy is observed, not just the self reaction. In other words, the rate of decay that is measured in the experiment is the effective self-reaction rate, k_{obs}, which receives contributions from Eqs. (2.23), (2.24), and (2.26). The relationship between k_{obs} and self-reaction rate k is documented in the literature [59]. To obtain the effective self-reaction rate, k_{obs}, of ethyl peroxy a departure is made from the pulsed experiments used to probe the first three aspects of ethyl peroxy and a cw-CRDS technique is implemented. The kinetics experiments are conducted in arm "A" of the 2λ system introduced Sect. 2.3.3.1. The present modifications to the apparatus [61] for this experiment are shown in Fig. 2.12. By adding a laser probe in arm "B", reaction rate constants between two reactive species may be measured. The ethyl peroxy radicals are generated by photolyzing 3-pentanone with 193 nm light from an excimer in the presence of oxygen with the photolysis laser pulse at $t = 15$ ms into the 50 ms experimental time frame between photolysis laser shots. (Pressures of precursor gases consisted of 0.2–0.6 torr 3-pentanone, 60.0 torr O_2, and 250.0 torr

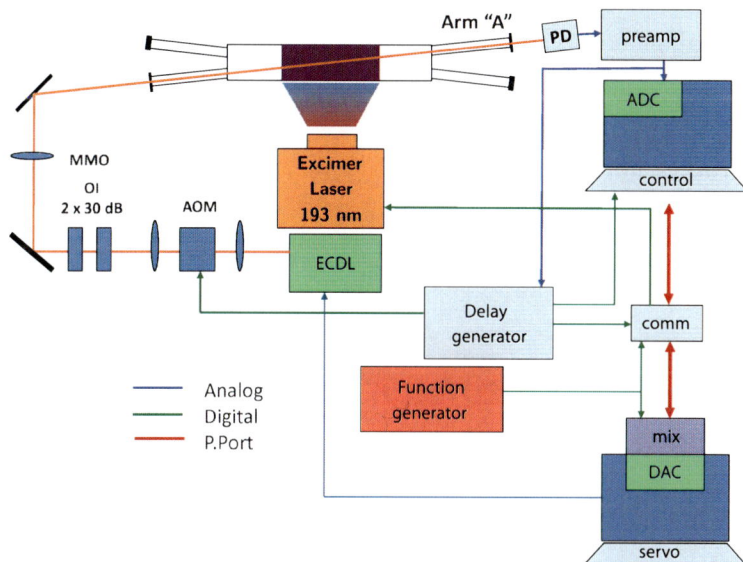

Fig. 2.12 cw-CRDS setup for measuring self-reaction rate of ethyl peroxy. Abbreviations: ECDL—external cavity diode laser, OI—optical isolators, AOM—acousto-optical modulator, MMO—mode matching optics

N_2.) To monitor the time decay of the ethyl peroxy radicals, the peak absorption for the origin of the G conformer at 7596 cm^{-1} is used. An external cavity diode laser is used to produce the radiation at 7596 cm^{-1} and is coupled to the TEM$_{00}$ mode of the cavity through mode-matching optics. Two 30-dB optical isolators are placed after the laser to prevent back reflection to the diode laser. The power output of the diode laser was maintained between 4.5–5.5 mW, with about 3 mW impinging on the CRDS mirror, in order to keep the laser operating with a single mode. There is also a piezo drive actuator driven by an arbitrary function generator that dithers the laser frequency slightly to keep the laser locked onto the TEM$_{00}$ mode of the cavity.

The time decay of the ethyl peroxy radicals monitored in 50 ms time intervals obtaining individual ringdown events every 250 µs. An example of the individual ringdown events recorded is shown in Fig. 2.13. In this particular time window the first two individual ringdown events are recorded in the initial 500 µs of the experiment. The ringdown times for the individual events are then determined and converted to an absorption (ppm/pass) to generate a concentration profile of ethyl peroxy with each individual ringdown event representing a single data point in the decay curve. Figure 2.14 shows the resulting kinetic decay curve for a 50 ms experiment including a photolysis laser shot. To obtain individual ringdown events, an acousto optic modulator is used to block the radiation coming into the ringdown cavity. Radiation from the diode laser is permitted to enter the CRDS cavity until a certain threshold voltage is reached on the photodiode. When the threshold voltage is attained, the photodiode sends a signal to switch the AOM on and block the radiation. The AOM is switched on for approximately 250 µs and a ringdown event is

Fig. 2.13 The individual
ringdown curves obtained in
the initial 500 μs of the 50 ms
kinetics experiment are
shown. Ringdown times are
determined and are converted
to absorbance values to plot
kinetic decay curve

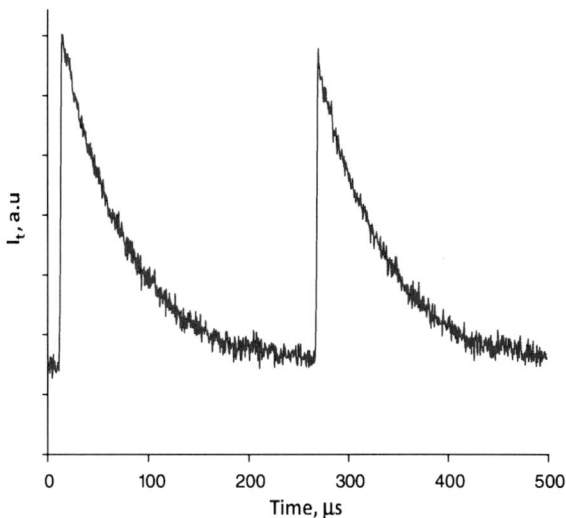

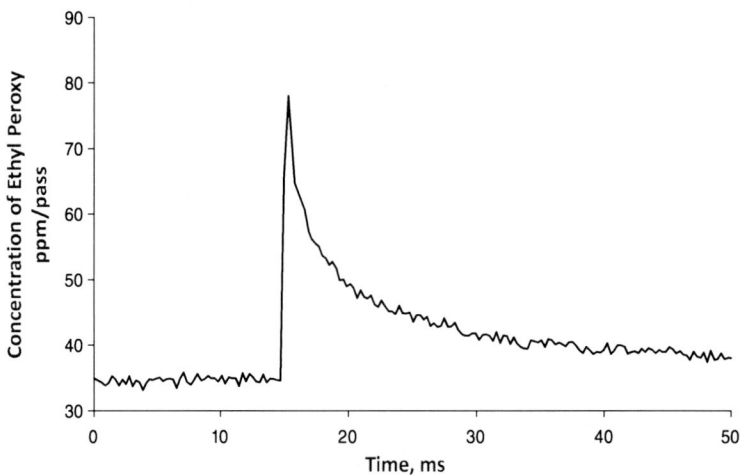

Fig. 2.14 Kinetic decay curve obtained for the self reaction of ethyl peroxy

recorded and stored in a file. At 250 μs the AOM is switched back off and radiation
is again allowed to enter the CRDS cell with the procedure repeating to obtain individual ringdown events. Previous studies by Atkinson and Hudgens [52] and Fenter
et al. [60] determined k_{obs} by measuring the absorption at a single delay time after
the photolysis pulse. One of the main sources of error associated with this technique
is that every data point is obtained with a different set of radicals, which can lead to
errors associated with variation in the initial radical concentration due to fluctuation
in the photolysis laser power and other variables. This source of error is effectively
removed in this experiment since all of the ringdowns and corresponding concen-

Table 2.3 Values of k_{obs} obtained from various experiments. Errors indicate 1σ uncertainties	k_{obs}, 10^{-13} cm^3/s	Reference
	0.966(44)	Ref. [61]
	1.08(34)	Ref. [59]
	1.29(7)	Ref. [60]

tration determinations are carried out on the same sample of radicals over a period of time.

2.3.4.2 Self-Reaction Kinetics: Observations and Results

To extract k_{obs} from the kinetic decay curve shown in Fig. 2.14, the simple second order rate law is used

$$\frac{dN_{C_2H_5O_2}(t, x)}{dt} = -2k_{obs} N^2_{C_2H_5O_2}(t, x) \qquad (2.28)$$

where k_{obs} accounts for the removal by self reaction and secondary chemistry and the number density of the ethyl peroxy radicals ($N_{C_2H_5O_2}$) is a function of time and position along the interrogating path. Since the photolysis beam is carefully shaped and the path length is not very long, it is reasonable to neglect the x dependence of the concentration. However, since the decay is relatively slow there is a contribution to the concentration decay of ethyl peroxy from the physical removal of radicals due to pumping out of the exhaust port. This means that on the time scale of the described experiment, Eqs. (2.23), (2.24), and (2.26) are not the only mechanisms by which the ethyl peroxy radical is removed and that the pumping out of the radicals needs to be considered as well. These effects can be well modeled and an excellent value of k_{obs} was determined [61]. This value is compared with previously reported values in Table 2.3. Since an independent determination of k_{obs} can be made every 50 ms, it is easy to average $\geq 10^3$ measurements to obtain an extremely precise value of k_{obs}.

2.4 Dynamics

In this chapter we have used a single example radical, ethyl peroxy, to illustrate different ringdown techniques that are available for the detection, spectral characterization, and monitoring of reactive intermediates. However, there are many other peroxy radicals that have been investigated by CRDS (for example, see Table 2.1) that were not discussed in this chapter and some of them illustrate other kinds of information obtainable from CRDS experiments on reactive intermediates. Room temperature, moderate resolution and jet cooled, high resolution CRDS experiments have been performed on a derivative of ethyl peroxy, the β-hydroxyethylperoxy (β-HEP) radical. Hydroxy peroxy radicals in general are formed [62–64] in the atmosphere by ·OH attack on alkenes to produce a hydroxy alkyl radical, followed by reaction with oxygen to produce the hydroxy peroxy radical. To generate β-HEP specifically, ethene is attacked by ·OH followed by a reaction with oxygen,

$$H_2C = CH_2 + \cdot OH \longrightarrow \cdot CH_2CH_2OH \tag{2.29}$$

$$\cdot CH_2CH_2OH + O_2 + M \longrightarrow \cdot OOCH_2CH_2OH \tag{2.30}$$

Hydroxy peroxy radicals can also be formed as part of the combustion process in motor vehicles as ethanol and other alcohols are added to automotive fuel to cut down on emissions generated by internal combustion engines. Under combustion conditions, hydroxy peroxy radicals are formed by the abstraction of an alkyl hydrogen with subsequent reaction of the radical with oxygen.

To generate β-HEP in both the room temperature and jet cooled apparatuses, iodoethanol can be used as a precursor [65],

$$IH_2CCH_2OH \longrightarrow \cdot CH_2CH_2OH + I\cdot \tag{2.31}$$

$$\cdot CH_2CH_2OH + O_2 + N_2 \longrightarrow \cdot OOCH_2CH_2OH + N_2 \tag{2.32}$$

In the room temperature apparatus the iodine atom is dissociated with 248 nm light from the photolysis excimer laser, with the spectrum obtained [65] shown in Fig. 2.15.

There are 13 conformers that are predicted to be stable minima on the potential energy surface, [65] however, only two are observed in the spectrum due to a combination of low oscillator strength and/or Boltzmann population at room temperature. The $G_1'G_2G_3$ and $G_1G_2G_3$ conformers, as defined by the $\angle OOCC$, $\angle OCCO$, and $\angle CCOH$ dihedral angles, respectively, are found to be carriers of the spectrum shown in Fig. 2.15. Overall, the β-HEP spectrum contains many of the same general features that unsubstituted peroxy radicals possess. The two broad bands appearing at the red end of the spectrum are the origin bands of the two conformers and roughly 900 cm^{-1} to the blue fundamental transitions due to O–O stretching appear for both conformers. There is also a very intense band appearing in the 7700 cm^{-1} region involving the $\angle CCOH$ torsion fundamentals of β-HEP with the first overtone of this band appearing at around 8100 cm^{-1}; additionally there are higher frequency combination bands appearing between 8600–8800 cm^{-1}.

In the jet cooled, high resolution apparatus the slit-jet discharge is used to dissociate the iodine atom from iodoethanol. A spectrum of the origin band for the $G_1G_2G_3$ conformer of β-HEP is shown [66] in Fig. 2.16. The top panel of Fig. 2.16, shows the spectrum of the fully protonated β-HEP while the bottom panel shows the spectrum of the singly deuterated analogue, ·OOCH$_2$CH$_2$OD. As demonstrated in Sect. 2.3.2, when using SRS the resolution of the jet-cooled high resolution apparatus is $\approx$250 MHz. Looking at the spectra of the protonated and singly deuterated β-HEP radical, one can see that there is no visible rotational structure in the

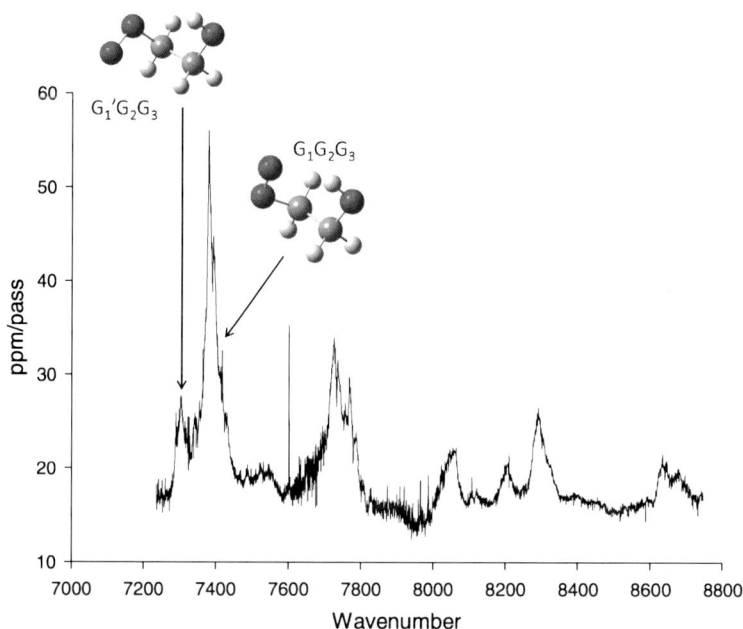

Fig. 2.15 Room temperature, moderate resolution spectrum of β-HEP radical. The origin bands of the two conformers present in spectrum are indicated by *arrows*

protonated analogue, but in the singly deuterated β-HEP there is clearly some rotational structure in the spectrum, as well as a small shift in the origin frequency due to isotopic substitution. By simulating the spectra, the experimental Lorentzian linewidths for the two isotopic analogues are determined to be much greater than the instrumental linewidth of 250 MHz with values of 7300 MHz and 3600 MHz determined for the protonated and singly deuterated β-HEP radicals, respectively. Chen et al. [66] studied this phenomenon extensively via isotopic substitution and determined it was related to an excited state dynamical effect along the reaction coordinate by which the hydrogen atom on the OH group is transferred to the peroxy moiety. In protonated β-HEP the excited state lifetime was determined to be 22 ps for this process, while in the singly deuterated isotopomer it was found to be 45 ps. When the deuterium is substituted for the hydrogen the rate of transfer slows down giving a longer excited state lifetime and a correspondingly smaller experimental linewidth. This experiment demonstrates the capability of high resolution CRDS to be used to monitor excited state dynamics.

When analyzing the high resolution spectra for both ethyl peroxy and β-HEP, the Lorentzian component of the linewidth was found to be significantly greater than the instrumental linewidth of 250 MHz, [33, 66] with the major contribution being due the dynamical effects which limited the excited state lifetime. While this observation allowed measurement of the dynamical process the downside is that lifetime broadening of linewidths can limit the resolution of the CRDS spectrum and the structural information obtainable from it.

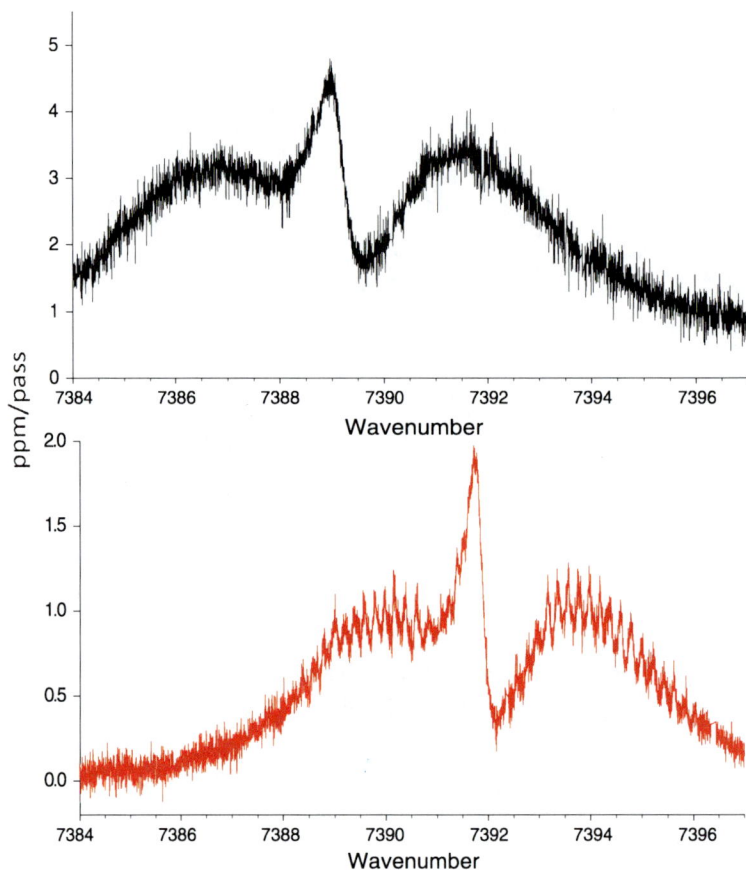

Fig. 2.16 Jet cooled, high resolution spectrum of $G_1G_2G_3$ conformer of β-HEP. The *top panel* (*black trace*) shows the high resolution spectrum of the fully protonated isotopologue, while the *bottom panel* (*red trace*) shows the singly deuterated version in which the hydroxyl hydrogen has been exchanged with a deuterium atom

While lifetime effects may limit rotational resolution for the jet-cooled ethyl peroxy and β-HEP spectra, their room temperature spectra are sufficiently resolved to allow the unambiguous assignment of the spectra to individual peroxy radicals and even specific conformers. However, as the peroxy radicals become larger in size there are other factors that can limit the spectral resolution and information obtainable. The molecule can assume a greater number of conformations and more conformers may be populated at room temperature and contribute to the spectrum, making explicit assignment of particular conformers difficult. This was observed in the study of the butyl peroxy radical family and also with pentyl peroxy radicals [67, 68]. In both of these cases, spectra were obtained of each of the individual isomers of butyl peroxy (1-butyl, 2-butyl, isobutyl, *t*-butyl) and of pentyl peroxy (1-pentyl, neopentyl, 2-methylbutyl, 3-methylbutyl, 3-pentyl,3-methyl-2-butyl, *t*-pentyl). In *t*-butyl peroxy only one conformer is expected to appear in the spectrum, but in the other three isomers of butyl peroxy there are numerous conformers present and individual conformer bands overlapped and could not be resolved with the room temperature, moderate resolution CRDS system. The same situation was observed for pentyl peroxy, with the one exception being neopentyl peroxy which only has two stable conformers. Both conformers were observed and uniquely assigned in that case. However, for the other four isomers of the pentyl peroxy family, spectra of the individual isomers was obtained, but conformer specific assignments proved to be difficult as there was too much spectral overlap present.

2.5 Conclusion

In this chapter the ethyl peroxy (and its derivative, β-HEP) radical have been used to illustrate the capabilities of CRDS for the study of reactive chemical intermediates both at room and low (jet-cooled) temperatures. It has been shown that the technique can identify the intermediate with respect to chemical species, and isomeric and conformeric structure. The room temperature CRDS spectral fingerprints can be used for detecting and monitoring these intermediates during chemical reactions, particularly atmospheric and combustion reactions. The jet-cooled spectra provide precise molecular parameters, e.g., rotational constants, spin-rotation tensor components, orientation of the dipole moment, etc. that can be used to benchmark various quantum chemical calculations. The jet-cooled spectra can also probe dynamical processing occurring in the picosecond regime. The fingerprint spectra, particularly the room temperature ones, provide a means not only for qualitatively identifying reactive intermediates but also to measure their concentrations and reaction rates. These tasks are most efficiently performed using a 2λ CRDS spectrometer.

Of course all techniques have limitations. For some reactive intermediates, dynamical effects may limit resolution and the precision of molecular parameters derived from the spectrum. While the CRDS techniques can provide very explicit species, isomer, and conformer identification for smaller organic radicals, eventually the molecules become too large to uniquely discriminate among species and structures. Nonetheless the rapid progress in the first decade of the 21st century of CRDS applications to reactive intermediates raises our expectations for similar such progress in the coming decade.

Acknowledgements   The authors acknowledge the financial support of this work by the US Department of Energy, via Grant DE-FG01-01ER14172. They also acknowledge helpful conversations with Dmitry Melnik.

References

1. C. Wang, A. Mbi, M. Shepherd, IEEE Sens. J. **10**, 54 (2010)
2. M. Snels, T. Venezia, L. Belfiore, Chem. Phys. Lett. **489**, 134 (2010)

3. H.D. Osthoff, S.S. Brown, T.B. Ryerson, T.J. Fortin, B.M. Lerner, E.J. Williams, A. Petters-son, T. Baynard, W.P. Dubé, S.J. Ciciora, A.R. Ravishankara, J. Geophys. Res. **111**, D12305(1) (2006)
4. S.S. Brown, H. Stark, A.R. Ravishankara, Appl. Phys. B **75**, 173 (2002)
5. M. Aldener, S.S. Browns, H. Stark, J.S. Daniel, A.R. Ravishankara, J. Mol. Spectrosc. **232**, 223 (2005)
6. D.S. Sayres, E.J. Moyer, T.F. Hanisco, J.M.St. Clair, F.N. Keutsch, A. O' Brien, N.T. Allen, L. Lapson, J.N. Demusz, M. Rivero, T. Martin, M. Greenberg, C. Tuozzolo, G.S. Engel, J.H. Kroll, J.B. Paul, J.G. Anderson, Rev. Sci. Instrum. **80**, 044102(1) (2009)
7. M.F. Witinski, D.S. Sayres, J.G. Anderson, Appl. Phys. B **102**, 375 (2011)
8. E.J. Moyer, D.S. Sayres, G.S. Engel, J.M.St. Clair, F.N. Keutsch, N.T. Allen, J.H. Kroll, J.G. Anderson, Appl. Phys. B **92**, 467 (2008)
9. P.B. Tarsa, A.D. Wist, P. Rabinowitz, K.K. Lehmann, Appl. Phys. Lett. **85**, 4523 (2004)
10. E.N. Sharp, P. Rupper, T.A. Miller, Phys. Chem. Chem. Phys. **10**, 3955 (2008)
11. H. Linnartrz, in *Cavity Ring-Down Spectroscopy: Techniques and Applications* (Wiley, New York, 2009)
12. H.D. Osthoff, M.J. Pilling, A.R. Ravishankara, S.S. Browns, Phys. Chem. Chem. Phys. **9**, 5785 (2007)
13. I. Glassman, *Combustion* (Academic Press, Orlando, 1987)
14. H.J. Curran, P. Gaffuri, W.J. Pitz, C.K. Westbrook, Combust. Flame **114**, 149 (1998)
15. A.R. Ravishankara, Proc. Natl. Acad. Sci. USA **106**, 13639 (2009)
16. O.J. Nielsen, T.J. Wallington, *Peroxyl Radicals* (Wiley, New York, 1997)
17. M.J. Pilling, I.W.M. Smith, in *Modern Gas Kinetics: Theory, Experiment, and Application* (Blackwell Scientific, Oxford, 1987), Chaps. C1 and C2
18. T.P. Murrells, M.E. Jenkin, S.J. Shalliker, G.D. Hayman, J. Chem. Soc. Faraday Trans. **87**, 2351 (1991)
19. T.J. Wallington, P. Dagaut, M.J. Kurylo, Chem. Rev. **92**, 667 (1992)
20. R. Atkinson, D.L. Baulch, R.A. Cox, R.F. Hampson, J.A. Kerr, J. Troe, J. Phys. Chem. Ref. Data **18**, 881 (1989)
21. E. Villenave, R. Lesclaux, J. Phys. Chem. **100**, 14372 (1996)
22. O.J. Nielsen, J. Sehested, S. Langer, E. Ljungstroem, I. Wangberg, Chem. Phys. Lett. **238**, 359 (1995)
23. J.A. Jafri, D.H. Phillips, J. Am. Chem. Soc. **112**, 2586 (1990)
24. S. Newman, I. Lane, A. Orr-Ewing, D. Newnham, J. Ballard, J. Chem. Phys. **110**, 10749 (1999)
25. P. Rupper, E.N. Sharp, G. Tarczay, T.A. Miller, J. Phys. Chem. A **111**, 832 (2007)
26. M. Ahmed, D. Blunt, D. Chen, A.G. Suits, J. Chem. Phys. **106**, 7617 (1997)
27. A.V. Baklanov, L.N. Krasnoperov, J. Phys. Chem. A **105**, 97 (2001)
28. H. Ito, Y. Nogata, S. Matsuzaki, A. Kuboyama, Bull. Chem. Soc. Jpn. **42**, 2453 (1969)
29. J.C. Rienstra-Kiracoffe, W.D. Allen, H.F. Schaefer III., J. Phys. Chem. A **104**, 9823 (2000)
30. H.E. Hunziker, H.R. Wendt, J. Chem. Phys. **64**, 3488 (1976)
31. S.J. Blanksby, T.M. Ramond, G.E. Davico, M.R. Nimlos, S. Kato, V.M. Bierbaum, W.C. Lineberger, G.B. Ellison, M. Okumura, J. Am. Chem. Soc. **123**, 9585 (2001)
32. V.L. Stakhursky, T.A. Miller, in *56th Annual OSU International Symposium on Molecular Spectroscopy*, Columbus, OH (2001)
33. G.M.P. Just, P. Rupper, T.A. Miller, W.L. Meerts, J. Chem. Phys. **131**, 184303 (2009)
34. E. Cromwell, T. Tricki, Y.T. Lee, A.H. Kung, Rev. Sci. Instrum. **60**, 2888 (1989)
35. W.F. Polik, D.R. Guyer, C.B. Moore, J. Chem. Phys. **92**, 3453 (1990)
36. C.K. Ni, A.H. Kung, Rev. Sci. Instrum. **71**, 3309 (2000)
37. P. Dupré, T.A. Miller, Rev. Sci. Instrum. **78**, 033102 (2007)
38. J.C. Barnes, N.P. Barnes, L.G. Wang, W. Edwards, Proc. IEEE **29**, 2684 (1993)
39. N.D. Finkelstein, W.L. Lempert, R.B. Miles, A. Finch, G.A. Rines, AIAA Paper Number 96-0177, 1 (1995)
40. T. Walther, M.P. Larsen, E.S. Fry, Appl. Opt. **40**, 3046 (2001)

41. S. Wu, P. Dupré, T.A. Miller, Phys. Chem. Chem. Phys. **8**, 1682 (2006)
42. C.M. Lovejoy, D.J. Nesbitt, J. Chem. Phys. **86**, 3151 (1987)
43. K.L. Busarow, B.A. Blake, K.B. Laughlin, R.C. Cohen, Y.T. Lee, R.J. Saykally, Chem. Phys. Lett. **141**, 2889 (1987)
44. D.T. Anderson, S. Davis, T.S. Zwier, D.J. Nesbitt, Chem. Phys. Lett. **258**, 207 (1996)
45. S. Davis, D.T. Anderson, G. Duxbury, D.J. Nesbitt, J. Chem. Phys. **107**, 5661 (1997)
46. K. Liu, R.S. Fellers, M.R. Viant, R.P. McLaughlin, M.G. Brown, R.J. Saykally, Rev. Sci. Instrum. **67**, 410 (1996)
47. S. Wu, P. Dupré, P. Rupper, T.A. Miller, J. Chem. Phys. **127**, 224305 (2007)
48. W.L. Meerts, M. Schmitt, Int. Rev. Phys. Chem. **25**, 353 (2006)
49. G. Myszkiewicz, W.L. Meerts, C. Ratzer, M. Schmitt, Chem. Phys. Chem. **6**, 2129 (2005)
50. W.L. Meerts, M. Schmitt, Int. Rev. Phys. Chem. **25**, 353 (2006)
51. C.-Y. Chung, C.-W. Cheng, Y.-P. Lee, H.-Y. Liao, E.N. Sharp, P. Rupper, T.A. Miller, J. Chem. Phys. **127**, 044311 (2007)
52. D.B. Atkinson, J.W. Hudgens, J. Phys. Chem. A **101**, 3901 (1997)
53. D.B. Atkinson, J.L. Spillman, J. Phys. Chem. A **106**, 8891 (2002)
54. H. Ismail, P.R. Abel, W.H. Green, A. Fahr, L.E. Jusinski, A.M. Knepp, J. Zador, G. Meloni, T.M. Selby, D.L. Osborn, C.A. Taatjes, J. Phys. Chem. A **113**, 1278 (2009)
55. C.A. Taatjes, J. Phys. Chem. A **110**, 4299 (2006)
56. W. Ludwig, B. Brandt, G. Frederichs, F. Temps, J. Phys. Chem. A **110**, 3330 (2006)
57. W.R. Bosenberg, D.R. Guyer, Appl. Phys. Lett. **61**, 387 (1992)
58. D. Melnik, R. Chhantyal-Pun, T.A. Miller, J. Phys. Chem. A **114**, 11583 (2010)
59. P.D. Lightfoot, R.A. Cox, J.N. Crowley, M. Destriau, G.D. Hayman, M.E. Jenkin, G.K. Moortgat, F. Zabel, Atmos. Environ. **26A**, 1805 (1992)
60. F.F. Fenter, V. Catoire, R. Lesclaux, P.D. Lightfoot, J. Phys. Chem. **97**, 3530 (1993)
61. D. Melnik, T.A. Miller, J. Chem. Phys. **139**, 0924201 (2013)
62. J.P. Senosiain, S.J. Klippenstein, J.A. Miller, J. Phys. Chem. A **110**, 6960 (2006)
63. S. Olivella, A. Solé, J. Phys. Chem. A **108**, 11651 (2004)
64. J.G. Calvert, R. Atkinson, J.A. Kerr, S. Madronich, G.K. Moortgat, T.J. Wallington, G. Yarwood, *The Mechanism of Atmospheric Oxidation of the Alkenes* (Oxford University Press, New York, 2000)
65. R. Chhantyal-Pun, N.D. Kline, P.S. Thomas, T.A. Miller, J. Phys. Chem. Lett. **1**, 1846 (2010)
66. M.-W. Chen, G.M.P. Just, T. Codd, T.A. Miller, J. Chem. Phys. **135**, 184304 (2011)
67. B.G. Glover, T.A. Miller, J. Phys. Chem. A **109**, 11191 (2005)
68. E.N. Sharp, P. Rupper, T.A. Miller, J. Phys. Chem. A **112**, 1445 (2008)
69. M.B. Pushkarsky, S.J. Zalyubovsky, T.A. Miller, J. Chem. Phys. **112**, 10695 (2000)
70. S.J. Zalyubovsky, D. Wang, T.A. Miller, Chem. Phys. Lett. **335**, 298 (2001)
71. G.M.P. Just, A.B. McCoy, T.A. Miller, J. Chem. Phys. **127**, 044310 (2007)
72. D. Melnik, P.S. Thomas, T.A. Miller, J. Phys. Chem. A **115**, 13931 (2011)
73. S.J. Zalyubovsky, B.G. Glover, T.A. Miller, C. Hayes, J.K. Merle, C.M. Hadad, J. Phys. Chem. A **109**, 1308 (2005)
74. G. Tarczay, S.J. Zalyubovsky, T.A. Miller, Chem. Phys. Lett. **406**, 81 (2005)
75. G.M.P. Just, P. Rupper, T.A. Miller, W.L. Meerts, Phys. Chem. Chem. Phys. **112**, 4773 (2010)
76. P.S. Thomas, R. Chhantyal-Pun, T.A. Miller, J. Phys. Chem. A **114**, 218 (2010)
77. P.S. Thomas, T.A. Miller, Chem. Phys. Lett. **514**, 196 (2011)
78. G.M.P. Just, E.N. Sharp, S.J. Zalyubovsky, T.A. Miller, Chem. Phys. Lett. **417**, 378 (2006)
79. P.S. Thomas, T.A. Miller, Chem. Phys. Lett. **491**, 123 (2010)
80. P.S. Thomas, N.D. Kline, T.A. Miller, J. Phys. Chem. A **114**, 12437 (2010)
81. N.D. Kline, T.A. Miller, Chem. Phys. Lett. **530**, 16 (2012)
82. S.J. Zalyubovsky, B.G. Glover, T.A. Miller, J. Phys. Chem. A **107**, 7704 (2003)

Chapter 3
Quantum Cascade Laser Based Chemical Sensing Using Optically Resonant Cavities

S. Welzel, R. Engeln, and J. Röpcke

Abstract Progress in the development of compact semiconductor-based mid-infrared light sources, more specifically of quantum cascade lasers (QCLs), has been astonishingly rapid in the 2 decades since their first realisation. Their performance makes them superior to conventional sources and has led to significant improvements and new developments in chemical sensing techniques encompassing cavity enhanced methods. The aim of this compilation is to provide an overview about useful combinations of QCLs with optical cavities and to highlight recent achievements thereby focussing on potential sensing applications.

Abbreviations

AS	absorption spectroscopy
CEA(S)	cavity enhanced absorption (spectroscopy)
CRD(S)	cavity ring-down (spectroscopy)
cw	continuous wave
DFB	distributed feedback
DFG	difference frequency generation
EC	external cavity
FC	(optical) frequency comb
FM	frequency modulation
FSR	free spectral range
FT(S)	Fourier-transform (spectroscopy)
FWHM	full-width at half maximum
HWHM	half width at half maximum

S. Welzel (✉) · R. Engeln
Eindhoven University of Technology, P.O. Box 513, 5600 MB Eindhoven, The Netherlands
e-mail: s.welzel@tue.nl

S. Welzel
Dutch Institute for Fundamental Energy Research (DIFFER), P.O. Box 1207, 3430 BE
Nieuwegein, The Netherlands
e-mail: welzel.s@freenet.de

J. Röpcke
INP Greifswald e.V., Felix-Hausdorff-Str. 2, 17489 Greifswald, Germany

G. Gagliardi, H.-P. Loock (eds.), *Cavity-Enhanced Spectroscopy and Sensing*,
Springer Series in Optical Sciences 179, DOI 10.1007/978-3-642-40003-2_3,
© Springer-Verlag Berlin Heidelberg 2014

ICL	interband cascade laser
ICLAS	intracavity laser absorption spectroscopy
ICO(S)	integrated output (spectroscopy)
IR	infrared
LN	liquid nitrogen
MDND	minimum detectable number density
MW	microwave
NEA	noise equivalent absorption
NICE-OHMS	noise-immune cavity enhanced optical heterodyne molecular spectroscopy
OA	off-axis
OF	optical feedback
OPO	optical parametric oscillator
p	pulsed
(QE-)PAS	(quartz-enhanced) photoacoustic spectroscopy
PP-LN/LT	periodically poled lithium niobate/lithium tantalite
PP-KTP/KTA	periodically poled potassium titanyl phosphate/potassium titanyl arsenate
PS	phase shift
QCL	quantum cascade laser
QPM	quasi phase matching
QW	quantum well
SNR	signal-to-noise ratio
TE	thermoelectrical(ly)
WM	wavelength modulation

3.1 Introduction to Molecular Absorption Spectroscopy

Solutions to present societal challenges like climate changes, the transition into a sustainable economy, or threats to homeland security often involve the identification, quantification and monitoring of molecular substances. High-resolution chemical sensing of small amounts of molecular species using laser-based absorption spectroscopy (AS), also known as trace gas sensing, is essential for such measurements. Scientifically, these studies of complex gas mixtures aim at scrutinising chemical, physical and biological processes. Chemical sensing applications are thereby as widespread as potential target molecules and encompass constituents of the atmosphere along with their isotopic ratios, potential marker molecules in breath gas monitoring, and stable or transient species in reactive environments such as combustion processes or electrical gas discharges [1–5].

Narrow line width tuneable laser sources of adequately high spectral output power are the key to achieve high selectivity and sensitivity, respectively. The narrow spectral width enables absorption features of different molecular species to be discriminated specifically at reduced sample pressures. Additionally, the absorption scales approximately with the ratio of absorption line width and laser spectral width [6]. A narrow laser line width is hence a critical parameter to realise a highly sensitive experiment, that is the quantification of a gas with very low and perhaps highly variable concentration. The requirement of high sensitivity along with high time-resolution, e.g., for the detection of radicals and reactive species in hot or turbulent gas flows, should be carefully distinguished from the need for high precision as often encountered in environmental sensing. In the latter case very small fractional changes in nearly constant levels of stable gases need to be precisely monitored with an accuracy clearly smaller than 1 % while sensitivity might be less important [7]. The important distinction between sensitivity and precision in gas-phase spectroscopy along with the robustness and complexity of the (optical) arrangement determines the selection of the specific absorption technique and the experimental equipment involved. These considerations are closely linked to the spectral discrimination and detection of light after propagation through the analyte.

In general, the absorption of radiation in a homogeneous sample can be described by the Beer-Lambert law of linear absorption which is

$$\ln\left(\frac{I_0(\nu)}{I(\nu)}\right) = k(\nu)L_{\text{eff}}.$$

(3.1)

$I_0(\nu)$ and $I(\nu)$ denote the incident and the transmitted light through the sample and the effective absorption length L_{eff} embodies all variants of absorption techniques, i.e. single-pass absorption, folded optical paths (multi-pass cells) or the transmission through optical cavities. The absorption coefficient, $k(\nu)$, is the sum of all individual absorption features, with absorption cross section $\sigma_i(\nu)$ and their corresponding number density, n_i (i.e., number of molecules per unit volume)

$$k(\nu) = \sum_i k_i(\nu) = \sum_i n_i \sigma_i(\nu).$$

(3.2)

The absorption cross section of a transition σ may also be expressed by a normalised line profile ϕ centred at ν_0 and the line strength, S, as proportionality factor which is frequently used in spectroscopic databases

$$\sigma(\nu - \nu_0) = S \cdot \phi(\nu - \nu_0).$$

(3.3)

The sensitivity of a measurement is typically evaluated by the signal-to-noise ratio, SNR, and can be estimated by transforming the absorption signal, $I_{\text{abs}}(\nu)$, into the transmitted intensity, $I(\nu)$, and assuming weak absorption in (3.1)

$$SNR = \frac{I_{\text{abs}}(\nu)}{\Delta I(\nu)} \approx \frac{I_0(\nu)}{\Delta I(\nu)} k(\nu) L_{\text{eff}}.$$

(3.4)

From Eq. (3.4) an expression for the minimum detectable absorption coefficient $k_{\min}$ or number density (MDND), $n_{\min}$, can be deduced for $SNR = 1$

$$n_{\min}\sigma = k_{\min} = \Delta k \approx \frac{\Delta I}{I_0} \cdot \frac{1}{L_{\text{eff}}}.$$

(3.5)

Fig. 3.1 Comparison of
selected absorption
spectroscopy (AS) techniques
according to their typically
achieved minimum detectable
absorption coefficients, k_{min},
their robustness and
complexity. Abbreviations
which are not explained in the
text can be found in the list of
abbreviations (adapted
from [17])

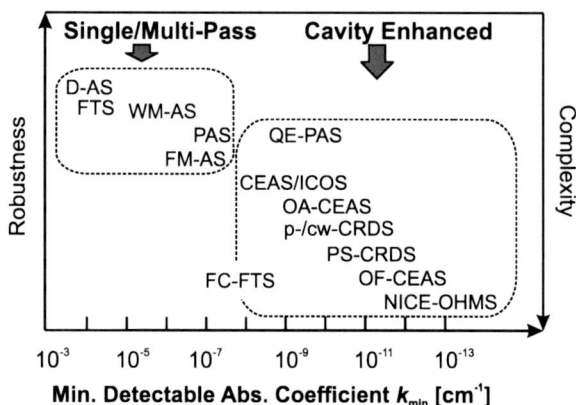

Table 3.1 Sensitivity comparison of different absorption techniques. The MDND (n_{min}) was estimated for a typical medium line strength of fundamental molecular transitions ($S = 10^{-20}$ cm/molecule) and 1 s integration times

Spectroscopic technique	$(\Delta I/I_0)_{min}$	k_{min} [cm^{-1}]	n_{min} [cm^{-3}]
Single-Pass (direct)	10^{-2}–10^{-4}	10^{-3}–10^{-5}	10^{14}
Single-Pass (averaging, WM, FM)	10^{-5}–10^{-7}	10^{-6}–10^{-8}	10^{13}–10^{12}
Multi-Pass (direct)	10^{-2}–10^{-4}	10^{-6}–10^{-8}	10^{12}
Multi-Pass (averaging, WM, FM)	10^{-5}–10^{-7}	10^{-9}–10^{-11}	10^{11}–10^{9}
CRDS (pulsed)	10^{-2}–10^{-3}	10^{-6}–10^{-10}	10^{10}
CRDS/CEAS (cw & variants)	10^{-3}–10^{-5}	10^{-8}–10^{-12}	10^{9}
NICE-OHMS	$>10^{-8}$	10^{-11}–10^{-14}	10^{9}–10^{6}

Consequently, a higher sensitivity is accomplished by (i) reducing the minimum detectable fractional absorbance, $\Delta I/I_0$, (ii) increasing the absorption path length L_{eff}, (iii) maximising the absorption cross section, or a useful combination of these methods.

A few approaches are collected in Fig. 3.1 and in Table 3.1. The discussion below is mainly limited to direct absorption, e.g., photoacoustic methods (PAS) are excluded. In a straightforward application of direct (laser) absorption spectroscopy (D-AS) fractional absorption as small as 10^{-4} can be achieved. This also holds for Fourier-Transform spectroscopy (FTS) based on broadband radiation sources. These techniques benefit from the multiplex or Fellgett advantage as the whole spectrum is recorded simultaneously. Spectral resolution is determined by the distance scanned by the moveable mirror in the Michelson interferometer. Thus, high spectral resolution as good as 0.001 cm^{-1} (30 MHz) is usually achieved at the expense of recording time [8]. More importantly, the étendue of standard FT spectrometers hampers their useful application to folded optical beam paths in order to increase L_{eff}. Recently, this has been overcome by spectrometers based on optical frequency combs (FCFTS) [9].

Further decrease in the noise level by a few orders of magnitude is facilitated by frequency-modulation (FM) or wavelength-modulation (WM) techniques although they may increase the complexity of the setup (Fig. 3.1). Long time averaging over minutes or averaging over thousands of rapidly acquired spectra are alternatively and successfully applied methods of reducing $\Delta I/I_0$ and are often accompanied by background subtraction. Mechanical fluctuations and optical fringes typically degrade the spectrometer performance leading to a practical lower limit of $\Delta I/I_0 \approx 10^{-6}$ for both averaging or modulation techniques [10–16]. Since the residual noise signal cannot arbitrarily be reduced the effective absorption path has to be extended to yield higher sensitivity. Folded optical lines of different geometries where the laser beam is reflected once by a retro-reflector up to several hundred times in sophisticated multi-pass arrangements enable an extension of L_{eff} up to hundreds of metres. Different types of multi-pass configurations are reported and are based on plane mirrors [18], spherical optics (e.g., White [19], Chernin [20] and Herriott type [21, 22] cells), or astigmatic mirrors [23, 24]. The interaction length can be further increased up to the kilometre range by employing optical cavities consisting of high reflectivity mirrors ($>99.9\,\%$) which enclose the sample volume. Similar to linear absorption techniques the relative measurement uncertainties in many variants of cavity ring-down (CRDS), cavity enhanced absorption (CEAS), and integrated cavity output (ICOS) spectroscopy can be as low as 10^{-4}, i.e. $k_{\min}$ is not much less than 10^{-10} cm^{-1} (Table 3.1). Another order of magnitude in sensitivity may be gained by frequency locking of the laser to the cavity [25]. Additional substantial sensitivity improvement is achieved by combining the basic ideas of FM techniques with the advantage of increased absorption lengths in optical resonators ($k_{\min} \sim 10^{-12}$ cm^{-1}). These modifications to the conventional CEAS approach and additional frequency locking of the resonator are known as noise-immune cavity enhanced optical heterodyne molecular spectroscopy (NICE-OHMS) providing record sensitivities (10^{-14} cm^{-1} or 10^{-9} fractional absorption) [26].

Details of the different cavity enhanced techniques encompassing optical feedback (OF) and resonant coupling of FCs with optical cavities resulting in highly sensitive high resolution measurements of broad spectral coverage, are discussed throughout individual chapters in this book as well as in several comprehensive reviews [25, 27–32]. The present compilation is intended to give an overview of the promising application of CEAS methods in conjunction with modern semiconductor based mid-infrared (mid-IR) laser sources, more precisely quantum cascade lasers (QCLs). Chemical sensing in the spectral range between 3 and 20 μm, often called the "molecular fingerprint region", is particularly desirable as it provides an elegant means to increase the absorption cross section σ in Eq. (3.5). The improvement in sensitivity is due to strong fundamental (ro-vibrational) molecular transitions in this spectral range. However, the majority of cavity based methods have used sources of radiation in the ultraviolet and visible regions. For many years the IR spectral range could not be employed either for CRDS or for CEAS techniques, because of the lack of suitable radiation sources with the required power and tunability, but this situation has now changed. Near-IR applications have profited from developments in telecommunications where cheap and compact light sources became available in the 1990s whereas similar lasers were not (yet) available in the mid-IR range. Progress in the development of QCLs has initiated a rapid trend away from niche sensing applications in research laboratories to field-deployable optical sensors. Additionally, the broader use of cavity based techniques in the mid-IR has so far not only been hindered by space or weight restrictions, but also by the need for cryogenic cooling of either the laser or (at least) the detector. Recent improvements of IR detectors have been equally rapid, specifically with respect to detectivity and time-response of thermoelectrically cooled devices [33, 34].

The present understanding of the excitation of optical resonators with radiation from QCLs in cavity enhanced sensing schemes is closely linked to the progress in realisation of different types and modes of operation of these lasers. In other words, the majority of CRDS and CEAS techniques have been used with QCLs since their first realisation in the mid 1990s and the aim of this chapter is to provide an overview about the rapidly growing number of chemical sensing applications. Therefore, firstly, basic properties of QCLs are summarised with respect to conventional mid-IR radiation sources (Sect. 3.2). Secondly, necessary formulae of different cavity enhanced approaches are collected in Sect. 3.3, while Sect. 3.4 is devoted to technical aspects that are specific to optical resonators combined with QCLs. Section 3.5 follows with selected applications and a comparison of QCL-based studies reported so far for the mid-IR.

3.2 Mid-Infrared Light Sources

3.2.1 Overview About Conventional Sources

An ideal compact light source for absorption spectroscopy should provide continuously tuneable radiation of narrow spectral width and relatively high spectral brightness enabling a selective identification of gas phase constituents and their quantification at a high signal-to-noise ratio (SNR) to be carried out. Since such a source still does not exist, a wide variety of light sources have been used so far (Fig. 3.2) and are briefly discussed following the classification of Tittel et al. [35]. They distinguish between direct generation of laser radiation and non-linear frequency down conversion of mainly near-IR (0.8–3 µm) pump sources. Apart from Tittel's review more details can be found elsewhere and in the wealth of references therein [36–39].

Line tuneable gas lasers have been used for many years, especially as long as the performance of semiconductor diode lasers in the mid-IR was not sufficient. Watt-level output power is easily achieved either in pulsed (p) or continuous wave (cw) mode. However, available wavelengths are limited to pressure broadened ro-vibrational laser transitions at 10/11 µm (CO_2 and isotopes), 5–8 µm (CO) and 3–5 µm (HF, HCl, HBr, N_2O, CO overtones) respectively [36]. Another important class

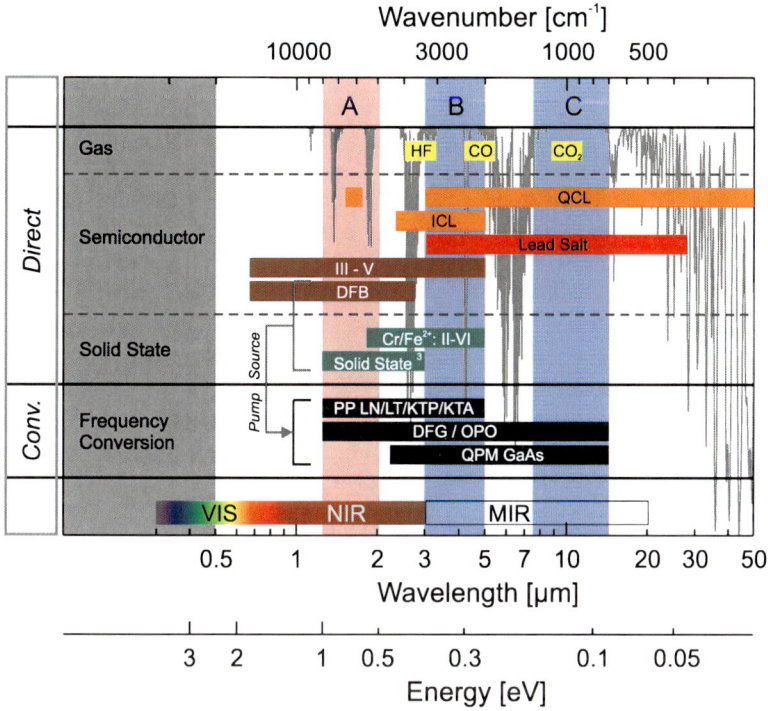

Fig. 3.2 Wavelength coverage of typical IR (laser) light sources. Radiation might be obtained from direct sources or by frequency conversion. Absorption features of atmospheric constituents (H_2O, CO_2; ~1 m absorption path, ambient conditions) are indicated (*grey*). The overtone region (A) and both atmospheric windows in the fingerprint region (B, C) are important for chemical sensing applications. For acronyms see text and list of abbreviations (Adapted from [35])

of direct sources is based on vibronic transitions occurring in doped crystalline and fibre lasers, e.g. Cr^{2+} and Fe^{2+} doped II–VI semiconductors (ZnSe, ZnS). Similarly, transitions of colour centres in alkali halides have been used to cover the wavelength range shorter than 4 µm. Common to all lasers is the requirement of optical pumping by means of adequately powerful sources resulting in output laser powers between tens of mW up to several W. Room temperature and continuous operation is now commonly achieved. In practise these types of lasers are mainly used as pump source for parametric frequency conversion by means of difference frequency generators (DFGs) and optical parametric oscillators (OPOs) rather than as primary mid-IR light sources.

DFGs convert a pump and an idler beam at different frequencies into a low frequency signal beam in single pass configuration. A resonant (i.e. cavity based) configuration is used for frequency-down conversion of a single pump laser into two output waves (idler and signal) which is referred to as OPO [35]. Both methods are based on non-linear materials and tuning is facilitated by either tuning the pump source or temperature tuning of the non-linear crystal. Since energy and momentum, i.e. frequency and phase, of the involved waves have to be conserved (phase matching condition), which is difficult to achieve with birefringent crystals in a preferred collinear beam configuration, quasi phase matching (QPM) has been introduced. QPM by applying external electric fields to periodic short regions of ferroelectric materials in non-linear crystals is observed, e.g., in periodically poled (PP) lithium niobate (LN), lithium tantalite (LT), potassium titanyl phosphate (KTP) and potassium titanyl arsenate (KTA) [35, 37]. The wavelength range up to 5 μm is covered by these materials. In order to generate longer wavelengths uncommon materials and pump sources are required [37]. DFGs typically provide μW–mW output power due to their limited conversion efficiency whereas tens of mW are observed—still often limited to pulsed mode—by using OPOs in combination with PP materials. The need for additional pump sources and sophisticated optical geometries to establish (only) the radiation source remain drawbacks of this approach.

Relatively compact semiconductor based lasers represent an emerging group of direct sources generating light due to stimulated emission across the band gap or quantised energy levels. While diode lasers provide single mode cw operation at room temperature with tens of mW output powers in the near-IR range [35, 40], their mid-IR counterparts, known as lead salt lasers, normally require cryogenic cooling to obtain less than a few mW multimode radiation. The recent advent of cascaded laser structures has made hundreds of mW single mode mid-IR radiation possible.

Diode lasers consist of semiconductor alloys forming a p-n junction where population inversion is accomplished by applying a forward bias to the device. The laser crystal is shaped into a (Fabry-Perot) optical cavity of ~0.5 mm length. Hence the multiple cavity modes are typically separated by several cm^{-1} (~100 GHz). The energy and thus the emitted wavelength is determined by the band gap energy of the semiconductor alloy. Lasing requires both electrons injected into the conduction band and holes from the valence band. The basic principle shown in Fig. 3.3 is referred to as double heterostructure diode laser. The potential well formed by alternating micrometer thick films of binary and ternary alloys serves (i) to better confine the charge carriers, and thus (ii) to improve the population inversion at room temperature, as well as (iii) to establish a better waveguide for the emitted radiation. Reducing the layer thickness down to nanometres leads to quantum wells (QWs) with discrete energy levels. Forthcoming progress in deposition technology being capable of producing multi-quantum well (MQW) structures and band gap engineering has led to the development of high performance diode lasers, mainly based on III–V compounds (i.e. GaAs, InP). These devices emit from the visible to the near-IR spectral range at room temperature. Auger recombination being the main loss mechanism at longer wavelengths degrades the performance of III–V based devices drastically. Consequently, mid-IR cw room temperature operation cannot be achieved routinely. Therefore cryogenically cooled lead salt lasers were the only alternative between 5–30 μm [38, 39]. Spectral tuning of diode lasers is accomplished by a temperature or charge carrier induced change of the refractive index of the gain medium [41]. The spectral coverage is not continuous and encompasses mode overlapping, mode hops or spectral gaps. Current tuning also comprises FM options which makes diode lasers superior to optically pumped mid-IR sources.

Fig. 3.3 Schematic band diagram of a heterostructure diode laser composed of materials with different band gap energies E_{gap}. Increased laser line widths ($\Delta \nu \sim \Delta E$) are caused by the curvature of the dispersion relation $E(\kappa_{||})$ of both bands

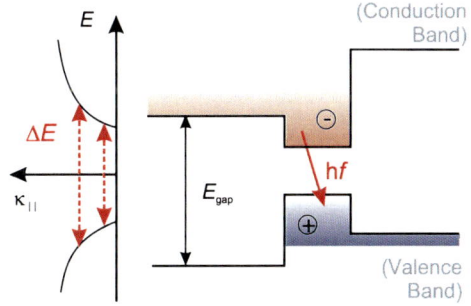

Fig. 3.4 Radiative transition (hf) in a QW as an example for a typical building block of (unipolar) QCLs. Reduced laser follow from the same curvature of the dispersion relation $E(\kappa_{||})$ of both quantum states. The band gap discontinuity (ΔE_{disc}) is also defined

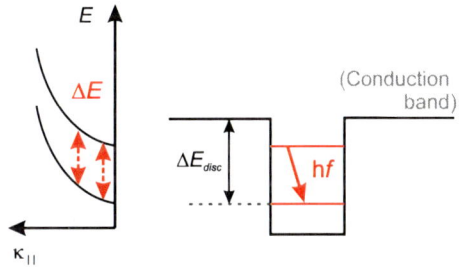

3.2.2 Quantum Cascade Lasers (QCLs)

3.2.2.1 Design Principles

Similar to QWs the radiative transition in QCLs is based on confined quantum states in stacks of semiconductor layers of nanometre thickness. However, in contrast to diode lasers, where electrons and holes are involved, QCLs are unipolar devices, i.e. only electrons are travelling between quantum states in the conduction band and lead to laser emission (Fig. 3.4). The energy of the radiation is determined by the separation of the energy levels and thus by the layer thickness providing a means of custom-tailoring the emission wavelength. In contrast to diode lasers, upper and lower laser level have now the same curvature of the dispersion relation (Figs. 3.3 and 3.4) yielding inherently narrower spectral widths of the transitions [42]. Electron transport is accomplished by applying an external electric field to a periodic structure of QWs (Fig. 3.5). Electrons are injected through specifically designed injector regions into the upper laser level followed by the radiative transition. The lower level is depopulated by tunneling through the potential well barrier which simultaneously serves as the injector of the next active region as illustrated in Fig. 3.5. A high tunneling probability is observed for the resonant case, i.e. subsequent lower and upper laser levels are aligned by the external field. Resonant tunneling accompanied by a negative differential resistance of the structure which could be used for light amplification was theoretically predicted in the early 1970s by Esaki [43], Kazarinov and Suris [44]. However, forthcoming progress in depositing such sophisticated thin film heterostructures was necessary to realise first devices in the

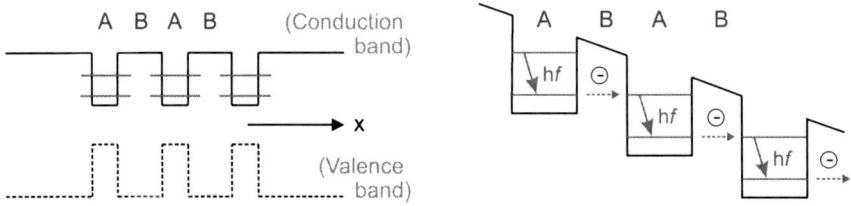

Fig. 3.5 *Left*: Band diagram of a multi-quantum well (MQW) structure without external bias. Confined quantum states of the active zone (A) and the tunneling barrier (B) are schematically shown. Space coordinate is symbolised by x. *Right*: Radiative transitions (hf) and resonant tunneling between cascaded active (A) and injector (B) regions in a biased and therefore aligned MQW structure

late 1980s [45, 46]. The potential of a new type of mid-IR light source based on linearly polarised intersubband emission in cascaded semiconductor superlattices was recognised [46–48] and introduced by Faist et al. in 1994 as QCL [49, 50]. Due to the cascaded structure of typically ∼30 active regions a single electron is repeatedly used yielding quantum efficiencies beyond 100 %. In this way, considerably higher output powers (mW–W) are obtained as compared to conventional lead salt lasers. An efficient population inversion in QCLs requires—amongst other measures—selective depopulation of the lower level which is mainly depleted by fast (non-radiative) phonon transitions. Consequently, the wall-plug efficiency, i.e. the portion of electrical input power that is converted into radiation, is rather low and QCLs are thus mainly producing heat. This also hampered cw lasing at room temperature for a long time and required pulsed operation for commercially available devices [51, 52]. Furthermore, the electrical input powers of QCLs are several Watts due to the relatively high compliance voltages which are necessary to align the involved energy levels. Wall-plug efficiency and thermal management, i.e. heat removal from the active zone, and thus the performance of QCLs have been subject to continuous improvement yielding now Watt-level cw output power at room temperature [53].

The picture of active zone and injector region in Figs. 3.4 and 3.5 is oversimplified. In general, both regions exhibit an intrinsic coupled QW or superlattice structure [49]. Radiative transitions may take place between coupled quantum states (intersubband) or adjacent minibands (interminiband) rather than single quantum states. Further design principles are detailed in several extensive reviews [7, 54–58]. The majority of QCLs consists of III–V compounds (e.g., GaInAs/AlInAs, GaAs/AlGaAs). Tailoring the emission of QCLs to shorter wavelengths is yet unsatisfactorily solved. Laser emission from GaAs/AlGaAs devices, which are often used for generating THz radiation, cannot fall below ∼8 μm, while 3.4 μm emission has been reported for strain-compensated GaInAs/AlInAs [59]. Consequently, strong C–H absorption features around 3 μm are difficult to access. The reason is the limited band gap discontinuity, ΔE_{disc} (Fig. 3.4), of the employed semiconductor compounds. Shorter wavelengths require an increased subband or miniband separation which has to be clearly smaller than the band gap discontinuity to avoid parasitic leakage from the upper laser level. Short wavelength QCLs may therefore be achieved with alloys of similar lattice constants but larger band gap differences than the aforementioned compounds [60]. Research is still in progress to tackle this issue by means of material engineering activities [61] and references therein. Alternatively, the integration of cascaded structures into bipolar devices and transitions across staggered quantum wells, also known as interband cascade lasers (ICLs), was proposed [62] and references therein.

Similar to diode lasers a QCL chip forms a Fabry-Perot cavity of a few millimetres length resulting in a multimode emission spectrum of ~ 1 cm^{-1} (30 GHz) mode spacing. Tuneable single mode operation is obtained (i) by means of distributed feedback (DFB) lasers, or (ii) in external cavity (EC) configuration. Particularly the highly developed processing technology for the initially employed III–V compounds enabled DFB gratings to be integrated into the QCL structure, more specifically in the waveguide structure [63]. Bragg reflection in complex- [63, 64] or index-coupled [65] gratings provide a means of observing single mode QCL emission by selecting one of the cavity modes. Spectral tuning is accomplished by temperature induced changes of the refractive index of the laser which tunes both the spectral gain and to a lesser extent the period of the Bragg grating [55]. Hence, the QCL current can only be used indirectly for sweeping the laser wavelength through internal laser heating. Additionally, Faist et al. found an inherent thermal drift in p-DFB-QCLs caused by the dissipated power in the active region which is also known as chirped pulse [63, 66]. The seed current for the laser generates heat during the nanosecond pulse which in turn causes a frequency decrease throughout the duration of the laser pulse. The total emission range of DFB-QCLs is typically limited to less than 7 cm^{-1} between -30 °C and $+30$ °C. Although an EC arrangement employing uncoated multimode p-QCLs has been successfully demonstrated [67], anti-reflection coated lasers are desirable for efficient spectral narrowing [68–70]. Temperature tuning as well as tilting the external grating yielded a spectral coverage of up to 20 % around the centre wavelength for a pulsed system and a ~ 8 % mode hop free tuning range for a cw laser. In common with the earlier commercialised DFB-QCLs, EC-QCLs are now routinely available and applied for spectroscopic purposes [71–76].

3.2.2.2 Spectroscopic Issues

The aforementioned design principles along with the observed radiation characteristics strongly suggest that QCLs are far more than a substitute to tuneable (lead salt) diode lasers [40, 77]. Therefore conventional principles and setups commonly used with diode lasers have to be carefully validated albeit a straightforward application to QCLs may appear obvious. Early experiments with p-QCLs combined short laser pulses (≤ 50 ns) with the conventional method for sweeping lead salt lasers by ramping a DC current [78–80]. Extensions such as the sweep integration method were adapted to spectrometers employing p-QCLs and later—with their increasing availability—cw-QCLs [14, 81–83]. While this approach yields excellent performance in conjunction with cw-lasers the situation changes considerably for p-QCLs.

In what follows, a few spectroscopic challenges connected to QCLs operated with short current pulses are detailed since these lasers appear perfectly suited for e.g., CRDS, where the incident radiation has to be interrupted to observe the light decay. Apart from pulse-to-pulse intensity fluctuations, a phenomenon well-known from other pulsed radiation sources, increased effective laser line widths of up to 1.2 GHz (0.04 cm^{-1}, FWHM) were reported as a major drawback of p-QCLs [77, 81, 84–86]. According to McCulloch et al. the spectral resolution of p-QCL spectrometers is determined both by the chirp rate and the pulse width of the lasers [87]. They estimated the effective QCL spectral width, and hence the spectral resolution $\Delta \nu$, of a pulsed spectrometer to be $\sim (C \cdot \alpha)^{1/2}$, where α is the spectral sweep rate (df/dt) and C a current pulse shape-dependent constant. Assuming a rectangular and a Gaussian time-window C equals 0.886 and 0.441, respectively. Hence, $C = 1$ may be considered as an upper limit. It should be noted that Mc-Culloch's estimate is true for an optimized pulse width $\Delta t = \Delta t_{opt}$ (best aperture time). In general, the spectral width $\Delta \nu$ is governed either by the uncertainty relation, $\Delta \nu \cdot \Delta t \geq C$, or by the frequency-down chirp of the laser, $\Delta \nu = \alpha \cdot \Delta t$. For extremely short pulses ($\Delta t = \Delta t_{short} < 5$ ns) the spectral bandwidth is Fourier transform limited $\Delta \nu_{short} \sim C/\Delta t_{short}$. For longer pulses $\Delta t = \Delta t_{long}$ the frequency chirp of the laser sets the fundamental limit $\Delta \nu_{long} = \alpha \cdot \Delta t_{long}$ and the spectral width clearly exceeds the theoretical value given by the uncertainty relation (i.e., $\Delta \nu_{long} \gg C/\Delta t_{long}$). McCulloch suggest therefore the best aperture time Δt_{opt} for which the Fourier transform limited bandwidth equals the frequency chirp (i.e., $\Delta \nu_{short} = \Delta \nu_{long}$) in order to estimate the spectral resolution of pulsed spectrometers. In practise, the pulse widths $t_{on} = \Delta t > \Delta t_{opt}$ are usually higher than this best pulse width. Hence, $(C \cdot \alpha)^{1/2}$ provides a lower limit approximation of the effective line width, since $\Delta \nu$ often remains chirp limited ($\sim \alpha \cdot t_{on}$).

Figure 3.6 depicts the spectral sweep rate α of 31 pulsed lasers, mainly commercially available and covering almost the entire mid-IR spectral range. Based on the recorded fringes of a germanium etalon, two different values were defined: an average chirp rate across a 100 ns interval, $\Delta f/dt_{100}$, (Fig. 3.6). Additionally, the chirp rate df/dt was determined and followed throughout the entire pulse length (Fig. 3.7). It is obvious from Fig. 3.6 that no general (average) chirp rate for pulsed QCLs exist. It strongly depends on each device and the thermal behaviour of the active laser zone. Nevertheless, a lower limit of the average chirp rate of about 90–150 MHz/ns (0.003–0.005 cm^{-1}/ns) can be estimated from this survey. Further generalisation is not feasible since the chirp rate is not a device constant. As demonstrated for the 4.5 μm laser (Fig. 3.6), which was operated at two different voltages, the chirp rate covers a certain range and is influenced by several parameters, such as the laser material properties, thermal management provided by the laser design, and the external operation conditions. The pulsed laser is in a thermal non-equilibrium state with a rapid increase of the core temperature during the pulse. Characteristic time scales are in the order of a few μs [88, 89] which in turn causes a temperature increase and thus a variable spectral sweep rate α throughout the laser

Fig. 3.6 Average spectral sweep rate $\Delta f/dt_{100}$ of pulsed QCLs emitting between 600 and 2250 cm^{-1}. Displayed values were averaged over a 100 ns interval. The *dashed line* represents a range of average chirp rate values for one specific laser achieved by adapting the operation conditions (suppliers: *open* and *full circles*—Alpes Lasers, *triangle down*—IAF Freiburg, *triangle up*—nanoplus, *triangle left*—Laser Comp., *triangle right*—Univ. of Sheffield)

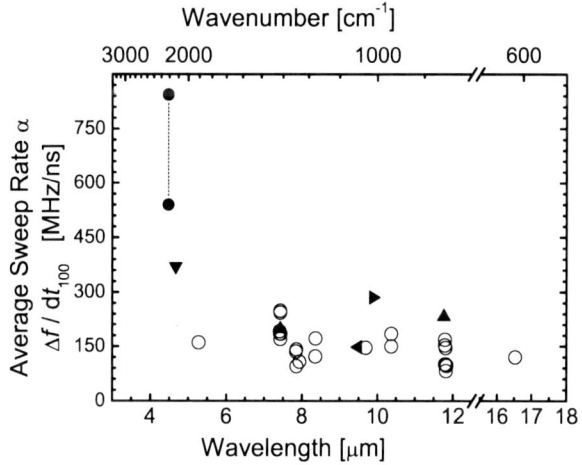

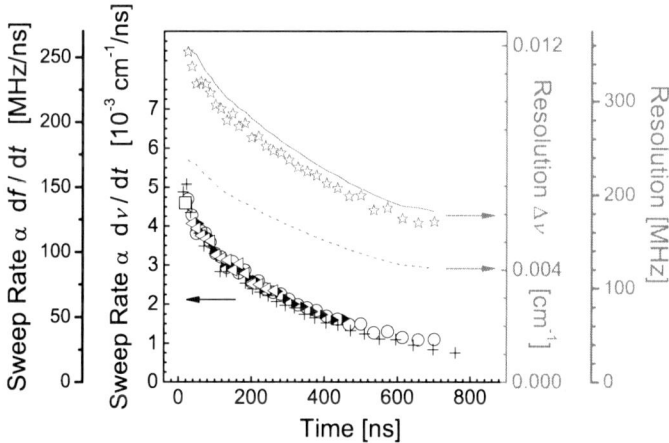

Fig. 3.7 Spectral sweep rates of a QCL (*black*) operated with different pulse widths (*square*—33 ns, *triangle left*—292 ns, *triangle right*—492 ns, *circle*—794 ns, +—997 ns) and the corresponding resolution in a spectrum (*grey*) following McCulloch's definition for three cases: the upper limit ($C = 1$, *solid line*), the lower limit (Gaussian time window, $C = 0.441$, *dashed line*) and the intermediate case (rectangular time window ($C = 0.886$, *star*)

pulse as shown in Fig. 3.7 for a p-QCL emitting at 8.5 μm. Although the device studied exhibits a moderate chirp rate, it is clear that for pulses shorter than 100 ns the resolution (based on McCulloch's estimate) cannot be better than 200–360 MHz (0.007–0.012 cm^{-1}).

Since at low pressure Doppler broadened absorption profiles typically show spectral widths of less than those mentioned previously, p-QCLs cannot be considered as narrow bandwidth mid-IR light source. The situation may be even worse because the laser in Fig. 3.7 was operated close to its threshold, i.e. at higher input power levels the temperature increase in the active zone and hence the chirp rate is higher which in turn would degrade the spectral resolution further. It is essential here to distinguish between different terms which can be found in the literature: the effective [81, 82, 90] or integrated [84] line width is connected with both the laser chirp and detection system. By contrast, the instantaneous [91] or intrinsic [92–94] line width of QCLs is theoretically only determined by the intersubband transition and the design of the QCL and was predicted to be ~15 kHz (5×10^{-7} cm^{-1}) [93]. Experiments with cw-DFB-QCLs yielded effective line widths (FWHM) of 24–40 MHz (0.0008–0.0013 cm^{-1}) indirectly determined from a fit to NO lines and using a standard power supply [82, 91]. Free-running cw-QCL line widths of 1–6 MHz (3–20×10^{-5} cm^{-1}) were found in combination with low-noise stabilised current sources from heterodyne experiments [94–96]. Although laser drivers are still the limiting factor, further improvements in current controller technology enabled QCL line widths of 150 kHz (5×10^{-7} cm^{-1}) to be obtained [93] whereas the theoretical prediction (12 kHz or 4×10^{-7} cm^{-1}) could be confirmed with frequency stabilisation employing a feedback-loop [97]. Recently, a sub-kHz intrinsic line-width has been reported [98]. As EC-QCLs are increasingly becoming available on a commercial basis, similar studies have been performed to assess their suitability for sensing applications. Effective line widths of the order of 30 MHz (0.001 cm^{-1}) were observed along with drift effects across spectral scans of up to 2.4 GHz (0.08 cm^{-1}) [73, 99].

3.3 Theoretical Considerations

This section summarises the basic equations which are necessary to analyse and discuss the recorded data. A more thorough introduction to different approaches and the assumptions lying behind the formulae which are in common use can be found e.g. in [27, 28]. Note that throughout this compilation wavenumbers (in cm^{-1}) are symbolised by v and are linked with the SI system by the frequency $f = vc$.

3.3.1 Cavity Ring-Down Spectroscopy (CRDS)

Following the description in [28] only a few restrictions are used for the ballistic assumption (or ping-pong model) to describe the behaviour of a laser pulse shorter than the round trip time in an optical resonator, namely that the reflectivity of the cavity mirrors is equal ($R_1 = R_2 \equiv R$) and scattering is neglected, i.e. the transmission $T \approx 1 - R$. The separation of the highly reflective mirrors is L, whereas the interaction length of the laser beam with the absorbing medium is d. The intensity I^μ leaking out of the cavity after μ round trips transforms readily into the well-known relationship for the cavity ring down intensity

$$I^\mu(v) = I^0(v) \cdot \exp\left(-\frac{t_\mu}{\tau(v)}\right), \tag{3.6}$$

where the time t_μ necessary for a cavity round trip is given by $t_\mu = (2\mu L)/c$ and c is the velocity of light. Although strictly speaking the ballistic assumption is only valid for short laser pulses exponential decays can be also observed for light pulses longer than the round trip time as long as the absorption line is broader than the spacing of the evolving cavity modes [100]. The decay time τ is defined by

$$\tau(v) = \frac{L}{c(k(v)d - \ln R)} \approx \frac{L}{c((1 - R) + k(v)d)}, \tag{3.7}$$

where the approximation assumes the typically high values for the mirror reflectivity, $R \rightarrow 1$. For an empty cavity ($k = 0$) and with $d = L$ (absorbing medium completely filling the cavity), an effective absorption path for the experiment $L_{\text{eff}} = c\tau_0 = L/(1 - R)$ can be defined. Transforming (3.7) yields the absorption coefficient:

$$n\sigma(v) = k(v) = \left(\frac{1}{\tau(v)} - \frac{1}{\tau_0(v)}\right)\frac{L}{dc}. \tag{3.8}$$

3.3.2 Cavity Enhanced Absorption Spectroscopy (CEAS/ICOS)

To describe the excitation of a cavity with continuous radiation rather than a short laser pulse, as introduced by Engeln et al. [101], the temporal development of the intensity inside the cavity needs to be considered. The steady state output I_{out} of the cavity is characterised both in time and amplitude by the cavity losses (R and k). Using the steady state output the ratio of the intensities for an empty (I_0) and a filled (I) cavity respectively can be defined as [102, 103]

$$\frac{I_0}{I} = \frac{I_{\text{out}}(k = 0)}{I_{\text{out}}} = 1 + GU, \tag{3.9}$$

which is valid for all R and k and where $G = R/(1 - R)$ and $U = (1 - \exp(-k(v)d))$ were introduced. In the weak absorption limit ($k \rightarrow 0$ and $R \rightarrow 1$) Eq. (3.9) becomes

$$n\sigma(v) = k(v) = \left(\frac{I_0(v)}{I(v)} - 1\right)\frac{1 - R}{d}. \tag{3.10}$$

By analogy with the Beer-Lambert law (3.1) assuming weak absorption the effective path length can thus be expressed as $L_{\text{eff}} = L/(1 - R)$ for $d = L$. On the other hand, if the natural logarithm is taken from (3.9) and plotted against the molecular number density n a linear relationship is expected, as long as a weak absorber is present in the cavity. In this case the expressions can be further simplified:

$$\ln\left(\frac{I_0(v)}{I(v)}\right) \approx GU \sim n. \tag{3.11}$$

Equation (3.11) fulfils therefore two functions. Firstly, it serves as a check on the validity of the weak absorption limit. Secondly, in the range where $\ln(I_0/I)$ is proportional to n, the slope of this linear relationship enables the mirror reflectivity R to be determined from a known concentration standard and hence provides a means for the absolute calibration of CEAS or ICOS.

ICOS was introduced for pulsed excitation of the cavity [104]. For data analysis there is in fact no difference from CEAS since Eq. (3.10) holds for cw as well as for pulsed excitation [103]. On the other hand, following the ping-pong model, the integrated cavity output I_{out} corresponds to an infinite sum over the intensity I^{μ} leaking out of the cavity. The sum over μ converges for the experimentally reasonable cases when $R \cdot \exp(-k(v)d) < 1$. By analogy to (3.9) the ratio of the time integrated signal of the cavity without ($k = 0$) and with the absorbing medium is

$$\frac{I_0}{I} = \frac{I_{out}(k = 0)}{I_{out}} = \frac{1 - (Re^{-k(v)d})^2}{(1 - R^2)e^{-k(v)d}}. \tag{3.12}$$

In the limits of weak absorber ($k \to 0$) and high reflectivity ($R \to 1$) Eq. (3.12) simplifies to

$$\frac{I_0(v)}{I(v)} \approx 1 + \frac{k(v)d}{1 - R}, \tag{3.13}$$

which is the standard formula used for ICOS, leading essentially to the same result as in the CEAS case embodied in Eq. (3.9). The terms CEAS and ICOS are used synonymously in the published literature and so they are here. However, ICOS will be used in the context of pulsed excitation and CEAS in the context of cw cavity excitation in this chapter.

3.3.3 Cavity Mode Structure

The eigenfrequencies of an optical cavity formed by 2 spherical mirrors with radii of curvature $r_1 = r_2 \equiv r$ separated by a distance L are

$$f_{q,mn} = \frac{c}{2L}\left(q + (n + m + 1)\frac{\theta}{2\pi}\right), \tag{3.14}$$

where $\theta = 2 \cdot \text{arccos}([g_1 g_2]^{1/2})$. The stability of a cavity, in other words the ability to confine the cavity modes during their round trips, is assessed by the geometrical cavity parameter $g_i = 1 - L/r_i$ and commonly expressed as $0 < g_1 g_2 < 1$ [27, 28]. The mode indices q, m and n characterise the electromagnetic field in the cavity and are referred to as longitudinal (q) and transverse (m, n) or TEM$_{mn}$ modes. The free spectral range (FSR) of a resonator is defined as the frequency difference between successive longitudinal modes of same m and n.

$$FSR = \frac{c}{2L}. \tag{3.15}$$

If the spherical resonator is aligned off-axis (OA) incident light will repeat after μ roundtrips as long as $L_{OA} = r \cdot [1 - \cos(\pi \cdot \kappa/\mu)]$ is fulfilled, where κ is an integer [21, 106]. This re-entrant condition for a "magic" cavity may be also expressed as $\theta = 2\pi \cdot \kappa/\mu$, so that the eigenfrequencies in Eq. (3.14) become [106]

$$f_{q,mn} = \frac{c}{2\mu L}(\mu q + (n + m + 1)\kappa). \qquad (3.16)$$

Equation (3.16) essentially expresses a "densification" of the longitudinal mode structure (in frequency space) in a cavity that is aligned off-axis and re-entrant after μ roundtrips. Such a cavity yields an effective FSR_{eff} [107]

$$FSR_{eff} = \frac{c}{2\mu L}. \qquad (3.17)$$

Additionally, if non-mode-matched conditions and thus transverse modes are considered, the cavity length provides an even denser structure as the FSR is divided into μ groups of degenerated and equidistant modes separated by $c\kappa/2\mu L$ [28, 106]. Practical approaches to reduce the FSR—which is highly desirable in CEAS as it may significantly improve the spectral resolution in the recorded spectra—are outlined in Sect. 3.4.2.

3.3.4 Sensitivity Considerations

For CRDS the uncertainty in the absorption coefficient Δk can be derived from Eq. (3.8) with $\tau \to \tau_0$

$$\Delta k = \frac{\Delta \tau}{\tau_o^2} \frac{L}{dc}. \qquad (3.18)$$

Conventionally the standard deviation of the ring down time is used as a measure of $\Delta \tau$. For CEAS and ICOS an analogous relationship can be found from (3.10) with $I \to I_0$

$$\Delta k = \frac{\Delta I}{I_0} \frac{1 - R}{d}, \qquad (3.19)$$

where ΔI represents the intensity fluctuations, either inherent to the light source or induced by mode fluctuations. In principle, the error in the absorption coefficient is even higher since the reflectivity of the mirrors R is not known, i.e. a calibration is required leading to a $\sqrt{2}$ times higher uncertainty in k [28]. In general, another quantity, namely the noise equivalent absorption (NEA), is used to estimate the detection limit

$$NEA = \Delta k \sqrt{2} f^{-1/2}, \qquad (3.20)$$

where f is the repetition rate of the measurements, i.e. the number of averaged scans in a 1 s interval. It should be noted that the NEA is not defined consistently throughout the literature. A useful comparison of expressions usually employed can be found in [105]. The *NEA* as given in (3.20) will be used in this chapter and would correspond to the "per scan" definition in [105].

An estimate of the MDND, n_{min}, is derived from the integrated absorption coefficient $K = nS$ that follows straightforwardly from spectral integration of the absorption coefficient as defined in Eq. (3.2). Under low (<500 Pa) and elevated pressure (>5000 Pa) conditions an analytic expression for K can be obtained by assuming a Gaussian or Lorentzian line profile for $\phi(\nu - \nu_0)$ in (3.3). In case the transmitted intensity at the peak position ν_0 can be evaluated, K becomes [40]

$$K = nS = \frac{1}{L_{eff}}\sqrt{\frac{\pi}{\ln 2}} \ln\left(\frac{I_0(\nu)}{I(\nu)}\right)\Bigg|_{\nu_0} \Delta\nu_D \quad (p < 500 \text{ Pa}), \quad \text{and} \quad (3.21)$$

$$K = nS = \frac{\pi}{L_{eff}} \ln\left(\frac{I_0(\nu)}{I(\nu)}\right)\Bigg|_{\nu_0} \Delta\nu_{col} \quad (p < 5000 \text{ Pa}). \quad (3.22)$$

The Doppler half width and the Lorentzian half width (in cm^{-1}, HWHM) are denoted by $\Delta\nu_D$ and $\Delta\nu_{col}$, respectively, and are given by [6]

$$\Delta\nu_D = \nu_0\sqrt{\frac{2k_B T \ln 2}{m_{molec}c^2}} = 3.58 \times 10^{-7}\sqrt{\frac{T}{M_{mol}}}\nu_0, \quad \text{and} \quad (3.23)$$

$$\Delta\nu_{col} \sim \gamma_p\frac{p}{\sqrt{T}}. \quad (3.24)$$

The molecular mass in (3.23) $m_{molec} = M_{mol}/N_A$ can also be expressed by the molar mass M_{mol} and the Avogadro number N_A. The gas temperature and Boltzmann constant are denoted by T and k_B. The pressure broadening coefficient, γ_p is usually provided in spectroscopic databases. Molecular ro-vibrational transitions exhibit pressure broadened half widths of a few MHz/mbar [6], e.g. ≤ 3 MHz (0.0001 cm^{-1}) at low pressures <500 Pa and ~ 1.5 GHz (0.05 cm^{-1}) at atmospheric pressure. If the peak-to-peak noise in a spectrum is considered as peak absorption, n_{min} is extracted by evaluating the expression $\ln(I_0/I)$ at ν_0 in Eqs. (3.21) and (3.22). Using Beer-Lambert law (3.1) along with (3.19) yields $\ln(I_0/I) \approx k_{min}L_{eff} \approx NEA \cdot L_{eff} \approx \Delta I/I_0$ and, hence, (3.21) and (3.22) can be transformed into an estimate of the MDND:

$$n_{min} \approx \frac{NEA}{S} \cdot \begin{cases} \Delta\nu_D \cdot \sqrt{\pi/\ln 2}, & p \leq 500 \text{ Pa}, \\ \Delta\nu_{col} \cdot \pi, & p \geq 5000 \text{ Pa}. \end{cases} \quad (3.25)$$

Although a numerical calculation would be required for the intermediate pressure range, it is clear from Eq. (3.25) that the low and high pressure cases are perfectly suited to estimate lower and upper limits for the MDND. In the previous discussion the instrumental broadening $\Delta\nu_{instr}$ was assumed to be substantially smaller than the main broadening contribution. Particularly at low pressure conditions the laser line width may be of the same order as the Doppler broadening. The convolution of absorption and instrumental profile, which is assumed to be Gaussian as well, yields the observed spectral width, $\Delta v_{obs} = (\Delta v_D^2 + \Delta v_{instr}^2)^{1/2}$, and replaces Δv_D in Eqs. (3.21) and (3.25).

3.4 QCL Based Excitation of Optical Cavities

This section is devoted to technical aspects linked with the combination of optical resonators and QCLs while the next section highlights a few selected applications. First, p-CRDS is scrutinised. Next, different excitation options of the resonator in CEAS are compared. Finally, OF-CEAS and the application of EC-QCLs are briefly addressed. Frequency locking in general and NICE-OHMS in particular are not discussed in what follows.

3.4.1 CRDS Using p-QCLs

As depicted in Fig. 3.8 the experimental arrangement using p-DFB-QCLs is straightforward [103, 108]. Welzel and co-workers employed a stable resonator formed by two high reflectivity mirrors of diameter 25.4 mm and $r = 1$ m radius of curvature. The mirrors also served to enclose a vacuum vessel which was made of standard vacuum components. Hence the vacuum cell determined the mirror separation of $L = 0.432$ m. Beam shaping optics (BSO) is commonly required (i) to collect the strongly divergent radiation from the QCL, and (ii) to reduce the beam diameter [103, 109]. In the example presented here the spherical resonator was aligned on-axis without additional mode-matching optics. Furthermore, the cavity length was not actively changed or dithered nor was the cavity locked to the illuminating light source. The radiation leaking out of the cavity was collected by an OA parabolic mirror of small f-number and directed to a liquid-nitrogen (LN) cooled detector (Judson Technologies, J15D12 series) and preamplifier with a response time of approximately 70 ns. Relative and absolute calibration was provided by recording the fringes of a germanium etalon and standard gas absorption spectra in a reference channel. The DFB-QCL was driven and frequency swept using an external trigger signal and a sub-threshold current ramp superimposed to the short laser pulses. Each laser pulse represents a spectral data point in the spectrum. The repetition frequency of the trigger was set to 10 kHz which enabled the decay transient—being of the order of less than 10 µs—to be followed completely and without any distortion. A fast digitising oscilloscope was used to acquire the ring down transients. The ring down transients were then processed according to Eqs. (3.7), (3.8) for CRDS and Eq. (3.13) for ICOS. Two p-DFB-QCLs (Alpes Lasers) were used for the experiments. One emitting between 1345 and 1352 cm^{-1} ($\sim$7.42 µm) almost coincided with the centre wavelength of the high reflectivity mirrors ($R = 99.84$ %), whereas the emission of the second laser between 1195 and 1200 cm^{-1} ($\sim$8.35 µm) coincided with the edge of the low loss range of the cavity mirrors. A pulse width of

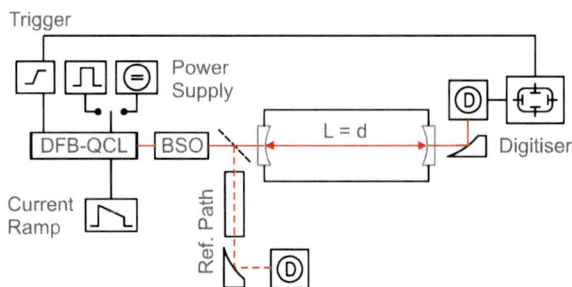

Fig. 3.8 Schematic diagram of the apparatus used for both CRDS and (on-axis) CEAS experiments using pulsed and cw laser sources, respectively. Beam shaping optics (BSO) provided efficient light transmission to the resonator. The detector (D) signals were recorded with data acquisition cards and synchronised with the current ramp applied to the DFB-QCLs [103]

60 ns was chosen for the 7.42 µm laser in order to inject enough power into the cavity. For the 8.35 µm laser a pulse width of 32 ns was adequate to detect a signal from the cavity. The pulse length was thereby chosen to be as short as possible while maintaining a reasonable cavity output [103].

The ICOS spectra appeared typically not to be as noisy as for the CRDS case (integration over the cavity output is in fact an averaging process). However, a thorough analysis of CH_4 and N_2O spectra revealed in both cases similar artefacts as observed by Sukhorukov et al. [110] although less dominant in their case: (i) absorption features are only slightly more pronounced that the scatter of the baseline data, (ii) the selectivity is decreased due to arbitrarily broadened and shifted absorption lines, and, most importantly, (iii) the retrieved absorption coefficients are a factor 10–25 smaller in comparison with theory. It was suggested to apply smaller QCL pulse widths to reduce the frequency-down chirp and the laser line width. However, even in the 8.35 µm region where the mirror reflectivity was lower and enabled the laser pulse width to be reduced to 32 ns while at the same time quadrupling the laser energy injected through the entrance mirror, a significantly improved *NEA* could not be achieved. On the other hand, the effective absorption path length fell to 150 m which is of the same order of magnitude as available from conventional Herriott-type long path cells. Figure 3.9 shows an example of a CRDS measurement performed with a gas mixture consisting of 1667 ppm N_2O in N_2 at a pressure of 100 mbar. An N_2O absorption feature is indicated by an unambiguous drop in the decay. Albeit this result is qualitatively "the best" obtained with this approach, it still falls short by at least a factor of 8 in the absorption coefficient.

It could be argued that the absorption features chosen as examples were far too strong for a standardisation test since saturation effects could arise thereby reducing the experimental absorption coefficients. In this case weaker absorption features would show a better agreement between experiment and calculation. However, considering e.g. the smaller N_2O lines which should appear at 1197.40 cm^{-1} and 1197.81 cm^{-1} (Fig. 3.9) it is clear that potential saturation effects cannot explain the observed effects. The small N_2O line at 1197.40 cm^{-1} appears above the noise

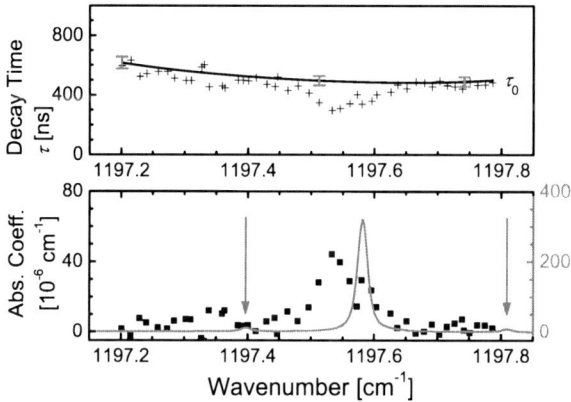

Fig. 3.9 CRDS results for 1667 ppm N_2O in N_2 at 100 mbar total pressure: decay time τ (*upper*) and the corresponding absorption coefficient k (*lower*). *Filled squares in the lower panel* are the experimental absorption points; *solid line* is the calculated spectrum. The *arrows* indicate two weak N_2O absorptions in the calculated spectrum. Note the 5 times scale magnification for the calculation (*right hand scale*) [103]

level and agrees with the calculation, but note that the latter is plotted with a times 5 scale magnification. The shortfall in k for weaker or stronger lines is very similar, i.e. the sensitivity is systematically limited. Optical feedback due to reflected light from the cavity mirrors does not provide an explanation for the line shift, because the p-DFB-QCL is already tuned out of resonance due to the laser chirp when the reflected light arrives at the laser, about 1–3 ns after the initial emission (for the actual distance between cavity and laser). Multimode behaviour of the lasers might be another explanation for apparent line broadening or line shifts. However, FT-IR spectra of the QCL emission showed no evidence for this hypothesis. Furthermore, the simultaneously recorded reference spectra were not affected in the same manner.

Provided that spectral averaging was carried out the achieved sensitivity for the p-CRDS measurements should be $4–6 \times 10^{-7}$ cm^{-1} $Hz^{-1/2}$. These values are approximately a factor of 10 better than estimated from similar experiments performed by Manne et al. [108] while for cw-CRDS with a cw-DFB-QCL a notably higher sensitivity was reported (4×10^{-9} cm^{-1} $Hz^{-1/2}$ [111]) also suggesting a systematic source of error with p-QCLs. For comparison, Menzel et al. used a Herriott cell type QCL spectrometer of path length 100 m. Their *NEA* for short time measurements ($\sim$1 s) is estimated to be 3×10^{-8} cm^{-1} $Hz^{-1/2}$ [112] and would therefore be superior to the high finesse cavity setup. Clearly, the obstacles might be overcome by calibration [108, 110], but in so doing one of the major advantages of CRDS, namely the calibration free measurement approach, is lost. Conversely this raises the question as to whether the combination of an optical cavity with a pulsed QCL in order to benefit from reduced sample volumes compared with multiple pass cells of the same effective length is desirable as a calibration free method.

In what follows it will be demonstrated that for a combination of p-QCLs with optical cavities both the spectral sweep rate α and the full frequency-down chirp are critical parameters. The consequences of such bandwidth effects for p-CRDS can be generalised. In order to increase the usable output signal of the cavity in a p-CRDS experiment the input can be increased either by modifying the driving voltage or the pulse width. Both factors significantly increase the full chirp of the p-QCL (cf. Sect. 3.2.2.2). From Fig. 3.6 it can be concluded that with small pulse widths (e.g. 32 ns, as used for the N_2O example) the chirped QCL pulse should nevertheless cover at least $\Delta f_{laser,tot} \approx 3$ GHz (0.1 cm^{-1}). This estimate implies a relatively slow sweep rate α of 100 MHz/ns (0.003 cm^{-1}/ns) which is even underestimated for such short laser pulses (Fig. 3.6). Additionally, the sweep rate would convert into an effective laser line width (Fig. 3.7) of $\Delta f_{aser,eff} \approx 300$ MHz (0.01 cm^{-1}). In other words, the total spectral coverage of the chirped pulse already exceeds a $\Delta f_{FWHM,line} = 900$ MHz (0.03 cm^{-1}) absorption feature, as might be observed at intermediate or atmospheric pressures, while the effective $\Delta f_{laser,eff}$ does not. The intrinsic QCL width might be even smaller and could be estimated to be $\Delta f_{laser,int} \approx 30$ MHz (0.001 cm^{-1}) which is certainly smaller than the (longitudinal) mode spacing of the cavity ($FSR = 300$ MHz or 0.01 cm^{-1}). This basically describes the ideal case for an optical resonator based spectroscopic experiment, where $\Delta f_{laser,int} < FSR < \Delta f_{FWHM,line}$ yielding single exponential decays. However, if now the chirped QCL pulse is taken into account the picture has to be modified to $\Delta f_{laser,int} < \Delta f_{laser,eff} \leq FSR < \Delta f_{FWHM,line} \ll \Delta f_{laser,tot}$. Although this situation has been partly considered by Sukhorukov et al. [110] and Silva et al. [113], the effect is usually underestimated. The effective laser line width $\Delta f_{laser,eff}$ is often assumed to give an adequate estimate of the number ϑ of affected cavity modes which is not the case for p-QCLs, since $\Delta f_{laser,tot}$ determines this number. The ratio between the absorption line width $\Delta \nu_{FWHM,line}$ and the number of excited modes times the FSR of the cavity, $\vartheta \cdot FSR$, is the important criterion. A detailed analysis of the specific 8.35 μm p-DFB-QCL in [103] revealed that a single laser pulse excited almost twice the number of cavity modes ($\Delta f_{laser,tot} = 4.8$ GHz or 0.16 cm^{-1}) than would be affected by a representative broad absorption feature at intermediate pressure ≥ 5000 Pa. This situation is illustrated in Fig. 3.10 which shows the mismatch between the number of (longitudinal) resonator modes covered a relatively broad absorption feature and those ones affected by the chirped laser pulse.

Laser bandwidth effects of this type have been described extensively by Hodges et al. [114] for a 3 GHz line width dye laser used to probe a 900 MHz (FWHM) absorption feature. They observed 5 to 8 times smaller absorptive losses compared to using a Ti:Al$_2$O$_3$ laser source having a bandwidth smaller than the absorption line. This reduction in sensitivity accords well with the approximately 10 times smaller absorption coefficients mentioned earlier. Berden et al. discussed laser bandwidth effects as the reason for non-exponential decays as well as underestimated absorption coefficients in a review [30]. They also pointed out that correct values might be extracted if the spectral density distribution of the light source is known. The investigation of individual pulses during a laser sweep, however, leads to the conclusion

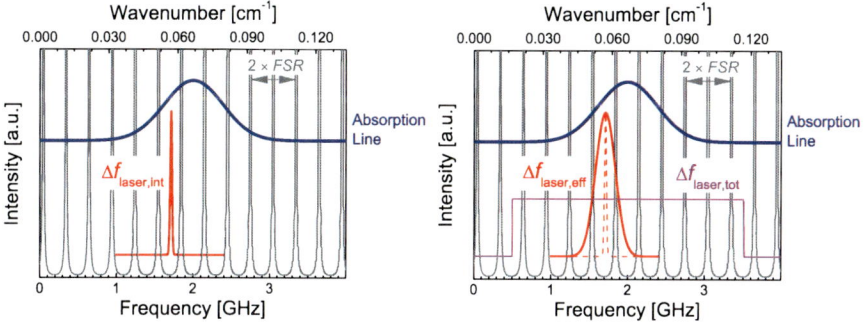

Fig. 3.10 Mode structure of an optical cavity of 0.5 m length (*FSR* = 300 MHz) calculated according to the equations given in [100] with respect to absorption line ($\Delta f_{\mathrm{FWHM,line}}$ = 900 MHz, *thick solid line*) and p-QCL widths as discussed in the text. *Left*: Idealised picture using the intrinsic QCL width ($\Delta f_{\mathrm{laser,int}}$ = 30 MHz). *Right*: Comparison of cavity modes affected by the intrinsic (*dotted line*) and effective p-QCL width ($\Delta f_{\mathrm{laser,eff}}$ = 300 MHz, thin *solid line*) along with the lower limit for a chirped laser pulse ($\Delta f_{\mathrm{laser,tot}} \geq 3.0$ GHz, *rectangularly shaped line*). Line profiles are stacked for clarity

that the determination of the spectral density distribution of QCL pulses is not an option due to the unconstant pulse width and intensity throughout the sweep [40, 77]. This would additionally be complicated by the variable chirp rate during an individual QCL pulse (Fig. 3.7)

The conclusion from this section is that an mid-IR source with much smaller effective line width or spectral coverage is needed to reach the ideal case of a high finesse cavity experiment. Since the frequency-down chirp is inherent to p-QCLs (i.e. also to p-EC-QCLs) the only solution appears to avoid rapid changes in the laser current that cause the spectral sweep, i.e. to use cw-QCLs. The potential of such a combination was first demonstrated with a cryogenically cooled cw-DFB-QCL [111]. Note, that—apart from higher investment costs for cw-DFB-QCLs—additional equipment such as acousto-optic modulators will be required to interrupt the laser injection into the cavity. Alternative approaches will be discussed in the next sub-section. It should be pointed out that the chirped laser pulse is counter productive in two ways that may be synergistic. Firstly, the chirp rate of at least 100 MHz/ns causes an insufficient cavity build-up because each cavity mode only coincides with the corresponding spectral part of the laser emission for a very short time. This corresponds in fact to a laser pulse at a fixed frequency shorter than the cavity round trip time propagating in the cavity. The throughput is then limited by the reflectivity of the mirrors [28]. Secondly, the full chirp of the QCL excites too many resonator modes as discussed above. In order to achieve a higher cavity output the pulse length or the operating voltage has to be raised which in turn increases the sweep rate and the full chirp respectively, so even more cavity modes are excited during one laser pulse while the interaction time may even be reduced.

3.4.2 CEAS Using cw-QCLs

3.4.2.1 On-Axis Excitation

The most straightforward implementation of a CEAS experiments is based on a quasi-resonant (on-axis) excitation of the cavity where specific mode-matching optics are obviated to enable transverse modes to be excited. The previously used arrangement as schematically shown in Fig. 3.8 was therefore equipped with cw-DFB-QCLs (Alpes Lasers) emitting between 1300 cm^{-1} and 1311 cm^{-1} ($\sim$7.65 μm). The cavity output signal was recorded with a LN cooled detector as well as with a thermoelectrically (TE) cooled detector. During a typical measurement the laser was driven with a constant DC current of 300–500 mA. An additional current ramp of <15 % of the DC level was impressed on it to sweep the laser frequency with about 1 kHz by $\leq$1 cm^{-1} (30 GHz) during the on-phase. The laser current was briefly reduced below its threshold value to record the offset of the zero detector signal in the off-phase. The spectra (on-phase) consisted of 800 points with an additional off-phase of 50 points. For a mirror distance of $L = 0.432$ m, used for the validation experiments, the sweep of the laser frequency corresponds to a scan over several tens of cavity modes. After approximately 14 μs a new cavity mode was excited which was sufficient to avoid an overlap between the decay events of different modes. On the other hand the sweep rate gave a full spectrum after slightly more than 1 ms (Fig. 3.11). A given number of spectra were then averaged and simultaneously fitted. In order to smooth the residual cavity mode noise on the baseline of the spectra, mainly caused by longitudinal modes, the whole cavity was mechanically destabilised, e.g. by an external vibrating source. Averaging over 5 s or more (i.e. >1000 single shot spectra) reduced the optically induced baseline modulation to below the noise level caused by all the data acquisition electronics. Due to instabilities in the commercial current source used for the cw-DFB-QCL the laser line widths (FWHM) of $\Delta f_{laser} < 40$ MHz (<0.0012 cm^{-1}), as suggested in Sect. 3.2.2.2, are expected to be a lower limit in practise. From fits to the spectra an upper limit (FWHM) of the observed instrumental broadening of $\Delta f_{obs} < 300$ MHz (0.010 cm^{-1}) was deduced. This reflects mainly the longitudinal mode spacing of the short cavity ($FSR = 360$ MHz or 0.012 cm^{-1}) rather than other effects. No additional optics were added to suppress possible optical feedback to the laser, since the reflected power was assumed to be negligible for the cw-DFB-QCLs given their small intrinsic line width and thus reduced susceptibility to reflected laser power [109].

Since the cavity mirrors itself served to enclose the evacuated cavity, which would be also the case for the intended application to plasma diagnostics (cf. Sect. 3.5.3), a mechanical dither of the mirror mounts by piezoelectric transducers could not be implemented. In order to improve the spectral resolution and sensitivity compared to what was reported in [103], firstly, the cavity length was increased to $L = 1.297$ m to reduce the FSR to 120 MHz (0.004 cm^{-1}) and to achieve a higher density of fundamental cavity modes. Secondly, the cw-DFB-QCL was externally modulated with a high frequency sinusodial signal at almost 200 kHz and out of

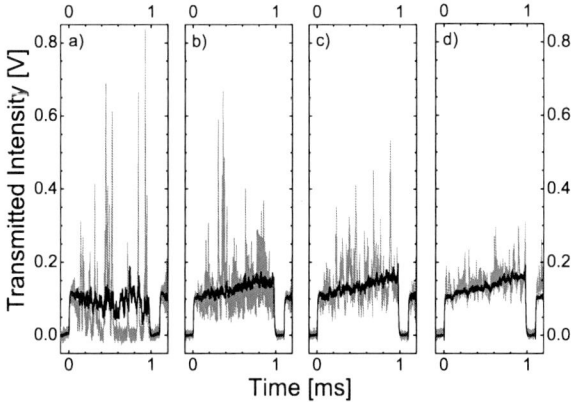

Fig. 3.11 Improvement of the residual baseline noise level by an electronically dithered current fed to a cw-DFB-QCL current. The influence is demonstrated for the single shot cavity transmission (*grey*) and the transmitted signal averaged over 50 laser sweep cycles (*black*). Detector signals correspond to an empty cavity (**a**) without a ramp or dither, (**b**) with an additional current ramp (17 % of DC level), (**c**) with a weakly dithered (0.5 % of DC level) current ramp, and (**d**) with a strongly dithered (2 % of DC level) current ramp applied

phase with the laser sweep rate. The current amplitude was usually below 1 % of the DC current. In this way, the resonance between the laser and (fundamental) cavity modes was electronically dithered. Although suddenly excited longitudinal modes capable of transmitting high intensities through the resonator are inherent to an on-axis alignment, their influence on the *SNR* can efficiently be reduced by averaging over theses events. First, dithering the current through the laser enables a better control of the randomised cavity excitation events than the mechanical destabilisation of the resonator used in the earlier validation measurements. Second, the laser emission frequency is now deliberately chirped which reduces the probability for resonant excitation and therefore decreases the mode noise level. On the other hand, several fundamental cavity modes may be excited simultaneously. In Fig. 3.11 the achieved transmission with an empty cavity is shown for 4 illustrative sample sweeps. If only a constant DC current is applied to the QCL (Fig. 3.11a) the transmitted signal is governed by occasionally occurring resonances of high intensity throughput which are more pronounced during the first half of the on-phase when the QCL temperature and hence the emission frequency stabilises. As soon as a current ramp is superimposed continuous transmission is observed leading to a smoothed average signal, albeit single resonance events may be still of high intensity (Fig. 3.11b). The influence of these spikes is reduced by dithering the QCL current (Fig. 3.11c and d). Although a high amplitude of the sinusoidal dither yields apparently a better *SNR* (Fig. 3.11d), the modulation amplitude had to be limited to <1 % of the DC current—resulting in an uncorrelated dither of slightly more than one cavity *FSR* per spectral data point (Fig. 3.11c)—to avoid asymmetric line shapes (skewing effect).

Spectra of N_2O from different mixtures in N_2 buffer gas were recorded and averaged over 20 s, in order to determine the effective reflectivity of the cavity mirrors, and provide a calibration. A fit of the transmission spectra between 1306.8 and 1306.9 cm^{-1} resulted in an effective reflectivity of $R = 99.96$ % for the cavity mirrors or an effective path length of $L_{eff} = 1080$ m in the case of the short cavity ($L = 0.432$ m) [103]. Since in this approach the excitation of high order transverse modes may occur which could have different diffraction losses this value of R should be regarded as an effective one [112]. Furthermore, even for absorptions exceeding 10 % a correction for non-linearity is recommended. For this purpose the validity of the weak absorption limit in Eq. (3.11) was assesed by plotting $\ln(I_0/I) \approx GU$ at the line maximum against the concentration of N_2O in the cavity. Generally, it can be concluded that the linear approximation is valid for $GU \leq 0.15$ whilst for absorption features bigger than 15 % a correction is essentially required. Equation (3.11) provides also a means to determine the effective mirror reflectivity from the slope of $\ln(I_0/I)$ as a function of N_2O density as long as the linear weak absorption regime is considered. Using this approach the above mentioned reflectivity was confirmed with an N_2O line at 1303.201 cm^{-1} ($R = 99.967$ %). This yielded an effective absorption path of $L_{eff} = 3900$ m for the extended cavity ($L = 1.297$ m).

The application of optical resonators in absorption spectroscopy under low pressure conditions might suffer from non-linear absorption, e.g., power saturation as observed for CRDS [115]. A "worst case" estimate—assuming perfect beam shaping and power injection to the TEM$_{00}$ cavity modes—suggests that the experimental conditions used in these experiments might be already sufficient to generate non-linear absorption effects [40]. Power saturation may be still negligible under the aforementioned conditions, since the laser is swept rapidly and hence excites higher order cavity modes. This would prevent a complete resonant intensity build-up inside the cavity as known from cw-CRDS. However, with forthcoming increase in QCL power levels and selective excitation of TEM$_{00}$ modes the situation may be considerably different as recently demonstrated in saturated absorption CRD experiments [116, 117].

Another artefact, namely a weak asymmetry in the absorption profiles, can be detected in spectra obtained by using a relatively rapid laser sweep [40, 103, 105, 118]. This skewing effect is observed when the laser scan frequency approaches the cavity time constant [105, 118] and has to be distinguished from other obstacles seen with QCLs, e.g. the "rapid passage" effect [87, 90]. While the latter is linked with the disturbed population transfer between ro-vibrational levels on the sub-nanosecond time-scale, the skew is connected to multiple incomplete excitation of resonator modes. Considering the previous experimental conditions, i.e. laser sweep of almost 0.8 cm^{-1} (24 GHz) in 1.1 ms (Figs. 3.11 and 3.12), the excitation of a neighbouring longitudinal cavity mode was estimated to occur after about five times the ring down time τ_0 which was assumed to be reasonably long to obviate overlapping ring-down events. However, such rapid sweep rates result in relatively short ring-up times (being the time when the QCL frequency coincides with a cavity mode, which was presently about 390 ns). Firstly, the cavity throughput becomes limited.

Fig. 3.12 CEAS transmission spectrum of a N_2O calibration mixture (1.23 ppm in N_2, 0.47 mbar) recorded with 1 s integration time. The N_2O line (*circle*, 1303.201 cm^{-1}) and both highlighted spectral positions (*triangle right*, 1303.40 cm^{-1} and *triangle left*, 1303.55 cm^{-1}) were used for further Allan plot analysis. Weak, unresolved N_2O features are indicated by *arrows*

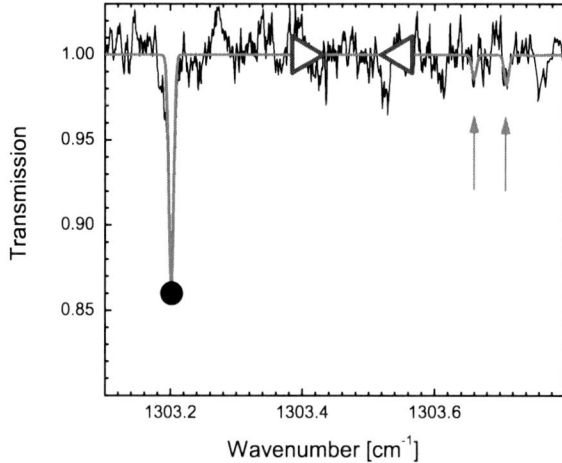

This has been modelled and discussed by Remy et al. for the analogous case of a fixed laser frequency and a moving cavity mirror [119]. Secondly, a ring-up time being short compared to the ring down event obviously leads to asymmetric line shapes. Any change of the absorption cross section inside the cavity, e.g. by pronounced absorption lines, is observed as an unambiguous intensity change within a very few spectral data points when the laser frequency approaches the absorption feature and as an additionally broadened wing when the laser frequency passed the line maximum. Since CEAS experiments require a calibration in any case, such effects are included in the effective reflectivity R or in L_{eff} [40, 105]. Nevertheless the sweep rate dependence should always be checked carefully and a compromise has to be found between achieving a high *SNR* using short integration times and reducing non-linear effects on the line shapes [118].

A N_2O detection limit (σ_A) of 96 ppb, 17 ppb and 12 ppb at 0.47 mbar was detected in Fig. 3.12 for integration times of 1 s, 30 s and 90 s respectively. The 1 s value is already interesting for time resolved measurements (e.g. plasma diagnostics, online breath gas analysis, cf. Sect. 3.5) whereas the long-time integration limits may be applicable to trace gas measurements. Averaging longer than 30 s is often limited in practise due to drift effects inherent to the system. Hence, the 90 s detection limit should be considered as a best case value of the current system. In order to facilitate general sensitivity conclusions for the CEAS system the (baseline) noise signal was analysed for the two spectral positions 1303.40 cm^{-1} and 1303.55 cm^{-1} by means of an Allan plot (Fig. 3.13). The Allan variance σ_A shows the same behaviour in both cases and is mainly determined by white noise. Minimum detectable fractional absorbance values, $\Delta I/I_0$, of 1.1×10^{-2}, 2×10^{-3}, and 1×10^{-3} were found for the three selected integration times. Since the system was aligned on-axis cavity mode noise inherent to CEAS or ICOS systems is still the main limiting factor [121]. In order to flatten the residual mode structure on the baseline a measurement interval of at least 5 s was necessary. After 30 s integration time small deviations of experimental data from the theoretical white

Fig. 3.13 Baseline signal detected at 1303.40 cm^{-1} (*upper left, triangle right* in Fig. 3.12) and 1303.55 cm^{-1} (*upper right, triangle left* in Fig. 3.12). *Lower*: Allan variance σ_A of the baseline noise for both *upper panels* (*symbols*) and the corresponding white noise trace (*grey line*). An example for an unstable experiment (data stream not shown) is also given (*dotted*)

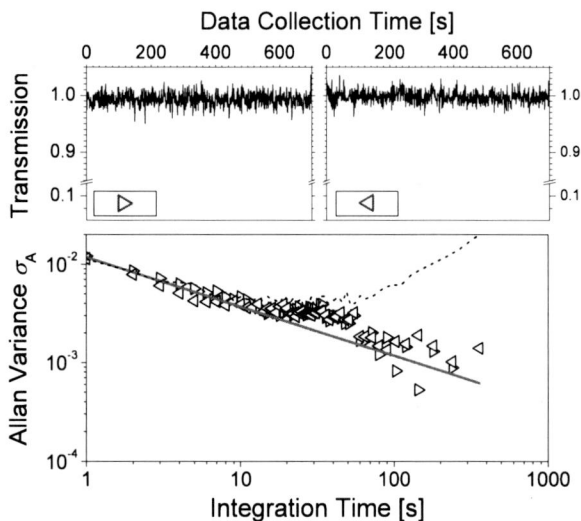

noise behaviour are observed which might be caused by drift effects. The reason for the drift can be directly (i.e. mechanically) and indirectly in nature. Particularly the latter case may hamper further averaging: a thermally induced drift of the laser intensity causes a gradual change of the background normalised transmission spectra. These drifts effects are visible as a strong increase of σ_A for integration times longer than 60 s (dotted line in Fig. 3.13). Additionally, the spectral position will shift and would require a correction or frequency locking methods.

The Allan variance σ_A of the baseline signal(s) can readily be converted into the *NEA* using Eqs. (3.19) and (3.20) which would be $\sim 4 \times 10^{-8}$ cm^{-1} Hz$^{-1/2}$ for the example discussed here. Furthermore, employing Eq. (3.25) the MDND ($n_{\min}$) can be inferred (for low pressure conditions). Note that the instrumental broadening $\Delta\nu_{\mathrm{obs}}$ should be used instead of the theoretical Doppler broadening $\Delta\nu_D$. In Table 3.2 the (anticipated) MDNDs are collected for several molecules of main interest in plasmas or atmospheric chemistry for three integration times. The given line strengths are typical values for absorption features in the mid-IR spectral range. Clearly for measurements at intermediate or atmospheric pressures sub-ppb levels could be measured although the lines would be broader and might be unresolved. Note that the detection limits in Table 3.2 were obtained with a LN cooled detector. If TE cooled detectors are applied the corresponding $\Delta I/I_0$ may increase by a factor of 2. The degradation in performance is mainly due to the lower sensitivity of TE cooled detectors and has been observed in linear QCL-based AS as well as with CEAS [103].

3.4.2.2 Off-Axis Excitation

Resonant coupling of the laser power to the lowest transverse electromagnetic mode (TEM$_{00}$) of a cavity, which is aligned on-axis with respect to the laser beam, re-

Table 3.2 Detection limits for molecules being detectable at ~ 1303 cm^{-1} (7.67 µm) with a CEAS system aligned on-axis. An instrumental broadening of 240 MHz (0.008 cm^{-1}, FWHM) and an absorption path of $L_{\text{eff}} \approx 4$ km was assumed to calculate the anticipated values ([a])

Species	Line strength S [cm/molecule]	Detection limits (MDND) n_{min} [cm^{-3}] Integration time		
		1 s	30 s	90 s
N$_2$O	1.5×10^{-19}	1.6×10^9	3×10^8	2×10^8
CH$_4$[a]	5.0×10^{-20}	4.7×10^9	9×10^8	5×10^8
C$_2$H$_2$[a]	1.0×10^{-19}	2.3×10^9	4×10^8	2×10^8
HNO$_3$[a]	1.0×10^{-20}	2.3×10^{10}	4.5×10^9	2.6×10^9
H$_2$O$_2$[a]	2.0×10^{-20}	1.2×10^{10}	2.2×10^9	1.3×10^9

mains the main challenge in CEAS (or ICOS) measurements. While such an mode-matched alignment is highly desired in cw-CRDS to retrieve single-exponential decays, implementations of CEAS seek to avoid even occasional excitation of fundamental modes as these modes transmit much more power through the resonator which can be found back as intensity noise in the recorded spectra [107]. Obviating mode-matching optics and the application of rapid laser scanning to excite higher order transverse modes (i.e. $m, n > 0$ in Eq. (3.14)) can partly solve the issue. However, the frequency sweep has to stay long enough in resonance with the cavity modes to enable a sufficient intensity build-up inside the cavity and to avoid skewing effects. At the same time the scan rate must be significantly higher than the jittering of the cavity modes to avoid large intensity fluctuations [101]. An alternative approach to generate a dense cavity mode structure was suggested by Paul et al. [122] and uses an off-axis alignment of the incident laser beam. In this way, a Herriott or Lissajous type pattern is generated. If the radiation fulfils the re-entrant condition after μ roundtrips (Eq. (3.16)) the effective FSR_{eff} is a factor $1/\mu$ smaller than under resonant excitation. On the other hand, the optical power is divided between all modes. Strictly speaking, the signal detected from the cavity is reduced by a factor μ. Hence, the SNR will be reduced by $\sqrt{\mu}$. Under idealised conditions this is compensated by averaging of μ times more spectral data points per laser sweep [106].

The spectrometer performance in OA-CEAS benefits from reduced optically induced noise levels compared to spectrometers using an experimentally slightly less demanding on-axis alignment. Additionally, the effective absorption length L_{eff} is increased for an identical cavity footprint L, i.e. similar to Herriott cell type spectrometers the gas sampling volumes remain small [123]. Although sensitivities down to the 10^{-10} cm^{-1} Hz$^{-1/2}$ range have been reported in the mid-IR spectral range, the entire systems were mainly based on LN cooled devices [105, 124]. The implementation is comparable to an on-axis experiment as shown in Fig. 3.14. Apart from highly reflective mirrors of increased diameter (up to 100 mm diameter) the reduced signal levels at the cavity output, are potential pitfalls of OA-CEAS detection schemes. By analogy to conventional multi-pass spectrometers interference effects due to overlapping laser spots on the mirror surfaces has to be considered during the design phase.

Fig. 3.14 Schematic diagram
of a typical arrangement used
for OA-CEAS experiments.
By analogy to Fig. 3.8 the
linearly polarised radiation
from a cw-DFB-QCL is
collimated, directed through a
waveplate (WP) to reduce
potential OF, and reduced in
diameter through a telescope.
A reference channel is used to
perform calibration.
(Reproduced with permission
from [107]. Copyright 2009
American Chemical Society)

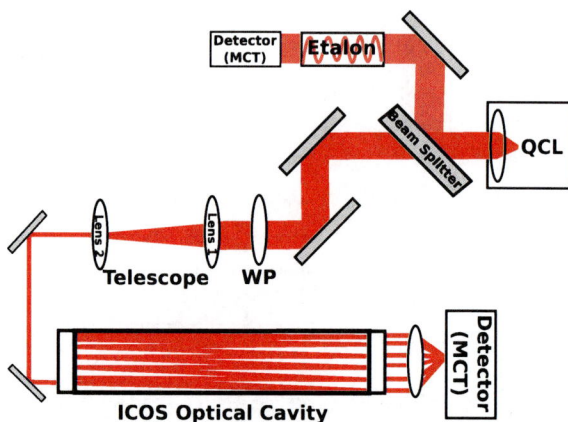

Figure 3.15 shows the consequences of an OA alignment for the sampling of a broad absorption line (900 MHz, 0.03 cm^{-1}, FWHM) as function of the excited modes in a stable resonator as introduced in Fig. 3.10 in Sect. 3.4.1. Note that by using cw-QCLs line width broadening effects due to the laser radiation are absent. In the upper panel the undesirable case of exploiting the longitudinal modes in on-axis configuration is depicted leading to poor spectral resolution. In the middle panel of Fig. 3.15 the case of $\mu = 4$ roundtrips in an OA-cavity is illustrated. Interestingly, a similar spectral resolution might be obtained if higher order transverse modes are efficiently excited in an on-axis experiment (i.e. $\mu = 1$, but $m, n \gg 0$). Finally, the lower panel shows the densification of the mode structure after 100 roundtrips which yields a FSR_{eff} significantly smaller than the cw-QCL line width (inset).

3.4.3 Optical Feedback to QCLs

A separate chapter will be concerned with OF techniques. Therefore, only a brief account on the specific challenges in conjunction with QCLs is given below. A typical experimental arrangement applying a "V-shaped" cavity is shown in Fig. 3.16. In common with cw-CRDS, OF-CEAS aims at complete, efficient and sequential excitation of longitudinal cavity modes while rapidly sweeping the laser frequency as typical for CEAS. Hence, fast digitising of ring-down transients can be obviated. Furthermore, optically induced noise caused by excitation of high-order transverse modes as known from on- or off-axis-CEAS is absent. Consequently, laser sweep rates can be reduced to avoid skew of absorption features and to obtain a sufficient cavity build-up time. Typical values are of the order of 1 ms per resonator mode. More importantly, the mirror size and hence the entire volume of the sampling cell can be smaller than required for OA-CEAS approaches. The attainable spectral resolution may be twice as high as in on-axis CEAS techniques using the same cavity base length L (due to 2 cavity arms of length L (Fig. 3.16)), but lower than in OA-CEAS (Fig. 3.15).

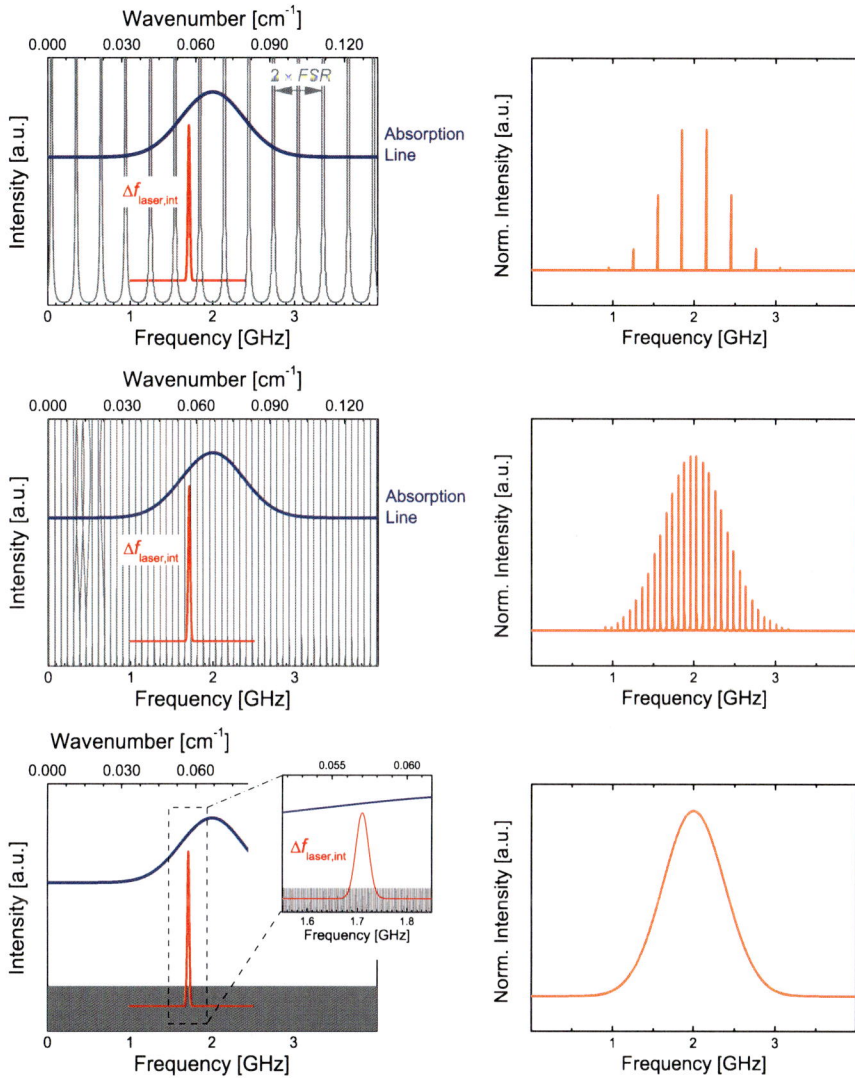

Fig. 3.15 *Left*: Mode structure of the same resonator of 0.5 m length as introduced in Fig. 3.10, now illustrating off-axis alignment cases. Absorption line ($\Delta f_{\text{FWHM,line}} = 900$ MHz, *thick solid line*) and intrinsic cw-QCL widths ($\Delta f_{\text{laser,int}} = 30$ MHz, *thin solid line*) are as shown in Fig. 3.10. *Right*: Resultant spectra are connected to the 3 distinct cases detailed in the text. *Upper*: Only longitudinal modes are excited. The *FSR* (Eq. (3.15)) is indicated. *Middle*: Off-axis alignment with $\mu = 4$, $\kappa = 1$ (Eq. (3.16)) yields a denser mode structure in frequency space and hence resembles the absorption profile more accurate. A similar result may be achieved in on-axis alignment and an efficient excitation of transverse modes. *Lower*: Typical off-axis alignment with $\mu = 100$, $\kappa = 1$. Not the change in cavity throughput (intensity) depicted in the left panel. Line profiles are stacked for clarity

Fig. 3.16 Schematic diagram of an OF-CEAS experiment employing a cw-DFB-QCL and a V-shaped cavity along with TE cooled detectors (PD). The OF phase is controlled by a piezoelectric translation (PZT)-mounted mirror [125]

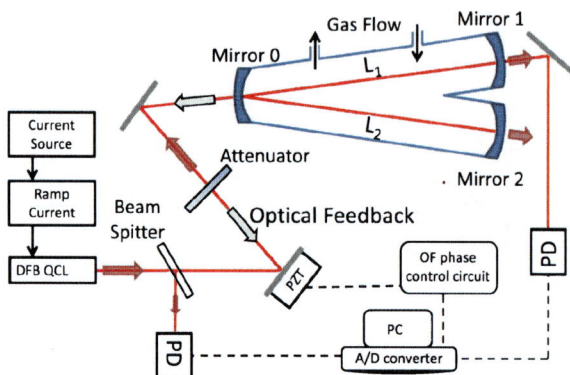

QCL-based OF-CEAS in the mid-IR has recently been demonstrated to be at least as sensitive (3×10^{-9} cm^{-1} Hz$^{-1/2}$) as in the near-IR, though the mirror quality in the latter spectral range is still superior to mid-IR cavity mirrors [109, 125]. It transpires that OF is challenging to achieve given the very low intrinsic line width of QCLs [109]. Additionally, in practise, a mid-IR beam, even if perfectly shaped, is more difficult to keep collimated as the divergence scales linearly with the wavelength of the radiation. In other words, mid-IR radiation diverges about an order of magnitude more rapidly in an optical setup than a beam in the visible spectral range. Therefore Hamilton and Orr-Ewing omitted the attenuator commonly used to control the OF (Fig. 3.16). On the other hand, Maison et al. discuss that, if the same feedback level as for DFB diode lasers can be obtained, QCLs may respond more pronounced yielding therefore an increase in power enhancement [125].

3.4.4 Absorption Spectroscopy Using EC-QCLs

The increasingly good performance of EC-QCLs capable of emitting single mode mid-IR radiation over broad spectral ranges ($\sim$100 cm^{-1}) has also triggered their use for sophisticated chemical sensing schemes. Since EC-QCLs inherently use a (Fabry-Perot) cavity to spectrally filter the radiation, intracavity laser absorption (ICLAS) may be performed [126, 127]. Such an approach combines the broad spectral gain of the laser chips with cavity enhanced sensing using very small cavity base lengths. Phillips and Taubman specifically analysed the changes in the compliance voltage of the EC-QCL to obtain a broadband spectrum of Freon-134a (Figs. 3.17 and 3.18, [126]) and reported a *NEA* of 2×10^{-6} cm^{-1} Hz$^{-1/2}$. Considering the short cavity base length the noise level is comparable to conventional absorption techniques. Further combination of high finesse optical cavities with EC-QCLs is forthcoming from the availability of cw-EC-QCLs. Early attempts with p-EC-QCLs [128] clearly suffer from the issues with chirped laser pulses as outlined for p-DFB-QCLs in Sect. 3.4.1. Recently, the successful implementation of EC-QCLs in a CRD and an OA-CEA spectrometer has been reported [129, 130].

Fig. 3.17 ICLAS using an EC-QCL and the changes in compliance voltage to record absorption spectra (adapted from [126])

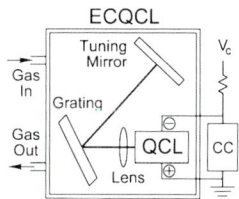

Fig. 3.18 Absorption spectrum of Freon-134a (upper) obtained while measuring the compliance voltage of an EC-QCL in ICLAS configuration. Database spectrum (lower) is plotted for comparison. The *inset* shows a calibration plot of compliance voltage versus absorption coefficient [126]

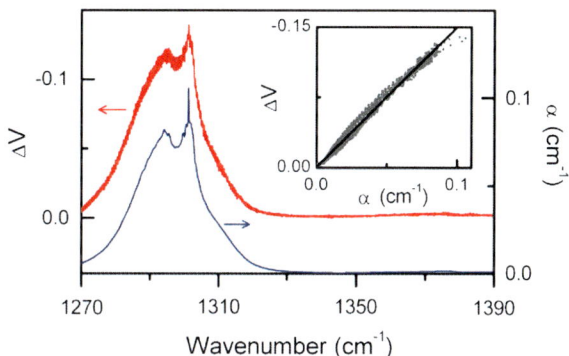

3.5 Applied Mid-Infrared Chemical Sensing Using Optical Cavities

3.5.1 Brief Historical Background

Early experiments in the mid-IR region were carried out with OPOs, Raman cells or shifters, CO_2 or CO lasers [131–142]. In all these cases sophisticated optical geometries were developed which were more suitable for the research laboratory than for field applications. To overcome the drawbacks of the bulky radiation sources, small semiconductor based lasers became increasingly used. In 2001 a mid-IR CRD spectrometer, using lead salt lasers, and the detection of CO was reported [119, 143]. More recently, an attempt of pulsed slit jet CRDS, using a 3 μm lead salt diode laser, has been published [144]. In both studies the technique suffered from low laser intensity and the beam quality. Continuous progress in non-linear frequency conversion techniques and available pump sources enabled OPO- or DFG-based cavity enhanced spectrometers to be designed for measuring specifically hydrocarbons in the 3 μm range [145, 146]. The advent of QCLs (and ICLs), which are increasingly becoming available on a commercial basis, has particularly renewed the interest in combining mid-IR laser sources with optical cavities. From application point of view two main objectives have become apparent: trace gas sensing applications (Sect. 3.5.2) seeking often for both high sensitivity along with high precision, and online or *in-situ* chemical sensing in reactive environments containing transient species (Sect. 3.5.3).

Fig. 3.19 Phosgene (10 ppm in N_2, 2500 Pa) survey spectrum recorded with CEAS by rapid laser sweeping and tuning the heat sink temperature (1 sweep range per colour) of a TE cooled cw-DFB-QCL. The dotted spectral range is detailed in Fig. 3.20. Note the spectral overlap with atmospheric H_2O (*top*)

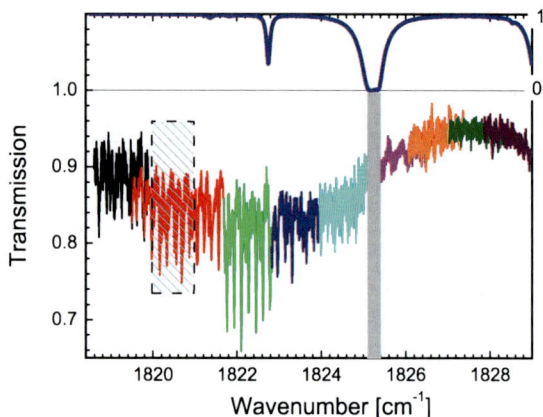

3.5.2 Trace Gas Sensing

3.5.2.1 Phosgene Detection Using CEAS

In many chemical sensing applications where toxic or explosive samples need to be analysed high sensitivity in conjunction with small sample volumes and short residence times are essential requirements. Although CEAS in the mid-IR fingerprint region appears ideally suited, chemically harsh environments may hamper the application of sophisticated optical arrangements where robust sensors rather than carefully aligned or mode-matched cavities are required. In order to detect and monitor trace levels of phosgene ($COCl_2$) below the indicative occupational exposure limit of 100 ppb for this highly toxic gas (as e.g. present in workplaces in chemical industry) a cavity-enhanced cw-DFB-QCL spectrometer was employed. On-axis alignment similar to the proof-of-principle study mentioned in Sect. 3.4.2.1 was applied. The fundamental bands of $COCl_2$ are usually located at wavelengths longer than 11 μm (<900 cm^{-1}) and are therefore difficult to measure using the emission of commercially available cw-DFB-QCLs. Thus the v_1 band was selected and was observed in a spectral window between 1816 cm^{-1} and 1829 cm^{-1} (Fig. 3.19). To reduce the residual noise structure on the baseline electronic dithering combined with rapid laser scanning was applied to a cavity of $L = 0.432$ m. Calibration of the mirror reflectivity suggests an effective absorption length of 280 m. Allan plot analysis yields a $NEA = 4 \times 10^{-8}$ cm^{-1} Hz$^{-1/2}$ corresponding to 50 ppb $COCl_2$ at reduced sample pressure (2500 Pa). Since an unstabilised cavity was used drift effects become dominant after 30 s integration time. The performance of the spectrometer is in good agreement with what was observed earlier for such an approach [103] and may be therefore considered as typical NEA value. Spectral congestion hinders the assignment of individual lines. The definition of an effective absorption cross section, k_{eff}, averaged across a certain spectral micro-window $\langle v \rangle$,

$$k_{eff} = n\sigma_{eff}(\langle v \rangle) = \frac{1}{L_{eff}} \ln\left(\frac{I_0(\langle v \rangle)}{I(\langle v \rangle)}\right), \tag{3.26}$$

Fig. 3.20 (Effective) absorption cross section of the ν_1 band of phosgene (*thin solid line*) in the spectral window between 1820 and 1821 cm^{-1} as measured by CEAS. Good agreement is observed, if spectral resolution is adjusted during post-processing (*thick solid line*) to literature data (*dotted line*)

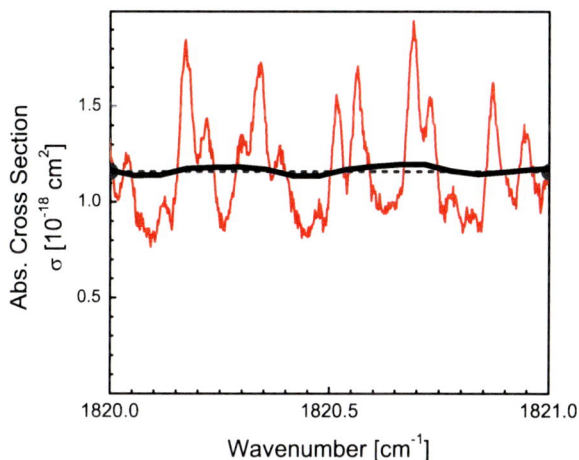

provides a means for quantitative measurements of COCl$_2$ samples. Figure 3.20 shows an example of the high-resolution effective absorption cross section of phosgene obtained by CEAS in comparison with the low-resolution cross section available in the PNNL-NWIR database [147]. Interestingly, the measured cross section, if artificially adjusted in spectral resolution, is in good agreement with the database value.

3.5.2.2 Atmospheric Sensing Using OF-CEAS

Optical sensors for atmospheric and environmental monitoring, e.g. of greenhouse gas levels, provide important input data to climate models. Field-deployable spectrometers should be capable of cryogenic free operation. The increased cavity transmission in OF-CEAS compared to OA-CEAS enables particularly TE cooled detectors to be used along with the advantage of small sampling volumes to keep sampling flow rates on a reasonably low level. The spectral resolution of such spectrometers is usually determined by the cavity *FSR*, e.g. $\approx$300 MHz (0.01 cm^{-1}) in a V-shaped cavity of $L \approx 70$ cm [109], which is sufficient at elevated sampling pressures with broad absorption features (Fig. 3.21). If optical feedback to the TE cooled cw-DFB-QCL is optimised *NEA* values as low as 3×10^{-9} cm^{-1} Hz$^{-1/2}$ can be achieved [125]. This value is about an order of magnitude better than what was reported for weak feedback cases [109] or for on-axis configurations as established for the aforementioned phosgene spectrometer.

3.5.2.3 OA-CEAS Instruments in Medical Science and In-Flight Studies

Real-time monitoring of marker molecules in breath gas analysis using cavity enhanced spectrometers is becoming a valuable alternative to conventional analytic

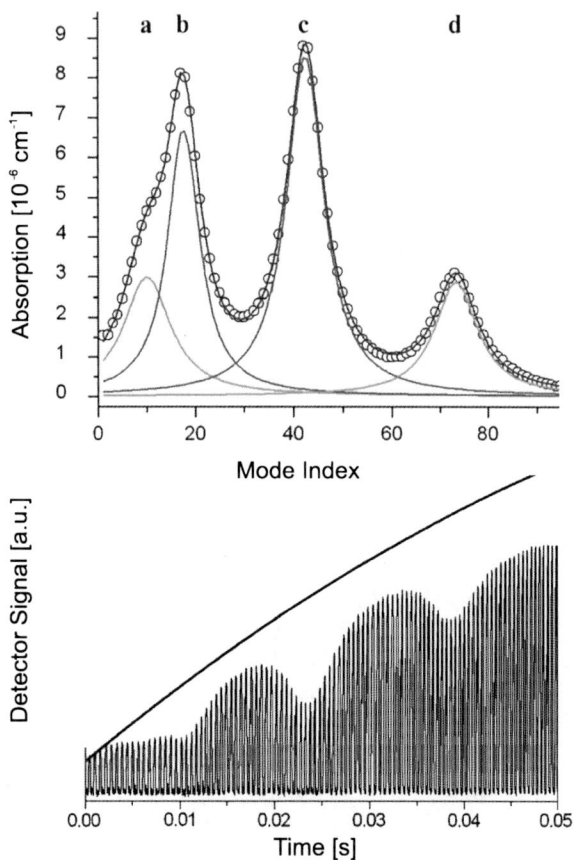

Fig. 3.21 *Lower*: Cavity output signal (*lower trace*) and the incident laser intensity (*upper trace*) measured during a single QCL current sweep centred around 1275 cm^{-1}. *Upper*: N$_2$O and CH$_4$ molecular absorptions ($a + d$) and ($b + c$) were measured in atmospheric air at 1 atm. The absorption features obviously reduce the cavity output signal (*lower pannel*) and yield the absorbance spectrum (*circles*) which was further analysed to establish an *NEA* of 5×10^{-8} cm^{-1} Hz$^{-1/2}$ [109]. (With kind permission from Springer Science and Business Media)

techniques such as mass spectrometry or gas chromatography. For example, nitric oxide is recognised as biomarker in asthma and other respiratory disorders. Ammonia can be used to examine the efficacy of renal dialysis or to diagnose liver disease while ethane may suggest oxidative stress. Elevated C_2H_6 levels are known from cigarette smokers [148]. Clinical breath analysis requires high sensitivity in order to measure sub-ppb levels of tracer molecules combined with high selectivity to distinguish constituents of exhaled breath and perhaps establish correlations between their temporal evolution. The latter fact also suggests relatively fast measurements within a few seconds rather than averaging for tens or hundreds of seconds to reach the Allan minimum of a spectrometer. Furthermore, LN free operation is desirable, but not yet entirely achieved as the majority of optical breath gas analysers is so far based on OA-CEAS (often referred to as OA-ICOS in literature [121, 148–150]). These instruments may achieve *NEA* values down to the 10^{-10} cm^{-1} Hz$^{-1/2}$ range, i.e. about an order of magnitude more sensitive than the previously discussed straightforward CEAS or OF-CEAS sensors of similar footprint. Apart from LN cooled

Fig. 3.22 Temporal evolution of C_2H_6 in exhaled breath measured with an OA-CEAS instrument based on an ICL 30 min after smoking a cigarette. CO and CO_2 concentrations were monitored by using commercial instruments [151]

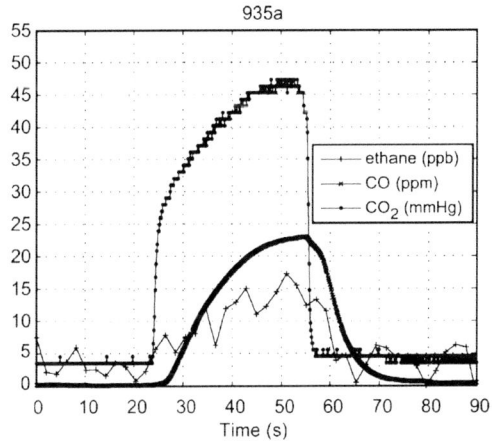

detectors, the target molecules such as ethane in the 3 μm C–H stretch region may (still) require LN cooled lasers [151]. So far, OA-CEAS instruments based on cascaded laser structures have been applied particularly for NO and C_2H_6 sensing in clinical environments [149, 151]. Figure 3.22 shows an example of an off-line C_2H_6 measurement, i.e. breath was first collected in a sampling bag and then transferred to the analyser.

Atmospheric monitoring as mentioned in Sect. 3.5.2.2 aims at measuring concentrations of stable species to establish incoming and outgoing fluxes, i.e. sources and sinks of these molecules. Correlations with isotopic signatures may help refining existing model assumptions for e.g. methane fluxes in the arctic region or water vapour levels in the tropical upper troposphere and lower stratosphere. The potential of cavity enhanced sensing has generated much interest since it may allow highly sensitive field measurements, even aboard aircraft, with robust optical alignment. Although sensitivity is an important criterion in these applications, high precision is even more essential to accurately determine isotopic ratios (X) of e.g. HDO/H_2O and $^{13}C/^{12}C$ which are often expressed as δX in per mille

$$\delta X = \left(\frac{(\text{isot.ratio})_{\text{sample}}}{(\text{isot.ratio})_{\text{stand.}}} - 1 \right) \times 1000. \qquad (3.27)$$

Typically, a precision better than a few per mille is needed for such measurements. Through careful design high finesse optical cavities in off-axis alignment can meet the performance requirements as demonstrated by the "Harvard ICOS isotope instrument" [105, 107, 118]. The spectrometer (Fig. 3.14), for which a *NEA* of 2×10^{-11} cm^{-1} Hz was reported, was optimised with respect to laser beam shaping, frequency sweep to minimise non-linear absorption behaviour (skew), sampling pressure and gas handling, respectively.

3.5.3 Reactive Gas Mixtures

3.5.3.1 Studies of Jet-Cooled Molecules

The combination of strong absorption features in the mid-IR fingerprint region with highly sensitive CEAS techniques is particularly appealing in chemical environments where both pressure and absorption path (i.e. the cavity base length) cannot arbitrarily chosen. Typical examples are process monitoring in low pressure electrical gas discharges, where the vacuum vessel determines the mirror separation, or fundamental studies of molecular beams expanding through a nozzle. The latter method is a valuable research tool to study the structure of transient molecules, particularly of those ones that cannot be observed by alternative methods [152]. Therefore cw-CRDS has been applied in research laboratories to study jet-cooled molecules or complexes in a pulsed slit jet source [144, 153] or a supersonic expansion source [152]. The first study was still employing lead salt lasers and the results suggested replacing the spectrometer by a QCL based OA-CEAS instrument with a sensitivity of 2×10^{-8} cm^{-1} Hz$^{-1/2}$ [153]. Recently CH_2Br_2 has been studied by cw-CRDS with a *NEA* of 5×10^{-8} cm^{-1} Hz$^{-1/2}$ [152]. The main limiting factor in these experiments is usually the performance of the IR detector since fast response times along with high detectivity are required to follow the ring-down transients. High detectivity suggests large area LN cooled detectors while the response time is inversely proportional to the detector area.

3.5.3.2 Plasma Diagnostics

Detection of transients and stable species and their time-resolved monitoring to scrutinise the kinetics in reactive environments has been an important field of application of mid-IR chemical sensing for years [4, 5]. Given their transient nature *ex-situ* sampling of radicals from chemically active gas phases, among them plasmas or combustion engines, is not an option. Hence, non-invasive optical spectroscopy has been applied to quantify species, e.g. using *in-situ* multi-pass optics. However, reactive environments also typically exhibit gradients in densities and temperatures along the line of sight. Increasing the sensitivity of mid-IR diagnostic methods is desirable either for lowering the limit of detection to levels where important transient species are expected to be produced in the plasma, or for obtaining better spatial resolution. In the latter case the line-of-sight is reduced and more localised measurements of molecular number densities become feasible. Both aims may be contradictory, because increased effective absorption path lengths L_{eff} are required to yield better detection limits. Increasing the effective absorption length by employing high finesse optical cavities can hence provide both increased sensitivities at inherently small base lengths.

An *in-situ* CEAS study applied to a microwave (MW) discharge was reported in [154]. A straightforward on-axis alignment was employed to detect CH_4 and especially HCN formed in a CH_4–Ar–N_2 plasma at 1304 cm^{-1}. The MW reactor was

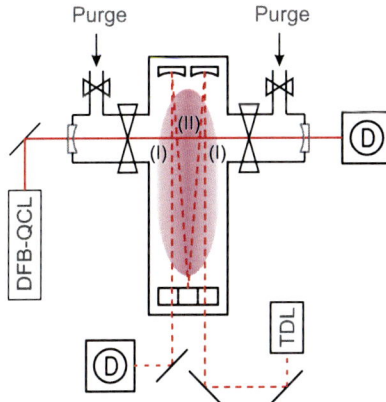

Fig. 3.23 Schematic diagram of the experimental setup used for CEAS with a MW discharge reactor. A TE cooled cw-DFB-QCL was guided through purged viewing apertures perpendicular to a lead salt laser (TDL) spectrometer aligned to accomplish 36 m absorption length within an *in-situ* White cell. The reactor clearance was about 0.2 m. The separation of cavity mirrors ($R = 99.965$ %) encompassing the purging equipment was 0.8 m. Cool (I) and hot (II) regions across the line of sight are indicated [154]

additionally equipped with a White type multi-pass cell of 36 m effective absorption path aligned perpendicular to the optical cavity (Fig. 3.23). A lead salt laser spectrometer coupled with the White cell provided complementary measurements around 1385 cm^{-1} which were used to validate the CEAS results. The reactor viewing apertures were purged with an Ar gas flow. Since the purge gas flow influences the effective absorption path for species produced in the plasma the flow rate has to be established in preliminary experiments. It was set to yield $d = 0.2$ m, i.e. to match the distance between the reactor walls. It should be noted that the mirror separation, $L = 0.8$ m, and the interaction length with the active medium (d in Eqs. (3.7)–(3.13) have to be carefully distinguished for these kind of applications. The effective reflectivity ($R = 99.965$ %) of the mirrors used in this study would yield almost 600 m effective absorption path.

The measurements performed in a CH_4–Ar–N_2 mixture were linked to earlier studies in the reactor. Analysing the $P5$ line of HCN, as measured by means of the complementary multi-pass lead salt laser spectrometer, yielded essentially the same HCN mixing ratios as observed in earlier studies. In these cases a homogeneous effective gas temperature of about 600 K along the line-of-sight (36 m) is assumed. The corresponding cavity transmission spectrum showing mainly HCN ($P34e$ and $P38$ lines) and CH_4 features is presented in Fig. 3.24. Two aspects of the CEAS transmission graph should be highlighted. Firstly, the (peak) ratio of the CH_4 lines—though already violating the weak absorption assumption and usually requiring a non-linear correction for quantitative spectroscopy—suggests a gas temperature of only ~400 K. Assuming that the viewing apertures are purged, this result is surprising, since most of the 0.2 m cavity base length, i.e., the reactor clearance, was supposed to be the active or hot part of the MW plasma. On the other hand,

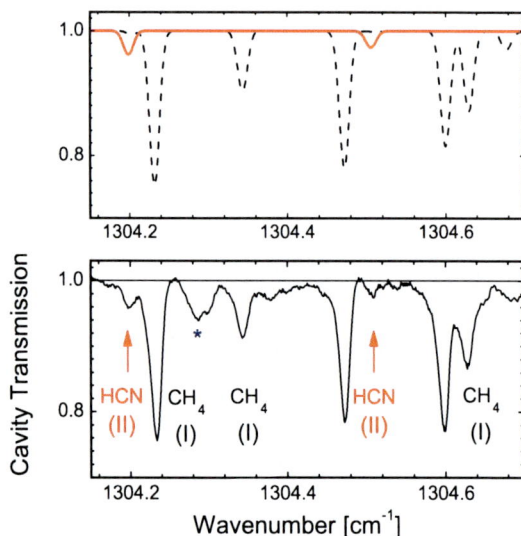

Fig. 3.24 *Upper*: Calculation of a transmission spectrum for CH$_4$ (400 K, *dotted trace*) and HCN (1200 K, *solid trace*). Since the gas temperature (gradient) and thus number density (gradients) are unknown this plot should be regarded as a guide to the eye for identifying the main features in the lower panel. *Lower*: Transmission spectrum obtained by *in-situ* CEAS. Molecular absorption lines of CH$_4$ and HCN were detected in a Ar(+ purge)/CH$_4$/N$_2$ plasma at 150 Pa in the MW reactor in Fig. 3.23. The HCN lines, $P34e$ and $P38$, are indicated by arrows. Potential H$_2$O absorption (leakage) is marked with * [154]

the HCN lines in the spectral range covered by the cw-DFB-QCL are too weak at low temperatures (i.e., 10^4 times weaker than the $P5$ line). However, HCN was still detected, although the gain in L_{eff} by CEAS cannot be higher than 15 compared with the lead salt spectrometer. Hence, the HCN produced in the plasma might be present in different rotationally and vibrationally excited states and a unique HCN number density cannot easily be extracted without detailed chemical modelling. Additionally, the observed CH$_4$ is very likely detected and formed only in the cool outer parts of the line-of-sight. In other words, a fit to the experimental results assuming a single gas temperature is not feasible. Secondly, Allan variance analysis (Fig. 3.25) suggests an Allan minimum after 4 s integration time which is significantly faster than for a spectrometer based on almost the same spectroscopic equipment (Sect. 3.4.2.1). This observation highlights the fact that time-constants in reactive environments (e.g. residence times, drift times, reaction times) are considerably shorter (microseconds–seconds) than what may be required for a cavity enhanced spectrometer to perform (slow) spectral sweeps and to achieve ultimate sensitivity after tens or hundreds of seconds.

Fig. 3.25 *Upper*: CH_4 peak absorption at 1303.71 cm^{-1} as a function of time in a CH_4–Ar discharge. *Lower*: Allan variance (*circles*) retrieved from the data stream plotted in the upper panel. The trend in white noise is indicated (*line*). Note the Allan minimum at around 4 s which appears significantly earlier (due to thermal drifts of the reactor wall) than in Fig. 3.13 employing a similar configuration [154]

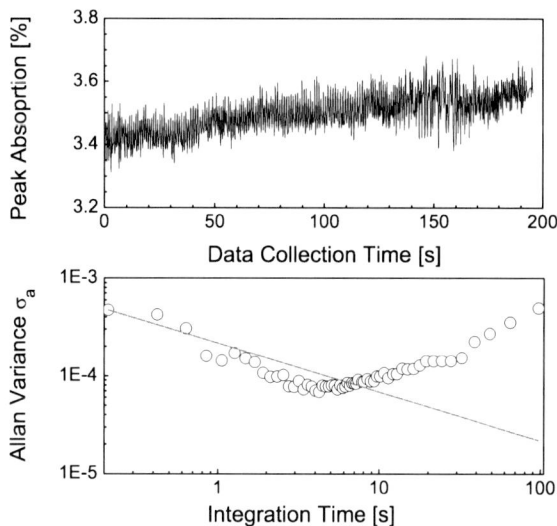

3.5.4 Overview of Cavity Enhanced mid-IR Spectrometers: System Performance

The preparation of a comprehensive compilation about (mid-IR) spectrometers based on cavity enhanced methods is challenging and particularly hampered by the inconsistent use of sensitivity parameters or other figures of merit for the instruments throughout the literature [105]. An extensive list of measured species, though already published in 2000, is provided in a review of Berden et al. [30]. Spectrometers based on QCLs are absent in this early overview since this new class of semiconductor laser just had become commercially available. Now, about a decade later, QCL have been employed with several cavity enhanced detection schemes to measure a wide variety of molecular species. A summary of the achievements is provided in Table 3.3 which is limited to high finesse cavities and QCLs as mid-IR light sources. In case the *NEA* or other parameters listed in Table 3.3 were not given in the corresponding publications, values were converted or estimated using the formulae given in Sect. 3.3.4. A comparison with linear AS using QCLs may be found in [103].

Using the *NEA* (per scan) to characterise and distinguish QCL based CEAS methods, the spectrometers reported in literature may be classified by 4 subgroups: CRD spectrometers employing cw-QCLs may achieve sensitivities in the 10^{-9} cm^{-1} $Hz^{-1/2}$ range. A straightforward application of CEA spectrometers in on-axis alignment typically yields a *NEA* of up to 10^{-8} cm^{-1} $Hz^{-1/2}$. Such a performance can be easily achieved under LN free conditions, i.e. using TE cooled lasers and detectors. Similarly, OF-CEAS instruments may entirely be composed of TE cooled devices and typically surpass the sensitivity of CEAS by one order of magnitude. Hence, OF-CEAS are of equally good performance as cw-QCL-CRD spectrometers. This may be explained by the commonly omitted cav-

Table 3.3 Intercomparison of QCL based spectrometers employing high finesse optical cavities sorted by method and QCL wavelength

Ref.	Method	λ [μm]	Gas	QCL	Operating temperature		L_{eff} [m]	NEA [cm^{-1} Hz$^{-1/2}$]	Year
					Laser	Detector			
[126]	ICLAS	7.5	CH_2FCF_3	EC	TE	I–V[a]	n.a.[b]	2×10^{-6}	2012
[108]	CRDS	10.3	NH_3	pulsed	TE	LN	240	5×10^{-6}[c]	2006
[111]	CRDS	8.5	NH_3	cw	LN	LN	280	3×10^{-9}[c]	2000
[152]	CRDS	8.4	CH_2Br_2	cw	LN	LN	3000	5×10^{-6}[c]	2010
[155]	CRDS	5.2	NO	cw	LN	LN	1050	6×10^{-10}[c]	2001
[129]	CRDS	6.1	NO_2	EC	TE	TE	2000	n.a.[b]	2010
[128]	CRDS	5.9–6.3	C_3H_6O, CH_3NO_2	EC	TE	LN	500	n.a.[b]	2012
[156]	CEAS (OA)	10.3	NH_3, C_2H_4	p	TE	LN	76	n.a.[b]	2010
[102]	ICOS (OA)	7.9	CH_4	cw	LN	LN	5560	2×10^{-9}	2002
[103]	CEAS	7.7	CH_4	cw	TE	TE	1080	2×10^{-7}	2008
[103]	CEAS	7.7	N_2O	cw	TE	TE	1080	2×10^{-7}	2008
[40]	CEAS	7.7	N_2O	cw	TE	LN	3900	4×10^{-8}	2009
[124]	ICOS (OA)[d]	7.7	CH_4	cw	TE	LN	2250	2×10^{-10}	2012
[118]	ICOS (OA)	7.7	$^{12,13}CH_4$	cw	TE	LN	2500	n.a.[b]	2011
[105]	ICOS (OA)	6.7	HDO	cw	LN	LN	4200	9×10^{-10}	2008
[107]	ICOS (OA)	6.7	$H_2^{16,18}O$, HDO	cw	LN	LN	4500	2×10^{-11}	2009
[153]	CEAS (OA)	6.7	NH_3	cw	LN	LN	n.a.[b]	2×10^{-8}	2009
[130]	ICOS (OA)	6.0	NO_2	EC	TE	TE	n.a.[b]	n.a.[b]	2011
[121]	ICOS (OA)	5.5	NO	cw	TE	LN	500	4×10^{-8}	2006
unpubl.	CEAS	5.5	$COCl_2$	cw	TE	LN	280	4×10^{-8}	2009

Table 3.3 (Continued)

Ref.	Method	λ [μm]	Gas	QCL	Operating temperature		L_{eff} [m]	NEA [cm^{-1} Hz$^{-1/2}$]	Year
					Laser	Detector			
[157]	ICOS (OA)[d]	5.5	NO	cw	TE	LN	700	$< 1 \times 10^{-7}$	2006
[112]	CEAS	5.2	NO	cw	LN	TE	670	3×10^{-6}	2001
[123]	ICOS (OA)[d]	5.2	NO	cw	LN	LN	75	2×10^{-7}	2004
[113]	ICOS	5.2	NO	pulsed	TE	LN	1500	6×10^{-8}	2005
[149]	ICOS (OA)	5.2	NO, CO$_2$	cw	LN	LN (?)	n.a.	$< 1 \times 10^{-8}$	2007
[158]	ICOS (OA)	3.5	H$_2$CO	cw[e]	LN	LN	83	5×10^{-7}	2006
[151]	ICOS (OA)	3.3	C$_2$H$_6$	cw (?)[e]	LN	LN	1350	$\sim 5 \times 10^{-9}$	2009
[109]	CEAS (OF)	7.8	N$_2$O, CH$_4$	cw	TE	TE	1045	5×10^{-8}	2011
[125]	CEAS (OF)	4.4	CO$_2$, N$_2$O	cw	TE	TE	2500	3×10^{-9}	2010
[159]	CEAS (locked)	8.7	N$_2$O	cw	LN (?)	n.a.[b]	n.a.[b]	8×10^{-10}	2004
[159]	NICE-OHMS	8.7	N$_2$O	cw	LN (?)	n.a.[b]	n.a.[b]	1×10^{-10}	2004

[a] Detected signal was the feedback to the compliance voltage of the QCL

[b] n.a. = not available in the reference

[c] NEA taken from comparison in [152]; note that these may not reflect "per scan" values

[d] WM techniques were also applied; the NEA given here is without WM

[e] ICL

ity mode noise in CRDS. For CEAS the minimum detectable absorption is typically limited to 10^{-2} to 10^{-3} due to incomplete averaging over the cavity resonances. In combination with wavelength modulation techniques this can be improved by a factor of about 5 at the expense of a more complex experimental layout and reduced laser sweep rates [123, 124, 157]. Not surprisingly the sensitivity of these systems cannot compete with the values obtained with sophisticated locked CEAS setups (8×10^{-10} cm^{-1} Hz$^{-1/2}$) or even with NICE-OHMS (9.7×10^{-11} cm^{-1} Hz$^{-1/2}$) [159]. QCL based spectrometers employing conventional long path cells accomplish sensitivities well below 1×10^{-7} cm^{-1} Hz$^{-1/2}$ [103] even down to the 10^{-10} cm^{-1} Hz$^{-1/2}$ range [24] and hence are still superior to short base length CEA spectrometers. The main reason is the residual mode noise which is absent in the spectra for those multi-pass spectrometers. However, the volume of such multi-pass cells covering effective path lengths up to 210 m is typically between 0.5 and 5 l. It is thus clearly bigger than sampling volumes of CEA spectrometers exploiting mainly fundamental or low-order transverse modes. Recently, the volume of Herriott cells has been further reduced [24]. Best system performance of field deployable instruments was reported for arrangements that combine off-axis alignment (as also used in Herriott-cells) with high finesse optical resonators. The *NEA* was of the order of 2×10^{-11} cm^{-1} Hz$^{-1/2}$ [105, 107]. Slightly less sensitive implementations of OA-CEAS may provide *NEA* values in the 10^{-9} cm^{-1} Hz$^{-1/2}$ range (Table 3.3).

3.6 Conclusions and Perspectives

Progress in the development of QCLs, has been astonishingly rapid in the 2 decades since their first realisation. The combination of optical resonators with compact mid-IR lasers thus seems to be ideally suited for decreasing the detection limit (Fig. 3.26) while reducing the necessary sampling volume and shrinking the apparatus dimensions in potentially field deployable systems at the same time. It transpires that p-QCLs cannot reasonably be used with high finesse optical resonators because the inherent frequency-down chirp hinders the intensity build-up inside the cavity.

Presently, the performance of linear multi-pass QCL spectrometers may be considered as good as accomplished by cavity enhanced instruments. The choice of an appropriate method depends on whether the important criterion is ultimate sensitivity or a more compact system. In the former case a multi-pass cell spectrometer would be preferable because of its better signal to noise characteristics but this configuration would exclude certain types of *in-situ* measurements or the detection of processes on short time scales (given the increased volumes of multi-pass cells). To achieve a compact system a small volume cavity based spectrometer employing cw-QCLs would be more appropriate. This configuration would also be of special interest for applications where the pressure cannot arbitrarily be chosen in order to adapt the absorption line width to the laser line width or instrumental broadening, e.g. in low pressure gas discharges. Moreover, for these applications *in-situ* measurements are essential because multi-pass cell sampling *ex-situ* is not an option.

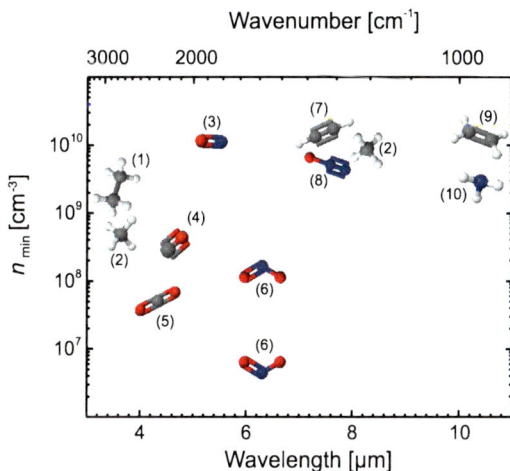

Fig. 3.26 Anticipated limits of detection for target molecules of high interest encompassing important chemical bonds among them C–H, C=H, C≡H, N–O, N=O, C=O, C≡O. The MDND (n_{min}) was estimated from corresponding spectrometers in Table 3.3 (assuming 5000 Pa, a pressure broadened line width (HWHM) of 600 MHz (0.02 cm^{-1}) and typical line strengths of the correpsonding species in Eq. (3.25)). Detection limits would usually improve by an order of magnitude, if low pressure conditions (~100 Pa) are applied. (1—C_2H_6, 2—CH_4, 3—NO, 4—CO, 5—CO_2, 6—NO_2, 7—C_2H_2, 8—N_2O, 9—C_2H_4, 10—NH_3)

Cavity mode injection noise persists for all CEAS variants without optical feedback, but can be significantly reduced in off-axis alignments. Although these configurations require dielectric mirrors of increased diameter progress in coating technology is envisaged which will push the reflectivity and thus he sensitivity further [146]. Additionally, QCL technology is progressing and seeks to close the gap of spectral coverage in the 3 μm range [61] and to improve the detectivity of TE cooled detectors and IR cameras [34]. The recent development of broadband EC-QCLs is particularly interesting for chemical sensing applications such as explosives detection or breath gas analysis where larger molecular (or organic) species are involved [72, 128].

Acknowledgements   The authors give sincere thanks to all present and former members of the laboratories involved in Greifswald, Eindhoven and Cambridge for permanent support and a stimulating working atmosphere. The authors are also indebted to Alpes Lasers for supporting the activities. This work was partly supported by the German Research Foundation (DFG) within the framework of the Collaborative Research Centre Transregio 24 'Fundamentals of Complex Plasmas'.

References

1. P. Hering, J.P. Lay, S. Stry (eds.), *Laser in Environmental and Life Science. Modern Analytical Methods* (Springer, Berlin, 2004). ISBN 978-3-540-40260-2

2. A. Amann, D. Smith (eds.), *Breath Analysis for Clinical Diagnosis and Therapeutic Monitoring* (World Scientific, Singapore, 2005). ISBN 981-256-284-2
3. M. Lackner, Rev. Chem. Eng. **23**, 65 (2007)
4. J. Röpcke, G. Lombardi, A. Rousseau, P.B. Davies, Plasma Sources Sci. Technol. **15**, S148 (2006)
5. J. Röpcke, P.B. Davies, N. Lang, A. Rousseau, S. Welzel, J. Phys. D, Appl. Phys. **45**, 423001 (2012)
6. W. Demtröder, *Laser Spectroscopy* (Springer, Berlin, 2003). ISBN 3540652256
7. R.F. Curl, F. Capasso, C. Gmachl, A.A. Kosterev, J.B. McManus, R. Lewicki, M. Pusharsky, G. Wysocki, F.K. Tittel, Chem. Phys. Lett. **487**, 1 (2010)
8. P. Crozet, A.J. Ross, M. Vervloet, Annu. Rep. Prog. Chem., Sect. C, Phys. Chem. **98**, 33 (2002)
9. F. Adler, P. Maslowski, A. Foltynowicz, K.C. Cossel, T.C. Briles, I. Hartl, J. Ye, Opt. Express **18**, 21861 (2010)
10. P. Werle, F. Slemr, M. Gehrtz, C. Bräuchle, Appl. Phys. B **49**, 99 (1989)
11. G. Friedrichs, Z. Phys. Chem. **222**, 1 (2008)
12. J. Reid, D. Labrie, Appl. Phys. B **26**, 203 (1981)
13. D.S. Bomse, A.C. Stanton, J.A. Silver, Appl. Opt. **31**, 718 (1992)
14. M.S. Zahniser, D.D. Nelson, J.B. McManus, P.L. Kebabian, Philos. Trans. R. Soc. Lond. A **351**, 371 (1995)
15. C.V. Horii, M.S. Zahniser, D.D. Nelson, J.B. McManus, S.C. Wofsy, Proc. SPIE **3758**, 152 (1999)
16. P. Werle, R. Mücke, F. Slemr, Appl. Phys. B **57**, 131 (1993)
17. F. Schmidt, *Laser-Based Absorption Spectrometry: Development of NICE-OHMS Towards Ultra-Sensitive Trace Species Detection* (Umea University, Umea, 2007)
18. H. Linnartz, Phys. Scr. **70**, C24 (2004)
19. J.U. White, J. Opt. Soc. Am. **32**, 285 (1942)
20. S.M. Chernin, E.G. Barskaya, Appl. Opt. **30**, 51 (1991)
21. D. Herriott, H. Kogelnik, R. Kompfner, Appl. Opt. **3**, 523 (1964)
22. D.R. Herriott, H.J. Schulte, Appl. Opt. **4**, 883 (1965)
23. J.B. McManus, P.L. Kebabian, M.S. Zahniser, Appl. Opt. **34**, 3336 (1995)
24. J.B. McManus, M.S. Zahniser, D.D. Nelson, Appl. Opt. **50**, A74 (2011)
25. B.A. Paldus, A.A. Kachanov, Can. J. Phys. **83**, 975 (2005)
26. J. Ye, L.S. Ma, J.L. Hall, J. Opt. Soc. Am. B **15**, 6 (1998)
27. G. Berden, R. Engeln (eds.), *Cavity Ring-Down Spectroscopy: Techniques and Applications* (Wiley-Blackwell, New York, 2009). ISBN 978-1-4051-7688-0
28. M. Mazurenka, A.J. Orr-Ewing, R. Peverall, G.A.D. Ritchie, Annu. Rep. Prog. Chem., Sect. C, Phys. Chem. **101**, 100 (2005)
29. G. Friedrichs, Z. Phys. Chem. **222**, 31 (2008)
30. G. Berden, R. Peeters, G. Meijer, Int. Rev. Phys. Chem. **19**, 565 (2000)
31. A. Foltynowicz, F.M. Schmidt, W. Ma, O. Axner, Appl. Phys. B **92**, 313 (2008)
32. M.J. Thorpe, F. Adler, K.C. Cossel, M.H.G. de Miranda, J. Ye, Chem. Phys. Lett. **468**, 1 (2009)
33. A. Rogalski, Opto-Electron. Rev. **16**, 458 (2008)
34. A. Rogalski, J. Antoszewski, L. Faraone, J. Appl. Phys. **105**, 091101 (2009)
35. F.K. Tittel, D. Richter, A. Fried, Top. Appl. Phys. **89**, 445 (2003)
36. R.F. Curl, F.K. Tittel, Annu. Rep. Prog. Chem., Sect. C, Phys. Chem. **98**, 219 (2002)
37. A. Godard, C. R. Phys. **8**, 1100 (2007)
38. W. Lei, C. Jagadish, J. Appl. Phys. **104**, 091101 (2008)
39. A. Joullié, P. Christol, C. R. Phys. **4**, 621 (2003)
40. S. Welzel, *New Enhanced Sensitivity Infrared Laser Spectroscopy Techniques Applied to Reactive Plasmas and Trace Gas Detection* (Logos, Berlin, 2009). ISBN 978-3-8325-2345-9
41. A.W. Mantz, Spectrochim. Acta A **51**, 2211 (1995)

42. J. Faist, C. Sirtori, F. Capasso, L. Pfeiffer, K.W. West, Appl. Phys. Lett. **64**, 872 (1994)
43. L. Esaki, R. Tsu, IBM J. Res. Dev. **14**, 61 (1970)
44. R.F. Kazarinov, R.A. Suris, Sov. Phys. Semicond. **5**, 707 (1971)
45. F. Capasso, K. Mohammed, A.Y. Cho, IEEE J. Quantum Electron. **22**, 1853 (1986)
46. P. Yuh, K.L. Wang, Appl. Phys. Lett. **51**, 1404 (1987)
47. M. Helm, P. England, E. Colas, F. DeRosa, S.J. Allen, Phys. Rev. Lett. **63**, 74 (1989)
48. J. Faist, F. Capasso, C. Sirtori, D. Sivco, A.K. Hutvhinson, S.G. Chu, A.Y. Cho, Appl. Phys. Lett. **64**, 1144 (1994)
49. J. Faist, F. Capasso, D.L. Sivco, C. Sirtori, A.L. Hutvhinson, A.Y. Cho, Science **264**, 553 (1994)
50. R. Tsu, Nature **369**, 442 (1994)
51. J. Faist, F. Capasso, C. Sirtori, D.L. Sivco, J.N. Baillargeon, A.L. Hutchinson, S.G. Chu, A.Y. Cho, Appl. Phys. Lett. **68**, 3680 (1996)
52. D. Hofstetter, M. Beck, T. Aellen, J. Faist, U. Oesterle, M. Ilegems, E. Gini, H. Melchior, Appl. Phys. Lett. **78**, 1964 (2001)
53. Y. Bai, S. Slivken, S. Kuboya, S.R. Darvish, M. Razeghi, Nat. Photonics **4**, 99 (2010)
54. F. Capasso, C. Gmachl, R. Paiella, A. Tredicucci, A.L. Hutchinson, D.L. Sivco, J.N. Baillargeon, A.Y. Cho, H.C. Liu, IEEE J. Sel. Top. Quantum Electron. **6**, 931 (2000)
55. C. Gmachl, F. Capasso, D.L. Sivco, A.Y. Cho, Rep. Prog. Phys. **64**, 1533 (2001)
56. J. Faist, D. Hofstetter, M. Beck, T. Aellen, M. Rochat, S. Blaser, IEEE J. Quantum Electron. **38**, 533 (2002)
57. F. Capasso, C. Gmachl, D.L. Sivco, A.Y. Cho, Phys. Today **55**, 34 (2002)
58. C. Sirtori, J. Nagle, C. R. Phys. **4**, 639 (2003)
59. J. Faist, F. Capasso, D.L. Sivco, A.L. Hutchinson, S.G. Chu, A.Y. Cho, Appl. Phys. Lett. **72**, 680 (1998)
60. I. Vurgaftman, J.R. Meyer, L.R. Ram-Mohan, J. Appl. Phys. **89**, 5851 (2001)
61. N. Bandyopadhyay, Y. Bai, S. Tsao, S. Nida, S. Slivken, M. Razeghi, Appl. Phys. Lett. **101**, 241110 (2012)
62. R.Q. Yang, J.L. Bradshaw, J.D. Bruno, J.T. Pham, D.E. Wortman, IEEE J. Quantum Electron. **38**, 559 (2002)
63. J. Faist, C. Gmachl, F. Capasso, C. Sirtori, D.L. Sivco, J.N. Baillargeon, A.Y. Cho, Appl. Phys. Lett. **70**, 2670 (1997)
64. J. Koeth, M. Fischer, M. Legge, J. Seufert, R. Werner, Photonik **36**, 1 (2005) (in German)
65. C. Gmachl, J. Faist, J.N. Baillargeon, F. Capasso, C. Sirtori, D.L. Sivco, S.G. Chu, A.Y. Cho, IEEE Photonics Technol. Lett. **9**, 1090 (1997)
66. E. Normand, G. Duxbury, N. Langford, Opt. Commun. **197**, 115 (2001)
67. G. Totschnig, F. Winter, V. Pustogov, J. Faist, A. Müller, Opt. Lett. **27**, 1788 (2002)
68. G.P. Luo, C. Peng, H.Q. Le, S.S. Pei, W.Y. Hwang, B. Ishaug, J. Um, J.N. Baillargeon, C.H. Lin, Appl. Phys. Lett. **78**, 2834 (2001)
69. G. Luo, C. Peng, H.Q. Le, S.S. Pei, H. Lee, W.Y. Hwang, B. Ishaug, J. Zheng, IEEE J. Quantum Electron. **38**, 486 (2002)
70. R. Maulini, M. Beck, J. Faist, E. Gini, Appl. Phys. Lett. **84**, 1659 (2004)
71. G. Wysocki, R. Lewicki, R.F. Curl, F.K. Tittel, L. Diehl, F. Capasso, M. Troccoli, G. Hofler, D. Bour, S. Corzine, R. Maulini, M. Giovannini, J. Faist, Appl. Phys. B **92**, 305 (2008)
72. A. Hugi, R. Maulini, J. Faist, Semicond. Sci. Technol. **25**, 083001 (2010)
73. D. Lopatik, N. Lang, U. Macherius, H. Zimmermann, J. Röpcke, Meas. Sci. Technol. **23**, 115501 (2012)
74. A. Karpf, G.N. Rao, Appl. Opt. **48**, 408 (2009)
75. A. Karpf, G.N. Rao, Appl. Opt. **49**, 1406 (2010)
76. K. Knabe, P.A. Williams, F.R. Giorgetta, M.B. Radunsky, C.M. Armacost, S. Crivello, N.R. Newbury, Opt. Express **21**, 1020 (2013)
77. S. Welzel, J. Röpcke, Appl. Phys. B **102**, 303 (2011)
78. K. Namjou, S. Cai, E.A. Whittaker, J. Faist, C. Gmachl, F. Capasso, D.L. Sivco, A.Y. Cho, Opt. Lett. **23**, 219 (1998)

79. A.A. Kosterev, F.K. Tittel, C. Gmachl, F. Capasso, D.L. Sivco, J.N. Baillargeon, A.L. Hutchinson, A.Y. Cho, Appl. Opt. **39**, 6866 (2000)
80. D.M. Sonnenfroh, W.T. Rawlins, M.G. Allen, C. Gmachl, F. Capasso, A.L. Hutchinson, D.L. Sivco, J.N. Baillargeon, A.Y. Cho, Appl. Opt. **40**, 812 (2001)
81. D.D. Nelson, J.H. Shorter, J.B. McManus, M.S. Zahniser, Appl. Phys. B **75**, 343 (2002)
82. J.B. McManus, D.D. Nelson, S.C. Herndon, J.H. Shorter, M.S. Zahniser, S. Blaser, L. Hvozdara, A. Muller, M. Giovannini, J. Faist, Appl. Phys. B **85**, 235 (2006)
83. B.H. Lee, E.C. Wood, M.S. Zahniser, J.B. McManus, D.D. Nelson, S.C. Herndon, G.W. Santoni, S.C. Wofsy, J.W. Munger, Appl. Phys. B **102**, 417 (2006)
84. C. Gmachl, F. Capasso, R. Köhler, A. Tredicucci, A.L. Hutchinson, D.L. Sivco, J.N. Baillargeon, A.Y. Cho, IEEE Circuits Devices Mag. **16**, 10 (2000)
85. Q. Shi, D.D. Nelson, J.B. McManus, M.S. Zahniser, M.E. Parrish, R.E. Baren, K.H. Shafer, C.N. Harward, Anal. Chem. **75**, 5180 (2003)
86. B. Grouiez, B. Parvitte, L. Joly, V. Zeninari, Opt. Lett. **34**, 181 (2009)
87. M.T. McCulloch, E.L. Normand, N. Langford, G. Duxbury, D.A. Newnham, J. Opt. Soc. Am. B **20**, 1761 (2003)
88. Y.G. Zhang, Y.J. He, A.Z. Li, Chin. Phys. Lett. **20**, 678 (2003)
89. M.S. Vitiello, G. Scamarcio, V. Spagnolo, Appl. Phys. Lett. **92**, 101116 (2008)
90. G. Duxbury, N. Langford, M.T. McCulloch, S. Wright, Chem. Soc. Rev. **34**, 921 (2005)
91. S.W. Sharpe, J.F. Kelly, J.S. Hartman, C. Gmachl, F. Capasso, D.L. Sivco, J.N. Baillargeon, A.Y. Cho, Opt. Lett. **23**, 1396 (1998)
92. C. Gmachl, F. Capasso, A. Tredicucci, D.L. Sivco, J.N. Baillargeon, A.L. Hutchinson, A.Y. Cho, Opt. Lett. **25**, 230 (2000)
93. T.L. Myers, R.M. Williams, M.S. Taubman, C. Gmachl, F. Capasso, D.L. Sivco, J.N. Baillargeon, A.Y. Cho, Opt. Lett. **27**, 170 (2002)
94. H. Ganser, B. Frech, A. Jentsch, M. Mürtz, C. Gmachl, F. Capasso, D.L. Sivco, J.N. Baillargeon, A.Y. Cho, W. Urban, Opt. Commun. **197**, 127 (2001)
95. D. Weidmann, L. Joly, V. Parpillon, D. Courtois, Y. Bonetti, T. Aellen, M. Beck, J. Faist, D. Hofstetter, Opt. Lett. **28**, 704 (2003)
96. F. Bielsa, A. Douillet, T. Valenzuela, J.P. Karr, L. Hilico, Opt. Lett. **32**, 1641 (2007)
97. R.M. Williams, J.F. Kelly, J.S. Hartman, S.W. Sharpe, M.S. Taubman, J.L. Hall, F. Capasso, C. Gmachl, D.L. Sivco, J.N. Baillargeon, A.Y. Cho, Opt. Lett. **24**, 1844 (1999)
98. S. Bartalini, S. Borri, P. Cancio, A. Castrillo, I. Galli, G. Giusfredi, D. Mazzotti, L. Gianfrani, P. De Natale, Phys. Rev. Lett. **104**, 083904 (2010)
99. V.L. Kasyutich, R.K. Ibrahim, P.A. Martin, Infrared Phys. Technol. **53**, 381 (2010)
100. P. Zalicki, R.N. Zare, J. Chem. Phys. **102**, 2708 (1995)
101. R. Engeln, G. Berden, R. Peeters, G. Meijer, Rev. Sci. Instrum. **69**, 3763 (1998)
102. J.B. Paul, J.J. Scherer, A. O'Keefe, L. Lapson, J.G. Anderson, C. Gmachl, F. Capasso, A.Y. Cho, Proc. SPIE **4577**, 1 (2002)
103. S. Welzel, G. Lombardi, P.B. Davies, R. Engeln, D.C. Schram, J. Röpcke, J. Appl. Phys. **104**, 093115 (2008)
104. A. O'Keefe, Chem. Phys. Lett. **293**, 331 (1998)
105. E.J. Moyer, D.S. Sayres, G.S. Engel, J.M.St. Clair, F.N. Keutsch, N.T. Allen, J.H. Kroll, J.G. Anderson, Appl. Phys. B **92**, 467 (2008)
106. J. Peltola, M. Vainio, V. Ulvila, M. Siltanen, M. Metsälä, L. Halonen, Appl. Phys. B **107**, 839 (2012)
107. D.S. Sayres, E.J. Moyer, T.F. Hanisco, J.M. St. Clair, F.N. Keutsch, A. O'Brien, N.T. Allen, L. Lapson, J.N. Demusz, M. Rivero, T. Martin, M. Greenberg, C. Tuozzolo, G.S. Engel, J.H. Kroll, J.B. Paul, J.G. Anderson, Rev. Sci. Instrum. **80**, 044102 (2009)
108. J. Manne, O. Sukhorukov, W. Jäger, J. Tulip, Appl. Opt. **45**, 9230 (2006)
109. D.J. Hamilton, A.J. Orr-Ewing, Appl. Phys. B **102**, 879 (2011)
110. O. Sukhorukov, A. Lytkine, J. Manne, J. Tulip, W. Jäger, Proc. SPIE **6127**, 61270A (2006)
111. B.A. Paldus, C.C. Harb, T.G. Spence, R.N. Zare, C. Gmachl, F. Capasso, D.L. Sivco, J.N. Baillargeon, A.L. Hutchinson, A.Y. Cho, Opt. Lett. **25**, 666 (2000)

112. L. Menzel, A.A. Kosterev, R.F. Curl, F.K. Tittel, C. Gmachl, F. Capasso, D.L. Sivco, J.N. Baillargon, A.L. Hutchinson, A.Y. Cho, W. Urban, Appl. Phys. B **72**, 859 (2001)
113. M.L. Silva, D.M. Sonnenfroh, D.I. Rosen, M.G. Allen, A. O'Keefe, Appl. Phys. B **81**, 705 (2005)
114. J.T. Hodges, J.P. Looney, R.D. van Zee, Appl. Opt. **35**, 4112 (1996)
115. I. Labazan, S. Rudic, S. Milosevic, Chem. Phys. Lett. **320**, 613 (2000)
116. G. Giusfredi, S. Bartalini, S. Borri, P. Cancio, I. Galli, D. Mazzotti, P. De Natale, Phys. Rev. Lett. **104**, 110801 (2010)
117. I. Galli, S. Bartalini, S. Borri, P. Cancio, D. Mazzotti, P. De Natale, G. Giusfredi, Phys. Rev. Lett. **107**, 270802 (2011)
118. M.F. Witinski, D.S. Sayres, J.G. Anderson, Appl. Phys. B **102**, 375 (2011)
119. J. Remy, M.M. Hemerik, G.M.W. Kroesen, W.W. Stoffels, IEEE Trans. Plasma Sci. **32**, 709 (2004)
120. D.D. Nelson, B. McManus, S. Urbanski, S. Herndon, M.S. Zahniser, Spectrochim. Acta A **60**, 3325 (2004)
121. M.R. McCurdy, Y.A. Bakhirkin, F.K. Tittel, Appl. Phys. B **85**, 445 (2006)
122. J.B. Paul, L. Lapson, J.G. Anderson, Appl. Opt. **40**, 4904 (2001)
123. Y.A. Bakhirkin, A.A. Kosterev, C. Roller, R.F. Curl, F.K. Tittel, Appl. Opt. **43**, 2257 (2004)
124. P. Malara, M.F. Witinski, F. Capasso, J.G. Anderson, P. De Natale, Appl. Phys. B **108**, 353 (2012)
125. G. Maisons, P. Gorrotxategi Carbajo, M. Carras, D. Romanini, Opt. Express **35**, 3607 (2010)
126. M.C. Phillips, M.S. Taubman, Opt. Lett. **37**, 2664 (2012)
127. G. Medhi, A.V. Muravjov, H. Saxen, C.J. Fredricksen, T. Brusentsova, R.E. Peale, O. Edwards, Proc. SPIE **8032**, 80320E (2011)
128. C.C. Harb, T.K. Boyson, A.G. Kallapur, I.R. Petersen, M.E. Calzada, T.G. Spence, K.P. Kirkbride, D.S. Moore, Opt. Express **20**, 15489 (2012)
129. G.N. Rao, A. Karpf, Appl. Opt. **49**, 4906 (2010)
130. G.N. Rao, A. Karpf, Appl. Opt. **50**, 1915 (2011)
131. J.J. Scherer, D. Voelkel, D.J. Rakestraw, J.B. Paul, C.P. Collier, R.J. Saykally, A. O'Keefe, Chem. Phys. Lett. **245**, 273 (1995)
132. J.J. Scherer, K.W. Aniolek, N.P. Cernansky, D.J. Rakestraw, J. Chem. Phys. **107**, 6196 (1997)
133. S. Wu, P. Dupré, T.A. Miller, Phys. Chem. Chem. Phys. **8**, 1682 (2006)
134. F. Ito, J. Chem. Phys. **124**, 054309 (2006)
135. J.B. Paul, R.J. Saykally, Anal. Chem. **69**, 287A (1997)
136. J.B. Paul, R.A. Provencal, C. Chapo, K. Roth, R. Casaes, R.J. Saykally, J. Phys. Chem. A **103**, 2972 (1999)
137. J.B. Paul, R.A. Provencal, C. Chapo, A. Petterson, R.J. Saykally, J. Chem. Phys. **109**, 10201 (1998)
138. R. Peeters, G. Berden, A. Olafsson, L.J.J. Laarhoven, G. Meijer, Chem. Phys. Lett. **337**, 231 (2001)
139. M. Mürtz, B. Freech, W. Urban, Appl. Phys. B **68**, 243 (1999)
140. D. Kleine, H. Dahnke, W. Urban, P. Hering, M. Mürtz, Opt. Lett. **25**, 1606 (2000)
141. H. Dahnke, G. von Basum, K. Kleinermanns, P. Hering, M. Mürtz, Appl. Phys. B **75**, 311 (2002)
142. D. Halmer, G. von Basum, P. Hering, M. Mürtz, Opt. Lett. **30**, 2314 (2005)
143. M.M. Hemerik, Ph.D. Thesis, Eindhoven University of Technology, 2001
144. W.S. Tam, I. Leonov, Y. Xub, Rev. Sci. Instrum. **77**, 063117 (2006)
145. D.D. Arslanov, S.M. Cristescu, F.J.M. Harren, Opt. Lett. **35**, 3300 (2010)
146. K.E. Whittaker, L. Ciaffoni, G. Hancock, R. Peverall, G.A.D. Ritchie, Appl. Phys. B **109**, 333 (2012)
147. S.W. Sharpe, T.J. Johnson, R.L. Sams, P.M. Chu, G.C. Rhoderick, P.A. Johnson, Appl. Spectrosc. **58**, 1452 (2004)

148. T.H. Risby, F.K. Tittel, Opt. Eng. **49**, 111123 (2010)
149. Y. Bakhirkin, G. Wysocki, F.K. Tittel, J. Biomed. Opt. **12**, 034034 (2007)
150. M.R. McCurdy, Y. Bakhirkin, G. Wysocki, R. Lewicki, F.K. Tittel, J. Breath Res. **1**, 014001 (2007)
151. K.R. Parameswaran, D.I. Rosen, M.G. Allen, A.M. Ganz, T.H. Risby, Appl. Opt. **48**, B73 (2009)
152. B.E. Brumfield, J.T. Stewart, S.L. Widicus Weaver, M.D. Escarra, S.S. Howard, C.F. Gmachl, B.J. McCall, Rev. Sci. Instrum. **81**, 063102 (2010)
153. Y. Xu, X. Liu, Z. Su, R.M. Kulkarni, W.S. Tam, C. Kang, I. Leonov, L. D'Agostino, Proc. SPIE **7222**, 722208 (2009)
154. S. Welzel, F. Hempel, M. Hübner, N. Lang, P.B. Davies, J. Röpcke, Sensors **10**, 6861 (2010)
155. A.A. Kosterev, A.L. Malinovsky, F.K. Tittel, C. Gmachl, F. Capasso, D.L. Sivco, J.N. Baillargeon, A.L. Hutchinson, A.Y. Cho, Appl. Opt. **40**, 5522 (2001)
156. J. Manne, A. Lim, W. Jäger, J. Tulip, Appl. Opt. **49**, 5302 (2010)
157. Y.A. Bakhirkin, A.A. Kosterev, R.F. Curl, F.K. Tittel, L. Hvodzdara, M. Giovannini, J. Faist, Appl. Phys. B **82**, 149 (2006)
158. J.H. Miller, Y.A. Bakhirkin, T. Ajtai, F.K. Tittel, C.J. Hill, R.Q. Yang, Appl. Phys. B **85**, 391 (2006)
159. M.S. Taubman, T.L. Myers, B.D. Cannon, R.M. Williams, Spectrochim. Acta, Part A, Mol. Biomol. Spectrosc. **60**, 3457 (2004)

Chapter 4
Saturated-Absorption Cavity Ring-Down (SCAR) for High-Sensitivity and High-Resolution Molecular Spectroscopy in the Mid IR

P. Cancio, I. Galli, S. Bartalini, G. Giusfredi, D. Mazzotti, and P. De Natale

Abstract A non-conventional cavity ring-down spectroscopic technique is described. When the light intensity is well above the saturation level for the molecular species inside a high-finesse cavity, each single cavity ring-down event simultaneously measures both the background losses from the cavity mirrors and the linear absorption from the gas. Such a differential scheme acting on very short time scales (a few tens of microseconds) can improve the sensitivity of conventional cavity ring-down by more than one order of magnitude, while achieving sub-Doppler resolution, if needed. Applications to optical detection of very rare molecular species like radiocarbon dioxide and resolved molecular hyperfine structure in $^{17}O^{12}C^{16}O$ are presented.

4.1 Sensitivity and Resolution Limits of Linear Cavity Ring-Down Molecular Spectroscopy with CW Laser Sources in the Mid IR

Since the first demonstrations of Cavity Ring Down (CRD) spectroscopy with CW laser sources as driving light (CW-CRD) [1–3], the improvements in terms of sensitivity and resolution with respect to pulsed CRD were quickly noticed. In CW-CRD only one longitudinal mode of the high-finesse non-degenerate Fabry-Pérot (F-P) cavity is excited in shorter build-up times and in a more efficient cavity coupling than pulsed version. In these conditions, a simple first order exponential decay is observed if the response of the intracavity medium is linear. This is because optical losses are determined by the well defined region of the cavity mirrors and of the intracavity sample, which is illuminated by that light mode. Indeed, lower shot-to-shot ring-down rate fluctuations than in pulsed CRD are measured for CW-CRD. As a consequence shot-noise-limited sensitivity can be in principle reached with this technique. In addition, the narrower linewidth of CW coherent sources allows residual Doppler and sub-Doppler resolution in CW-CRD spectroscopy [4–6].

P. Cancio (✉) · I. Galli · S. Bartalini · G. Giusfredi · D. Mazzotti · P. De Natale
Istituto Nazionale di Ottica, Consiglio Nazionale delle Ricerche, Via N. Carrara 1,
50019 Sesto Fiorentino, Italy
e-mail: pablo.canciopastor@ino.it

G. Gagliardi, H.-P. Loock (eds.), *Cavity-Enhanced Spectroscopy and Sensing*,
Springer Series in Optical Sciences 179, DOI 10.1007/978-3-642-40003-2_4,
© Springer-Verlag Berlin Heidelberg 2014

Moreover, practical advantages in terms of power consumption, user-friendly operation, reduced size and cost give a further appeal to CW-CRD with respect to its pulsed version.

Despite the lack of CW laser sources and highly reflective mirrors for mid-IR wavelengths, CW-CRD spectroscopy is appealing in this spectral region [5, 7, 8] because there lie the strongest ro-vibrational transitions for the simplest molecules, belonging to the so-called fingerprint region [9]. Nowadays, CW-CRD spectroscopy in the mid IR is a well assessed technique for trace gas sensing [10–12] thanks to the technological development of new tunable CW coherent sources in this spectral region [13–20]. In these CRD experiments, linear optical absorption from the sample was only considered. Non-linear molecular absorption, which can occur in CW-CRD due to the strong optical field stored in the cavity, has been exploited to increase spectroscopic resolution [4–6], but only recently with the development of SCAR [21] a boost in sensitivity has also been achieved. Combination of SCAR with Difference-Frequency-Generated (DFG) IR radiation [16, 17], frequency stabilized and controlled against an Optical Frequency Comb (OFC) [22], has pushed CW-CRD to the new limits in terms of sensitivity [23] and resolution [21].

Apart from the technical simplicity, allowing extreme sensitivity with a short averaging time, the increased popularity and applicability of CRD spectroscopy is due to its being a quantitative method, allowing direct and accurate determination of the absolute value of the absorption coefficient (α) of the targeted transition, with no need for complex calibration routines. Thus, a very low quantity of absorbers can be accurately measured, if the transition strength is known independently. In conventional CRD, the linear absorption loss rate $c\alpha$ is added to other cavity losses and it is measured from the increase of the cavity decay rate with respect to the non-absorbing sample situation. Hence, accurate measurements require an independent determination of the empty-cavity loss spectrum to know the zero absorption baseline [24]. The ultimate sensitivity limit does not only depend on the precision in measuring the empty-cavity decay rate γ_c but also on its repeatability and stability. Coating non-uniformity, surface defects and dirty spots of the portion of the cavity mirrors illuminated by the laser light, laser alignment fluctuations, excitation of high-order transverse modes nearly degenerate with longitudinal ones, fluctuations in the injection power and field shape (excess noise), optical fringes, and finite extinction ratio of the CW light are some of the factors which worsen the shot-to-shot γ_c fluctuations, and hence prevent possible shot-noise-limited sensitivity in conventional CW-CRD spectroscopy. Indeed, we have noticed such worse γ_c stability for CW-CRD spectroscopic measurements performed around 4.5 μm [11].

In these experiments we use an OFC-assisted DFG IR coherent source [16] (see Sect. 4.3.1 for details) coupled in a 1-m-long high-finesse ($\sim$11500) F-P cavity. A double-pass acusto-optic modulator (AOM) is used to quickly switch off the input IR light when a threshold coupling level is reached, and thus the transmission cavity decay is detected by a N_2-cooled InSb detector during 100 μs. Quick switch between three laser frequencies at intervals of 450 MHz (three cavity Free Spectral Ranges (FSR)) is allowed by using other two double-pass AOM's. The absolute frequency stability of the DFG source and the slow cavity-length modulation around

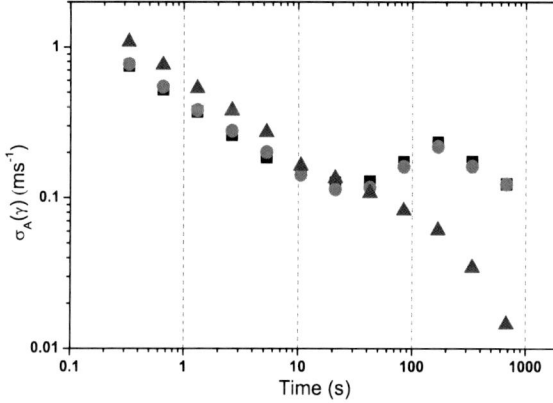

Fig. 4.1 Allan deviation of the empty-cavity decay rate at 4.5 μm for two laser frequencies, three cavity FSR apart (450 MHz) (*black squares* and *red circles*), and for its difference (*blue triangles*). Each exponential fitted signal is the result of an average of 8 CRD decays recorded in an acquisition time of 330 ms (Color figure online)

laser resonance limits to about 42 ms the shot-to-shot acquisition time. The standard error on γ_c, 4×10^{-2} ms^{-1} achieved in a single shot (i.e. at $T = 100$ μs), yields an absorption sensitivity of 1.3×10^{-11} cm^{-1} Hz$^{-1/2}$, only about twice worse than the shot-noise-limited minimum detectable absorption, taking into account the experimental parameters. Such sensitivity worsen very quickly, as shown in Fig. 4.1. There, the Allan deviation of consecutive γ_c measurements is plotted for two different DFG IR frequencies 450 MHz apart (black square and red circle points). Each γ_c value is the result of an average of 8 consecutive decays which limits the shot-to-shot acquisition time to 330 ms. As we can see, the γ_c uncertainty is increased by about 25 times with respect to the one at 100 μs, and consequently the sensitivity is decreased by the same quantity. If the noise sources of such instability are white, long term average could again improve the sensitivity, as shown in Fig. 4.1 up to 20 s of acquisition time with almost one order of magnitude of improvement. Longer averaging does not produce gain in terms of precision due to long-term γ_c fluctuations.

We noticed the same variance behavior for both frequencies which excludes dependence of such fluctuations on optical frequency. An usual strategy of CRD spectroscopy is the use of consecutive γ_c measurements at different laser frequencies, one in resonance with the targeted absorption transition and one out of resonance, in order to directly determine the sample absorption contribution by difference. At the same time, frequency independent γ_c instabilities could be canceled out or minimized in such difference. Indeed, the Allan deviation in this case (blue triangle points in Fig. 4.1) shows such improvement beyond 20 s acquisition time. Unfortunately, low repeatability of the measured γ_c prevents sensitivity increase by using long-term averages. In Fig. 4.2, we show the repeatability of the measured γ_c during different days at three different laser frequencies (black squares, red circles and blue triangles) at 450 MHz frequency intervals. In this case, each plotted point is the

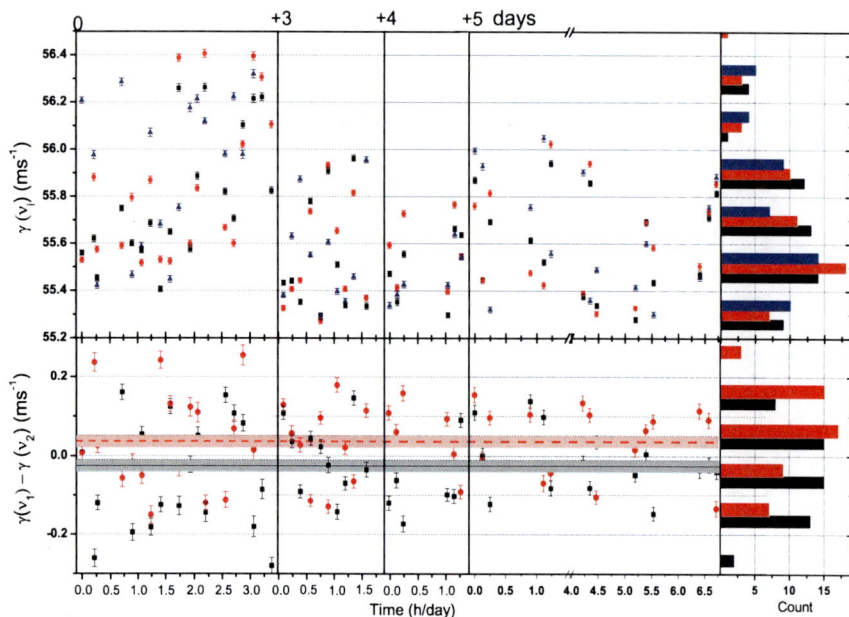

Fig. 4.2 *On the top*: repeatability of the empty-cavity decay rate at 4.5 μm for three laser frequencies, at intervals of three cavity FSR (450 MHz) (0 MHz: *black squares*, +450 MHz: *red circles* and −450 MHz: *blue triangles*). On the bottom: repeatability of the difference of γ at different frequencies (0–(−450) MHz: *black squares*, 0–450 MHz: *red circles*). *Horizontal bars* with the average values and uncertainty are shown in this case. Histogram distribution for each case is shown in *right-side graphs*. Each point is the average of 500 consecutive decay-rate values

average of the 500 consecutive γ_c values. The standard deviation with respect to the weighted mean γ_c value at each frequency is three times worse than the uncertainty of the single point. Moreover, the non-Gaussian statistical distribution (see top bar graph on right of Fig. 4.2) prevents further improvement even when averaging a large number of measurements. If we examine the repeatability of the differences between γ_c at different optical frequencies (see bottom graph of Fig. 4.2), we note that a Gaussian distribution is restored, as expected, but the weighted mean values differs by more than one standard deviation. Even more, these values differ from zero contrary to what one might expect for the difference of decay rates of cavity modes only few FSR apart. As a consequence, such differential method cannot be applied to remove the empty cavity contribution, and the baseline spectrum must be accurately measured, at least at the repeatability limit. Determination of this frequency-dependent baseline can be even more dramatic when a background from other absorbing substances present in the sample is added. The sensitivity limits due to uncorrelated fluctuations of the cavity decay rate, evidenced by these results, suggested us to follow a totally different approach to decouple other cavity losses from the targeted absorption loss in the shortest possible time, i.e. in a single CRD

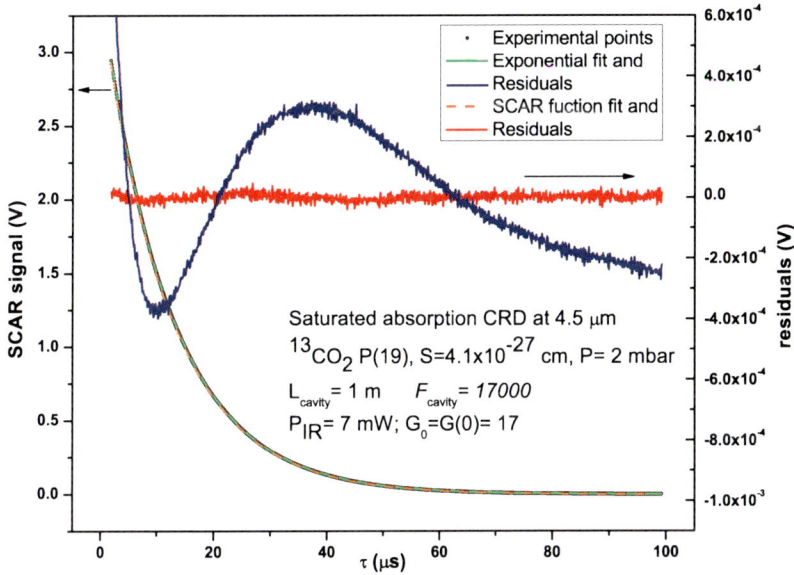

Fig. 4.3 Typical SCAR decay for a saturating DFG light resonant with a CO_2 transitions around 4.5 μm and a Fabry-Perot cavity of finesse about 11000. Exponential (*green trace*) and SCAR (Eq. (4.7), *orange trace*) decay fit functions and respective residual plots (*blue* and *red traces*, respectively) are shown. *Left vertical scale* applies to decay signal and fit functions, and *right vertical scale* to residual plots

event, approaching, in this way, the ultimate shot-noise-limited sensitivity. SCAR spectroscopy does that.

4.2 SCAR Technique

From the experimental point of view, SCAR does not differ from other conventional CRD experiments. Laser light is coupled to a high-finesse F-P cavity up to a threshold level, and then it is quickly switched off a cavity resonance. Transmitted ring-down light is detected and the decay rate is measured. If a molecule inside of the cavity absorbs coupled light, the loss rate due to such absorption is measured as an increase of the empty cavity loss rate. What is different in SCAR is that saturation effects of the molecular absorption induce a deviation of the ring-down signal from the perfectly exponential behavior, as expected for linear intracavity losses. Indeed, as represented in Fig. 4.3, when the ring-down signal that contains saturated molecular absorption is fitted to an exponential function, the residuals differ strongly from a flat behavior. Instead, a fit to a decay function which takes into account non-linear absorption effects, as the one of Eq. (4.7), gives a flat residual plot, hence explaining the physical effect of the saturated molecular transition.

Saturation effects in CRD spectroscopy have been analyzed both for pulsed and CW experiments. In the formers, theoretical mean-field analyses on dynamic absorption saturation were performed a few years ago, either considering inhomogeneous broadening of the transition [25] or not [26]. Instead, for CW CRD experiments, non-linear absorption effects were discussed in two specific regimes. When the ring-down time is much shorter than the population and coherence relaxation times, absorption loss keeps constant at the saturated value existing at the CRD start time and the decay will be simply exponential [5, 6]. In this case, the decay-time variation due to saturated absorption is smaller than in non-saturated one, depending on the saturation level, and it could be even negligible for strong saturation regime. In the opposite regime (the so-called "adiabatic" approximation), the saturation level of the optical transitions reaches the steady state at each point of the ring-down profile and a more complete and complex theoretical model will be needed to fit the non-exponential decay curves [4, 21, 23]. Here, we describe the model corresponding to the "adiabatic" regime, because in this case SCAR spectroscopy is effectively exploited to improve the sensitivity of conventional CRD. In addition, the equations resulting from this treatment can be easily deduced for the non-adiabatic regime by considering a constant saturation parameter during the decay.

For the analysis, we consider two different cases, depending on whether the molecular transition is homogeneously or inhomogeneously broadened. We are assuming that the gas interacts with intracavity radiation in a TEM_{00} mode with a time-dependent intensity I and power P given by the following expressions:

$$I(\rho, t) = I_0(t)e^{-2(\rho/w)^2}, \qquad P(t) = \frac{\pi w^2}{2}I_0(t), \qquad (4.1)$$

where $\rho = \sqrt{x^2 + y^2}$ is the radial coordinate, $I_0(t) \equiv I(\rho = 0, t)$ is the peak intensity on the cavity axis z, and w is the beam waist, assumed to be constant along z. We also define several physical quantities related with saturated absorption of a two-level transition:

$$I_S = ch\frac{\Gamma_{\parallel}\Gamma_{\perp}k^3}{A}, \quad \text{sat. intensity} \qquad (4.2)$$

$$P_S = \frac{\pi w^2}{2}I_S, \quad \text{sat. power} \qquad (4.3)$$

$$G(t; \Delta\nu) = I(t)\frac{g(\Delta\nu)}{I_S} = P(t)\frac{g(\Delta\nu)}{P_S} = P(t)U_S(\Delta\nu), \quad \text{sat. function} \quad (4.4)$$

where A is the Einstein coefficient of the transition, k is the wave vector, $2\pi\Gamma_{\parallel}$ is the population relaxation rate and $2\pi\Gamma_{\perp}$ is the homogeneous dephasing rate. In Eq. (4.4), $g(\Delta\nu)$ is the homogeneous line profile normalized to one on the peak, and $\Delta\nu$ is the laser frequency detuning from resonance. In the following analysis, we are neglecting the effects of the standing-wave light field inside the cavity and we are spatially averaging the different saturation levels of molecules interacting with light in node and antinode positions.

4.2.1 Inhomogeneously Broadened Transitions

In the presence of inhomogeneous broadening due to a thermal Gaussian distribution of molecular velocities, the absorption coefficient, affected by saturation, obeys the following equation [27]:

$$\alpha(\rho, t; \Delta v) = \frac{\alpha_0 g_D(\Delta v)}{\sqrt{1 + I(\rho, t)/I_S}} \tag{4.5}$$

where α_0 is the non-saturated value of the absorption coefficient and $g_D(\Delta v)$ is the Doppler profile due to the molecular velocity distribution. In this case, I_S is the saturation intensity defined in Eq. (4.2), which is constant for each velocity class if the Doppler linewidth is larger than the natural linewidth.

Intracavity power attenuation due to gas absorption, along the z axis, can be expressed as

$$\frac{dP}{dz}(t) = -2\pi \int_0^\infty \alpha(\rho, t; \Delta v) I(\rho, t) \rho d\rho$$
$$= -\alpha_0 g_D(\Delta v) \frac{2P(t)}{1 + \sqrt{1 + P(t)/P_S}} \tag{4.6}$$

where the spatial integration over the transverse beam profile corresponds to a "local" approximation. That is correct at pressure regimes where the diffusion time of molecules is longer than the population relaxation time of molecular levels. If we combine this absorption loss mechanism, which obeys the Beer-Lambert law, with mirror losses, and by using the speed of light relation $\frac{d}{dz} = \frac{1}{c}\frac{d}{dt}$, we get the following rate equation for the saturation function:

$$\frac{dG}{dt}(t) = -\gamma_c G(t) - \gamma_g \frac{2G(t)}{1 + \sqrt{1 + G(t)}}, \quad G(0) = G_0 \tag{4.7}$$

where $\gamma_c = c(1 - R)/L_c$ (R is the mirror reflectivity and L_c is the cavity length) and $\gamma_g = c\alpha_0 g_D(\Delta v)$ are the cavity and molecular absorption decay rates, respectively. In Eq. (4.7), $G_0 = P_0/P_S$ is the saturation parameter value when the SCAR decay event starts, triggered by a threshold level reached on the F-P transmitted signal.

Unfortunately, an analytical solution $G = G(t)$ of this differential equation is not possible even by explicitly inverting the solution $t = t(G)$. Alternatively, a numerical integration of Eq. (4.7) can be performed by using the 4th-order Runge-Kutta algorithm (RK4) [28] as a step of the fitting procedure. Since in real CRD molecular spectroscopy experiments $\gamma_g < \gamma_c$ and the dynamic range of the decay curve exceeds several decades, we can factorize G in the following way:

$$G(t) = G_0 e^{-\gamma_c t} f(t; G_0, \gamma_c, \gamma_g), \tag{4.8}$$

and a direct substitution in Eq. (4.7) gives that the function f obeys the following differential equation:

$$\frac{df}{dt}(t) = -\gamma_g \frac{2f(t)}{1 + \sqrt{1 + G_0 e^{-\gamma_c t} f(t)}}, \quad f(0) = 1. \tag{4.9}$$

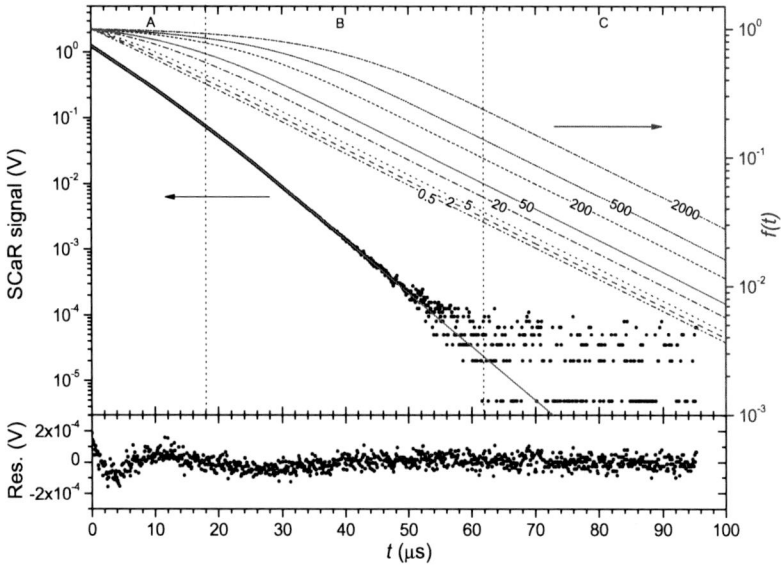

Fig. 4.4 Saturated decay functions f, simulated for different values of G_0 (labeling curves) are plotted with the *right vertical scale*. Experimental data points, fit curve and bottom residuals are plotted with the *left scale*. Experimental SCAR decay was recorded by averaging 3072 decay signals measured at a fixed frequency close to the center of the $(03^31$–$03^30)$ $R(50)$ transition of the $^{12}C^{16}O_2$ (line-center frequency 2343.663 cm^{-1}, line-strength $S = 7.86 \times 10^{-24}$ cm) with cavity length of 1 m, and gas temperature and pressure of 296 K and 50 μbar, respectively. The SCAR fit was used to determine $\gamma_g = 50$ ms^{-1} and $\gamma_c = 130$ ms^{-1}, kept fixed in the plotted simulations (Color figure online)

Numerical integration of Eq. (4.9) and G factorization in Eq. (4.8) can be used for a better computing precision of the fitted parameters G_0, γ_c and γ_g. In addition, the function f gives a rough and intuitive explanation for the intrinsic ability of the SCAR technique to improve sensitivity by distinguishing between gas absorption losses and cavity ones. In Fig. 4.4, simulated $f(t)$ functions are plotted for different G_0 values when γ_c and γ_g are kept at a fixed value in all cases. The value of such parameters were obtained by fitting an experimental SCAR decay event with $G_0 = 50$. As depicted in Fig. 4.4, three consecutive time intervals can be recognized in the SCAR signal: zones A, B, and C carry information, respectively, on γ_c, $\gamma_c + \gamma_g$ and the detection offset, which in the IR region is due to thermal background (whose fitted value has been subtracted from the signal in Fig. 4.4). The A–B transition is marked by a slope change in the SCAR signal, which needs the condition $G_0 \gg 1$ to be well observable. As a consequence, a large detection dynamics is needed to measure the transition from strong-saturation to linear-absorption regime in the SCAR decay, before it falls below the noise level (B–C transition). The best choice for G_0 should give such three zones with similar durations.

More important, as evidenced by Eq. (4.7) (and by Eq. (4.12) for homogeneous broadening case), both losses due to gas absorption γ_g and other cavity losses γ_c

are measured simultaneously and independently in time intervals corresponding to each single CRD event, which allows the SCAR technique to achieve very high sensitivity in trace gas detection.

4.2.2 Homogeneously Broadened Transitions

At high gas pressures the target transition is homogeneously broadened, and the saturated absorption coefficient can be approximated by:

$$\alpha(\rho,t;\Delta v) = \frac{\alpha_0 g_L(\Delta v)}{1 + \frac{I(\rho,t)}{I_S} g_L(\Delta v)} \tag{4.10}$$

where α_0 is the non-saturated absorption at resonance and $g_L(\Delta v)$ is the Lorenztian profile, normalized to 1 on the peak, due to molecular collisions.

As in the inhomogeneous case, the molecular absorption attenuates the intracavity power along the absorption path, with the following law:

$$\frac{dP}{dz}(t) = -2\pi \int_0^\infty \alpha(\rho,t;\Delta v) I(\rho,t)\rho d\rho$$

$$= -\alpha_0 g_L(\Delta v) \ln\left[1 + \frac{P(t)}{P_S} g_L(\Delta v)\right], \tag{4.11}$$

where a "local" approximation was again used in the spatial integration over the transverse beam profile. Taking into account also other cavity losses γ_c and following a similar procedure to that used in the previous section, we get the following rate equation for the SCAR decay:

$$\frac{dG}{dt}(t) = -\gamma_c G(t) - \gamma_g \ln\left[1 + G(t)\right], \tag{4.12}$$

where $\gamma_g = c\alpha_0 g_L(\Delta v)$ and the saturation parameter at the start of the decay depends on the detuning:

$$G(t;\Delta v) = \frac{P(t)}{P_S} g_L(\Delta v), \qquad G(0;\Delta v) = \frac{P(0)}{P_S} g_L(\Delta v). \tag{4.13}$$

Equation (4.12) can be used to fit the SCAR decay for homogeneously-broadened transitions, by following a similar numerical integration procedure described in the previous section. In this regime, we note that stronger saturation parameters are needed to get the sensitivity benefits of the SCAR technique even in the tails of the line profile.

Finally, an hybrid regime where homogeneous broadening is comparable with the inhomogeneous one can be analyzed by using the same equations as described in this section, but with the Lorentz line profile $g_L(\Delta v)$ replaced by a Voigt profile $g_V(\Delta v)$.

4.3 Mid-IR CW Laser Sources for SCAR Spectroscopy

For SCAR spectroscopy, the laser source must be coupled to the cavity in the most efficient way, so that the intracavity laser power at the start of the cavity decay, $P(0)$, is higher than the saturation power of the targeted molecular transition (see Sect. 4.2). Independently of the power enhancement factor of the cavity due to its high-finesse, the requirement for the incident laser power at the cavity input is more relaxed for TEM_{00} and narrow-linewidth laser sources. On the other hand, this technique, like other molecular spectroscopy ones, requires laser sources tunable in a wide spectral range, as well as absolute frequency calibration of their frequency. Moreover, laser frequency synthesis could be a useful tool to improve sensitive detection of molecules with SCAR, by implementing long-time absorption average (see Sect. 4.5). Differently from the visible and near-IR regions, where laser narrowing techniques and absolute frequency referencing by using OFCs are well established, a considerable technological effort has been made in the last decade to transfer such characteristics to CW laser sources in the mid IR [15–17, 19, 20, 29–34]. In this section, we give an overview of such sources, paying attention to features useful for SCAR spectroscopy.

4.3.1 Non-linear Frequency Down Converted Coherent Sources

DFG in non-linear crystals from near-IR CW lasers has proven to be a low-noise [35], well suited tool for high-resolution [36] and high-sensitivity [11] molecular spectroscopy in the mid IR. Sub-Doppler spectra with absolute frequency scales have been observed [37] using such a DFG source, linked to an OFC. The DFG wide tuning range is only limited by the tunability of the pump/signal sources and by the transparency of the non-linear medium. Single-pass conversion efficiency is one of its main limitations for high-power demanding applications, even when periodically-poled (PP) non-linear crystals are used. Indeed, the power of a few hundreds of μW generated with a power-boosted single-pass DFG source [35] could be not enough to perform SCAR spectroscopy of CO_2 transitions around 4.5 μm, when its linewidth is larger than the cavity linewidth (that translates in lower coupling efficiency). To overcome such limit, we have developed a technique to control the DFG frequency against a visible/near-IR OFC [16], which not only traces the mid-IR frequency against the primary frequency standard, but also narrows the DFG intrinsic linewidth down to 10 Hz.

Our single-pass DFG uses an Extended-Cavity-Diode-Laser (ECDL) around 860 nm and a fiber-amplified Nd:YAG laser around 1064 nm as pump and signal sources, respectively, for the non-linear process in a PP-Lithium Niobate crystal. The pump and signal lasers are beaten against a self-referenced OFC which operates in the 500–1100 nm range. The beatings are used to get an effective phase lock of the ECDL frequency to the Nd:YAG one by use of a direct-digital-synthesis (DDS) technique [38], without any phase-noise contribution from the OFC. As a consequence, the DFG linewidth is a given fraction of the signal laser one. Taken into account the spectral purity of the Nd:YAG laser, the DFG intrinsic linewidth was narrowed down to a white-noise limit of 10 Hz with a residual jitter of 1 kHz in the 100 ms time scale. Similar results could be achieved with other noisier lasers, frequency narrowed by using well established techniques in the visible/near-IR range. In addition, a low-bandwidth frequency lock of the signal laser to the OFC allowed absolute frequency tracing of the DFG radiation. In these conditions, for a 1-m-long Fabry-Perot cavity with ZnSe plano-concave mirrors (100 ppm absorption and scattering and 170 ppm transmission losses, $F \sim 11500$), a coupling efficiency of 83 % was achieved, a bit less than the expected 86 %, due to slight deviations from a perfect TEM_{00} profile of this DFG source. Nevertheless, intracavity powers in excess of 10 mW were achieved with few μW of incident DFG radiation, enough to strongly saturate CO_2 transitions around 4.5 μm at low pressure conditions.

When high input powers are needed, good alternatives are pump-enhanced intracavity DFG [17] or Optical Parametric Oscillator (OPO) sources [20]. In the former case, DFG IR radiation is generated inside the cavity of a Ti:Sapphire laser injected by an ECDL around 860 nm, enormously increasing the pump power. A fiber amplified Nd:YAG laser is coupled to the intracavity PP-Lithium Niobate non-linear crystal acting as signal laser for the DFG process. Since the pump-master and signal lasers are basically the same as in the single-pass DFG, OFC narrowing and absolute frequency control of the generated IR radiation was allowed by the above described technique, but with an output power larger than 30 mW at the transparency edge of the non-linear crystal ($\sim$4.5 μm). In addition, the DFG tunability can be extended, exploiting the wide emission range of the Ti:Sapphire laser. Moreover, the excellent TEM_{00} mode of the generated IR, thanks to the optical beam profile quality provided by the Ti:Sapphire cavity, allowed to couple 86 % of the input power to the same F-P cavity used for the single-pass DFG. As a consequence, higher intracavity power was provided at the CRD start allowing, for example, strong saturation of CO_2 v_3-band ro-vibrational transitions at higher pressures (see Sect. 4.5).

Since their first implementation, OPOs have provided high mid-IR output powers, but only recently high-resolution and precise molecular spectroscopy applications have been demonstrated [20]. An OFC-referenced pump source was used in such mid IR OPO to get both IR narrowing and absolute frequency traceability. Differently from the DFG sources described above, OFC phase noise was present in the generated radiation, which could be avoided by narrowing the pump laser with high-finesse F-P cavities, length-controlled against the OFC. Nevertheless, sub-Doppler spectroscopy of the hyperfine structure of the CH_3I molecule around 3 μm was performed, demonstrating the useful features for application to SCAR.

4.3.2 Quantum Cascade Lasers

Quantum cascade lasers (QCLs) emitting directly in the mid-IR region are ideal laser sources for spectroscopic applications with molecular targets [19, 39, 40] due

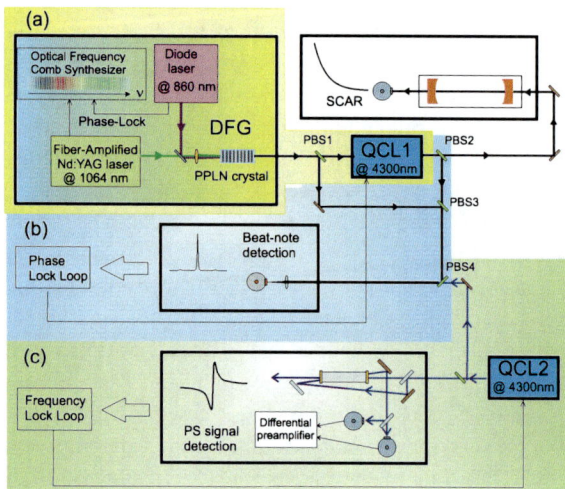

Fig. 4.5 Narrowing and frequency control schemes of a CW-QCL (QCL1) in the mid IR: (**a**) Optical injection locking [31] (*yellow box*); (**b**) Phase-locking [34] to an OFC-assisted DFG IR source (*blue box*); (**c**) phase-locking to a second QCL (QCL2) frequency locked and narrowed to a CO_2 molecular transition [33]. Polarization of the different IR sources must be controlled at the polarizing beam-splitters (PBSs) to switch from one configuration to the other. Application of such QCL to SCAR spectroscopy is also shown

to their room-temperature operation, compactness and low cost. High output powers of QCLs (from mW up to W level [41]), their intrinsic spectral purity, at the hundreds of Hz level [42–44], and their frequency control against visible/near-IR OFCs [29–32], give them appeal to be used for SCAR spectroscopy. Nevertheless, a control of the excess frequency noise at Fourier frequencies within bandwidth of few hundred of kHz [40, 45] must be achieved to perform efficient coupling of QCLs to a high-finesse IR F-P. We have proposed and realized three possible approaches [31, 33, 34], depicted schematically in Fig. 4.5.

The first one is based on controlling the QCL frequency jitter by optical injection-locking from a frequency stable IR source. Our single-pass OFC-assisted DFG source (see Sect. 4.3.1) was used as a master laser for such optical injection at 4.67 μm. As a result, the room-temperature QCL slave source was frequency narrowed close to the level of the master one, with an absolute frequency traceability to the primary frequency standard, and with higher power (few mW) available for CRD or SCAR spectroscopy.

The second approach is based on phase-locking the QCL frequency to a stable IR oscillator. Again, we used our single-pass OFC-assisted DFG source as a master oscillator for such lock [34]. In this way, the QCL frequency was stabilized and narrowed within a 450 kHz bandwidth, acquiring the same spectral characteristics of the DFG source, but with boosted power. Alternative option is to upconvert the QCL frequency by using non-linear sum-frequency generation, and to directly compare the generated frequency with an OFC [32]. Such beatnote is used to phase-lock the QCL frequency to the OFC.

Recently, we have proposed to frequency lock the QCL to a narrow sub-Doppler molecular absorption. We have narrowed down to a sub-kHz level the linewidth of a room-temperature QCL at 4.3 µm, by frequency locking it to a saturation dip of a CO_2 transition [33] recorded with Doppler-free polarization spectroscopy [40]. Moreover, such transition is a secondary frequency standard to fix the QCL frequency. Such molecular-dip-locked QCL can be used as a master oscillator for the phase-locking scheme described above, instead of the DFG source, lowering the cost and increasing the compactness.

4.4 Sub-Doppler SCAR Molecular Spectroscopy

As noted in Sect. 4.1, one of the benefits of SCAR is to resolve the sub-Doppler structure of an inhomogeneously-broadened molecular transition. Molecules that simultaneously interact with the forward and backward laser light between cavity mirrors (i.e. molecules that have no axial velocity component along the direction of the laser beam), undergo a doubled saturation effect with respect to molecules of other velocity classes. As a consequence, a decrease of the absorption coefficient is observed in the center of the inhomogeneous Doppler profile of the transition and it is referred to as a Lamb dip. Indeed, observation of such features in hot ro-vibrational transitions of NO_2 molecule around 800 nm was the first evidence of saturated-absorption effects in CW-CRD spectroscopy [4]. In that experiment, in which adiabatic conditions were satisfied, a model like that described in Sect. 4.2 was not developed to fit the observed non-exponential decays. Each CRD decay was divided in four consecutive time intervals, and each one was exponentially fitted while the laser was scanned across Doppler broadened transitions. As a consequence, evident Lamb dips were observed for the line profiles corresponding to the first decay interval, which almost disappeared in the line profiles resulting from the fits of the last interval of the CRD decays.

The first systematic study of Lamb dips observed by CRD was performed on sub-Doppler transitions of the ro-vibrational band of ethylene (C_2H_4) around 941.8 cm^{-1} [5]. In such experiment, microwave side-bands of a CO_2 laser at 10.6 µm were coupled to a medium-finesse cavity ($F \sim 617$), interacting with C_2H_4 molecules at low pressure in a saturated-absorption regime. Since the non-adiabatic saturation regime applies to that case, the variation of the decay rate was measured when the side-band was scanned across the Doppler-broadened line. As a consequence of saturation, that variation was lower around the line-center, and thus dips with homogeneous line-shape were observed. Even if sub-Doppler signals are analyzed by using the steady-state approach of optical saturation of a two level system in the plane-wave limit, we note that Eq. (4.7) could be used for that purpose in the approximation $G(t) = G_0$ during all decay time. Lamb dips in H_2O transitions around 925 nm were measured by using frequency-stabilized CRD spectroscopy [6].

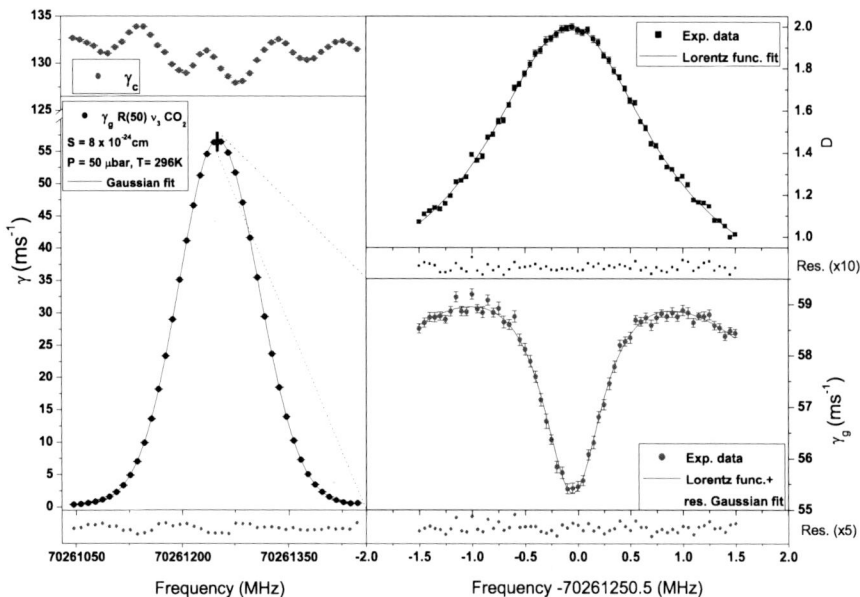

Fig. 4.6 Saturation dip of the $(03^31–03^30)$ $R(50)$ transition of $^{16}C^{12}O_2$ molecule at 2343.663 cm^{-1} recorded with an absorption cavity length of 1 m. For these spectra, the single-pass DFG source and the F–P cavity described in Sect. 4.3.1 were used. On the *left graph*, a Doppler broadened spectrum of the gas loss rate, γ_g (*black points*) and cavity loss rate, γ_c (*red points*). On the *right-bottom graph*, a 3 MHz zoom of γ_g around line center, where the saturation dip is observable. On the *right-top graph*, D parameter in the same 3 MHz scan is depicted. Fits to expected functions and residuals are shown in all cases. The fitted line centers of γ_g and D sub-Doppler spectra were in agreement, considering their uncertainties (Color figure online)

In that case, exponential fits of the CRD decays were used to measure a decrease of the absorption rate in the Doppler line profile around the line center. Such dips were used as frequency markers for line-shape analysis.

In the case of adiabatic saturation regime, we have demonstrated that sub-Doppler resolution can be achieved with SCAR spectroscopy [21]. At the center of the Doppler-broadened transitions, the saturation parameter G is twice its value at detuned frequencies. Indeed, a saturation dip can be measured in this case by using fitted values of either G_0 or γ_g as shown in Fig. 4.6. In the former case, a new dimensionless parameter D is defined:

$$D = G_0 \frac{P_S}{P_0}, \tag{4.14}$$

which varies from 1 to 2 when the saturation is doubled for molecules that simultaneously interact with forward and backward laser beams. The sub-Doppler resolution of the SCAR technique was used to resolve and measure the hyperfine structure of the $(00^01–00^00)$ $R(0)$ transition ($S = 1.25 \times 10^{-22}$ cm) of the $^{17}O^{12}C^{16}O$ molecule at 2340.76 cm^{-1}. The measured absolute frequencies of the three hyperfine components were used to determine the line-center frequency with an accuracy of 5×10^{-11} and to improve the value of the electric-quadrupole coupling constant in the excited ro-vibrational state of this molecule. In this experiment, a natural abundance (7.5×10^{-4}) $^{17}O^{12}C^{16}O$ gas sample at very low pressure ($\sim$2 µbar) was used. These conditions did not satisfy the local assumption of the model described in Sect. 4.2, because the population relaxation time is dominated by the time of interaction of molecules with light, τ_t (transit time across the cavity mode). Instead, a mean-field approximation of the spatial integration over the transverse beam profile in Eq. (4.6) was made. Even more, the adiabatic approximation $\tau_t \ll 1/\gamma_c$ is at the validity edge. Nevertheless, the quality of the obtained results certified the potential of SCAR spectroscopy to observe sub-Doppler structures for a very low quantity of molecular absorbers, and to analyze them with a very simplified model.

4.5 Ultra High-Sensitivity SCAR Molecular Spectroscopy: Application to Detection of Very Rare Species

Trace gas detection by optical spectroscopy plays a key role in applications that demand quantitative measurements of extremely small amounts of molecular gases. Combination of CW-CRD spectroscopy and strong absorption of the allowed molecular transitions in the mid IR can fulfill such requirement with a compact and low-cost optical technology. As noted in Sect. 4.1, despite its simplicity, CW-CRD spectroscopy is a method to accurately measure absorption coefficients of molecular transitions without any need for complex calibration routines. Nevertheless, state-of-the-art conventional CW-CRD spectrometers in the mid IR [10, 11] do not reach minimum detectable absorptions below 10^{-9} cm^{-1} Hz$^{-1/2}$ level, partly due to the empty-cavity decay rate fluctuations and also to the technical difficulty to get a higher cavity finesse in this spectral region. The SCAR technique improves by more than one order of magnitude such limit, thanks to a sample absorption measurement which is independent from the other cavity losses during the same cavity decay event. In the left graph of Fig. 4.6 the sensitivity benefit added by the SCAR technique is graphically described. The large γ_c fluctuations (red points in the top graph on the left) are removed from the baseline background of the target CO_2 transition (γ_g black points in Fig. 4.6), as confirmed by the good agreement between the experimental γ_g values and the expected Gaussian fit. Moreover, the measured spectral area of the absorption agrees with the value calculated by taking into account the tabulated linestrength for this transition [9] and the gas pressure.

In order to compare and rank different trace gas sensors based on different spectroscopic techniques, spectral ranges, sources, and detectors, the minimum detectable pressure of a given molecular species (in units of mbar Hz$^{-1/2}$) can be used. Figure 4.7 shows a collection of the best ever published sensitivities, labeled by molecular species, wavelength, and technique [10, 46–53]. From such a comparison, CO_2 detection with the SCAR technique results to be the most sensitive, with detected molecular gas pressures, in a 1-Hz bandwidth, of a few tens of femtobar.

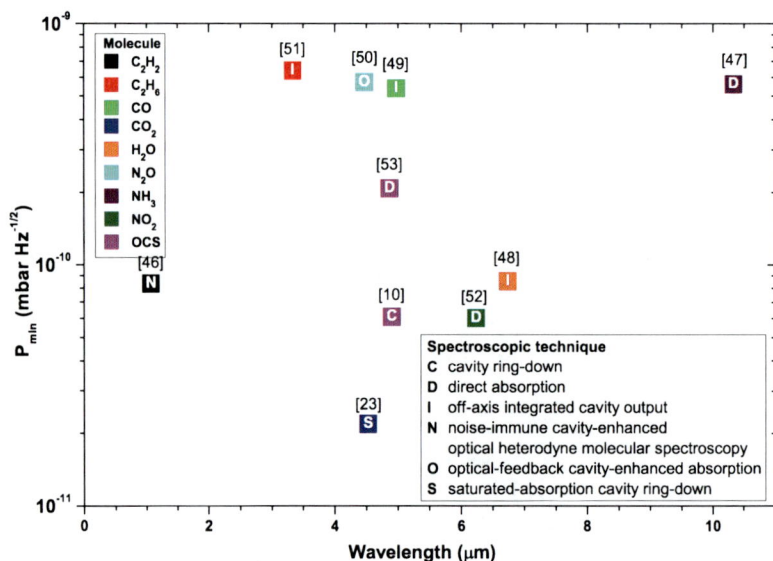

Fig. 4.7 Trace gas sensitivity ranking graph with the best values achieved by using the molecular absorption spectroscopy in the IR spectral range. The minimum detectable pressures (up to 10^{-9} mbar Hz$^{-1/2}$) are related to 9 different molecular species (*color*) and 6 different laser absorption techniques (*letter*). For all these data, the noise-equivalent absorbance for the strongest transition of the detected molecule at the given wavelength was calculated. References are given in *brackets*

Such sensitivity could be applied to the detection of very low abundant or rare molecular species. This is the case for radiocarbon (^{14}C)-containing molecules, such as ^{14}C^{16}O$_2$ [17, 54]. Radiocarbon is a very elusive atom, its concentration with respect to the most abundant ^{12}C isotope being about one part per trillion (10^{-12}). Notwithstanding, for longer than 60 years radiocarbon has been the "natural clock" for dating organic-matter-containing samples. Recently, we have performed optical detection of radiocarbon concentrations at a 43 parts-per-quadrillion level by using SCAR spectroscopy on ^{14}C^{16}O$_2$. In this experiment, the (00^01-00^00) P(20) ro-vibrational transition of ^{14}C^{16}O$_2$ around 4.5 μm was targeted with a SCAR instrument that uses the DFG IR source [17] and the high-finesse ($F \sim 11500$) F–P described elsewhere in this chapter. CO$_2$ gas samples at pressures of about 12 mbar and a temperature of 195 K are used to get the best trade-off between minimum carbon content and maximum sensitivity, also minimizing interference effects from absorptions of other CO$_2$ isotopologues present in the sample. The ^{14}C^{16}O$_2$ concentration is determined by measuring the spectral area of the P(20) line recorded by scanning the DFG frequency by 750 MHz across line center, and by using the following equation:

$$\frac{P_0}{P} = \frac{1}{n_s} \frac{P_s}{P} \frac{T}{T_s} \frac{1}{S} \int \frac{\gamma_g(v; P_i, T)}{c} dv, \qquad (4.15)$$

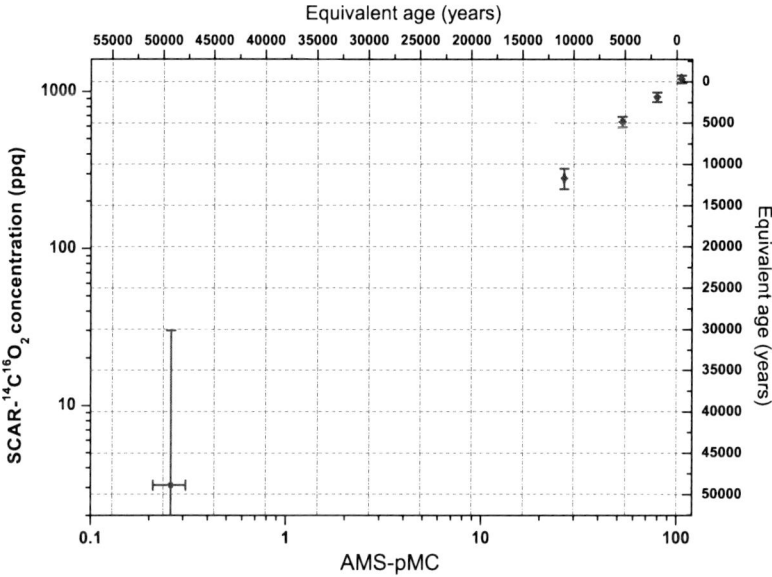

Fig. 4.8 AMS-SCAR intercalibration curve for radiocarbon detection. Radiocarbon-dioxide concentration measured by SCAR spectroscopy vs. pMC measured by AMS in CO_2 gas samples ranging from ancient to 2010. In the *top/right scales*, equivalent age is given in terms of radiocarbon content with respect to modern standard (year 1950). Such age must be corrected for possible fluctuations of the ^{14}C concentration during different epochs to get the real age of the sample (Color figure online)

with P total CO_2 pressure, P_i partial pressures of all CO_2 isotopologue contained in the sample (P_0 for the targeted one, $^{14}C^{16}O_2$ in this case), T gas temperature, S linestrength of the targeted transition ($P(20)$ in this case), and n_s the Loschmidt density ($\sim 2.5 \times 10^{19}$ cm^{-3}) which is the density of an ideal gas at standard thermodynamic conditions: $P_s = 1$ atm and $T_s = 296$ K.

The $^{14}C^{16}O_2$ SCAR spectroscopy table-top setup represents a valid analysis method of ^{14}C-marked samples, alternative to accelerator mass spectrometry (AMS), which is the standard method presently used for radiocarbon-based applications, such as biomedical, earth-atmosphere monitoring and dating [55, 56]. Indeed, we have recently calibrated the radiocarbon SCAR concentrations against AMS by measuring with both methods CO_2 samples from present natural abundance to totally depleted ^{14}C content (see Fig. 4.8) [54]. From this comparison, the precision for modern radiocarbon content of the SCAR technique is 2 pMC (percent of Modern Carbon), or an equivalent age of 32000 years with respect to modern epoch (1950), instead of 0.3 pMC (or 50000 years) achieved by AMS. In addition, a reduction of needed carbon mass and measurement time for the SCAR technique is mandatory to be competitivwe with AMS, at least for dating. On the contrary, SCAR offers a more compact, less expensive and low-maintenance equipment without degradation of the dated gas sample. Furthermore, SCAR has a large dynamic range: a perfectly linear behavior was measured from present natural abundance up to 6000 times higher.

Apart from ultrasensitive trace gas sensing, SCAR technique could provide new insights into elusive quantum mechanical effects encoded in molecules, such as parity violation due to weak interactions [57], violation of the Fermi-Bose statistics and/or the symmetrization postulate of quantum mechanics [58], time variation of fundamental physical constants, e.g., the proton-to-electron mass ratio [59].

References

1. R. Engeln, G. Helden, G. Berden, G. Meijer, Phase shift cavity ring down absorption spectroscopy. Chem. Phys. Lett. **262**, 105–109 (1996)
2. D. Romanini, A.A. Kachanov, F. Stoeckel, Diode laser cavity ring down spectroscopy. Chem. Phys. Lett. **270**, 538–545 (1997)
3. B.A. Paldus, J.J.S. Harris, J. Martin, J. Xie, R.N. Zare, Laser diode cavity ring-down spectroscopy using acousto-optic modulator stabilization. J. Appl. Phys. **82**, 3199 (1997)
4. D. Romanini, P. Dupre, R. Jost, Non-linear effects by continuous wave cavity ringdown spectroscopy in jet-cooled NO_2. Vib. Spectrosc. **19**, 93 (1999)
5. C.R. Bucher, K.K. Lehmann, D.F. Plusquellic, G.T. Fraser, Doppler-free nonlinear absorption in ethylene by use of continuous-wave cavity ringdown spectroscopy. Appl. Opt. **39**, 3154 (2000)
6. D. Lisak, J.T. Hodges, R. Ciuryło, Comparison of semiclassical line-shape models to rovibrational H_2O spectra measured by frequency-stabilized cavity ring-down spectroscopy. Phys. Rev. A **73**, 012507 (2006)
7. J.J. Scherer, D. Voelkel, D.J. Rakestraw, J.B. Paul, C.P. Collier, R.J. Saykally, A. O'Keefe, Infrared cavity ringdown laser-absorption spectroscopy (IR-CRLAS). Chem. Phys. Lett. **245**, 273–280 (1995)
8. M. Muertz, B. Frech, W. Urban, High-resolution cavity leak-out absorption spectroscopy in the 10 μm region. Appl. Phys. B **69**, 243–249 (1999)
9. Harvard-Smithsonian Center for Astrophysics, The HITRAN database (2009). http://www.cfa.harvard.edu/hitran
10. D. Halmer, G. von Basum, P. Hering, M. Murtz, Mid-infrared cavity leak-out spectroscopy for ultrasensitive detection of carbonyl sulfide. Opt. Lett. **30**, 2314 (2005)
11. D. Mazzotti, P. Cancio, A. Castrillo, I. Galli, G. Giusfredi, P. De Natale, A comb-referenced difference frequency spectrometer for cavity ring-down spectroscopy in the 4.25-μm region. J. Opt. A **8**, S490–S493 (2006)
12. I. Galli, P. Cancio, G. Di Lonardo, L. Fusina, G. Giusfredi, D. Mazzotti, F. Tamassia, P. De Natale, The v_3 band of $^{14}C^{16}O_2$ molecule measured by optical-frequency-comb-assisted cavity ring-down spectroscopy. Mol. Phys. **109**, 2267–2272 (2011)
13. D. Mazzotti, P. De Natale, G. Giusfredi, C. Fort, J.A. Mitchell, L.W. Hollberg, Difference-frequency generation in PPLN at 4.25 μm: an analysis of sensitivity limits for DFG spectrometers. Appl. Phys. B **70**, 747–750 (2000)
14. P. Maddaloni, G. Gagliardi, P. Malara, P. De Natale, A 3.5-mW continuous-wave difference-frequency source around 3 μm for sub-Doppler molecular spectroscopy. Appl. Phys. B **80**, 141–145 (2005)
15. E.V. Kovalchuk, T. Schuldt, A. Peters, Combination of a continuous-wave optical parametric oscillator and a femtosecond frequency comb for optical frequency metrology. Opt. Lett. **30**, 3141–3143 (2005)
16. I. Galli, S. Bartalini, P. Cancio, G. Giusfredi, D. Mazzotti, P. De Natale, Ultra-stable, widely tunable and absolutely linked mid-IR coherent source. Opt. Express **17**, 9582–9587 (2009)

17. I. Galli, S. Bartalini, S. Borri, P. Cancio, G. Giusfredi, D. Mazzotti, P. De Natale, Ti:sapphire laser intracavity difference-frequency generation of 30 mW cw radiation around 4.5 μm. Opt. Lett. **35**, 3616–3618 (2010)

18. S. Borri, S. Bartalini, P. Cancio, I. Galli, G. Giusfredi, D. Mazzotti, P. De Natale, Quantum cascade lasers for high-resolution spectroscopy. Opt. Eng. **49**, 111122 (2010)

19. P. Cancio, S. Bartalini, S. Borri, I. Galli, G. Gagliardi, G. Giusfredi, P. Maddaloni, P. Malara, D. Mazzotti, P. De Natale, Frequency-comb-referenced mid-IR sources for next-generation environmental sensors. Appl. Phys. B **102**, 255–269 (2011)

20. I. Ricciardi, E. De Tommasi, P. Maddaloni, S. Mosca, A. Rocco, J.-J. Zondy, M. De Rosa, P. De Natale, Frequency-comb-referenced singly-resonant OPO for sub-Doppler spectroscopy. Opt. Express **20**, 9178–9186 (2012)

21. G. Giusfredi, S. Bartalini, S. Borri, P. Cancio, I. Galli, D. Mazzotti, P. De Natale, Saturated-absorption cavity ring-down spectroscopy. Phys. Rev. Lett. **104**, 110801 (2010)

22. P. Maddaloni, P. Cancio, P. De Natale, Optical comb generators for laser frequency measurement. Meas. Sci. Technol. **20**, 052001 (2009)

23. I. Galli, S. Bartalini, S. Borri, P. Cancio, D. Mazzotti, P. De Natale, G. Giusfredi, Molecular gas sensing below parts per trillion: radiocarbon-dioxide optical detection. Phys. Rev. Lett. **107**, 270802 (2011)

24. D. Romanini, A.A. Kachanov, E. Stoeckel, Cavity ringdown spectroscopy: broad band absolute absorption measurements. Chem. Phys. Lett. **270**, 546–550 (1997)

25. J.Y. Lee, J.W. Hahn, Theoretical analysis on the dynamic absorption saturation in pulsed cavity ringdown spectroscopy. Appl. Phys. B **79**, 653 (2004)

26. S.S. Brown, H. Stark, A.R. Ravishankara, Cavity ring-down spectroscopy for atmospheric trace gas detection: application to the nitrate radical (NO_3). Appl. Phys. B **75**, 173 (2002)

27. W. Demtroder, *Laser Spectroscopy*. Advanced Texts in Physics (Springer, New York, 2003)

28. E.W. Weisstein, Gill's method. From MathWorld—a Wolfram web resource. http://mathworld. wolfram.com/GillsMethod.html

29. S. Bartalini, P. Cancio, G. Giusfredi, D. Mazzotti, P. De Natale, S. Borri, I. Galli, T. Leveque, L. Gianfrani, Frequency-comb-referenced quantum-cascade laser at 4.4 μm. Opt. Lett. **32**, 988–990 (2007)

30. S. Borri, S. Bartalini, I. Galli, P. Cancio, G. Giusfredi, D. Mazzotti, A. Castrillo, L. Gianfrani, P. De Natale, Lamb-dip-locked quantum cascade laser for comb-referenced IR absolute frequency measurements. Opt. Express **16**, 11637–11646 (2008)

31. S. Borri, I. Galli, F. Cappelli, A. Bismuto, S. Bartalini, P. Cancio, G. Giusfredi, D. Mazzotti, J. Faist, P. De Natale, Direct link of a mid-infrared QCL to a frequency comb by optical injection. Opt. Lett. **37**, 1011–1013 (2012)

32. A.A. Mills, D. Gatti, J. Jiang, C. Mohr, W. Mefford, L. Gianfrani, M. Fermann, I. Hartl, M. Marangoni, Coherent phase lock of a 9 μm quantum cascade laser to a 2 μm thulium optical frequency comb. Opt. Lett. **37**, 4083–4085 (2012)

33. F. Cappelli, I. Galli, S. Borri, G. Giusfredi, P. Cancio, D. Mazzotti, A. Montori, N. Akikusa, M. Yamanishi, S. Bartalini, P. De Natale, Sub-kilohertz linewidth room-temperature mid-IR quantum cascade laser using a molecular sub-Doppler reference. Opt. Lett. **37**, 4811 (2012)

34. I. Galli et al., Comb-assisted sub-kilohertz linewidth quantum cascade laser for high-precision mid-IR spectroscopy. Appl. Phys. Lett. **103**, 12117 (2013). doi:10.1063/1.4799284

35. S. Borri, P. Cancio, P. De Natale, G. Giusfredi, D. Mazzotti, F. Tamassia, Power-boosted difference frequency source for high-resolution infrared spectroscopy. Appl. Phys. B **76**, 437–477 (2003)

36. D. Mazzotti, S. Borri, P. Cancio, G. Giusfredi, P. De Natale, Low-power Lamb-dip spectroscopy of very weak CO_2 transitions around 4.25 μm. Opt. Lett. **27**, 1256–1258 (2002)

37. D. Mazzotti, P. Cancio, G. Giusfredi, P. De Natale, M. Prevedelli, Frequency-comb-based absolute frequency measurements in the mid-IR with a difference-frequency spectrometer. Opt. Lett. **30**, 997–999 (2005)

38. H.R. Telle, B. Lipphardt, J. Stenger, Kerr-lens mode-locked lasers as transfer oscillators for optical frequency measurements. Appl. Phys. B **74**, 1–6 (2002)

39. S. Borri, S. Bartalini, P. De Natale, M. Inguscio, C. Gmachl, F. Capasso, D.L. Sivco, A.Y. Cho, Frequency modulation spectroscopy by means of quantum-cascade lasers. Appl. Phys. B **85**, 223–2229 (2006)

40. S. Bartalini, S. Borri, P. De Natale, Doppler-free polarization spectroscopy with a quantum cascade laser at 4.3 μm. Opt. Express **17**, 7440–7449 (2009)

41. A. Lyakh, R. Maulini, A. Tsekoun, R. Go, C. Pflügl, L. Diehl, Q.J. Wang, F. Capasso, C. Kumar, N. Patel, 3 W continuous-wave room temperature single-facet emission from quantum cascade lasers based on nonresonant extraction design approach. Appl. Phys. Lett. **95**, 141113 (2009)

42. S. Bartalini, S. Borri, P. Cancio, A. Castrillo, I. Galli, G. Giusfredi, D. Mazzotti, L. Gianfrani, P. De Natale, Observing the intrinsic linewidth of a quantum-cascade laser: beyond the Schawlow-Townes limit. Phys. Rev. Lett. **104**, 083904 (2010)

43. S. Bartalini, S. Borri, I. Galli, G. Giusfredi, D. Mazzotti, T. Edamura, N. Akikusa, M. Yamanishi, P. De Natale, Measuring frequency noise and intrinsic linewidth of a room-temperature DFB quantum cascade laser. Opt. Express **19**, 17996–18003 (2011)

44. M.S. Vitiello, L. Consolino, S. Bartalini, A. Taschin, M. Inguscio, P. De Natale, Quantum-limited frequency fluctuations in a terahertz laser. Nat. Photonics **6**, 525–528 (2012)

45. S. Borri, S. Bartalini, P. Cancio, I. Galli, G. Giusfredi, D. Mazzotti, M. Yamanishi, P. De Natale, Frequency-noise dynamics of mid-infrared quantum cascade lasers. IEEE J. Quantum Electron. **47**, 984–988 (2011)

46. J. Ye, L.-S. Ma, J.L. Hall, Ultrasensitive detections in atomic and molecular physics: demonstration in molecular overtone spectroscopy. J. Opt. Soc. Am. B **15**, 6 (1998)

47. J.B. McManus, J.H. Shorter, D.D. Nelson, M.S. Zahniser, D.E. Glenn, R.M. McGovern, Pulsed quantum cascade laser instrument with compact design for rapid, high sensitivity measurements of trace gases in air. Appl. Phys. B **92**, 387 (2008)

48. J.E.J. Moyer et al., Design considerations in high-sensitivity off-axis integrated cavity output spectroscopy. Appl. Phys. B **92**, 467 (2008)

49. M. Sowa, M. Murtz, P. Hering, Mid-infrared laser spectroscopy for online analysis of exhaled CO. J. Breath Res. **4**, 047101 (2010)

50. G. Maisons, P. Gorrotxategi Carbajo, M. Carras, D. Romanini, Optical-feedback cavity-enhanced absorption spectroscopy with a quantum cascade laser. Opt. Lett. **35**, 3607 (2010)

51. D.D. Arslanov, S.M. Cristescu, F.J.M. Harren, Optical parametric oscillator based off-axis integrated cavity output spectroscopy for rapid chemical sensing. Opt. Lett. **35**, 3300 (2010)

52. B.H. Lee et al., Simultaneous measurements of atmospheric HONO and NO_2 via absorption spectroscopy using tunable mid-infrared continuous-wave quantum cascade lasers. Appl. Phys. B **102**, 417 (2011)

53. J.B. McManus, M.S. Zahniser, D.D. Nelson, Dual quantum cascade laser trace gas instrument with astigmatic Herriott cell at high pass number. Appl. Opt. **50**, A74 (2011)

54. I. Galli, S. Bartalini, P. Cancio, P. De Natale, D. Mazzotti, G. Giusfredi, M.E. Fedi, P.A. Mando, Optical detection of radiocarbon dioxide: first results and AMS intercomparison. Radiocarbon **55**(2–3), 213–223 (2013)

55. R.N. Zare, Analytical chemistry: ultrasensitive radiocarbon detection. Nature **482**, 312–313 (2012)

56. D. Mazzotti, S. Bartalini, S. Borri, P. Cancio, I. Galli, G. Giusfredi, P. De Natale, All-optical radiocarbon dating. Opt. Photonics News **23**(12), 52 (2012)

57. Ch. Daussy, T. Marrel, A. Amy-Klein, C.T. Nguyen, Ch.J. Bordé, Ch. Chardonnet, Limit on the parity nonconserving energy difference between the enantiomers of a chiral molecule by laser spectroscopy. Phys. Rev. Lett. **83**, 1554 (1999)

58. D. Mazzotti, P. Cancio, G. Giusfredi, M. Inguscio, P. De Natale, Search for exchange-antisymmetric states for spin-0 particles at the 10^{-11} level. Phys. Rev. Lett. **86**, 1919–1922 (2001)

59. A. Shelkovnikov, R.J. Butcher, C. Chardonnet, A. Amy-Klein, Stability of the proton-to-electron mass ratio. Phys. Rev. Lett. **100**, 150801 (2008)

Chapter 5
Cavity Enhanced Absorption Spectroscopy with Optical Feedback

Jérôme Morville, Daniele Romanini, and Erik Kerstel

Abstract In general, Cavity-Enhanced Absorption Spectroscopy (CEAS) suffers from inefficient and noisy cavity injection when using an unstabilized laser source with a linewidth exceeding that of the cavity resonances. The solution is to tightly frequency lock the laser to the cavity resonance, and drastically reduce its emission linewidth. This has been possible using either the Pound-Drever-Hall technique, giving the NICE-OHMS scheme, or Optical Feedback (OF). Here we review the OF-CEAS method that has the advantage of allowing simple self-locking of the laser to successive cavity modes during a frequency scan. OF-CEAS also produces a stronger linewidth reduction compared to an all-electronic locking loop, and works well with semiconductor lasers possessing a broad emission spectrum modulated by high-frequency phase noise that are difficult or impossible to lock using electronic control. The efficient laser linewidth narrowing and the locking onto successive $TEM_{0,0}$ cavity modes provides spectra with both a high S/N and an intrinsically precise and linear frequency scale over the short time of a single laser frequency scan.

5.1 Introduction

In this chapter we present absorption spectroscopy techniques that rely on self-locking of the laser to an optical cavity containing the sample under study. Following the success of Cavity Ring-Down Spectroscopy using CW lasers (CW-CRDS), the principle of laser self-locking to a cavity by providing spectrally filtered optical feedback (OF) was introduced to mitigate the problem of injecting the spectrally fluctuating emission of a diode laser into the narrow cavity resonances. The cavity-

J. Morville (✉)
Institut Lumière Matière, UMR 5306, Université Lyon 1, CNRS, Université de Lyon,
69622 Villeurbanne cedex, France
e-mail: jerome.morville@univ-lyon1.fr

D. Romanini · E. Kerstel
Laboratoire Interdisciplinaire de Physique, UMR 5588, Université J. Fourier (Grenoble I), CNRS,
38402 Saint-Martin d'Hères, France

G. Gagliardi, H.-P. Loock (eds.), *Cavity-Enhanced Spectroscopy and Sensing*,
Springer Series in Optical Sciences 179, DOI 10.1007/978-3-642-40003-2_5,
© Springer-Verlag Berlin Heidelberg 2014

enhanced spectrum obtained using OF can be recorded in either of two ways. The first involves a ring-down measurement, based on recording the light decay following the abrupt interruption of the cavity injection. This is then repeated for different laser wavelengths in order to construct the sample absorption spectrum. The second involves recording the light intensity at the cavity output during a laser frequency scan. In the first case the technique is generally referred to as Optical Feedback Cavity Ring-Down Spectroscopy (OF-CRDS), whereas the second technique corresponds to the narrow interpretation of the term Optical-Feedback Cavity-Enhanced Absorption Spectroscopy (OF-CEAS). The term OF-CEAS is sometimes (but not here) used in a wider sense, as to also include OF-CRDS (a similar remark was made concerning CRDS and CEAS in the introduction chapter of this book). While OF-CEAS was introduced almost a decade ago [1], after some developments involving OF-CRDS [2–7], previous experience with OF from an optical cavity to a diode laser existed with applications in atomic physics, rather than in molecular spectroscopy (see e.g. [8, 9]).

Although it may be usual for a review to begin with a historical overview, we have decided to first present a basic description of the experimental implementation of OF-CEAS (Sect. 5.2) in its currently most common configuration, before proceeding with the theoretical foundation of OF-CEAS (Sect. 5.3). This approach should facilitate the reading of more fundamental aspects of laser physics. The theoretical section, all along illustrated with experimental results, identifies the constraints posed by OF-CEAS on the laser source and the type of cavity, and allows the reader to appreciate the different experimental implementations of OF-CEAS described in Sect. 5.4. Section 5.5 presents technical aspect of the technique, with a close look at the calibration procedure that enables the measurement of absolute cavity losses. Typical OF-CEAS performance figures are presented in Sect. 5.6, and limitations and prospects are discussed. An historical overview is then given in Sect. 5.7. We conclude with an overview, as complete as possible, of OF-CRDS and OF-CEAS applications.

5.2 An Intuitive Picture of the Technique

Inefficient cavity injection is one of the major problems in CEAS/CRDS, due to the fact that the laser linewidth is usually much larger than the cavity mode width (typically in the kHz range). Concretely, we will consider a typical semiconductor (SC) laser with a Distributed Feed-Back grating in the laser structure (DFB diode laser or Quantum Cascade Laser, QCL) stabilizing its single mode emission, yielding a short term (ms) linewidth of around 1 MHz. This leads to an injection efficiency that is orders of magnitude lower than for an ideal monochromatic laser, and a very noisy cavity output signal when the laser goes through resonance with one of the cavity modes, for example when frequency-tuned by a ramp applied to its injection current (see Fig. 5.1). As discussed in the introduction chapter, the laser linewidth ultimately results from an instantaneous monochromatic wave emission suffering

Fig. 5.1 Typical cavity transmission signals when tuning a free-running (no optical feedback) DFB diode laser (linewidth $\Delta \nu_{las} \sim 1$ MHz) over a cavity mode (linewidth $\Delta \nu_{cav} \sim 10$ kHz), for two different scanning speeds η expressed in unit of $\Delta \nu_{cav}/\tau_{RD}$

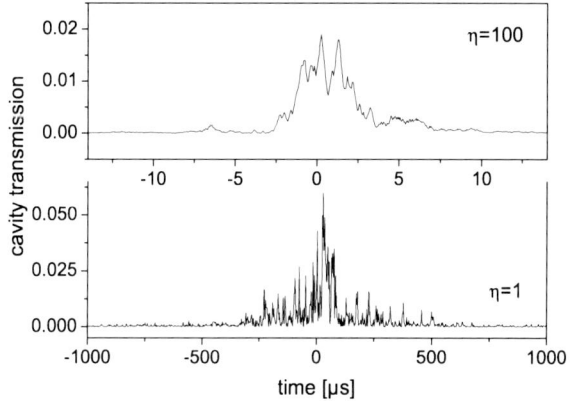

random phase fluctuations due to spontaneous emission, which is an unavoidable quantum effect intrinsic to laser emission. The result of these random phase jumps is that over a millisecond time scale, the laser emission frequency traces out a power spectral density with a characteristic Lorentzian profile. For a DFB diode laser or QCL, typical widths are measured in MHz. However, for a QCL this value may, in practice, not correspond to the intrinsic spontaneous emission effect (which in this case results in a linewidth as low as a few tens of kHz), but rather to a non-negligible injection current noise.[1]

OF-CEAS exploits the sensitivity of SC lasers to OF in order to optimize the injection of a high-finesse cavity by returning to the laser light that has been spectrally filtered by the optical cavity. Without loss of generality, we will consider the experimental scheme depicted in Fig. 5.2, where a V-shaped cavity made of three, usually identical, concave mirrors is used to produce OF at resonance (i.e., when the laser emission frequency coincides with a cavity mode). Other cavity schemes have been used for OF-CEAS and will be presented in detail in Sect. 5.4.2. The important point is that no light is able to return to the laser directly from the input cavity mirror. This is of course in contrast to the case of on-axis injection of a two-mirror cavity, in which case there is always feedback to the laser, whether on or off resonance (actually, the OF decreases as the cavity becomes transmissive). The cavity mirrors are labeled M_a for the folding mirror at the apex of the cavity, and M_1 and M_2 for the two end mirrors. The laser output is focused by a single (aspheric) lens to have its waist coincide with one of the two waists centered between the cavity

---

[1]The Schawlow-Townes limit [10] expresses the laser linewidth when only spontaneous emission takes place. Its value scales with the inverse of the laser cavity photon lifetime, itself proportional to the laser length. For semiconductor lasers, this limit is increased by a factor ranging from 2 to 10, due to a coupling between gain and phase in the laser medium [11, 12]. QCL cavities are typically ten times longer than those of diode lasers and the coupling is expected to be practically zero, resulting in a smaller intrinsic linewidth. However, compared to diode lasers, QCLs have a similar current tuning coefficient but typically operate at an injection current level one order of magnitude higher, thus require a comparably higher fractional current stability. For instance, a current noise on the order of 10^{-6} is needed to obtain a 1 MHz laser linewidth.

Fig. 5.2 Schematic of a
typical OF-CEAS
experimental setup

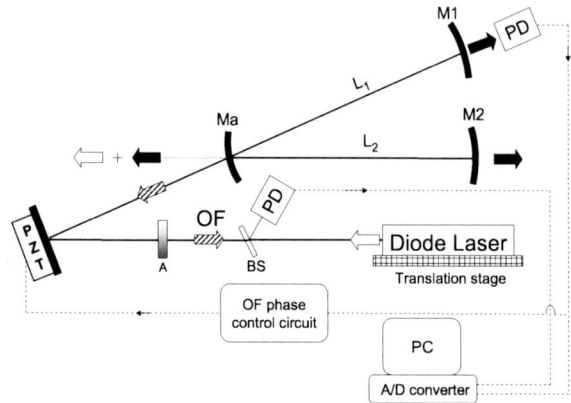

mirrors. A reference detector monitors the laser power, while a signal detector behind one of the cavity mirrors records the cavity output. As the laser is frequency
tuned and approaches a cavity resonance, its radiation can build up inside the cavity,
and a fraction of the intracavity field returns to the laser. An adjustable attenuator is
used to reduce the OF level experienced by the laser, while a piezo-mounted steering mirror allows precise adjustment of the laser-to-cavity distance, and therewith
the phase of the OF signal. If specific conditions (further discussed below) concerning the phase and intensity of the OF beam are respected, the SC laser will lock its
frequency to the cavity resonance and its emission linewidth will collapse to below
the resonance width. This spectral collapse is intuitively understood by considering
that the OF signal coming from inside the cavity is spectrally filtered by the cavity
resonance profile on each passage of photons traveling back and forth between the
laser and the cavity during the period that the OF locking is active and before a
steady state is reached.

The frequency locking mechanism that is thus established allows all the laser
power to be admitted inside the resonance width and at the same time increases
the residence time of the laser at resonance: As long as laser tuning is not too fast
a complete build-up of the intracavity field is attained. In particular, this injection
scheme eliminates the above mentioned amplitude noise that accompanies the passage through resonance of a free-running (optically isolated) laser. It is important to
understand that with OF-locking the laser frequency does not come to a full stop,
but its tuning speed is strongly reduced so that the laser remains much longer in resonance, as shown in Fig. 5.3 [1]. This figure will be discussed in detail further down.
For the moment we mention that the slowing down of the tuning is proportional to
the OF level, which can be adjusted by the attenuator placed between the laser and
the cavity. An ideal level of OF is reached when the slow-down factor is equal to the
cavity finesse, so that tuning across a resonance takes as long as it would take the
free-running laser to tune over a full cavity free spectral range (FSR). We will refer
to the spectral width of the locking region as "locking range". In order to obtain
the "true" cavity mode profile at this reduced speed, an adiabatic condition must
be respected: In practice one finds that in order to obtain "adiabatic" transmission

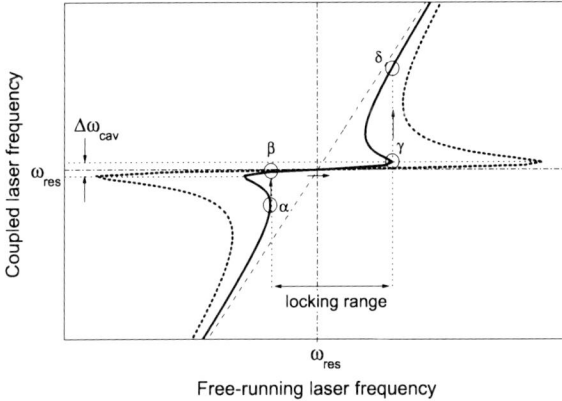

Fig. 5.3 Coupled laser frequency tuning in the vicinity of a cavity mode resonance for two levels of optical feedback (*solid* and *dot lines* for weak and strong feedback rates, respectively) as a function of the free laser frequency (proportional to injection current). Arrows indicate the path ($\alpha \to \beta \to \gamma \to \delta$) followed by the coupled frequency when the free laser frequency is tuned (by an injection current ramp), and show the slowing down of the coupled laser tuning close to a resonance. The feedback phase must be optimized to obtain this behavior

profiles the time for tuning across the resonance (with frequency locking) should be some 50 to 100 times longer than the cavity response time. Faster tuning induces visible distortion and fluctuation of these profiles (after having accounted for the different time scale associated with slower locked-frequency tuning). For a cavity with a 20 μs ring-down, the tuning speed should then be of the order of one FSR per ms considering the optimal locking range of 1 FSR. With a 50 cm V-shaped cavity the FSR is 150 MHz, such that it should take approximately 200 ms to cover the 200 FSRs that correspond to 1 cm^{-1}, the typical current scan range of a SC laser.

Another important requirement is that the OF wave returning to the laser arrives in phase with the laser field, in order to obtain constructive interference in the laser cavity. Normally, this would require that the distance between laser and cavity be adjusted continuously while the laser scans over successive cavity resonances. It is, however, possible to choose a laser-to-cavity distance such that the OF phase remains the same for all these resonances. Indeed each cavity standing wave has nodes on the two cavity output mirrors. This standing wave is composed of a travelling wave propagating back and forth between these end mirrors, changing sign upon reflection, which produces the nodal surfaces coinciding with the mirrors surfaces. It is easy to see that the fraction of this wave propagating out of the cavity through M_a and in the direction of the laser, will then have a fixed phase relationship with the laser field if the laser-cavity distance is chosen to be equal to the length of the other cavity arm relative to the arm to which the laser beam is aligned. A more rigorous insight into this point is provided later. Once the above condition is realized, a piezo-mounted mirror between laser and cavity enables the adjustment of the phase to obtain optimal injection of all resonances and requires only sub-wavelength adjustments during the laser scan, mainly to compensate for thermal drift and vibrations. A control signal for the piezo-mounted mirror is obtained by a real-time (analogue or digital [13]) analysis of the mode profiles in transmission, and the goal is to stabilize symmetric-looking profiles.

In the most practical and common situation considered here, with the laser-cavity distance matching the length of the cavity arms, OF-CEAS produces spectra sampled at all the longitudinal cavity modes inside the laser scan (with frequency interval given by the cavity FSR). A minor disadvantage of this scheme is that cavity modes for which the round trip cavity length is an even or an odd multiple of the wavelength (from here, named even and odd cavity modes), experience a slightly different reflection coefficient at the folding mirror [1]. This results in a slight and constant offset between the spectra registered by the odd and even modes that, however, is easily included in the data analysis procedure.

With minimal spatial mode matching, this scheme enables the successive injection of all longitudinal cavity modes that are within the laser frequency tuning range. Indeed, a welcome advantage of OF-CEAS is that it suffices to align the laser beam onto the optical axis of one of the cavity arms, and focus it to a point somewhere inside the cavity, to obtain OF predominantly from the $TEM_{0,0}$ mode, which then produces OF locking preferentially. One factor helping in this automatic mode matched injection, is that the locking range can be adjusted to be close to the cavity FSR. Then, higher-order transverse modes do not get a chance to be injected, since the laser jumps from one to the next longitudinal cavity mode rather than scanning continuously the whole cavity FSR. The above is illustrated in Fig. 5.4: In (a) the laser does not receive OF from the cavity (an optical isolator is placed between laser and cavity). Linear tuning of the laser frequency results in a periodic re-occurrence of resonances over one cavity FSR with the cavity mode structure that appears very noisy. The average intensity transmitted by the cavity is also orders of magnitude lower than in the case of Fig. 5.4(b), in which the laser receives in-phase OF, forcing it to lock to successive longitudinal modes of the cavity, and to simultaneously narrow its emission bandwidth to below that of the cavity resonance. As is shown, higher order transverse modes do not become excited in this case. We will see that there are other factors that also favor this "automatic mode matching".

An OF-CEAS spectrum is finally obtained by considering maxima of the successive $TEM_{0,0}$ cavity mode profiles, determined by a software routine that takes as input the ratio of the cavity output signal over the reference signal. Since the signal is determined at the cavity mode profile maxima, instead of by integration of the entire profile, the cavity enhancement factor equals $2\mathcal{F}/\pi$ (as explained in the introduction chapter to this book). The $TEM_{0,0}$ mode maxima are then determined with a typical S/N in excess of 1000, without any need for averaging over several frequency scans. For a 20 μs ring-down time as assumed above, this S/N corresponds to a noise equivalent absorption of about 10^{-9}/cm obtained over 200 spectral points from the acquisition of a single 200 ms scan, which in other CEAS schemes may be attained only after seconds to minutes averaging.

Thus, OF efficiently solves the problem of light injection into a high-finesse optical cavity. The alternative would be to force the laser frequency to match that of a cavity mode by an electronic feedback loop acting on the laser injection current,

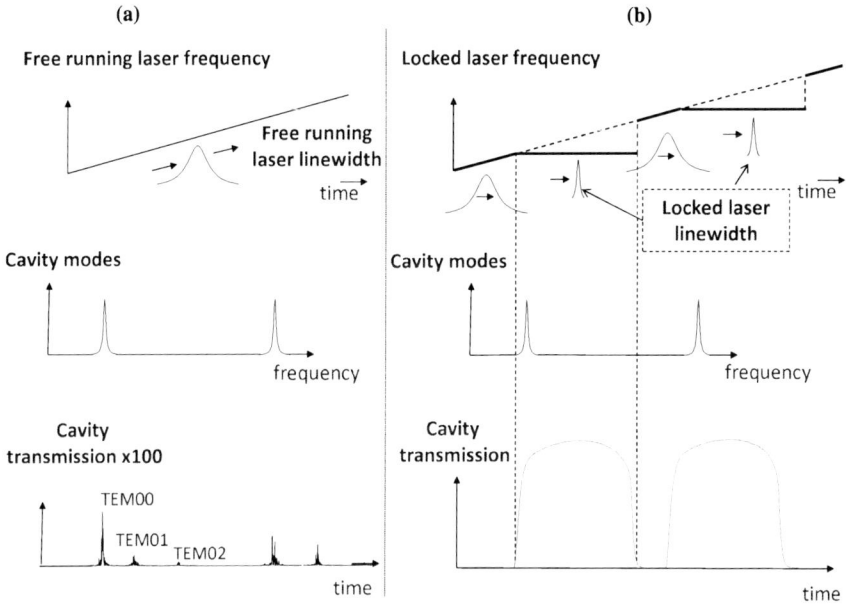

Fig. 5.4 Injection of a high-finesse optical cavity without (**a**) and with (**b**) optical feedback. For simplicity only the longitudinal cavity modes are schematically shown in the "cavity modes" central panes. *On the right*, the feedback rate is adjusted to have a locking range equal to a cavity FSR, which does not allow higher-order transverse mode to get injected

which is the only frequency control parameter in the system with a sufficiently high bandwidth. Still, in order to squeeze the emission of a SC laser and keep it inside a narrow cavity resonance, a servo loop with high bandwidth would be required, given the wide noise spectrum extending up to GHz frequencies for some of these lasers [14]. Even the most advanced of the electronic locking schemes, such as the one introduced by Pound, Drever and Hall [15]), have a hard time to narrow a diode laser linewidth to much below 10 kHz [16], whereas the highest finesse cavities exhibit modes with a width of the order of a few hundred Hz. As recently demonstrated [17], laser linewidth narrowing using the PDH scheme is more effective with a QCL than with a diode laser. Even then, it is necessary to force the locked laser to quickly jump between successive cavity modes and to subsequently re-establish the electronic lock.

On the other hand, even though OF-CEAS also requires an active control for the optical feedback phase, a bandwidth of 100 Hz is sufficient [18]. As will become clear later, reasonably small residual fluctuations in the OF-phase do affect the shape, and in particular the symmetry of the cavity transmission profiles. However, the maxima of these profiles continue to coincide with those of the true cavity resonances, such that the CEAS spectrum remains unperturbed.

As a last important aspect we mention that OF-CEAS provides an intrinsically precise frequency scale by sampling a molecular absorption spectrum on the uniform frequency grid of the cavity modes. At the timescale of spectra acquisition ($\sim$100 ms), the cavity frequency comb stability is mainly affected by high-frequency acoustic and mechanical vibrations. Using a stiff cavity construction this jitter of the cavity mode frequencies can be reduced to sub-MHz levels (corresponding to fluctuations of a few nm for a 50 cm cavity length at $\lambda = 1$ μm). Thus, while laser linewidth narrowing provides low amplitude noise at the cavity output, frequency locking to successive cavity modes comes along with a low level of frequency noise, and both aspects contribute to a high quality of the absorption spectra.

5.3 Theoretical Foundations of OF-CEAS

In this section, we give the theoretical ingredients enabling the understanding of the particular cavity transmission patterns induced by OF. A rigorous treatment should take into account the dynamic behavior of the coupled system composed of the SC laser and the external optical cavity, considering that a current ramp is applied to the SC laser injection current to induce frequency tuning. This interaction is rather complex, as it is described by differential equations including an integral term due to the cavity memory of past field evolution. However, a detailed understanding can still be obtained by considering only stationary solutions of the coupled system. In this limit, the reasonable requirement is that the injection current variations (or, equivalently, the free-laser frequency variations) should belong to the adiabatic regime. The tuning speed should then stay below $\Delta \nu_{cav}/\tau_{RD}$, such that the time of passage over a cavity resonance (FWHM $\Delta \nu_{cav}$) is longer than the cavity response time (the ring-down time τ_{RD}). This would, in turn, impose very long acquisition times. Experimental observations show that cavity transmission patterns obtained with frequency scanning speeds several orders of magnitude higher than this theoretical limit still agree perfectly with patterns calculated from the stationary model, in the adiabatic limit. In Sect. 5.3.6 we will give a simple interpretation of this welcome behavior.

5.3.1 Optical Feedback: Model Equations

In order to interpret specific effects induced by the resonant OF, it is useful to identify the coupled system as a single laser with an output coupler formed by the combination of the laser chip facet and the cavity response function in reflection (see Fig. 5.5). The cavity system has an amplitude and phase defined by the cavity OF transfer function $h_{OF}(\omega)$. Therewith, when the free running laser frequency ω_{free} is far from a cavity resonance the effective output coupler reflectivity is given by the facet only, since OF from the cavity is absent. On the other hand, close to a resonance the effective amplitude reflectivity increases if the OF phase is "constructive". Consequently, the losses of the effective laser decrease, which favors emission at the cavity resonant frequency. This is the effect of frequency locking induced by the res-

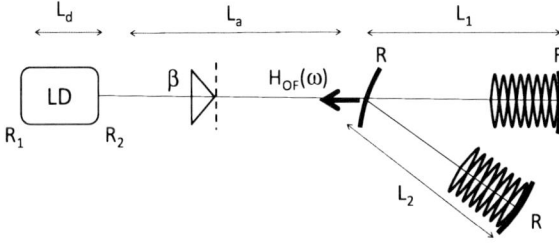

Fig. 5.5 Schematic of the coupled system for a V-shaped cavity. All quantities are expressed in relation to the laser intensity. The intensity transfer function for the feedback from the cavity is thus given by $H_{OF}(\omega) = |h_{OF}(\omega)|^2$ where $h_{OF}(\omega)$ is the field (amplitude) transfer function

onant feedback. However, the locking depends on how the reflected wave interferes in the gain medium with the outgoing wave, and therefore the sum of the phase of $h_{OF}(\omega)$ and the phase accumulated throughout the propagation between the laser and the cavity will also determine how the effective laser losses are affected.

The stationary solutions of the coupled system could be obtained by writing the lasing condition of the coupled laser. However, it is preferable to start with the isolated LD. The field after one LD cavity round trip should be preserved, which can be written in the following manner:

$$\sqrt{\mathcal{R}_1 \mathcal{R}_2} \exp(-ik2L_d) = 1, \tag{5.1}$$

where $\mathcal{R}_{1,2}$ are the diode facet reflection coefficients supposed equal in the following, L_d its length and k the complex wave vector:

$$k = n_{\text{free}} \frac{\omega_{\text{free}}}{c} + i\frac{1}{2}(g_{\text{free}} - a), \tag{5.2}$$

where ω_{free}, g_{free} and n_{free} are the laser stationary frequency, gain, and refractive index without optical feedback, while a stands for the distributed losses of the gain medium.

The OF from the cavity modifies the output facet reflectivity of the diode and a new lasing condition is obtained as:

$$\sqrt{\mathcal{R}_1 \mathcal{R}_{2\text{eff}}(\omega_L)} \exp\left(-i\frac{\omega_L}{c} 2n_L L_d\right) \exp\big((g_L - a)L_d\big) = 1, \tag{5.3}$$

where the frequency ω_L, the gain g_L and the refractive index n_L are the stationary values of the coupled (locked) laser and $\mathcal{R}_{2\text{eff}}(\omega_L)$ is the effective reflection coefficient. In order to work in the low coupling regime, an optical system is placed between the laser and the cavity to attenuate the intensity fraction returning to the laser, by a factor $\beta < 1$. The effective facet reflectivity has thus the following expression:

$$\sqrt{\mathcal{R}_{2\text{eff}}} = \sqrt{\mathcal{R}_2}\left(1 - \sqrt{\beta}\frac{1 - \mathcal{R}_2}{\sqrt{\mathcal{R}_2}} h_{OF}(\omega) \exp\left(-i\frac{\omega}{c} 2L_a\right)\right), \tag{5.4}$$

corresponding to the sum of the field reflected by the facet and the field coming from the cavity with the complex amplitude $h_{OF}(\omega)$ attenuated by $\sqrt{\beta}$, phase shifted by propagation over the length L_a from the laser to the cavity input mirror and finally transmitted by the facet.

Inserting Eq. (5.4) in the lasing condition Eq. (5.3), and developing the gain to first order, $g_L = g_{free} + \Delta g$, one obtains the product $(1 - \epsilon_1)(1 - \epsilon_2)(1 - \epsilon_3) = 1$ where:

$$\epsilon_1 = \sqrt{\beta}\frac{1 - \mathcal{R}_2}{\sqrt{\mathcal{R}_2}}h_{OF}(\omega_L)\exp\left(-i\frac{\omega_L}{c}2L_a\right)$$

$$\epsilon_2 = (n_L\omega_L - n_{free}\omega_{free})\frac{2L_d}{c}$$

$$\epsilon_3 = \Delta g L_d$$

are all small quantities in case of weak OF rates. This justifies keeping only the first order of the product. Expressing the refractive index as $n_L = n_{free} + \Delta n$ and introducing the Henry factor [19], which expresses the dependence of the SC refractive index variation with the gain variation[2], $\alpha_H = -2\frac{\omega}{c}\frac{\partial n}{\partial g}$, then separating the real and imaginary parts, one obtains general expressions for the stationary shift of the laser gain and of the lasing frequency in the presence of weak OF:

$$\Delta g = -\sqrt{\beta}\frac{n_{free}}{c\tau_d}\left[P(\omega_L)\cos\left(\frac{\omega_L}{c}2L_a\right) + Q(\omega_L)\sin\left(\frac{\omega_L}{c}2L_a\right)\right]$$

$$\omega_L - \omega_{free} = -\frac{\sqrt{\beta(1 + \alpha_H^2)}}{2\tau_d} \tag{5.5}$$

$$\times\left[P(\omega_L)\sin\left(\frac{\omega_L}{c}2L_a + \theta\right) - Q(\omega_L)\cos\left(\frac{\omega_L}{c}2L_a + \theta\right)\right],$$

where $\theta = \arctan(\alpha_H)$, and $\tau_d = (n_{free}L_d/c)\sqrt{\mathcal{R}_2}/(1 - \mathcal{R}_2)$ is the photon lifetime of the isolated laser diode. $P(\omega)$ and $Q(\omega)$ are respectively the real and imaginary part of the field transfer function $h_{OF}(\omega)$. Expressions Eq. (5.5) formalize what has been mentioned above: in addition to the amplitude response associated with a resonance, the role of the phase associated to the dispersive response is also crucial. It should also be noted that the role of the phase acquired over the cavity-to-laser distance is expressed by the cosine and sine terms in these equations.

All things being equal, considering the factors in front of the square brackets in Eq. (5.5), we see that the gain change and frequency shift (thus the locking range) both increase with β and decrease with the isolated laser photon lifetime τ_d. On the other hand, the Henry factor α_H has a positive impact only on the frequency shift, where it also induces a phase shift θ relative to the gain response as a function of the laser to cavity distance L_a.

The expressions given by Eq. (5.5) are general for any kind of feedback generated by an optical system with a linear response, allowing for the use of the transfer function in reflection $h_{OF}(\omega)$. In particular, they can be used with a simple reflector (mirror), characterized by its amplitude reflection coefficient $\mathcal{R}$, in which case

---

[2]The Henry factor and its physical origin are discussed in Sect. 5.4.1.

$h_{OF}(\omega) = \sqrt{\mathcal{R}}$. In the case of a V-shaped cavity with equal mirror coefficients $\mathcal{R}$, the expression of the field transfer function is given by:

$$h_{OF}(\omega) = \frac{(1 - \mathcal{R})\sqrt{\mathcal{R}}\exp(-i\frac{\omega}{c}2L_1)}{1 - \mathcal{R}^2 \exp(-i\frac{\omega}{c}2(L_1 + L_2))}, \tag{5.6}$$

where $L_{1,2}$ are the lengths of the two arms of the V-cavity, L_1 being the arm along which the laser beam is aligned. The exponential phase factor in the numerator of $h_{OF}(\omega)$ can be grouped with the exponential phase factor of ϵ_1, by defining a new transfer function taking its origin after one round trip in the arm L_1:

$$h'_{OF}(\omega) = \frac{(1 - \mathcal{R})\sqrt{\mathcal{R}}}{1 - \mathcal{R}^2 \exp(-i\frac{\omega}{c}2(L_1 + L_2))}, \tag{5.7}$$

which allows to write the gain and frequency shift as modified expressions:

$$\Delta g = -\sqrt{\beta}\frac{n_{free}}{c\tau_d}\left[P'(\omega_L)\cos\left(\frac{\omega_L}{c}2(L_a + L_1)\right) \right.$$
$$\left. + Q'(\omega_L)\sin\left(\frac{\omega_L}{c}2(L_a + L_1)\right) \right], \tag{5.8}$$

$$\omega_L - \omega_{free} = -\frac{\sqrt{\beta(1 + \alpha_H^2)}}{2\tau_d}\left[P'(\omega_L), \sin\left(\frac{\omega_L}{c}2(L_a + L_1) + \theta\right) \right.$$
$$\left. - Q'(\omega_L)\cos\left(\frac{\omega_L}{c}2(L_a + L_1) + \theta\right) \right], \tag{5.9}$$

where $h'_{OF}(\omega) = P'(\omega) + iQ'(\omega)$ for the modified transfer function. This expression clearly shows that the effect of the phase acquired during propagation should take into account the first round trip in the cavity arm L_1. Also, when the laser frequency is outside a resonance, P' and Q' are both close to zero and the laser frequency equals the free laser frequency, irrespective of the values of L_1 and L_a.

As explained before, cavity modes are standing-waves with nodes on mirrors M_1 and M_2. The same phase condition for all cavity modes is obtained when the OF wave that propagates out of M_a towards the laser presents an integer number of cycles counting from the node at M_1 all the way to the laser output facet. This condition is met if the laser-to-cavity distance L_a matches a length $L_2 + k(L_1 + L_2)$ where k is an integer ≥ 0. Indeed, the arguments of the sine and cosine functions in Eq. (5.9) will then become synchronized to the argument of the exponential in the denominator of the Airy function in Eq. (5.7), which generates the comb of cavity resonances.

Before discussing the consequence of Eq. (5.9), it is convenient to introduce here the "feedback rate" of the coupled system. This is commonly defined as the optical power returning back to the laser and coupled to the laser active region divided by the emitted power. It is proportional to the attenuation factor β previously defined, squared because of the double pass, and to the square of the spatial coupling coefficient for the intensity ϵ_{00}, which is itself the square modulus of the superposition

integral of the laser beam field with the $TEM_{0,0}$ cavity field [20]. The square originates from the forward coupling of the laser beam to the cavity mode followed by
the backward coupling of the cavity OF beam to the laser mode, which give the same
superposition integral. Lastly, the feedback rate must include the square modulus of
the maximum of the field transfer function, $H_{OF,max} = |h_{OF,max}|^2$, which depends
on the cavity geometry and its losses and sets the maximum optical power that can
return back to the laser. These three independent factors define the feedback rate as
$\beta' = \epsilon_{00}^2 H_{OF,max}\beta$ which is in practice adjustable by acting on the attenuation factor
β. It should be noted that accurate measurement of all these factors is not trivial.

5.3.2 The Locked Frequency Behavior and the Resulting Cavity Transmission Pattern

To aid the discussion of the laser behavior as it scans through a resonance at frequency ω_{res}, we plot in Fig. 5.6(a) the real and imaginary parts of the transfer function h'_{OF} for a lossless V-shaped cavity.[3] In addition we represent in Fig. 5.6(b) the
stationary solution Eq. (5.9) as a plot of the coupled laser frequency ω_L against the
free laser frequency ω_{free}.[4] Since ω_{free} is proportional (almost linearly) to the SC
laser injection current, this is the free driving parameter when considering the evolution of the system and of the locked laser frequency ω_L. If we consider applying
an increasing linear current ramp, the horizontal axis is then equivalent to a time
axis.

To start with, we have chosen L_a satisfying the condition $\omega_{res}2(L_a + L_1)/c =
-\theta$, which makes the sine term in Eq. (5.9) to vanish. It is then the imaginary part of
h'_{OF} that determines the behavior of the coupled frequency, as shown in Fig. 5.6(b).
Its dispersion-like shape generates regions where three mathematical solutions coex-
ist for a given ω_{free}. With reference to this figure, if we consider an increasing ω_{free},
then ω_L evolves first along a positive slope section before entering a region where
two other solutions are possible. Still, the system cannot jump to any of these other
solutions, since that would correspond to a spontaneous change of the intracavity
field frequency: A long-lived intracavity field (even if in the wing of the cavity resonance) tends to maintain a given frequency in the coupled system. Frequency jumps
can only occur at the turning points, where the system passes from three solutions
back to only one available solution: At these turning points, if the driving parameter
ω_{free} moves towards the point where 2 solutions collapse, then the system frequency
ω_L may perform a forced jump if it was previously evolving according to one of the
disappearing solutions. Inspection of the figure and considering also the case of a
decreasing ω_L ramp, will confirm that actually only the positive slope regions are

---

[3]In this case, it is shown that the maximum cavity transmission $H_{OF,max}$ is equal to 0.25 which
corresponds well to the sum $P'^2 + Q'^2$ at exact resonance.

[4]In practice it is trivial to plot ω_{free} as a function of ω_L and exchange axes in the plot.

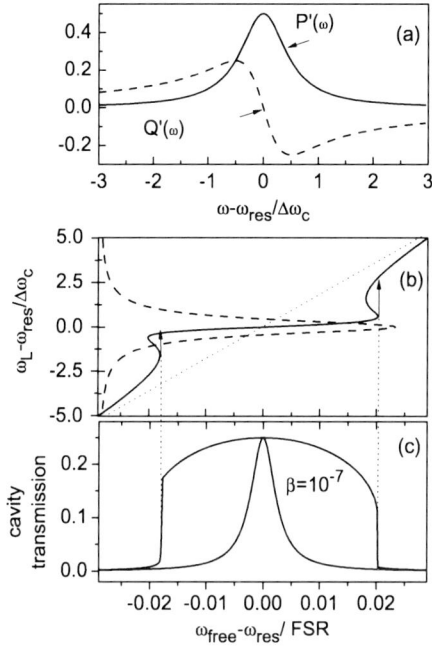

Fig. 5.6 (a) Real and imaginary part, $P'(\omega)$ and $Q'(\omega)$, of the feedback transfer function $h'_{OF}(\omega)$ for a lossless V-shaped cavity around a cavity resonance ω_{res} of linewidth $\Delta\omega_c$. (b) Stationary solutions of the coupled laser frequency as a function of the free-running laser frequency ω_{free} (expressed in units of FSR, equal to 150 MHz for $L_1 = L_2 = 50$ cm in our simulation). For this we use Eq. (5.9) with only the dispersive part $Q'(\omega)$ given the choice of L_a such that $2\omega_{res}(L_a + L_1)/c = -\theta$. Simulation is for a typical near Infra-Red (IR) DFB diode laser ($\mathcal{R}_2 = 0.25$ and $L_d = 500$ µm) with a multiple-quantum-wells gain medium ($n_{free} = 3.5$ and $\alpha = 2$), and for an attenuation factor $\beta = 10^{-7}$ resulting in a weak optical feedback rate $\beta' = 2.5 \times 10^{-8}$. In order to appreciate the details of the coupled frequency behavior, a moderate cavity finesse of 200 is used. *Arrows* indicate laser frequency jumps when ω_{free} is tuned towards higher values. In order to locate the coupled frequency relative to the resonance, the resonance profile is also shown (*dashed line*), but for convenience referenced to the coupled laser frequency on the *vertical axis*. (c) Corresponding modified cavity transmission figure, compared to the normal transmission profile in the absence of feedback (Lorentzian-like narrower profile)

accessible, while the negative slope regions are unreachable and represent unphysical solutions. It is also easy to see that the regions with three solutions generate a hysteresis effect, with a different ω_L evolution depending on whether ω_{free} is tuned towards lower or higher frequencies. This also produces an asymmetric cavity transmission mode profile as visible in Fig. 5.6(c) for the case of a positive going ramp, since the locked laser frequency jumps directly close to the center of the resonance profile, which is also plotted (dashed line) as a function of ω_L in (b) and is thus referenced to the vertical axis.

As a consequence, in response to a large free laser frequency excursion, the locked laser frequency scans only a small spectral range lying inside the cavity mode linewidth. Then, the laser frequency exits the resonance with a sudden jump to a value equal to the free running frequency (the 1:1 dotted reference line in the figure). Neglecting for the moment the OF effect on the laser output power, the cavity transmission plotted in Fig. 5.6(c) will reflect this laser frequency evolution. In particular, the strong reduction of the scanning speed corresponds to an apparent broadening the transmission profile of the mode. It should also be mentioned that, given the hysteresis, a scan in the opposite direction will not produce the same profile.

These simulations are based on typical DFB multiple-quantum-wells near-IR laser parameters reported in the caption of Fig. 5.6. For clarity, the coupled laser frequency as a function of the free-laser frequency following (Eq. (5.9)) is computed with a moderate cavity finesse ($\mathcal{F} \sim 200$). As shown by the behavior of the cavity transmission, in this case a feedback rate as small as 2.5×10^{-8} is already sufficient to broaden the transmission peak and give a locking range of more than 4 % of the FSR. As Eq. (5.9) shows, using a cavity with a much higher finesse ($\mathcal{F} \sim 20000$) does not change the locking range, whereas increasing the feedback rate to around 2.5×10^{-5} (keeping all other parameters fixed), will increase the locking range to roughly one FSR, i.e. ~ 150 MHz for a 50 cm base-length V-shaped cavity.

5.3.3 Optical Feedback Phase and the Cavity Transmission Beating Patterns

In the previous paragraph, the value of the propagation term $\omega_{res} 2(L_a + L_1)/c$ of the OF phase was adjusted in order to null the real part of the field transfer function for a given cavity resonance. As we have seen, this is a favorable case, as the transmitted mode profile in time will appear as a horizontal zoom-in around the top of that cavity resonance spectral response. But, in general, if this is the case for a given mode, adjacent modes may have different phases resulting in different frequency behaviors and different transmission profiles.

The propagation phase associated with the n-th cavity mode with resonant frequency $\omega_n = n\pi c/(L_1 + L_2)$ is given by $\phi_n = n2\pi(L_1 + L_a)/(L_1 + L_2)$ which implies a fixed phase increment of $2\pi(L_1 + L_a)/(L_1 + L_2)$ between successive resonances. Considering a sequence of cavity modes starting from n, there exists a mode m such that $\phi_m - \phi_n$ becomes larger than or equal to 2π. The cavity transmission over successive cavity modes thus exhibits a quasi-periodic beating structure imposed by the value of L_a (and strictly periodic if $(\phi_m - \phi_n)[2\pi] = 0$). If $L_a = (q/N)(L_1 + L_2) - L_1$, where q/N is an irreducible ratio of integer numbers, the transmission profiles of a succession of modes will present a beating structure with period $N = m - n$ (expressed in number of modes). In Fig. 5.7 we plot a measured pattern of cavity modes where eight beating periods are scanned. A zoom over one period shows the different mode profiles obtained for successive OF phase conditions. A simulated profile of the same region obtained with the same parameters given previously is also shown. The OF phase condition of the central mode corre-

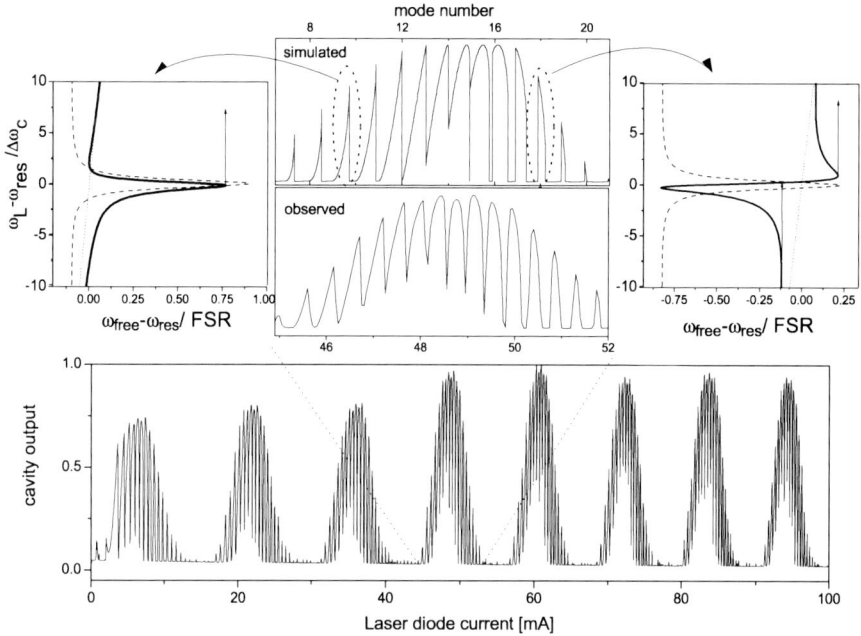

Fig. 5.7 *Bottom*: Cavity output signal recorded by applying a linear current ramp to a near-IR DFB diode laser. The laser-to-cavity distance L_a is such that the OF phase changes when successive cavity modes are excited. *Top*: The central figure is a zoom of a single period of the beating pattern, compared to simulated profiles. The whole sequence of mode transmission profiles is well reproduced by a simulation based on the adiabatic model with the same parameters as in Fig. 5.6, except for a higher feedback rate $\beta' = 2.5 \times 10^{-5}$. Side panels show the frequency behavior of the numerical solution of Eq. (5.9) for two modes for which the OF phase is not optimal. As in Fig. 5.6, the resonance profile is also shown (*dashed line*) as a function of the coupled laser frequency on the *vertical axis*

sponds roughly to the previously determined optimal condition $\phi_n = -\theta$, where the locked frequency has a symmetric trajectory around the exact resonance, determined by the imaginary part of the transfer function (plotted in Fig. 5.6(a)). The transmission profile in the presence of OF is, however, not symmetric, given the hysteresis effect shown in Figs. 5.6(b) and (c). Away from this central mode, the stationary solution of the laser frequency is determined by both the real and imaginary parts of the transfer function, and the peaked real part pushes the locked frequency trajectory away from the center of the resonance, in a direction determined by the relative sign of the sine and cosine terms in Eq. (5.9). Sufficiently far away from the center (see inset of Fig. 5.7), the resonance frequency does not even fall inside the small spectral range covered by the locked frequency, the transmission maxima are reduced in amplitude, and the transmission profiles become peaked and strongly asymmetric. For phase conditions where the real part of the transfer function dominates in Eq. (5.9), the laser frequency is then pushed away from the resonance, producing a repulsive OF and resulting in zero cavity transmission.

Fig. 5.8 Measured cavity transmission as a function of time during a linear frequency scan of an ECDL in the Littmann configuration. Here L_a is adjusted to have the same optimal OF phase for all modes. The mode intensity variations correspond to an absorption line by the sample gas present inside the cavity. Effects of OF are also visible on a simultaneous recording of the laser power and of a low-finesse solid-etalon trace, resulting here in a step-sine behavior

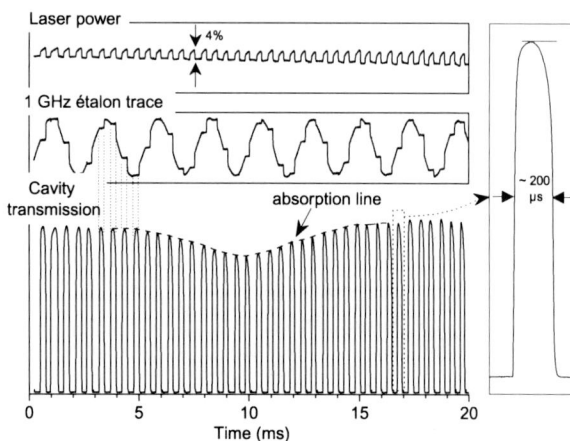

5.3.4 The OF-CEAS Operating Conditions

In order to obtain OF-CEAS spectra, L_a must then be adjusted to have the same OF phase for all successive cavity resonances in the laser scan, for instance by choosing $L_a = L_2$. To this end the laser-to-cavity distance is first adjusted roughly, e.g. by a delay line, until the transmission profiles for all resonances appear to have the same shape. The precision of this adjustment must be such that the phase error from the first to the last resonance in the scan is small enough to induce an almost undetectable change in the corresponding transmission profiles. For a typical OF-CEAS setup with a 1 cm^{-1} laser scan and a 50 cm cavity, the precision of this adjustment is of the order of 100 μm. Subsequent, sub-wavelength fine-adjustment of the optical path length, e.g. using a piezo-actuated translating mirror in the optical path between laser and cavity, then insures that the OF phase remains at the value that yields symmetric mode profiles for all resonances. An example of cavity transmission measured under these conditions is given in Fig. 5.8. The amplitudes of the transmitted peaks clearly correspond to what could be obtained in the ideal case of a monochromatic wave injecting a high finesse cavity as described in the introduction of this book. In addition, the locking onto each resonance results in wide and symmetric top sections of the transmission profiles, as described in the previous section, which allows an easy determination of the maximum of each peak. In particular, absorption lines are directly observable by their effect on the mode amplitudes.

Also clearly visible in Fig. 5.8 are the variations of the laser power induced by the OF, as expected from the gain modification in Eq. (5.9). This effect is particularly marked here given the use of an External-Cavity Diode Laser (ECDL). Due to this effect, the cavity output should be normalized by the incident power in order to obtain a correct estimate of the cavity transmission (the transfer function) and hence quantitative absorption measurements. It should be underlined that for modes falling inside an absorption line, the OF level decreases together with the cavity output, thus the effect on the laser power also decreases. For this reason, it is generally not possible to approximate the normalization function by a low-order polynomial.

We have derived the model equations accounting for the behavior of the stationary laser frequency and the peculiar features of cavity transmission patterns (asymmetric profiles and OF phase beatings). Both Fig. 5.7 and Fig. 5.8 include experimental observations. Clearly, a good agreement is obtained with the model equations despite the fact that the laser linewidth and the relatively high frequency scanning speed are, as remarked earlier (see Sect. 5.3), outside the validity range of the model. However, the observed agreement justifies *a posteriori* the validity of the adiabatic limit, and the low noise level on the profiles recorded at cavity output is an empirical confirmation that the laser spectrum is narrowing down well below the resonance width.

5.3.5 Locked-Laser Linewidth and Optical Feedback Phase Tolerance

A rigorous frequency noise analysis of the coupled laser is quite complex and analytical expressions are derived only for particular values of the OF phase. The interested reader will find all derivations in [21]. Notably, it is shown that for the in-phase optimal condition $\phi_n = -\theta$, the flat spectral density of the laser frequency noise, obtained when only spontaneous emission is considered, is strongly reduced over the cavity bandwidth. More specifically, the reduction factor inside the cavity bandwidth is given by the square of the reduced slope $p = d\omega_L/d\omega_{\text{free}}$. This derivative is obtained using expression Eq. (5.9). It is then shown that the laser linewidth is reduced by the same factor such that the locked laser linewidth is given by:

$$\Delta \nu_{\text{locked}} = \frac{\Delta \nu_{\text{free}}}{\beta'(1 + \alpha_H^2)} \left(\frac{\tau_d}{\tau_{\text{RD}}} \right)^2, \tag{5.10}$$

for a high finesse cavity. This actually holds only with white noise sources (flat noise spectral density), such as spontaneous emission. However, for a more realistic situation where $1/f$ noise is present, it is shown that the square-dependence reduces to a linear dependence on τ_d/τ_{RD}, and the locked laser linewidth is given by:

$$[\Delta \nu_{\text{locked}}]_{1/f} = \frac{\Delta \nu_{\text{free}}}{\sqrt{\beta'(1 + \alpha_H^2)}} \frac{\tau_d}{\tau_{\text{RD}}}, \tag{5.11}$$

where $\Delta \nu_{\text{free}}$ is the free laser linewidth. The linewidth reduction factor in this last case is equal to the slope reduction factor defining the locked-laser tuning speed relative to the free tuning speed: This also acts to reduce all spectral fluctuations, even of technical origin, e.g. those induced by injection current noise. Applying the parameter values used in Fig. 5.6 ($\tau_d \sim 3$ ps, $\alpha_H = 2$) to Eq. (5.11), but with feedback rate $\beta = 10^{-4}$, a reduction of five orders of magnitude is obtained with a cavity possessing a ring-down time of 10 μs. Starting with a 1 MHz linewidth (typical for a DFB diode laser), a 10 Hz locked laser linewidth should result. Of course, this linewidth is relative to the cavity resonance, which has a width of 15 kHz and should also be stabilized, actively or passively.

In the end, the relevant point is that the locked laser linewidth is easily orders of magnitude below the cavity resonance linewidth. Moreover, cavity transmission patterns measured when the laser-to-cavity distance is such that the OF phase condition changes as a function of the mode number, agree extremely well with simulations. This demonstrates that the linewidth narrowing is sufficient to produce an essentially monochromatic laser as soon as the OF phase value enables a non-zero cavity injection.

Finally, we note that our model calculations confirm that OF-CEAS is tolerant of small OF phase fluctuations, typically up to about $2\pi/10$, which do not have an impact on the CEAS spectra. As Fig. 5.7 shows, for one full phase beating period, 3 out of 30 modes reach the maximum transmission level.

5.3.6 The High Frequency Scanning Speed of OF-CEAS

We already remarked that OF allows to achieve full excitation of a cavity resonance within a surprisingly short time scale. As shown in Fig. 5.8, low-noise mode profiles are acquired in less than 1 ms, under typical conditions. If we consider instead a perfectly monochromatic laser continuously scanning over one FSR during the same time interval Δt, in the absence of OF its emitted field would interact with the cavity mode during a time $\Delta t/\mathcal{F}$. For a typical cavity finesse $\mathcal{F} = 10^4$ this time equals 100 ns, corresponding to a few tens of cavity round-trips, allowing for a very partial cavity field build-up, which occurs over the same time as a ring-down event. Comparing the scanning speed W to the adiabatic scanning speed defined in Sect. 5.3, we find $W \approx 100 \times \Delta\nu_{\text{cav}}/\tau_{\text{RD}}$, which places the situation in the impulsive regime of cavity excitation. Turning to the particular situation with feedback, let us first recall the different timescales of the coupled system. Starting with the laser alone, the gain medium polarization relaxes on the timescale of intra-band relaxation ($\sim 10^{-13}$ s), the inverted population relaxes on the timescale of the carrier lifetime ($\sim 10^{-9}$ s) and the photon lifetime equals a few times 10^{-12} s for a diode laser or a QCL. A coupled system is then formed by adding a cavity with a photon lifetime of tens of μs. It is thus clear that any variation of the intracavity field is adiabatically followed by the laser dynamics. Moreover, the shortest delay of the feedback field (the time for light to make a single cavity round-trip and then to return to the laser) is already an order of magnitude longer than the longest timescale for the isolated laser dynamics. We can thus expect the laser to follow any cavity field variation even on the shortest timescale. It appears that a reasonable condition to define the onset of OF locking, is that the re-injected field amplitude exceeds the incoherent effect of spontaneous emission which fundamentally drives laser field fluctuations and is responsible for its intrinsic linewidth (in the Schawlow–Townes limit).

Cavity excitation by a passage through resonance in the presence of OF may then be described as follows. The instantaneous laser frequency, subject to fluctuations and noise defining its free linewidth $\Delta\nu_{\text{free}}$, is initially out of resonance and scanned at a rate W. At a given time, the laser field distribution starts to overlap with the resonance and to build-up a small intracavity field (as a random process). A fraction of the intracavity buildup field feeds back to the laser and is re-amplified. If the build-up is sufficient to obtain an OF amplitude inside the laser cavity which surpasses the spontaneous emission in the laser mode, OF starts affecting the laser emission frequency, inducing linewidth narrowing. These effects increase the time interval during which the laser field is able to contribute to the cavity field buildup, and improve the laser field injection in the cavity, resulting in a self-amplifying process as the frequency scanning slows down more and more, while frequency fluctuations reduce which in turn strengthens the build-up and the locking effect. This finally converges to a stationary regime in which the previously introduced slope $p = d\omega_L/d\omega_{\text{free}}$ defines the reduced scanning speed W/p and linewidth $\Delta\nu_{\text{free}}/p$.

We can understand now why the maximum frequency scanning rate at which cavity modes may become fully excited by OF locking does not correspond to the adiabatic scanning speed $\Delta\nu_{\text{cav}}/\tau_{\text{RD}}$. A faster critical scanning speed W_c exists, for which it is no longer possible to buildup a sufficient field in the cavity for starting up the dynamic transition to the coupled laser regime. Above this limit, no locking occurs, the laser is undisturbed, and light transmitted by the cavity has the same low intensity and temporal profile as described earlier and shown in Fig. 5.4. In practice, a transient regime occurs for $W \sim W_c$, where the OF locking works occasionally, as the onset of locking depends on a cavity injection which may randomly attain the buildup threshold level, due to the preexistent fluctuations of the laser field. This regime is quite easy to be observed experimentally either by decreasing β or by increasing W. For a given laser sensitivity, depending on the photon lifetime τ_d and Henry factor α_H, the critical frequency scanning speed W_c increases with the feedback rate β'. In Fig. 5.9, we report measurements of the injection efficiency (maximum cavity output at resonance, for an optimized OF phase, normalized to maximum cavity output for low W) as a function of W for different feedback rates, clearly illustrating this dependency.

In a more general way, Fig. 5.9 summarizes the relevance of using resonant feedback to inject all longitudinal cavity modes in rapid succession.

First of all, the linewidth narrowing enables to fully excite the cavity mode and to use its transfer function to describe its amplitude as it would be obtained with an ideal monochromatic laser. Thus, losses present inside the cavity may be determined quantitatively by their influence on the mode amplitude.

Second, the time of scanning is (to first order) not limited by the cavity response time, which would lead to a prohibitive acquisition time. For instance, without OF, a cavity with a ring-down time τ_{RD} around 20 µs and a FSR of 150 MHz (finesse $\mathcal{F} = 2 \times 10^4$), injected by an ideal monochromatic laser, requires a minimum acquisition time $T_{\text{acq}} = N \cdot \text{FSR}/(\Delta\nu_{\text{cav}}/\tau_{\text{RD}}) = 2\pi N \cdot \text{FSR} \cdot \tau_{\text{RD}}^2$, or several tens of seconds to cover $N = 100$ cavity modes. With resonant feedback, the acquisition time is reduced by several orders of magnitude, depending on the laser sensitivity and its robustness to OF. For a given laser, the locking range is adjusted by the feedback rate to optimally approach one FSR, while the free scanning speed can be set

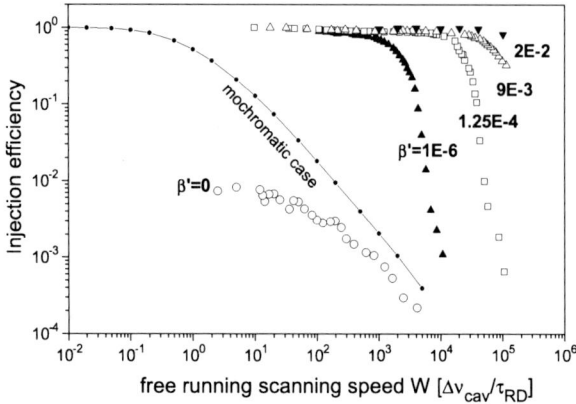

Fig. 5.9 Measured injection efficiency as a function of scanning speed for different feedback rates (reproduced with permission from [1]). Maximum cavity mode transmissions are recorded with the optimal OF phase and a given feedback rate β'. When the scanning speed approaches, and eventually surpasses, the critical scanning speed, the maxima are reduced and start fluctuating in intensity, since the time available for the initial intracavity field build-up, itself subjected to the free laser frequency fluctuations, becomes the limiting factor. For this reason, the figure is based on averaged peak intensity values. The theoretical curve for the monochromatic laser is also shown, together with an experimental curve ($\beta' = 0$) obtained without OF and the same laser, having $\Delta \nu_{\text{free}} = 1$ MHz

to a value just below the critical scanning speed. With a DFB diode laser, for example, a locking range of about one FSR implies a feedback rate of around 10^{-4}, and the acquisition time required to cover $N = 100$ modes is only about 100 ms. As discussed in the next section, some lasers do not manifest sufficient sensitivity to OF as needed to extend the locking range over one FSR or are not sufficiently robust to OF to tolerate such level of feedback rate while maintaining single frequency emission, the locking range and scanning speed must consequently be kept at a lower level than optimal.

Last, in addition to the effect of locking range discussed in Sect. 5.2, the dependency of the critical scanning speed with the feedback rate β' also enables to selectively excite fundamental $TEM_{0,0}$ modes with no particular attention to the mode-matching. As long as the laser beam is aligned onto the cavity optical axis and focused at the cavity waist, the intensity spatial coupling coefficient onto the $TEM_{0,0}$ mode, ϵ_{00}, dominates over high order mode coupling coefficients. As this coefficient participates (squared) to the feedback rate (see end of Sect. 5.3.1), the critical scanning speed for the high order transverse modes is always below the actual scanning speed and excitation with resonant OF can not operate. The fundamental $TEM_{0,0}$ modes are then selectively excited even in case of reduced locking range or for modes close to the center of a strong absorption line (which also have a reduced locking range induced by the reduced $H_{\text{OF,max}}$). This situation is somewhat idealized since a significantly lower scanning speed is required in practice to obtain mode transmission profiles which are really close to those of the adiabatic model. High order transverse cavity modes could then appear in correspondence of a strong

Fig. 5.10 Recorded cavity output obtained with an ECDL in Littrow configuration. A low-finesse solid-etalon trace is also shown, making clearly visible the step-by-step behavior of the laser frequency spending all the scanning time on successive cavity mode frequencies. Reproduced with permission from [22]

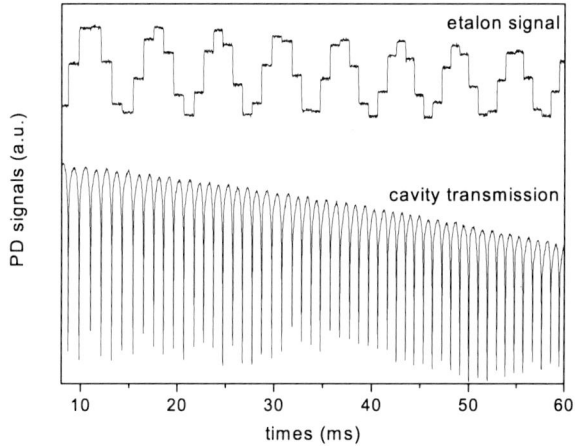

absorption line if mode matching is too coarse. However, their amplitude may be easily kept low by optimizing the on-axis alignment of the laser beam, and they can be selectively neglected by using a threshold scheme in the analysis software.

In Fig. 5.10 an experimental registration of the cavity output is shown, together with a low-finesse etalon trace. The regular sine trace from a continuous, smooth laser scan, normally obtained with such an etalon, is here replaced by a step-sine trace, showing that practically most of the time the laser is effectively injecting cavity modes, in a rapid succession.

5.4 Implementation

5.4.1 Different Laser Sources for OF-CEAS

The use of optical feedback in the OF-CEAS technique limits the choice of laser sources. In addition, the range of optical feedback rates accessible with a given cavity is a function of its geometry and its losses (see Sect. 5.4.2). With typical V-shaped cavities, OF rates up to several percent are easily accessible. This corresponds to no attenuation ($\beta = 1$) and roughly to the higher limit of the low coupling regime for "OF-robust" lasers like DFB diodes lasers or QCLs. It is often seen that even lower OF rates drive a laser out of single mode emission and eventually into a chaotic behavior. A laser ideally suited for OF-CEAS should have an OF sensitivity enabling locking ranges up to one cavity FSR (of the order of one hundred to several hundred MHz) with accessible OF rates, while maintaining stable, single mode operation. At the same time, the laser output power should ideally be affected as little as possible.

With regard to Eqs. (5.9), the laser sensitivity is governed by both the laser photon lifetime τ_d and the Henry factor $\alpha_H = -2\frac{\omega}{c}\frac{\partial n}{\partial g}$, describing the degree of coupling between gain and refractive index variations. This amplitude-phase coupling depends on the specific energy bands involved in the laser transition. In a SC laser, optical transitions take place between carrier energy levels whose non-uniform density induces an asymmetric shape of the real part of the susceptibility (the gain) as a function of frequency. Thus, any small change of carrier density induces both vertical and horizontal shifts of the gain bands [12]. One consequence is that the lasing frequency, determined by the maximum gain amplitude, varies with carrier fluctuations induced by spontaneous emission. This explains why SC lasers exhibit an increased linewidth, roughly equal to $(1 + \alpha_H^2)$ times the Schawlow–Townes linewidth limit [23, 24]. For this reason, the Henry factor is also known as the linewidth enhancement factor. For a diode laser with an hetero-junction gain, typical values of α_H lie between 3 and 8 [25]. Values in the range of 1 to 3 are obtained with multiple-quantum-wells gain structures [26], normally used for DFB diode lasers. In QCLs, transitions occur between discrete states, the gain has a symmetric shape, and expected values for α_H are close to zero. However, while some authors confirm extremely low values [27, 28], others measure values as low as -0.5 and up to 3 for high injection currents [29]. To conclude, the range of variation of the Henry factor for SC laser roughly span over one order of magnitude whereas laser photon lifetimes for general laser system easily range from ps to hundreds of ns. The predominant parameter affecting sensitivity to OF is thus the photon lifetime and up to now, only mm or sub-mm range laser cavity lengths with high output coupling factors have been used for OF-CEAS.

The second crucial laser property for OF-CEAS application is its robustness to OF which is governed by the spectral selectivity of its single-mode emission. Single mode emission means that almost all the optical power is closely concentrated around the carrier frequency, defining the laser linewidth. However, some residual optical power is still present far away from the emission line and may then overlap with different cavity resonances. This usually negligible effect could be amplified by the laser gain if: (i) the feedback rate (at the first order equal for all cavity resonances) is sufficiently high and (ii) the laser spectral selectivity sufficiently low. The parasitic OF could then drive the laser emission out of its principal emitting mode, creating instable frequency jumps and ultimately collapse of coherence. When looking at the cavity output, the first sign indicating the route to instability appears on the top of the transmitted mode profile where a large dip occurs when the optical power starts to leak out of the main optical carrier frequency.

First applications of OF coupling for laser frequency stabilization employed double hetero-structure Fabry-Perot type diode lasers, demonstrating an extremely high OF sensitivity. However, their single-mode emission, when it exists (photorefractive effects must counteract spatial gain hole-burning), is generally not sufficiently robust for most OF-CEAS applications. In fact, the first OF-CEAS demonstration used a DFB diode laser [1] for which the spectral selectivity of the guided gain, index modulated structure, reinforces the single-mode emission. These lasers are extremely robust with respect to OF, with rates up to 10^{-2} still not inducing instabilities. Observed locking ranges extending over almost 500 MHz (more than three cavity FSRs) are easily observed. Laser power variations induced by OF are generally below the 1 % level.

ECDLs have also been successfully employed, with both Littrow [30] and Litmann-Metcalf geometries [31]. Their longer photon lifetime reduces sensitivity to OF and their single-mode robustness appears to be lower. This is particularly evident in the case of a Litmann-Metcalf configuration, in which case locking ranges of more than 100 MHz are hardly ever reached before instabilities appear. For this last type of source the reduced OF sensitivity in the frequency domain is also accompanied by a larger OF effect on the output power. As evidenced by Fig. 5.8, power variations of several percent are observed.

With respect to application of Vertical-External-Cavity Surface-Emitting Lasers (VECSEL) to OF-CEAS [32], the low output coupler transmission ($\sim 1\%$) may be compensated by a short laser cavity (275 μm in air), which maintains a short photon lifetime in the laser cavity. However, locking ranges are still limited to several tens of MHz.

Finally, QCL lasers have a cavity length in the millimeter range. This, together with an output coupling made by facet cleaving, guarantees extremely short photon lifetimes, making these lasers well suited for OF-CEAS. Three applications have already been reported [33–35]. Although those studies used a V-shaped cavity, they reported significantly different laser behavior in terms of sensitivity to OF and the magnitude of the OF effect on intensity. Contrary to [33], in [34] no attenuator was placed between the laser and the cavity to control the locking range. At the same time, the power effect was observed to be in the percent range in [34], whereas it surpassed ten percent in [33]. These observations appear to call for more investigations relatively to the response of a QCL to resonant optical feedback.

5.4.2 OF-CEAS Schemes

Since the earliest developments, several OF-CEAS schemes have been developed combining different laser sources (as already discussed in Sect. 5.4.1) with different cavity geometries. Here we present a brief review of these cavity geometries. All have in common an ability to produce selective optical feedback at resonance. However, they differ by several important parameters relating to the laser-cavity coupling. The maximum feedback rate they can produce ($\beta'_{max} = H_{OF,max}$ when $\beta = \epsilon_{00}^2 = 1$) determines, together with the laser sensitivity to OF, whether a particular laser source may be employed or not. Their polarization eigenstates determine the acceptable incident laser polarization(s). Also, this may have impact on the way the feedback rate β' is adjusted in practice. Lastly, some specific artifacts, discussed below, induced by the presence of a standing wave in a linear or folded cavity, will not be present in the case of a traveling-wave ring-shaped cavity.

The first cavity employed for OF-CEAS was a V-shaped cavity formed by three mirrors, as in Fig. 5.11(a). Here, injection occurs through the folding (apex) mirror, so that the direct reflection from the incident beam is not coupled back to the laser. On the other hand, one of the "exit ports" from the cavity is in the direction of the incident beam and generates the resonant OF. It is easily shown that with four exit

a) The V-shaped cavity

$\beta = P_{OF}/P_{inc}$

b) The Brewster angle cavity

c) The linear cavity with residual mirror birefringence

d) The ring cavity

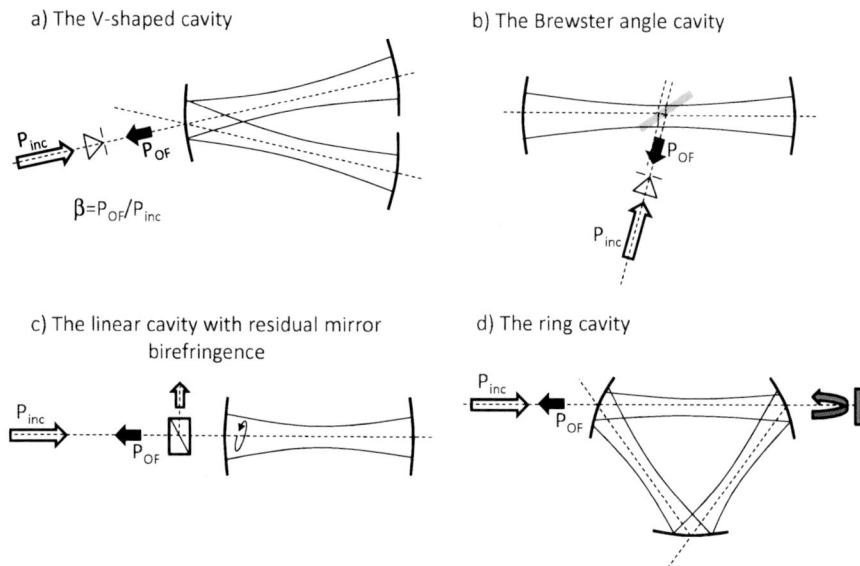

Fig. 5.11 Schematic drawing of the four different cavity geometries used with OF cavity injection. The *black arrows* indicate the cavity exiting port creating the feedback beam. See text for details

ports through which light can leak out of the cavity, the maximum transmission of one port is 25 % for a lossless, V-shaped cavity with equal mirrors (see Eq. (5.16) in Sect. 5.5). However, typical values for a real, high-finesse cavity are closer to 5 %, which corresponds to the common case of mirrors that have equal loss and transmission coefficients. Furthermore, as we discussed in Sect. 5.3.1, the squared value of the mode-matching coefficient also contributes to β'. Taking all together, maximum OF rates obtained with a V-shaped cavity are typically in the range of 10^{-2}, and a variable attenuator placed before the cavity is generally the preferred method of adjusting β'. This has the drawback of lowering the incident power and hence the level of useful signal for the cavity transmission measurement, but may have the advantage to reduce the danger of power saturation of the molecular transitions, which is generally a nuisance and would be favored by the large intra-cavity power buildup allowed by OF injection. If desired, however, it is possible to exploit most of the laser power for cavity injection by using a more complex optical attenuation scheme. For instance, one could employ an optical isolator (a 45° Faraday rotator sandwiched between two crossed polarizers) with an adjustable output polarizer. This would allow an adjustable fraction of the OF field to be returned to the laser with little effect on the forward transmission level.

Whatever the scheme of feedback rate adjustment, the polarization of the incident light should match the polarization eigenstate of the cavity modes. If we consider the V-shaped cavity, the non-normal reflection at the folding mirror induces slightly different phase shifts for the s-polarization and the p-polarization. This breaks the polarization degeneracy present in a linear cavity and induces a slight spectral shift of the polarization modes. As the reflection coefficient too depends on the polarization orientation, each family of polarization modes is associated with a different cavity finesse.

Even when a V-shaped cavity is injected using either vertical or horizontal linear polarization, another effect modulates the $TEM_{0,0}$ mode losses. Let us consider a cavity $TEM_{0,0}$ mode with its transverse Gaussian profile. This standing wave, when it reflects at the folding mirror, produces vertical fringes on the mirror surface (we suppose the cavity laying in the horizontal plane). The period of these fringes is given by the inverse of the projection of the field wave vector k along the mirror plane, and thus is rather large when the V-cavity angle is small. For a cavity with a half-angle of $1°$, the fringe period would be on the order of 100 µm for near-IR wavelengths. Small mirror defects or dust particles will then induce more or less losses depending on whether they are located near a fringe a maximum field line or a nodal line (zero field). The mode losses will then contain a term given by the sum of losses by all localized surface defects weighted by the fringe pattern. The observed even-odd loss modulation is then associated with the fact that the fringe pattern translates horizontally by a quarter of a period when going from a mode to the next, so that nodal lines exchange with maximum-field lines. In practice it is easily observed that deposition of dust on the folding mirror strongly increases the even-odd effect while also producing an overall decrease of the cavity finesse. This even-odd loss modulation is rather constant over a laser tuning range, only if the cavity arms are exactly matched, since a change in wavelength will increase the fringe spacing while maintaining the central fringe position on the mirror surface. If these have different lengths, the fringe pattern will then shift as a function of wavelength producing a sinusoidal modulation of the even-odd modes loss. Its period, in units of mode number, is given by $(L_1 - L_2)/(L_1 + L_2)$. Most often, considering the small spectral scan achievable by current tuning of a SC laser, the even-odd difference can be included as a linear sloping baseline in a fitting routine for the spectral line profiles. Signals shown in Fig. 5.10 were obtained with a V-shaped cavity for which the odd and even modes families are barely distinguishable one from another.

A Brewster-plate cavity has also been used with OF-CEAS. It consists of a linear two-mirror cavity, containing a superpolished flat glass plate at its center, oriented close to the Brewster angle with respect to the cavity axis. The cavity is injected trough the weak reflection obtained on this plate when the incident light is p-polarized. The plate produces 4 output beams, one returning to the laser and creating the desired OF as illustrated in Fig. 5.11(b). For given mirror coefficients and given Brewster-plate losses, the signal level obtained through the mirror, the maximum OF level ($H_{OF,max}$) and the cavity finesse are all a function of the incident angle over the Brewster plate. In the study reported in [31], cavity finesses range from 10^3 to 10^4 and OF values are at the percent level. Here, polarization modes are highly discriminated by the near-Brewster reflection. Also, and contrary to the case of the V-shaped cavity, the cavity modes are entirely free of astigmatism. However, the standing wave, in combination with the intra-cavity plate surfaces, still introduces an even-odd like mode structure. Its pattern and dependence on the plate position inside the cavity may be quite complex because of the two plate interfaces. However by using a thin microscope slide (0.15 mum typically), which turns out to possess highly polished surfaces, it is possible to reduce this loss pattern to the noise level of a single OF-CEAS scan [31]. We have noted that the advantage of this cavity is the ability to modify the finesse by a simple adjustment of the plate orientation, which allows widely increasing the dynamic range of OF-CEAS measurements. On the other hand, the use of a thin microscope cover plate inside the cavity complicates the setup and makes it more sensitive to vibrations and temperature drift.

Mirror birefringence was also exploited to produce a resonant OF [36]. Birefringence results in a modified polarization state after cavity excitation which can be used to feed back selectively a small fraction of the resonant field against the strong background of the directly reflected linearly polarized input field. The advantage is that a simple linear 2 mirrors cavity configuration can be used, as depicted in Fig. 5.11(c). The mirror birefringence is principally linear and originates from the stress acquired during the coating process. Typical values of the phase shift at reflection between the birefringence axes are in the μrad range and can be easily increased by adding controlled mechanical stress to the mirror substrate. As long as the mirrors phase shifts are small, by rotating the mirrors it is possible to control the orientation and strength of the resulting round-trip phase, and therefore the orientation of the polarization eigenmodes, which are almost linearly polarized (and orthogonal). However, the principle of operation of this scheme requires that the resulting cavity phase shift does not exceed a critical value $\phi_c = 2\pi/\mathcal{F}$, in order not to split the polarization modes. Under this condition, and by placing an optical system in front of the cavity, such as a polarizer followed by a 45° Faraday rotator, or a quarter-wave plate (this enables to filter out the direct input mirror reflection, while any other polarization is partly transmitted), the maximum OF level $H_{OF,max}$ is given by $(\phi_c \mathcal{F}/2\pi)^2$. For a cavity with a finesse of 10^4 and $\phi_c \simeq 10^{-5}$ rad, a feedback rate of 2.5×10^{-4} may be attained. Given the use of only 2 mirrors, this scheme is not affected by the even-odd modes loss structure, but the low level of feedback rate limits its application to lasers that are extremely sensitive to OF, such as DFB diode lasers. Lastly, the optical system used to filter out the direct input mirror reflection is the critical part of the scheme, as it must be selective at a sub-ppm level. This requires high quality optical components and high stability of the opto-mechanical setup, which constitutes the main drawback of this approach.

More recently, in order to avoid the even-odd modes loss structure, Hamilton and co-workers [37] used a ring cavity composed of three mirrors with the same radius of curvature, as depicted in Fig. 5.11(d). In principle, as a traveling wave is injected inside the ring and detected in transmission, no spatially selective losses can be produced. In order to create the desired OF, a mirror placed behind one of the cavity exiting ports reflects back the transmitted beam, which re-enters the cavity and creates new transmitted beams, one of is returning back to the laser. Adjustment of the feedback rate is done through the slight misalignment of the returning beam. It is straightforward to show that for equal mirrors $H_{max} = (\mathcal{T}/(1 - \mathcal{R}^{3/2}))^2$ (which has a maximum value of $\sim 4/9$ in the limit of negligible losses). It is interesting to note that with this approach, the transfer function of the feedback field is not proportional to the transfer function describing a single exit port, but rather is proportional to the square of this expression. This point has not been discussed by the authors, but could reveal interesting potential for laser stabilization, as the linewidth reducing factor is now given by the square of the slope, even in the case of $1/f$ noise. A second consequence of the consecutive passage is that the photon dynamics of the feedback field may be longer than the single cavity photon lifetime. However, absorption lines by the sample will have a stronger negative impact on the available OF level, probably inducing a limited dynamic range for this configuration. Lastly, the even-odd modes loss could re-emerge if small imperfections are present somewhere on the mirrors surfaces probed by the laser beam. This is potentially sufficient to excite the counter-propagating wave [38] and to create a standing wave, as the scattering imposes a phase relation between the two traveling waves. For sufficient scattering a strong coupling takes place between the two traveling waves, which would cause a frequency splitting of the cavity resonances. However, such effect was found to be quite small by Hamilton et al., and with careful cleaning of the mirrors its contribution could be kept negligible [37].

5.5 Absorption Scale Calibration by a Single Ring-Down

We present a simple and direct calibration procedure used to convert normalized transmission signals into absorption spectra, following the derivation introduced in 2006 by Kerstel et al. [39]. An experimental verification of its validity was provided by Motto-Ross et al. [40]. Although the derivation is specific for a V-shaped cavity, we will see that its adaptation to other cavity geometries is straightforward. A major advantage of this approach is that it provides an absolute estimation of the total cavity losses including the molecular absorption, without the need for an independent measurement of the empty cavity losses (i.e. without absorption by the sample).

Let us consider the transmission function of the V-shaped cavity setup of Fig. 5.2 with arms of length L_1 and L_2, in the case of slow passage through resonance by a very narrow bandwidth laser. As explained above, these conditions are fulfilled in the case of OF-CEAS, because the optically self-locked laser linewidth narrows to below the width of the cavity resonance, while its emission is forced to scan slowly through the peak of the resonance. The resulting transmission profile would have the shape of an Airy function in the case of a laser that is optically isolated from the cavity (see, e.g., the Introductory chapter), but is altered by optical feedback as shown in Fig. 5.6 and as discussed in Sect. 5.3.2. The important point is that controlling the optical feedback phase insures that the transmission profile corresponds to a zoomed section of the Airy function around its maximum, such that the measured peak transmission corresponds to the maximum transmission of the resonance.

The peak transmission at resonance is easily obtained by squaring the sum of the amplitudes of the constructively interfering electric fields after each passage at the output mirror (M_1 in Fig. 5.2):

$$H_{\max} = \left| \frac{E_{out}}{E_{in}} \right|^2 = \left(\mathbf{t}^2 a_1 \sum_{n=0}^{\infty} (a_1^2 a_2^2 \mathbf{r}^4)^n \right)^2$$

$$= \left(\frac{\mathbf{t}^2 a_1}{1 - a_1^2 a_2^2 \mathbf{r}^4} \right)^2. \tag{5.12}$$

Here $\mathbf{t} = \sqrt{t_a t_1}$, where t_a and t_1 are, respectively, the field transmission coefficients of the apex mirror M_a and of the output mirror, while $\mathbf{r}$ collects the field reflection coefficients over one round-trip, $\mathbf{r} = (r_1 r_2 r_a^2)^{1/4}$. The single pass field transmissions by the sample in the two arms of the V-cavity are given by a_1 and a_2, and can be expressed in terms of the absorption coefficient α. The latter is defined for the field intensity, whose one pass transmission would be $\exp(-\alpha L_i)$, and can be applied to the field amplitude by taking the square root of this expression, while we will neglect the corresponding imaginary part that contributes only to dispersion:

$$a_i = e^{-\alpha L_i/2} \quad (i = 1, 2). \tag{5.13}$$

This is analogous to Eq. (8) obtained for a linear cavity in the introduction chapter.

Likewise, if we neglect reflection phase change, the field transmission and reflection coefficients of the mirrors can be replaced by the corresponding intensity coefficients ("averaged" over one round-trip) in the following manner:

$$\mathcal{T} = \mathbf{t}^2 = \sqrt{\mathcal{T}_a \mathcal{T}_1},$$
$$\mathcal{R} = \mathbf{r}^2 = \sqrt{\mathcal{R}_a} \sqrt[4]{\mathcal{R}_1 \mathcal{R}_2}; . \tag{5.14}$$

We should note that the reflectivity of the three mirrors usually does not differ noticeably, since in most instances they are taken from the same production batch, and because the angle between the two cavity arms is small enough to not appreciably affect the reflectivity of M_a.

We thus obtain for the peak of the transmission transfer function for a V-shaped cavity:

$$H(\alpha) = \left(\frac{\mathcal{T} e^{-\alpha L_1/2}}{1 - \mathcal{R}^2 e^{-\alpha(L_1+L_2)}} \right)^2. \tag{5.15}$$

Conservation of energy then requires that $\mathcal{L} + \mathcal{R} + \mathcal{T} = 1$, where $\mathcal{L}$ represents the combined losses (scattering and absorption) of the mirrors. It is now straightforward to show, by setting α and $\mathcal{L}$ equal to zero in Eq. (5.15), that the maximum transmission for a completely lossless cavity reaches 25 %, reflecting the fact that light will leak out of the cavity in four different directions:

$$H(\alpha \to 0, \mathcal{L} + \mathcal{T} \ll 1) \approx \frac{1}{4} \frac{1}{(1 + \frac{\mathcal{L}}{\mathcal{T}})^2}. \tag{5.16}$$

As discussed in the introduction chapter, a derivation based on the summation of the intensities would miss the resonance effect and yield a much lower cavity transmission of only $\mathcal{T}/4$, corresponding to the case of a broadband incoherent light source or that of integrated cavity enhanced absorption spectroscopy. Also, as other cavity enhanced techniques, OF-CEAS benefits from an increased dynamic range compared to traditional direct absorption, as is seen by comparing Eq. (5.15) to the Beer-Lambert absorption, as a function of α, for the same effective absorption path length.

Direct inversion of Eq. (5.15), in order to determine the intracavity absorption coefficient α from the measured intensity ratio $H(\alpha)$, requires knowledge of the mirror transmission and reflection coefficients. In addition, it is difficult to measure the value of $H(\alpha)$ accurately since it depends on the ratio of photodiode signals, one monitoring the spatially filtered cavity output, the other monitoring a fraction of the laser intensity at cavity input from a beamsplitter. This would require not only accurate knowledge of the response of both detectors, but also the efficiency of the coupling of the laser mode to the cavity $TEM_{0,0}$ mode, as well as the exact beam splitting ratio. We will show here that knowledge of all these parameters is conveniently replaced by a single ring-down measurement, carried out once per spectral scan. This measurement, obtained in the presence of sample absorption, suffices to calibrate the absorption scale with a high degree of accuracy.

To this end we approximate the exponential factor in the numerator of Eq. (5.15) by unity, which allows to extract an approximation of α as a function of the other parameters, in particular the measured H_k, where the index k is for the specific cavity mode that we consider:

$$\tilde{\alpha}(H_k) = \frac{1}{L_1 + L_2}\left(\ln(\mathcal{R}^2) + \ln\left(1 - \frac{T}{\sqrt{H_k}}\right)\right). \tag{5.17}$$

Other approximate expression can be derived, but having numerically compared different possibilities we believe this one is the simplest and most effective.

The first r.h.s. term of Eq. (5.17) is recognized as the empty-cavity baseline absorption $\alpha_0 = \gamma_0/c$, where γ_0 is the empty cavity loss rate, which can be obtained by considering that the exponential decay of intracavity light must amount to $\mathcal{R}^2$ (as defined above for a V-shaped cavity) after one pass in the cavity:

$$\exp(-\gamma_0 t)|_{(t=\frac{L_1+L_2}{c})} = \mathcal{R}^2. \tag{5.18}$$

Thus Eq. (5.17) can be rewritten as:

$$\tilde{\alpha}(H_k) = -\frac{\gamma_0}{c} - \frac{1}{L_1 + L_2}\ln\left(1 - \frac{T}{\sqrt{H_k}}\right). \tag{5.19}$$

On the other hand, the absorption coefficient of mode k can be written in a highly recognizable way to anyone familiar with CRDS:

$$\alpha_k = \frac{1}{c}\left(\frac{1}{\tau_k} - \frac{1}{\tau_0}\right) = \frac{1}{c}(\gamma_k - \gamma_0), \tag{5.20}$$

where $\tau_0 = 1/\gamma_0$ is the ring-down time for the empty cavity, while $\tau_k = 1/\gamma_k$ is the ring-down time (the inverse of the corresponding loss rate) for mode k. This can be determined experimentally by performing a ring-down measurement following an abrupt termination of the laser once the maximum injection for mode k has been reached, which is necessary for the simultaneous measurement of H_k.

Comparing the last two equations gives:

$$\frac{\gamma_k}{c} = \frac{-1}{L_1 + L_2} \ln\left(1 - \frac{T}{\sqrt{H_k}}\right) \simeq \frac{1}{L_1 + L_2} \frac{T}{\sqrt{H_k}}, \qquad (5.21)$$

which is generally a very good approximation since $T/\sqrt{H} \ll 1$, considering that large sample absorptions will be avoided such that H_k should normally be larger than 1 %, while T is typically smaller than 100 ppm.

This last approximation allows writing:

$$\gamma_k \sqrt{H_k} = \frac{cT}{L_1 + L_2} = \gamma_m \sqrt{H_m}, \qquad (5.22)$$

where we observe that over a small laser scan we can indeed consider T as a constant and write the last equality, as the expression is valid for any mode m. Thus, after γ_k has been measured, together with all H_m in the laser scan including $m = k$, this simple expression permits to calculate directly the γ_m values for all cavity modes. This way an OF-CEAS spectrum composed of the values H_m, can be "calibrated", i.e. converted to values $\gamma_m = c\alpha_m$ (which include the baseline γ_0) in absolute absorption units.

As mentioned before, this expression can be expected to be very precise in the limit of $T \ll \sqrt{H}$, but we should also underline that the result is almost exact for mode number k for which the ring-down measurement is performed. This has been experimentally verified by comparing the measured ratio H_k/H_m with a registration of the ring-down rate γ_m for all modes in the spectral range being considered [40].

We should also note that the presence of an alternance of intensities between "even" and "odd" longitudinal cavity modes, explained elsewhere in this chapter, does not conflict with this normalization procedure. This effect is due to variations of intracavity losses occurring on the input mirror surface, which can be considered equivalent to sample absorption. The same can be stated concerning broad band intracavity losses produced, e.g., by particles or by Rayleigh scattering.

In order to investigate the accuracy of the normalization approach we compare the result of Eq. (5.22) to the exact result of Eq. (5.15) for a model case, where we assume to know all parameter values (T, L_1, L_2, etc.). Although the latter can be solved analytically in the case that $L_1 = L_2 = L$ (which requires solving a fourth order polynomial in $x = \exp(-\alpha L)$), it is straightforward to solve the equation numerically, as was done here to produce Fig. 5.12. This figure shows the relative error $\tilde{\alpha}/\alpha - 1$ of the total absorption (with baseline) for a number of scenarios: low and high mirror reflectivity/transmission, and low and high absorption at the ring-down mode number k. The high-reflectivity mirror case ($T = 5$ ppm and $\mathcal{L} = 3$ ppm) corresponds to a cavity with 130 µs ring-down time, whereas the (relatively!) low-reflectivity mirrors ($T = 50$ ppm and $\mathcal{L} = 35$ ppm) give a ring-down time of 20 µs (the cavity arm length equals 50 cm in both cases). For each case, the fractional error is determined for two different choices of the mode for which the ring-down normalization is carried out. In the one case the normalization is carried out for a mode at which the intracavity absorption results in a cavity transmission of 50 % of the maximum value, in the other case the cavity transmission at the ring-down

Fig. 5.12 The relative error of the absorption over baseline as determined by the procedure described in this section. The error is calculated for two different combinations of mirror transmission and losses, corresponding to cavities with 20 μs and 130 μs ring-down time, and for two different choices of the transmission of the cavity mode used for the normalization ring-down measurement. Interestingly, the relative error is smaller over a broader dynamic range for the higher finesse cavity. See the text for details

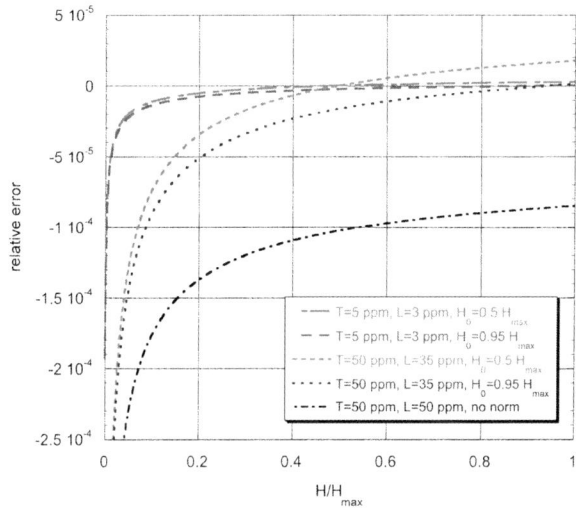

normalization mode equals 95 %. For comparison, the relative error has also been calculated for the case that the approximate Eq. (5.17) is used with the true values of $\mathcal{T}$ and $\mathcal{R}$ (assuming that one could determine these independently). It is clear that the normalization procedure simply shifts this error curve to cross zero at the value used for H_k.

Figure 5.12 also shows that even for a typical OF-CEAS cavity with a ring-down time of ~ 20 μs, the relative error associated with the normalization of the absorption scale is smaller than 1 part in 10^4, provided that the transmission of the cavity remains higher than 10 % of its maximum value ($H > 0.1$). This value of transmission corresponds to an absorption coefficient of $\sim 4 \times 10^{-6}$/cm or, considering the typical noise floor below 10^{-10}/cm, a dynamic range of 4 decades. The analysis also shows that there is no obvious advantage of performing the ring-down normalization on a cavity mode that shows a high transmission (close to the empty cavity limit), i.e., away from strong intra-cavity absorption features, other than that a long ring-down time recording will be composed of more samples with larger signal levels, and thus allow for the ring-down time to be determined with a higher signal-to-noise ratio. In the particular case where isotope ratios are determined by directly comparing absorption line intensities of different isotopologues within a single spectral recording, the error associated with the absorption scale normalization can be safely ignored if the sample used as reference material has similar concentration, thus similar intensities for the absorption lines. But also in the case of trace gas concentration measurements, this source of error is usually negligibly small, and could potentially be further reduced by about one order of magnitude with a simple linear correction based on the calculated error.

Finally, the derivation presented here is specifically for the case of a V-shaped cavity. However, it is straightforward to apply the same analysis to other cavity geometries by replacing Eq. (5.15) with the transmission function of the cavity geometry in question. Indeed, the experimental validation presented in [40] was done with a Brewster cavity injection scheme.

5.6 Typical Performance of OF-CEAS

In this section we present examples of data obtained with OF-CEAS and discuss its performance in terms of typical sensitivity, acquisition time and dynamic range. As already pointed out, uncertainties related to both the frequency and the intensity axis contribute to sensitivity and accuracy of acquired spectra. We will consider different sources of noise limiting the ultimate sensitivity and indicate the direction for future improvement. As it is an important issue for the precision of the frequency scale, we also discuss the effect of intracavity dispersion induced by molecular absorption itself.

The principal feature of the OF-CEAS technique is its ability to obtain high sensitivity spectra with a linear frequency scale in a fraction of a second. Typical duration for a tuning range covering approximately hundred cavity modes is 100 ms and for a cavity length $L = L_1 + L_2$ generally not exceeding one meter, the tuned spectral range is a few tens of GHz. The minimum detectable absorption coefficient $\delta\alpha_{min}$ is for a single spectrum given by the smallest measurable normalized intensity variation $\delta H/H$ divided by the cavity-enhanced effective absorption length at resonance (see Eq. (17) of the Introduction chapter), which can be written as:

$$\delta\alpha_{min} = \frac{\pi}{2\mathcal{F}L} \frac{\delta H}{H}. \tag{5.23}$$

This $\delta\alpha_{min}$ should correspond to the baseline noise of a single spectrum when care is taken to maintain parasitic optical fringes below this level. As already pointed out, the smoothed cavity mode profile and the high signal level, both induced by the optical self-locking, enable the recording of the normalized amplitude with an uncertainty at the per mille level. For a cavity of finesse 10 000 and length 1 m, $\delta\alpha_{min}$ is about 10^{-9}/cm. If this noise level is further normalized by accounting for the single point acquisition rate, we deduce a figure of merit around 3×10^{-11}/cm/$\sqrt{Hz}$.[5] Indeed, such a performance for a single spectrum acquisition have been reached with a DFB diode laser in the near-IR [36], with an ECDL in the visible [40], and recently with a QCL in the mid-IR [35].

When considering averaging of several spectra, a more refined expression is required. As the product $\mathcal{F}L$ is generally determined for each single spectrum, its uncertainty should be taken into account. Following our calibration procedure (Sect. 5.5), which uses the simultaneous measurement of the ring-down time and

---

[5]This "figure of merit" is discussed in the introduction chapter (Sect. 1.3) together with detection-limit-related issues. It is not to be considered as a true detection limit since $1/f$ and fringe noise are not taken into account in its definition.

the normalized intensity of a single cavity mode, a basic model for the minimum detectable absorption is given by:

$$\delta\alpha_{min} = \frac{\pi}{2\mathcal{F}L}\sqrt{\left(\frac{\delta\tau_{RD}}{\tau_{RD}}\right)^2 + 2\left(\frac{\delta H}{H}\right)^2},\qquad(5.24)$$

where uncertainties of the ring-down time and of the normalized intensities of different modes (according to Eq. (5.22)) are all assumed to be uncorrelated. The full power build-up offered by the self-locking effect also makes that an uncertainty on the ring-down time determination at the per mille level may be achieved. This multiplies the previous limit expressed by 5.23 by a factor smaller than 2. In other terms, the shot-to-shot variation of a single spectral point (on the spectrum baseline), will have a standard deviation given by expression 5.24. For spectral points close to an absorption line $\alpha(\nu)$, a contribution to shot-to-shot intensity variations also comes from fluctuations of the cavity frequency comb. These are translated into amplitude fluctuations proportionally to the slope on the sides of the absorption line. Applying some basic algebra to Eq. (5.22) and still considering uncorrelated noise, this adds to the root of Eq. (5.24) the square of the term $\frac{2}{\alpha(\nu)+\gamma_k/c}\frac{d\alpha}{d\nu}\delta\nu$, where γ_k is the loss rate of the mode used for calibration, $\delta\nu$ is the root mean square value of the mode frequency fluctuation (at the time scale of a single spectrum acquisition) and $\frac{d\alpha}{d\nu}$ is the derivative of the absorption line.

As long as white noise sources are dominant in the experiment, we expect to increase the sensitivity by averaging spectra. For a typical OF-CEAS system, this is true, at best, up to a few seconds, with about hundred spectra averaged before reaching an ultimate sensitivity of typically a few times 10^{-10}/cm (for a ring-down of around 10 μs). At that point, the noise on the averaged spectrum converges to a fixed pattern, which may contain a few recognizable optical interference fringes (regular oscillations), which can be associated to etaloning caused by diffused light on specific optical surfaces. This fixed pattern noise evolves slowly with further averaging and has rms fluctuations across spectral datapoints which do not decrease. The origin of this fixed pattern noise is still at present unclear, but it is universally present in all OF-CEAS setups that the authors have experience with.

Figure 5.13 provides an example of a OF-CEAS spectrum of air at ambient conditions, with the oxygen B-band $^R Q_1$ line at 14 532.900 cm^{-1}. This is obtained with an ECDL and a Brewster cavity with a ring-down time of 12 μs. 50 spectra are averaged over 5 s and result in a baseline rms noise of 3×10^{-10}/cm, as shown in the figure. This is determined by calculating the standard deviation (on a baseline section) of the residuals of a multi-Lorentzian line fit which includes the main strong line (7×10^{-6}/cm peak absorption) plus other extremely weak lines ($\sim 10^{-9}$/cm). The lower panel of the figure displays the relative absorption uncertainty for each cavity mode (from the shot-to-shot rms variations divided by the square root of the number of averaged spectra), indicating a relative baseline noise level just above 4×10^{-5}.[6] In absolute value this corresponds to a $\delta\alpha_{min}$ of about 10^{-10}/cm, 3 times

---

[6] We recall that the absorption baseline is not zero and corresponds to the empty cavity losses.

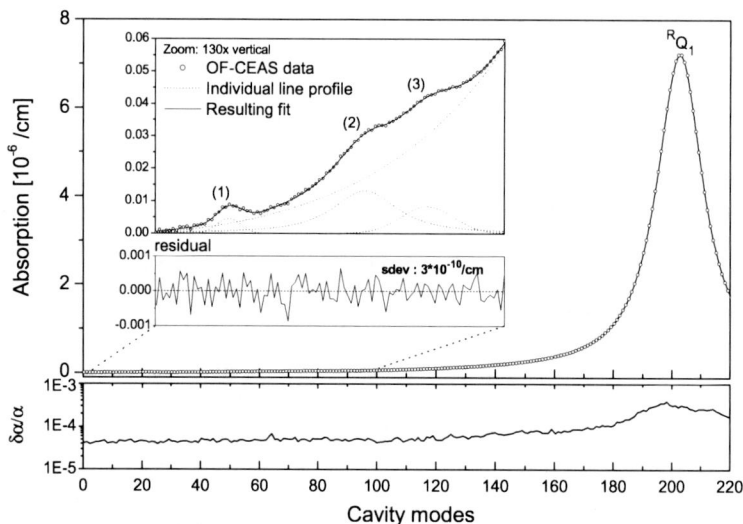

Fig. 5.13 OF-CEAS spectra of the oxygen B-band $^R Q_1$ line at $14\,532.900$ cm^{-1} obtained with an ECDL and the Brewster-angle injection scheme. This spectra covers more than 200 modes corresponding to a wavenumber range of 1.35 cm^{-1}. A single mode acquisition rate of 2 kHz enables to record single spectra at 10 Hz. Here, 50 spectra are averaged in 5 s. In a vertical zoom of the spectrum, shown in the *inset*, three weak transitions in the far wing of the $^R Q_1$ line are visible. Two transitions labeled as (2) and (3) belong to the third most abundant oxygen isotope $^{16}O^{17}O$. The transition labeled as (1) is the $^R Q_{39}$ line. The rms of the fit residuals gives a minimum absorption coefficient of 3×10^{-10}/cm. The *lower panel* shows the loss rate fractional uncertainty $\delta\alpha/\alpha = \delta H/H$ of each cavity mode. Fringe effects appearing with averaging have not been corrected here

lower than the above fixed-pattern baseline noise of an averaged spectrum. We see therefore that this pattern noise is limiting the quality of OF-CEAS spectra after just a few seconds averaging.

Nevertheless, considering the averaged spectrum noise level relative to the strong line intensity, the above example illustrates that a dynamic range of a few time 10^4 is readily attained. Also, as expected, an increase of the relative loss rate uncertainty due to insufficient cavity stability is observed for cavity modes close to the absorption line (bottom trace in Fig. 5.13). However, the signal reduction induced by absorption also contributes to this increase, which explains why the uncertainty does not drop at line center (zero spectral slope). With a noise analysis as a function of the mode position relative to the absorption line, the cavity stability can be deduced. In the work presented in [40], where a non-monolithic Brewster-angle cavity is used, a RMS value of $\delta\nu$ at the MHz level at the time scale of a single spectrum acquisition (100 ms) is deduced. On the other hand, OF-CEAS systems with monolithic cavities as described in [41] do not manifest excess noise on the wings of strong absorption lines, indicating negligible fluctuations of the mode positions over a single laser scan.

Let us consider now the shot-noise limit for OF-CEAS absorption measurements. This is expressed by:

$$\delta\alpha_{min}^{SN} = \frac{\pi}{\mathcal{F}L}\sqrt{\frac{eB}{\eta P}}, \tag{5.25}$$

where e [C] is the electron charge; B and η are the bandwidth [Hz] and responsivity [A/W] of the detector, respectively; P [W] is the optical power received by the photodetectors. This expression takes into account the incident power normalization, and for simplicity it is assumed that both the signal and the reference detector receive equal power. The latter is a reasonable hypothesis as usually the fraction of laser power used for the normalization corresponds well to a cavity mode transmission of several percent. With typically 10 μW on the photodetectors, a bandwidth around 1 MHz as needed to record ring-down times in the several μs range, and a detector responsivity of 0.5 A/W, $\delta\alpha_{min}^{SN}$ is estimated to be a few times 10^{-10}/cm. This is close to the performance described in Fig. 5.13 obtained however after averaging and limited by fringes and fixed pattern noise. Therefore we can state that, on one hand, improvement on photodetectors down to the shot noise level should allow achieving the current ultimate noise level with a single OF-CEAS scan. On the other hand, further improvement can only be expected after overcoming the parasitic optical fringes and pattern noise effects. With respect to spectral fringes caused by scattering from optical surfaces (for more details see Sect. 3 of the introduction chapter), we should underline that, thanks to the linear frequency scale of OF-CEAS and its fast acquisition time, these appear as perfectly periodic sinusoidal perturbations of the spectral baseline. In more conventional spectroscopy techniques, periodic fringes are obtained only after linearization of the frequency scale of acquired spectra, which is often a complex and imperfect procedure. Thanks to this, after a preliminary fit of a OF-CEAS spectrum, fringe periods are readily determined from the positions of peaks present in the Fourier transform of the fit residuals. A refined fit may be then executed with inclusion of sine functions of the predetermined periods in the fitting routine, which enables automatic subtraction of the fringes from the absorption spectrum. The price to pay is an increase in the number of fitting parameters and a (usually modest) increase of the indetermination associated with the resulting fit parameters values.

Improvements to the technique would also include the active stabilization of the cavity length and the temperature of the whole optical system. This allows having stable fringes and pattern noise and to obtain reproducible spectral fit results on the long term, which, combined with calibration procedures (periodic injection of reference gas samples) may still deliver high accuracy concentration or isotope ratio measurements. Of course, choosing high quality optical surfaces and assuring a careful alignment and choice of the distances separating the optical components would also help reducing the effect of optical fringes. Further gain in sensitivity is also possible by using higher reflectivity cavity mirrors. Nowadays, a cavity finesse in excess of 10^5 is easily available at certain wavelengths (e.g. the telecom range [42]), thanks to advances in dielectric mirror coatings.

Fig. 5.14 *Top panel*: Absorption spectrum described by a Lorentzian profile at atmospheric conditions (HWHM = Γ = 1 GHz) and the corresponding variation of the sample refractive index. A maximum absorption coefficient value of 10^{-5}/cm corresponds to a maximum refractive index variation of almost 8×10^{-11}. *Middle panel*: The cavity mode frequency shifts. The maximum absolute shift corresponds to the maximum absorption. As a consequence, a spectral dependency of the cavity FSR is produced (*right scale*). *Lower panel*: Deviation of the measured absorption profile if the mode spacing is assumed to be a constant FSR (taken in the absence of absorption)

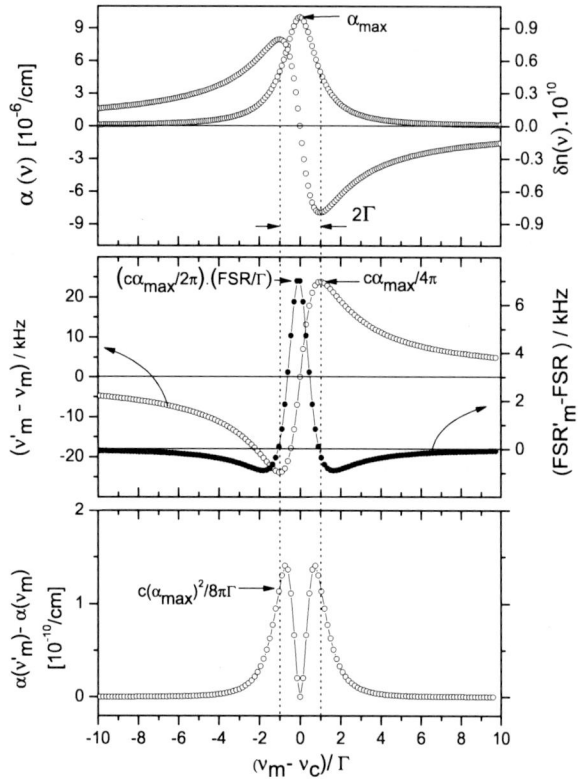

As a last issue relevant to precision of OF-CEAS measurements, it is interesting to consider the effect of cavity mode frequency shift induced by dispersion intrinsically associated with molecular absorption. Indeed, the resonant absorption of light is accomplished with a phase delay that leads to a dispersive perturbation of the refractive index of gas sample. Considering the absorption coefficient of a single absorption line described by a Lorentzian profile $\alpha(\nu - \nu_c)$ where ν_c is the transition frequency (see top panel of Fig. 5.14), the refractive index variation is given by [12]:

$$\delta n(\nu) = -\frac{c}{2\pi\nu}\frac{\nu - \nu_c}{\Gamma}\alpha(\nu - \nu_c), \tag{5.26}$$

where Γ is the Half Width at Half Maximum (HWHM) of the Lorentzian profile. The cavity mode frequency of order m is now given by:

$$\nu_m' = \nu_m\left(1 - \frac{\delta n(\nu_m')}{n_{nr}}\right), \tag{5.27}$$

where $\nu_m = mc/2n_{nr}L$ is the empty cavity mode frequency and n_{nr} is the non-resonant refractive index of the gas sample. This implicit equation for ν_m' is numerically solved and the difference with the empty cavity mode frequency is plotted in the middle panel of Fig. 5.14 for an absorption coefficient of 10^{-5}/cm at the line center. This maximum value α_{max} is approximately equal to the largest absorption that the typical cavity considered here can tolerate, and still produce a reasonable level of transmission and feedback. Mode frequency shifts are largest at 1 HWHM from line center, reaching a few tens of kHz. This is sufficiently small to replace v'_m by v_m in the argument of $\delta n(v)$ in Eq. (5.27). In that case, it is easy to show that the largest frequency shift is given by $(v' - v)_{max} = c\alpha_{max}/4\pi$ and depends only on α_{max}. However, if a constant frequency increment is assumed between OF-CEAS spectral data points, given by the cavity FSR (in the absence of dispersion effects), it is pertinent to consider the variation of the FSR induced by absorption, $FSR_m - FSR = (v'_{m+1} - v'_m) - c/2n_{nr}L$. As expected, the middle panel of Fig. 5.14 shows that this variation is largest at the transition frequency where $\delta n(v)$ has the largest slope. Using the same approximation, and in the restricted case $FSR \ll \Gamma$, the largest FSR variation is given by $\Delta FSR_{max} = (\frac{FSR}{2\pi}) \cdot (\frac{c\alpha_{max}}{\Gamma})$, which depends both on α_{max} and Γ. Even with $\alpha_{max} = 10^{-5}$/cm and a HWHM of 1 GHz, the largest variation is just a few kHz. If datapoints in spectra acquired mode-by-mode are assumed to have a constant spacing, the effect induces a defect on the intensity of the absorption line, which is plotted for our example in the bottom panel of Fig. 5.14. We see then, that a detection limit below 10^{-10}/cm around the absorption line would be required to resolve this effect. In other words, a signal-to-noise ratio (S/N) of 10^5 is necessary. As confirmed by Fig. 5.14, the largest deviation of the residuals appears at about 1 HWHM from the line center, which enables to derive a simple analytical expression for it. Indeed, $\alpha(v'_m) - \alpha(v_m) = \frac{\partial \alpha}{\partial v}(v'_m - v_m)$ and $(\alpha(v'_m) - \alpha(v_m))_{max} \simeq \frac{\partial \alpha}{\partial v}|v'_m - v_m|_{max}$, which gives at the 1 HWHM position: $\Delta\alpha_{max} = c(\alpha_{max})^2/8\pi\Gamma$. In particular, this expression shows that the largest deviation of the residuals scales with the square of the absorption coefficient. A quadratically lower detection limit of 10^{-12}/cm would occur for a more typical absorption coefficient of 10^{-6}/cm.

To conclude, we see that the current performance of OF-CEAS is not far from the level that would enable the observations of this small effect. Using a narrower absorption line at lower sample pressure (close to the Doppler limit) would allow to make the effect stronger. However, in order to observe this effect a high-accuracy knowledge of the absorption line profile must be provided. Conversely, the equations above would allow correcting OF-CEAS spectral profiles in case high precision measurements are needed.

5.7 Precursory Works on OF-Locking of a Semiconductor Laser to an Optical Cavity

5.7.1 Laser Stabilization

The first application of OF-locking of a SC laser to an optical cavity appears to have occurred in atomic physics [8, 43]. The goal was to obtain a spectrally narrow and stable laser emission without a high bandwidth electronic feedback loop. It was discovered that with a low level of optical feedback coming from the resonant field of an optical cavity a SC laser is able to self-lock to the cavity resonance, even as its injection current would be driving its free running frequency away from the resonance. As it was readily noticed, this locking effect is accompanied by a dramatic reduction of the laser linewidth and of its relative intensity noise. The locking range could reach hundreds of MHz and be controlled by the feedback phase and level. By synchronized sweeping of the laser current and the reference cavity length, all the while controlling the feedback phase, a continuous narrow-linewidth scan of the laser frequency could be obtained over small spectral windows ($\sim$1 GHz). The potential of this experimental approach led to several studies, providing a solid theoretical basis for this coupled optical system [21, 44–47]. Even though the main motivation was to obtain a stabilized laser, it was also recognized that the power buildup inside the reference cavity could be exploited for atomic beam collimation [48] or intracavity frequency doubling [38].

In the aforementioned studies, several optical cavity geometries were investigated. The first demonstration of Dahmani et al. [8] used a confocal cavity with a tilted injection providing the resonant optical feedback from an intracavity stable V-shaped light trajectory (produced as a superposition of frequency-degenerate transverse cavity modes, as discussed in the introductory chapter of this book). Shortly thereafter, Li et al. [47] used a V-shaped cavity as described in the previous section. This eliminates the need for a precise adjustment of the cavity length imposed by the confocal condition. Tanner et al. [48] used the small mirror birefringence and a polarization filter to discriminate the direct input mirror reflection from the resonant field. In the work of Hemmerich et al. [38], OF was obtained in a ring-cavity configuration: The parasitic diffusion on optical surfaces is sufficient to weakly excite the counter-propagating mode in a ring-cavity. This fact is actually not as surprising as it may seem at first, considering that intracavity scattering is enhanced by the large intracavity field and constitutes a light source internal to the cavity, directly seeding the counter-propagating cavity mode.

In all these applications, the stabilized laser was not intended, nor readily exploitable, for molecular spectroscopy, which requires a wide and fast tunability. In addition, stiff and short cavities with moderate finesse (about 100) were employed. Indeed, OF locking drives the laser linewidth to orders of magnitude below the width of the cavity resonance, thus a moderate finesse is already sufficient to attain sub-kHz laser emission.

5.7.2 Optical-Feedback Cavity Ring-Down Spectroscopy

Historically, before it could be established that OF is exploitable to obtain CEAS spectra injecting longitudinal cavity modes in rapid succession, another, experimentally simpler configuration was introduced. This uses ring-down events to measure the intracavity sample absorption [7, 49]. In this early implementation of OF-CRDS, a ring-down signal is obtained by applying a square current pulse to the SC laser. The pulse instantaneously induces laser emission as soon as the lasing threshold is exceeded, but the thermal equilibrium of the laser chip is abruptly modified. The initial heat-up of the laser induces a fast chirp in the emitted wavelength before a steady state with a constant temperature is reached. In other words, the abrupt laser current injection initiates a fast frequency sweep which progressively slows down. The entire frequency range spanned by this process depends on the current step, but can be easily on the order of ten GHz or more, or about hundred cavity FSRs. At the beginning of the pulse, the scanning speed is so high that the optical feedback (resulting from the extremely low transient intracavity build up) is not sufficient to affect the laser emission, and no light is detected behind the cavity. As the scanning speed reaches sufficiently low values to make the optical feedback operate, signals appear behind the cavity corresponding to the cavity modes which start to appear "enlarged" by the frequency locking effect. By adjusting a feedback rate in such a way that the locking range exceeds the free spectral range, the cavity is almost always injected at the end of the pulse. The abrupt current drop at pulse end is thus likely followed by a ring-down event at the frequency of the last injected mode. To construct a ring-down spectrum, it is sufficient to scan the current pulse amplitude. In order to correctly map a spectral window, the pulse duration and its amplitude increment are roughly optimized. For a typical feedback rate of 10^{-4} used with a DFB diode laser, the pulse duration is typically a few hundred μs, giving a ring-down acquisition rate in the kHz range.

This OF-CRDS scheme is simple and robust since no special adjustment nor active control of the laser-cavity distance is required. Actually, vibrations are welcome since they favor shot-to-shot variations of the OF phase at the end of the laser pulse. This induces an averaging over the selection of the last cavity mode to which the laser is locked when the ring-down event occurs. If mode matching is not good, this last mode may even fluctuate among different transverse modes, which require different OF phases for optimal laser locking. As it is well known that the ring-down time is different for different transverse modes, it is clear that this OF-CRDS scheme not only provides spectra with a frequency uncertainty of more than one cavity FSR, but can easily suffer of an excess amplitude noise associated with bad selectivity in transverse mode excitation.

Nonetheless, as mentioned before, the simplicity of this OF-CRDS scheme, its insensitivity to vibrations, and the high ring-down acquisition rate have led to the development and successful application of low cost transportable trace gas analyzers. Detection of methane, water vapor, and hydrogen fluoride has been demonstrated both in the laboratory and in field measurements [7, 49]. A typical absorption baseline noise of a few 10^{-8}/cm may be obtained on a single spectrum, and averaging over a few minutes enables a sub-10^{-9}/cm absorption limit with a ring-down time just above 10 μs. More recently, the high acquisition rate has been exploited for aerosols extinction measurements [50, 51] and precise measurement of an ultra-high mirror reflectivity was shown to be possible with an equally simple optical layout [52] (see next section).

5.8 Applications

OF-CEAS has been used in a number of different applications that will be reviewed in this section. We group them by applications area: trace gas detection, isotope ratio analyses, aerosol studies, and finally other applications such as mirror reflectivity measurements. At the same time, different techniques will be looked over: OF-CRDS and OF-CEAS; near-IR DFB diode lasers and ECDLs, as well as mid-IR laser sources; photoacoustic detection; V-shaped, birefringent linear, Brewster-angle, and ring cavity configurations.

5.8.1 Trace Gas Detection

The first spectroscopic applications of OF-CRDS were the detection of trace gases in the near-IR region of the spectrum. Romanini and co-workers demonstrated sensitive detection of H_2O and HF near 1.3 μm and CH_4 at 1.65 μm with at the time a performance only slightly inferior to that of CW-CRDS, but with a greatly simplified and lower cost experimental setup [7, 49, 53]. In 2005 the group demonstrated OF-CEAS with a noise equivalent absorption sensitivity (NEAS) of 5×10^{-10} cm$^{-1}/\sqrt{Hz}$ for a spectral recording over ~ 1 cm^{-1}. This was for a cavity with a ring-down time of 17 μs, yielding a sub-ppb detection limit for HF at 1.312 μm in an open V-shaped cavity [1]. Courtillot et al. extended the OF-CEAS method to ECDLs and to the blue region of the optical spectrum, by measuring NO_2 at 411 nm with an estimated detection limit below 1 ppb (with few seconds averaging) [30].

Measurements of atmospheric methane concentrations using an OF-CEAS instrument on board of NASA's DC-8 research platform served to demonstrate the relative insensitivity of the technique to environmental factors, in particular vibrations, but also changing ambient temperature and pressure [18]. Operating at 1.6 μm, the airborne spectrometer was able to resolve fast (0.3 s, limited by the gas cell exchange time) methane concentration changes of 1 ppb. This instrument was not temperature controlled, still its stability was established to be about 20 ppb (pk-pk) during a 10-days long direct comparison with a half-hourly automatically calibrated gas chromatograph.

An extension of the technique towards the mid-IR was first demonstrated using a DFB diode laser emitting near 2.33 μm for the detection of CO and CH_4 in fumarolic gases (in the presence of about 98 % CO_2) [41]. The instrument achieved a relatively poor NEAS of 3×10^{-9} cm$^{-1}/\sqrt{Hz}$, attributed partly to a short cavity ring-down time τ_0 of 6.5 μs, partly to residual optical interference fringes. An improved version of the instrument was applied to exhaled breath analysis for CO and CH_4, as well as NH_3, in a hospital setting [54]. With better mirrors ($\tau_0 = 30$ μs) and fringe suppression, a tenfold reduction in NEAS was realized in this case.

A rather different approach to the same multiple gas analysis was taken by Cermak et al. [32], who combined the OF-CEAS V-shaped cavity with a VECSEL operating near 2.33 μm. The short and stiff external cavity design of the optically pumped laser enabled broad tuning over as much as 50 cm^{-1} with low frequency noise, resulting in a spectrometer that exhibited a NEAS comparable to that of more common near-IR DFB diode laser OF-CEAS setups.

The high potential of combining OF-CEAS with thermo-electrically cooled DFB QCLs was illustrated by Maisons et al. [33] and by Hamilton et al. [34]. The first study used a 4.46 μm-QCL to construct a spectrometer that achieved a NEAS of 3×10^{-9} cm$^{-1}/\sqrt{Hz}$ for the registration of a ~ 1 cm^{-1} wide spectrum of diluted (1:25) atmospheric air. The corresponding detection sensitivities were 35 pptv for N_2O and 15 ppbv for CO_2 with $\tau_0 = 8.5$ μs. The second group used a QCL operating near 7.78 μm for the simultaneous detection of CH_4 and N_2O with a spectrometer NEAS of 5.5×10^{-8} cm$^{-1}/\sqrt{Hz}$ and $\tau_0 = 3.5$ μs. These experiments serve to highlight the high cavity throughput advantage in the mid-IR of OF-CEAS over alternative CEAS-based techniques in this part of the optical spectrum, considering that the QCL used in this study exhibited a modest output power level of, respectively, 15 and 4 mW maximum (comparable to near-IR DFB diode laser output powers), while the detector sensitivities in the mid-IR are typically two to three orders of magnitude lower than those used in the near-IR. As mirror quality continues to improve and cavity transmission to decrease, this is likely to become an even larger advantage (see more recent results in [35]).

Motto-Ros et al. investigated the B-band of molecular oxygen at 688 nm using a novel optical layout consisting of a linear cavity with a microscope cover plate at a near-to-Brewster angle with respect to the cavity symmetry axis and approximately centered between the end-mirrors [40]. The Brewster plate serves to couple the laser light into the cavity, while simultaneously enabling an adjustment of the cavity finesse. This in turn offers the possibility to increase the dynamic range of the technique by up to one order of magnitude. Conceivably, the microscope plate also provides a convenient way to introduce a liquid or solid absorber, ad- or ab-sorbed on its surface.

Another interesting optical layout was investigated by Hamilton et al. [37]. They used a three-mirror ring cavity to record weak transitions of two molecular oxygen isotopologues in the A-band near 762 nm, demonstrating a NEAS of $\sim 2 \times 10^{-9}$ cm$^{-1}/\sqrt{Hz}$. Potential advantages of the ring configuration are the ability to easily modulate the optical feedback level by means of a retro-reflecting mirror placed after one of the cavity high-reflectivity mirrors, without introducing an attenuating optical element in the beam path between laser and cavity, and the absence of the even odd mode structure.

Yet another interesting development for trace gas detection has been the combination of OF-CEAS with photoacoustic detection, demonstrated with the sensitive detection of high overtones of the water molecule near 635 nm [55]. The NEAS scales linearly with laser power and in this study was demonstrated to reach of 4.4×10^{-9} cm$^{-1}/\sqrt{Hz}$ for 1.7 mW of incoming laser power. The technique shares the principal advantages (such as its zero-base-line character), as well as disadvantages (among others, the dependence of its sensitivity on the gas matrix composition), with more common photoacoustic implementations.

5.8.2 Isotope Ratio Analyses

Isotope ratio measurements are particularly prone to benefit from a sensitive technique as CEAS to access the generally weak absorption lines of the isotopologues (see [56] and [57] for an introduction and review of the application of laser-based techniques in this field of research). We have therefore been an early adopter of OF-CEAS for water isotope ratio measurements. In early 2004 a spectrometer flew on the NASA DC-8 (together with the above mentioned CH_4 analyzer), demonstrating the feasibility of measuring isotope ratios with an airborne instrument operating in the near-IR and with good precision, even at the low water concentration encountered at an altitude of 13 km in the upper troposphere. This first water isotope spectrometer was characterized by a NEAS of 4×10^{-10} cm$^{-1}/\sqrt{Hz}$ for a spectral recording over 200 modes ($\sim$1 cm^{-1}), a τ_0 of 20 µs, and a cavity base length of 50 cm. It measured the $^2H/^1H$, $^{17}O/^{16}O$, and $^{18}O/^{16}O$ isotopic ratios with a precision of 9 ‰, 3 ‰, and 1 ‰ after averaging over 30 s [39]. A second generation of this spectrometer was flown on the European M55-Geophysica high-altitude research aircraft during the 2006 AMMA/SCOUT-03 (African Monsoon Multidisciplinary Analysis/Stratospheric Climate links with emphasis on the Upper Troposphere and lower stratosphere) campaign in Burkina Faso [58, 59]. The same instrument was used to measure near-surface tropospheric water moisture at two orders of magnitude higher water mixing ratios [60]. More recently we have carried out high-precision isotope measurements of moisture at the Antarctic station of Troll with a spectrometer much improved in terms of cavity temperature and pressure stabilization, and with an increased empty cavity ring-down τ_0 of 130 µs. [61]

Wehr et al. have used a DFB diode laser at 2.007 µm to access the $\nu_1 + 2\nu_2^0 + \nu_3$ combination band of CO_2 [62]. This band is almost two orders of magnitude stronger than the $\nu_1 + 4\nu_2^0 + \nu_3$ combination band at 1.6 µm. Despite this, the demonstrated performance on the $^{13}CO_2/^{12}CO_2$ isotopic ratio determination was somewhat disappointing and attributed to mechanical instability in the setup and a relatively strong residual optical fringe due to the use of a flat, non-anti-reflection coated mirror substrate.

Currently, in the Grenoble laboratory, efforts are underway to miniaturize the water isotope spectrometer for it to be embarked on a novel ice-core drilling device to be deployed in 2015 at Dome Concordia in central Antarctica. The goal is to measure in-situ the isotopic composition of the melted ice, while the probe drills down to bedrock at about 3 km under the ice surface in one Antarctic summer season ($\sim$70 days). At the same time, a QCL-equipped OF-CEAS system is being developed for the accurate measurement of carbon dioxide isotopologues recovered from tiny air bubbles contained in ice-cores that have been drilled in the traditional manner. The sensitivity boost required to deal with the very small sample size is to come from exciting strong CO_2 isotopologue transitions in the 4.35 µm region, while maintaining a small cavity volume of about 20 mL.

5.8.3 Aerosol Studies

Orr-Ewing and co-workers realized the potential of OF-CEAS for aerosol measurements, in particular its capability to perform CRDS measurements with a high repetition rate (1.25 kHz). This enabled multiple measurements of the extinction by single spherical polymer beads, revealing the dependence of light scattering on the position of a particle within the laser beam. A model is proposed to explain quantitatively this phenomenon [50]. Further studies show that the linear relation between the extinction coefficient and its variance, predicted by a Poisson statistical model, allows for the determination of the extinction cross-section without prior knowledge of the particle number density [51].

5.8.4 Other Applications

Morville and Romanini demonstrated that the weak birefringence of high-reflectivity dielectric mirrors can be measured with good spatial resolution and a sensitivity better than 10^{-7} radians by mapping the optical feedback rate as a function of the mirror rotation around the cavity symmetry axis [36]. More recently[63] a similar scheme was used to obtain the static Kerr effect in a gas by lock-in detection of the birefringence induced by a modulated transverse electric field. The state-of-the-art detection limit of 3×10^{-13} radians was attained by averaging shot-noise-limited measurements as long as one hour.

Gong and colleagues [64] used OF with a very simple linear cavity optical layout in combination with a spectrally wide, multi-mode Fabry-Perot diode laser, to measure an ultra-high mirror reflectivity with a precision of 0.3 ppm. This presents a gain in sensitivity of two-orders of magnitude compared with previous phase-shift CRDS mirror reflectivity measurements, at a significantly reduced cost. In this particular application, the direct OF from the mirror back surface is acceptable as the resulting perturbed laser frequency remains within the high-reflectivity optical bandwidth of the mirrors. More recently Tong-Kai et al. repeated the same experiment, but with both a linear and a V-shaped cavity configuration, and found the two to give comparable results [65].

5.9 Conclusions

In this chapter we have presented in detail the technique of cavity enhanced absorption spectroscopy exploiting selective optical feedback from the cavity resonances to the laser. The technique has several advantages over CEAS schemes that rely on an electronic control of the laser emission frequency to achieve sufficient spectral overlap between the laser emission and a cavity resonance. Its implementation is straightforward and requires few optical components. The spectral narrowing and frequency locking to $TEM_{0,0}$ cavity modes induced by the optical feedback achieve a higher level of performance than accessible with complex, high-performance electronic feedback circuitry. This results in a fast and complete light injection of the cavity on a well-defined grid of equally spaced frequencies, and provides clear advantages in terms of acquisition time and signal-to-noise ratio of the measurement of absorption lines as both absorption and frequency scales are precisely defined. Although the technique requires that the laser exhibits a certain level of sensitivity to optical feedback, it has been shown to work with a large variety of modern semi-conductor lasers, including DFB diode lasers, ECDLs, VECSELs, and QCLs, together spanning the spectral range from 400 nm to beyond 8 μm. Admittedly, successful application of the technique requires a good understanding of its underlying physical principles, which are arguably more complex than those of competing techniques.

We have started the chapter with an intuitive description of the OF-CEAS technique in order to ease the reader through the following more detailed description of the time and frequency behavior of a coupled diode-laser-cavity system, which is reasonably simple in the quasi stationary approximation. Model equations have been derived and used to analyze the cavity transmission when optical feedback modifies the laser frequency during a scan of the free laser frequency (or the injection current). In return, this analysis clearly identifies the configuration space of the system parameters (laser-cavity distance, feedback rate and feedback phase, frequency scanning speed) and their limits imposed by the cavity geometry and losses, and by the laser sensitivity and robustness to optical feedback.

After this general outline of the OF-CEAS working principle, we have discussed various implementations reported in the literature up to date, including the different laser sources as well as the cavity configurations that have been successfully applied to OF-CEAS.

We dedicated an entire section to a calibration procedure that we developed, based on the measurement of a single ring-down event during the laser scan, which confers to the technique the advantage of absolute loss measurements.

We then presented typical performance arising from this procedure when combined with the property of equally spaced frequency data points provided by the cavity comb sampling of OF-CEAS spectra. Limiting factors on both scales of a spectrum (absorption and frequency) have been discussed including perspectives of improvement. Absorption baseline noise on single spectra covering 1 cm^{-1} acquired in 100 ms is presently around 10^{-9}/cm, as obtained with cavities of moderate finesse ($\mathcal{F} \sim 10\,000$) from the blue range to the IR range using DFB multiple-quantum-wells diode lasers, ECDLs, and DFB room-temperature QCLs.

We have provided a brief historical overview of developments leading to the currently most common implementation of OF-CEAS. The last section is devoted to the different fields of application of OF-CEAS, from trace to isotope ratio analysis, to birefringence, aerosols, or mirror reflectivity measurements.

References

1. J. Morville, S. Kassi, M. Chenevier, D. Romanini, Appl. Phys. B, Lasers Opt. **80**(8), 1027 (2005). doi:10.1007/s00340-005-1828-z. http://www.springerlink.com/index/10.1007/s00340-005-1828-z

2. A. Kachanov, D. Romanini, M. Chenevier, A. Garnache, F. Stoeckel, in *Part of SPIE Conference on Air Monitoring and Detection of Chemical and Biological Agents II*, Boston, MA, 19–22 September 1999. Proc. SPIE, vol. 3855 (1999), p. 51

3. D. Romanini, A.A. Kachanov, J. Morville, M. Chenevier, in *Part of SPIE EUROPTO Conference on Environmental Sensing and Applications, CLEO-Europe*, Munich, 14–17 June 1999. Proc. SPIE, vol. 3821 (1999), p. 94

4. J. Morville, D. Romanini, Appl. Phys. B **74**, 495 (2002)

5. J. Morville, D. Romanini, M. Chenevier, J. Phys. IV **12**, 389 (2002)

6. J. Morville, M. Chenevier, A.A. Kachanov, D. Romanini, in *Optical Spectroscopic Techniques, Remote Sensing, and Instrumentation for Atmospheric and Space Research IV*. Proc. SPIE, vol. 4485 (2002), p. 236

7. J. Morville, D. Romanini, A.A. Kachanov, M. Chenevier, Appl. Phys. B, Lasers Opt. **78**(3–4), 465 (2004). doi:10.1007/s00340-003-1363-8. http://www.springerlink.com/openurl.asp?genre=article&id=doi:10.1007/s00340-003-1363-8

8. B. Dahmani, L.W. Hollberg, R.E. Drullinger, Opt. Lett. **12**(11), 876 (1987). doi:10.1364/OL.12.000876. http://www.ncbi.nlm.nih.gov/pubmed/19741901

9. H. Patrick, C.E. Wieman, Rev. Sci. Instrum. **62**(11), 2593 (1991). doi:10.1063/1.1142236

10. A.L. Schawlow, C.H. Townes, Phys. Rev. **112**, 1940 (1958)

11. C. Henry, R. Logan, K. Bertness, J. Appl. Phys. **52**(7), 4457 (1981)

12. A. Yariv, *Quantum Electronics*, 3rd edn. (Wiley, New York, 1989)

13. J.C. Habig, J. Nadolny, J. Meinen, H. Saathoff, T. Leisner, Appl. Phys. B, Lasers Opt. **106**(2), 491 (2011). doi:10.1007/s00340-011-4804-9. http://www.springerlink.com/index/10.1007/s00340-011-4804-9

14. C.E. Wieman, L.W. Hollberg, Rev. Sci. Instrum. **62**(1), 1 (1991)

15. R.W.P. Drever, J.L. Hall, F.V. Kowalski, J. Hough, G.M. Ford, A.J. Munley, H. Ward, Appl. Phys. B, Lasers Opt. **31**(2), 97 (1983). doi:10.1007/BF00702605. http://www.springerlink.com/index/10.1007/BF00702605

16. D.A. Long, A. Cygan, R.D. van Zee, M. Okumura, C.E. Miller, D. Lisak, J.T. Hodges, Chem. Phys. Lett. **536**, 1 (2012). doi:10.1016/j.cplett.2012.03.035. http://linkinghub.elsevier.com/retrieve/pii/S0009261412003466

17. M.S. Taubman, T.L. Myers, B.D. Cannon, R.M. Williams, Spectrochim. Acta, Part A, Mol. Biomol. Spectrosc. **60**(14), 3457 (2004). doi:10.1016/j.saa.2003.12.057. http://www.ncbi.nlm.nih.gov/pubmed/15561632

18. D. Romanini, M. Chenevier, S. Kassi, M. Schmidt, C. Valant, M. Ramonet, J. Lopez, H.J. Jost, Appl. Phys. B, Lasers Opt. **83**(4), 659 (2006). doi:10.1007/s00340-006-2177-2. http://www.springerlink.com/index/10.1007/s00340-006-2177-2

19. C.H. Henry, IEEE J. Quantum Electron. **18**(2), 259 (1982). doi:10.1109/JQE.1982.1071522. http://ieeexplore.ieee.org/lpdocs/epic03/wrapper.htm?arnumber=1071522

20. K.K. Lehmann, D. Romanini, J. Chem. Phys. **105**(23), 10263 (1996). doi:10.1063/1.472955. http://link.aip.org/link/JCPSA6/v105/i23/p10263/s1&Agg=doi

21. P. Laurent, A. Clairon, C. Breant, IEEE J. Quantum Electron. **25**(6), 1131 (1989)

22. I. Courtillot, J. Morville, V. Motto-Ros, D. Romanini, Appl. Phys. B **85**, 407 (2006)

23. C.H. Henry, J. Appl. Phys. **52**(7), 4457 (1981). doi:10.1063/1.329371. http://link.aip.org/link/?JAP/52/4457/1&Agg=doi

24. C.A. Green, N.K. Dutta, W. Watson, Appl. Phys. Lett. **50**(20), 1409 (1987). doi:10.1063/1.97836. http://link.aip.org/link/APPLAB/v50/i20/p1409/s1&Agg=doi

25. K. Petermann, *Laser Diode Modulation and Noise* (Kluwer Scientific, Tokyo, 1991)

26. Y. Arakawa, A. Yariv, IEEE J. Quantum Electron. **QE-21**(10), 1666 (1985)

27. T. Aellen, R. Maulini, R. Terazzi, N. Hoyler, M. Giovannini, J. Faist, Appl. Phys. Lett. **89**, 091121 (2006)
28. S. Bartalini, S. Borri, P. Cancio, A. Castrillo, I. Galli, G. Giusfredi, D. Mazzotti, L. Gianfrani, P.D. Natale, Phys. Rev. Lett. **104**, 083904 (2010)
29. J. von Staden, T. Gensty, W. Elsäßer, G. Giuliani, C. Mann, Opt. Lett. **31**(17), 2574 (2006). doi:10.1364/OL.31.002574. http://ol.osa.org/abstract.cfm?URI=ol-31-17-2574
30. I. Courtillot, J. Morville, V. Motto-Ros, D. Romanini, Appl. Phys. B, Lasers Opt. **85**(2–3), 407 (2006). doi:10.1007/s00340-006-2354-3. http://www.springerlink.com/index/10.1007/s00340-006-2354-3
31. V. Motto-Ros, J. Morville, P. Rairoux, Appl. Phys. B, Lasers Opt. **87**(3), 531 (2007). doi:10.1007/s00340-007-2618-6. http://www.springerlink.com/index/10.1007/s00340-007-2618-6
32. P. Cermak, M. Triki, A. Garnache, L. Cerutti, D. Romanini, IEEE Photonics Technol. Lett. **22**(21), 1607 (2010). doi:10.1109/LPT.2010.2075922. http://ieeexplore.ieee.org/lpdocs/epic03/wrapper.htm?arnumber=5570899
33. G. Maisons, P. Gorrotxategi-Carbajo, M. Carras, D. Romanini, Opt. Lett. **35**(21), 3607 (2010)
34. D.J. Hamilton, A.J. Orr-Ewing, Appl. Phys. B, Lasers Opt. **102**(4), 879 (2010). doi:10.1007/s00340-010-4259-4. http://www.springerlink.com/index/10.1007/s00340-010-4259-4
35. P. Gorrotxategi-Carbajo, E. Fasci, I. Ventrillard, M. Carras, G. Maisons, D. Romanini, Appl. Phys. B, 309–314 (2013)
36. J. Morville, D. Romanini, Appl. Phys. B, Lasers Opt. **74**(6), 495 (2002). doi:10.1007/s003400200854. http://www.springerlink.com/Index/10.1007/s003400200854
37. D.J. Hamilton, M.G.D. Nix, S.G. Baran, G. Hancock, A.J. Orr-Ewing, Appl. Phys. B, Lasers Opt. **100**(2), 233 (2009). doi:10.1007/s00340-009-3811-6. http://www.springerlink.com/index/10.1007/s00340-009-3811-6
38. A. Hemmerich, D.H. McIntyre, C. Zimmermann, T.W. Hänsch, Opt. Lett. **15**(7), 372 (1990). doi:10.1364/OL.15.000372. http://www.opticsinfobase.org/abstract.cfm?URI=ol-15-7-372
39. E.R.T. Kerstel, R.Q. Iannone, M. Chenevier, S. Kassi, H.J. Jost, D. Romanini, Appl. Phys. B, Lasers Opt. **85**(2–3), 397 (2006). doi:10.1007/s00340-006-2356-1. http://www.springerlink.com/index/10.1007/s00340-006-2356-1
40. V. Motto-Ros, M. Durand, J. Morville, Appl. Phys. B, Lasers Opt. **91**(1), 203 (2008). doi:10.1007/s00340-008-2950-5. http://www.springerlink.com/index/10.1007/s00340-008-2950-5
41. S. Kassi, M. Chenevier, L. Gianfrani, A. Salhi, Y. Rouillard, A. Ouvrard, D. Romanini, Opt. Express **14**(23), 11442 (2006)
42. S. Kassi, A. Campargue, J. Chem. Phys. **137**(23), 234201 (2012). doi:10.1063/1.4769974. http://www.ncbi.nlm.nih.gov/pubmed/23267478
43. L.W. Hollberg, M. Ohtsu, Appl. Phys. Lett. **53**(11), 944 (1988). doi:10.1063/1.100077. http://link.aip.org/link/APPLAB/v53/i11/p944/s1&Agg=doi
44. A. Clairon, B. Dahmani, P. Laurent, C. Breant, in *Proceeding of EFTF88, Second European Frequency and Time Forum*, Neuchatel, Switzerland (1988), p. 537
45. H. Li, N.B. Abraham, Appl. Phys. Lett. **53**(23), 2257 (1988). doi:10.1063/1.100271. http://link.aip.org/link/APPLAB/v53/i23/p2257/s1&Agg=doi
46. H. Li, N.B. Abraham, IEEE J. Quantum Electron. **25**(8), 1782 (1989). doi:10.1109/3.34036. http://ieeexplore.ieee.org/lpdocs/epic03/wrapper.htm?arnumber=34036
47. H. Li, H.R. Telle, IEEE J. Quantum Electron. **25**(3), 257 (1989). doi:10.1109/3.18538. http://ieeexplore.ieee.org/lpdocs/epic03/wrapper.htm?arnumber=18538
48. C.E. Tanner, B.P. Masterson, C.E. Wieman, Opt. Lett. **13**(5), 357 (1988). doi:10.1364/OL.13.000357. http://www.opticsinfobase.org/abstract.cfm?URI=ol-13-5-357
49. J. Morville, M. Chenevier, A.A. Kachanov, D. Romanini, in *Proceedings of SPIE*, vol. 4485, ed. by A.M. Larar, M.G. Mlynczak (2002), pp. 236–243. doi:10.1117/12.454256
50. T.J. Butler, J.L. Miller, A.J. Orr-Ewing, J. Chem. Phys. **126**, 174302 (2007). doi:10.1063/1.2723735
51. T.J.A. Butler, D. Mellon, J. Kim, J. Litman, A.J. Orr-Ewing, J. Phys. Chem. A **113**(16), 3963 (2009). doi:10.1021/jp810310b

52. Y. Gong, B. Li, Y. Han, Appl. Phys. B, Lasers Opt. **93**(2–3), 355 (2008),. doi:10.1007/s00340-008-3247-4. http://www.springerlink.com/index/10.1007/s00340-008-3247-4

53. D. Romanini, A.A. Kachanov, J. Morville, M. Chenevier, in *Proc. SPIE*, vol. 3821 (1999), p. 94. doi:10.1117/12.364170. http://scholar.google.com/scholar?hl=en&btnG=Search&q=intitle:Measurement+of+trace+gases+by+diode+laser+cavity+ringdown+spectroscopy+.#0

54. I. Ventrillard, T. Gonthiez, C. Clerici, D. Romanini, J. Biomed. Opt. **14**(6), 64026 (2009). doi:10.1117/1.3269677. http://www.ncbi.nlm.nih.gov/pubmed/20059264

55. M. Hippler, C. Mohr, K.A. Keen, E.D. McNaghten, J. Chem. Phys. **133**(4), 44308 (2010). doi:10.1063/1.3461061. http://www.ncbi.nlm.nih.gov/pubmed/20687651

56. E.R.T. Kerstel, in *Handbook of Stable Isotope Analytical Techniques*, vol. 1, ed. by P.A. De Groot (Elsevier, Amsterdam, 2004), pp. 759–787

57. E.R.T. Kerstel, L. Gianfrani, Appl. Phys. B, Lasers Opt. **92**(3), 439 (2008). doi:10.1007/s00340-008-3128-x. <GotoISI>://000258703600017

58. R.Q. Iannone, S. Kassi, H.J. Jost, M. Chenevier, D. Romanini, H.A.J. Meijer, S. Dhaniyala, M. Snels, E.R.T. Kerstel, Isot. Environ. Health Stud. **45**(4), 303 (2009). doi:10.1080/10256010903172715. http://www.ncbi.nlm.nih.gov/pubmed/19670069

59. F. Cairo, J.P. Pommereau, K.S. Law, H. Schlager, A. Garnier, F. Fierli, M. Ern, M. Streibel, S. Arabas, S. Borrmann, J.J. Berthelier, C. Blom, T. Christensen, F. D'Amato, G. Di Donfrancesco, T. Deshler, A. Diedhiou, G. Durry, O. Engelsen, F. Goutail, N.R.P. Harris, E.R.T. Kerstel, S. Khaykin, P. Konopka, A. Kylling, N. Larsen, T. Lebel, X. Liu, A.R. MacKenzie, J. Nielsen, A. Oulanowski, D.J. Parker, J. Pelon, J. Polcher, J.A. Pyle, F. Ravegnani, E.D. Rivière, A.D. Robinson, T. Röckmann, C. Schiller, F. Simões, L. Stefanutti, F. Stroh, L. Some, P. Siegmund, N. Sitnikov, J.P. Vernier, C.M. Volk, C. Voigt, M. von Hobe, S. Viciani, V. Yushkov, Atmos. Chem. Phys. Discuss. **9**(5), 19713 (2009)

60. R.Q. Iannone, D. Romanini, O. Cattani, H.A.J. Meijer, E.R.T. Kerstel, J. Geophys. Res. **115**(D10), 1 (2010). doi:10.1029/2009JD012895. http://www.agu.org/pubs/crossref/2010/2009JD012895.shtml

61. J. Ladsberg, E. Kerstel, D. Romanini, Appl. Phys. Lett. (2013 prepared for submission)

62. R. Wehr, S. Kassi, D. Romanini, L. Gianfrani, Appl. Phys. B, Lasers Opt. **92**(3), 459 (2008). doi:10.1007/s00340-008-3086-3. http://www.springerlink.com/index/10.1007/s00340-008-3086-3

63. M. Durand, J. Morville, D. Romanini, Phys. Rev. A **82**(3), 031803(R) (2010). doi:10.1103/PhysRevA.82.031803. http://link.aps.org/doi/10.1103/PhysRevA.82.031803

64. Y. Gong, B. Li, Appl. Opt. **47**(21), 3860 (2008). http://www.ncbi.nlm.nih.gov/pubmed/18641755

65. Z. Tong-Kai, Q. Zhe-Chao, H. Yan-Ling, L. Bin-Cheng, Chin. Phys. Lett. **27**(10), 100701 (2010). doi:10.1088/0256-307X/27/10/100701

Chapter 6
NICE-OHMS—Frequency Modulation Cavity-Enhanced Spectroscopy—Principles and Performance

Ove Axner, Patrick Ehlers, Aleksandra Foltynowicz, Isak Silander, and Junyang Wang

Abstract Noise-immune cavity-enhanced optical heterodyne molecular spectroscopy (NICE-OHMS) is a sensitive technique for detection of molecular species in gas phase. It is based on a combination of frequency modulation for reduction of noise and cavity enhancement for prolongation of the interaction length between the light and a sample. It is capable of both Doppler-broadened and sub-Doppler detection with absorption sensitivity down to the 10^{-12} and 10^{-14} $Hz^{-1/2}\,cm^{-1}$ range, respectively. This chapter provides a thorough description of the basic principles and the performance of the technique.

6.1 Introduction

Noise-immune cavity-enhanced optical heterodyne molecular spectroscopy (NICE-OHMS) is a powerful technique for detection of molecular compounds in gas phase that is based on a successful combination of two concepts: frequency modulation spectroscopy (FMS) for reduction of noise, and cavity enhancement, for prolongation of the interaction length between the light and the sample [1–5]. Choosing the modulation frequency equal to the free spectral range (FSR) of the cavity allows FMS to be performed inside the cavity. In addition, all spectral components of the FM triplet are transmitted through the cavity in an identical manner, whereby they are affected by any frequency-to-amplitude noise conversion in the same way. This implies that the amplitude noise cancels in the FM signal transmitted through the cavity, which is referred to as "noise immunity" [2, 4]. As a result, NICE-OHMS allows FMS to be performed as if the cavity would not be present, yet fully benefiting from the prolonged interaction length inside the resonator.

NICE-OHMS has the ability to detect a wide variety of signals. Owing to the use of FMS, it can measure both absorption and dispersion signals. Due to the presence of high intensity counter-propagating beams inside the cavity, the technique can detect both sub-Doppler (sD) and Doppler-broadened (Db) signals. Finally, in order

---

O. Axner (✉) · P. Ehlers · A. Foltynowicz · I. Silander · J. Wang
Department of Physics, Umeå University, 901 87 Umeå, Sweden
e-mail: ove.axner@physics.umu.se

G. Gagliardi, H.-P. Loock (eds.), *Cavity-Enhanced Spectroscopy and Sensing*,
Springer Series in Optical Sciences 179, DOI 10.1007/978-3-642-40003-2_6,
© Springer-Verlag Berlin Heidelberg 2014

to reduce any remaining low-frequency noise or to remove the Db background on which the sD signals reside, an additional layer of wavelength modulation (wm) can be applied [5].

The technique was originally developed in the late 1990s by John Hall, Jun Ye, and Long-Sheng Ma at JILA, Boulder, CO, for frequency standard applications [1–3]. In its original realization, using a well-stabilized fixed-frequency Nd:YAG laser and a cavity with a finesse of 100 000, a detection sensitivity down to astonishing relative absorption of 5×10^{-13} at 1 s (corresponding to a length-normalized absorption of 1×10^{-14} cm^{-1}) was demonstrated for sD detection of C_2HD at 1064 nm [3]. This is better than any other laser-based absorption technique has ever demonstrated.

In the following years, several attempts were made to implement the technique with the use of tunable lasers, mostly in the near-infrared wavelength region, for a variety of applications. NICE-OHMS has been used for sD spectroscopy of molecular overtone bands in CH_4 [6] and CH_3I [7], Db detection of weak magnetic dipole transitions of O_2 [8], Db detection of the sixth overtone band of NO [9, 10], sD detection of N_2O in the mid-infrared [11], Db measurement of ultraweak transitions in the visible region of molecular oxygen [12, 13], Db and sD detection of C_2H_2 [14–16], Db detection of CH_4 [17] and the HO_2 radical [18], and for velocity modulation spectroscopy of a fast molecular ion beam [19, 20]. For a detailed description of various applications of NICE-OHMS, see Chap. 7 of this book [21]. These works demonstrated detection sensitivities typically in the 10^{-9}–10^{-10} range (corresponding to 10^{-10}–10^{-11} cm^{-1} Hz$^{-1/2}$), which are in pair with or better than the best other cavity-enhanced techniques for sensitive trace gas analysis [22, 23]. There are several reasons why these sensitivities are inferior to that of the first NICE-OHMS demonstration. First, tunable lasers have larger linewidth than fixed-frequency solid-state lasers, which makes locking them to a narrow cavity mode more demanding; second, the finesse of the cavities is most often in the 10^3–10^4 range, one-to-two orders of magnitude below that of the first realization; and third, when the technique is used for detection of Db signals it is affected by background signals to a larger extent than when sD signals are detected.

The biggest technological hurdle of NICE-OHMS is the requirement of a rather tight lock of the laser frequency to a mode of the cavity. Since the linewidth of tunable lasers is often larger than the width of a transmission mode of a high finesse cavity, high servo bandwidths are needed to efficiently couple the laser light into the cavity. Using a narrow linewidth Er-doped fiber laser (linewidth below 1 kHz on sub-ms timescales), which facilitates locking, we have demonstrated in a series of papers [14–16, 24–28] that NICE-OHMS can be realized in a compact and robust setup capable of hours of uninterrupted operation. The detection sensitivity of fiber-laser-based NICE-OHMS (FLB-NICE-OHMS) is currently 5.6×10^{-12} cm^{-1} Hz$^{-1/2}$ (corresponding to a single-pass absorption of 7.2×10^{-11} at 10 s) [28], which is the best so far achieved for Db detection by NICE-OHMS. FLB-NICE-OHMS [16, 24–28], together with distributed-feedback-laser (DFB) based NICE-OHMS [29–32], have been used to assess a number of properties of the technique, e.g. the effect of optical saturation on both the Db and the sD signals [24, 25, 33], to validate theoretical descriptions of both frequency-modulated and wavelength-modulated Db and sD NICE-OHMS signals under a variety of conditions [24–26, 30–33], and for the establishment of optimum detection conditions [16, 27].

In this chapter we describe the basic principles and performance of the NICE-OHMS technique. The theoretical description of NICE-OHMS signals is given in Sect. 6.2. Since NICE-OHMS is based on frequency modulation spectroscopy, we start this section by recalling the basics of this technique. We then discuss the various line shapes of Db and sD NICE-OHMS signals that appear under different conditions. In Sect. 6.3 we describe a typical experimental setup as well as give a more detailed description of the FLB-NICE-OHMS system. The performance, sensitivity, and some practical limitations of the technique are discussed in Sect. 6.4. We summarize, conclude, and present a future outlook in Sect. 6.5.

6.2 Theory—NICE-OHMS Analytical Signals

6.2.1 Frequency Modulation Spectroscopy

In FMS the phase of the electric field is modulated at a radio frequency, v_m, with an amplitude represented by a modulation index, β, according to

$$\tilde{\mathbf{E}}^{fm}(v_c, t) = \frac{E_0}{2}\hat{\boldsymbol{\varepsilon}}e^{i[2\pi v_c t + \beta \sin(2\pi v_m t)]} + \text{c.c.}, \tag{6.1}$$

where E_0 is the field amplitude, $\hat{\boldsymbol{\varepsilon}}$ is a unit vector representing the direction of field polarization, v_c is the carrier frequency, and c.c. stands for the complex conjugate. For small modulation indices ($\beta < 1$) this field consists mainly of a triplet comprising a carrier and two sidebands separated by the modulation frequency, whose amplitudes are given by Bessel functions of order j, $J_j(\beta)$, and can be written as [34]

$$\tilde{\mathbf{E}}^{fm}(v_c, t) = \frac{E_0}{2}\hat{\boldsymbol{\varepsilon}}\left[J_0(\beta) + J_1(\beta)e^{i2\pi v_m t} - J_1(\beta)e^{-i2\pi v_m t}\right]e^{i2\pi v_c t} + \text{c.c.} \tag{6.2}$$

The signal is given by the component of the intensity at the modulation frequency. For a purely frequency modulated electric field, i.e. a fully balanced triplet, the beat signal of the lower sideband and the carrier cancels that of the upper sideband and the carrier. This implies that the intensity does not contain any component at the modulation frequency. Thus, there is no FM signal in the absence of analyte, which is the main advantage of FMS.

When the FM triplet passes through a gaseous sample in the vicinity of a molecular transition the balance of the triplet is compromised. The transmitted electric field becomes

$$\tilde{\mathbf{E}}_T^{fm}(v_c, t) = \frac{E_0}{2}\hat{\boldsymbol{\varepsilon}}\left[\tilde{T}_0(v_c)J_0(\beta) + \tilde{T}_1(v_c)J_1(\beta)e^{i2\pi v_m t}\right.$$
$$\left. - \tilde{T}_{-1}(v_c)J_1(\beta)e^{-i2\pi v_m t}\right]e^{i2\pi v_c t} + \text{c.c.}, \tag{6.3}$$

where $\tilde{T}_j(v_c) = \exp[-\delta_j(v_c) - i\phi_j(v_c)]$ is the complex transmission function of the analyte, where in turn $\delta_j(v_c)$ and $\phi_j(v_c)$ are the amplitude attenuation and the optical phase shift of the electrical field at the frequency $v_j = v_c + jv_m$, where $j = 0, \pm 1$, respectively. The intensity of a frequency modulated field transmitted through a sample contains a term oscillating at the modulation frequency, which, in the limit of low absorption (i.e. for $|\delta_0 + \delta_{\pm 1}| \ll 1$ and $|\phi_0 - \phi_{\pm 1}| \ll 1$), can be written as

$$I_T^{fm}(v_d, t) = 2I_0 J_0(\beta) J_1(\beta)$$
$$\times \left\{ \left[\phi_{-1}(v_d) - 2\phi_0(v_d) + \phi_1(v_d) \right] \sin(2\pi v_m t) \right.$$
$$\left. + \left[\delta_{-1}(v_d) - \delta_1(v_d) \right] \cos(2\pi v_m t) \right\}, \tag{6.4}$$

where I_0 (W/m^2) is the intensity of light in the absence of a sample and where we have introduced v_d as the detuning of the laser frequency from the transition center frequency v_0, i.e. as $v_d = v_0 - v_c$. The sine term, which is in-phase with the modulation [see Eq. (6.1)] and therefore referred to as the in-phase signal, is proportional to the difference between the phase shift experienced by the carrier and the average phase shift experienced by the sidebands, while the cosine term (the out-of-phase signal) is proportional to the difference of the attenuation experienced by the two sidebands.

The FMS signal is obtained by demodulation of the detector signal at the modulation frequency at a given detection phase θ_{fm}, which yields a DC signal

$$S_T^{fm}(v_d, \theta_{fm}) = \eta_{fm} P_0 J_0(\beta) J_1(\beta)$$
$$\times \left\{ \left[\phi_{-1}(v_d) - 2\phi_0(v_d) + \phi_1(v_d) \right] \cos \theta_{fm} \right.$$
$$\left. + \left[\delta_{-1}(v_d) - \delta_1(v_d) \right] \sin \theta_{fm} \right\}, \tag{6.5}$$

where P_0 (W) is the optical power incident on the detector and η_{fm} (V/W) is an instrumentation factor including the detector responsivity and amplifier gain.

6.2.2 Doppler-Broadened NICE-OHMS

In NICE-OHMS the FM triplet is coupled to three modes of an external cavity by making the modulation frequency equal to an integer multiple of the FSR of the cavity, v_{FSR}, most often as $v_m = v_{FSR}$, as is schematically shown in Fig. 6.1.

When the single-pass absorption of the analyte is much smaller than the intracavity losses the influence of the analyte on the properties of the cavity is negligible. This is valid when $\alpha_0 L \ll \pi/(2F)$, where α_0 is the absorption coefficient per unit length (cm^{-1}), L the cavity length (cm), and F the finesse of the cavity, given by $\pi \sqrt{R}/(1 - R)$, where R is the mirror reflectivity [3]. The cavity finesse can then be assumed constant and the shift of the cavity mode frequencies due to dispersion of the transition can be neglected. Under this condition, the NICE-OHMS signal is given by the ordinary FMS signal, Eq. (6.5), with the field attenuation and the phase shift increased by the cavity-enhancement factor $2F/\pi$.

Fig. 6.1 Schematic illustration of the incoupling of the FM triplet into an external high finesse cavity in NICE-OHMS

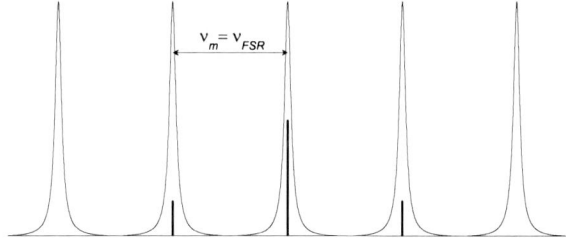

Fig. 6.1 Schematic illustration of the incoupling of the FM triplet into an external high finesse cavity in NICE-OHMS

6.2.2.1 General Expressions

The attenuation and phase shift of an electric field due to an analyte can be expressed in terms of the number density of molecules, n_A (molecules/cm^3), and an area-normalized absorption line shape function, χ_j^{abs}, and its dispersion counterpart, χ_j^{disp} (both in cm), as [33]

$$\delta_j(\nu_d, G) = \frac{\hat{S} n_A L}{2} \chi_j^{abs}(\nu_d, G) \tag{6.6}$$

and

$$\phi_j(\nu_d, G) = \frac{\hat{S} n_A L}{2} \chi_j^{disp}(\nu_d, G), \tag{6.7}$$

where G is the optical saturation parameter and $\hat{S}$ is the integrated molecular line strength [cm^{-1}/(molecule cm^{-2})] [35]. The Db NICE-OHMS signal, S^{Db}, can therefore be written as

$$S^{Db}(\nu_d, \theta_{fm}, G) = \eta_{fm} \frac{F}{\pi} P_0 J_0(\beta) J_1(\beta) \hat{S} n_A L$$
$$\times \left\{ \left[\chi_{-1}^{disp}(\nu_d, G) - 2\chi_0^{disp}(\nu_d, G) + \chi_1^{disp}(\nu_d, G) \right] \cos\theta_{fm} \right.$$
$$\left. + \left[\chi_{-1}^{abs}(\nu_d, G) - \chi_1^{abs}(\nu_d, G) \right] \sin\theta_{fm} \right\}. \tag{6.8}$$

The entity within the curly brackets is the *Db NICE-OHMS line shape function*, $\chi_{NO}^{Db}(\nu_d, \theta_{fm}, G)$, which, in turn, is composed of a *Db dispersion* and *absorption NICE-OHMS line shape function*, $\chi_{NO}^{Db,disp}(\nu_d, G)$ and $\chi_{NO}^{Db,abs}(\nu_d, G)$, obtained for detection phases θ_{fm} of 0 and $\pi/2$, respectively. The product $\hat{S} n_A$ can be expressed alternatively as $S c_{rel} p$, where S is the integrated gas line strength (cm^{-2}/atm), related to $\hat{S}$ through the ideal gas law, i.e. as $S = n_0(T_0/T)\hat{S}$, where n_0 is the Loschmidt's number, 2.667×10^{19} cm^{-3}/atm, and T_0 is 273.15 K, c_{rel} is the dimensionless relative concentration of absorbers, and p is the total pressure (atm). It is moreover useful to define the *Db NICE-OHMS signal strength* (V), S_0^{Db}, as the product of the factors multiplying the NICE-OHMS line shape function in Eq. (6.8) and the peak value of the unsaturated absorption line shape function, $\chi^{abs}(0,0)$, i.e.

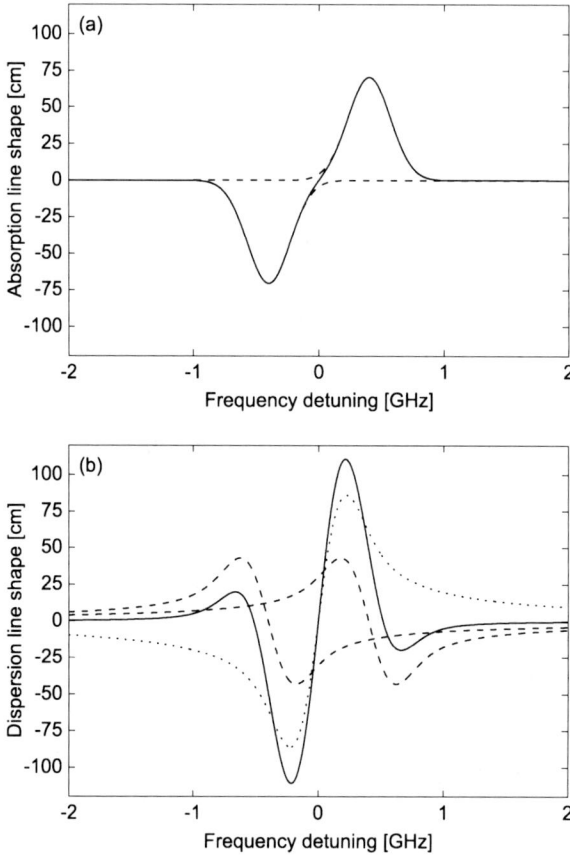

Fig. 6.2 Absorption (**a**) and dispersion (**b**) NICE-OHMS line shape functions (solid curves) in the Doppler limit for a Doppler width of 200 MHz and a modulation frequency of 400 MHz, with the sideband and carrier contributions indicated separately as dashed and dotted curves, respectively [53]

as

$$S_0^{Db} = \eta_{fm} \frac{F}{\pi} P_0 J_0(\beta) J_1(\beta) \alpha_0 L, \tag{6.9}$$

where $\alpha_0 L = S c_{rel} p L \chi^{abs}(0,0)$ is the single-pass on-resonance absorption inside the cavity. This signal strength is independent of detection phase and it thus allows determination of the concentration of the analyte without the prior knowledge of θ_{fm} by performing a fit of Eq. (6.8) to the experimental signal.

Figure 6.2 shows an example of a Db absorption and dispersion NICE-OHMS line shape function (solid curves), as well as the individual line shape functions which they are composed of (dashed or dotted curves), in the Doppler limit for a case with the modulation frequency equal to twice the Doppler width. In this particular case, the Db absorption NICE-OHMS line shape function consists of two almost separated absorption line shapes with opposite signs, which implies that its peak-to-peak value is $2\chi_1^{abs}(0, G)$. The Db dispersion NICE-OHMS line shape function, on the other hand, is given by three partly overlapping dispersion line shape functions, with a peak-to-peak value slightly above $3\chi_1^{abs}(0, G)$. In general, the peak-to-peak value of the dispersion line shape function depends more strongly on the ratio of the FSR and the line width than the absorption function.

6.2.2.2 Voigt Line Shapes

In order to ensure high accuracy and precision in the assessment of gas concentrations or line strengths, and to determine the optimum detection conditions of the technique, it is of importance to model the line shape functions correctly. The most commonly used line shape function is the Voigt profile, which takes the two main broadening mechanisms (Doppler and pressure broadening) into account by a convolution of a Gaussian and a Lorentzian function. The absorption and dispersion Voigt line shape functions, $\chi_{V,j}^{abs}$ and $\chi_{V,j}^{disp}$, can be expressed in terms of the real and the imaginary parts of the complex error function of a complex argument, $W(x + iy)$, as [33]

$$\chi_{V,j}^{abs}(x, y, G) = \frac{c}{\sqrt{\pi}\Gamma_D'} \frac{1}{\sqrt{1 + G_j}} \text{Re}\left[W(x_j + iy_j)\right] \tag{6.10}$$

and

$$\chi_{V,j}^{disp}(x, y, G) = -\frac{c}{\sqrt{\pi}\Gamma_D'} \text{Im}\left[W(x_j + iy_j)\right], \tag{6.11}$$

where x is the Doppler-width-normalized frequency detuning of the carrier, given by ν_d/Γ_D', and y is the unsaturated Voigt parameter, given by Γ_L/Γ_D', where, in turn, Γ_D' is given by $\Gamma_D/\sqrt{\ln 2}$, where Γ_D is the Doppler width (HWHM) of the transition (Hz) and Γ_L is the HWHM homogenous linewidth (Hz). Moreover, c is the speed of light (cm/s), while G_j, x_j, and y_j are the mode-specific (for mode j) degree of saturation, the Doppler-width-normalized frequency detuning, given by $(\nu_d + j\nu_m)/\Gamma_D'$, and the saturated Voigt parameter, given by $\sqrt{1 + G_j}y$, respectively.

Figure 6.3 displays a set of simulations of Db absorption and dispersion NICE-OHMS line shape functions in the Voigt regime for four different unsaturated Voigt parameters (for y ranging from 0 to 0.1) and various degrees of saturation for a case with a modulation frequency of $1.6\Gamma_D$ and a modulation index of 0.36 [33]. The figure shows that the absorption and dispersion line shape functions behave differently under optically saturated conditions. For $y > 0$ they are both affected by saturation broadening of the homogeneous linewidth in the error function. However, the magnitude of the absorption line shape is additionally decreased by a factor of $\sqrt{1 + G_1}$. This demonstrates that optically saturated absorption and dispersion signals are not related to each other by the Kramers–Kronig relations [36, 37]. It is worth noting that since the absorption signal does not contain contribution from the

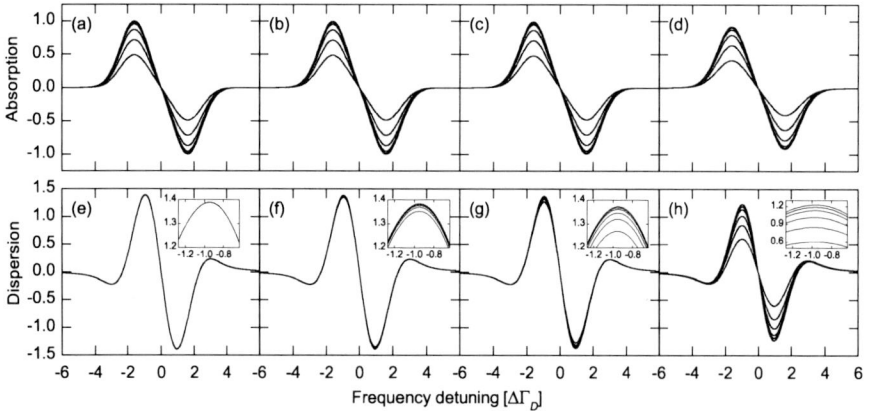

Fig. 6.3 Doppler-broadened absorption NICE-OHMS line shape functions [(**a**)–(**d**)] and their dispersion counterparts [(**e**)–(**h**)] in the Voigt regime (in terms of $\chi_G^{abs}(0,0)$). The four pairs of panels, (**a**) and (**e**), (**b**) and (**f**), (**c**) and (**g**), and (**d**) and (**h**), represent $y = 0$, 0.003, 0.01, and 0.1, respectively. Each panel displays six curves, corresponding to saturation parameters of the carrier, G_0, of 0, 1, 3, 10, 30, and 100 counted from the uppermost curve in each panel. The inset in each lower panel shows a zoom of the dispersion signal line shape around the positive peak [33]. Reproduced with permission from Optical Society of America

carrier, it is affected solely by the optical saturation of the sidebands [see Eqs. (6.8) and (6.10)], which carry only a fraction of the total optical power. This, together with the rather weak dependence of the dispersion signal on optical saturation, implies that Db NICE-OHMS is affected significantly less by optical saturation than other cavity-enhanced absorption techniques [24].

6.2.2.3 The Doppler Limit—Gaussian Line Shapes

Under low pressure conditions, when the collision broadening is insignificant (i.e. $\Gamma_L \ll \Gamma_D$, whereby $y \ll 1$), the line shapes simplify to a Gaussian absorption line shape function and its dispersion counterpart, which can be expressed as [33]

$$\chi_{G,j}^{abs}(x,G) = \frac{c}{\sqrt{\pi}\,\Gamma_D'}\frac{1}{\sqrt{1+G_j}}e^{-x_j^2} \tag{6.12}$$

and

$$\chi_{G,j}^{disp}(x) = -\frac{c}{\sqrt{\pi}\,\Gamma_D'}\frac{2}{\sqrt{\pi}}e^{-x_j^2}\int_0^{x_j}e^{s^2}ds. \tag{6.13}$$

This shows that in the Doppler limit the shapes of the absorption and dispersion NICE-OHMS line shape functions are unaffected by optical saturation. However, the amplitude of the absorption line shape function decreases as $1/\sqrt{1+G_1}$, while that of the dispersion function is independent of optical saturation [as can be seen from panels (a) and (e) in Fig. 6.3]. This implies that although the transition is

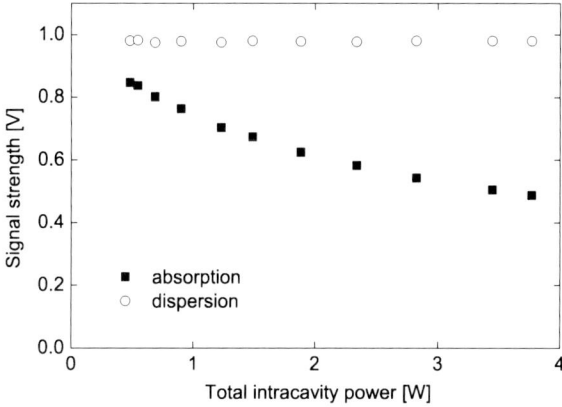

Fig. 6.4 The saturated absorption and dispersion NICE-OHMS signal strengths in the Doppler limit from 1000 ppm of C_2H_2 at 10 mTorr of N_2 as a function of intracavity power measured at the $P_e(11)$ transition at 1531.588 nm [24]. The measurements were performed at different intracavity powers, while the power impinging on the detector was held constant. Reproduced with permission from Optical Society of America

optically saturated it is possible to express the Db NICE-OHMS signal in terms of the *unsaturated* Db NICE-OHMS line shape function, $\chi_{NO}^{Db}(\nu_d, \theta_{fm}, G = 0)$. The signal strength will then be dissimilar for the two detection phases. The dispersion signal strength, $S_{0,disp}^{Db}$, is given by Eq. (6.9) with $\chi^{abs}(0, 0)$ equal to $c/(\sqrt{\pi}\,\Gamma_D')$, whereas the absorption signal strength, $S_{0,abs}^{Db}$, is related to $S_{0,disp}^{Db}$ by

$$S_{0,abs}^{Db} = \frac{1}{\sqrt{1 + G_1}} S_{0,disp}^{Db}. \qquad (6.14)$$

Since the ratio of these signal strengths is given by $1/\sqrt{1 + G_1}$, measuring them provides a convenient means of assessing the degree of optical saturation in the Doppler limit in NICE-OHMS [24].

This is illustrated in Fig. 6.4, which displays the absorption and dispersion NICE-OHMS signal strength from an acetylene transition at a pressure of 10 mTorr as a function of optical power inside a cavity (while the power incident on the detector, P_0, was kept constant). The dispersion signal strength is constant, while the absorption signal strength decreases with increasing intracavity power as predicted by Eq. (6.14). The reason for the insensitivity of the dispersion NICE-OHMS signal in the Doppler limit to optical saturation is that the molecules with an axial velocity higher than that corresponding to the center of the Bennet hole contribute to the dispersion signal in opposite manner to those with a correspondingly lower axial velocity, thus canceling each other's contributions [33].

6.2.2.4 Collision Dominated Regime—Lorentzian Line Shapes

Under high pressure conditions, i.e., in the collision dominated regime, where $\Gamma_L \gg \Gamma_D$, the line shapes have a Lorentzian form, which can be written as [38–40]

$$\chi_{L,j}^{abs}(\nu_d, \Gamma_L, G) = \frac{c}{\pi} \frac{1}{\sqrt{1+G_j}} \frac{\Gamma_L'(G_j)}{(\nu_d + j\nu_m)^2 + \Gamma_L'^2(G_j)} \tag{6.15}$$

and

$$\chi_{L,j}^{disp}(\nu_d, \Gamma_L, G) = -\frac{c}{\pi} \frac{\nu_d + j\nu_m}{(\nu_d + j\nu_m)^2 + \Gamma_L'^2(G_j)}, \tag{6.16}$$

where $\Gamma_L'(G_j)$ is the saturated HWHM homogenous linewidth, given by $\sqrt{1+G_j}\Gamma_L$. Also in this regime the absorption and dispersion line shape functions are affected by optical saturation in dissimilar ways; although their linewidth is broadened by a factor of $\sqrt{1+G_j}$ their peak values are reduced by a factor of $(1+G_j)$ and $\sqrt{1+G_j}$, respectively. However, since NICE-OHMS has so far not been performed under collision dominated conditions, these line shapes have not yet been verified.

6.2.2.5 Line Shapes Beyond the Voigt Profile—Dicke Narrowing and Speed Dependent Effects

The use of the Voigt line shape function implies tacitly that the collision and Doppler broadening processes are assumed to be independent. However, as has been demonstrated repeatedly in the literature, this is not always an appropriate assumption.

When the mean free path of the molecules in the sample is of the same order of magnitude as (or shorter than) the wavelength of the transition, the increased collision rate can affect the velocity distribution of the molecules, resulting in a narrower Doppler profile, an effect referred to as Dicke narrowing [41]. The two most common models for Dicke narrowing in absorption spectrometry are the so-called soft and hard collision models (for collision partners lighter and heavier than the analyte, respectively), proposed by Galatry [42] and Rautian and Sobel'man [43], respectively.

For the soft collision model (SCM) the absorption and dispersion line shape functions can be written as [30]

$$\chi_{SCM,j}^{abs} = \frac{c}{\pi\Gamma_D'} \text{Re}\left[\frac{1}{(1/2z) + y - ix_j} \times M\left(1; 1 + \frac{1}{2z^2} + \frac{y - ix_j}{z}; \frac{1}{2z^2}\right)\right] \tag{6.17}$$

and

$$\chi_{SCM,j}^{disp} = -\frac{c}{\pi\Gamma_D'} \text{Im}\left[\frac{1}{(1/2z) + y - ix_j} \times M\left(1; 1 + \frac{1}{2z^2} + \frac{y - ix_j}{z}; \frac{1}{2z^2}\right)\right], \tag{6.18}$$

where $M(a; b; c)$ is a confluent hypergeometric function. The dimensionless parameter z is defined as $z = \beta_{soft}/\Gamma'_D$ where β_{soft} is the effective frequency of velocity-changing collisions, given by $\beta_{soft} = \beta^0_{soft}p$, where, in turn, β^0_{soft} is the collisional narrowing coefficient for soft collisions.

For the hard collision model (HCM), the corresponding line shape functions can be written as [30]

$$\chi^{abs}_{HCM,j} = \frac{c}{\sqrt{\pi}\Gamma'_D}\text{Re}\left[\frac{W[x_j + i(y + \zeta)]}{1 - \sqrt{\pi}\zeta W[x_j + i(y + \zeta)]}\right] \quad (6.19)$$

and

$$\chi^{disp}_{HCM,j} = -\frac{c}{\sqrt{\pi}\Gamma'_D}\text{Im}\left[\frac{W[x_j + i(y + \zeta)]}{1 - \sqrt{\pi}\zeta W[x_j + i(y + \zeta)]}\right], \quad (6.20)$$

where ζ is a dimensionless parameter defined as $\zeta = \beta_{hard}/\Gamma'_D$ where β_{hard} is the total collision frequency, given by $\beta^0_{hard}p$, where β^0_{hard} is the collisional narrowing coefficient for hard collisions.

Another deviation from Voigt line shapes comes from the fact that the relaxation rates (and thereby the collision widths) are velocity dependent. This implies that not all velocity classes of molecules have the same homogenous width [44]. This phenomenon gives rise to a narrowing of the collision broadening and is referred to as speed-dependent effects (SDEs).

An absorption and a dispersion line shape function that take SDEs into account [based upon the so-called speed-dependent Voigt (SDV) model] can be written as [31]

$$\chi^{abs}_{SDV,j} = \frac{c}{\Gamma'_D}\frac{1}{\sqrt{\pi}}\text{Re}\left[W(\zeta_{1,j}) - W(\zeta_{2,j})\right] \quad (6.21)$$

and

$$\chi^{disp}_{SDV,j} = -\frac{c}{\Gamma'_D}\frac{1}{\sqrt{\pi}}\text{Im}\left[W(\zeta_{1,j}) - W(\zeta_{2,j})\right], \quad (6.22)$$

where the real and imaginary parts of the complex arguments of the error functions can be written as

$$\text{Re}(\zeta_{1,j}) = \text{Re}(\zeta_{2,j}) = -\text{sign}(\beta_j)\sqrt{\sqrt{(\varepsilon^2 + \alpha)^2 + \beta_j^2} - (\varepsilon^2 + \alpha)}/\sqrt{2} \quad (6.23)$$

and

$$\text{Im}(\zeta_{1/2,j}) = \sqrt{\sqrt{(\varepsilon^2 + \alpha)^2 + \beta_j^2} + (\varepsilon^2 + \alpha)} \mp \varepsilon/\sqrt{2}, \quad (6.24)$$

where $\text{Im}(\zeta_{1,j})$ and $\text{Im}(\zeta_{2,j})$ correspond to the upper and lower sign in front of ε, respectively, $\alpha = \Gamma_L/\Gamma_{SD} - 3/2$, $\beta_j = (\nu_d + j\nu_m)/\Gamma_{SD}$, and $\varepsilon = \Gamma'_D/(2\Gamma_{SD})$, where in turn Γ_{SD} is a parameter that represents the speed-dependence of the collision width, and is referred to as the speed-dependent (SD) collision width, which also

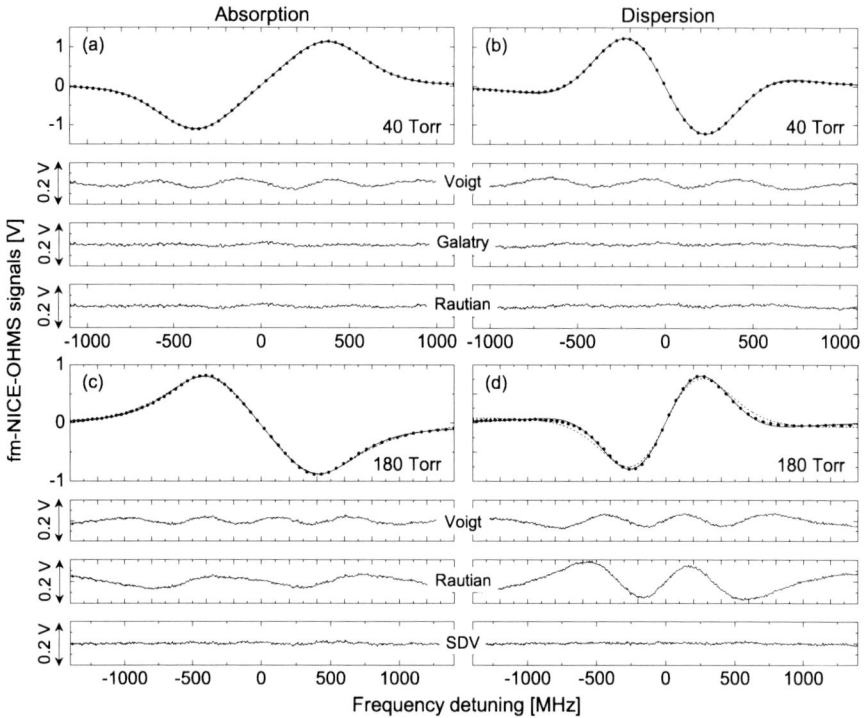

Fig. 6.5 *Upper row*: absorption (**a**) and dispersion (**b**) NICE-OHMS signals from the isolated $P_e(33)$ transition of acetylene at 1552.946 nm taken at a pressure of 40 Torr in N_2 [30]. The individual markers in the upper windows represent measurements whereas the three curves (overlapping to a large extent) show the best fits of NICE-OHMS signals based on the Voigt lineshape, the SCM (Galatry), and the HCM (Rautian). The residuals of the fits of each line shape are given in the lower windows, as marked. *Lower row*: absorption (**c**) and dispersion (**d**) NICE-OHMS signals from the same transition taken at a pressure of 180 Torr [32]. The three curves show the best fits of NICE-OHMS signals based on the Voigt, the HCM (Rautian), and the SDV models. The residuals of the fits of each line shape are given in the lower windows, as marked. Reproduced with permission from Optical Society of America

can be written as a product of a speed-dependent collision coefficient, Γ_{SD}^0, and the gas pressure, p.

The effects of Dicke narrowing appear at lower pressures than the SDEs. The upper row of panels in Fig. 6.5 shows Db absorption and dispersion NICE-OHMS signals from an isolated line of acetylene in N_2 in the 1.5 µm region measured at a pressure of 40 Torr [30]. The figure shows that when the expression for the NICE-OHMS signal based upon the Voigt line shape functions is fitted to the signals, systematic structures in the residuals appear both in the absorption and the dispersion modes of detection. On the other hand, both models for Dicke narrowing, (i.e. soft and hard collisions) reproduce the detected signals adequately.

The lower row of panels in Fig. 6.5 displays absorption and dispersion NICE-OHMS signals from the same transition taken at a higher pressure (180 Torr) [32].

The residuals in these panels show that both the ordinary Voigt and the Rautian profiles are inadequate to model the signals correctly in this pressure region, and that, for the latter, the discrepancy is significantly stronger for dispersion than absorption. On the other hand, the fits from the SDV line shape model provide virtually structureless residuals, which demonstrate that these line shape functions can satisfactory model NICE-OHMS signals in this pressure range. It should be noted that the actual pressure ranges of validity of Dicke narrowing and SDEs depend on the transition as well as the species (both the analyte and the perturber).

6.2.2.6 Wavelength-Modulated Doppler-Broadened NICE-OHMS

In order to remove low frequency noise in the NICE-OHMS signal an additional dither can be applied to the laser frequency via modulation of the cavity length. The signal is then demodulated at the dither frequency yielding a so-called wavelength-modulated (wm) NICE-OHMS signal, which is given by the first even Fourier coefficient of the signal [45, 46]. An example of an absorption wm-NICE-OHMS signal in the Voigt regime is shown in Fig. 6.6. The wm-NICE-OHMS signal can be written in terms of the Fourier coefficients of the individual wavelength-modulated absorption and dispersion line shape functions, $viz.$ as

$$
S_1^{Db,wm}(\nu_d, \nu_a, \theta_{fm}, G)
$$

$$
= \eta_{wm}\eta_{fm}\frac{F}{\pi}P_0 J_0(\beta)J_1(\beta)\hat{S}n_A L
$$

$$
\times \left\{ \left[\chi_{-1,1}^{disp}(\nu_d, \nu_a, G) - 2\chi_{0,1}^{disp}(\nu_d, \nu_a, G) + \chi_{1,1}^{disp}(\nu_d, \nu_a, G) \right] \cos\theta_{fm} \right.
$$

$$
\left. + \left[\chi_{-1,1}^{abs}(\nu_d, \nu_a, G) - \chi_{1,1}^{abs}(\nu_d, \nu_a, G) \right] \sin\theta_{fm} \right\}, \tag{6.25}
$$

where the second subscript '1' indicates the first (even) Fourier coefficient of each wavelength-modulated line shape function, ν_a is the modulation amplitude, and η_{wm} an instrumentation factor. The first Fourier coefficient of the wavelength modulated line shape function is defined as

$$
\chi_{j,1}^{abs/disp}(\nu_d, \nu_a, G) \equiv \frac{2}{\tau}\int_0^\tau \chi_j^{abs/disp}\left[\nu_d + \nu_a\cos(2\pi f_{wm}t), G\right]\cos(2\pi f_{wm}t)dt, \tag{6.26}
$$

where τ is the integration time, given by the inverse of the wavelength modulation (dither) frequency, f_{wm}, or an integer multiple thereof.

To allow real time data analysis, and for convenience, analytical expressions for the Fourier coefficients of the most common line shape functions have been derived or alternative numerical calculation procedures have been developed. There exists no analytical solution for these coefficients for the Voigt line shape function, wherefore they have to be calculated numerically. However, calculation of these using the general definition of Eq. (6.26) is time consuming due to the existence of nested integrals, one for the Voigt function and one for the wavelength modulation (in the

Fig. 6.6 An absorption
wm-NICE-OHMS signal for
the P7P7 transition of the hot
band of oxygen at 772 nm
(*dots*) measured at a pressure
of 397.4 mbar with the fit of
the expected line shape (*solid
curve*) and the corresponding
residual to the fit (*bottom
graph*) [13]. Reproduced with
permission from Optical
Society of America

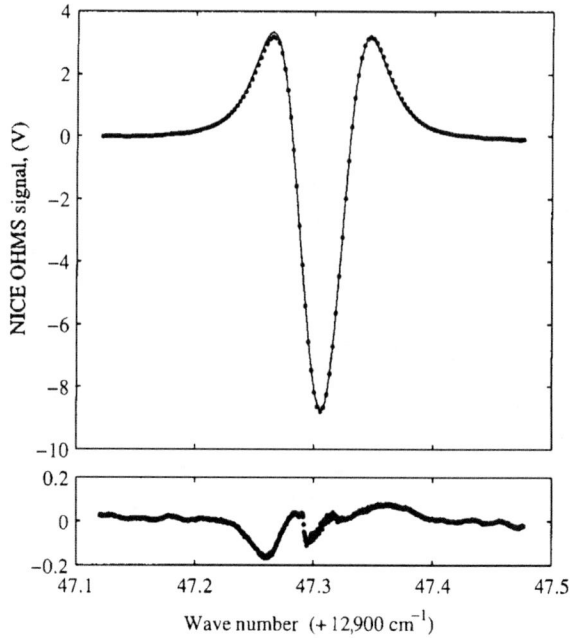

order of seconds for a single line shape function). Instead, it was recently shown
that the first even Fourier coefficient of a modulated (unsaturated) Voigt line shape
function can be calculated significantly faster (in the order of ms) as a convolution
of the first even Fourier coefficient of a modulated (unsaturated) Lorentzian line
shape function, $\chi_{L,j,1}^{abs}(\bar{v}_d, \bar{v}_a)$, and an area-normalized velocity-distribution function giving rise to the Doppler broadening, $f_v(\bar{v}_d)$, which formally can be written
as [47]

$$\chi_{V,j,1}^{abs}(\bar{v}_d, \bar{v}_a) = \chi_{L,j,1}^{abs} \otimes f_v(\bar{v}_d), \tag{6.27}$$

where $\chi_{L,j,1}^{abs}$ is given by Eq. (6.28) below with $G = 0$, and $f_v(\bar{v}_d)$ is given by
$y/\sqrt{\pi} \exp(-y^2\bar{v}_d^2)$, $\bar{v}_d$ and $\bar{v}_a$ are the homogeneous width-normalized detuning and
the corresponding modulation amplitude, given by v_d/Γ_L and v_a/Γ_L, respectively.
Although not formally proven in Ref. [47], the extension of the expression above to
account for optical saturation should be straightforward. Moreover, the coefficient
for the modulated Voigt dispersion line shape function, i.e. $\chi_{V,j,1}^{disp}$, is assumed to be
given by a corresponding expression [48].

The strength and line shape of wm-NICE-OHMS signals in the Doppler limit
have been meticulously scrutinized by Foltynowicz et al. [26]. An analytical expression exists only for the Fourier coefficients of the Gaussian absorption line
shape function, i.e. of Eq. (6.12). However, this expression is given in terms of a
sum of polynomials, see Eqs. (7) or (8) in Ref. [49], whose convergence is rather
poor when the transition is over-modulated, whereby it is not always expedient to
use. The reader is referred to Ref. [49] for details.

The first even Fourier coefficients of a modulated area-normalized absorption and the corresponding dispersion Lorentzian line shape functions can be written in a succinct manner, *viz.* as [50, 51]

$$\chi_{L,j,1}^{abs}(\bar{\nu}_d, \bar{\nu}_a, G) = \frac{c}{\pi \Gamma_L'(G_j)} \frac{1}{\sqrt{1+G_j}} \frac{2}{\bar{\nu}_{a,j}}$$
$$\times \left[\frac{-\bar{\nu}_{d,j}\sqrt{R_j + M_j} + \text{sign}(\bar{\nu}_{d,j})\sqrt{R_j - M_j}}{\sqrt{2}R_j} \right] \quad (6.28)$$

and

$$\chi_{L,j,1}^{disp}(\bar{\nu}_d, \bar{\nu}_a, G) = \frac{c}{\pi \Gamma_L'(G_j)} \frac{2}{\bar{\nu}_{a,j}}$$
$$\times \left[-1 + \frac{\sqrt{R_j + M_j} + \bar{\nu}_{d,j}\text{sign}(\bar{\nu}_{d,j})\sqrt{R_j - M_j}}{\sqrt{2}R_j} \right], \quad (6.29)$$

respectively, where $R_j = \sqrt{M_j^2 + 4\bar{\nu}_{d,j}^2}$ and $M_j = 1 + \bar{\nu}_{a,j}^2 - \bar{\nu}_{d,j}^2$, where, in turn, $\bar{\nu}_{a,j}$ and $\bar{\nu}_{d,j}$ are the saturated width-normalized modulation amplitude and detuning for mode j, given by $\nu_a/\Gamma_L'(G_j)$ and $(\nu_d + j\nu_m)/\Gamma_L'(G_j)$, respectively. The factors $c/[\pi \Gamma_L'(G_j)] \cdot 1/\sqrt{1+G_j}$ and $c/[\pi \Gamma_L'(G_j)]$ convert the Fourier coefficients of peak-normalized line shape functions (used in the original derivations in [50] and [51]) to an area-normalized absorption line shape function and its dispersion counterpart, respectively, and take simultaneously optical saturation into account. The expression for the first Fourier coefficient of the dispersion Lorentzian line shape function is particularly useful to model sub-Doppler signals.

6.2.3 Sub-Doppler NICE-OHMS

6.2.3.1 Origin and Properties

Sub-Doppler signals can be observed with NICE-OHMS due to the presence of high intensity counter-propagating waves inside the cavity. As shown in Fig. 6.7, as the FM triplet is scanned across a transition there are nine occasions at which two counter-propagating waves interact with the same velocity group of molecules, occurring at five different detunings. Figure 6.8 then shows how the sub-Doppler absorption and dispersion NICE-OHMS signals appear on top of the Doppler-broadened signals.

As is illustrated in Figs. 6.7(a) and (e), when the detuning of the carrier from the transition center is equal to (±) the modulation frequency, two of the sidebands propagating in opposite directions interact with the same group of molecules, namely those with a zero velocity component along the optical axis. These interac-

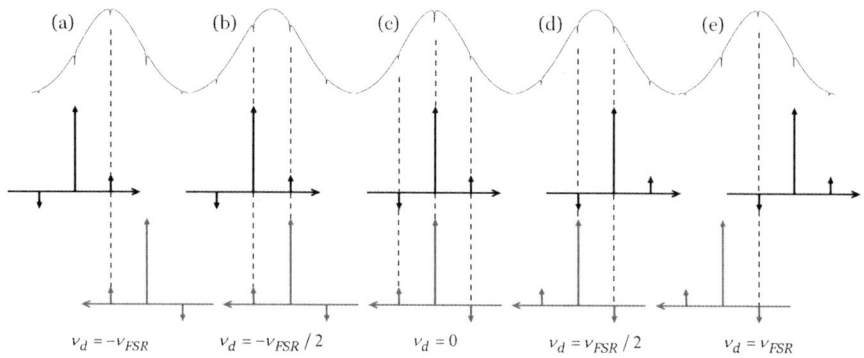

Fig. 6.7 Schematic illustration of spectral hole burning in a Doppler-broadened medium with a Gaussian velocity distribution of molecules (*uppermost row*) by two counter-propagating FM triplets, shown in *black* (positive direction, *second row*) and *gray* (negative direction, *lowermost row*). The x-*axis* of the distributions in the upper row represents velocity, whereas those of the other rows indicate frequency. The *five panels* correspond to the five detunings of the carrier from the center of the transition at which various components of two counter-propagating FM triplets interact with a common velocity group of molecules. Each mode interacts with the group of molecules that has a velocity $v = \lambda(\pm v_d + jv_m)$ where the upper and lower signs correspond to the electrical fields that propagate in the positive and negative directions, respectively. The velocity groups of molecules that are addressed simultaneously by a negatively and positively propagating field, which are indicated with dashed lines, are those that contribute to the sD signal [53]

tions give rise to the outermost (the weakest) sD signals displayed in Fig. 6.8 (at around ±400 MHz). When the detuning of the carrier is equal to (±) half the modulation frequency, the carrier going in the positive direction interacts with the same group of molecules as one of the sidebands going in the negative direction, and, simultaneously, a sideband going in the positive direction interacts with the molecules addressed by the carrier propagating in the negative direction, as shown in the panels (b) and (d) in Fig. 6.7. This gives rise to the sD signals that appear at around ±200 MHz in Fig. 6.8. Finally, as shown in Fig. 6.7(c), when the carrier is tuned to molecular resonance, the two counter-propagating carriers interact with molecules with a zero velocity component along the optical axis, while simultaneously the upper and lower sidebands propagating in the opposite directions interact with the groups of molecules that have a velocity component along the optical axis equal to $\pm v_{FSR}\lambda$. This case gives rise to the largest sD signal, appearing at zero detuning in dispersion in Fig. 6.8(b). The absence of signal in the absorption phase is due to a generic insensitivity of FMS to the attenuation of the carrier and the fact that the sideband-sideband interactions on resonance cancel (they have equal magnitude but opposite sign). All this implies that when the transition is optically saturated by the carrier as well as the sidebands five sD signals appear on top of the Db signal at dispersion phase, and four at absorption phase.

Fig. 6.8 Absorption (**a**) and dispersion (**b**) NICE-OHMS signals from 500 ppm of C_2H_2 in 20 mTorr of N_2 measured at the $P_e(11)$ transition at 1531.588 nm with an intracavity power of 4.6 W, corresponding to a degree of saturation of 50 and 1.5 for the carrier and the sidebands, respectively [16]. Reproduced with permission from Optical Society of America

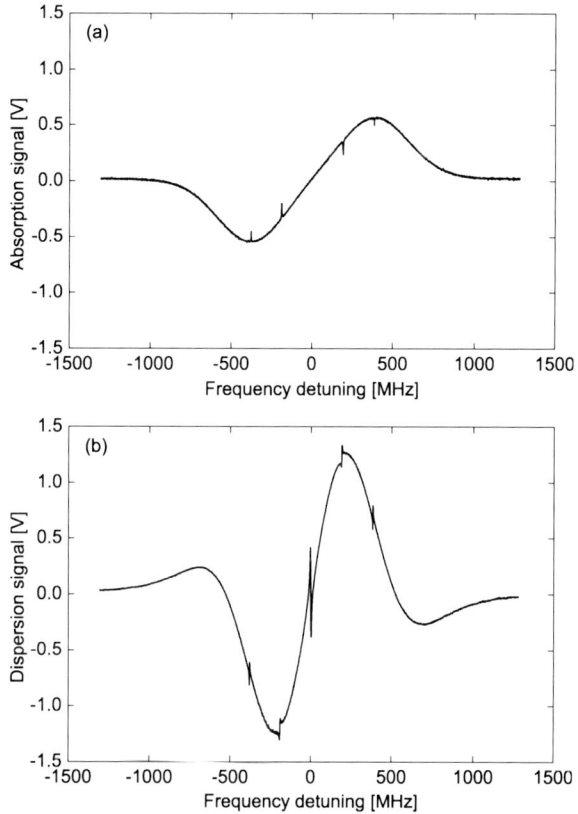

6.2.3.2 The Center sD Dispersion Signal—Line Shape and Signal Strength

The center dispersion sD signal is of most interest for practical applications, since it is the largest, it resides on top of a nearly linear background, and it appears at zero detuning (i.e. in the center of the transition). Moreover, the shape of this signal does not change with detection phase since it does not exist in absorption. This also implies that it can be used to find the correct electronic phases for detection of pure absorption and dispersion signals.

The center sD dispersion signal, $S^{sD}(\nu_d, G_0)$, can be expressed in terms of an sD optical phase shift, $\phi_{00}(\nu_d, G_0)$, as [25]

$$S^{sD}(\nu_d, G_0) = -\eta_{fm} \frac{4F}{\pi} P_0 J_0(\beta_1) J_1(\beta_1) \phi_{00}(\nu_d, G_0), \qquad (6.30)$$

where G_0 is the saturation parameter of the carrier. The line shape of the sD optical phase shift cannot be written in a simple analytical form, but can be well approximated up to high degrees of saturation by an unsaturated peak-to-peak-normalized Lorentzian dispersion function, $(\pi \Gamma_L/c)\chi_{L,0}^{disp}$, where $\chi_{L,0}^{disp}$ is given by Eq. (6.16)

Fig. 6.9 Peak-to-peak
sub-Doppler optical phase
shift in terms of the
absorption coefficient, i.e.,
$\Phi(G_0)$, as a function of the
degree of saturation induced
by the carrier. The *solid
markers* show experimental
data whereas the *solid curve*
displays the theoretical
dependence for a beam with a
Gaussian intensity
distribution, i.e. Eq. (6.31).
Data taken from Ref. [25]

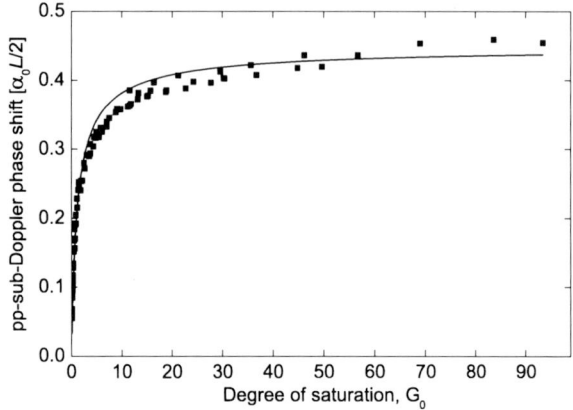

with $G_0 = 0$ [25, 52]. For a beam with a Gaussian-shaped intensity distribution, the peak-to-peak value of the sD optical phase shift, $\phi_{00}^{pp}(G_0)$, can be expressed in terms of the single-pass absorption as

$$\phi_{00}^{pp}(G_0) = \Phi(G_0)\frac{\alpha_0 L}{2} = 0.45\frac{\alpha_0 L}{2}\frac{8}{w^2}\int_0^\infty \frac{G_0 e^{-4(r/w)^2}}{1 + 2G_0 e^{-2(r/w)^2}}r\,dr, \qquad (6.31)$$

where w is the radius of a Gaussian beam inside the cavity. The dimensionless function $\Phi(G_0)$ is plotted in Fig. 6.9 as a function of the saturation parameter (solid curve), together with experimental data (individual markers), for a particular transition in C_2H_2 around 1.5 μm. The figure shows that the peak-to-peak sub-Doppler optical phase shift increases monotonically with the degree of saturation towards a value of $0.45\alpha_0 L/2$, which is in contradiction to predictions in early NICE-OHMS publications [1, 3].

Making use of the definition of $\Phi(G_0)$, the center sD dispersion NICE-OHMS signal can be written as [53]

$$S^{sD}(\nu_d, G_0) = S_0^{sD}(G_0)\frac{\pi\Gamma_L}{c}\chi_{L,0}^{disp}(\nu_d, \Gamma_L, 0), \qquad (6.32)$$

where the sD NICE-OHMS signal strength, $S_0^{sD}(G_0)$, is given by

$$S_0^{sD}(G_0) = -\eta_{fm}\frac{2F}{\pi}P_0 J_0(\beta)J_1(\beta)\alpha_0 L\Phi(G_0). \qquad (6.33)$$

Two such signals from C_2H_2, on top of the Db signal, for two different intracavity pressures, are illustrated in Fig. 6.10(a).

The sD NICE-OHMS signal strength, $S_0^{sD}(G_0)$, also can be expressed in terms of the unsaturated Db NICE-OHMS signal strength, S_0^{Db}, as $2\Phi(G_0)S_0^{Db}$. This provides a means to relate the maximum peak-to-peak values of the sD and the Db dispersion NICE-OHMS signals to each other. The former is directly given

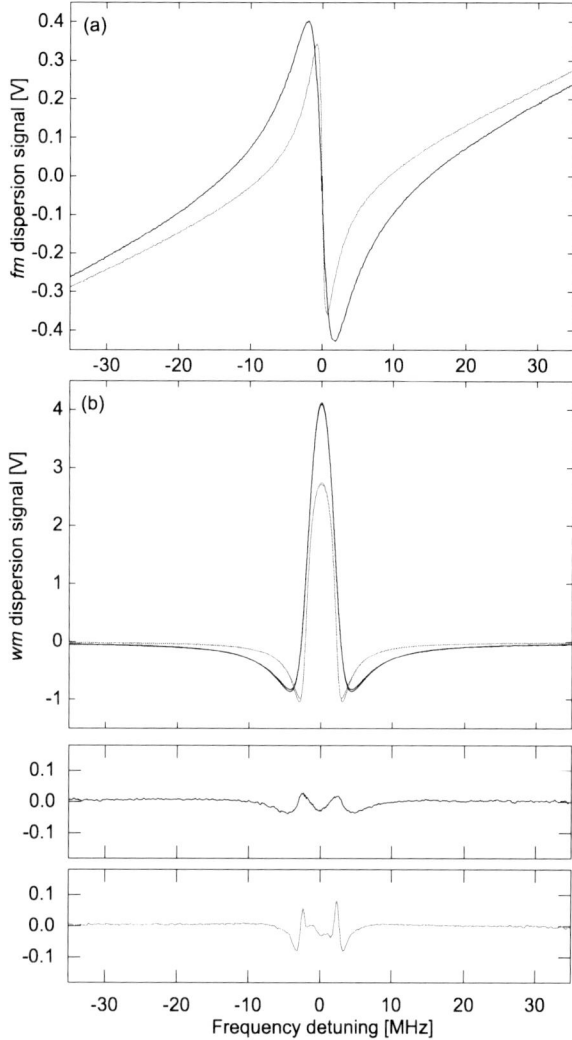

Fig. 6.10 Sub-Doppler dispersion NICE-OHMS signals from 10 μTorr of C_2H_2 in 10 mTorr of N_2 measured at the $P_e(11)$ transition at 1531.588 nm with two intracavity powers: 4.1 W (*black curves*, larger signal) and 0.49 W (*gray curve*, smaller signals) without (**a**) and with (**b**) *wm* dither (the latter with a modulation amplitude of 2.3 MHz). In panel (**b**), fits of Eq. (6.34) are also shown, with residuals displayed below, where the upper and lower residual correspond to the higher and lower intracavity power, respectively [16]. Reproduced with permission from Optical Society of America

by $S_0^{sD}(G_0)$, whereas the latter is given by $S_0^{Db} \chi_{NO,pp}^{Db,disp}(G)/\chi^{abs}(0,0)$, where pp indicates the peak-to-peak value. This implies that the peak-to-peak value of the central sD NICE-OHMS signal is a fraction $2\Phi(G_0)\chi^{abs}(0,0)/\chi_{NO,pp}^{Db,disp}(G)$ of the corresponding Db dispersion signal. For example, for a degree of saturation around 10–50, $2\Phi(G_0) \approx 0.8$ (Fig. 6.9), while $\chi_{NO,pp}^{Db,disp}(G)/\chi^{abs}(0,0)$ is slightly above 3 for a modulation frequency equal to twice the Doppler width (Fig. 6.2). This implies that for relatively large degrees of saturation, the peak-to-peak value of the sD NICE-OHMS signal is expected to be around $1/3.5$ of that of the Db signal. This agrees well with the experimental data presented in Fig. 6.8(b).

6.2.3.3 Wavelength-Modulated Sub-Doppler NICE-OHMS

In order to remove the linear slope of the Db signal and to eliminate any possible $1/f$ type of noise, the sD dispersion signal is most often measured with a wm-dither. Making use of the nomenclature from above, the wm-sD NICE-OHMS signal can then be written as

$$S_1^{sD,wm}(\nu_d, \nu_a, G_0) = \eta_{wm} S_0^{sD}(G_0) \frac{\pi \Gamma_L}{c} \chi_{L,0,1}^{disp}(\nu_d, \nu_a, 0), \qquad (6.34)$$

where $\bar{\chi}_{L,0,1}^{disp}$ is the first (even) Fourier coefficient of an unsaturated Lorentzian dispersion line shape function, which is given by Eq. (6.29) with $G = 0$. Some typical wm-sD NICE-OHMS signals from C_2H_2 in N_2 together with corresponding fits are shown in Fig. 6.10(b).

6.3 Experimental Implementation

6.3.1 Generic Setup

6.3.1.1 Principles

As was discussed above, the core of NICE-OHMS is the combination of an external cavity with frequency modulation. A typical NICE-OHMS set up is schematically illustrated in Fig. 6.11.

The frequency of the laser is locked to a mode of a high finesse cavity by the Pound-Drever-Hall (PDH) technique [54–56]. This incorporates a modulation of the laser frequency at a frequency, ν_{PDH}, that is larger than the linewidth of the cavity modes and different from the FSR of the cavity, ν_{FSR}. When the carrier is close to resonance, this choice of modulation frequency ensures that the PDH sidebands are reflected by the cavity and that their phase relation to the carrier provides instantaneous information about the frequency of the laser with respect to that of the cavity mode. A PDH error signal is derived by in-phase demodulation at ν_{PDH} of the light reflected by the cavity. This signal is then processed in a servo control unit and fed back to the laser frequency actuator. The modulation for the PDH locking, which often is in the low MHz range, is either performed by modulating the laser source directly or by the use of an external electro-optic modulator (EOM).

To provide sidebands for the FMS, the laser frequency is simultaneously modulated at a frequency, ν_m, most often by the use of an EOM. Since the sidebands should be transmitted through the cavity (as was shown in Fig. 6.1), the modulation frequency needs to be equal to (a multiple of) the FSR of the cavity. Inside the cavity, which contains the analyte, the three modes will experience different attenuation and phase shift. This introduces an imbalance in the triplet that gives rise to a modulation of the intensity at the beat frequency. The NICE-OHMS signal is obtained

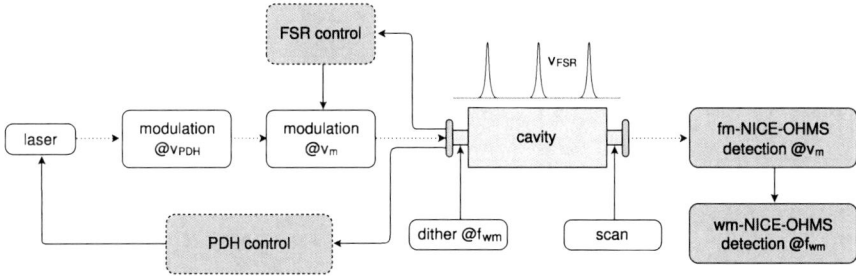

Fig. 6.11 Schematic illustration of a typical NICE-OHMS setup. The modulations for the PDH locking as well as the frequency modulation are often produced externally, e.g. with an electro-optic modulator. The cavity reflected light is demodulated at ν_{PDH} to provide an error signal for the locking of the laser to the cavity by the PDH technique and at $\nu_m \pm \nu_{PDH}$ for locking of the modulation frequency ν_m to the FSR of the cavity, ν_{FSR}, by the deVoe-Brewer technique. The length of the cavity is controlled by one or two piezo-electric actuators, to allow for scanning over a molecular transition and dithering at frequency f_{wm} for wavelength modulation. The NICE-OHMS signal is measured in transmission by phase-sensitive detection at ν_m. In the case of wm, it is additionally demodulated at f_{wm}

by phase-sensitive detection of the intensity of light transmitted through the cavity at ν_m.

Since the laser is locked to the cavity, a scan across a resonance is achieved by altering the length of the cavity. However, this affects the FSR as well, and to sustain noise immunity during the scan the modulation frequency needs to follow the cavity FSR. This is of particular importance for large alterations of the cavity length, which take place for Db detection. A common means of locking the modulation frequency to the ν_{FSR} is the so-called deVoe-Brewer method [57]. In this technique, an error signal, used for correcting the modulation frequency, is derived from the reflected light at either of the frequencies $\nu_{FSR} \pm \nu_{PDH}$. The feedback is sent to the frequency source, usually a voltage controlled oscillator (VCO). By locking the laser frequency to a cavity mode and the modulation frequency to the FSR of the cavity, a constant incoupling of the triplet to the cavity during a scan is achieved.

Finally, to increase the signal-to-noise ratio in the system, and to reduce the background signal in sD detection, the cavity length can additionally be dithered and the NICE-OHMS signal demodulated at the dither frequency.

6.3.1.2 Lasers, Cavities, and Locking

NICE-OHMS has so far been realized with a variety of lasers: Nd:YAG [2, 3], Ti:sapphire [3, 10, 19], external cavity diode laser (ECDL) [6, 8, 12, 13, 17], quantum cascade (QC) [11], distributed feedback (DFB) [29], and fiber laser (FL) [14, 15], and most recently with a cw optical parametric oscillator (OPO) [58]. While some lasers are tunable by current and temperature only, others have additional tuning stages, e.g. a PZT control of the resonator length or a fiber stretcher.

Due to their construction, the free-running line widths of these lasers range from a few kHz to a few MHz.

Most realizations of NICE-OHMS have been made using a linear (two-mirror) Fabry-Pérot (FP) cavity, see e.g. [2, 8, 14, 29], although also a ring-resonator has been used [17]. A linear FP cavity allows for counter-propagating modes of light, which is necessary to obtain sub-Doppler signals. However, the light reflected from such a cavity is always perpendicular to the reflecting surface of the mirrors, which can, by multiple reflections to other optical elements in the system, produce background signals (etalons) that hamper detection. A ring-resonator circumvents this problem because the light impinges non-perpendicularly on the cavity mirrors. However, a drawback of this type of cavity is that the light propagates in only one direction, which prevents sub-Doppler detection.

The finesse of the cavities is usually in the range of a few hundreds to a few tens of thousands. This implies for a cavity with a typical length of a few tens of cm that the mode width is typically in the tens of kHz range. In order to couple all laser light into a cavity mode, the bandwidth of the PDH servo loop needs to be larger than the line width of the laser. Moreover, to provide a tight lock, the gain at low frequencies should be as high as possible [56]. However, the maximum achievable bandwidth is limited by the phase response of the system consisting of the cavity, the laser frequency actuator, and the electronics. The transfer function of the cavity can be modeled as that of a low-pass filter with a corner frequency equal to the cavity line width, while the transfer function of the laser depends on the laser and its frequency actuator and needs to be experimentally assessed. If the bandwidth of the latter is insufficient, it can be necessary to employ an external acousto-optic modulator (AOM) or an EOM to achieve sufficiently good locking. The servo electronics can be a conventional P-I-D controller, although in most cases electronics with tailored transfer functions perform better.

The laser light has also to be spatially matched to the TEM_{00} mode of the cavity. This implies that the phase fronts of the Gaussian shaped laser beam should be matched to the curvature of the mirrors of the cavity. This is typically achieved by the use of one or two lenses prior to the cavity. Since cavities often are made with lengths of a few tens of cm and mirrors with radii of curvature larger than this, the Rayleigh range of the Gaussian light field is of similar length as the cavity. This implies that the beam diameters on the mirrors are of similar size as that in the focus of the beam, usually on the order of a mm or below, and that it suffices if the cavity is constructed using mirrors with small diameters, e.g. half an inch.

Matching both the frequency and the spatial distribution of the beam to that of the cavity modes usually ensures a high incoupling efficiency for cavities with low or medium finesse (below 10^4). However, it is of importance to consider also the impedance matching of the cavity, especially for cavities with high finesse. The cavity is impedance matched when $t_1 = t_2 + l_1 + l_2$, where t_i and l_i are the transmission and loss of mirror i, respectively [3]. This condition, which can be fulfilled for a pair of mirrors with dissimilar reflectivities, provides the lowest (zero) power reflected on resonance and thereby the highest intracavity power, which is of importance for sub-Doppler detection. This implies that the highest transmission through

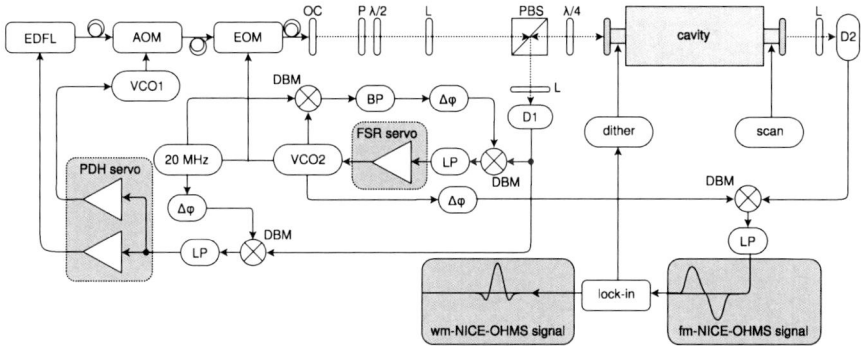

Fig. 6.12 Schematic illustration of an FLB-NICE-OHMS instrumentation. *Solid lines* with circles: optical fibers; dotted lines: free-space light path; OC: output coupler; P: polarizer; $\lambda/2$: half-wave plate; L: lens; PBS: polarizing beam splitter cube; $\lambda/4$: quarter-wave plate; D1: reflection detector with RF-amplification at 20 and 381 MHz; D2: transmission detector with RF-amplification at 381 MHz; DBM: double-balanced mixer; LP: low-pass filter; 20 MHz: fixed 20 MHz source for PDH and FSR locking; VCO1: voltage controlled 110 MHz source; VCO2: 381 MHz source, BP: band-pass filter @ 360 MHz; $\Delta\varphi$: phase shifter; lock-in: lock-in amplifier

a cavity with a given finesse, which is given by $t_1 t_2 (F/\pi)^2$, is not obtained for an impedance matched cavity. The optimum choice of cavity mirrors depends on the type of application.

6.3.2 Fiber-Laser-Based NICE-OHMS

The compact and robust NICE-OHMS system that has provided the so far highest (best) detection sensitivity for Db detection is realized around a DFB-laser pumped Er-doped fiber laser (EDFL). The long term stability of the system allowed for proper investigation of several features of the NICE-OHMS technique [14–16, 24–28, 53], some of which are discussed in Sects. 6.2 and 6.4, wherefore it is presented here in more detail.

The system, which is schematically illustrated in Fig. 6.12 and pictured in Fig. 6.13, is based on an EDFL that is locked to an optical cavity. The particular EDFL (Koheras, Adjustik E15) has a free running line width of 1 kHz over 120 µs. It can be operated at wavelengths from 1530.8 to 1531.8 nm and provides a fast tuning by the use of a PZT fiber stretcher with a maximum tuning range of 3 GHz and a modulation bandwidth of around 30 kHz. The plano-concave cavity consists of two mirrors separated by 39.4 cm, yielding an FSR of 381 MHz, and has a finesse of 5700. The mirrors are attached to a spacer (Zerodur, Schott AG) through two ring piezo-electric transducers (PZT) which jointly act as a gas chamber.

The light from the laser is sent through a fiber-coupled AOM used for laser frequency stabilization. The first-order output, shifted by 110 MHz, is connected to the input of a fiber-coupled EOM used for frequency modulation. The fiber-coupled

Fig. 6.13 The FLB-NICE-OHMS instrumentation realized at Umeå University [16, 24–28]. Laser light comes in a fiber (*green, in the foreground*) from an EDFL placed outside the picture and passes an AOM (square aluminum box, *red on top*) before it goes to the EOM (brass colored item in the *blue open box*). The light continues to the output coupler after which it passes a polarizer and a half-wave plate. It is redirected by a mirror before it passes a lens and impinges upon a polarizing beam splitter cube and continues, via a second mirror, through a quarter-wave plate, to the cavity (solid block of glass) onto which two cylindrical PZT actuators and mirrors are glued. The NICE-OHMS signal is measured in transmission by the leftmost detector. The reflected light, used for PDH and deVoe-Brewer locking, is detected by the rightmost detector. The detector in the background is optionally used for active control of the EOM. ©Chemical Biological Centre KBC, Umeå University, picture by Johan Gunséus, Synk

EOM is based on a small crystal with a waveguide, wherefore it is broadband (up to several GHz) and has a low π-voltage (around a few Volt). This allows for simultaneous modulation at both ν_{PDH} and ν_{FSR}. The modulated light is sent out into free space by an output coupler that provides a beam with a diameter of around 1.3 mm in its focus, which, in turn, is spatially mode-matched to the optical cavity by a single lens with a focal length of 1000 mm. Before impinging upon the cavity, the beam, whose polarization is cleaned and aligned by a polarizer and a half-wave plate, is directed through a polarizing beam splitter cube and a quarter-wave plate for pick-up of the cavity-reflected light used for locking. The cavity reflected and transmitted light is focused onto two high bandwidth (1 GHz) photo-detectors (D1 and D2, respectively). The laser is swept over a molecular transition by scanning the length of the cavity with the help of any of the two cavity PZTs, typically at a rate of 1 Hz. The frequency scale is calibrated and linearized after signal acquisition with the cavity modes acting as frequency markers, having the laser-cavity lock turned off and the scan of the cavity unchanged.

For the PDH locking scheme, a 20 MHz signal from a fixed-frequency source is fed to the EOM producing a modulation with a modulation index of 0.2. The PDH error signal is obtained by demodulating the cavity-reflected light, detected by D1, at the modulation frequency. Home-made servo electronics send slow corrections, up to 100 Hz, to the internal PZT-controlled fiber stretcher of the EDFL, whereas fast corrections, in the frequency range from 100 Hz to 100 kHz, are sent to the AOM driven by a voltage controlled oscillator (VCO1). This design ensures that the large optical frequency deviations, e.g. those from the scan of the cavity length, are corrected by the PZT tuning stage, keeping the AOM within its working range. Double integration in the PZT part of the servo provides a gain level of around 100 dB at 1 Hz and guarantees a tight laser-cavity lock at the scanning frequency. It should be noted that the system can be operated also without the AOM, although with reduced detection sensitivity as is discussed further in Sect. 6.4. In this case, all feedback is sent to the PZT tuning stage of the laser. The achievable bandwidth is then limited to 10 kHz by a resonance in the EDFL piezo at around 30 kHz, but still sufficient to keep the narrow line width EDFL locked to a cavity mode.

In order to produce a NICE-OHMS signal, the EOM is fed with an RF signal generated by a voltage controlled oscillator, VCO2, at a frequency equal to the FSR of the cavity, i.e. at 381 MHz, yielding a modulation index of 0.36. This frequency is locked to the FSR of the cavity by the deVoe-Brewer method, employing an error signal extracted from the beat note between the two modulation frequencies in cavity reflection at 361 MHz. The RF output of the transmission detector, D2, is amplified, demodulated with a reference signal from VCO2 by a double balanced mixer (DBM), and low-pass-filtered, to give the NICE-OHMS signal. For best performance of the FSR lock, the offset of the deVoe-Brewer error signal, which determines the locking point, has to be precisely adjusted by minimizing the noise in (unfiltered) absorption NICE-OHMS signal [53].

For wm-detection of NICE-OHMS signals, the cavity length is additionally dithered with a sinusoidal signal from a lock-in amplifier (for Db detection at 20 Hz and for sD at 125 Hz) while the output of the DBM is sent to a lock-in amplifier for demodulation at the first harmonic. The dither frequency should preferably be as high as possible, while still assuring a tight laser-cavity lock. It is beneficial to add a resonant gain circuit to the PDH servo at the dither frequency, which is less troublesome and more efficient than a feed forward to the laser frequency control, which requires a careful adjustment of amplitude and phase. In order to maximize the wm-signal, for sD detection the dither amplitude should be chosen as $1.3\Gamma_L$ [16], while for DB detection the optimum value is dependent on the FM detection phase [26]. The non-linearity and hysteresis in the wm-dither and the scan are held down by sending the two modulations to separate cavity PZTs.

To reach the highest detection sensitivities (by time averaging) and to allow for accurate calibration-free concentration assessments under long measurement campaigns, an UHV vacuum system is used, based upon conflate (CF) flanges, that provide low leakage rates (down to 10^{-8} mTorr L/s per flange coupling), and thereby a good long-term stability of the gas density. In its present configuration, the system shows a total cavity leakage of only 3 mTorr per day. The system is additionally equipped with an oil-free scroll vacuum pump and a turbo-molecular pump for fast and convenient gas handling.

6.4 Performance

6.4.1 Concentration, Pressure, and Power Dependence of the Analytical Signal

The NICE-OHMS signal depends on a number of parameters of which the analyte concentration, total pressure, and optical power inside the cavity are of particular importance. Although each specific system will have its own characteristics, the general dependences of the signals on these parameters are common to all of them.

6.4.1.1 Doppler-Broadened NICE-OHMS

Doppler-broadened NICE-OHMS signals have been measured at pressures ranging from a few mTorr [24] up to almost an atmosphere (900 mbar) [12]. At a constant pressure, in the absence of optical saturation, the signal strength, as defined by Eq. (6.9), is independent of detection phase and linear with analyte concentration, provided that the single-pass absorption is significantly lower than the empty cavity losses, $\alpha_0 L \ll \pi/(2F)$ [14]. Figure 6.14(a) shows the signal strength as a function of single-pass absorption inside a cavity with a finesse of 1400. As is shown by the inset, the signal strength is linear with the absorption for relative absorption up to around 10^{-4}, which is a fraction of the intracavity losses that for this cavity were 1.1×10^{-3}. However, for higher absorption (above 10^{-4}), the signal strength rolls off, as is shown by the second order polynomial fit. This indicates that there is an upper limit of the linear dynamic range of the technique. However, this does not pose an upper limit to the applicability of the technique; it can still be used for samples with higher absorbance provided the non-linear behavior of the signal strength with respect to absorption is taken into account.

In the Doppler limit, i.e. at low intracavity pressures, the transition is often affected by optical saturation whereby the absorption and dispersion NICE-OHMS signal strengths become dissimilar, as predicted by Eq. (6.14). This is experimentally illustrated in Fig. 6.14(b), which shows the saturated absorption and dispersion NICE-OHMS signal strength from 50 ppm of acetylene in nitrogen measured in a cavity with a finesse of 4800 for an intracavity power of 3.8 W as a function of intracavity pressure [24]. Although the dispersion signal strength shows a perfectly linear behavior with pressure, the absorption signal does not. The absorption signal strength (solid squares) is lower than the dispersion (open circles) by the factor $\sqrt{1 + G_1}$, which, in turn, increases with decreasing pressure due to the increase of optical saturation. As mentioned before, the ratio of the two signal strengths provides a convenient means of determining the degree of saturation induced by the sidebands. In the absence of optical saturation in the Doppler limit both the signal strengths and the peak-to-peak values of the NICE-OHMS signal are linear with pressure.

At higher pressures, for which the pressure broadening starts to influence the line shapes, the peak-to-peak values of the Db NICE-OHMS signals are no longer

Fig. 6.14 (**a**) The unsaturated NICE-OHMS signal strength (*markers*) as a function of single-pass absorption of C_2H_2 in N_2 measured at the $P_e(11)$ transition at 1531.588 nm in a cavity with a finesse of 1400 and a length of 37.8 cm with a polynomial fit (*solid curves*). The *inset* shows a linear fit to the data for the lowest relative absorption [14]. (**b**) Pressure dependence of the NICE-OHMS signal strength in the Doppler limit for absorption (*solid squares*) and dispersion phase (*open circles*) for the same transition in C_2H_2 and 50 ppm concentration measured in a cavity with a finesse of 4800 and intracavity power of 3.8 W [24]. The solid curve shows a linear fit to the dispersion data. (**c**) Pressure dependence of the peak-to-peak value of the absorption (*solid squares*) and dispersion (*open circles*) NICE-OHMS signal in the pressure range from 25 to 250 Torr measured on the $P_f(30)$ line of acetylene at 1550.786 nm using 1000 ppm of acetylene in N_2 [29], together with fits of the predicted pressure dependence. Reproduced with permission from Optical Society of America

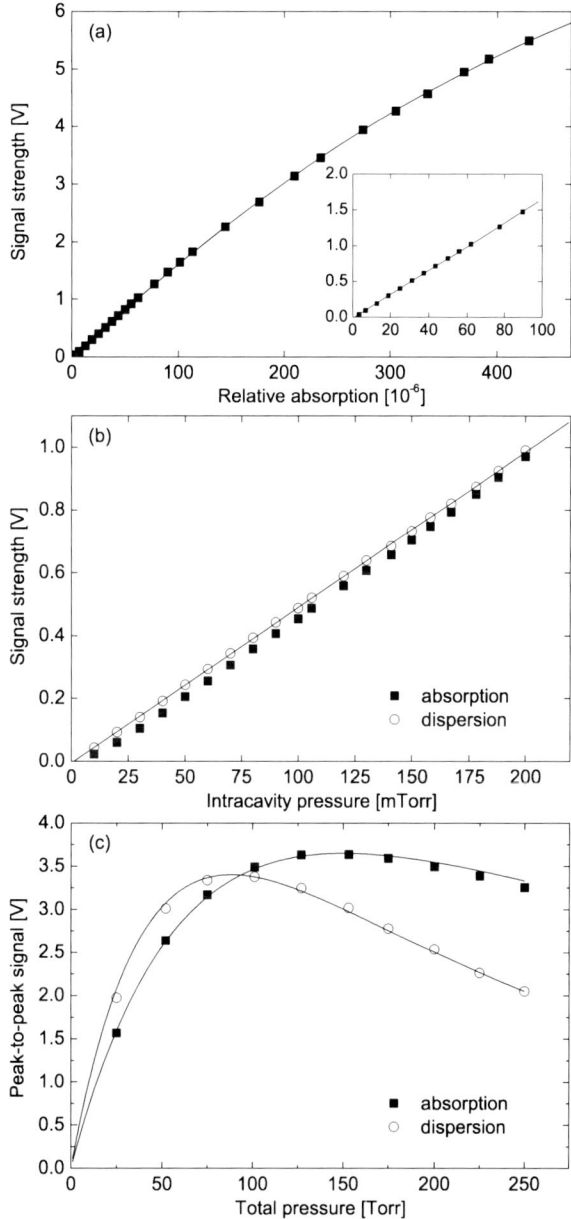

linear with pressure. This is shown in Fig. 6.14(c), which displays the peak-to-peak absorption and dispersion NICE-OHMS signals from a constant concentration of acetylene (1000 ppm) measured at pressures up to 250 Torr [29]. The peak-to-peak values of the signals increase with increasing pressure, reach a maximum, and then decrease for higher pressures. This is caused by a combination of two effects: (i) the ordinary pressure broadening phenomenon, which implies that the peak value of the absorption signal is independent of pressure in the pressure dominated regime [since the increase in number density is canceled by the decrease in peak value of the Lorentzian lineshape, as shown by Eq. (6.15)]; and (ii) the fact that the broadening causes the individual line shape functions to overlap and cancel to a larger degree. Since the degree of overlap is dissimilar in the absorption and dispersion NICE-OHMS line shape functions, as was shown by Fig. 6.2, the pressure at which the peak-to-peak value reaches a maximum, referred to as the optimum pressure, is dissimilar for the two detection phases and depends on experimental parameters such as the FSR of the cavity and the Doppler- and pressure-broadened widths of the transition addressed.

6.4.1.2 Sub-Doppler NICE-OHMS

Similar to the case for Doppler-broadened signals, for a given pressure and provided the single-pass absorption is lower than the intracavity losses, the sub-Doppler NICE-OHMS signal strength is linear with analyte concentration. This is illustrated in Fig. 6.15(a), which displays the dependence of the sD signal strength, as defined in Eq. (6.33), on the acetylene concentration for three different total pressures and a constant intracavity power [16]. The figure shows that for the two lowermost pressures, 10 and 100 mTorr, the dependence is linear, while the upper part of the curve for the highest pressure, 500 mTorr, bends over due to the fact that the intracavity absorption starts to rival the losses in the cavity.

Since the degree of saturation depends strongly on the total pressure inside the cavity, the sD signal strength has, in general, a non-linear dependence on intracavity pressure. Figure 6.15(b) shows the dependence of the sD NICE-OHMS signal strength on pressure for a constant acetylene concentration and four different intracavity powers. For each intracavity power, the signal strength increases with pressure at low pressures, and reaches a maximum before it decreases for larger pressures. The roll-off and decrease of the signal with pressure originates from the fact that the degree of saturation decreases with increasing pressure. The figure also illustrates that both the signal strength and the sD optimum pressure, at which the highest signal-to-noise ratio is expected, increase with intracavity power. Sub-Doppler NICE-OHMS has been thoroughly characterized by Ye et al. [1–3] and Foltynowicz et al. [16].

6.4.2 Noise and Background Signals

The detection sensitivity of a spectroscopic system is defined as the lowest absorption or concentration of a given species that can be detected. It is in general limited by the noise and the presence of background signals. To fully take advantage of the extraordinary detection sensitivity of NICE-OHMS the sources of these must therefore be understood and, if possible, reduced or eliminated.

Fig. 6.15 (a) Concentration dependence of the wm-sD NICE-OHMS signal strength at three different intracavity pressures (10, 50, and 500 mTorr) for an intracavity power of 4.45 W, together with linear and polynomial fits [16] (**b**). Pressure dependence of the wm-sD NICE-OHMS signal strength for the $P_e(11)C_2H_2$ transition at 1531.588 for a concentration of 20 ppm and four different intracavity powers. Reproduced with permission from Optical Society of America

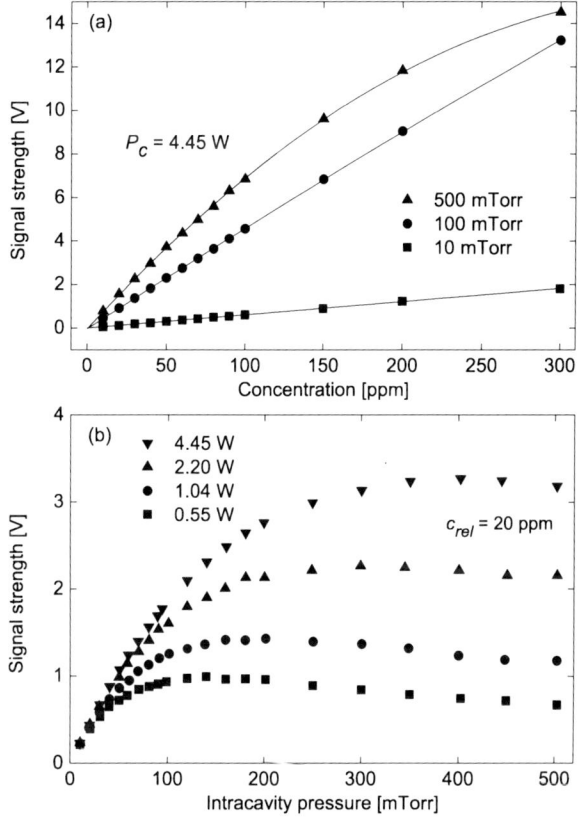

Fig. 6.15 (a) Concentration dependence of the wm-sD NICE-OHMS signal strength at three different intracavity pressures (10, 50, and 500 mTorr) for an intracavity power of 4.45 W, together with linear and polynomial fits [16] (**b**). Pressure dependence of the wm-sD NICE-OHMS signal strength for the $P_e(11)C_2H_2$ transition at 1531.588 for a concentration of 20 ppm and four different intracavity powers. Reproduced with permission from Optical Society of America

6.4.2.1 Shot Noise

Shot noise, which originates from the discrete nature of photons, is the fundamental limitation of all optical techniques, including NICE-OHMS. An expression for the shot noise equivalent absorption in NICE-OHMS was derived by Ye et al. [2]. However, it did not take into account the NICE-OHMS line shape function. For Db NICE-OHMS, the shot noise equivalent absorption, defined as the single-pass absorption that provides a signal with a peak value that is equal to one standard deviation of the shot noise, is given by

$$\alpha_0 L|_{SN}^{Db} = \frac{\pi}{2F} \sqrt{\frac{e \Delta f}{\eta_c P_0}} \frac{1}{J_0(\beta) J_1(\beta)} \frac{\chi^{abs}(0,0)}{\max[\chi_{NO}^{Db}(\nu_d, \theta_{fm}, G)]}, \qquad (6.35)$$

where e is the elementary charge (C), Δf is the detection bandwidth (Hz), η_c is the current response of the detector (A/W), and χ_{NO}^{Db} is the Db NICE-OHMS lineshape function, as defined by Eq. (6.8). For a typical case with a cavity with a finesse of 10^4, a detector with a current response of 1 A/W, a detected power of 1 mW, a detection bandwidth of 1 Hz, and a modulation index of 0.4, the shot noise equivalent absorption is 10^{-11}. For a transition with an integrated molecular line strength of 10^{-20} cm^{-1}/(molecule cm^{-2}) (corresponding to an integrated gas line strength of around 0.25 cm^{-2}/atm at room temperature), and a peak value of the unsaturated absorption line shape function, $\chi^{abs}(0,0)$, of 30 cm at a pressure of 0.1 atm, measured in a cavity with a length of 40 cm this corresponds to a concentration of absorbers of around 0.3 ppt (3×10^{-13}). The use of longer integration times, a higher cavity finesse, or a transition with larger line strength, decreases this number even further. This clearly shows the enormous potential of Db NICE-OHMS.

For sD NICE-OHMS, the corresponding shot noise equivalent absorption is given by

$$\alpha_0 L|_{SN}^{sD} = \frac{\pi}{2F} \sqrt{\frac{e\Delta f}{\eta_c P_0}} \frac{1}{J_0(\beta)J_1(\beta)} \frac{2}{2\Phi(G_0)}, \qquad (6.36)$$

where the factor of 2 in the numerator originates from a ratio $c/(\pi \Gamma_L)/$ $\max[\chi_{L,0}^{disp}(\nu_d, \Gamma_L, 0)]$, i.e. the inverse of half of the peak-to-peak-normalized Lorentzian dispersion function, where the latter, by definition, is unity. For the same experimental conditions, and for a well-saturated transition, this limit is about 3.5 times higher than that for Db detection. This shows that also sD NICE-OHMS can achieve exceptionally high detection sensitivities. A detection sensitivity close to the shot noise limit has to date been achieved only for sD detection by Ye et al. [2, 3] and by Ishibashi et al. [6]. As is further discussed below, for Db detection the best detection sensitivity is still a few times above the shot noise limit [28].

6.4.2.2 Background Signals—the Source of Noise

The noise-immune property of NICE-OHMS originates from the fact that the technique is ideally background free. However, in reality, there exist various sources of background signals through which noise can couple in and reduce the sensitivity of the technique. In FM spectroscopy background signals originate in general from an unbalance in the triplet in the absence of analyte, which can have several causes. Three of these are of particular importance for NICE-OHMS: residual intensity modulation of the laser, polarization-dependent dispersion in birefringent components, and wavelength-dependent transmission of the optical system, the latter predominantly caused by multiple reflections between optical components. Drift, fluctuation, or noise in any component which these background signals depend on gives rise to noise in the NICE-OHMS signal.

As an example, consider a NICE-OHMS system with a background signal corresponding to a single-pass absorption on the 10^{-5} level that is stable over short time scales. For short integration times the main noise contribution is usually from the intensity noise that is caused by the frequency-to-amplitude noise conversion, which often can be on the 10^{-2} level. Since the background signal is proportional to the power, it will in this case couple in noise on the 10^{-7} level, which for a cavity with a cavity enhancement factor, $2F/\pi$, of 10^3 corresponds to a single-pass absorption of around 10^{-10}. Additional drifts and noise in other components can reduce the detection sensitivity further. Hence, the detection sensitivity of NICE-OHMS depends to a large degree on the ability to reduce the background signals.

6.4.2.3 Background Signals from Residual Amplitude Modulation

When a laser is directly frequency modulated (e.g. a semiconductor laser though its injection current) an associated intensity modulation in the laser, also referred to as residual amplitude modulation (RAM), causes an unbalance in the triplet. The sidebands caused by amplitude modulation are in phase with each other, contrary to those created by a pure frequency modulation process, which are fully out of phase. As a result, the balance of the triplet is compromised, and a background signal appears. Any noise in the detected laser intensity that is within the detection bandwidth will then contribute to noise in the NICE-OHMS signal. The use of an external modulator, i.e. an EOM, for frequency modulation reduces this problem.

6.4.2.4 Background Signals from Birefringent Components

A triplet can also become unbalanced when it propagates through birefringent components, e.g. polarization maintaining (PM) fibers or electro-optic crystals, if its polarization axis is not fully aligned along any of the principal axes of the components. In such system linearly polarized light will exit as elliptically polarized because of the dissimilar indices of refraction (dispersion) for the different directions of polarization (n_e and n_o). However, since the optical path difference along the two polarization directions, given by $(n_e - n_o)L_b$, where L_b is the length of the birefringent component, gives rise to dissimilar phase shifts for the three components of the triplet, the degree of ellipticity is different for the carrier and the sidebands. This implies that a triplet propagating in a birefringent component becomes unbalanced and gives rise to background signals at all polarization directions except those along the optical axes of the material [59].

A related effect takes place as a consequence of the modulation process in an EOM. In an EOM the modulation takes place mainly along one axis of the crystal wherefore sidebands will likewise be produced mostly with their polarization along this axis. This implies that the ellipticity of the carrier is much higher than that of the sidebands, which gives rise to background signals. Since the absorption signal is not sensitive to the attenuation of the carrier, this type of background signal is much smaller at the absorption than the dispersion phase [59].

Wong and Hall showed that background signals from free space EOMs can be eliminated by a careful alignment of a polarizer before or after the EOM, so as to extract only the part of the light that carries a balanced triplet [60]. However, this approach is not applicable to fiber-coupled EOMs due to inherent manufacturing misalignments between the PM fibers and the EOM waveguide. For such EOMs, a couple of alternative approaches have been developed. One is to make use of lithium niobate EOMs with a waveguide created by a proton exchange (PE) process, which supports propagation only along one polarization axis [27, 59]. No matter how large the misalignment between the PM fibers and the electro-optic crystal is, such an EOM will not produce an unbalanced triplet. Another approach, which can be applied to EOMs with titanium-diffused (TD) waveguides, is also based upon the work of Wong and Hall for FMS [60] and utilizes a DC voltage feedback to the EOM to balance the relative phase of the triplet components. The error signal is derived from a part of the light sampled prior to the cavity [27, 59], so the advantage of this technique is that the active feedback reduces all types of background signals created in the optical system up to the sampling point, including any possible etalons (see below). On the other hand, the use of an EOM with a PE waveguide has the advantage of simplicity.

A comparison of typical NICE-OHMS signals measured with the FLB-NICE-OHMS system with EOMs in three different configurations is shown in Fig. 6.16. The three panels (a), (b), and (c) correspond to an EOM with a PE waveguide, and an EOM with a TD waveguide with and without active feedback, respectively. As is shown in panel (b), there is virtually no background signal when an EOM stabilized by active feedback is used. Panel (c) reveals that the same EOM can produce a substantial background signal when it is free-running because of the strong temperature dependence of the EOM crystal. The use of an EOM with a PE waveguide gives rise to a low but constant background signal as shown in panel (a).

It is clear from the figure that the in-coupling of intensity noise is proportional to the magnitude of the background signal; the noise in panel (c) is significantly larger than that in the other two panels. Hence, NICE-OHMS systems incorporating fiber-coupled EOMs should either make use of PE-EOM or TD-EOM with feedback [27, 59].

6.4.2.5 Background Signals from Etalons

Background signals originating from multiple reflections between various surfaces of optical components, so-called etalons, are common in all types of optical systems used for absorption spectrometry, including FMS. They have a strong wavelength and temperature dependence that limits the detection sensitivity in the system. The use of antireflection-coated or wedged components placed at non-perpendicular directions with respect to the beam reduces etalon effects but does not eliminate them completely.

To further reduce etalon effects it is in FMS possible to separate the optical components by so-called etalon-immune distances (EID). In the same way as the modulation frequency is matched to the FSR of the cavity so that the FM triplet can pass through the cavity undisturbed, it is possible to "hide" an etalon from detection by choosing the distances between the optical components so that the various modes of the FM triplet are affected identically also by the etalon. This takes place if the length of the etalon is an integer multiple of the length of the cavity, L_c, since then

Fig. 6.16
Doppler-broadened
NICE-OHMS signals from
10 ppm of acetylene in
180 mTorr of N_2 measured
with the FLB-NICE-OHMS
system using: (**a**) an EOM
with a PE waveguide; (**b**) an
EOM with a TD waveguide
with active DC voltage
feedback; and (**c**) 3 h after
turning the feedback off [27].
Reproduced with permission
from Optical Society of
America

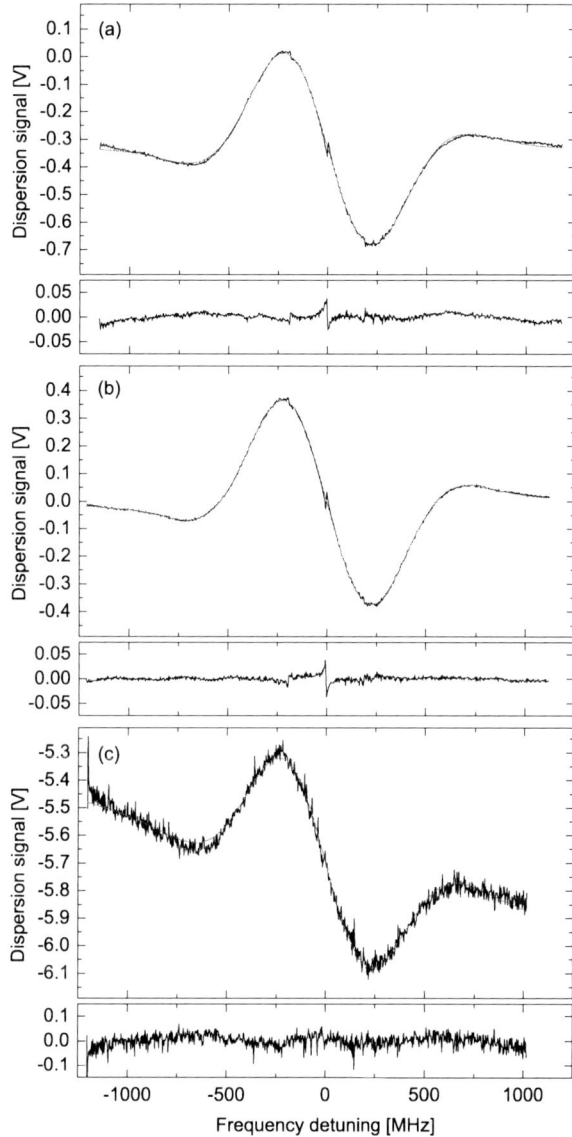

the frequency modulation of the light also will match the FSR of the etalon [61]. Moreover, since the absorption signal is given solely by the attenuation of the two sidebands, which are separated by twice the modulation frequency, background signals from etalons created by distances of half the cavity length will also evade detection in the absorption mode. This is exemplified in Fig. 6.17, which shows the magnitude of an etalon background signal at the absorption and dispersion detection phases as a function of distance between the reflecting surfaces, calculated using

Fig. 6.17 Normalized
amplitude of etalon
background signals for etalon
lengths up to one cavity
length. The *solid* and *dotted*
curves correspond to
dispersion and absorption,
respectively

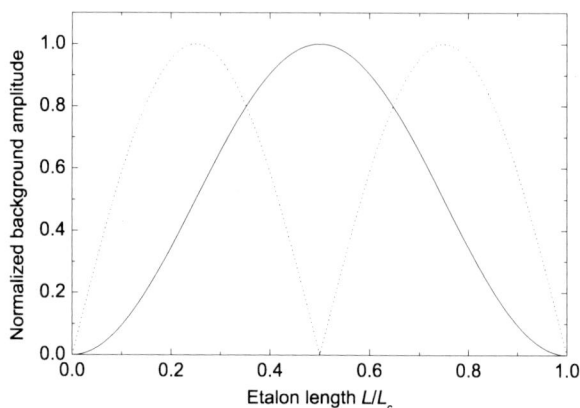

Eq. (6.3) and the transfer function of an etalon. The figure also illustrates that the dispersion background signals for lengths close to the EID, including short lengths (i.e. for $L \ll L_c$), are smaller than those in absorption. Systems with EID implemented, or with optical components placed in groups separated by EID, will therefore have lower background signals from etalons in dispersion than in absorption phase.

6.4.3 Detection Sensitivity

To separate the response from the analyte from that of other concomitant species or background signals and to accurately assess the signal strength, the signal in NICE-OHMS is most often acquired by scanning the laser across the transition of the analyte. For Db-NICE-OHMS the scan range is usually chosen a few times the Doppler width, as seen in Fig. 6.16, while in sD-NICE-OHMS it is only a fraction of this, as is illustrated in Fig. 6.10. An expression for the NICE-OHMS signal [Eq. (6.8) for Db and Eq. (6.32) for sD detection] based upon an appropriate line shape function is then fitted to the experimental data. The signal strengths obtained from such fits can then be converted to a suitable entity, e.g. absorption, $\alpha_0 L$, absorption per unit length, α_0, or species concentration, c_{rel}.

The performance of the system in terms of detection sensitivity and stability is often assessed by the means of the Allan variance [62]. Figure 6.18 shows the Allan deviation (defined as the square-root of the Allan variance, in terms of absorption) of different sets of Db dispersion measurements performed with the FLB-NICE-OHMS setup described in Sect. 6.3.2 on the $P_e(11)$ transition of C_2H_2 with an analyte concentration of 10 ppm in N_2 at 150 mTorr. The Allan deviation for short integration times decreases with the square root of the measurement time and represents the level of white noise in the system. The increasing deviation for long

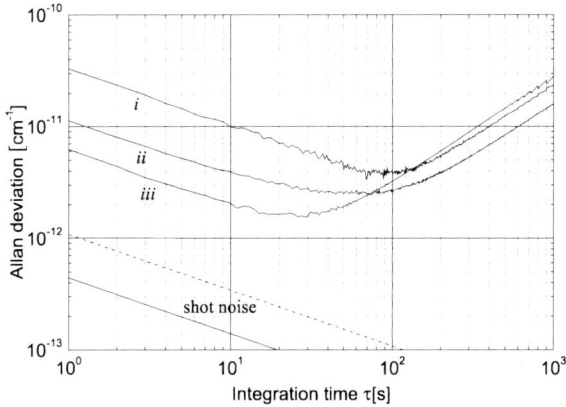

Fig. 6.18 Allan deviation of absorption measured with the FLB-NICE-OHMS setup with different noise reduction schemes; *i.* with a fiber-coupled EOM with a PE waveguide [28], *ii.* with a fiber-coupled EOM with a TD waveguide and active feedback [27], and *iii* with a fiber-coupled EOM with a PE waveguide and an AOM to improve the laser-cavity lock [28]. All data taken with optical components mounted at etalon immune distances whenever possible. The *solid* and *dashed straight lines* show the shot-noise-equivalent absorption evaluated at signal maximum and from a scan, respectively

integration times corresponds to the types of noise that have a $1/f^b$ ($b > 1$) dependence, i.e. drifts. It is customary to define the optimum integration time for a system as the time for which the Allan deviation has its minimum. Note though that the Allan deviation is equal to the standard deviation only as long as it shows white-noise behavior [62].

In Fig. 6.18 the curves *i* and *ii* correspond to measurements with a fiber-coupled EOM with a PE waveguide and with a TD waveguide and an active voltage feedback, corresponding to the signals displayed in Figs. 6.16(a) and (b), respectively. In both cases the EDFL was locked to the cavity solely with the use of the PZT fiber stretcher. The use of active feedback reduces the background signals and thereby the in-coupling of intensity noise, which is clearly seen by comparing the two curves at short integration times. Curve *iii* corresponds to measurements with a fiber-coupled EOM with a PE waveguide and a laser-cavity lock employing an AOM (as shown in Fig. 6.12) [28]. Since the curves *i* and *iii* were taken with the same EOM, the difference in white noise level in the two Allan plots is attributed to the reduction of the frequency noise that is caused by the improvement in the laser-to-cavity lock by the AOM, which, in turn, reduces the transmitted intensity noise and thereby the noise in the background signals.

For longer integration times the sensitivity is limited mainly by drifting etalon signals. As discussed above, active feedback to the EOM reduces all unbalance in the triplet that is created before the cavity, including that from etalons, which shows up as a reduced Allan deviation for longer integration times in curve *ii*. Hence, while the system with a PE-EOM and AOM provides the best short term stability (curve *iii*), that with a TD-EOM with an active feedback (curve *ii*) shows the best long term stability.

Figure 6.18 also shows that the Allan deviations are a few times above the shot noise limit, given by Eq. (6.35) and indicated by the solid line. However, it should be noted that this shot noise limit represents that measured under the conditions for which Eq. (6.35) is valid, i.e. on the peak of the signal during the entire measurement process. The data giving rise to the curves *i* to *iii* were, in fact, taken during repeated frequency scans over the transition, and the signal was extracted only over a part of the scan. Moreover, scans solely in one direction were analyzed. This implies that the signal was effectively measured only over a fraction of the entire scan, in this particular case about a sixth of the inverse scanning rate. The actual shot noise limit, to which the three measurement curves should be compared, is therefore higher than that given by Eq. (6.35), roughly by a factor of $\sqrt{6} \approx 2.5$. This level is marked by the dashed straight line in Fig. 6.18. This implies that curve *iii* effectively has a noise that is around 6 times above the shot noise limit.

In conclusion, the Allan deviations shown in Fig. 6.18 indicate that the FLB-NICE-OHMS system exhibits a white-noise behavior to at least 10 s, which for the best configuration (*iii*) gives an instrument with a minimum detectable Db absorption per unit length of 1.8×10^{-12} cm^{-1} and a single-pass absorption of 7.2×10^{-11} (both at 10 s). The latter number corresponds, for the $P_e(11)$ transition, to a relative concentration of acetylene in N_2 of 4 ppt.

6.5 Summary, Conclusions, and Future Outlook

NICE-OHMS is a highly sensitive spectroscopic technique for detection of molecular compounds in gas phase that has a number of unique properties of which the most important are the following: (i) it utilizes an external cavity for prolongation of the interaction length with the sample, (ii) it makes use of FMS for reduction of noise, (iii) it benefits from an immunity to frequency-to-amplitude noise conversion that affects many other cavity-enhanced techniques, (iv) it can provide both Doppler-broadened and sub-Doppler signals, (v) it can detect signals at both absorption and dispersion phase, or at any combination thereof, and (vi) it can provide continuous scans over molecular transitions. The combination of these features makes NICE-OHMS truly exclusive and it provides, among other things, high signal-to-noise ratios and thereby an exceptional detection sensitivity, in some cases close to the shot noise limit.

The basic properties of the technique indicate that it is a powerful alternative to other cavity-enhanced techniques, e.g. cavity ring down spectroscopy (CRDS), integrated cavity output spectroscopy (ICOS), or optical-feedback cavity-enhanced absorption spectrometry. The restricted number of publications about NICE-OHMS (presently about 30) as compared to, e.g. CRDS ($>10^3$), implies that the technique is still under development and that its full potential for practical applications has not yet been explored. However, substantial work regarding the assessment of basic properties of the technique has already been pursued [14–16, 24–28]. For example, the appropriate line shape functions on which the technique relies under various conditions have been derived and experimentally confirmed. Most lately, the influence of Dicke narrowing and SDEs on dispersion was established [30–32]. This has provided a basis for the assessment of the dependence of the NICE-OHMS signal on physical entities such as optical power (and thereby optical saturation) [16, 24, 25, 33], analyte concentration [14], and total gas pressure [29]. It has been shown, for example, that the technique is less influenced by optical saturation than other cavity-enhanced techniques [24, 33], and that for a given system, and a given mode of operation, there is an optimum pressure for which the NICE-OHMS signal is maximal [29].

The preferred mode of operation for trace gas detection depends on the particular application. An advantage of the Doppler-broadened detection is that it does not require optical saturation and that it therefore can be performed under any pressure. Its principles are nowadays also well understood. Although the sub-Doppler signal requires optical saturation, which precludes detection under the highest pressures, an evident advantage of the sub-Doppler mode of detection is that it removes virtually all spectral interferences. Moreover, although the full applicability for quantitative trace gas applications has still to be assessed, it can be concluded that this mode of detection is expected to find use for analytical applications if one is willing to sacrifice some of the detection sensitivity for an improved spectral selectivity.

Since its original realization for sD detection for frequency standard applications [1–3], NICE-OHMS has mainly been used in its Db mode of operation for fundamental investigations, e.g. high-precision spectroscopic investigations of weak molecular transitions [12, 19, 20], for assessments of line strengths [10, 12], and studies of line shapes [30–32], and for characterization of its applicability for chemical analysis and trace gas detection [11, 14–18, 28, 53, 58]. It has achieved a detection sensitivity for Db detection of 5.6×10^{-12} cm^{-1} Hz$^{-1/2}$ (corresponding to a relative single-pass absorption of 7.2×10^{-11} at 10 s) [28], a few times above the shot noise limit, which, for the targeted C_2H_2 line [$P_e(11)$ at 1531.588 nm], corresponds to a relative concentration of acetylene in N_2 of 4 ppt. It has a large dynamic range in terms of absorption, which is linear over at least five decades, and it can be used for sensitive and accurate assessments of gases at virtually any pressure (up to 900 mbar [12]). All this indicates that the technique has a high potential to be applied to trace gas analysis. The sD mode of operation has so far been only sparsely used, primarily for high-precision spectroscopic investigations of weak molecular transitions [6, 7, 63].

Until now, NICE-OHMS has been based predominantly on lasers operating in the near-infrared region, in which most molecules only have weaker overtone transitions. To fully benefit from the extraordinary detection sensitivity and obtain low concentration detection limits for a large variety of molecules, the technique should be realized in the mid-infrared region, where molecules have their strongest fundamental vibrations. There exist, so far, only two types of implementations of mid-infrared NICE-OHMS, one based on a QC laser [11] and another on a cw-OPO [64].

The future performance and applicability of the technique depends therefore on the ability of the community of scientists to incorporate new laser sources, predominantly in the mid-IR region. Moreover, more work is needed to develop sturdier and more compact NICE-OHMS systems, capable of stable and unmanned operation. Recent work has shown that the NICE-OHMS principle can be implemented in a compact way that facilitates the use of the technique [14, 15, 27–29, 65]. For example, the construction of a fully fiber-based system, of which several important advances recently have been taken [65], would be an important step towards these goals.

All this demonstrates that NICE-OHMS is indeed a powerful technique for detection of molecules in gas phase. When the technique has been fully realized in the mid-IR wavelength region it will presumably provide detection sensitivities in the low or even sub ppt ranges for a large number of molecules, including a variety of hydrocarbons. This will not only allow for highly sensitive trace gas analysis, it will also permit accurate isotopological assessments of gases that exist only in low concentrations. There is no doubt that the present knowledge about the potential and limitations of the technique, together with the rapid development of new lasers and electro-optical and optical components, will open up the possibility for future applications of NICE-OHMS in the field of chemical sensing and trace gas detection, e.g. regarding medicine (breath analysis), biology, ecology or atmospheric (climate) research.

Acknowledgements   This work was supported by the Swedish Research Council under the projects 621-2008-3674 and 621-2011-4216. The authors would also like to acknowledge the Kempe foundations, the Carl Trygger's foundation, and Stiftelsen J. Gust. Richert for support.

References

1. J. Ye, Ultrasensitive high resolution laser spectroscopy and its application to optical frequency standards. Ph.D. thesis, University of Colorado, Boulder, CO, 1997
2. J. Ye, L.S. Ma, J.L. Hall, Ultrasensitive detections in atomic and molecular physics: demonstration in molecular overtone spectroscopy. J. Opt. Soc. Am. B **15**, 6–15 (1998)
3. L.S. Ma, J. Ye, P. Dube, J.L. Hall, Ultrasensitive frequency-modulation spectroscopy enhanced by a high-finesse optical cavity: theory and application to overtone transitions of C_2H_2 and C_2HD. J. Opt. Soc. Am. B **16**, 2255–2268 (1999)
4. J. Ye, J.L. Hall, Absorption detection at the quantum limit: probing high-finesse cavities with modulation techniques, in *Cavity-Enhanced Spectroscopies*, ed. by R.D. van Zee, J.P. Looney (Academic Press, New York, 2002), pp. 83–127
5. A. Foltynowicz, F.M. Schmidt, W. Ma, O. Axner, Noise-immune cavity-enhanced optical heterodyne molecular spectroscopy: current status and future potential. Appl. Phys. B **92**, 313–326 (2008)
6. C. Ishibashi, H. Sasada, Highly sensitive cavity-enhanced sub-Doppler spectroscopy of a molecular overtone band with a 1.66 μm tunable diode laser. Jpn. J. Appl. Phys. 1(**38**), 920–922 (1999)
7. C. Ishibashi, H. Sasada, Near-infrared laser spectrometer with sub-Doppler resolution, high sensitivity, and wide tunability: a case study in the 1.65-μm region of CH_3I spectrum. J. Mol. Spectrosc. **200**, 147–149 (2000)

8. L. Gianfrani, R.W. Fox, L. Hollberg, Cavity-enhanced absorption spectroscopy of molecular oxygen. J. Opt. Soc. Am. B **16**, 2247–2254 (1999)
9. J. Bood, A. McIlroy, D.L. Osborn, Cavity-enhanced frequency modulation absorption spectroscopy of the sixth overtone band of nitric oxide. Proc. SPIE **4962**, 89–100 (2003)
10. J. Bood, A. McIlroy, D.L. Osborn, Measurement of the sixth overtone band of nitric oxide, and its dipole moment function, using cavity-enhanced frequency modulation spectroscopy. J. Chem. Phys. **124**, 084311 (2006)
11. M.S. Taubman, T.L. Myers, B.D. Cannon, R.M. Williams, Stabilization, injection and control of quantum cascade lasers, and their application to chemical sensing in the infrared. Spectrochim. Acta A **60**, 3457–3468 (2004)
12. N.J. van Leeuwen, H.G. Kjaergaard, D.L. Howard, A.C. Wilson, Measurement of ultraweak transitions in the visible region of molecular oxygen. J. Mol. Spectrosc. **228**, 83–91 (2004)
13. N.J. van Leeuwen, A.C. Wilson, Measurement of pressure-broadened, ultraweak transitions with noise-immune cavity-enhanced optical heterodyne molecular spectroscopy. J. Opt. Soc. Am. B **21**, 1713–1721 (2004)
14. F.M. Schmidt, A. Foltynowicz, W.G. Ma, O. Axner, Fiber-laser-based noise-immune cavity-enhanced optical heterodyne molecular spectrometry for Doppler-broadened detection of C_2H_2 in the parts per trillion range. J. Opt. Soc. Am. B **24**, 1392–1405 (2007)
15. F.M. Schmidt, A. Foltynowicz, W.G. Ma, T. Lock, O. Axner, Doppler-broadened fiber-laser-based NICE-OHMS—improved detectability. Opt. Express **15**, 10822–10831 (2007)
16. A. Foltynowicz, W.G. Ma, O. Axner, Characterization of fiber-laser-based sub-Doppler NICE-OHMS for quantitative trace gas detection. Opt. Express **16**, 14689–14702 (2008)
17. C.L. Bell, G. Hancock, R. Peverall, G.A.D. Ritchie, J.H. van Helden, N.J. van Leeuwen, Characterization of an external cavity diode laser based ring cavity NICE-OHMS system. Opt. Express **17**, 9834–9839 (2009)
18. C.L. Bell, J.P.H. van Helden, T.P.J. Blaikie, G. Hancock, N.J. van Leeuwen, R. Peverall, G.A.D. Ritchie, Noise-immune cavity-enhanced optical heterodyne detection of HO_2 in the near-infrared range. J. Phys. Chem. A **116**, 5090–5099 (2012)
19. B.M. Siller, M.W. Porambo, A.A. Mills, B.J. McCall, Noise immune cavity enhanced optical heterodyne velocity modulation spectroscopy. Opt. Express **19**, 24822–24827 (2011)
20. A.A. Mills, B.M. Siller, M.W. Porambo, M. Perera, H. Kreckel, B.J. McCall, Ultra-sensitive high-precision spectroscopy of a fast molecular ion beam. J. Chem. Phys. **135**, 224201 (2011)
21. B. McCall, B.M. Siller, Applications of NICE-OHMS to molecular spectroscopy, in *Cavity-enhanced Spectroscopy and Sensing*, ed. by G. Gianluca, H.P. Loock (Springer, Berlin, 2013)
22. J. Ye, T.W. Lynn (eds.), *Applications of Optical Cavities in Modern Atomic, Molecular, and Optical Physics*. Advances in Atomic Molecular, and Optical Physics, vol. 49, pp. 1–83 (2003)
23. B.A. Paldus, A.A. Kachanov, An historical overview of cavity-enhanced methods. Can. J. Phys. **83**, 975–999 (2005)
24. A. Foltynowicz, W.G. Ma, F.M. Schmidt, O. Axner, Doppler-broadened noise-immune cavity-enhanced optical heterodyne molecular spectrometry signals from optically saturated transitions under low pressure conditions. J. Opt. Soc. Am. B **25**, 1156–1165 (2008)
25. O. Axner, W.G. Ma, A. Foltynowicz, Sub-Doppler dispersion and noise-immune cavity-enhanced optical heterodyne molecular spectroscopy revised. J. Opt. Soc. Am. B **25**, 1166–1177 (2008)
26. A. Foltynowicz, W.G. Ma, F.M. Schmidt, O. Axner, Wavelength-modulated noise-immune cavity-enhanced optical heterodyne molecular spectroscopy signal line shapes in the Doppler limit. J. Opt. Soc. Am. B **26**, 1384–1394 (2009)
27. A. Foltynowicz, I. Silander, O. Axner, Reduction of background signals from fiber-based NICE-OHMS. J. Opt. Soc. Am. B **28**, 2797–2805 (2011)
28. P. Ehlers, I. Silander, J. Wang, O. Axner, Fiber-laser-based noise-immune cavity-enhanced optical heterodyne molecular spectrometry instrumentation for Doppler-broadened detection in the 10^{-12} cm^{-1} Hz$^{-1/2}$ region. J. Opt. Soc. Am. B **29**, 1305–1315 (2012)

29. A. Foltynowicz, J.Y. Wang, P. Ehlers, O. Axner, Distributed-feedback-laser-based NICE-OHMS in the pressure-broadened regime. Opt. Express **18**, 18580–18591 (2010)
30. J.Y. Wang, P. Ehlers, I. Silander, O. Axner, Dicke narrowing in the dispersion mode of detection and in noise-immune c avity-enhanced optical heterodyne molecular spectroscopy—theory and experimental verification. J. Opt. Soc. Am. B **28**, 2390–2401 (2011)
31. J.Y. Wang, P. Ehlers, I. Silander, J. Westberg, O. Axner, Speed-dependent Voigt dispersion lineshape function—applicable to techniques measuring dispersion signals. J. Opt. Soc. Am. B **29**, 2971–2979 (2012)
32. J.Y. Wang, P. Ehlers, I. Silander, O. Axner, Speed-dependent effects in dispersion mode of detection and in noise-immune cavity-enhanced optical heterodyne molecular spectrometry—experimental demonstation and validation of predicted line shape. J. Opt. Soc. Am. B **29**, 2980–2989 (2012)
33. W.G. Ma, A. Foltynowicz, O. Axner, Theoretical description of Doppler-broadened noise-immune cavity-enhanced optical heterodyne molecular spectroscopy under optically saturated conditions. J. Opt. Soc. Am. B **25**, 1144–1155 (2008)
34. G.C. Bjorklund, Frequency-modulation spectroscopy: a new method for measuring weak absorptions and dispersions. Opt. Lett. **5**, 15–17 (1980)
35. L.S. Rothman, C.P. Rinsland, A. Goldman, S.T. Massie, D.P. Edwards, J.M. Flaud, A. Perrin, C. Camy-Peyret, V. Dana, J.Y. Mandin, J. Schroeder, A. McCann, R.R. Gamache, R.B. Wattson, K. Yoshino, K.V. Chance, K.W. Jucks, L.R. Brown, V. Nemtchinov, P. Varanasi, The HITRAN molecular spectroscopic database and HAWKS (HITRAN Atmospheric Workstation): 1996 edition. J. Quant. Spectrosc. Radiat. Transf. **60**, 665–710 (1998)
36. H.A. Kramers, La diffusion de la lumiere par les atomes. Atti. Congr. Int. Fis. Como. **2**, 545–557 (1927)
37. R.L. Kronig, On the theory of dispersion of X-rays. J. Opt. Soc. Am. **12**, 545–556 (1926)
38. M. Sargent III., M. Scully, W.E. Lamb Jr., *Laser Physics* (Addison-Wesley, Reading, 1974)
39. W. Demtröder, *Laser Spectroscopy*, 2nd edn. (Springer, Berlin, 1996)
40. A.G. Marshall, R.E. Bruce, Dispersion versus absorption (DISPA) lineshape analysis—effects of saturation, adjacent peaks, and simultaneous distribution in peak width and position. J. Magn. Res. **39**, 47–54 (1980)
41. R.H. Dicke, The effect of collisions upon the Doppler width of spectral lines. Phys. Rev. **89**, 472–473 (1953)
42. L. Galatry, Simultaneous effect of Doppler and foreign gas broadening on spectral lines. Phys. Rev. **122**, 1218–1223 (1961)
43. S.G. Rautian, I.I. Sobelman, Effect of collisions on Doppler broadening of spectral lines. Sov. Phys. Usp. **9**, 701–716 (1967)
44. P.R. Berman, Speed-dependent collisional width and shift parameters in spectral profiles. J. Quant. Spectrosc. Radiat. Transf. **12**, 1331–1342 (1972)
45. P. Kluczynski, O. Axner, Theoretical description based on Fourier analysis of wavelength-modulation spectrometry in terms of analytical and background signals. Appl. Opt. **38**, 5803–5815 (1999)
46. P. Kluczynski, J. Gustafsson, A.M. Lindberg, O. Axner, Wavelength modulation absorption spectrometry—an extensive scrutiny of the generation of signals. Spectrochim. Acta B **56**, 1277–1354 (2001)
47. J. Westberg, J.Y. Wang, O. Axner, Fast and non-approximate methodology for calculation of wavelength-modulated Voigt lineshape functions suitable for real-time curve fitting. J. Quant. Spectrosc. Radiat. Transf. **113**, 2049–2057 (2012)
48. J. Westberg, J. Wang, O. Axner, Methodology for fast curve fitting to modulated Voigt dispersion lineshape functions. J. Quant. Spectrosc. Radiat. Transf. (2012). doi:10.1016/j.jqsrt.2013.08.008
49. P. Kluczynski, A.M. Lindberg, O. Axner, Wavelength modulation diode laser absorption signals from Doppler broadened absorption profiles. J. Quant. Spectrosc. Radiat. Transf. **83**, 345–360 (2004)

50. O. Axner, P. Kluczynski, A.M. Lindberg, A general non-complex analytical expression for the nth Fourier component of a wavelength-modulated Lorentzian lineshape function. J. Quant. Spectrosc. Radiat. Transf. **68**, 299–317 (2001)

51. J. Westberg, P. Kluczynski, S. Lundqvist, O. Axner, Analytical expression for the nth Fourier coefficient of a modulated Lorentzian dispersion lineshape function. J. Quant. Spectrosc. Radiat. Transf. **112**, 1443–1449 (2011)

52. C.J. Borde, J.L. Hall, C.V. Kunasz, D.G. Hummer, Saturated absorption-line shape—calculation of transit-time broadening by a perturbation approach. Phys. Rev. A **14**, 236–263 (1976)

53. A. Foltynowicz, Fiber-laser-based noise-immune cavity-enhanced optical heterodyne molecular spectrometry. Ph.D. thesis, Umeå University, Umeå, Sweden, 2009

54. R.W.P. Drever, J.L. Hall, F.V. Kowalski, J. Hough, G.M. Ford, A.J. Munley, H. Ward, Laser phase and frequency stabilization using an optical resonator. Appl. Phys. B **31**, 97–105 (1983)

55. E.D. Black, An introduction to Pound-Drever-Hall laser frequency stabilization. Am. J. Phys. **69**, 79–87 (2001)

56. R.W. Fox, C.W. Oates, L.W. Hollberg, Stabilizing diode lasers to high-finesse cavities, in *Cavity-Enhanced Spectroscopies*, ed. by R.D. van Zee, J.P. Looney (Elsevier Science, New York, 2002)

57. R.G. DeVoe, R.G. Brewer, Laser frequency division and stabilization. Phys. Rev. A **30**, 2827–2829 (1984)

58. K.N. Crabtree, J.N. Hodges, B.M. Siller, A.J. Perry, J.E. Kelly, P.A. Jenkins II., B.J. McCall, Sub-Doppler mid-infrared spectroscopy of molecular ions. Chem. Phys. Lett. **551**, 1–6 (2012)

59. I. Silander, P. Ehlers, J. Wang, O. Axner, Frequency modulation background signals from fiber-based electro optic modulators are caused by crosstalk. J. Opt. Soc. Am. B **29**, 916–923 (2012)

60. N.C. Wong, J.L. Hall, Servo control of amplitude-modulation in frequency-modulation spectroscopy—demonstration of shot-noise-limited detection. J. Opt. Soc. Am. B **2**, 1527–1533 (1985)

61. E.A. Whittaker, M. Gehrtz, G.C. Bjorklund, Residual amplitude-modulation in laser electro-optic phase modulation. J. Opt. Soc. Am. B **2**, 1320–1326 (1985)

62. P. Werle, R. Mucke, F. Slemr, The limits of signal averaging in atmospheric trace-gas monitoring by tunable diode-laser absorption spectroscopy (TDLAS). Appl. Phys. B **57**, 131–139 (1993)

63. C. Ishibashi, M. Kourogi, K. Imai, B. Widiyatmoko, A. Onae, H. Sasada, Absolute frequency measurement of the saturated absorption lines of methane in the 1.66 μm region. Opt. Commun. **161**, 223–226 (1999)

64. K.N. Crabtree, J.N. Hodges, B.M. Siller, A.J. Perry, J.E. Kelly, P.A. Jenkins II., B.J. McCall, Sub-Doppler mid-infrared spectroscopy of molecular ions. Chem. Phys. Lett. **551**, 1–6 (2012). doi:10.1016/j.cplett.2012.1009.1015

65. P. Ehlers, I. Silander, J. Wang, O. Axner, Fiber-laser-based NICE-OHMS incorporating a fiber-coupled optical circulator (2013 submitted for publication)

Chapter 7
Applications of NICE-OHMS to Molecular Spectroscopy

Brian M. Siller and Benjamin J. McCall

Abstract This chapter briefly discusses the general operation of the technique of Noise Immune Cavity Enhanced Optical Heterodyne Molecular Spectroscopy (NICE-OHMS), then goes on to describe the various laser systems that it has been implemented with and the molecules it has been used to study. The relative strengths and weaknesses of each of the laser systems, both in the near- and mid-infrared spectral regions, are highlighted. The molecules that have been studied with NICE-OHMS are described, with particular focus given to those systems that differ in some way from the 'generic' NICE-OHMS setup, and those spectra from which new scientific information was extracted.

7.1 Introduction

Noise Immune Cavity Enhanced Optical Heterodyne Molecular Spectroscopy (NICE-OHMS) is the most sensitive direct absorption technique, as it combines the advantages of cavity enhancement and heterodyne detection for very long effective path lengths through samples and (typically) near shot noise limited detection. Since its first demonstration by Ye et al. [1] in 1998, it has been implemented with a variety of laser systems to study a number of different molecules and extract various information from the obtained spectra.

The technique of NICE-OHMS is discussed thoroughly in another chapter in this book, so only a brief description will be given here. A generic NICE-OHMS experimental setup is shown in Fig. 7.1. The two distinguishing features of NICE-OHMS are an optical cavity and heterodyne detection, with the heterodyne sidebands coupled into separate cavity modes. The optical cavity provides path length enhancement by a factor of $2 \times F/\pi$ compared to a single-pass setup, where F is the cavity

B.M. Siller (✉)
Tiger Optics, 250 Titus Avenue, Warrington, PA 18976, USA
e-mail: bsiller@tigeroptics.com

B.J. McCall
Department of Chemistry, University of Illinois, 600 South Mathews Avenue, Urbana, IL 61801, USA
e-mail: bjmccall@illinois.edu

G. Gagliardi, H.-P. Loock (eds.), *Cavity-Enhanced Spectroscopy and Sensing*,
Springer Series in Optical Sciences 179, DOI 10.1007/978-3-642-40003-2_7,
© Springer-Verlag Berlin Heidelberg 2014

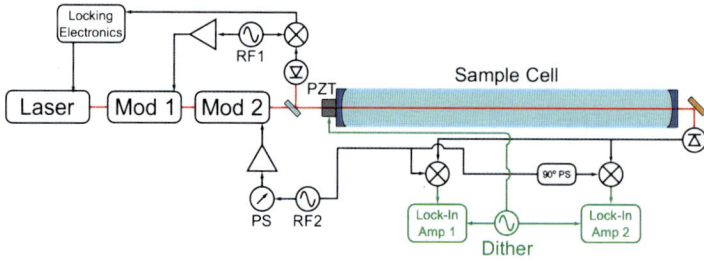

Fig. 7.1 A generic NICE-OHMS experimental layout. The green components are utilized only in wavelength modulated (wm-NICE-OHMS) setups. For fm-NICE-OHMS, the final experimental signals are taken directly from the mixer outputs. Alternatively, in velocity modulation setups, as described in Sects. 7.3.2.2 and 7.3.2.3, the 'dither' is applied to the discharge voltage (not pictured here) across the cell, inducing an alternating Doppler shift of the ions within the cell

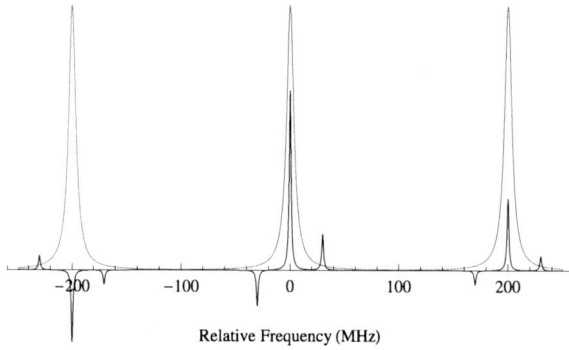

Fig. 7.2 The spectrum of a laser modulated for use in a NICE-OHMS setup overlaid with three optical cavity modes. Note the two sets of sidebands on the laser, in this example spaced at 30 MHz for cavity locking and 200 MHz for heterodyne detection. The second set of sidebands must be spaced at an integer multiple of the cavity FSR. They are most commonly spaced at a single FSR, but can be spaced at two [5] or even nine [6] or more times the free spectral range

finesse, which has ranged from 120 [2] to 100,000 [1] in the various NICE-OHMS implementations.

The laser is modulated at some multiple of the cavity free spectral range (FSR), effectively creating a set of heterodyne sidebands that can be coupled into separate cavity modes, as shown in Fig. 7.2. A second, typically weaker, set of sidebands is added to the laser frequency to enable locking of the laser frequency to the cavity length to constantly keep the carrier frequency on resonance with one of the cavity modes using the Pound-Drever-Hall method [3]. In some NICE-OHMS setups, these two sets of sidebands are also used for locking the sideband spacing to the cavity FSR using the DeVoe-Brewer method [4], to avoid frequency mismatch induced by the cavity FSR changing as its length is scanned.

The frequency modulation is typically applied using a pair of electro-optic modulators (EOM), as in [1], though both sets of sidebands can be applied by a single

EOM if it is capable of simultaneously modulating at two different frequencies with appropriate modulation depths [7]. Or, if the laser frequency can be modulated directly at sufficiently high frequencies, the modulation can be applied to the laser, as in [8].

7.2 Laser Systems

NICE-OHMS has been implemented with a fairly wide variety of laser systems, both in the near- and mid-infrared spectral regions.

7.2.1 Near-Infrared

To date, the vast majority of NICE-OHMS experimental setups have worked in the near-IR spectral region. Near-infrared optical components are generally less expensive, more readily available, and have better performance than the corresponding components designed for the mid-infrared, in part due to the large amount of research and development invested in the field by the telecommunications industry. Component selection is particularly important when it comes to the optical components that need to work at radio frequencies of $\gtrsim 100$ MHz, namely a high-speed detector and a method of modulating the laser frequency for heterodyne spectroscopy. Dielectric coatings for cavity mirrors also tend to be better-developed and higher performing in the near-infrared than in the mid-infrared, with losses of up to an order of magnitude lower than the best mid-infrared coatings available.

One disadvantage of working in the near-IR is that no fundamental vibrational bands for molecules lie in this region, so NICE-OHMS spectrometers in this region need to observe overtones or combination bands, which tend to be much weaker than fundamental bands. Since NICE-OHMS is such a sensitive technique, these detections are still possible, but it makes the detection limit in terms of quantity of analyte required for detection significantly larger than it would be for a spectrometer of the same sensitivity observing fundamental band transitions.

7.2.1.1 Neodymium:YAG

NICE-OHMS was first demonstrated by Ye et al. with a Neodymium:YAG (Nd:YAG) laser, and this first demonstration remains to this day the most sensitive NICE-OHMS implementation yet recorded, with a detection sensitivity of 1×10^{-14} cm^{-1} Hz$^{-1/2}$ [1, 9]. One of the scans from this setup is shown in Fig. 7.3, demonstrating the signal-to-noise advantage of NICE-OHMS compared to cavity enhanced absorption spectroscopy (CEAS). This extreme sensitivity was obtained through a combination of very high finesse (100,000) cavity, with a very stable,

Fig. 7.3 The initial
demonstration of
NICE-OHMS by Ye et al.,
from Ref. [1]. Both scans
were collected with
wavelength-modulated
detection of a Lamb dip of
acetylene. The top scan is a
direct DC detection of the
dither signal, while the
bottom utilized NICE-OHMS
using the same cavity and
dither. Fourteen years later,
this spectrometer still retains
the record for NICE-OHMS
detection, with a sensitivity of
5×10^{-13}

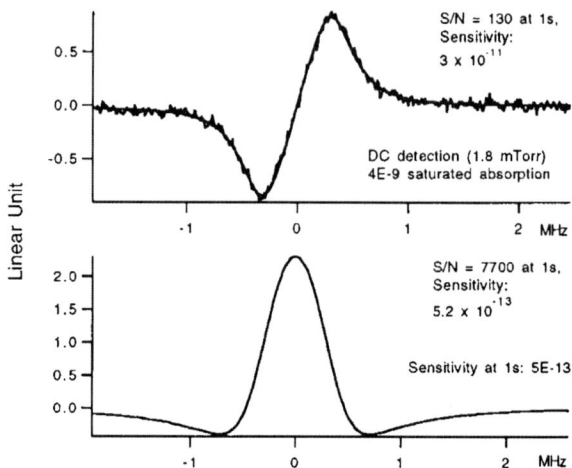

well-locked laser (down to 1 mHz relative frequency), and a sensitive detection system that allowed for an observed noise level that was within a factor of 1.4 of the shot noise limit. Although an impressive feat, not much work has been done with Nd:YAG laser systems since, primarily because they are not very widely tunable and thus can't be used to observe a very wide variety of chemical species.

7.2.1.2 Ytterbium:YAG

Ye et al. have also implemented NICE-OHMS with a Ytterbium:YAG (Yb:YAG) laser at 1030 nm [10]. The primary purpose of this work was stabilization of the laser frequency at both short and long timescales, since no precision reference existed at the Yb:YAG wavelength. To that end, a very narrow transition is needed at an absolute frequency. In this work, the R(29) transition of the $3\nu_3$ band of acetylene was used as the absolute frequency reference.

The Yb:YAG beam was sent through a double-pass acousto-optic modulator (AOM), which shifted the frequency of the beam by $\sim$160 MHz without changing the pointing of the beam as the AOM modulation frequency changed. The laser was locked to a stable optical cavity of finesse 75,000 using the Pound-Drever-Hall method. Most of the locking corrections up to 150 kHz control bandwidth were sent to the voltage-controlled oscillator (VCO) that controlled the AOM frequency, while the very slow ($\lesssim 1$ Hz) corrections were sent to the Yb:YAG temperature controller. Using these two controls, the absolute laser frequency was stabilized to the cavity to within $\sim$1 kHz relative linewidth, limited primarily by vibrations in the cavity.

For long-term stabilization against drift, a small dither was added to the cavity length, which the laser frequency followed, and a second-harmonic wm-NICE-OHMS signal of the central dispersion Lamb dip was used as an error signal, as its lineshape is antisymmetric. For optimal stabilization, the error signal should have as high a S/N and as narrow a linewidth as possible. In this work, a S/N of 900 was obtained from the sensitivity of 7×10^{-11} at 1 s averaging, and the full width at half maximum (FWHM) linewidth of the signal was ~500 kHz, limited by a convolution of transit-time, pressure broadening, and the applied dither. The drift-correction error signal was processed and sent to the piezo-electric transducer on which one of the cavity mirrors was mounted to control the cavity length to prevent the laser frequency from drifting by any more than ~1 kHz over long timescales.

7.2.1.3 Titanium:Sapphire

Shortly after its initial demonstration with a Nd:YAG, NICE-OHMS was implemented with a Titanium:Sapphire (Ti:Sapph) laser near 790 nm by Ma et al., and its performance was directly compared to that of the Nd:YAG setup [9]. The Ti:Sapph system has the advantage of being much more broadly tunable than the Nd:YAG one, which makes it a much more versatile tool for molecular spectroscopy. Because of this versatility, several other research groups have since implemented Ti:Sapph-based NICE-OHMS spectrometers at wavelengths ranging from 730 nm to 930 nm [6, 11].

The biggest disadvantage of the Ti:Sapph-based system compared to the Nd:YAG one is that the Ti:Sapph has a significantly broader free-running linewidth and cannot be stabilized as tightly to the optical cavity. To extend the bandwidth of the laser-cavity lock, double-pass AOMs have been used to provide a higher-frequency transducer to the system [6, 9]. To extend the bandwidth even further, a third EOM can be used with a sweeping applied voltage to change the laser frequency even faster than the AOM allows, since the EOM is not limited by the propagation time of an acoustic wave through a crystal, as an AOM is [9]. With both of these extra frequency transducers, Ma et al. were able to stabilize the Ti:Sapph laser to a cavity of finesse 17,000 to within a relative linewidth of 400 mHz. This is more than satisfactory for performing NICE-OHMS, but is still a factor of ~300 lower in sensitivity than they were able to achieve with the Nd:YAG setup and cavity with 100,000 finesse.

7.2.1.4 External Cavity Diode Lasers

NICE-OHMS has also been implemented with different types of diode lasers, including external-cavity diode lasers (ECDL) and distributed feedback (DFB) diode lasers. Both of these types of diode lasers have the advantages of being relatively inexpensive and much more broadly tunable than the relatively fixed-frequency YAG-based lasers.

For example, Bell et al. built an ECDL-based NICE-OHMS spectrometer for studying the HO_2 radical (see Sect. 7.3.2.1), and also used it to study CH_4 and CO_2 as diagnostic tests of their system [5]. In working with their system, they found that the sensitivity varied significantly with wavelength; they attributed this to the response of the laser varying as the wavelength was tuned, particularly with respect to the locking corrections being sent to the laser current. At 6596 cm^{-1}, where they optimized the system, they could achieve a sensitivity of 3×10^{-11} cm^{-1} with wm-

NICE-OHMS, but tuning the laser by $\sim$40 cm^{-1} to the red limited their sensitivity to 2×10^{-10} cm^{-1}. This is still very sensitive in absolute terms, but is nearly an order of magnitude less sensitive than their optimal value; this shows that optimizing a NICE-OHMS system under a particular set of conditions does not necessarily optimize it over a broad range of conditions.

Gianfrani et al. used a similar ECDL-based setup to study molecular oxygen (see Sect. 7.3.1.5) [12]. One unique aspect of their setup was the way in which they locked the heterodyne frequency to the cavity FSR. Rather than using the common DeVoe-Brewer method, they applied a small dither to the RF frequency at 70 kHz, demodulated the cavity back-reflection signal at that frequency, and used the resulting signal as an error signal for locking the sideband spacing to exactly the cavity FSR with a locking bandwidth of $\sim$10 kHz. They also found that the locking corrections being sent to the laser injection current produced significant intensity noise, so to combat this effect, they used an AOM in another feedback loop to keep the laser power as constant as possible. They did this because while in principle, the noise-immune property of NICE-OHMS prevents laser intensity noise from contributing to the net spectroscopic signal, in practice, residual amplitude modulation (RAM) is always present, and can cause laser intensity noise to couple through the entire detection train and into the final signal. This effect of RAM often makes it worth taking the time to clean up whatever intensity noise there is, rather than relying on the noise-immune property to maximize the sensitivity of the instrument. The intensity stabilizer not only improved their sensitivity, but also allowed them to collect broad scans (up to 8 GHz wide) with very flat baselines.

7.2.1.5 Distributed Feedback Diode Lasers

NICE-OHMS has also been implemented with a distributed feedback (DFB) diode laser [13]. Compared to ECDLs, DFB lasers are more robust, since they don't rely on an external grating that is susceptible to mechanical vibrations, so they have greater potential for use in more robust spectrometers. The DFB laser used in this work has the additional advantage of having a fiber-coupled output, which makes it easier to use the fiber-coupled acousto-optic and electro-optic modulators that were used for modulation and locking. Compared to free-space components, fiber components tend to be significantly easier to align and use, and often have better performance than their free-space counterparts.

7.2.1.6 Fiber Lasers

Fiber lasers offer excellent frequency stability and mode structure, which makes them well-suited for efficient coupling into and locking to optical cavities. Since the laser is fiber-coupled to start with, and exits the fiber in free space before entering the optical cavity, the instrument designer has a choice of using either fiber or free-space components for laser frequency modulation and control. Fiber EOMs are particular attractive for NICE-OHMS since they are much more efficient (lower half-wave voltage) compared to free space EOMs without the need for a resonant electronic circuit to amplify effective RF voltages. This allows for the laser frequency to be modulated at essentially any frequency up to several GHz with almost arbitrary depth of modulation. It is also possible to apply both of the modulation signals needed (for locking and for heterodyne detection) to a single EOM by using a simple RF combiner. Schmidt et al. demonstrated NICE-OHMS with an erbium-doped fiber laser, and observed acetylene with a sensitivity of 2.4×10^{-9} cm^{-1}, which was about a factor of 1000 above the shot noise limit [7]. In a follow-up paper in which they used a different EOM with shorter fibers and used an active temperature controller for the temperature of the EOM and fibers, they improved the sensitivity to 5×10^{-11} cm^{-1} Hz$^{-1/2}$, which is within a factor of 26 of the shot noise limit [14].

7.2.2 Mid-Infrared

NICE-OHMS hasn't been implemented nearly as often in the mid-infrared as it has in the near-infrared, in part because high performance mid-infrared components are not very readily available, but the mid-infrared does offer the significant advantage of being the region of fundamental vibrational modes of many molecules. In the decade following the initial discovery of NICE-OHMS, only a single demonstration was done at wavelengths beyond 2 μm, which used a quantum cascade laser (QCL) at ∼8.5 μm. Recently, mid-infrared systems have been implemented using nonlinear processes to frequency-shift near-IR lasers into the mid-IR. This allows for all laser processing to be done on the near-IR systems, meaning the only specialized mid-IR components required are cavity mirrors with appropriate coatings and fast detectors.

7.2.2.1 Quantum Cascade Lasers

Quantum cascade lasers (QCL) offer the advantage of being available at wavelengths ranging from 3.5 to 20 μm, a range inaccessible by most laser systems. Taubman et al. demonstrated a QCL-based NICE-OHMS spectrometer at ∼8.5 μm with a sensitivity of 9.7×10^{-11} cm^{-1} Hz$^{-1/2}$ [8]. They had two major difficulties in setting up this system: modulating the QCL, and detecting the resulting heterodyne beat signal, both of which typically need to operate with bandwidths of hundreds of MHz.

Frequency modulating the QCL was accomplished by modulating its injection current. They found that the modulation efficiency fell proportional to $1/f$, where f is the modulation frequency, up to ∼100 MHz, and dropped proportional to $1/f^2$ above 100 MHz. They did, however find a resonance in their QCL at 387.5 MHz where the modulation efficiency was significantly higher than nearby frequencies, so they chose this as their heterodyne frequency and designed their optical cavity length to match. They didn't offer a physical explanation for the resonance, but they note that resonant frequencies vary between different lasers, even for lasers on the same chip. This paper also describes a scheme for modulating a QCL with an injection-locking scheme using two lasers, a master and a slave, and they showed that this method reduced the observed RAM level by 49 dB compared to the current modulation scheme that was used for spectroscopy. Although this modulation scheme has not yet been used in a NICE-OHMS system, it is promising for future work.

The other challenge they faced was finding and characterizing a detector fast enough for optimal heterodyne detection. They used a mercury-cadmium-telluride (MCT) detector, and to determine its frequency response, they used two separate QCLs, tuned them to slightly different frequencies, and combined them onto the detector element. This gave them a heterodyne beat that should have constant amplitude over the $\sim$800 MHz that they tuned the frequency difference. By recording the detector output signal versus the heterodyne frequency, they found that the net detector signal at 387 MHz was attenuated by 35 dB compared to the DC response, and they attribute this detector inefficiency as the reason for their detection sensitivity being an order of magnitude above the shot noise limit.

One of the limitations of working with QCL systems is the limited tunability, which can be anywhere from $\sim$20 cm^{-1} (for stand-alone QCL systems such as the one used for NICE-OHMS) to $\sim$200 cm^{-1} (for external-cavity systems).

7.2.2.2 Difference Frequency Generation

Recently, NICE-OHMS has been implemented with a broadly tunable mid-IR source through difference frequency generation (DFG) [15]. The DFG process works by combining two lasers within a nonlinear material to produce a beam whose frequency is the difference of the two input lasers' frequencies. This experimental setup was based around a fixed-frequency Nd:YAG laser and a tunable Ti:Sapph laser, which were combined in a periodically poled lithium niobate (PPLN) crystal. Because near-IR components typically have better performance, lower prices, and are more readily available than their mid-IR counterparts, all laser frequency control was done on the near-IR pump lasers before generating the mid-IR DFG, so the only mid-IR specific components that were needed were the detectors for cavity transmission and back-reflection, and the cavity mirrors, which were specified for the 3.0–3.4 μm range. The 2.8–4.8 μm tuning range of the laser system, limited by the poling periods of the PPLN crystal, is a particularly attractive range for fundamental vibrational modes of many molecules.

To minimize the effects of frequency-dependent RAM and etalon effects on the ultimate signal from the instrument, the two EOMs were both placed on the fixed-frequency Nd:YAG rather than the tunable Ti:Sapph. The system remains sensitive to any etalons on the mid-IR beam after the PPLN crystal, but it is insensitive to those effects on either of the two near-IR beams.

This DFG system was used to acquire spectra of methane with both fm- and wm-NICE-OHMS. The fm-NICE-OHMS setup was used to acquire Doppler-broadened scans, and its sensitivity of 2×10^{-7} cm^{-1} Hz$^{-1/2}$ was limited primarily by etalons in the mid-IR beam path, particularly between the PPLN crystal and the cavity input mirror and between the cavity output mirror and the detector. In the wm-NICE-OHMS setup, a 50 Hz dither was added with a 1.7 MHz peak-to-peak modulation to observe just the sub-Doppler features of methane. In this configuration, the sensitivity was over an order of magnitude better: 6×10^{-9} cm^{-1} Hz$^{-1/2}$, approximately a factor of 60 above the shot noise limit.

7.2.2.3 Optical Parametric Oscillators

Recently, NICE-OHMS has also been implemented with an optical parametric oscillator (OPO), which relies on a nonlinear process similar to that used in the DFG system [2]. A 1064 nm ytterbium-doped fiber laser was used as the seed. This beam was passed through a fiber EOM that applies both the locking and heterodyne sidebands before the beam is sent to a fiber amplifier, which amplifies the total laser power to $\sim$10 W to be used as the pump of the OPO.

The OPO consists of a fan PPLN, which enables continuous tuning of the poling periods across its range by translating the crystal, and a singly resonant cavity, that is resonant with just the signal beam of the OPO. The signal is tunable from 1.5–1.6 μm, while the idler is tunable from 3.2–3.9 μm. Because the locking and heterodyne sidebands on the pump beam are not spaced at an exact multiple of the FSR of the OPO signal cavity, the signal beam remains a single frequency. Due to conservation of energy, this means that the sidebands get transferred entirely to the idler beam.

The NICE-OHMS cavity used in this work had a fairly low finesse, $\sim$120, but with the idler power of $\sim$1 W, there was more than enough intracavity power to enable sub-Doppler spectroscopy of molecular ions. Velocity modulation was coupled with NICE-OHMS in a technique referred to as Noise Immune Cavity Enhanced Optical Heterodyne Velocity Modulation Spectroscopy (NICE-OHVMS). Not only does velocity modulation help to combat some of the difficulties associated with DC detection (e.g. RAM and etalons), but it also affords discrimination of ionic signals from neutral ones through phase sensitive detection. After RF demodulation to separate absorption from dispersion followed by a second level of demodulation at twice the plasma drive frequency, a total of four data channels were collected simultaneously. There was some significant fringing in the signals from three of the detection channels, and the quietest of the channels exhibited a sensitivity of 3.9×10^{-9} cm^{-1}.

7.3 Molecules

Several different molecules have been observed with the NICE-OHMS spectrometers described Sect. 7.2. Many of these have been chosen to demonstrate and optimize spectrometers because they have bands that coincide with the spectral coverage of the laser systems used, while others were observed based on scientific interest and NICE-OHMS was the technique of choice due to its sensitivity and resolution. This section is intended to give an overview of the molecules observed with NICE-OHMS to date, briefly mentioning the molecules that were used as tests to demonstrate the capabilities of instruments, while discussing more in depth the studies from which new molecular information was extracted from NICE-OHMS spectra.

7.3.1 Stable Neutral Molecules

The vast majority of NICE-OHMS papers that have been published to date have demonstrated the detection of stable neutral molecules. Typically, these spectrometers have a static sample cell, often made of Invar, Zerodur, or some other material with a low coefficient of thermal expansion, with the cavity mirrors permanently affixed to the cell, and one mirror mounted on a piezo-electric transducer.

7.3.1.1 Acetylene (C_2H_2)

For many of the papers whose primary purpose was to demonstrate the technique of NICE-OHMS and to characterize and optimize the various aspects of the technique, acetylene and its isotopologues have been favorite targets [1, 7, 16–18]. Several vibrational combination bands have been observed at wavelengths ranging from 730 nm to 1530 nm.

7.3.1.2 Methane (CH_4)

Bell et al. observed an unassigned methane transition at 6610.06 cm^{-1} as a diagnostic of their NICE-OHMS spectrometer described in Sect. 7.3.2.1, recording the pressure broadening in helium to verify the linearity of their spectrometer [5]. Their pressure-broadening coefficient of 1.5 ± 0.1 MHz/Torr (HWHM) agreed well with the literature values for the $2\nu_3$ band of methane at 1.65 μm.

Ishibashi et al. observed several lines of the $2\nu_3$ band of methane with their ECDL-based wm-NICE-OHMS spectrometer with sub-Doppler resolution [19]. They also observed several lines of $^{13}CH_4$, and used their acquired spectra to determine that their sensitivity was 9.5×10^{-11} cm^{-1}, which is within a factor of 2.6 of the shot noise limit, and that their spectral resolution was 320 kHz, limited primarily by transit time, but with a small contribution from residual frequency noise in the system.

Porambo et al. used methane to demonstrate the capabilities of their DFG-based NICE-OHMS spectrometer [15]. They performed both Doppler-broadened scans with fm-NICE-OHMS, and sub-Doppler scans with wm-NICE-OHMS. They observed several transitions of the ν_3 fundamental band of CH_4, and found that they could obtain a sensitivity of $\sim 6 \times 10^{-9}$ cm^{-1} Hz$^{-1/2}$.

7.3.1.3 Methyl Iodide (CH$_3$I)

Ishibashi et al. also used their ECDL-based NICE-OHMS setup to observe the $2\nu_4$ band of CH$_3$I centered around 1.65 μm [20]. They observed a total of 56 rovibrational transitions from the P, Q, and R branches with $\sim$1 MHz resolution. This allowed them to achieve full resolution of the electric quadrupole hyperfine components for the P and R branch lines, while the Q branch hyperfine components were partially blended. Because the hyperfine splitting pattern differs for different spectral lines, they found that hyperfine resolution was useful for assigning their acquired spectra.

7.3.1.4 Nitric Oxide (NO)

Bood et al. studied the sixth overtone band of nitric oxide near 797 nm in order to determine its transition dipole moment [17]. This high overtone band is too weak for most direct absorption techniques, which is why NICE-OHMS was chosen. They observed a total of 15 rovibrational transitions of the $^2\Pi_{1/2}$–$^2\Pi_{1/2}$ sub-band of this vibrational band at a pressure of 75 Torr, and from their data, they extracted absolute intensities of individual lines, the vibrational transition dipole moment, and Herman-Wallis coefficients. Comparing the transition dipole moment of 3.09 μD for this band to the lower overtone bands indicated a significant influence from anharmonicity at the $v = 7$ vibrational level. They also parametrized the electric dipole moment of NO for bond lengths ranging between 0.91 and 1.74 Å more accurately than had previously been done.

7.3.1.5 Molecular Oxygen (O$_2$)

Gianfrani et al. used an ECDL and a cavity of finesse 6,000 to perform NICE-OHMS of molecular oxygen with a sensitivity of 6.9×10^{-11} cm^{-1} Hz$^{-1/2}$, a factor of 30 above the shot noise limit [12]. They studied weak magnetic-dipole transitions of the $b^1\Sigma_g^+(v' = 0) \leftarrow X^3\Sigma_g^-(v'' = 0)$ band near 762 nm. They started by characterizing their spectrometer in both fm- and wm-NICE-OHMS modes of operation using the $^PQ(13)$ line of $^{16}O_2$ with 50 mTorr of pressure, looking at both lineshapes and sensitivity compared to direct CEAS detection.

They then went on to look for the $^{16}O_2$ $^PP(12)$ forbidden transition. As a consequence of the symmetrization postulate of quantum mechanics, transitions starting from even rotational quantum numbers are forbidden, so they used their NICE-OHMS spectrometer to look for one of these forbidden transitions as a test of this postulate. They saw no trace of a signal at up to 200 Torr of pressure, so based on their calculated sensitivity, they could set the upper limit on the violation of the symmetrization postulate for O$_2$ at 5×10^{-8}, which is about an order of magnitude lower than the previous upper limit.

Fig. 7.4 NICE-OHMS signal from Gianfrani et al. [12] of the $^PQ(13)$ line of $^{16}O^{17}O$ in its natural abundance (0.037 %) at 764.489 nm for four different pressures ranging from 660 to 1.3×10^4 Pa

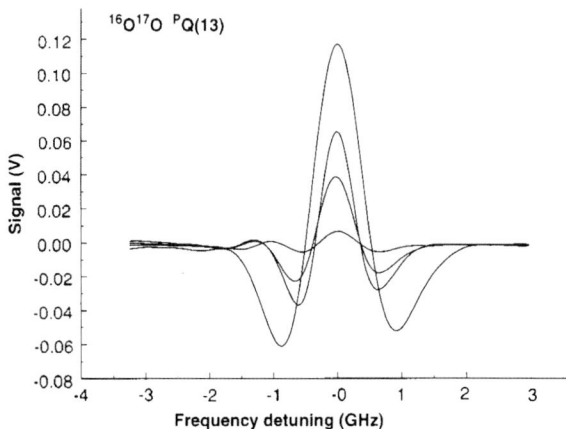

As shown in Fig. 7.4, they also observed some weak lines of $^{16}O^{18}O$ and $^{16}O^{17}O$ in their natural abundances (0.2 % and 0.037 %, respectively), and noted some nonlinearity with pressure that they attributed to two factors: the onset of pressure broadening at higher pressures, and the decreasing modulation depth of their constant-amplitude dither as linewidths broadened at higher pressures.

7.3.1.6 Nitrous Oxide (N₂O)

Taubman et al. characterized their QCL-based NICE-OHMS spectrometer with a nitrous oxide line at 1174.9515 cm^{-1} [8]. They used the observed spectra to determine that their experimental sensitivity was 9.72×10^{-11} cm^{-1} Hz$^{-1/2}$, and compared the obtained NICE-OHMS signals to those obtained with CEAS, showing the drastic noise reduction enabled by the technique.

Bell et al. performed spectroscopy of the R28(e) transition of the ν_3 band of N₂O as another diagnostic of their spectrometer described in Sect. 7.3.2.1 [5]. They found the pressure-broadening coefficient of 2.2 ± 0.1 MHz to be in good agreement with the literature value of the P(26) transition of the same band, and furthermore verified the linearity of their detection system, as indicated by the linear fit of the linewidth versus pressure plot.

7.3.1.7 Carbon Dioxide (CO₂)

Bell et al. also performed spectroscopy on the $\nu_1 + \nu_2 + 2\nu_3$ band of carbon dioxide at 6646.58 cm^{-1} as a diagnostic test of their ECDL-based NICE-OHMS spectrometer whose primary purpose was spectroscopy of HO₂ radical (see Sect. 7.3.2.1). They achieved a sensitivity of 2×10^{-10} cm^{-1} in 10 s of averaging, which was two orders of magnitude more sensitive than their CEAS setup with the same cavity.

7.3.2 Radicals and Ions

More recently, NICE-OHMS has been used to observe HO_2, N_2^+, and H_3^+. The sensitivity of NICE-OHMS is very well suited to detection of these species, since under the conditions used to generate these species, radicals and ions are often orders of magnitude less abundant than their precursor molecules.

7.3.2.1 Hydroperoxyl Radical (HO_2)

The HO_2 radical, which is of interest to atmospheric chemistry, has recently been studied using NICE-OHMS by Bell et al. [5] NICE-OHMS was chosen as the optimal technique due to its species-specific detection (as opposed to some other indirect techniques that have been used for this molecule), its capability of extracting absolute number densities from spectra, and its extremely high sensitivity. The first vibrational overtone of the OH stretch ($2\nu_1$) band, centered at 6649 cm^{-1}, was studied in this work, as it falls within the range of the ECDL used (1480–1540 nm).

Both fm- and wm-NICE-OHMS were performed. Heterodyne sidebands were spaced at ~219 MHz, twice the cavity's FSR, and this frequency was locked using the DeVoe-Brewer method. For the wavelength-modulated work, a 60 Hz dither was applied to one of the cavity mirrors to induce a frequency dither with 100 MHz amplitude. By performing cavity-enhanced absorption spectroscopy (CEAS) of a known transition of methane at 6595.90 cm^{-1} within their spectrometer, the authors determined that their cavity finesse was 2100 ± 100.

Within the vacuum chamber, 2 cm above the cavity, five UV lamps were used for photolysis of Cl_2. The generated Cl atoms then reacted with methanol to form CH_2OH, which then reacted with O_2 to form HO_2 and formaldehyde. The authors performed a detailed chemical analysis of all reactions to take place within the chamber to predict the abundance of HO_2 as well as that of any potentially interfering species.

Two transitions of the first vibrational overtone of the OH stretch ($2\nu_1$) were studied: the $^qP_1(12)$ transition at 6623.32 cm^{-1}, and the $^qP_2(10)$ transition at 6623.57 cm^{-1}, as shown in Fig. 7.5. fm-NICE-OHMS was performed, and background scans were collected and subtracted by turning off the UV lamps for ~10 s and repeating the scans. A sensitivity of 1.8×10^{-9} cm^{-1} was achieved, which corresponded to a minimum detectable concentration of ~4×10^{10} radicals/cm^3.

Kinetic studies were performed by turning off either the chlorine gas flow or the UV lamps, and observing the rate at which the observed signal decayed. The observed signal loss rate combined with kinetic modeling provided further evidence that the observed lines were, in fact, from the HO_2 radical and not some other species in the sample cell. The authors also used the decay rate to estimate the rate at which HO_2 is broken down by the walls of the chamber.

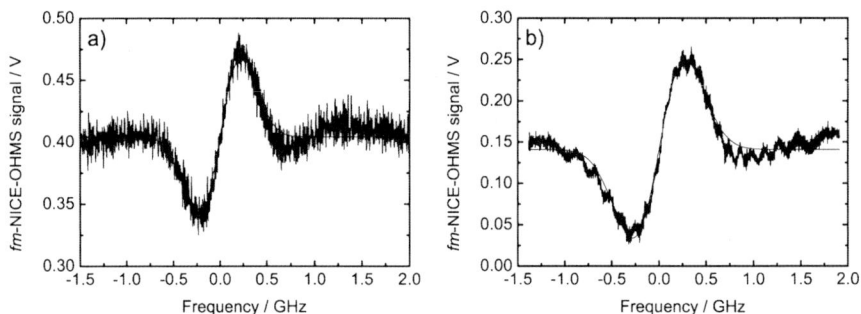

Fig. 7.5 fm-NICE-OHMS spectra of HO_2 with fits from Bell et al. [5] (**a**) $^qP_1(12)$ transition at 6623.32 cm^{-1}, (**b**) $^qP_2(10)$ transition at 6623.57 cm^{-1}

7.3.2.2 Molecular Nitrogen Cation (N_2^+)

N_2^+ has recently been studied with NICE-OHMS in both positive column [6] and ion beam [21] experiments. Both of these experiments relied on the same Ti:Sapph laser system, and because the cavity mirrors were physically separated from the sample cells, the positive column cell and the ion beam chamber could be moved in and out of the cavity without the need for a full optical realignment.

The cavity in both experiments had a finesse of ~ 300 and a free spectral range of ~ 113 MHz, and both experiments observed several transitions in the $v = 1 \leftarrow 0$ band of the Meinel system ($A^2\Pi_u$–$X^2\Sigma_u^+$) of N_2^+. Both also used a form of velocity modulation in addition to the usual NICE-OHMS heterodyne modulation.

In the positive column work, a plasma discharge cell was placed within the cavity, and light was coupled through Brewster windows mounted on either side of the cell. Two different heterodyne configurations were used: one with sidebands spaced at 1.02 GHz, 9 times the cavity free spectral range, and the other with sidebands spaced at 113 MHz, a single cavity free spectral range. A sample 9FSR scan is shown in Fig. 7.6.

The plasma discharge voltage was modulated at 40 kHz, and the net signal was demodulated at twice that frequency to extract both the velocity- and concentration-modulated components of the ion signals, as well as the concentration-modulated signals of any excited neutral species. The 9-FSR setup was used to collect wide Doppler broadened scans of N_2^+ and N_2^* (an electronically excited state of neutral N_2), while the 1-FSR setup was used to primarily to collect scans of the sub-Doppler features at the center of the N_2^+ lineshapes. The Lamb dips were found to have much steeper pressure broadening (~ 8 MHz/Torr) than that of typical neutral molecules, as well as an extrapolated zero-pressure linewidth of ~ 32 MHz, which isn't fully understood at this point.

Sub-Doppler scans were calibrated with an optical frequency comb, and with the extreme absolute accuracy afforded by comb calibration and the precision afforded by sub-Doppler resolution, line centers were determined with an absolute accuracy of ~ 300 kHz, which is approximately two orders of magnitude more pre-

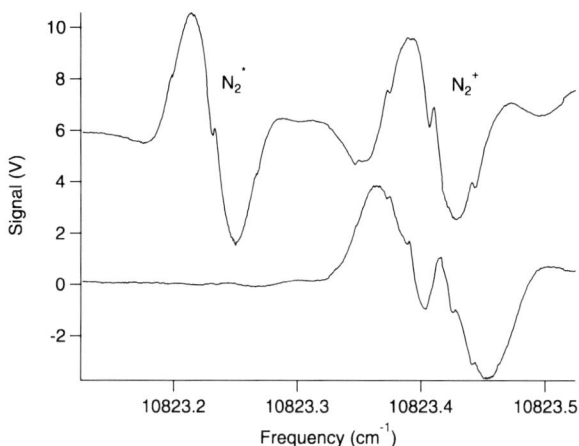

Fig. 7.6 A signal from the NICE-OHVMS system with a nitrogen plasma. The two traces are from the X and Y outputs of the lock-in amplifier. For this scan, the heterodyne sidebands were spaced at ~1 GHz, so Lamb dips are spaced at ~500 MHz on top of the Doppler profiles. Two spectral lines are shown, one from N_2^+ that is both concentration- and velocity-modulated, and the other from N_2^* that is just concentration-modulated. The RF detection phase was tuned to show primarily the dispersion signal, as evidenced by the strong central Lamb dips, and the plasma detection phase was tuned to isolate all of the N_2^* in a single detection phase

cise than traditional Doppler-broadened, wavemeter-calibrated velocity modulation spectroscopy.

For the ion beam work, the plasma discharge cell was removed from the optical cavity, and was replaced by a large vacuum chamber containing the ion beam and its associated ion optics. To avoid vibrations from the turbo pumps coupling into the cavity mirrors (and thus making locking more difficult), the ion beam chamber was mechanically separated from the optical setup, resting directly on the lab floor while all of the optics, including the cavity mirrors, were mounted on a floated optics table. Again, light was coupled through Brewster windows that were mounted on the sides of the chamber, and the ion beam within the chamber was made collinear with the intracavity laser beam by moving the chamber (for coarse control) and steering the ion beam by tuning voltages on the various ion optics (for fine control).

The ions were extracted from the source and accelerated through a 3.8 kV potential drop before being steered through a metal tube along the collinearity path of the ion and laser beams. Velocity modulation was accomplished by applying a 4 kHz 2 V peak-to-peak square wave voltage to the metal drift tube. The NICE-OHMS signal was then demodulated at this modulation frequency to extract the ion signal. Because the modulation voltage required in this setup was quite small, very little electrical interference was introduced by it, as opposed to the positive column cell that used kV-level modulation voltages. Also, the modulation is completely independent of the laser, unlike wm-NICE-OHMS, which can still be somewhat sensitive to RAM and etalons. This allowed this spectrometer to obtain a noise-equivalent absorption of $\sim 2 \times 10^{-11}$ cm^{-1} Hz$^{-1/2}$, within a factor of 1.5 of the shot noise limit.

The lines observed by this instrument are Doppler-shifted by ~ 6 cm^{-1} from their rest frequencies, and the linewidths obtained are narrowed through kinematic compression to ~ 120 MHz, which was limited by the beam energy spread of the ions extracted from the source into the beam. With frequency comb calibration, transition rest frequencies were determined to within ~ 8 MHz of those determined by the NICE-OHVMS work [6], which measured transition rest frequencies directly. The accuracy was limited by the asymmetry observed in the lineshapes, which is thought to be caused by imperfect alignment of the two beams and the beam energy stability over time.

7.3.2.3 Trihydrogen Cation (H_3^+)

The technique of NICE-OHVMS, which combines NICE-OHMS with velocity modulation spectroscopy within a positive column discharge cell, has been extended into the mid-infrared using an OPO, as described in Sect. 7.2.2.3, and its capabilities were demonstrated by observing H_3^+ in a liquid nitrogen cooled discharge cell, which lowered the rotational temperature of the ions to ~ 300 K, compared to the 600–700 K temperatures that are typical for air-cooled cells [2].

Like the N_2^+ work, the detector signal was demodulated twice, first at the heterodyne frequency with two RF mixers, then at twice the plasma frequency using a pair of dual-channel lock-in amplifiers. Since the technique of NICE-OHVMS is sensitive to both concentration- and velocity-modulated signals, all four detection channels had some signal, and these signals were not completely separable from one another. This makes fitting the Doppler profiles of the acquired signals difficult, so it was not attempted in this work.

Because Lamb dips probe only the zero-velocity population, velocity modulation of the overall ion population does not have the same effect on the signal as it does for the Doppler profile. Rather, the zero-velocity population increases and decreases throughout the cycles of the plasma discharge, so both velocity- and concentration-modulated signals appear as concentration modulation when just the zero-velocity population of the ions is considered.

Both the R(1, 0) and R(1, 1)^u lines of H_3^+, which are separated by ~ 0.3 cm^{-1}, were collected in a continuous scan. The overall continuous tuning range of this system is ~ 8 cm^{-1}, limited by the tuning range of the fiber seed laser, and the overall wavelength coverage of the OPO system is 3.2–3.9 μm, though with additional OPO modules, the tuning range could be extended to 2.2–4.6 μm, limited by the transparency of lithium niobate.

Finer resolution scans were collected of just the R(1, 0) line, and the four data channels acquired from each scan were fit simultaneously to find a linecenter. Although the accuracy of linecenter determination was limited to ~ 100 MHz by the wavemeter calibration, the precision of the fit was found to be ~ 70 kHz, which represents the ultimate limit that could be obtained if one were to calibrate the spectra with a more accurate method, e.g. with an optical frequency comb. The Lamb dips were found to each be $\sim$110 MHz wide (FWHM), so each individual Lamb dip was not resolvable, since the Lamb dips are spaced by FSR/2, $\sim$40 MHz.

7.4 Future Prospects

Even though NICE-OHMS has been implemented by a number of research groups over the past 15 years, we have still only begun to scratch the surface of what NICE-OHMS makes possible. The recent developments extending NICE-OHMS into the mid-infrared hold promise to enable the detection of strong fundamental bands of a greater variety of molecules. There has also been a good deal of work attempting to make NICE-OHMS a more robust technique, one that has the potential in being deployed in more robust instruments to observe trace gases in a wider variety of environments rather than being confined to a laboratory setting.

References

1. J. Ye, L.S. Ma, J.L. Hall, J. Opt. Soc. Am. B **15**(1), 6 (1998). doi:10.1364/JOSAB.15.000006. http://josab.osa.org/abstract.cfm?URI=josab-15-1-6
2. K.N. Crabtree, J.N. Hodges, B.M. Siller, A.J. Perry, J.E. Kelly, P.A. Jenkins, B.J. McCall, Chem. Phys. Lett. (2012). doi:10.1016/j.cplett.2012.09.015. http://www.sciencedirect.com/science/article/pii/S0009261412010597
3. R.W.P. Drever, J.L. Hall, F.V. Kowalski, J. Hough, G.M. Ford, A.J. Munley, H. Ward, Appl. Phys. B, Lasers Opt. **31**, 97 (1983). doi:10.1007/BF00702605. http://dx.doi.org/10.1007/BF00702605
4. R.G. DeVoe, R.G. Brewer, Phys. Rev. A **30**, 2827 (1984). doi:10.1103/PhysRevA.30.2827. http://link.aps.org/doi/10.1103/PhysRevA.30.2827
5. C.L. Bell, J.P.H. van Helden, T.P.J. Blaikie, G. Hancock, N.J. van Leeuwen, R. Peverall, G.A.D. Ritchie, J. Phys. Chem. A **116**(21), 5090 (2012). doi:10.1021/jp301038r
6. B.M. Siller, M.W. Porambo, A.A. Mills, B.J. McCall, Opt. Express **19**(24), 24822 (2011)
7. F.M. Schmidt, A. Foltynowicz, W. Ma, O. Axner, J. Opt. Soc. Am. B **24**(6), 1392 (2007). doi:10.1364/JOSAB.24.001392
8. M.S. Taubman, T.L. Myers, B.D. Cannon, R.M. Williams, Spectrochim. Acta, Part A, Mol. Biomol. Spectrosc. **60**(14), 3457 (2004). doi:10.1016/j.saa.2003.12.057. http://www.sciencedirect.com/science/article/pii/S1386142504001258
9. L. Ma, J. Ye, P. Dube, J. Hall, J. Opt. Soc. Am. B **16**(12), 2255 (1999). doi:10.1364/JOSAB.16.002255
10. J. Ye, L.S. Ma, J.L. Hall, J. Opt. Soc. Am. B **17**(6), 927 (2000). doi:10.1364/JOSAB.17.000927. http://josab.osa.org/abstract.cfm?URI=josab-17-6-927
11. A. Foltynowicz, F.M. Schmidt, W. Ma, O. Axner, Appl. Phys. B, Lasers Opt. **92**(3), 313 (2008). doi:10.1007/s00340-008-3126-z
12. L. Gianfrani, R.W. Fox, L. Hollberg, J. Opt. Soc. Am. B **16**(12), 2247 (1999). doi:10.1364/JOSAB.16.002247. http://josab.osa.org/abstract.cfm?URI=josab-16-12-2247
13. A. Foltynowicz, J. Wang, P. Ehlers, O. Axner, Opt. Express **18**(18), 18580 (2010)
14. A. Foltynowicz, W. Ma, O. Axner, Opt. Express **16**(19), 14689 (2008). http://www.opticsexpress.org/viewmedia.cfm?uri=oe-16-19-14689

15. M.W. Porambo, B.M. Siller, J.M. Pearson, B. McCall, Opt. Lett. **37**, 4422 (2012).
    doi:10.1364/OL.37.004422
16. F.M. Schmidt, A. Foltynowicz, W. Ma, T. Lock, O. Axner, Opt. Express **15**(17), 10822 (2007).
    doi:10.1364/OE.15.010822
17. J. Bood, A. McIlroy, D.L. Osborn, J. Chem. Phys. **124**(8), 084311 (2006). doi:10.1063/
    1.2170090. http://link.aip.org/link/?JCP/124/084311/1
18. W. Ma, A. Foltynowicz, O. Axner, J. Opt. Soc. Am. B **25**(7), 1144 (2008). doi:10.1364/
    JOSAB.25.001144
19. C. Ishibashi, H. Sasada, Jpn. J. Appl. Phys. **38**(Part 1, No. 2A), 920 (1999). doi:10.1143/
    JJAP.38.920. http://jjap.jsap.jp/link?JJAP/38/920/
20. C. Ishibashi, H. Sasada, J. Mol. Spectrosc. **200**(1), 147 (2000)
21. A.A. Mills, B.M. Siller, M.W. Porambo, M. Perera, H. Kreckel, B.J. McCall, J. Chem. Phys.
    135(22), 224201 (2011). doi:10.1063/1.3665925

Chapter 8
Cavity-Enhanced Direct Frequency Comb Spectroscopy

P. Masłowski, K.C. Cossel, A. Foltynowicz, and J. Ye

Abstract In less than fifteen years since the development of the first optical frequency comb (OFC), the device has revolutionized numerous research fields. In spectroscopy, the unique properties of the OFC spectrum enable simultaneous acquisition of broadband spectra while also providing high spectral resolution. Due to the regular structure of its spectrum, an OFC can be efficiently coupled to an optical enhancement cavity, resulting in vastly increased effective interaction length with the sample and absorption sensitivities as low as 1.3×10^{-11} $cm^{-1}\,Hz^{-1/2}$ per spectral element. This technique, called cavity-enhanced direct frequency comb spectroscopy (CE-DFCS), provides ultra-sensitive absorption measurements simultaneously over a wide spectral range and with acquisition times shorter than a second.

This chapter introduces the main ideas behind CE-DFCS including properties of various comb sources, methods of coupling and locking the OFC to the enhancement cavity, and schemes for broadband, simultaneous detection. Examples of experimental implementations are given, and a survey of applications taking advantage of the rapid, massively parallel acquisition is presented.

8.1 Introduction

An optical frequency comb (OFC) [1–8] can be defined in the frequency domain as a series of narrow, equally spaced lines with a known, and controllable, optical frequency and a well-defined phase relationship between them. A frequency comb

P. Masłowski (✉)
Institute of Physics, Faculty of Physics, Astronomy and Informatics, Nicolaus Copernicus University, Grudziadzka 5, 87-100 Torun, Poland
e-mail: pima@fizyka.umk.pl

K.C. Cossel · J. Ye
JILA, National Institute of Standards and Technology and University of Colorado, Department of Physics, University of Colorado, Boulder, CO 80309-0440, USA

A. Foltynowicz
Department of Physics, Umeå University, 901 87 Umeå, Sweden

G. Gagliardi, H.-P. Loock (eds.), *Cavity-Enhanced Spectroscopy and Sensing*,   271
Springer Series in Optical Sciences 179, DOI 10.1007/978-3-642-40003-2_8,
© Springer-Verlag Berlin Heidelberg 2014

in the spectral domain originates from an interference of a train of equally spaced pulses in the time domain. The frequency spacing between adjacent lines—called comb modes—in the frequency domain is given by the inverse of the time between two consecutive pulses. This frequency is called the repetition rate (f_{rep}) and is determined by the optical path-length inside the laser resonator. The inverse of the duration of each pulse sets the frequency domain spectral bandwidth. There exists one additional degree of freedom, which is the pulse-to-pulse phase shift of the electric-field carrier wave relative to the pulse envelope (the carrier-envelope phase shift, $\Delta\varphi_{CE}$). This shifts the comb modes in the frequency domain such that the frequency of the mth mode is given by $v_m = mf_{rep} + f_0$. Here m is an integer mode number, and f_0 is the offset frequency (also called the carrier-envelope offset frequency) due to the carrier-envelope phase shift, given by $f_0 = f_{rep}\Delta\varphi_{CE}/(2\pi)$.

Optical frequency combs were initially developed for frequency and time metrology [6, 7]. Due to both the uniformity of the frequency comb mode spacing at the 10^{-19} level [9] and the possibility to stabilize individual modes to below 10^{-15} fractional frequency stability [4, 10–12], they have indeed revolutionized the field. Today, OFCs allow precise and yet relatively easy metrology of optical frequencies without the need for complicated frequency chains [3, 4] and enable the transfer of frequency stability and phase coherence over long distances [13–17] or to different spectral regions [11, 18]. In a relatively short period of time, frequency combs have found numerous applications in a variety of scientific fields [8, 19] such as astronomy [20–24], bioengineering [25, 26], and microwave generation for communication systems, deep space navigation and novel imaging [27, 28]. They have helped discover properties of physical objects ranging in size from atoms [29–31] to planets and planetary systems [23].

With the capability for precise measurement of the optical frequencies provided by the OFCs, it was natural to consider their application to laser spectroscopy [29]. Continuous wave (cw-) laser spectroscopy referenced to a frequency comb provides significantly increased frequency precision and accuracy [1, 32–37], promising exciting improvements in the determination of fundamental constants [29, 33, 38–41]. However, to fully exploit the unique properties of the comb spectrum, an OFC can be used directly as the light source for spectroscopy, in an approach called direct frequency comb spectroscopy (DFCS), first introduced in Ref. [30]. In this technique, precise control of the frequency comb structure is used to simultaneously interrogate a multitude of atomic or molecular levels or to study time-dependent quantum coherence [30, 31, 42]. The frequency comb spectrum provides several benefits for spectroscopic applications when compared to more commonly used sources such as incoherent (thermal or LED) light or cw lasers [43–45]. The broad emission spectrum (broader than emission spectra of some incoherent sources) allows simultaneous acquisition over thousands of channels, facilitating the measurement of different spectroscopic features at the same time. The main advantage lies in the fact that this broadband operation is combined with high spectral resolution, fundamentally limited only by the linewidth of a single comb tooth, which can be as narrow as below 1 Hz [10, 11]. Comb spectroscopy thus maintains the high resolution of cw-laser-based spectroscopy, while removing the need for time-consuming, and often troublesome, frequency scanning.

Due to the spatial coherence of the frequency comb sources, the detection sensitivity for DFCS can be significantly enhanced by extending the path length of the optical beam through the absorbing sample. Absorption cells of moderate lengths with geometrical designs providing tens to hundreds of passes are widely used in cw-laser spectroscopy [46, 47] and can be used with DFCS as well [48, 49], enabling broadband, parallel signal acquisition with moderate sensitivity. One advantage is that the metallic mirror coatings allow the entire OFC spectrum to be used for spectroscopic interrogation [48]. In addition, since no resonance condition exists, these systems allow straightforward optical coupling of the comb spectrum. These features make multipass cells suitable for future commercial spectrometers with intermediate sensitivities based on broadband femtosecond laser sources.

For more demanding applications such as breath analysis, atmospheric research, or detection of hazardous substances, the required concentration detection limits are often lower than those obtainable with multipass cells. A further increase of the interaction length between the laser beam and a gaseous sample can be obtained with a high-finesse optical enhancement cavity. The interaction length inside the cavity can be increased by a factor of tens-of-thousands, providing superb absorption sensitivity. In the 1990's researchers working on cavity ring down spectroscopy started to experiment with coupling of picosecond lasers into passive optical cavities [50]. The very limited spectral coverage of a picosecond laser made it unnecessary to consider the carrier-envelope offset frequency of a comb (a concept that was matured only after the invention of optical frequency combs at the turn of the century), and hence only the matching between the repetition frequency and the cavity free spectral range (FSR) was considered. When this approach was naturally extended to femtosecond mode-locked lasers, termed mode-locked laser cavity-enhanced absorption spectroscopy (ML-CEAS) [51], the authors operationally found a "magic point"—the cavity length for which the coupling was most efficient (i.e. where f_{rep} and the FSR were matched). Rigorous analysis of comb-cavity coupling, which included the concept of the adjustment of the carrier-envelope offset frequency, was first provided in Ref. [52]. This knowledge led to the demonstration in 2006 of cavity-enhanced direct frequency comb spectroscopy (CE-DFCS) [53], which delivers sensitivities beyond the capabilities of traditional Fourier-transform spectroscopy (FTS)—the workhorse of broadband molecular spectroscopy. The potential of CE-DFCS has already been demonstrated for detection of molecular species important for environmental research [48, 53–55], monitoring of production processes [56], and human breath analysis [57], as well as for analysis of the energy level structure of exotic molecules [58, 59], and tomography of supersonic jet expansion [60]. The number of applications is constantly growing, triggered by the continuing development of frequency comb sources covering new spectral domains ranging from the extreme ultraviolet (EUV) [61–64], through the UV, visible, and extending to the mid-infrared (mid-IR) [65–70].

The aim of this chapter is to introduce the reader to the rapidly evolving technique of CE-DFCS and to review the existing approaches and experimental setups. Various OFC sources are presented, including the most common Ti:sapphire and fiber-based lasers, as well as those based on nonlinear frequency conversion working in various spectral ranges. In addition, a comparison of laser sources used for spectroscopic applications is given. Next, schemes for coupling and locking an OFC to an enhancement cavity are discussed. A variety of broadband detection approaches are introduced along with examples of the existing experimental configurations. Finally, a few CE-DFCS applications are presented to demonstrate the unique capabilities of this broadband spectroscopic technique.

8.2 Frequency Comb Sources

Frequency combs can be generated by a variety of laser sources, which can be divided into three general categories: mode-locked lasers, indirect sources, and cw-laser based sources. Mode-locked lasers, which emit trains of ultrashort pulses, can produce frequency combs directly at the output of the laser when their frequency domain structure is controlled and stabilized [1–3, 71, 72]. Such precise control has only been reliably accomplished in a small subset of all mode-locked lasers, and we focus our attention on the most common of those gain media, namely: Ti^{3+}:sapphire, Yb^{3+}:fiber, Er^{3+}:fiber, Tm^{3+}:fiber, and Cr^{2+}:ZnSe. Indirect sources use nonlinear optical effects to modify or shift the spectrum of a mode-locked laser. Cw-laser based sources use nonlinear or electro-optical materials to generate multiple sidebands from a cw laser. Below we provide an overview of how each of these types of sources work and briefly discuss some common features and their relative advantages and disadvantages.

The first consideration when choosing a comb source suitable for a particular application is usually its spectral coverage. For molecular spectroscopy, the strongest transitions, and thus the highest detection sensitivity, exist in the visible to ultraviolet (electronic transitions) and in the mid-infrared (fundamental vibrational modes) wavelength ranges. Many interesting atomic transitions, such as the 1S–2S transition in He or He-like ions or nuclear transitions in Th or U, occur below 200 nm. This has pushed the development of comb sources towards all of these spectral regions. Other important factors to consider when choosing a comb source are the available bandwidth, output power, and repetition rate. A broad spectral bandwidth is desired for multi-species detection; however, for a given laser power and repetition rate, a broader bandwidth results in a lower power per comb line, potentially decreasing the detection sensitivity, depending on the efficiency of the read-out method. Higher repetition rate provides higher power per comb line, but if the spacing of the comb lines is larger than the width of individual spectral features, the comb lines must be scanned to cover the full spectrum. Finally, for applications outside of spectroscopy labs, the robustness and portability of the system also need to be considered. These trade-offs have led to the development of a wide range of comb sources as illustrated in Fig. 8.1. This figure shows that it is now possible to find a comb source at any wavelength between the deep-UV and the mid-IR.

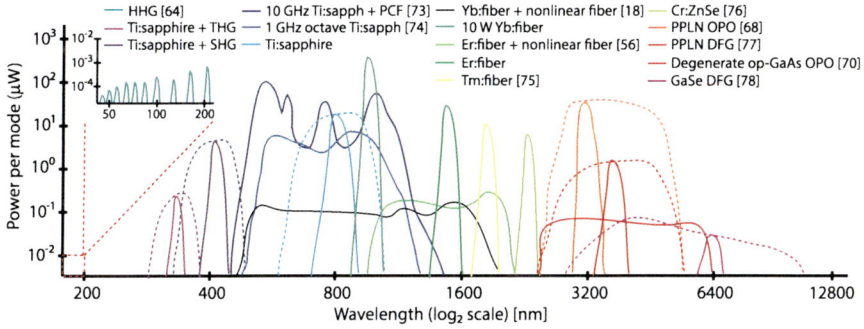

Fig. 8.1 Comparison of the operating spectral range and the output power of different comb sources. *Dashed curves* indicate the tuning range for tunable sources. Descriptions of the systems can be found in the text. HHG—high harmonic generation; THG—third harmonic generation; SHG—second-harmonic generation; PCF—photonic crystal fiber; PPLN—periodically-poled lithium niobate; OPO—optical parametric oscillator; DFG—difference frequency generation; op-GaAs—orientation-patterned gallium arsenide; GaSe—gallium selenide

8.2.1 Mode-Locked Lasers

Over the past few decades mode-locked operation has been achieved in many different gain media, including solid-state lasers, dye lasers, fiber lasers, and diode lasers. Several mechanisms exist for mode-locking these lasers [79–81], and they can be broadly divided into active or passive methods.

Active Mode-Locking Active mode-locking is obtained by forcing multiple modes of a laser resonator to lase with a well-defined phase relation between them [81]. This can be accomplished with an intra-cavity electro-optic or acousto-optic modulator that is driven at f_{rep} (i.e., at the desired comb mode spacing, determined by the optical path-length of the laser cavity). Thus, laser light at one mode will have sidebands located at adjacent modes. These sidebands will experience gain and will produce their own sidebands, which results in a cascaded generation of laser modes with fixed phases. While active mode-locking is very robust, it cannot produce extremely short pulses due to the limited bandwidth of active modulators. Because of this limitation, most comb sources rely on some form of passive mode-locking instead of, or in addition to, active mode-locking.

Passive Mode-Locking: Saturable Absorber Passive mode-locking is achieved by modifying the temporal response of the laser cavity to favor pulse formation over cw lasing. One way to accomplish this is to incorporate an intra-cavity absorbing medium with an absorption coefficient given by

$$\alpha = \frac{\alpha_0}{1 + \frac{I}{I_s}}, \tag{8.1}$$

where α_0 is the zero-power absorption coefficient, I is the intra-cavity intensity, and I_s is the saturation intensity. The intensity-dependent absorption has the effect of reducing the power of the leading edge of a pulse relative to the peak of the pulse. The carrier lifetime of common saturable absorbers is in the range of picoseconds to nanoseconds, which would limit the achievable pulse duration to this time scale; however, when combined with gain saturation to reduce the gain for the trailing edge of the pulse, it is possible to generate shorter pulses.

The most common saturable absorbers are semiconductor-based quantum wells (such as GaAs or InGaAsP) grown on the surface of Bragg reflecting mirrors [82]. These are called either saturable Bragg reflectors (SBRs) or semiconductor saturable absorber mirrors (SESAMs) and are typically used as an end-mirror (at the focus of a lens) in the laser cavity. To overcome some of the limitations (such as carrier lifetime) of semiconductor-based saturable absorbers and to extend the applicability to other wavelength regions, saturable absorbers based on materials such as carbon nanotubes and graphene have been demonstrated [83–85].

Passive Mode-Locking: Kerr Effect   An optical field (E) induces a polarization (P) in a material given by $P = \chi_e E$, where χ_e is the material susceptibility [86]. For strong fields, χ_e exhibits a nonlinear behavior $\chi_e \approx \chi^{(1)} + \chi^{(2)} E + \chi^{(3)} E^2 + \cdots$. For a material with inversion symmetry (and for all amorphous materials), $\chi^{(2)} = 0$. Using $I = (c n_0 \varepsilon_0 |E|^2)/2$ and $n^2 = 1 + \chi_e$, where c is the speed of light, n_0 is the refractive index in vacuum and ε_0 is the vacuum permittivity, the refractive index is given by $n \approx n_0 + n_2 I$ for small $\chi^{(3)}$. The nonlinear index n_2 is proportional to $\chi^{(3)}$. This (instantaneous) modification of the index of refraction as a function of intensity is known as the Kerr effect.

In solid-state lasers such as Ti:sapphire [87, 88] and Cr:ZnSe [89–91], the Kerr effect results in a nonlinear phase that varies as a function of the radial position across the beam profile [81, 92]

$$\phi_{nl}(r, t) = \left(\frac{2\pi}{\lambda}\right) n_2 d I(t) e^{-\left(\frac{2r^2}{w_o^2}\right)} \approx \left(\frac{2\pi}{\lambda}\right) n_2 d I(t) \left(1 - 2\frac{r^2}{w_o^2}\right), \qquad (8.2)$$

for a thin material of thickness d and a Gaussian beam of waist w_o. This parabolic phase front results in an effective lens of focal length

$$f = \frac{w_o^2}{4 n_2 d I_0}, \qquad (8.3)$$

where I_0 is the peak pulse intensity. This dynamic lens can be used as an effective saturable absorber for mode-locking (called Kerr-lens mode-locking or KLM) in several ways. First, with an addition of a geometrical aperture inside the laser cavity, the transmission through the aperture will increase with more dynamic lensing, so the net gain will be higher for shorter pulses. Even without a hard aperture, the presence of the Kerr lens modifies the cavity parameters. Thus a cavity near or beyond the stability edge for cw operation can be made more stable for pulsed operation. KLM lasers may not be self-starting, but typically a small perturbation is sufficient to initiate mode-locking. Because the Kerr effect is instantaneous, the pulse duration is only limited by intra-cavity dispersion or fundamentally by the gain bandwidth of the crystal. In general, KLM lasers provide the shortest pulse durations achievable directly from the gain medium.

The Kerr effect also occurs in Yb- and Er-doped gain fibers where it results in a nonlinear rotation of elliptically polarized light. This polarization rotation can be used in a ring cavity with polarization selective elements to achieve mode-locking by increasing the transmission for pulsed light through an intra-cavity polarizer [93]. Polarization-rotation mode-locking can be used to create all-fiber femtosecond lasers with no free-space sections, providing reliable, self-starting mode-locked operation.

8.2.2 Indirect Sources

The ability to produce ultrashort pulses from mode-locked lasers results in high peak intensity per pulse. This enables efficient use of nonlinear effects in many optical materials to broaden or shift the spectrum of a mode-locked laser. Frequency combs can thus be generated in spectral regions that are impossible or difficult to access with cw lasers, such as the extreme ultraviolet (below 100 nm) or the mid-infrared (2–10 µm). In addition, frequency combs can simultaneously cover multiple octaves of spectral bandwidth using nonlinear optical effects, far exceeding the tuning range of any cw lasers, while still maintaining the same high-resolution capability. This flexibility makes indirect comb sources well suited for new applications in spectroscopy.

Many optical crystals do not possess inversion symmetry and they therefore exhibit $\chi^{(2)}$ nonlinearity, which can be used for second-harmonic generation (SHG), sum-frequency generation (SFG), difference-frequency generation (DFG), and parametric generation [86]. Phase matching between three optical frequencies must be satisfied in all of these processes; this can be accomplished by using different axes of a birefringent crystal and tuning the input angle and polarization relative to the crystal axes. Alternatively, a crystal can be periodically poled such that the phase walk-off is periodically corrected, thus maintaining phase matching over many wavelengths. Periodically-poled materials can have "fan-out" poling-periods that vary linearly across one dimension of the crystal so that the operating (phase-matched) wavelength can be tuned by translating the crystal. In addition, longer periodically-poled crystals can be used without severely limiting the angular acceptance range (and thus the phase-matched spectral bandwidth), resulting in higher conversion efficiencies.

Large nonlinear effects can also be obtained by tight confinement of an optical field in small-core optical fibers [94, 95] (such as photonic crystal fiber [96, 97], microstructure fiber [98, 99], and highly nonlinear fiber [100]). These fibers provide not only high peak intensities due to the small mode size but also long interaction lengths for nonlinear effects to accumulate. In addition, the dispersion profile of a fiber can be tailored by adjusting the mode size and by varying dopants, which provides more control over the existing nonlinear effects. For these reasons, fiber-based coherent supercontinuum sources can be designed to cover over 1.5 octaves of spectral bandwidth [18] or to produce light in specific spectral regions. Currently, fibers with large nonlinearities are readily available throughout the visible and near-infrared wavelength ranges, enabling comb generation from 400 nm to 2100 nm, and have recently been demonstrated in the mid-infrared range [101–106]. The spectrum obtained from a nonlinear fiber results from a complex interplay of multiple $\chi^{(3)}$ processes including self-phase modulation, cross-phase modulation, stimulated Raman scattering, and four-wave mixing [94, 95, 107, 108]. Because of this complexity, fiber supercontinuum sources are often modeled numerically using a generalized nonlinear Schrödinger equation [94, 101].

Frequency combs operating at very short wavelengths (below 100 nm) can only be produced through the process of high-harmonic generation (HHG). With extremely high peak electric fields, like those attainable at a tight focus in an optical cavity, the perturbative expansion to the polarizability breaks down and the medium (e.g., a jet of a noble gas) is ionized. The electron is accelerated away from the atom for some period of time until the electric field reverses and accelerates the electron back toward the atom. The electron can then recombine with the atom, emitting all of the excess energy gained from the field in high-order (odd) harmonics of the original laser frequency [109–113]. The HHG process has been used to produce combs in the vacuum ultra-violet (VUV, 100 to 200 nm) and extreme ultra-violet (XUV, below 100 nm) [61, 62, 114] with over 100 µW of average power per harmonic order [115]. Very recently, high-resolution spectroscopy with an XUV comb was demonstrated, conclusively demonstrating the coherence of the HHG process [64].

8.2.3 Other Types of OFC Sources

Other types of comb sources, including these based on continuous-wave lasers, have not seen many applications in spectroscopy and are therefore only mentioned here briefly. It is possible to make a frequency comb simply by applying a strong frequency modulation to the carrier frequency, which creates sidebands at harmonics of the modulation frequency [116–119]. With very strong modulation it is possible to put optical power into high-order sidebands, resulting in a comb in the frequency domain. However, in practice, the achievable spectral bandwidth is small and thus does not provide much of an advantage over cw laser spectroscopy.

Frequency combs can also be generated by driving Raman transitions between rotational levels in molecules using two (possibly pulsed) laser frequencies. With enough Raman gain, it is possible to cascade this process and produce an octave-spanning comb [120]. These combs have only been demonstrated with large mode spacing (given by the rotational spacing of the molecules used), which potentially limits their usefulness for spectroscopy.

Recently, frequency combs based on parametric frequency conversion of cw-lasers in microresonators have been demonstrated [121–123]. When a cw laser is injected into a high-finesse microresonator, the high intra-cavity intensity results in cascaded four-wave mixing and can generate a comb with a repetition rate set by the free spectral range of the microresonator at the pump wavelength. These sources show some interesting potential for spectroscopic applications due to their compact size and inherent simplicity; however, they are currently limited to repetition rates above 20 GHz. In addition, while mode-locking of microresonator sources has been demonstrated very recently [124, 125], the inherent noise properties of the generated comb are not yet completely understood or controllable [126, 127].

8.2.4 Typical Comb Sources

Ti:Sapphire   The first realization of a fully-stabilized frequency comb was based on a mode-locked Ti:sapphire laser. These lasers, which are still widely used as comb sources, are usually Kerr-lens mode-locked. Their extremely broad gain bandwidth enables generation of ultrashort (10 fs) pulses and correspondingly large spectral bandwidth (covering about 700 nm to 1050 nm) directly from the laser. It is even possible to generate octave-spanning spectra directly from the laser with intracavity self-phase modulation [74, 128]. Meanwhile, systems offering longer pulses and thus narrower bandwidth, but widely tunable in the entire gain range, are readily available. Ti:sapphire combs can be made with repetition rates ranging from below 100 MHz up to 10 GHz [129], providing large flexibility for different applications. By the use of SHG [55] or supercontinuum fibers [96, 97] they provide high power tunable combs in the visible to UV ranges. Unfortunately, the free-space cavity of a Ti:Sapphire laser limits the robustness of this type of comb. In addition, the pump laser is still expensive and relatively bulky, further limiting the field applicability of the system.

Yb:Fiber   Yb:fiber lasers [79] produce combs directly in the 1000 to 1100 nm spectral region. These lasers are typically mode-locked using a saturable absorber and are limited in spectral bandwidth (and thus in pulse duration, which is limited to about 80 fs) by the gain bandwidth of the fiber. A common configuration, illustrated in Fig. 8.2(a), consists of a linear cavity with a pump diode coupled to the gain fiber via a wavelength-division multiplexer (WDM), a fiber Bragg grating (FBG) acting as one cavity mirror (the output coupler), and a short free-space section containing a waveplate, focusing lens and saturable Bragg reflector. This design is very robust since the pump is entirely fiber-coupled and the cavity consists mostly of fiber. Additionally, the pump diodes are compact and fairly inexpensive. It is also possible to use only polarization rotation mode-locking; however, in this case free-space intracavity gratings are required for dispersion compensation. Yb:fiber combs have been built with repetition rates up to 1 GHz [131]. Yb:fiber based amplifiers are excellent for power scaling due to the high doping possible in large mode area fibers. In fact, chirped-pulse amplifiers have been used to produce a comb with 80 W average power at f_{rep} equal to 150 MHz [130]. One drawback of Yb:fiber lasers is that, due to the narrow gain bandwidth, they are not very tunable without external spectral broadening; however, the high power enables efficient nonlinear optics.

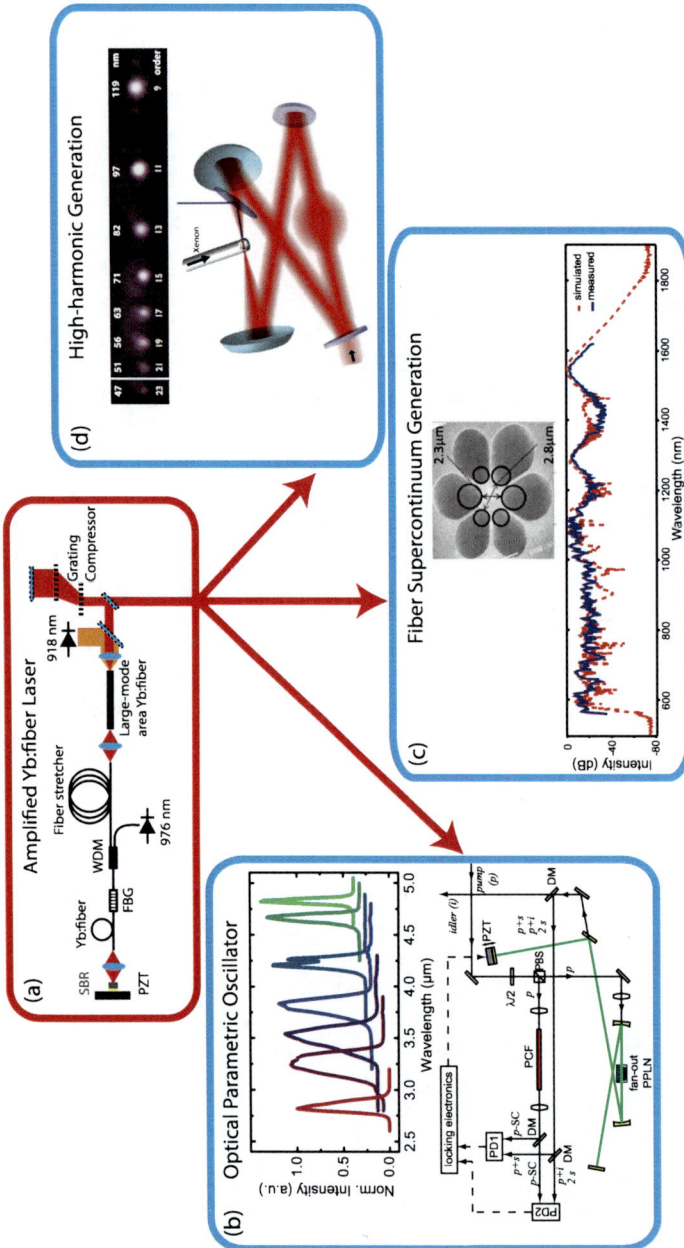

Fig. 8.2 Frequency comb sources at different spectral regions based on a Yb:fiber laser at 1.06 µm. (a) A linear Yb:fiber oscillator and chirped-pulse amplifier [11, 130] (WDM—wavelength-division multiplexer; SBR—saturable Bragg reflector; FBG—fiber Bragg grating; PZT—piezo-electric transducer). This type of system can be used to produce (b) a tunable comb in the mid-IR with an optical parametric oscillator [68] (DM—dichroic mirror, PBS—polarizing beamsplitter, PD—photodetector, PCF—photonic crystal fiber, SC—supercontinuum), (c) a broadband comb in the visible and near-IR from nonlinear fiber [18], and (d) an XUV comb using cavity-enhanced high-harmonic generation [64]

For example, Fig. 8.2 shows how a high average power Yb:fiber comb has been used to generate frequency combs in different spectral ranges via various nonlinear processes: a comb in the mid-infrared with an optical parametric oscillator (OPO) [Fig. 8.2(b)] [68], a comb spanning over 1.5 octaves in the near-infrared by the use of a highly nonlinear fiber [Fig. 8.2(c)] [18], and a comb in the XUV down to 50 nm with high-harmonic generation [Fig. 8.2(d)] [64].

Er:Fiber Er:fiber lasers mode-locked (typically) using nonlinear polarization rotation [132, 133] have become very popular for several reasons. First, since they produce combs near 1550 nm, they can take advantage of telecommunication technology and are inexpensive and fairly easy to build. Second, they can be made entirely out of fiber, without any free-space sections, which makes them very robust and portable [134] as demonstrated in a drop tower experiment with a deceleration of 50 g [135]. It is also possible to split the comb output into multiple branches and amplify each branch separately without loss of coherence [136], thus providing a large amount of flexibility. With highly nonlinear fiber, Er:fiber lasers can cover a wavelength range from 1000 nm to over 2100 nm [56]. When combined with SHG, they can be used to provide tunable combs throughout the visible region [137]. They have also been used for mid-infrared comb generation using either DFG [138] or OPO [139]. However, the output power of amplified Er:fiber lasers is currently limited to typically 500 mW per branch. Also, their spectral coverage without broadening or frequency conversion is not ideal for spectroscopy. Repetition rates up to 1 GHz have been demonstrated [140], typically limited by the fiber length needed to obtain sufficient gain and compensate for dispersion.

Tm:Fiber and Cr:ZnSe Two newly developed mode-locked lasers have pushed the operating wavelengths toward the mid-infrared. Tm:fiber combs [75], which operate at 2 to 2.1 μm, function in many ways similarly to Yb:fiber lasers, sharing many of their advantages and disadvantages. Currently, they can provide about 1 W of power at repetition rates up to about 100 MHz, although there is no foreseeable reason that those could not be increased. Instead of directly mode-locking a Tm:fiber laser, it is also possible to seed a Tm:fiber amplifier with the spectrally shifted output of an Er:fiber comb [141]. The primary advantage of Tm:fiber systems is that they allow the use of new nonlinear crystals with transmission windows starting above 1.6 μm for frequency conversion farther to the mid-IR [70]. The second new system is a Cr:ZnSe mode-locked laser operating around 2.5 μm [91], which is similar in many aspects to Ti:sapphire lasers. The fractional gain bandwidth for Cr:ZnSe is even larger than that of Ti:sapphire, which provides the potential for ultrashort pulse generation in the mid-IR and thus a broad spectral bandwidth directly from the laser. Currently, however, challenges such as cavity-dispersion control [76] and stability have limited the use of Cr:ZnSe lasers.

DFG and OPO The most common approaches for producing mid-infrared frequency combs beyond 3 μm are either DFG or parametric generation (using an OPO). DFG combs have been demonstrated using the spectrum generated directly from a Ti:sapphire laser [142]; however, the achievable powers are very low. More power can be obtained by using two synchronized Ti:sapphire lasers [143], but this is experimentally challenging. Multi-branch Er:fiber lasers enable mW-level tunable DFG: one branch is equipped with a nonlinear fiber to provide tunable frequency-shifted light and the second branch provides high-power, unshifted light [138, 144]. Up to 100 mW of DFG-based mid-IR light has recently been achieved using a fan-out periodically-poled crystal and a Yb:fiber laser [145]. In this case, some of the light from the fiber laser was sent through a nonlinear fiber and the red-shifted Raman soliton was mixed with the remaining unshifted pump light to generate the difference frequency. DFG systems are convenient and compact, but the power limitations can hinder some applications.

Higher power is possible with an OPO in which the signal and/or idler light produced by parametric generation is resonant with a cavity containing a nonlinear crystal. This greatly increases the conversion efficiency but also adds some complexity. For comb generation the cavity is pumped synchronously, i.e. the OPO cavity free spectral range matches an integer multiple of the pump-laser repetition rate. In addition, the cavity length must be actively controlled since it sets the f_0 of the generated comb [68, 146, 147]. Using a fan-out periodically-poled lithium niobate (PPLN) crystal, a comb has been demonstrated with a simultaneous bandwidth of up to 200 nm, a center wavelength tunable from 2.8 μm to 4.8 μm, and an output power up to 1.5 W at 136 MHz repetition rate [Fig. 8.2 (b)] [68]. It is also possible to construct an OPO with degenerate signal and idler frequencies (a "divide-by-two" system), which has been used to produce near-octave spanning spectral bandwidth around 3.3 μm [69, 148]. The long wavelength limit to the attainable spectral range is set by the absorption edge of the nonlinear crystals. To reach longer wavelengths crystals other than lithium niobate must be used. These are usually angled-tuned crystals such as AgGaSe$_2$ [149]. Recently, periodic patterning of GaAs has been developed, which could enable significantly higher OPO output powers. To use GaAs for mid-infrared generation, the pump wavelength must be above about 1.6 μm, thus the increased interest in Tm:fiber systems. Very recently, an octave-spanning mid-infrared spectrum covering up to 6.1 μm was demonstrated with a Tm:fiber laser and a degenerate OPO using orientation-patterned GaAs [70], although the average power was only about 30 mW.

8.3 Comb-Cavity Coupling

Efficient coupling of the frequency-comb light into a cavity is possible because the frequency-domain spectrum of Fabry-Perot cavity and that of an OFC are similar in structure. However, while the comb line spacing is constant, the optical cavity resonances are separated from each other by a frequency-dependent free spectral range (*FSR*) [150], given by

$$FSR(\omega_0) = \frac{c}{2L + c\frac{\delta\phi}{\delta\omega}|_{\omega_0}}, \tag{8.4}$$

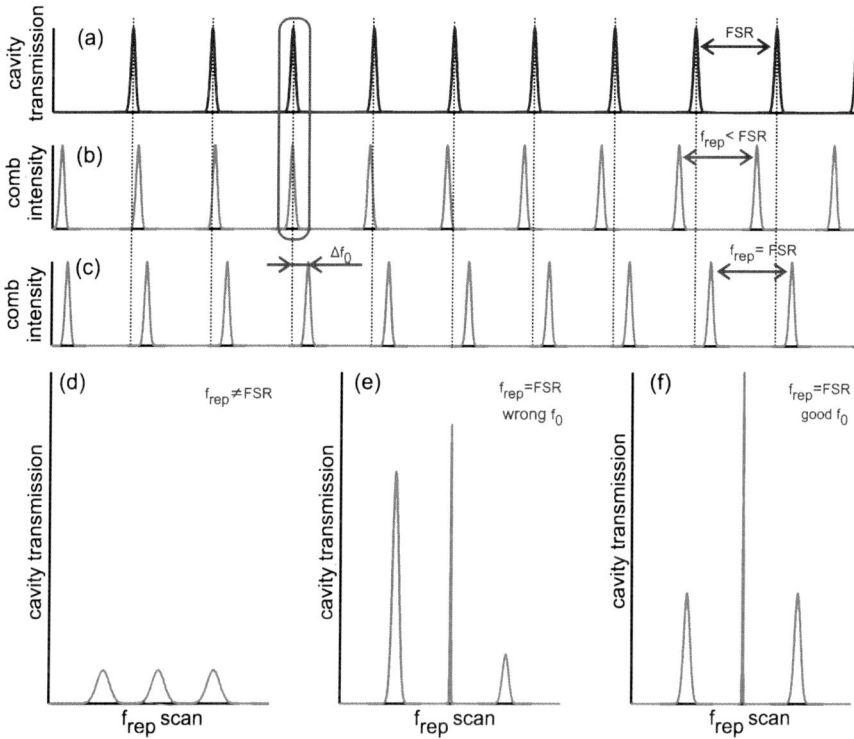

Fig. 8.3 Optimization of comb-cavity coupling. The optical cavity modes are shown in (**a**), the OFC spectrum with f_{rep} not matched to the cavity *FSR* in (**b**), and the comb spectrum with the correct f_{rep} but incorrect value of f_0 in (**c**). In (**d**)–(**f**), the cavity transmission signal during a scan of f_{rep} around different central values of f_{rep} and for different values of f_0 is presented. (**d**) As indicated by the common box in (**a**) and (**b**), even for $f_{rep} \neq FSR$ there is coincidental overlap between some of the cavity and the OFC modes, resulting in low and broad transmission peaks. (**e**) When f_{rep} is scanned around the value matched to the *FSR* but the f_0 is not optimized, the heights of the two side peaks are asymmetric, while (**f**) when f_0 is correct the height of the main transmission peak is maximal and the heights of the two neighboring peaks are equal

where c is speed of light, L is the length of the cavity, and $\frac{\delta\phi}{\delta\omega}$ represents the intra-cavity dispersion, typically originating mostly from the cavity mirror coatings. This dispersion term means that the separation of consecutive cavity resonant modes is approximately constant only in a limited frequency range. Thus, some care must be taken to minimize the cavity dispersion at the spectral region of interest, usually by operating as close as possible to the maximum of the mirror reflectivity.

An analysis of the comb-cavity coupling [52] shows that optimum coupling requires matching both the spacing (set by f_{rep}) and absolute frequencies (f_0) of the comb lines with corresponding cavity modes over as wide spectral range as possible. This implies that the repetition rate of the OFC must be matched to the cavity *FSR* (or an integer multiple or sub-harmonic thereof) at the spectral region of interest. As shown in Figs. 8.3(a)–(b), an incorrect repetition rate results in an overall mismatch between the comb lines and cavity resonances. Optimizing f_{rep} can be accomplished by changing the laser oscillator cavity length, for example with a piezo-electric transducer (PZT) that moves a mirror in the free-space part of the laser cavity (in a Ti:Sapphire or fiber lasers) or by stretching a piece of fiber in all-fiber laser sources [151]. Once the f_{rep} is matched to the cavity FSR as shown in Fig. 8.3(c), the f_0 of the comb also has to be chosen appropriately so that the resonances of the cavity and comb lines coincide, making the value of Δf_0 in Fig. 8.3(c) equal to zero. Tuning of f_0 is realized by modifying the dispersion of the laser cavity either by introducing an additional optical component such as a prism or by changing the power of the pump laser (which couples to the cavity dispersion through the nonlinear index of refraction of the gain medium) [5, 152].

The proper values of f_{rep} and f_0 can be determined while sweeping either the repetition rate of the laser or the length of the enhancement cavity and observing the cavity transmission signal, such as presented in Figs. 8.3(d)–(f). As the central value of f_{rep} is tuned close to the correct value, the intensity of the observed peaks increases and their width decreases [Figs. 8.3(d) to (e)], since more comb lines come on resonance with cavity modes simultaneously [44]. After f_{rep} is matched to the cavity FSR, the value of f_0 is optimized by comparing the height of the transmission peaks next to the most prominent peak. The correct value of f_0 is found when the height of the center peak is maximized and the two neighboring peaks have the same height [cf. Figs. 8.3(e) and (f)].

Once the f_{rep} and f_0 are optimized, these values must also be maintained over an extended period of time. This can be achieved in two different ways: by tightly locking the comb and cavity modes in order to ensure a constant transmission through the cavity, or by modulation of either one of the comb parameters or the cavity FSR so that laser light is periodically coupled into the laser cavity (the swept coupling scheme). Each of these schemes has advantages and limitations as explained below.

8.3.1 Comb-Cavity Coupling—Tight Locking Scheme

In the tight locking scheme the match between the comb modes and cavity resonances is actively maintained with feedback. The error signal for stabilization can be generated with the Pound-Drever-Hall (PDH) [153] method, the Hänsch-Couillaud method [154], or with a dither lock, where the comb modes are dithered around the cavity modes (or vice-versa) with amplitude smaller or comparable to the full width at half maximum (FWHM) cavity linewidth.

One way to achieve a tight lock is to stabilize the two degrees of freedom of the frequency comb to an external frequency reference and to implement a feedback loop that actively controls the cavity FSR to match it to f_{rep}. A difficulty arises due to the fact that for different cavity conditions (e.g. at different intra-cavity pressures) the optimum comb f_0 takes different values, so the locking electronics must be able to follow these changes. Another limitation arises from the fact that the length of linear cavities is often controlled with a large-travel-range PZT. This, combined with the large mass of the cavity mirrors, results in a PZT response bandwidth usually limited to a few hundred Hz, which is not fast enough to remove high-frequency (acoustic) noise, although in principle this can be addressed by designing a cavity that can accommodate a lower-mass mirror.

An alternative scheme is to lock the comb to the resonances of the enhancement cavity by stabilizing the two degrees of freedom of the comb with two servo loops—one for f_{rep} and another for f_0. The error signals for each of the feedback loops are derived from different parts of the cavity reflection spectrum. This can be accomplished by applying a modified PDH detection scheme, using separate photodiodes to detect different parts of the reflected laser spectrum after diffraction from a grating. A detection scheme using three photodiodes was implemented in Ref. [152]: here the PDH error signal from the center of the reflected spectrum was used to control f_{rep} and the difference between two error signals from opposite ends of the reflection bandwidth was used to control f_0. A similar setup using only two error signals obtained in reflection was used in later experiments [45, 155] where an Er:fiber comb spanning from 1510 to 1610 nm was tightly locked to an enhancement cavity with a finesse of 8000. This enabled a 50-nm transmission bandwidth centered at 1530 nm, limited by the dispersion of the dielectric mirror coatings, which were designed for peak reflectivity at 1600 nm.

The two-point locking scheme was also used in the experiment shown in Fig. 8.4, where a mid-infrared OPO was tightly locked to a cavity with a peak finesse of 3800 at 3.8 μm and a Fourier transform spectrometer was used for sensitive detection of C_2H_2, NO_2 and H_2O_2 [156]. A portion of the reflected beam from the cavity was dispersed with a diffraction grating, and two widely separated spectral regions were detected with two photodiodes, PD1 and PD2. The PDH error signal from PD1 was transformed by a proportional-integral (PI) filter with a corner frequency of 30 kHz and sent to a fast PZT controlling the length of the Yb:fiber pump-laser cavity, effectively changing f_{rep}. The output signal of the PI filter was further integrated and sent to a fiber stretcher in the Yb:fiber oscillator, cancelling low frequency drifts of the repetition rate. The second error signal, obtained from PD2, passed through a second PI filter with a corner frequency of 30 kHz and was sent to a fast PZT in the OPO cavity, effectively changing f_0 of the mid-IR comb [147, 157].

Figures 8.5(a) and 8.5(b) show the transmission spectra of the cavity measured with the Fourier-transform spectrometer at two different wavelength regions (solid blue curves). The locking points used by the f_{rep} and the f_0 feedback loops are marked in red and black, respectively. For comparison, the dashed blue lines show the transmission curves with only one locking point used (i.e., stabilizing only f_{rep} and with f_0 deliberately mismatched to reduce the transmitted bandwidth) as the locking point is tuned across the spectrum. As seen by comparison of the solid blue curve with the peaks of the dashed curves, the two-point locking enables broadband cavity-comb coupling with almost the entire incident OPO spectrum transmitted. In the wavelength region shown in panel (b) the losses on the edges of the

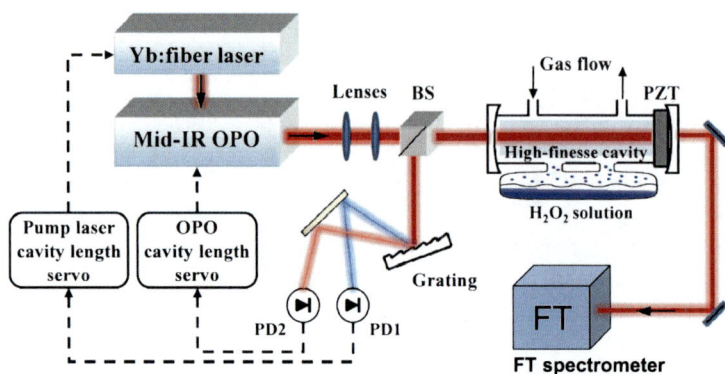

Fig. 8.4 Schematic of a CE-DFCS setup at 3.8 μm using two-point cavity locking. The optical parametric oscillator (OPO) is synchronously pumped by a femtosecond Yb:fiber laser and delivers an OFC tunable from 2.8 to 4.8 μm (see Fig. 8.2). The mid-IR comb is locked to a high-finesse cavity containing the gas sample by the use of a two-point tight locking scheme. In this scheme, the cavity reflected light is isolated with a beam splitter (BS) and sent onto a diffraction grating, which disperses the spectrum. Two distantly separated parts of the spectrum are measured on two photodetectors (PD1 and PD2) in order to create PDH error signals at the two wavelengths. The feedback signals are sent to the pump laser (mostly influencing f_{rep}) and the OPO cavity length (f_0). One mirror of the enhancement cavity is mounted on a PZT to allow adjustment of the cavity length. The light transmitted through the cavity is coupled into a fast-scanning Fourier-transform (FT) spectrometer (from Ref. [156]. With kind permission from Springer Science+Business Media: Appl. Phys. B, Cavity-enhanced optical frequency comb spectroscopy in the mid-infrared application to trace detection of hydrogen peroxide, vol. 110, year 2013, p. 163, A. Foltynowicz et al., Fig. 2a)

comb spectrum were more severe because of the larger mirror dispersion around 3900 nm [farther from the peak cavity finesse, plotted in Fig. 8.5(c)]. In particular, less light is coupled into the cavity at longer wavelengths (i.e., past 3950 nm) where the *FSR* of the cavity changes more rapidly. The larger dispersion also necessitated choosing the two locking points closer to each other. For this particular cavity, the transmission bandwidth with two-point locking ranged from 150 nm at 3700 nm to 70 nm at 3900 nm, which allowed simultaneous recording of the entire band of N_2O [Fig. 8.5(d)].

The two-point locking scheme enables broadband, tight locking of the comb modes to the cavity resonances, maximizing the enhancement factor and providing constant, high-power signal in cavity transmission. It can therefore be efficiently combined with Fourier-transform-based detection methods. On the other hand, this scheme requires more complicated locking electronics compared to the swept coupling scheme and also requires higher-bandwidth servo actuators in the laser systems. Moreover, residual frequency noise is translated into amplitude noise (FM-to-AM conversion) by the cavity resonances, which must be actively removed to reach high signal-to-noise ratio [155], as will be discussed later.

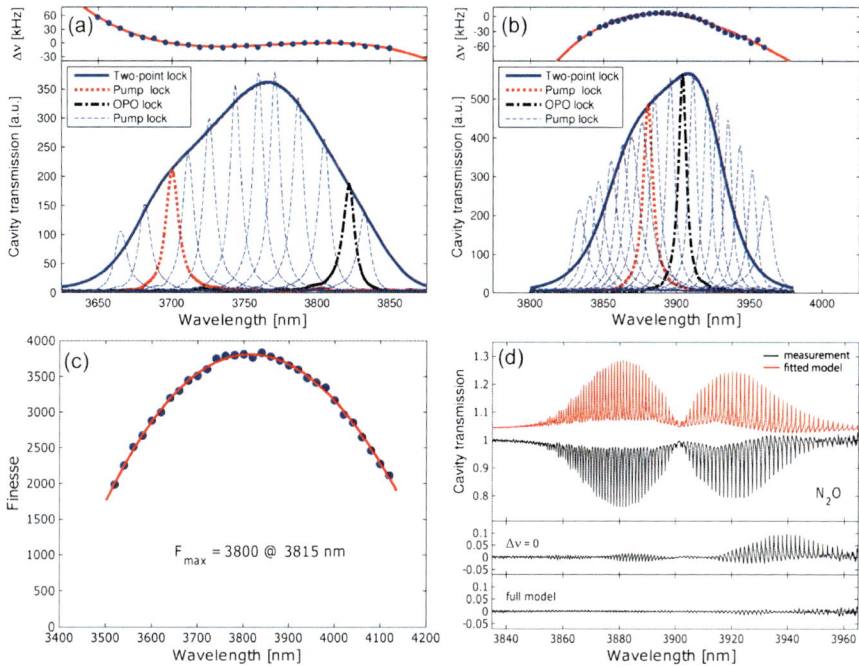

Fig. 8.5 Characterization of the two-point tight lock between a mid-IR comb and a cavity. (**a**) and (**b**) show the comb transmission through the cavity when both feedback loops are on (*solid blue*) for a center OPO wavelength of 3770 and 3920 nm, respectively. The two locking points are shown by the *red* (pump laser lock) and *black* (OPO lock) *dotted curves* when they are individually activated. The cavity transmission spectra with only one locking point tuned across the OPO spectrum are shown as *dashed blue curves*. The frequency mismatch between cavity and comb modes ($\Delta \nu$) when both loops are on is shown in the *upper part of the panels*. (**c**) The wavelength-dependent finesse of the cavity. (**d**) The spectrum of 0.8 ppm of nitrous oxide at the pressure of 760 Torr of nitrogen at room temperature recorded in 1 s (*black*) and the fitted model spectrum (*red*). The *lower parts* of (**d**) show fit residuals using two models, the first one ignoring the mismatch between comb modes and cavity modes and the second one based on the full model of Eq. (8.5), which improves the fit significantly (from Ref. [156]. With kind permission from Springer Science+Business Media: Appl. Phys. B, Cavity-enhanced optical frequency comb spectroscopy in the mid-infrared application to trace detection of hydrogen peroxide, vol. 110, year 2013, p. 163, A. Foltynowicz et al., Figs. 3, 4, 5)

8.3.2 Comb-Cavity Coupling—Swept Coupling Scheme

The walk-off between the comb lines and the cavity modes due to dispersion in the mirror coatings is the main factor limiting the spectral bandwidth that can be achieved for most CE-DFCS experiments with the tight comb-cavity lock. Moreover, the bandwidth limitation is more severe when a higher finesse cavity is used, resulting in a compromise between the sensitivity enhancement factor and the simultaneously usable spectral bandwidth. The negative effects caused by cavity dispersion can be reduced by special low-dispersion cavity designs (including prism cavities based on total internal reflection) [158]. Alternatively, the swept coupling scheme allows the entire spectral bandwidth of a comb to be transmitted through the cavity, as demonstrated both in early applications of ML-CEAS [51] as well as in CE-DFCS experiments [43]. In this scheme, the comb modes are periodically swept across the cavity modes (or vice versa) with the amplitude of the sweep larger than the FWHM linewidth of the cavity resonance (in contrast to dither lock), resulting in periodic transmission peaks. By stabilizing the time interval between the transmission peaks (with feedback to either the comb or the cavity), it is possible to eliminate slow drifts between the cavity and comb modes. The dispersion problem is largely avoided as the corresponding mode pairs come into resonance rapidly one after another, thus when averaging over a sweep, the full bandwidth of the source spectrum is transmitted [43]. In addition, the bandwidth required of the locking electronics is much lower than in the tight locking scheme, since low-bandwidth feedback is sufficient to stabilize the time between transmission peaks. Another advantage is that the swept coupling scheme efficiently reduces amplitude noise in cavity transmission caused by low-frequency mechanical noise and by cavity FM-to-AM noise conversion.

The main disadvantage of this method is the reduction of the useful experimental duty cycle and transmitted power arising from the limited time over which the comb modes are on resonance with the cavity. A more subtle effect is connected with the enhancement factor of the absorption signal. The enhancement of the interaction length is related to the cavity finesse, F, which is proportional to the inverse of the cavity losses. For fast sweeps, the time that the comb lines spend on resonance becomes only a fraction of the cavity lifetime. This results in an enhancement factor that is closer to F/π for rapid sweeps instead of $2F/\pi$ for the tight locking case [43, 55], which decreases the absorption sensitivity assuming the same fractional noise levels. The fact that the enhancement factor is dependent on the sweep parameters implies that a calibration of the effective length enhancement can be necessary for intermediate sweep speeds, although this is not absolutely necessary for higher sweep speeds as the enhancement factor approaches F/π. While the swept coupling scheme can be combined with Fourier-transform-based detection methods, the performance is worse than with a tight lock [159], due to the fact that the amplitude modulation of the transmission signal leads to a disturbance of the detected interference signals.

The swept cavity-comb coupling scheme has been used in ML-CEAS and CE-DFCS multiple times [44, 51, 57], and its robustness and reliability has been demonstrated by its application for trace gas detection in the north coast of France [55, 160] and in the Antarctic [161]. Amplitude noise reduction down to the shot-noise level has also been demonstrated [162].

An example of an experimental setup using the swept coupling scheme and a VIPA spectrometer (Sect. 8.4.2) is shown in Fig. 8.6. The FSR of the enhancement cavity, with a finesse of 28000, was matched to the f_{rep} of the Er:fiber comb. To couple the frequency comb to the cavity, the positions of the frequency comb lines were modulated with respect to the cavity resonances at a frequency of 1.5 kHz by sweeping f_{rep}. The sweep amplitude was chosen such that the comb lines were on

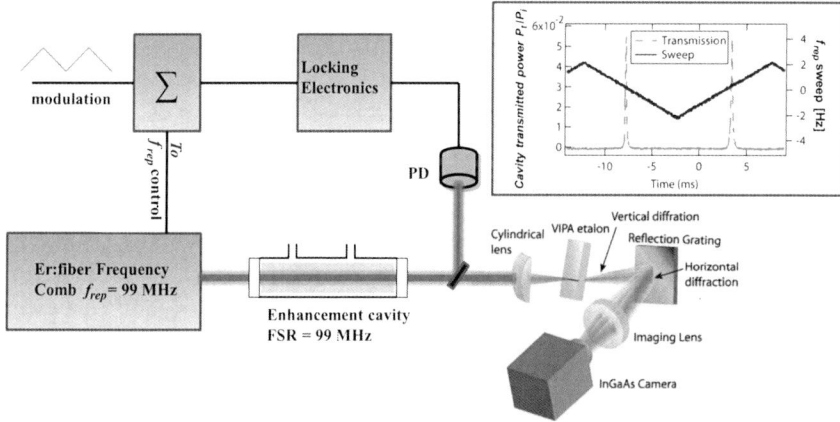

Fig. 8.6 Schematic of a CE-DFCS experimental setup with swept coupling scheme working in the 1.5–1.7 μm range. The repetition rate of the Er:fiber frequency comb was modulated, and a small part of the light transmitted through the cavity was focused on a photodetector (PD), whose output signal was used to generate the feedback signal applied to the f_{rep} control of the Er:fiber frequency comb ($\sum$—summing box). The *framed inset* shows the PD signal for a 100 nm incident spectrum while f_{rep} is scanned. The light transmitted through the cavity was analyzed with a VIPA spectrometer [43]

resonance with the cavity for 1/3 of the modulation period. The inset in Fig. 8.6 shows the cavity transmission measured by the photodetector in front of the VIPA spectrometer when f_{rep} was swept with a larger amplitude. Due to low cavity dispersion and optimized cavity-comb coupling, the cavity transmission signal on the photodiode was sharply peaked. The time between the successive transmission signals was used to produce an error signal. This error signal, after integration, was fed back to the laser repetition rate control, which allowed stable, long-term operation and transmission of the entire laser bandwidth through the cavity [43].

8.3.3 Effect of Comb-Cavity Resonance Mismatch on the Observed Line Shape

In the tight locking scheme, the centers of the cavity and comb modes are not perfectly matched over the entire transmission bandwidth, as shown in the upper panels of Figs. 8.5(a) and (b). This mismatch causes a distortion of the line shapes in the observed absorption spectrum, which can be seen in Fig. 8.5(d). This effect, visible in numerous experiments, must be taken into account in the data analysis to accurately determine the molecular concentrations.

The intensity of a single comb line at a frequency v transmitted through a cavity containing an absorbing sample is given by [155, 156]

$$I_t(v) = I_0(v) \frac{t^2(v)e^{-2\delta(v)L}}{1 + r^2(v)e^{-4\delta(v)L} - 2r(v)e^{-2\delta(v)L}\cos[2\phi(v)L + \varphi(v)]}, \quad (8.5)$$

where $I_0(v)$ is the intensity of the comb line incident on the cavity, $t(v)$ and $r(v)$ are the frequency-dependent intensity transmission and reflection coefficients of the cavity mirrors, respectively, $\delta(v)$ and $\phi(v)$ are the attenuation and phase shift of the electric field per unit length due to the analyte, and $\varphi(v)$ is the round-trip phase shift in the cavity, given by $4\pi v n L/c$, where n is the refractive index of the buffer gas. If there is a mismatch of Δv between a comb mode and the center of the nearest cavity mode, the round-trip intra-cavity phase shift is equal to $\varphi(\Delta v) = 2q\pi + 2\pi \Delta v/FSR$, where q is an integer mode number. At frequencies close to molecular transitions, the molecular dispersion adds an additional shift to the cavity modes. This results in the characteristic overshoots observed in the absorption line shapes, easily visible in the upper fit residuals in Fig. 8.5(d). In addition to the lineshape asymmetry, the peak absorption is reduced, which can lead to underestimation of the molecular concentrations. The model of Eq. (8.5), when applied to data in Fig. 8.5(d), is able to significantly decrease the discrepancies between the measured and theoretical spectra, as presented in the bottom fit residuals in Fig. 8.5(d).

8.4 Detection Methods

The frequency comb light transmitted through the enhancement cavity contains broadband spectroscopic information about the interrogated system; however, this broadband information must be detected to be of use. In some DFCS measurements, a single atomic or molecular transition, well separated from other spectroscopic features, served as an optical frequency discriminator by only interacting with a single comb tooth (e.g., [64]), removing the need for a broadband, parallel acquisition system. For sensing applications with dense spectra, the capability to make simultaneous measurements across tens of thousands of parallel channels with high spectral resolution is required. Thus the applied detection techniques have to satisfy two seemingly contradicting requirements—broadband operation and high spectral resolution—while also providing high sensitivity for absorption. Although the perfect experimental setup with unlimited spectral bandwidth and resolution does not exist, several detection schemes allowing multiplexed acquisition have been demonstrated and are summarized in Fig. 8.7.

In the initial work on ML-CEAS, the output of the enhancement cavity was dispersed with a grating spectrograph and detected with a detector array [Fig. 8.7(a)], and the FM-to-AM noise conversion due to the cavity was circumvented with the swept coupling scheme (Sect. 8.3.2) [51]. The sensitivity was increased in later experiments with active feedback to the cavity length and by using higher finesse

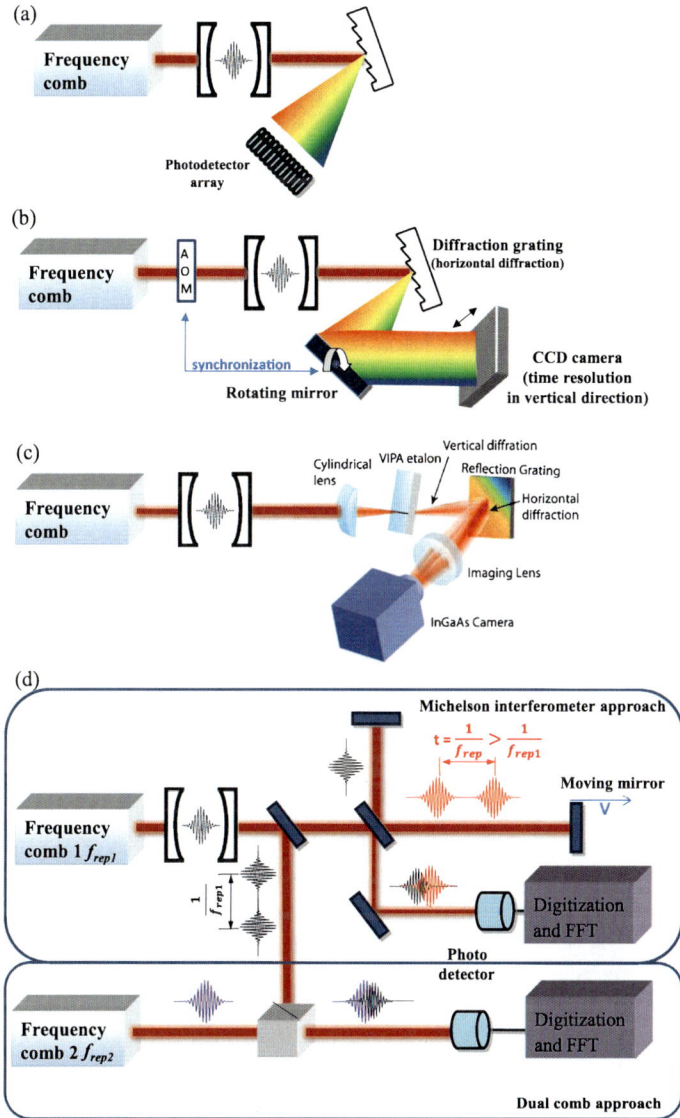

Fig. 8.7 Detection schemes for CE-DFCS. (**a**) A grating spectrometer equipped with a detector array. (**b**) Broadband cavity ringdown setup. A grating provides spectral dispersion in the horizontal direction and a rotating mirror creates the time scale in the vertical direction for a two-dimensional detector array. An acousto-optic modulator (AOM) is used to turn off the laser beam to initiate the ringdown. (**c**) A VIPA spectrometer. The cavity output is dispersed vertically with a high-resolution VIPA etalon and is cross-dispersed with a diffraction grating. (**d**) Femtosecond Fourier-transform spectrometer. In the Michelson interferometer approach, the cavity output is analyzed by an interferometer with a fast scanning mirror in one or both arms. This effectively changes the repetition rate of the OFC. The pulses of the combs from the two interferometer arms with different f_{rep} interfere on the detector. In the dual comb approach, the output of the enhancement cavity beats with another OFC with $f_{rep2} \neq f_{rep1}$ on a photodetector

cavities [54, 55], eventually reaching shot-noise limited performance [162]. This method, especially in combination with the swept coupling scheme, is fast and reliable and does not require any complicated locking schemes; however, due to the coupling scheme used, the transmitted power is significantly reduced thus reducing the shot-noise limited sensitivity. On the other hand, it is difficult to use this detection method with tight locking due to amplitude noise on the transmitted light. Also, the spectral resolution is limited by the grating spectrometer (typically to 10 GHz or above), which usually does not allow single comb modes to be resolved.

One way to avoid the FM-to-AM noise conversion introduced by the high-finesse cavity is to detect cavity-ringdown signals [Fig. 8.7(b), see Sect. 8.4.1], which are inherently more immune to intensity noise. To combine cavity-ringdown with CE-DFCS a two-dimensional detector is used [53]: one axis of the detector corresponds to time and the other corresponds to wavelength. The wavelength dispersion is obtained with a diffraction grating, resulting in the same limited spectral resolution. The temporal axis is generated with a rotating mirror to deflect the dispersed light or by clocking the readout of a CCD camera. The small number of resolved spectral elements, limited by the size of the detector array, has reduced the applicability of this readout method.

Higher spectral resolution and more spectral elements can be obtained with a virtually-imaged phased array (VIPA) etalon [163] combined with a grating cross-disperser and a two-dimensional detector array—schematically shown in Fig. 8.7(c) (Sect. 8.4.2). The resolution of a VIPA-based spectrometer can be 1 GHz or below, which can enable single-comb mode resolution [58, 145, 164]. As in the one-dimensional system, it is difficult to use tight locking due to FM-to-AM noise conversion; however, in combination with the swept coupling scheme [43, 44], this method provides high-resolution spectra with high sensitivity [57, 58].

It is also possible to employ Fourier-transform spectroscopy (FTS) techniques with CE-DFCS [Fig. 8.7(d)]. These spectrometers are based either on a Michelson interferometer with a mechanical translation stage [165] (Sect. 8.4.3) or dual comb spectroscopy [166, 167]. They allow measurements in the spectral range where a VIPA etalon is not readily available and also do not suffer from the trade-off between resolution and bandwidth due to the limited detector array sizes [168]. To deliver a good performance, FTS requires a continuous transmission signal and thus a tight comb-cavity lock. This can lead to increased amplitude noise if the comb modes are not stabilized to better than a small fraction of cavity resonance linewidth. An efficient scheme to circumvent this problem employing active cancelation of amplitude noise down to the quantum limit has recently been demonstrated [155] and provides long-term, uninterrupted operation. Commercial FTIRs have been used with the swept-coupling scheme as well, but this combination tends to result in decreased absorption sensitivity and lower quality of spectra [159, 169].

Another way to avoid the frequency resolution limitation of a grating spectrograph is to use a Vernier spectrometer [170], in which the cavity FSR is purposefully mismatched from the comb f_{rep} so that the cavity only transmits comb teeth separated by $k f_{rep}$ (k is an integer depending on the FSR). This is illustrated in Figs. 8.8(a)–(b) for the specific case where the cavity FSR is equal to $4 f_{rep}/5$. As

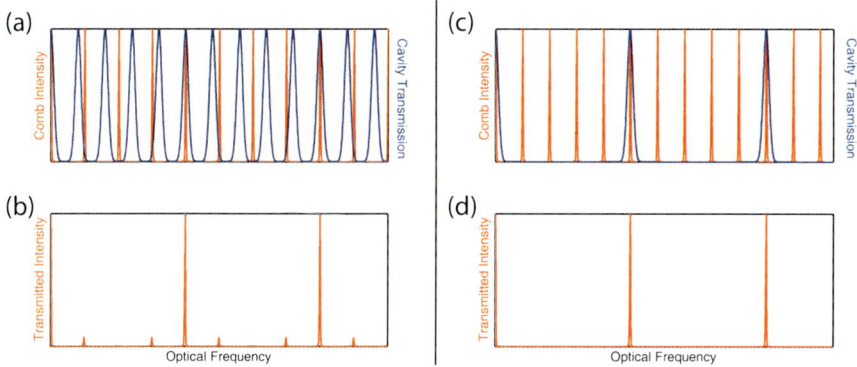

Fig. 8.8 Comb filtering using Vernier spectroscopy or a filter cavity. (**a**) Cavity transmission spectrum (*blue*) with $FSR = 200$ MHz and spectrum of an OFC (*red*) with repetition rate of 250 MHz ($FSR/f_{rep} = 4/5$), (**b**) transmitted comb spectrum (*red*) in which the comb spacing, effectively quadrupled, equals to $4f_{rep} = 1000$ MHz. The residual transmission between the main peaks is largely suppressed. (**c–d**) Cavity filtering by setting $FSR = 4f_{rep}$. The transmitted comb spectrum (**d**) again has spacing equal to 1000 MHz

a result, in the frequency domain, the separation of the transmitted comb teeth is increased by a factor of four; with a different cavity length the separation could be significantly higher. A similar result can be obtained using a filter cavity with $FSR = 4f_{rep}$ to filter the comb, as shown in Figs. 8.8(c)–(d). In this way, it is possible to increase the separation of the transmitted comb teeth to be resolvable by the grating. Thus an experimental setup similar to the one illustrated in Fig. 8.7(a) can be used, with a detector array recording a single comb line on each pixel. However, a major drawback is that due to the filtering effect more scanning is required to record a full spectrum. In addition, the swept-coupling scheme is challenging to use with the Vernier scheme, while with the filter cavity approach, the interaction length with the sample is significantly decreased (due to the shorter cavity length) or two cavities are required. The approach of comb filtering has found application in calibration of astronomical spectrographs with the frequency comb, where a mode spacing of tens of GHz is desired [171–173].

8.4.1 Broadband Cavity Ringdown Spectroscopy

In cavity ringdown spectroscopy (CRDS) [174], the decay rate of the light leaking out of the enhancement cavity is measured after removal of the input light; thus it is less sensitive to intrinsic intensity noise caused by FM-to-AM noise conversion. Unfortunately, multichannel detectors typically only offer readout rates from Hz to tens-of-kHz, depending on the size of the detectors, which is too slow for broadband CRDS. This problem was circumvented with an approach called broadband ring-down spectral photography [175, 176] where a rotating mirror was used

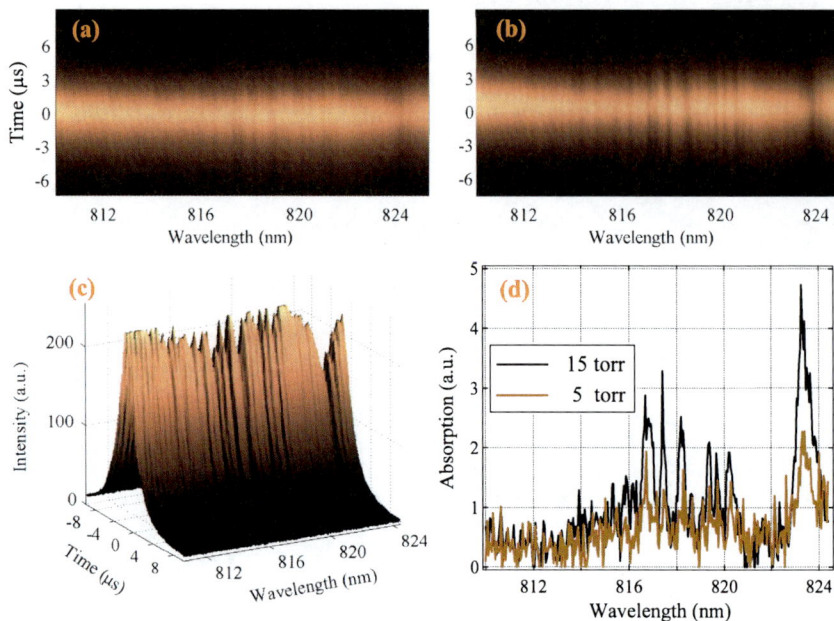

Fig. 8.9 Results from broadband CRDS. (**a**) and (**b**) Images of broadband ringdown spectra of the H_2O overtone spectrum at pressures of 5 and 15 Torr, respectively, acquired in 30 µs. In (**c**) a surface plot of 15 averaged images is plotted, showing the lower buildup power and faster decays in spectral regions containing H_2O absorption lines. (**d**) The resulting spectra of H_2O at pressures of 15 and 5 Torr (from Ref. [53]. Reprinted with permission from AAAS)

to create time axis in the plane of a two-dimensional camera. A similar result was achieved with no moving parts by clocking the readout of a CCD array [177]. CRDS was for the first time applied to measurements with an optical frequency comb in Ref. [53]. The experimental setup, shown schematically in Fig. 8.7(b), was based on a Ti:sapphire frequency comb centered at 800 nm, which allowed access to overtone spectra of C_2H_2, H_2O, NH_3, and forbidden electronic transitions of O_2. The comb light was coupled to an enhancement cavity with a finesse of 4500. The cavity output was dispersed horizontally by a grating and focused at the imaging plane of a 2D camera. A rotating mirror placed between the imaging lens and the camera moved the beam vertically along the detector array for temporal resolution. An AOM in front of the the cavity was used to rapidly switch off the comb light and start the decay event. The resulting images are presented in Fig. 8.9. The system allowed simultaneous measurement of 340 ringdown events with 25 GHz spectral resolution, reaching an absorption sensitivity of 2.5×10^{-10} cm^{-1} $Hz^{-1/2}$. A typical ringdown spectrum covering a 15 nm bandwidth was acquired in 30 µs.

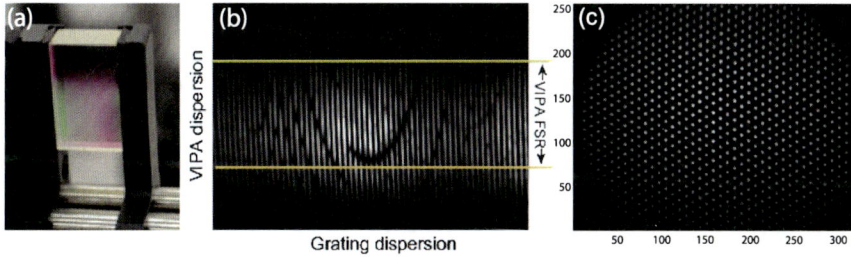

Fig. 8.10 (**a**) Virtually-imaged phased array (VIPA) etalon. The brighter stripe in the lower part is the entrance slit. (**b**) An image from a VIPA spectrometer working around 1600 nm. *Dark spots* indicate CO_2 absorption features; *yellow lines* indicate one VIPA *FSR*, which is the area with unique spectral information (from Ref. [43]). (**c**) The comb-mode resolved image from a VIPA spectrometer at 3800 nm. The separation between the comb lines is 2 GHz (obtained by cavity filtration of a lower repetition rate comb); the single OFC modes are clearly resolved (from Ref. [145])

8.4.2 VIPA Spectrometer

A VIPA, or virtually-imaged phased array [163, 178, 179], is a rectangular etalon plate [Fig. 8.10(a)] with a high reflectivity coating on the entrance surface (reflectivity >99.9 %) and a lower reflectivity back surface (reflectivity typically about 97 %). Additionally, an antireflection-coated entrance stripe is placed at the bottom of the entrance surface [visible in Fig. 8.10(a)] to allow light, which is focused in one dimension, to be coupled into the device. In contrast to an ordinary Fabry-Perot etalon, the VIPA etalon is tilted at an angle; because of this and the focusing of the incoming light, it transmits all the incident wavelengths. At the output it generates a pattern of overlapping mode orders that repeat at the etalon *FSR*—typically 50 to 100 GHz, depending on the thickness of the plate. The main advantage of a VIPA is the extremely high angular dispersion, up to 30–40 times higher than a traditional (first-order) diffraction grating. To spectrally resolve overlapped orders a cross-dispersion grating with a resolution better than the VIPA *FSR* is used. After the grating the light is imaged on a two-dimensional camera.

An example of the resulting image is shown in Fig. 8.10(b). The vertical stripes are single mode orders of the VIPA, which are dispersed in the horizontal direction by the lower-resolution grating. The yellow horizontal lines show the range of one VIPA *FSR* on the camera, where the spectral information is unique. Thus, a single column within this region contains spectral information from one VIPA *FSR*. By reading the pixels column-by-column a traditional one-dimensional spectrum can be created. The dark spots in the image represent CO_2 absorption signals for individual ro-vibrational transitions around 1600 nm.

The resolution of the VIPA spectrometer is limited by the finesse and *FSR* of the etalon. The finesse is determined both by the reflectivity of the coatings and by the number of round-trip passages within the tilted etalon. Resolution of 500 MHz has been demonstrated with VIPA *FSR*s of 25 to 100 GHz. This is sufficient to resolve single modes of high repetition rate combs, as demonstrated with Ti:sapphire combs

[58, 164]. Another approach to achieve comb mode resolution with lower repetition rate lasers is to use an intermediate filter cavity. This is shown in Fig. 8.10(c) for a filtered comb with 2 GHz mode spacing; here, the filter cavity was set to transmit every 15th comb tooth of a femtosecond mid-infrared OPO [145].

One advantage of VIPA spectrometers is the short integration time for a single image, on the order of tens to hundreds of μs [145], making them an ideal tool for studies of transient events. The fact that the setup does not involve any moving parts and is compact makes this technique particularly attractive for field-deployable devices. In addition, the approach is compatible with the swept coupling scheme, which reduces the overall complexity of the experimental setup. On the other hand, since the VIPA is an etalon the resulting fringe pattern is sensitive to optical alignment and to mechanical vibrations. Furthermore, calibration of the frequency axis needs to be carefully checked since the pixel-to-frequency mapping is nonlinear. VIPA etalons tend to be relatively expensive and are available only in limited spectral regions (a single etalon typically covers about 200 nm in the visible to near-IR ranges). Recently, a VIPA covering more than 1 μm in the important mid-IR region around 4 μm has been demonstrated [145].

8.4.3 Fourier-Transform Spectroscopy

Fourier-transform spectroscopy (FTS) based on interferometry with an incoherent light source is a workhorse for analytical chemistry and molecular science, providing broad bandwidth and relatively high spectral resolution [180]. It utilizes extremely broadband light sources between 2 and 50000 cm^{-1} and can reach resolutions of about 0.001 cm^{-1} (30 MHz, although resolutions of 1–10 GHz are more typical). The main shortcoming of this method is the long acquisition time, associated with the use of thermal light sources and the need for long averaging times to obtain high sensitivity. The lack of spatial and spectral coherence of thermal sources makes it very challenging to use long optical path length (required for high resolution) or to efficiently couple light into an enhancement cavity for high sensitivity [181].

Replacing the thermal source of an existing Fourier-transform (FT) spectrometer with an optical frequency comb offers an instantaneous increase in spectral brightness and thus reduces averaging times significantly. The coherent frequency comb also allows an efficient combination of cavity-enhanced spectroscopy and FTS. The schematic of a Michelson interferometer-based CE-DFCS setup is shown in Fig. 8.7(d): the comb is split in two arms of the Michelson interferometer with one arm scanning at a speed v (in practice both arms can be scanned simultaneously in opposite directions). The resulting time-domain interference signal is monitored by photodetectors at an output port of the interferometer. This interferogram is then Fourier transformed to obtain the frequency-domain spectrum, as in traditional FTS. When the spectrometer is capable of resolving each OFC mode, the actual resolution of each of spectral element is given by the comb-mode linewidth. This value is much

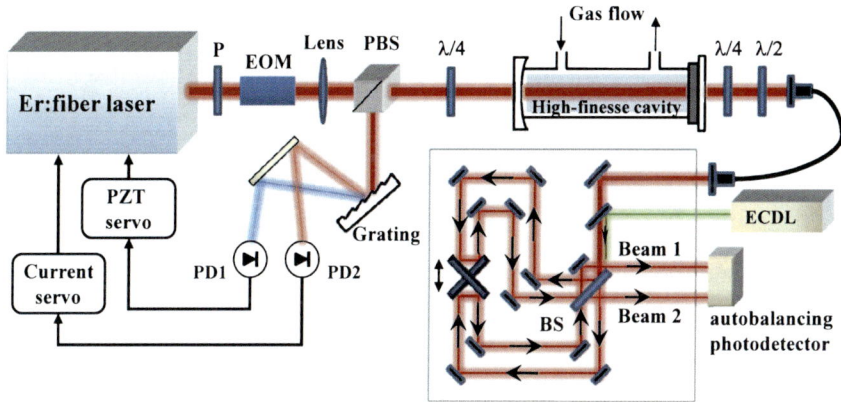

Fig. 8.11 Experimental setup for CE-DFCS near 1550 nm with FTS-based readout. An Er:fiber femtosecond laser is tightly locked to a high-finesse optical cavity containing a gas sample using a two-point PDH lock. An electro-optic modulator (EOM) is used to phase modulate the comb light at 14 MHz, and the cavity reflected light is dispersed by a grating and measured with two photodetectors (PD1 and PD2) in order to create error signals at two different wavelengths. The feedback is sent to the pump diode current controller and to a PZT inside the laser cavity. The cavity transmitted light is coupled through a polarization-maintaining fiber into a fast-scanning Fourier-transform spectrometer. The two outputs of the interferometer (beams 1 and 2) are incident on two photodiodes of an auto-balanced photodetector. The beam of a cw 780 nm external cavity diode laser (ECDL), used for frequency calibration, co-propagates with the frequency comb beam and is monitored with a separate detector (not shown). P—polarizer, (P)BS—(polarizing) beam splitter, $\lambda/4$—quarter wave plate, $\lambda/2$—half wave plate (from Ref. [155]. Reprinted figure with permission from A. Foltynowicz et al., Phys. Rev. Lett. 107, 233002. Copyright (2011) by the American Physical Society)

lower than the OFC mode spacing and improves on the resolution of traditional FTS setups by many orders of magnitude [182].

Michelson interferometer-based FTS has been applied as a detection method for DFCS [48, 67, 165, 183] as well as for CE-DFCS, either with a tight lock between the cavity and optical frequency comb [45, 155, 156] or with a swept coupling scheme [159, 169, 184, 185]. Figure 8.11 shows an FTS-based CE-DFCS system developed at JILA using a 250-MHz Er:fiber comb covering from 1510 to 1610 nm. The laser was locked to a 60-cm long enhancement cavity with a finesse of 8000 using the two-point locking scheme described in Sect. 8.3.1. The wavelengths for the locking points were chosen to assure high cavity transmission in the spectral range where strong acetylene absorption occurs (1530–1540 nm). The output light from the cavity was sent into a Michelson interferometer equipped with a fast-scanning delay stage on which two retro-reflectors were mounted. To increase the optical path difference corresponding to the movement of this stage, the lengths of both inter-ferometer arms were scanned simultaneously but in opposite directions (Fig. 8.11). Light from a stable 780-nm cw laser co-propagated with the comb light and was used as a reference for frequency calibration. The comb interferograms at both out-put ports of the Michelson interferometer were monitored with an auto-balanced In-

Fig. 8.12 Signals recorded by an FTS-based CE-DFCS setup. (**a**) Time-domain interferogram with auto-balanced detection disabled (*red*) and enabled (*blue*). (**b**) Frequency-domain signals with auto-balanced detection disabled and enabled (*red* and *blue*, respectively), with an expanded viewed of the region of interest, revealing acetylene absorption, in (**c**) (from Ref. [155]. Reprinted figure with permission from A. Foltynowicz et al., Phys. Rev. Lett. 107, 233002. Copyright (2011) by the American Physical Society)

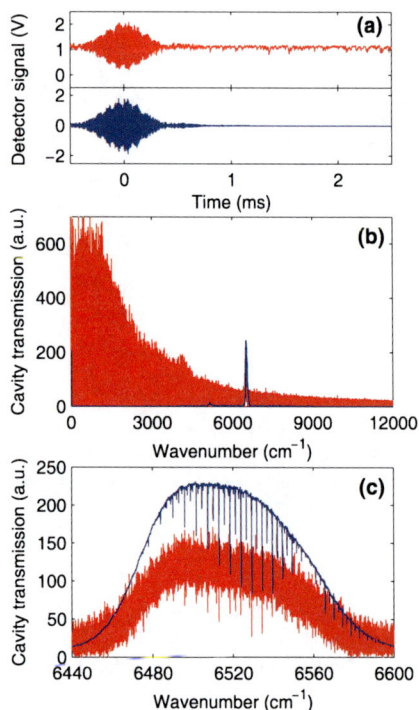

GaAs detector [186], which subtracts the photo-currents from the two photodiodes and uses active feedback to keep the DC currents equal. This subtraction doubles the useful signal compared to a single interferometer output (due to the fact that the two output signals of the interferometer are complementary [187]) while removing common mode amplitude noise, mainly caused by FM-to-AM conversion by the enhancement cavity. Both interferograms (from the comb and the reference cw laser) were digitized by a two-channel analog-to-digital converter at 1 Msample/s with 22-bit amplitude resolution.

Figure 8.12 shows two time-domain interferograms and the resulting spectrum of acetylene recorded with this system. The signal obtained with one output of the interferometer blocked (corresponding to a single detector setup) is shown in red in panel (a), while the blue trace shows the signal recorded using the auto-balanced detection. Significant removal of amplitude noise is clearly visible, especially in the baseline at times greater than 1 ms. The use of auto-balanced detection decreased the noise at the interferogram fringe frequency (corresponding to 6500 cm^{-1}) by a factor of 600, down to the shot noise limit, significantly increasing the signal-to-noise ratio of the resulting spectrum, as shown in panels (b) and (c). The acquisition time of one scan with a spectral resolution of 380 MHz was 3 s and a shot-noise limited sensitivity of 1.4×10^{-9} cm^{-1} was obtained in the normalized spectrum, corresponding to 3.4×10^{-11} $cm^{-1}\,Hz^{-1/2}$ per spectral element.

In general the repetition rate of the pulse train reflected from a moving mirror in a Michelson interferometer arm is Doppler shifted in comparison to the initial repetition rate of the femtosecond laser. Therefore, the interference at the output port can be seen as that between two frequency combs with slightly different repetition rates ($\Delta f_{\text{rep}} = 2\frac{v}{c} f_{\text{rep}}$). This makes it equivalent to the dual-comb spectroscopy technique initially proposed in Ref. [166] and demonstrated by several groups [49, 65, 167, 168, 188, 189]. In this technique the heterodyne beat note between two frequency combs with different repetition rates creates an interferogram in the time domain [see Fig. 8.7(d)—dual comb approach box]. In most implementations of the dual comb technique, Δf_{rep} is set in the 0.2–4 kHz range, which assures short acquisition times of a single interferogram. Since the dual comb technique does not use any mechanical moving parts and the footprint of the detection system is compact, it is potentially suitable for field applications; however, the need for two fully stabilized femtosecond sources (as the performance depends strongly on the stability of the mismatch of the repetition frequencies) significantly increases the cost and complexity of the system. Referencing schemes using cw lasers have recently been demonstrated [190–192], which can lead to a relaxation of the requirements for repetition rate stabilities.

8.4.4 Sensitivity of CE-DFCS Detection Methods

The sensitivity of a detection system can be determined from the relative noise, σ, on the baseline of the spectrum. The minimum detectable absorption coefficient can be defined as

$$\alpha_{\text{min}} = [\sigma/L_{\text{eff}}] \left[\text{cm}^{-1}\right], \tag{8.6}$$

where L_{eff} is the effective interaction path length with the sample. For a tight comb-cavity lock, L_{eff} is equal to $2FL/\pi$, where L is the physical cavity length (assuming a linear cavity); in the case of cavity ringdown, L_{eff} is reduced to FL/π. In intermediate cases, i.e., for the swept coupling scheme, L_{eff} takes a value between those two limiting cases [43]. When the system is white noise limited, α_{min} averages down with the square root of the number of samples, so the minimum detectable absorption coefficient normalized to 1-s acquisition time is used for comparison between systems:

$$\alpha_{\text{min}}^{1\,s} = [\sigma/L_{\text{eff}}]\sqrt{T} \left[\text{cm}^{-1}\,\text{Hz}^{-1/2}\right], \tag{8.7}$$

where T is the acquisition time of the spectrum. Note that this formula is valid only if the system is white-noise limited on the time scale of a second or longer. For broadband systems such as DFCS, a figure-of-merit including the number of simultaneously resolved spectral elements, M, can be used. This absorption sensitivity per spectral element, $\alpha_{\text{min}}^{\text{DFCS}}$, is defined as

$$\alpha_{\text{min}}^{\text{DFCS}} = [\sigma/L_{\text{eff}}]\frac{\sqrt{T}}{\sqrt{M}} \left[\text{cm}^{-1}\,\text{Hz}^{-1/2}\right]. \tag{8.8}$$

This value can be interpreted as the performance in terms of absorption sensitivity that a cw-laser system would need, including scanning time, to match the performance of the broadband system.

8.5 Applications

The powerful features of CE-DFCS make the technique attractive for numerous scientific applications. While a number of proof-of-principle experiments demonstrating the general capabilities of the technique have been reported, the list of systems applied to specific scientific problems is still relatively short [26, 48, 55, 57–60, 156, 160, 161]. A few of these are described in several extensive reviews [43, 44, 193]. In this section we present five systems developed at JILA with specific applications in mind and operating in different spectral regions spanning from the visible to the mid-IR. Combined they demonstrate the potential of applying CE-DFCS to a wide range of scientific fields. The first three examples illustrate broadband and ultra-sensitive detection of multiple molecular species using enhancement cavities: breath analysis using a VIPA spectrometer at 1.6 μm; semiconductor manufacturing gas impurity analysis with a broadband VIPA spectrometer at 1.75–2 μm; detection of hydrogen peroxide for medical applications using comb-FTS working at 3.74 μm. Two final examples show how resolving individual comb modes enables ultrahigh resolution with many simultaneous detection channels over a broad spectral bandwidth, which can then be used for rapid spectroscopy of cooled molecular beams and for velocity-modulation spectroscopy of molecular ions.

8.5.1 Breath Analysis

The idea of disease diagnosis based on the 'smell' of human breath can be traced back to Hippocrates. It has an enormous potential as a noninvasive, inherently-safe and low-cost method for disease detection, preventive medicine, and metabolic status monitoring. However, despite some progress, breath analysis is still not an established diagnosis tool, with the exceptions of NO detection (for asthma), and to some extent CO detection and $^{13}C/^{12}C$ ratio measurements, which have been used in a few clinical applications. A main issue preventing breath analysis from becoming a widespread diagnostic tool is the large number of volatile organic compounds present in human breath [194, 195]. The CE-DFCS technique has the potential to address this issue due to its capability of simultaneous multispecies detection.

The first test of the applicability of CE-DFCS for analysis of breath samples was conducted by Thorpe et al. [57]. The experimental setup with breath sampling is shown in Fig. 8.13(a) and was based on a 100-MHz Er:fiber OFC and a VIPA spectrometer (see Sect. 8.4.2). The comb light was coupled into an enhancement cavity with an FSR of 100 MHz and a finesse of 28000 using the swept coupling scheme (see Sect. 8.3.2). The laser generated a 100-nm-wide spectrum centered at 1.55 μm and could be shifted to the 1.6–1.7 μm region by Raman shifting in the amplifier, which also increased the average power to 300 mW.

The first experiment focused on the detection of CO and CO_2 near 1565 nm, as marked in Fig. 8.13(b). Breath samples were measured from two different

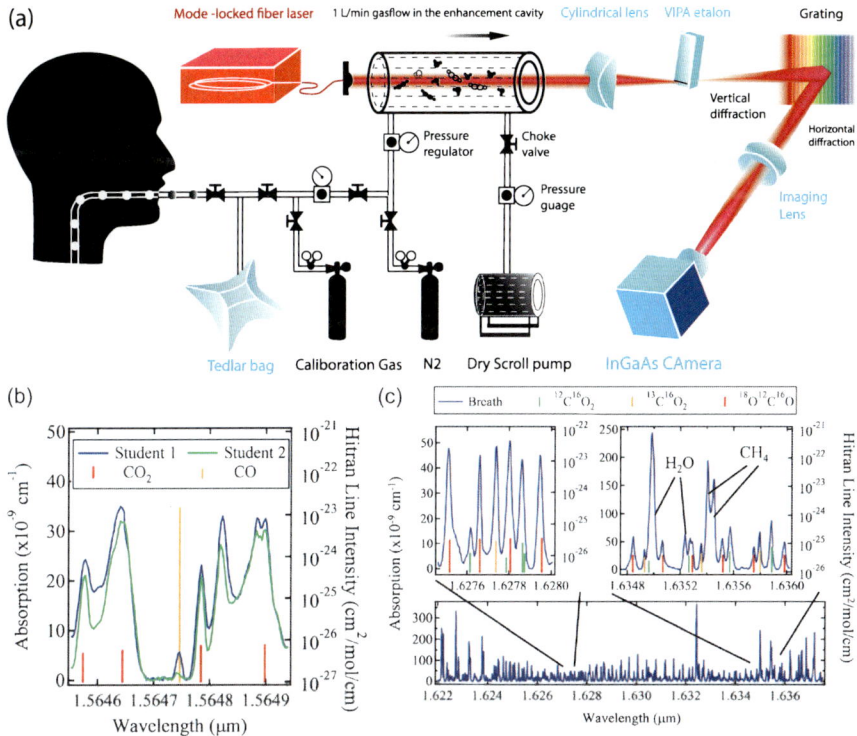

Fig. 8.13 Breath analysis with CE-DFCS around 1.6 μm. (**a**) The experimental setup including the breath sample handling system. (**b**) Zoomed-in view of two patients' breath spectra, showing measured absorption lines of CO_2 and CO (continuous spectrum) and the positions of CO_2 and CO lines from HITRAN (*red* and *yellow bars*, respectively). (**c**) A breath spectrum between 1.622 μm and 1.638 μm. Several windows in this region contain spectroscopic features of three isotopologues of CO_2, with nearly equal absorption strengths. Two zoomed-in spectral windows with line positions and intensities of relevant transitions are shown in the upper plots. Besides CO_2 peaks, strong absorption features of H_2O and CH_4 are detected in the measured spectral range (from Ref. [57])

students—one smoker (blue curve) and one non-smoker (green curve). The smoking student had a cigarette 15 minutes before the start of the test. The average concentration of CO for the smoker was determined to be 6.5 ppm—five times the value for the non-smoker. These results agree with previous studies of the CO level in a smoker's breath, ranging from 1 to 68 ppm [196]. In addition, a large change of the CO_2 concentration was observed as a function of time for which the patient held his breath before the sample was acquired, whereas the CO concentration was not influenced by the hold time.

The second test focused on the $^{13}C/^{12}C$ isotope ratio by comparing the concentrations of $^{13}CO_2$ and $^{12}CO_2$. This isotope ratio is of medical interest since it can provide information about liver function, bacterial overgrowth, or pancreatic function [197]; the main interest here was on the isotope ratio as a test for *helicobacter pylori* infection, which is a common cause of ulcers [198]. The infection usually does not cause any visible symptoms, and it is estimated that two-thirds of the world population is infected with *helicobacter pylori* [199], so there is a need for a quick, non-invasive detection method allowing for screening of this infection in a large population.

The lower panel of Fig. 8.13(c) shows the absorption spectrum of breath from a healthy patient centered around 1.63 μm. In the wavelength range shown, there are 78 lines from $^{12}CO_2$ and 29 from $^{13}CO_2$. The two zoomed-in panels show that even in the presence of water and methane, the absorption lines of $^{12}CO_2$ and $^{13}CO_2$ are clearly visible. Using five lines for each isotope, an isotope ratio of $\delta^{13}C$ of -28.1 ± 4.1 was determined ($\delta^{13}C$ is defined as the difference in the isotope abundance from natural abundance in parts per thousand). Similar precision was obtained for $\delta^{18}O$, derived from the intensities of ^{18}OCO lines also present in this wavelength range. The precision of these measurements can be improved by including a larger number of lines in the multiline fit, with no cost in the measurement time. Since a precision of 1 part per thousand is more than suitable for a *helicobacter pylori* test [197], this experiment demonstrates the ability to simultaneously record multiple stable isotope ratios at clinically relevant levels.

With small changes to the experimental setup, for example by using a multi-branch laser, it would be possible to combine these two tests into a single measurement. The potential to identify a large number of compounds in a single sample of gaseous mixture is important for medical applications where samples can be challenging or time consuming to obtain. Mid-IR frequency combs provide access to many more molecular species of interest and may allow construction of a bench-top device capable of detecting bio-markers of multiple diseases in a single, quick test.

8.5.2 Trace Water in Arsine Vapor

Another application of CE-DFCS is monitoring of trace impurities in process gasses, such as silane (SiH_4), germane (GeH_4), phosphine (PH_3), and arsine (AsH_3), used in the semiconductor industry. Trace levels of contaminants (down to the tens-of-ppb level) in these process gasses have a direct negative impact on the performance and lifetime of the final products owing to the incorporation of donor or acceptor impurity sites in the semiconductor. Some of the main impurities of concern include oxygen, carbon dioxide, hydrocarbons such as methane and ethane, hydrogen sulfide, and water vapor [200–202]. Currently, different systems are used to monitor each contaminant, many of which are bulky or very slow. In particular, on-line monitoring of water contamination of the process gasses is important, since this contaminant is difficult to remove completely due to its low vapor pressure.

Laser-based gas monitoring has obvious advantages—it allows real-time analysis and unambiguous identification of molecules. The main problem with the detection of these impurities using absorption spectroscopy is the dense spectrum of the process gasses themselves, especially since these gases are at nearly 100 % purity, so the contaminant absorption is extremely weak compared to the absorption background of the process gas. In addition, the possibility for rapid, simultaneous monitoring of multiple contaminants would be desirable. All these issues can be addressed with broadband CE-DFCS to achieve the required detection limits.

A CE-DFCS system focusing on detecting ppb levels of water contamination in arsine gas was constructed [56]. Due to the density of the arsine spectrum, the system was designed to work in the 1.75–1.95 μm wavelength region, where the arsine spectrum was previously unexplored but was expected to have a transparency window. To obtain an optical frequency comb in this wavelength range, the spectrum of an Er:fiber mode-locked laser, which provided 150 mW of power at a 250-MHz repetition rate, was amplified to 400 mW in an Er:fiber amplifier and was subsequently spectrally broadened in a 6-cm-long piece of highly-nonlinear silica fiber. The spectrum covered 1.2–2.1 μm and provided 17 mW of average power in a 40-nm bandwidth around 1.86 μm. The light was coupled into a 60-cm long enhancement cavity with a finesse of 30000 using the swept coupling scheme (see Sect. 8.3.2) [43]. In this case, f_{rep} was modulated at 7.5 kHz with a sweep amplitude for one comb tooth of 150 kHz in the optical domain. Slow feedback to the cavity PZT maintained the high transmission of the cavity for extended periods of time. A VIPA spectrometer [described in Sect. 8.4.2 and shown schematically in Fig. 8.7(c)] was used for detection, providing a spectral resolution of 900 MHz (0.031 cm^{-1}) and a simultaneous spectral coverage of 20 nm (50 cm^{-1}), limited by the size of the camera. Over 2000 spectral channels were collected simultaneously within 150 ms of integration time for a single image; typically 20 images were averaged per single spectrum.

Figure 8.14 shows the spectrum of 160 Torr of arsine gas contaminated with 1.27 ppm of water. The measured absorption spectrum is shown negative in black, with the water line strengths and positions from HITRAN [211] shown positive in blue. Even though the absorption of arsine gas increases rapidly close to the edges of the measured spectrum, a good transparency window does exist and allows for clear identification of water lines, visible in the two zoomed-in panels. The absorption sensitivity for water in arsine was determined to be 2.4×10^{-8} cm^{-1}, with a total integration time of 600 s. The sensitivity is limited by the background arsine absorption and by the low gas-switching speed as the initial gas handling system was not built for this application. This sensitivity corresponds to a minimum detectable absorption of 31 ppb for the trace amount of water in pure arsine. In summary, the system determined the required transmission window of arsine absorption and reached a detection limit suitable for the semiconductor industry.

8.5.3 Trace Detection of Hydrogen Peroxide

Another case where the absorption spectrum of a target molecule overlaps with spectra of more strongly absorbing or more abundant species is the detection of hydrogen peroxide (H_2O_2)—a molecule of interest for medical human breath analysis as well as atmospheric chemistry. In medicine, H_2O_2 is a potential marker of oxidative stress in the lungs, which can be connected to diseases such as asthma,

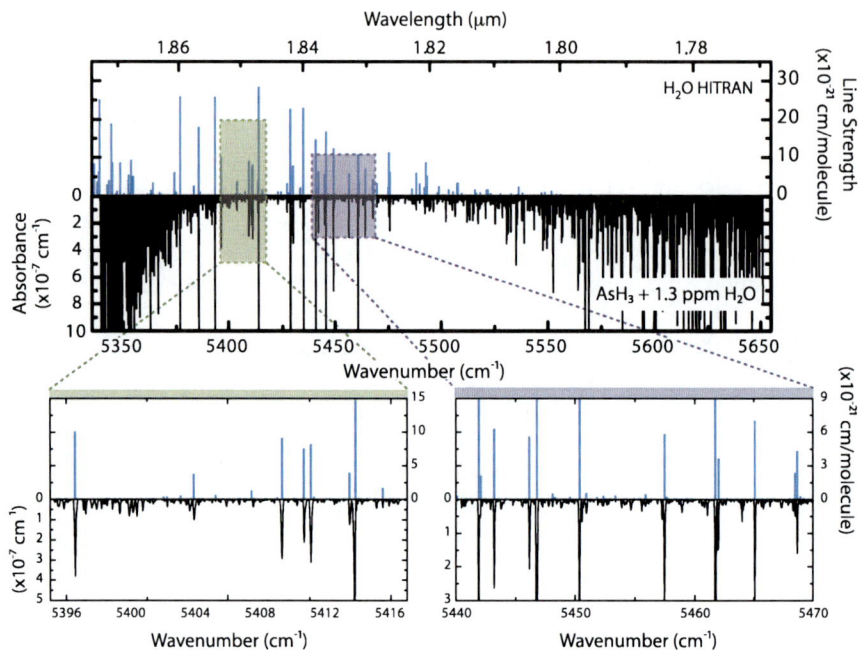

Fig. 8.14 Detection of water impurities in arsine process gas. The measured absorption spectrum of 1.27 ppm of water vapor in arsine is plotted negative (*black*) and the strengths of water lines taken from HITRAN are plotted positive (*blue*). The arsine absorption continues to increase both above 5650 cm^{-1} and below 5350 cm^{-1}. The *two bottom panels* with expanded views show easily resolvable water lines even in a strongly absorbing background gas (from Ref. [56]. With kind permission from Springer Science+Business Media: Appl. Phys. B, Analysis of trace impurities in semiconductor gas via cavity-enhanced direct frequency comb spectroscopy, vol. 100, year 2010, p. 917, K.C. Cossel et al., Fig. 4)

chronic obstructive pulmonary disease (COPD), or acute respiratory distress syndrome (ARDS) [203–205]. In atmospheric chemistry, H_2O_2 has a significant role as a stratospheric reservoir for HO_x [206] and is associated with biomass burning [207, 208]. However, the detection of this molecule is difficult since it is highly reactive and its fundamental ν_1 and ν_5 vibration bands overlap with strong absorption bands of water [209], which is usually the main interfering species, particularly in breath analysis (where water is present at a concentration of a few percent). The traditional approach used in breath analysis relies on water removal from the sample by filtering or on the use of exhaled breath condensate, both of which exclude real time operation. To test the possibility of real time monitoring of H_2O_2 concentration in gas samples with high water background, the H_2O_2 $\nu_2 + \nu_6$ intercombination band was targeted with a mid-infrared CE-DFCS system. This band, at 3.76 μm, is far detuned from all strong water absorption features and yet it is strong enough to provide sufficient detection limits [209, 210].

The experimental setup, shown in Fig. 8.6, was based on the tunable OPO comb and Fourier-transform spectrometer with auto-balanced detection described in

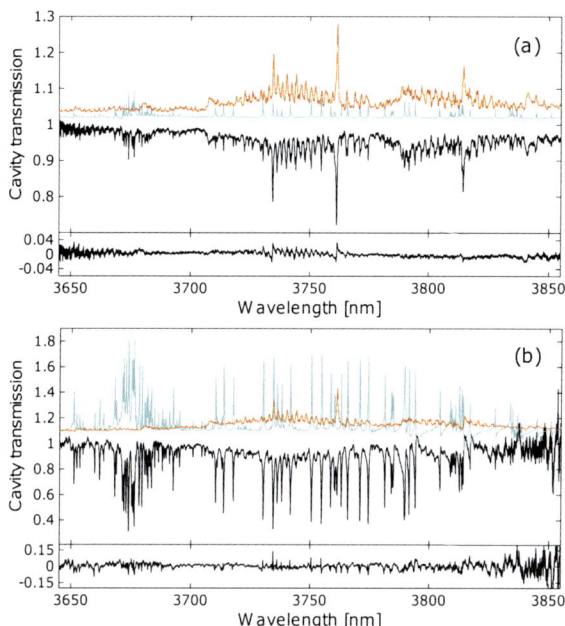

Fig. 8.15 Spectra of H_2O_2 solution vapor measured with mid-infrared CE-DFCS. In both panels the experimental spectra are shown in *black*, while the fit spectra of water and hydrogen peroxide in *blue* and *red*, respectively, inverted and offset for clarity. The *lower plots in each panel* display the fit residuals. (**a**) Spectrum taken at a pressure 630 Torr and room temperature with a small amount of N_2 flowing through the cavity spacer tube. The concentration values returned by the fit were 645 ppm for water and 3.6 ppm for H_2O_2. (**b**) Spectrum taken in 630 Torr of air and at a temperature of 37 °C with no N_2 flow through the cavity spacer tube. The concentration values returned by the fit were 1.2 % for water and 5 ppm for H_2O_2 (from Ref. [156]. With kind permission from Springer Science+Business Media: Appl. Phys. B, Cavity-enhanced optical frequency comb spectroscopy in the mid-infrared application to trace detection of hydrogen peroxide, vol. 110, year 2013, p. 163, A. Foltynowicz et al., Fig. 6)

Sects. 8.2.2 and 8.4.3, respectively. The enhancement cavity was constructed with two mirrors on ZnSe substrates and had a peak finesse of 3800 at 3.8 μm. The cavity length was chosen to be 54.7 cm so that the *FSR* was twice the comb repetition rate. The cavity was equipped with a glass container in which a liquid sample could be introduced, as well as a gas inlet and outlet allowing for constant flow of gaseous samples. This container was filled with a solution of 27 % of H_2O_2 in water to create a gaseous sample of hydrogen peroxide in the cavity. The surface of the peroxide solution was 6 cm below the cavity axis. The frequency comb was locked to the cavity with a two-point PDH lock (Sect. 8.3.1).

Figure 8.15(a) shows the normalized spectrum (in black) of the hydrogen peroxide vapor at a pressure of 630 Torr and at room temperature. To reduce the amount of water in the sample, a small flow of N_2 was maintained though the cavity during the measurement. The sum of model spectra of H_2O (from HITRAN [211]) and H_2O_2 (from [209]) was fit to the data; the constituent spectra from the fit are shown in color (blue for H_2O and red for H_2O_2, inverted and offset for clarity). A sensitivity of 5.4×10^{-9} cm^{-1} Hz$^{-1/2}$ was reached, corresponding to 6.9×10^{-11} cm^{-1} Hz$^{-1/2}$ per spectral element for 6000 resolved elements. This resulted in a noise equivalent concentration detection limit for H_2O_2 in the absence of water of 8 ppb for 1 s averaging time.

To simulate the conditions of human breath analysis, the container and the cavity cell were heated to a temperature of 37 °C and the N_2 purge was turned off. To avoid water condensation on the cavity mirror surfaces, the mirrors were heated to around 40 °C. Under these conditions, water absorption dominated the measured spectrum, as can be seen in Fig. 8.15(b); however, it was still possible to observe absorption due to hydrogen peroxide. The concentrations obtained from the fit are 5 ppm of H_2O_2 and 1.2 % of H_2O. It can be seen that the discrepancies in the center of the fit are larger than those for the low concentration case, suggesting that water was affecting the detection limit of H_2O_2.

To estimate the effects of water on the detection limit of hydrogen peroxide, pure deionized water was introduced into the container and the spectra of its vapor in air were recorded. The temperature and pressure conditions were the same as those for the hydrogen peroxide measurement at elevated temperature shown in Fig. 8.15(b). A model including both H_2O and H_2O_2 was fit to 100 consecutively measured spectra and a mean H_2O and H_2O_2 concentrations of 2.83 % and 75 ppb were obtained, respectively, with a standard deviation of 130 ppb in the second case, which implies a detection limit of 130 ppb for H_2O_2 in the presence of almost 3 % of water. This value is significantly higher than that expected from the noise level; however, for the detection of a weak absorption signal out of overlapping strong absorption background of other constituents, the detection limit for the weak component is set by the quality of the model spectra of the strong components rather than by the detection noise. To remove this limitation, the spectra of the strong components need to be precisely determined under the relevant experimental conditions, including effects not described by the Voigt profile such as Dicke narrowing or the speed dependence of both pressure broadening and shifting [212, 213], which are not included in the HITRAN database. Thus, systematic analysis of the line shapes for the overlapping water band will be critical for further performance improvements.

This experiment was the first demonstration of CE-DFCS in the mid-IR region and confirmed that it is capable of efficient detection of H_2O_2 at the sub-ppm level even in the presence of a large amount of water, without the need for water removal. The data obtained with low water concentration [Fig. 8.15(a)] suggest that the CE-DFCS technique allows detection of H_2O_2 using the $\nu_2 + \nu_6$ band down to the low ppb level. While this detection limit is not as low as that of a cw-laser based system used for atmospheric research [214], CE-DFCS offers the capability for simultaneous detection of additional molecular species of interest such as methane, acetylene, nitrous oxide, and formaldehyde, in a single system.

It is interesting to compare the performance of this mid-IR CE-DFCS system to that based on the same OPO and FTS and employing a multipass cell instead of the cavity [48]. The absorption sensitivity obtained with the Herriot cell is significantly lower (3.8×10^{-8} cm^{-1} Hz$^{-1/2}$ per spectral element), mostly due to the

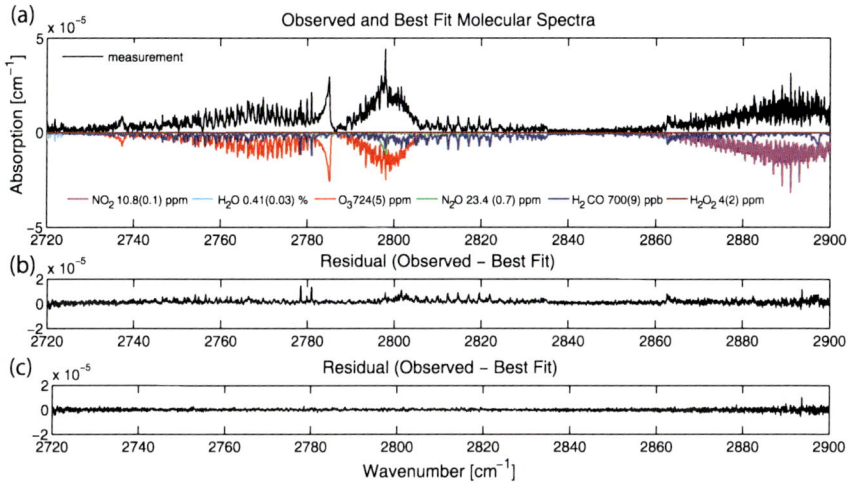

Fig. 8.16 (**a**) The absorption spectrum of the plasma effluent (*black*) at a pressure of 620 Torr measured around 3.6 μm (2800 cm^{-1}), recorded at 28 s after the plasma is turned on. The individual molecular spectra (*magenta*—nitrogen dioxide, *red*—ozone, *grey*—water, *green*—nitric oxide, *blue*—formaldehyde, *brown*—hydrogen peroxide), whose sum is fit to the measured data, are plotted negative for clarity. The legend lists the fit concentrations with 1σ uncertainties. (**b**) The residual of the fit for the case when the spectrum of formaldehyde is not included in the model. (**c**) The residual of the fit when a model including all the mixture components is applied (from Ref. [26])

difference in effective path length (36 m inside the cell, compared to 1.3 km in the cavity), and the lack of autobalanced detection in the experiment with the multipass cell. However, the multipass cell allows acquisition of spectra over the entire tuning range (2.8–4.6 μm) of the OPO, as was demonstrated in Ref. [48], while the cavity effectively limits the useful bandwidth, as shown in Fig. 8.5.

The mid-IR system with multipass cell was used in an experiment that nicely demonstrates the general capabilities of DFCS. It was employed for measurements of hydrogen-peroxide-enhanced non-thermal plasma effluent, created in a device made for disinfection purposes in various medical settings [26]. The multipass cell was connected to the plasma device, taking the place of a disinfection chamber and the device ran in a closed-loop configuration maintaining a constant flow of disinfection gas mixture through the cell. The mixture was expected to be a combination of constituents such as O_3, N_2O, NO_2, NO, OH^* and H_2O_2: the latter was added separately to enhance the disinfection performance as confirmed by separate tests with bacteria [26]. To optimize the disinfection efficiency, quantitative information on the concentration levels of individual constituents under different plasma operating conditions was needed.

Figure 8.16 shows an example of the experimental spectrum of plasma effluent at a pressure of 620 Torr. Panel (a) shows the measured absorption spectrum (black) and the model spectra of the individual constituents (plotted negative for clarity, in color), whose sum is fit to the experimental data. Panel (b) plots the fit residuals from the model including all expected species visible in the spectral window (O_3, N_2O, NO_2 and H_2O_2). The structure in the residual indicates that the assumed model of the spectrum did not account for all the absorbing species produced by the medical device. Indeed, the fit significantly improves after adding H_2CO (initially not expected in the mixture) spectrum to the model, as shown by the residuals in panel (c). The obtained values of the single components concentrations with 1σ uncertainties are shown in panel (a).

The broadband coverage of DFCS enables precise determination of concentrations of multiple molecular species simultaneously without *a priori* knowledge of all constituents. Such capability is desirable for many scientific applications, including atmospheric research and breath analysis, and is difficult to realize with cw-laser based systems; even if multiple cw lasers would be used, their frequencies would be set to transitions of constituents expected to be present in the mixture, precluding the observation of unexpected features. The fact that the full spectrum is acquired simultaneously eliminates drifts across the spectrum due to the scanning time of cw lasers. Although the presented example does not employ a cavity, a similar application with a cavity-based system is straightforward.

8.5.4 Comb-Mode Resolved Spectroscopy

The examples described above focus on spectroscopic applications of CE-DFCS without the need to resolve individual comb lines. Below we present two different applications demonstrating the advantages of simultaneously acquiring spectra with comb-mode resolution. In the first example, the enhanced resolution permits detailed examinations of both translational and rotational cooling in a supersonic molecular beam expansion. In the second example, modulation and lock-in detection are introduced to CE-DFCS, allowing high-resolution velocity-modulation spectroscopy of molecular ions to be performed on many comb lines simultaneously.

Spectroscopy of Cold Molecules   Pulsed supersonic expansions are widely used in physical chemistry and molecular physics to obtain internal cooling of molecules without the need for cryogenic systems. The cooling dynamics in such systems can be complex and depend not only on the species to be cooled and the nozzle geometry, but also on the carrier gas species, backing pressure, and background pressure. Thus, the cooling in a supersonic expansion is often both spatially dependent and internal state dependent (i.e., the rotational, vibrational, and electronic temperatures are not in equilibrium).

Rapid tomography of a pulsed supersonic expansion of 2 % acetylene seeded in argon has been demonstrated using CE-DFCS [60]. A cavity with an *FSR* of 700 MHz and a finesse of 6300, centered at the 1.53 μm ($v_1 + v_3$) acetylene band, was aligned perpendicular to the molecular jet direction. Thus, the cavity mode measured a small, cylindrical region of the molecular beam expansion, perpendicular to the beam propagation direction. The expansion nozzle position was translatable *in situ* in three dimensions, which results in the cavity mode sampling different regions of the expansion. The light from an Er:fiber frequency comb ($f_{rep} = 100$ MHz) was coupled into the cavity using the swept coupling scheme. The cavity increased the comb mode spacing to the cavity FSR, so that only one comb mode interacted with a given molecular ro-vibrational line (with approximately 300 MHz linewidth) at a time. This provides single comb-mode resolution since only one ro-vibrational line occurred within the VIPA spectrometer resolution. The cavity length was actively stabilized to an iodine-stabilized reference laser and was scanned over one FSR by shifting the reference laser frequency.

The comb light transmitted through the cavity was recorded using a VIPA spectrometer (see Sect. 8.4.2). The camera integration time was synchronized with the gas pulse to enable gating of specific regions of the pulse. To create a full tomographic reconstruction of the expansion, a spectrum at a given nozzle position was measured by stepping the cavity length over one FSR; this process was then repeated for different transverse and longitudinal positions of the nozzle. Each spectrum contains a wealth of information: first, by integrating over the width of each ro-vibrational line, a rotational distribution integrated along the cavity axis was obtained. By using measurements at different heights, and relying on the cylindrical symmetry of the expansion, a radial density and temperature profile of the expansion could be derived. In addition, individual ro-vibrational line shapes, as shown in Fig. 8.17(a), could be analyzed in detail. These Doppler-broadened line shapes provided information about transverse-velocity and rotational-state dependent effects. For example, in the shaded regions, an increased population in lower rotational states relative to higher rotational states was observed, which suggests that the cooling was optimal off the molecular beam axis (i.e., in regions with non-zero transverse velocity), potentially due to on-axis clustering.

This experiment illustrates how CE-DFCS combines the seemingly incompatible capabilities of wide spectral bandwidth, high sensitivity, sub-millisecond temporal resolution, and high spectral resolution, all in one experimental platform. In this case, the wide, simultaneous spectral bandwidth along with the temporal resolution provided information about the rotational cooling as a function of the spatial and the temporal positions in the gas pulse. The single comb mode resolution enabled simultaneous comparison of the line shapes of many ro-vibrational lines. This makes CE-DFCS a potentially powerful tool for studying state-dependent collisions and reactions in applications such as crossed molecular beams, trapped cold and ultracold molecules, and studies of unstable intermediates in chemical reactions.

Velocity Modulation Spectroscopy Broadband spectroscopy of molecular ions has many applications in areas ranging from fundamental physics to physical chemistry and astrochemistry. In astrochemistry, for example, it is believed that many of the currently unidentified visible to near-IR absorption bands, collectively termed the diffuse interstellar bands (DIBs), may be due to interstellar absorption from molecular ions [215, 216]. Molecular ions also hold potential for precision measurements [217–221] such as testing QED calculations, tests of time variation of the electron-to-proton mass ratio and the fine structure constant, or measurement of the permanent electric dipole moment of the electron (eEDM) due to the long coherence times possible with trapped ions.

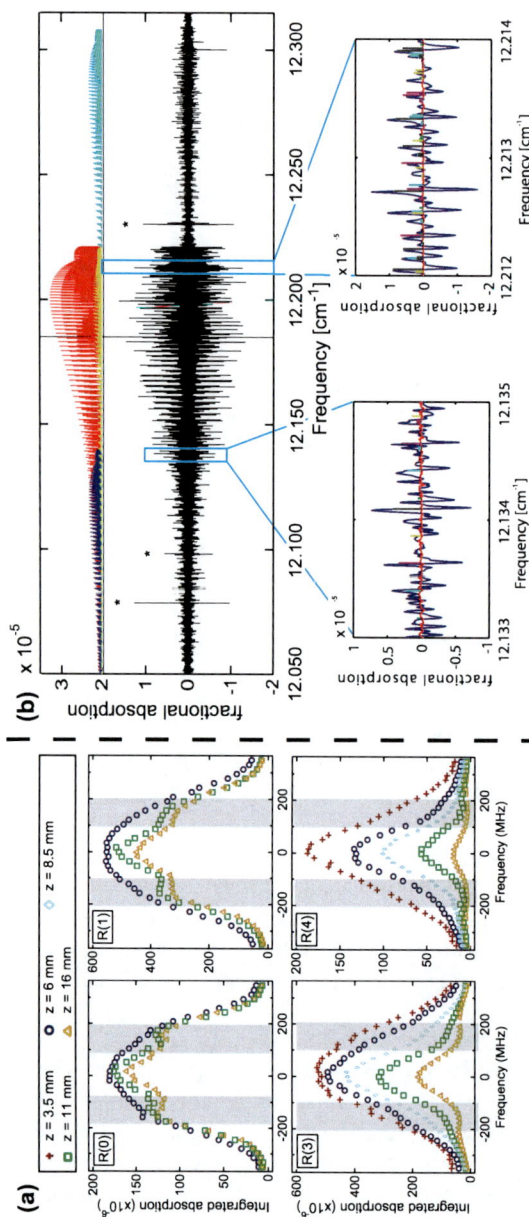

Fig. 8.17 Comb-mode resolved CE-DFCS. (**a**) Line shapes of four ro-vibrational lines in the R-branch of the acetylene $\nu_1 + \nu_3$ band recorded in a pulsed supersonic expansion as a function of distance from the nozzle. Notice that lower J lines (*upper two plots*) show an increase in population in the shaded regions compared with the higher J lines (*two lower plots*) [60]. (**b**) The spectrum of HfF$^+$ recorded with CE-DFCS combined with velocity modulation spectroscopy. Fits to four different ro-vibronic bands are shown above the experimentally obtained spectrum. The *insets* show the high-resolution spectra, with individual isotopes (marked in different colors) clearly resolved

One of the challenges in using molecular ions for precision measurement is a lack of available spectroscopic data. For example, in the JILA eEDM experiment multiple states of HfF^+ or ThF^+ [222] will be needed for state preparation and spin readout; thus a method for broad survey spectroscopy to characterize the electronic structure of these ions was needed. Pulsed-field-ionization zero-kinetic-energy measurements (PFI-ZEKE) recently provided information about the lowest lying states of HfF^+ [223] and ThF^+ [224]. Theoretical calculations exist for the states of HfF^+, but due to the challenges of including so many electrons in the calculations, the uncertainties were at least a thousand wavenumbers [225].

Recently, CE-DFCS has been incorporated into velocity-modulation spectroscopy (VMS) to enable broadband, ion-specific detection [58, 59]. In this experiment, a 3-GHz repetition rate Ti:sapphire comb was locked (using a single-point tight lock) to a ring cavity with an *FSR* equal to $1/25$ of f_{rep} and a finesse of about 100. The cavity contained an AC discharge cell that both produced the ions and modulated their drift velocity, resulting in a modulated absorption signal. The transmitted comb light was comb-mode resolved by the VIPA spectrometer and imaged onto a lock-in camera, which simultaneously demodulated each pixel (and thus each comb tooth) at the discharge frequency. Polarization optics were used to rapidly switch the direction of light propagation through the cavity for noise reduction.

About 1500 channels, spanning 150 cm^{-1}, were recorded simultaneously. One comb tooth was locked to a stable cw Ti:sapphire laser, which provided an absolute frequency reference in our spectrum. For each measurement, images for each direction of propagation were averaged and subtracted. Additionally, the power in each comb tooth was measured by applying a calibrated amplitude modulation to the laser. To fully sample the spectrum, 30 measurements with the cw laser stepped over 3 GHz were interleaved. This resulted in a spectrum that covered 150 cm^{-1} sampled with a step-size of 100 MHz. The spectrum was then interpolated onto a fixed 0.001 cm^{-1} grid to allow scans to be easily averaged or combined. One full scan lasted about 30 minutes and resulted in a single-pass fractional absorption sensitivity of 3×10^{-7}. Since one scan contained 45000 channels, this equated to a sensitivity of 4×10^{-8} $Hz^{-1/2}$ per spectral element.

The top panel of Fig. 8.17(b) shows part of the HfF^+ spectrum acquired using the comb-VMS system, with the different observed ro-vibronic bands shown above in different colors. The spectrum centered around 12200 cm^{-1} is extremely congested due to the presence of many bands, each with five isotopes, and the high temperature of the oven, which results in observed J values up to about 70. The dynamic range of frequency-comb VMS is demonstrated by our ability to identify two overlapping vibronic bands (shown in red and green) despite an order of magnitude difference in line strengths and an offset in the band origin of only about 1 cm^{-1}. Since wide spectral windows are acquired simultaneously when using the comb, relative line strengths within the region are not influenced by variability in oven and discharge conditions, which significantly helped to disentangle the bands. The total spectral coverage of the system is extendable using nonlinear optics or different comb sources as discussed earlier. This technique is thus a powerful system for accurate survey spectroscopy of many molecular ion species over a wide spectral range.

8.6 Summary

This chapter discusses the principles of the CE-DFCS technique and reviews existing frequency comb sources ranging from the XUV to the mid-IR. Methods of coupling and locking frequency combs to an enhancement cavity are presented, as are different read-out systems capable of simultaneous acquisition of thousands of channels. Combined, these techniques represent the current state of the art of CE-DFCS; however, the rapid progress in this field guarantees improvements in spectral coverage, absorption sensitivity, and acquisition rate. Development of new frequency comb sources and improved properties of their spectra will enable access to molecular and atomic systems in previously unaccessible spectral regions.

An overview of various systems and applications of the technique was presented to highlight the different combinations of broadband coverage, high resolution, and ultra-high sensitivity provided by CE-DFCS. In a number of research areas, such as breath analysis, trace detection of contaminants in industrial gases, or atmospheric chemistry, the enormous potential is just starting to be recognized. We are confident that new windows of opportunity will arise in other fields, including precision spectroscopy, studies of molecular interaction and reaction dynamics, and coherent control of atomic or molecular systems.

Acknowledgements We gratefully thank many of our colleagues who have contributed to the work described in this review. They are F. Adler, T. Allison, T. Ban, C. Benko, B. Bjork, T. Briles, A. Cingöz, E. Cornell, S. Diddams, M. Fermann, A. Fleisher, M. Golkowski, J. Hall, I. Hartl, L.-S. Ma, M. Martin, J. Repine, A. Ruehl, L. Sinclair, M. Thorpe, and D. Yost. The development of CE-DFCS has been supported by NIST, AFOSR, DARPA, NSF, DTRA, and Agilent.

References

1. S.A. Diddams et al., Direct link between microwave and optical frequencies with a 300 THz femtosecond laser comb. Phys. Rev. Lett. **84**(22), 5102–5105 (2000)
2. R. Holzwarth et al., Optical frequency synthesizer for precision spectroscopy. Phys. Rev. Lett. **85**(11), 2264–2267 (2000)
3. T. Udem, R. Holzwarth, T.W. Hansch, Optical frequency metrology. Nature **416**(6877), 233–237 (2002)
4. S.T. Cundiff, J. Ye, Colloquium: femtosecond optical frequency combs. Rev. Mod. Phys. **75**(1), 325–342 (2003)
5. J. Ye, S.T. Cundiff (eds.), *Femtosecond Optical Frequency Comb: Principle, Operation, and Applications* (Springer, Berlin, 2004)
6. T.W. Hansch, Nobel lecture: passion for precision. Rev. Mod. Phys. **78**(4), 1297–1309 (2006)
7. J.L. Hall, Nobel lecture: defining and measuring optical frequencies. Rev. Mod. Phys. **78**(4), 1279–1295 (2006)
8. S.A. Diddams, The evolving optical frequency comb. J. Opt. Soc. Am. B, Opt. Phys. **27**(11), B51–B62 (2010)
9. L.S. Ma et al., Optical frequency synthesis and comparison with uncertainty at the 10^{-19} level. Science **303**(5665), 1843–1845 (2004)
10. A. Bartels et al., Stabilization of femtosecond laser frequency combs with subhertz residual linewidths. Opt. Lett. **29**(10), 1081–1083 (2004)

11. T.R. Schibli et al., Optical frequency comb with submillihertz linewidth and more than 10 W average power. Nat. Photonics **2**(6), 355–359 (2008)

12. M.J. Martin et al., Testing ultrafast mode-locking at microhertz relative optical linewidth. Opt. Express **17**(2), 558–568 (2009)

13. S.M. Foreman et al., Coherent optical phase transfer over a 32-km fiber with 1 s instability at 10^{-17}. Phys. Rev. Lett. **99**(15), 153601 (2007)

14. S.M. Foreman et al., Remote transfer of ultrastable frequency references via fiber networks. Rev. Sci. Instrum. **78**(2), 021101 (2007)

15. A.D. Ludlow et al., Sr lattice clock at 1×10^{-16} fractional uncertainty by remote optical evaluation with a Ca clock. Science **319**(5871), 1805–1808 (2008)

16. K. Predehl et al., A 920-kilometer optical fiber link for frequency metrology at the 19th decimal place. Science **336**(6080), 441–444 (2012)

17. O. Lopez et al., Ultra-stable long distance optical frequency distribution using the Internet fiber network. Opt. Express **20**(21), 23518–23526 (2012)

18. A. Ruehl et al., Ultrabroadband coherent supercontinuum frequency comb. Phys. Rev. A **84**(1), 011806(R) (2011)

19. N.R. Newbury, Searching for applications with a fine-tooth comb. Nat. Photonics **5**(4), 186–188 (2011)

20. T. Steinmetz et al., Laser frequency combs for astronomical observations. Science **321**(5894), 1335–1337 (2008)

21. D.A. Braje et al., Astronomical spectrograph calibration with broad-spectrum frequency combs. Eur. Phys. J. D **48**(1), 57–66 (2008)

22. G.G. Ycas et al., Demonstration of on-sky calibration of astronomical spectra using a 25 GHz near-IR laser frequency comb. Opt. Express **20**(6), 6631–6643 (2012)

23. T. Wilken et al., A spectrograph for exoplanet observations calibrated at the centimetre-per-second level. Nature **485**(7400), 611–614 (2012)

24. D.F. Phillips et al., Calibration of an astrophysical spectrograph below 1 m/s using a laser frequency comb. Opt. Express **20**(13), 13711–13726 (2012)

25. S.J. Lee et al., Ultrahigh scanning speed optical coherence tomography using optical frequency comb generators. Jpn. J. Appl. Phys. **40**(8B), L878–L880 (2001)

26. M. Golkowski et al., Hydrogen-peroxide-enhanced nonthermal plasma effluent for biomedical applications. IEEE Trans. Plasma Sci. **40**(8), 1984–1991 (2012)

27. A. Bartels et al., Femtosecond-laser-based synthesis of ultrastable microwave signals from optical frequency references. Opt. Lett. **30**(6), 667–669 (2005)

28. T.M. Fortier et al., Generation of ultrastable microwaves via optical frequency division. Nat. Photonics **5**(7), 425–429 (2011)

29. T. Udem et al., Absolute optical frequency measurement of the cesium D1 line with a mode-locked laser. Phys. Rev. Lett. **82**(18), 3568–3571 (1999)

30. A. Marian et al., United time-frequency spectroscopy for dynamics and global structure. Science **306**(5704), 2063–2068 (2004)

31. M.C. Stowe et al., High resolution atomic coherent control via spectral phase manipulation of an optical frequency comb. Phys. Rev. Lett. **96**(15), 153001 (2006)

32. J. Ye et al., Accuracy comparison of absolute optical frequency measurement between harmonic-generation synthesis and a frequency-division femtosecond comb. Phys. Rev. Lett. **85**(18), 3797–3800 (2000)

33. T.W. Hansch et al., Precision spectroscopy of hydrogen and femtosecond laser frequency combs. Philos. Trans. R. Soc., Math. Phys. Eng. Sci. **363**(1834), 2155–2163 (2005)

34. G. Galzerano et al., Absolute frequency measurement of a water-stabilized diode laser at 1.384 µm by means of a fiber frequency comb. Appl. Phys. B, Lasers Opt. **102**(4), 725–729 (2011)

35. I. Galli et al., Molecular gas sensing below parts per trillion: radiocarbon-dioxide optical detection. Phys. Rev. Lett. **107**(27), 270802 (2011)

36. J. Domyslawska et al., Cavity ring-down spectroscopy of the oxygen B-band with absolute frequency reference to the optical frequency comb. J. Chem. Phys. **136**(2), 024201 (2012)

37. I. Ricciardi et al., Frequency-comb-referenced singly-resonant OPO for sub-Doppler spectroscopy. Opt. Express **20**(8), 9178–9186 (2012)
38. M. Fischer et al., New limits on the drift of fundamental constants from laboratory measurements. Phys. Rev. Lett. **92**(23), 230802 (2004)
39. G. Casa et al., Primary gas thermometry by means of laser-absorption spectroscopy: determination of the Boltzmann constant. Phys. Rev. Lett. **100**(20), 200801 (2008)
40. C.G. Parthey et al., Improved measurement of the hydrogen 1S–2S transition frequency. Phys. Rev. Lett. **107**(20), 203001 (2011)
41. C. Lemarchand et al., Progress towards an accurate determination of the Boltzmann constant by Doppler spectroscopy. New J. Phys. **13**, 073028 (2011)
42. A. Marian et al., Direct frequency comb measurements of absolute optical frequencies and population transfer dynamics. Phys. Rev. Lett. **95**(2), 023001 (2005)
43. M.J. Thorpe, J. Ye, Cavity-enhanced direct frequency comb spectroscopy. Appl. Phys. B, Lasers Opt. **91**(3–4), 397–414 (2008)
44. F. Adler et al., Cavity-enhanced direct frequency comb spectroscopy: technology and applications. Annu. Rev. Anal. Chem. **3**(3), 175–205 (2010)
45. A. Foltynowicz et al., Optical frequency comb spectroscopy. Faraday Discuss. **150**, 23–31 (2011)
46. J.U. White, Long optical paths of large aperture. J. Opt. Soc. Am. **32**(5), 285–288 (1942)
47. D.R. Herriott, H.J. Schulte, Folded optical delay lines. Appl. Opt. **4**(8), 883 (1965)
48. F. Adler et al., Mid-infrared Fourier transform spectroscopy with a broadband frequency comb. Opt. Express **18**(21), 21861–21872 (2010)
49. A.M. Zolot et al., Direct-comb molecular spectroscopy with accurate, resolved comb teeth over 43 THz. Opt. Lett. **37**(4), 638–640 (2012)
50. E.R. Crosson et al., Pulse-stacked cavity ring-down spectroscopy. Rev. Sci. Instrum. **70**(1), 4–10 (1999)
51. T. Gherman, D. Romanini, Mode-locked cavity-enhanced absorption spectroscopy. Opt. Express **10**(19), 1033–1042 (2002)
52. R.J. Jones, J. Ye, Femtosecond pulse amplification by coherent addition in a passive optical cavity. Opt. Lett. **27**(20), 1848–1850 (2002)
53. M.J. Thorpe et al., Broadband cavity ringdown spectroscopy for sensitive and rapid molecular detection. Science **311**(5767), 1595–1599 (2006)
54. G. Mejean, S. Kassi, D. Romanini, Measurement of reactive atmospheric species by ultraviolet cavity-enhanced spectroscopy with a mode-locked femtosecond laser. Opt. Lett. **33**(11), 1231–1233 (2008)
55. R. Grilli et al., Trace measurement of BrO at the ppt level by a transportable mode-locked frequency-doubled cavity-enhanced spectrometer. Appl. Phys. B, Lasers Opt. **107**(1), 205–212 (2012)
56. K.C. Cossel et al., Analysis of trace impurities in semiconductor gas via cavity-enhanced direct frequency comb spectroscopy. Appl. Phys. B, Lasers Opt. **100**(4), 917–924 (2010)
57. M.J. Thorpe et al., Cavity-enhanced optical frequency comb spectroscopy: application to human breath analysis. Opt. Express **16**(4), 2387–2397 (2008)
58. L.C. Sinclair et al., Frequency comb velocity-modulation spectroscopy. Phys. Rev. Lett. **107**(9), 093002 (2011)
59. K.C. Cossel et al., Broadband velocity modulation spectroscopy of HfF$^+$: towards a measurement of the electron electric dipole moment. Chem. Phys. Lett. **546**, 1–11 (2012)
60. M.J. Thorpe et al., Tomography of a supersonically cooled molecular jet using cavity-enhanced direct frequency comb spectroscopy. Chem. Phys. Lett. **468**(1–3), 1–8 (2009)
61. R.J. Jones et al., Phase-coherent frequency combs in the vacuum ultraviolet via high-harmonic generation inside a femtosecond enhancement cavity. Phys. Rev. Lett. **94**(19), 193201 (2005)
62. C. Gohle et al., A frequency comb in the extreme ultraviolet. Nature **436**(7048), 234–237 (2005)

63. A. Ozawa et al., High harmonic frequency combs for high resolution spectroscopy. Phys. Rev. Lett. **100**(25), 253901 (2008)
64. A. Cingoz et al., Direct frequency comb spectroscopy in the extreme ultraviolet. Nature **482**(7383), 68–71 (2012)
65. F. Keilmann, C. Gohle, R. Holzwarth, Time-domain mid-infrared frequency-comb spectrometer. Opt. Lett. **29**(13), 1542–1544 (2004)
66. K.A. Tillman et al., Mid-infrared absorption spectroscopy of methane using a broadband femtosecond optical parametric oscillator based on aperiodically poled lithium niobate. J. Opt. A, Pure Appl. Opt. **7**(6), S408–S414 (2005)
67. E. Sorokin et al., Sensitive multiplex spectroscopy in the molecular fingerprint 2.4 µm region with a Cr(2+): ZnSe femtosecond laser. Opt. Express **15**(25), 16540–16545 (2007)
68. F. Adler et al., Phase-stabilized, 1.5 W frequency comb at 2.8–4.8 µm. Opt. Lett. **34**(9), 1330–1332 (2009)
69. N. Leindecker et al., Broadband degenerate OPO for mid-infrared frequency comb generation. Opt. Express **19**(7), 6304–6310 (2011)
70. N. Leindecker et al., Octave-spanning ultrafast OPO with 2.6–6.1 µm instantaneous bandwidth pumped by femtosecond Tm-fiber laser. Opt. Express **20**(7), 7046–7053 (2012)
71. T. Udem et al., Accurate measurement of large optical frequency differences with a mode-locked laser. Opt. Lett. **24**(13), 881–883 (1999)
72. D.J. Jones et al., Carrier-envelope phase control of femtosecond mode-locked lasers and direct optical frequency synthesis. Science **288**(5466), 635–639 (2000)
73. A. Bartels, D. Heinecke, S.A. Diddams, 10-GHz self-referenced optical frequency comb. Science **326**(5953), 681 (2009)
74. T.M. Fortier, A. Bartels, S.A. Diddams, Octave-spanning Ti:sapphire laser with a repetition rate >1 GHz for optical frequency measurements and comparisons. Opt. Lett. **31**(7), 1011–1013 (2006)
75. J. Jiang et al., Fully stabilized, self-referenced thulium fiber frequency comb, in *The European Conference on Lasers and Electro-Optics* (Optical Society of America, Washington, 2011)
76. M.N. Cizmeciyan et al., Operation of femtosecond Kerr-lens mode-locked Cr:ZnSe lasers with different dispersion compensation methods. Appl. Phys. B, Lasers Opt. **106**(4), 887–892 (2012)
77. T.W. Neely, T.A. Johnson, S.A. Diddams, High-power broadband laser source tunable from 3.0 µm to 4.4 µm based on a femtosecond Yb:fiber oscillator. Opt. Lett. **36**(20), 4020–4022 (2011)
78. A. Ruehl et al., Widely-tunable mid-infrared frequency comb source based on difference frequency generation. Opt. Lett. **37**(12), 2232–2234 (2012)
79. M.E. Fermann, A. Galvanauskas, G. Sucha (eds.), *Ultrafast Lasers: Technology and Applications* (CRC Press, Boca Raton, 2002)
80. U. Keller, Recent developments in compact ultrafast lasers. Nature **424**(6950), 831–838 (2003)
81. J.-C. Diels, W. Rudolph, *Ultrashort Laser Pulse Phenomena*, 2nd edn. (Academic Press, New York, 2006)
82. U. Keller et al., Semiconductor saturable absorber mirrors (SESAM's) for femtosecond to nanosecond pulse generation in solid-state lasers. IEEE J. Sel. Top. Quantum Electron. **2**(3), 435–453 (1996)
83. C.C. Lee et al., Ultra-short optical pulse generation with single-layer graphene. J. Nonlinear Opt. Phys. Mater. **19**(4), 767–771 (2010)
84. H.J. Kim et al., High-performance laser mode-locker with glass-hosted SWNTs realized by room-temperature aerosol deposition. Opt. Express **19**(5), 4762–4767 (2011)
85. A. Schmidt et al., 175 fs Tm:Lu(2)O(3) laser at 2.07 µm mode-locked using single-walled carbon nanotubes. Opt. Express **20**(5), 5313–5318 (2012)
86. R.W. Boyd, *Nonlinear Optics*, 3rd edn. (Academic Press, New York, 2008)

87. D.E. Spence, P.N. Kean, W. Sibbett, 60-fsec pulse generation from a self-mode-locked Ti-sapphire laser. Opt. Lett. **16**(1), 42–44 (1991)
88. U. Keller et al., Femtosecond pulses from a continuously self-starting passively mode-locked Ti-sapphire laser. Opt. Lett. **16**(13), 1022–1024 (1991)
89. I.T. Sorokina, K.L. Vodopyanov (eds.), *Solid-State Mid-Infrared Laser Sources*. Topics in Applied Physics (Springer, Berlin, 2003)
90. M. Ebrahim-Zadeh, I.T. Sorokina (eds.), *Mid-Infrared Coherent Sources and Applications*. NATO Science for Peace and Security Series B: Physics and Biophysics. (Springer, Dordrecht, 2008)
91. M.N. Cizmeciyan et al., Kerr-lens mode-locked femtosecond Cr(2+):ZnSe laser at 2420 nm. Opt. Lett. **34**(20), 3056–3058 (2009)
92. F. Salin, J. Squier, M. Piche, Mode-locking of Ti-Al(2)O(3) lasers and self-focusing—a Gaussian approximation. Opt. Lett. **16**(21), 1674–1676 (1991)
93. H.A. Haus, E.P. Ippen, K. Tamura, Additive-pulse modelocking in fiber lasers. IEEE J. Quantum Electron. **30**(1), 200–208 (1994)
94. G. Agrawal, in *Nonlinear Fiber Optics*, 4th edn. (2006)
95. G. Genty, S. Coen, J.M. Dudley, Fiber supercontinuum sources (invited). J. Opt. Soc. Am. B, Opt. Phys. **24**(8), 1771–1785 (2007)
96. P. Russell, Photonic crystal fibers. Science **299**(5605), 358–362 (2003)
97. J.C. Knight, Photonic crystal fibres. Nature **424**(6950), 847–851 (2003)
98. L. Dong, B.K. Thomas, L.B. Fu, Highly nonlinear silica suspended core fibers. Opt. Express **16**(21), 16423–16430 (2008)
99. L.B. Fu, B.K. Thomas, L. Dong, Efficient supercontinuum generations in silica suspended core fibers. Opt. Express **16**(24), 19629–19642 (2008)
100. T. Okuno et al., Silica-based functional fibers with enhanced nonlinearity and their applications. IEEE J. Sel. Top. Quantum Electron. **5**(5), 1385–1391 (1999)
101. J.M. Dudley, S. Coen, Numerical simulations and coherence properties of supercontinuum generation in photonic crystal and tapered optical fibers. IEEE J. Sel. Top. Quantum Electron. **8**(3), 651–659 (2002)
102. P. Domachuk et al., Over 4000 nm bandwidth of mid-IR supercontinuum generation in sub-centimeter segments of highly nonlinear tellurite PCFs. Opt. Express **16**(10), 7161–7168 (2008)
103. J.S. Sanghera, L.B. Shaw, I.D. Aggarwal, Chalcogenide glass-fiber-based mid-IR sources and applications. IEEE J. Sel. Top. Quantum Electron. **15**(1), 114–119 (2009)
104. R. Cherif et al., Highly nonlinear As_2Se_3-based chalcogenide photonic crystal fiber for mid-infrared supercontinuum generation. Opt. Eng. **49**(9), 095002 (2010)
105. W.Q. Zhang et al., Fabrication and supercontinuum generation in dispersion flattened bismuth microstructured optical fiber. Opt. Express **19**(22), 21135–21144 (2011)
106. A. Marandi et al., Mid-infrared supercontinuum generation in tapered chalcogenide fiber for producing octave-spanning frequency comb around 3 μm. Opt. Express **20**(22), 24218–24225 (2012)
107. J.M. Dudley, G. Genty, S. Coen, Supercontinuum generation in photonic crystal fiber. Rev. Mod. Phys. **78**(4), 1135–1184 (2006)
108. J.M. Dudley, J.R. Taylor (eds.), *Supercontinuum Generation in Optical Fibers* (Cambridge University Press, Cambridge, 2010)
109. J.L. Krause, K.J. Schafer, K.C. Kulander, High-order harmonic-generation from atoms and ions in the high-intensity regime. Phys. Rev. Lett. **68**(24), 3535–3538 (1992)
110. A. Lhuillier, P. Balcou, High-order harmonic-generation in rare-gases with a 1-ps 1053-nm laser. Phys. Rev. Lett. **70**(6), 774–777 (1993)
111. P.B. Corkum, Plasma perspective on strong-field multiphoton ionization. Phys. Rev. Lett. **71**(13), 1994–1997 (1993)
112. M. Lewenstein et al., Theory of high-harmonic generation by low-frequency laser fields. Phys. Rev. A **49**(3), 2117–2132 (1994)

113. T. Popmintchev et al., The attosecond nonlinear optics of bright coherent X-ray generation. Nat. Photonics **4**(12), 822–832 (2010)
114. A.K. Mills et al., XUV frequency combs via femtosecond enhancement cavities. J. Phys. B, At. Mol. Opt. Phys. **45**(14), 142001 (2012)
115. D.C. Yost et al., Power optimization of XUV frequency combs for spectroscopy applications. Opt. Express **19**(23), 23483–23493 (2011)
116. M. Kourogi, K. Nakagawa, M. Ohtsu, Wide-span optical frequency comb generator for accurate optical frequency difference measurement. IEEE J. Quantum Electron. **29**(10), 2693–2701 (1993)
117. M. Kourogi, T. Enami, M. Ohtsu, A monolithic optical frequency comb generator. IEEE Photonics Technol. Lett. **6**(2), 214–217 (1994)
118. J. Ye et al., Highly selective terahertz optical frequency comb generator. Opt. Lett. **22**(5), 301–303 (1997)
119. S.A. Diddams et al., Broadband optical frequency comb generation with a phase-modulated parametric oscillator. Opt. Lett. **24**(23), 1747–1749 (1999)
120. T. Suzuki, M. Hirai, M. Katsuragawa, Octave-spanning Raman comb with carrier envelope offset control. Phys. Rev. Lett. **101**(24), 243602 (2008)
121. P. Del'Haye et al., Full stabilization of a microresonator-based optical frequency comb. Phys. Rev. Lett. **101**(5), 053903 (2008)
122. P. Del'Haye et al., Octave spanning tunable frequency comb from a microresonator. Phys. Rev. Lett. **107**(6), 063901 (2011)
123. T.J. Kippenberg, R. Holzwarth, S.A. Diddams, Microresonator-based optical frequency combs. Science **332**(6029), 555–559 (2011)
124. T.B.V. Herr, M.L. Gorodetsky, T.J. Kippenberg, Soliton mode-locking in optical microresonators. arXiv:1211.0733 (2012)
125. K. Saha et al., Modelocking and femtosecond pulse generation in chip-based frequency combs. Opt. Express **21**(1), 1335–1343 (2013)
126. S.B. Papp, S.A. Diddams, Spectral and temporal characterization of a fused-quartz-microresonator optical frequency comb. Phys. Rev. A **84**(5), 053833 (2011)
127. T. Herr et al., Universal formation dynamics and noise of Kerr-frequency combs in microresonators. Nat. Photonics **6**(7), 480–487 (2012)
128. L. Matos et al., Direct frequency comb generation from an octave-spanning, prismless Ti:sapphire laser. Opt. Lett. **29**(14), 1683–1685 (2004)
129. A. Bartels, D. Heinecke, S.A. Diddams, Passively mode-locked 10 GHz femtosecond Ti:sapphire laser. Opt. Lett. **33**(16), 1905–1907 (2008)
130. A. Ruehl et al., 80 W, 120 fs Yb-fiber frequency comb. Opt. Lett. **35**(18), 3015–3017 (2010)
131. I. Hartl et al., *Fully Stabilized GHz Yb-Fiber Laser Frequency Comb*. Advanced Solid State Phonics (Optical Society of America, Washington, 2009)
132. J. Rauschenberger et al., Control of the frequency comb from a mode-locked erbium-doped fiber laser. Opt. Express **10**(24), 1404–1410 (2002)
133. B.R. Washburn et al., Phase-locked, erbium-fiber-laser-based frequency comb in the near infrared. Opt. Lett. **29**(3), 250–252 (2004)
134. B.R. Walton et al., Transportable optical frequency comb based on a mode-locked fibre laser. IET Optoelectron. **2**(5), 182–187 (2008)
135. S. Herrmann et al., Atom optical experiments in the drop tower: a pathfinder for space based precision measurements, in *38th COSPAR Scientific Assembly* (2010)
136. F. Adler et al., Attosecond relative timing jitter and 13 fs tunable pulses from a two-branch Er:fiber laser. Opt. Lett. **32**(24), 3504–3506 (2007)
137. K. Moutzouris et al., Multimilliwatt ultrashort pulses continuously tunable in the visible from a compact fiber source. Opt. Lett. **31**(8), 1148–1150 (2006)
138. A. Gambetta, R. Ramponi, M. Marangoni, Mid-infrared optical combs from a compact amplified Er-doped fiber oscillator. Opt. Lett. **33**(22), 2671–2673 (2008)
139. N. Coluccelli et al., 250-MHz synchronously pumped optical parametric oscillator at 2.25–2.6 μm and 4.1–4.9 μm. Opt. Express **20**(20), 22042–22047 (2012)

140. D. Chao et al., Self-referenced erbium fiber laser frequency comb at a GHz repetition rate—OSA technical digest, in *Optical Fiber Communication Conference OW1C.2*, (2012)

141. F. Adler, S.A. Diddams, High-power, hybrid Er:fiber/Tm:fiber frequency comb source in the 2 μm wavelength region. Opt. Lett. **37**(9), 1400–1402 (2012)

142. R.A. Kaindl et al., Broadband phase-matched difference frequency mixing of femtosecond pulses in GaSe: experiment and theory. Appl. Phys. Lett. **75**(8), 1060–1062 (1999)

143. S.M. Foreman, D.J. Jones, J. Ye, Flexible and rapidly configurable femtosecond pulse generation in the mid-IR. Opt. Lett. **28**(5), 370–372 (2003)

144. C. Erny et al., Mid-infrared difference-frequency generation of ultrashort pulses tunable between 3.2 and 4.8 μm from a compact fiber source. Opt. Lett. **32**(9), 1138–1140 (2007)

145. L. Nugent-Glandorf et al., Mid-infrared virtually imaged phased array spectrometer for rapid and broadband trace gas detection. Opt. Lett. **37**(15), 3285–3287 (2012)

146. J.H. Sun, B.J.S. Gale, D.T. Reid, Composite frequency comb spanning 0.4–2.4 μm from a phase-controlled femtosecond Ti:sapphire laser and synchronously pumped optical parametric oscillator. Opt. Lett. **32**(11), 1414–1416 (2007)

147. D.T. Reid, B.J.S. Gale, J. Sun, Frequency comb generation and carrier-envelope phase control in femtosecond optical parametric oscillators. Laser Phys. **18**(2), 87–103 (2008)

148. S.T. Wong, K.L. Vodopyanov, R.L. Byer, Self-phase-locked divide-by-2 optical parametric oscillator as a broadband frequency comb source. J. Opt. Soc. Am. B, Opt. Phys. **27**(5), 876–882 (2010)

149. S. Marzenell, R. Beigang, R. Wallenstein, Synchronously pumped femtosecond optical parametric oscillator based on AgGaSe$_2$ tunable from 2 μm to 8 μm. Appl. Phys. B, Lasers Opt. **69**(5–6), 423–428 (1999)

150. E. Hecht, *Optics*, 3rd edn. (Addison-Wesley, Reading, 1998)

151. I. Hartl et al., Cavity-enhanced similariton Yb-fiber laser frequency comb: 3×10^{14} W cm^{-2} peak intensity at 136 MHz. Opt. Lett. **32**(19), 2870–2872 (2007)

152. R.J. Jones, I. Thomann, J. Ye, Precision stabilization of femtosecond lasers to high-finesse optical cavities. Phys. Rev. A **69**(5), 051803 (2004)

153. R.W.P. Drever et al., Laser phase and frequency stabilization using an optical-resonator. Appl. Phys., B Photophys. Laser Chem. **31**(2), 97–105 (1983)

154. T.W. Hansch, B. Couillaud, Laser frequency stabilization by polarization spectroscopy of a reflecting reference cavity. Opt. Commun. **35**(3), 441–444 (1980)

155. A. Foltynowicz et al., Quantum-noise-limited optical frequency comb spectroscopy. Phys. Rev. Lett. **107**(23), 233002 (2011)

156. A. Foltynowicz et al., Cavity-enhanced optical frequency comb spectroscopy in the mid-infrared application to trace detection of hydrogen peroxide. Appl. Phys. B **110**(2), 163–175 (2013)

157. R. Gebs et al., 1-GHz repetition rate femtosecond OPO with stabilized offset between signal and idler frequency combs. Opt. Express **16**(8), 5397–5405 (2008)

158. K.K. Lehmann, P.S. Johnston, P. Rabinowitz, Brewster angle prism retroreflectors for cavity enhanced spectroscopy. Appl. Opt. **48**(16), 2966–2978 (2009)

159. S. Kassi et al., Demonstration of cavity enhanced FTIR spectroscopy using a femtosecond laser absorption source. Spectrochim. Acta, Part A, Mol. Biomol. Spectrosc. **75**(1), 142–145 (2010)

160. R. Grilli et al., Frequency comb based spectrometer for in situ and real time measurements of IO, BrO, NO$_2$, and H$_2$CO at pptv and ppqv levels. Environ. Sci. Technol. **46**(19), 10704–10710 (2012)

161. R. Grilli et al., First investigations of IO, BrO, and NO$_2$ summer atmospheric levels at a coastal East Antarctic site using mode-locked cavity enhanced absorption spectroscopy. Geophys. Res. Lett. **40**, 1–6 (2013)

162. R. Grilli et al., Cavity-enhanced multiplexed comb spectroscopy down to the photon shot noise. Phys. Rev. A **85**(5), 051804 (2012)

163. M. Shirasaki, Large angular dispersion by a virtually imaged phased array and its application to a wavelength demultiplexer. Opt. Lett. **21**(5), 366–368 (1996)

164. S.A. Diddams, L. Hollberg, V. Mbele, Molecular fingerprinting with the resolved modes of a femtosecond laser frequency comb. Nature **445**(7128), 627–630 (2007)
165. J. Mandon, G. Guelachvili, N. Picque, Fourier transform spectroscopy with a laser frequency comb. Nat. Photonics **3**(2), 99–102 (2009)
166. S. Schiller, Spectrometry with frequency combs. Opt. Lett. **27**(9), 766–768 (2002)
167. I. Coddington, W.C. Swann, N.R. Newbury, Coherent multiheterodyne spectroscopy using stabilized optical frequency combs. Phys. Rev. Lett. **100**(1), 013902 (2008)
168. B. Bernhardt et al., Cavity-enhanced dual-comb spectroscopy. Nat. Photonics **4**(1), 55–57 (2010)
169. X.D.D. Vaernewijck et al., Cavity enhanced FTIR spectroscopy using a femto OPO absorption source. Mol. Phys. **109**(17–18), 2173–2179 (2011)
170. C. Gohle et al., Frequency comb vernier spectroscopy for broadband, high-resolution, high-sensitivity absorption and dispersion spectra. Phys. Rev. Lett. **99**(26), 263902 (2007)
171. T. Steinmetz et al., Laser frequency combs for astronomical observations. Science **321**(5894), 1335–1337 (2008)
172. T. Steinmetz et al., Fabry-Perot filter cavities for wide-spaced frequency combs with large spectral bandwidth. Appl. Phys. B, Lasers Opt. **96**(2–3), 251–256 (2009)
173. F. Quinlan et al., A 12.5 GHz-spaced optical frequency comb spanning >400 nm for near-infrared astronomical spectrograph calibration. Rev. Sci. Instrum. **81**(6), 063105 (2010)
174. A. Okeefe, D.A.G. Deacon, Cavity ring-down optical spectrometer for absorption-measurements using pulsed laser sources. Rev. Sci. Instrum. **59**(12), 2544–2551 (1988)
175. J.J. Scherer, Ringdown spectral photography. Chem. Phys. Lett. **292**(1–2), 143–153 (1998)
176. J.J. Scherer et al., Broadband ringdown spectral photography. Appl. Opt. **40**(36), 6725–6732 (2001)
177. S.M. Ball et al., Broadband cavity ringdown spectroscopy of the NO_3 radical. Chem. Phys. Lett. **342**(1–2), 113–120 (2001)
178. S.J. Xiao, A.M. Weiner, 2-D wavelength demultiplexer with potential for ≥1000 channels in the C-band. Opt. Express **12**(13), 2895–2902 (2004)
179. S.X. Wang, S.J. Xiao, A.M. Weiner, Broadband, high spectral resolution 2-D wavelength-parallel polarimeter for dense WDM systems. Opt. Express **13**(23), 9374–9380 (2005)
180. P.R. Griffiths, J.A. de Haseth, *Fourier Transform Infrared Spectrometry* (Wiley, Hoboken, 2007)
181. A.A. Ruth, J. Orphal, S.E. Fiedler, Fourier-transform cavity-enhanced absorption spectroscopy using an incoherent broadband light source. Appl. Opt. **46**(17), 3611–3616 (2007)
182. P. Balling et al., Length and refractive index measurement by Fourier transform interferometry and frequency comb spectroscopy. Meas. Sci. Technol. **23**(9), 094001 (2012)
183. J. Mandon et al., Femtosecond laser Fourier transform absorption spectroscopy. Opt. Lett. **32**(12), 1677–1679 (2007)
184. X.D.D. Vaernewijck, S. Kassi, M. Herman, (OCO)-O-17-C-12-O-17 and (OCO)-O-18-C-12-O-17 overtone spectroscopy in the 1.64 µm region. Chem. Phys. Lett. **514**(1–3), 29–31 (2011)
185. X.D.D. Vaernewijck, S. Kassi, M. Herman, (OCO)-O-17-C-12-O-17 and (OCO)-O-18-C-12-O-17 spectroscopy in the 1.6 µm region. Mol. Phys. **110**(21–22), 2665–2671 (2012)
186. P.C.D. Hobbs, Ultrasensitive laser measurements without tears. Appl. Opt. **36**(4), 903–920 (1997)
187. S.A. Davis, M. Abrams, J. Brault, *Fourier Transform Spectrometry* (Academic Press, San Diego, 2001), p. 262
188. A. Schliesser et al., Frequency-comb infrared spectrometer for rapid, remote chemical sensing. Opt. Express **13**(22), 9029–9038 (2005)
189. N.R. Newbury, I. Coddington, W. Swann, Sensitivity of coherent dual-comb spectroscopy. Opt. Express **18**(8), 7929–7945 (2010)
190. T. Ideguchi et al., Adaptive dual-comb spectroscopy in the green region. Opt. Lett. **37**(23), 4847–4849 (2012)

191. S. Boudreau, J. Genest, Referenced passive spectroscopy using dual frequency combs. Opt. Express **20**(7), 7375–7387 (2012)

192. J.D. Deschenes, P. Giaccari, J. Genest, Optical referencing technique with CW lasers as intermediate oscillators for continuous full delay range frequency comb interferometry. Opt. Express **18**(22), 23358–23370 (2010)

193. A. Schliesser, N. Picque, T.W. Hansch, Mid-infrared frequency combs. Nat. Photonics **6**(7), 440–449 (2012)

194. T.H. Risby, S.F. Solga, Current status of clinical breath analysis. Appl. Phys. B, Lasers Opt. **85**(2–3), 421–426 (2006)

195. L. Pauling et al., Quantitative analysis of urine vapor and breath by gas-liquid partition chromatography. Proc. Natl. Acad. Sci. USA **68**(10), 2374–2376 (1971)

196. A.J. Cunnington, P. Hormbrey, Breath analysis to detect recent exposure to carbon monoxide. Postgrad. Med. J. **78**(918), 233–237 (2002)

197. E.R. Crosson et al., Stable isotope ratios using cavity ring-down spectroscopy: determination of C-13/C-12 for carbon dioxide in human breath. Anal. Chem. **74**(9), 2003–2007 (2002)

198. B.J. Marshall, J.R. Warren, Unidentified curved bacilli in the stomach of patients with gastritis and peptic ulceration. Lancet **1**(8390), 1311–1315 (1984)

199. Helicobacter pylori: fact sheet for health care providers. Center for Disease Control, Atlanta, GA (1998)

200. S.Y. Lehman, K.A. Bertness, J.T. Hodges, Detection of trace water in phosphine with cavity ring-down spectroscopy. J. Cryst. Growth **250**(1–2), 262–268 (2003)

201. J. Feng, R. Clement, M. Raynor, Characterization of high-purity arsine and gallium arsenide epilayers grown by MOCVD. J. Cryst. Growth **310**(23), 4780–4785 (2008)

202. H.H. Funke et al., Techniques for the measurement of trace moisture in high-purity electronic specialty gases. Rev. Sci. Instrum. **74**(9), 3909–3933 (2003)

203. I. Horvath et al., Combined use of exhaled hydrogen peroxide and nitric oxide in monitoring asthma. Am. J. Respir. Crit. Care Med. **158**(4), 1042–1046 (1998)

204. S.A. Kharitonov, P.J. Barnes, Exhaled biomarkers. Chest **130**(5), 1541–1546 (2006)

205. S.A. Kharitonov, P.J. Barnes, Exhaled markers of pulmonary disease. Am. J. Respir. Crit. Care Med. **163**(7), 1693–1722 (2001)

206. P.S. Connell, D.J. Wuebbles, J.S. Chang, Stratospheric hydrogen-peroxide—the relationship of theory and observation. J. Geophys. Res., Atmos. **90**(Nd6), 10726–10732 (1985)

207. J.A. Snow et al., Hydrogen peroxide, methyl hydroperoxide, and formaldehyde over North America and the North Atlantic. J. Geophys. Res., Atmos. **112**(D12), D12s07 (2007)

208. C.P. Rinsland et al., Detection of elevated tropospheric hydrogen peroxide (H_2O_2) mixing ratios in atmospheric chemistry experiment (ACE) subtropical infrared solar occultation spectra. J. Quant. Spectrosc. Radiat. Transf. **107**(2), 340–348 (2007)

209. T.J. Johnson et al., Absolute integrated intensities of vapor-phase hydrogen peroxide ($H(2)O(2)$) in the mid-infrared at atmospheric pressure. Anal. Bioanal. Chem. **395**(2), 377–386 (2009)

210. J.D. Rogers, Calculation of absolute infrared intensities of binary overtone, combination, and difference bands of hydrogen-peroxide. J. Phys. Chem. **88**(3), 526–530 (1984)

211. L.S. Rothman et al., The HITRAN 2008 molecular spectroscopic database. J. Quant. Spectrosc. Radiat. Transf. **110**(9–10), 533–572 (2009)

212. R. Ciurylo, Shapes of pressure- and Doppler-broadened spectral lines in the core and near wings. Phys. Rev. A **58**(2), 1029–1039 (1998)

213. J.-M. Hartmann, C. Boulet, D. Robert, *Collisional Effects on Molecular Spectra: Laboratory Experiments and Model, Consequences for Applications* (Elsevier, Amsterdam, 2008)

214. T.A. Staffelbach et al., Comparison of hydroperoxide measurements made during the Mauna Loa observatory photochemistry experiment 2. J. Geophys. Res., Atmos. **101**(D9), 14729–14739 (1996)

215. P.J. Sarre, The diffuse interstellar bands: a major problem in astronomical spectroscopy. J. Mol. Spectrosc. **238**(1), 1–10 (2006)

216. T.P. Snow, V.M. Bierbaum, Ion chemistry in the interstellar medium. Annu. Rev. Anal. Chem. **1**, 229–259 (2008)
217. S. Schiller, V. Korobov, Tests of time independence of the electron and nuclear masses with ultracold molecules. Phys. Rev. A **71**(3), 032505 (2005)
218. E.R. Meyer, J.L. Bohn, M.P. Deskevich, Candidate molecular ions for an electron electric dipole moment experiment. Phys. Rev. A **73**(6), 062108 (2006)
219. J.C.J. Koelemeij et al., Vibrational spectroscopy of HD^+ with 2-ppb accuracy. Phys. Rev. Lett. **98**(17), 173002 (2007)
220. L.V. Skripnikov et al., On the search for time variation in the fine-structure constant: ab initio calculation of HfF^+. JETP Lett. **88**(9), 578–581 (2008)
221. K. Beloy et al., Rotational spectrum of the molecular ion NH^+ as a probe for alpha and $m(e)/m(p)$ variation. Phys. Rev. A **83**(6), 062514 (2011)
222. A.E. Leanhardt et al., High-resolution spectroscopy on trapped molecular ions in rotating electric fields: a new approach for measuring the electron electric dipole moment. J. Mol. Spectrosc. **270**(1), 1–25 (2011)
223. B.J. Barker et al., Communication: spectroscopic measurements for HfF^+ of relevance to the investigation of fundamental constants. J. Chem. Phys. **134**(20), 201102 (2011)
224. B.J. Barker et al., Spectroscopic investigations of ThF and ThF^+. J. Chem. Phys. **136**(10), 104305 (2012)
225. A.N. Petrov, N.S. Mosyagin, A.V. Titov, Theoretical study of low-lying electronic terms and transition moments for HfF^+ for the electron electric-dipole-moment search. Phys. Rev. A **79**(1), 012505 (2009)

Chapter 9
Whispering Gallery Mode Biomolecular Sensors

Yuqiang Wu and Frank Vollmer

Abstract Optical resonator-based biosensors are emerging as one of the most sensitive microsystem biodetection technology that boasts all of the capabilities for a next-generation lab-on-chip device: label-free detection down to single molecules, multiplexed sensing capability, operation in aqueous environment as well as cost-effective integration on microchips. A scholarly introduction to the emerging field of whispering gallery mode resonator-based biosensors is given and their current applications are reviewed.

9.1 Nano-Biotechnology: Sensors Interface the Molecular World

The ability to fabricate devices with micro-to nanoscale precision using the tools of nanotechnology is revolutionizing biotechnology and is spawning the new field of nanobiotechnology. In nanobiotechnology, researchers harness the physics and utilize engineering of nanoscale devices to interface and probe the molecular world. Here, the fabrication of miniature biosensor elements is particularly important since the increase in surface to volume ratio of micro-to nanoscale devices can dramatically boost their sensitivity, for example of mechanical [1, 2], electrical [3, 4], photonic and plasmonic [5–7] biosensors. This unprecedented detection capability of nanobiotechnology-enabled sensors is not only revolutionizing clinical diagnostics [8] but is also opening the door for detailed explorations of the molecular world, the world of biomolecules, single molecules, molecular machines, and their interactions [9]. In fact the development of some of these nanobiosensors has already enabled detection down to single virus particles and even single molecules [3, 5, 10, 11].

For all of the proposed biosensing applications it is most important to achieve highly sensitive detection capability in *aqueous* solution since almost all biological and clinical samples are water-based. Furthermore, biomolecules function only in

Y. Wu (✉) · F. Vollmer
Laboratory of Nanophotonics & Biosensing, Max Planck Institute for the Science of Light,
G. Scharowksy Str. 1 / Bau 24, 91058 Erlangen, Germany
e-mail: yuqiang.wu@mpl.mpg.de

G. Gagliardi, H.-P. Loock (eds.), *Cavity-Enhanced Spectroscopy and Sensing*,
Springer Series in Optical Sciences 179, DOI 10.1007/978-3-642-40003-2_9,
© Springer-Verlag Berlin Heidelberg 2014

nanobiotechnology virus sensors
based on

optical resonator   mechanical resonator   nanowire

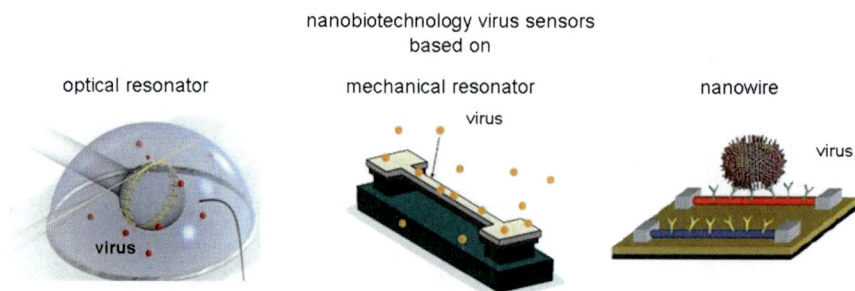

Fig. 9.1 Nanobiotechnology-enabled single virus detectors. From left to right: optical resonator, nanomechanical resonator and nanowire sensor. Adapted from [10, 11, 13]

Table 9.1 Overview of the capabilities of nanobiotechnology-enabled sensors: optical resonator, nanomechanical resonator and nanowire biosensors. MW = molecular weight

	Optical resonator	Mechanical resonator	Nanowire sensor
transduction scheme	optical frequency shift [12]	mechanical frequency shift [11]	conductance change [3]
sensitive to	polarizability, proportional to MW [14, 15]	MW [2]	charge [3]
operation in water	yes [12, 15, 16]	limited [17]	yes [18]
detection by molecular recognition	yes [15, 19, 20]	possible [17]	yes [18]
single molecule detection capability	possible [12]	demonstrated in vacuum [11]	possible [21]
single virus detection	demonstrated for Influenza A virus [10]	demonstrated for virus in vacuum [22]	demonstrated for Influenza A virus [13]
microfluidic integration	yes [23]	limited [17]	yes [21]
multiplexing	yes [24]	yes [2]	yes [3]
fabrication on chip	bottom-up [15] as well as top down [25]	top-down [2]	bottom-up [3]

aqueous environments and also clinical diagnostic assays rely on specific molecular recognition *in solution*. It is equally important to achieve sensitive detection *in real-time*, *without the need for labels*, on fully integrated and automated chip-scale devices.

Optical resonator biosensors, also called optical microcavities, are emerging as one of the most sensitive micro-/nanosystems biodetection technology that boasts all of the desired capabilities [12]: label-free detection, potential single molecule sensitivity, multiplexed sensing capability, operation in aqueous environment and cost-effective integration on microchips. Figure 9.1 and Table 9.1 compare optical resonator biosensors with other nanobiotechnology-enabled sensors that have been developed, in this example, for label-free single virus detection.

The chapter is organized as follows: in Sect. 9.2, a brief introduction to optical microcavities and their sensing principle will be given. In Sect. 9.3, we will present results for detecting protein and Influenza A virus and demonstrate the multiplexed sensing capability. We will also highlight the recent development for achieving single molecule sensitivity in a hybrid photonic-plasmonic sensor. An outlook for this type of sensor will be given at the end of the chapter.

9.2 Whispering Gallery Mode Resonator Biosensors

9.2.1 Whispering Gallery Mode

Perhaps the simplest example for an optical microcavity is a glass microbead, about ~ 100 μm in diameter, which is used to trap a light ray by total internal reflection (TIR). TIR is a well known optical effect that is observed for light reflected at the surface of a glass prism if the incident angle θ is larger than the critical angle $\theta_c = \arcsin(n_{air}/n_{glass})$, see Fig. 9.2 left. The idea of continuous TIR of waves was first considered to explain the observation of acoustical phenomenon in whispering galleries. There are over ten acoustical whispering galleries around the world, the oldest of which is in China (1400 BC), the largest of which is in St. Peters, Vatican City and the most famous of which is St. Paul's Cathedral in London, see Fig. 9.2 right. In a whispering gallery, sound waves that bounce off a curved surface with minimal diffraction are efficiently reflected so that they strike the wall again at the same angle and, thereby, travel along the surface of the wall. For the case of the microsphere, multiple TIRs guide the light around the sphere's circumference so that the returning light wave starts to interfere with itself, resulting in a traveling wave called 'Whispering Gallery Mode (WGM)' that is turning inside the cavity like a 'gear'. Light can only remain confined on this trajectory if the interference is constructive which means that an exact integer number N of wavelengths λ have to fit on the closed optical path.

The electromagnetic field in a microsphere can be described by a Helmholtz equation $\nabla^2 E + k^2 E = 0$, where E is the transverse electric (TE) or transverse magnetic (TM) component of the light field and $k = \omega/c$ is the wave number. Here, we only consider the TE modes and a similar process can be used to study TM modes. The TE modes can be written in spherical coordinates as solutions to

$$\frac{1}{r^2}\frac{\partial}{\partial r}\left(r^2\frac{\partial E}{\partial r}\right) + \frac{1}{r^2\sin\theta}\frac{\partial}{\partial\theta}\left(\sin\theta\frac{\partial E}{\partial\theta}\right) + \frac{1}{r^2\sin^2\theta}\frac{\partial^2 E}{\partial\phi} + k^2 E = 0 \quad (9.1)$$

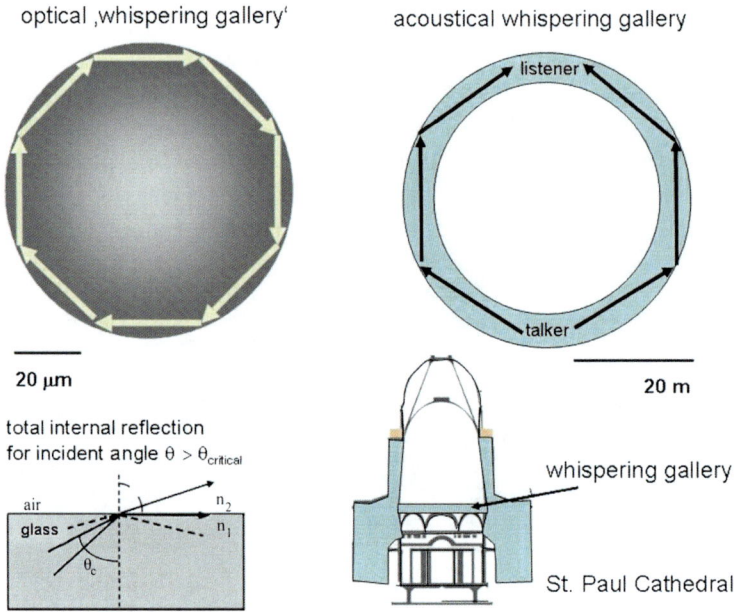

Fig. 9.2 *Left*: Total-internal reflection of light in a microsphere and at a prism. *Right*: Acoustic 'Whispering Gallery' in St. Paul's Cathedral

Equation (9.1) can be solved by a separation of variables, $E = \Psi_r(r)\Psi_\phi(\phi)\Psi_\theta(\theta)$. The solutions are given as

$$
\begin{cases}
\Psi_r = \begin{cases} A[j_l(kr)], & \text{for } r \leq R_0, \\ B\exp[-\alpha(r-R_0)], & \text{for } r > R_0, \end{cases} \\
\Psi_\phi(\phi) = N_\phi \exp(im\phi), \\
\Psi_\theta(\theta) = N_\theta P_l^m(\cos\theta),
\end{cases}
\tag{9.2}
$$

where A and B are constants determined by the boundary conditions at the surface and the normalization condition, N_φ and N_θ are normalization constants, R_0 is the radius of the sphere, j_l is the spherical Bessel function of the first kind, $P_m^l(\cos\theta)$ is an associated Legendre polynomial, $\alpha = \sqrt{\frac{l(l+1)}{R_0^2} - k^2}$, and $l \ (= 0, 1, 2, \ldots)$ and $m \ (= -l, l-1, \ldots, 0, l+1, l)$ are the angular and azimuthal mode numbers. They describe the field intensity distribution in the polar direction and the number of maxima in the sinusoidal variation of the field intensity in the azimuthal direction, respectively. Figure 9.3(A) shows an example for the radial field distribution and Fig. 9.3(B) shows the polar field. More detailed explanation of the physics of WGM can be found in [26].

Each solution to Eq. (9.1) is associated with a specific resonance frequency or wavelength of this respective WGM mode which is characterized by its radial (n),

Fig. 9.3 (**A**) Radial field distribution of WGMs for $m = 85$ and radial number $n = 1$. (**B**) Polar field distribution of WGMs with the same mode numbers (listed on the *upper left corner*). The *dashed red/white line* is the air/glass interface (Color figure online)

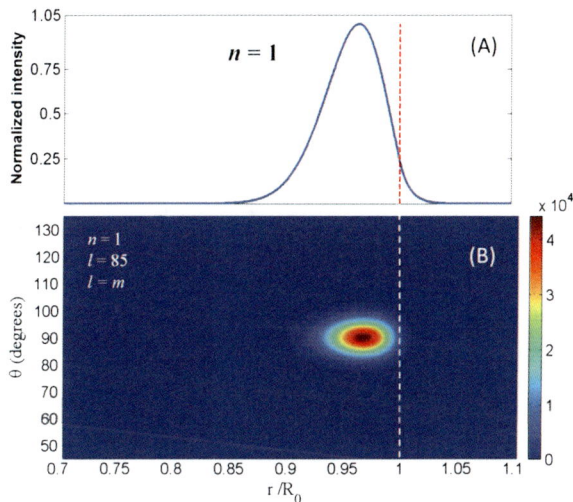

angular (l) and azimuthal (m) numbers. From Fig. 9.3, one can see that the length of the WGMs evanescent tail extending from the silica sphere is small, ~300 nm in aqueous solution. The attachment of even smaller biomolecules (several nanometers in diameter) within the evanescent field will cause a resonance wavelength red shift. The WGM red shift occurs since the biomolecule becomes polarized (at optical frequency) once attached within the evanescent field. Effectively, part of the light field is then pulled towards the outside of the microsphere, increasing the optical path length upon biomolecular binding. The details of this reactive sensing mechanism will be introduced in the following.

Other examples of optical resonators are silicon ring resonators [25, 27–30], toroidal glass resonators [31–35], and hollow glass capillaries [29, 36–39]. The different geometries and materials have all their unique advantages in biosensing: microspheres are extremely simple to fabricate, exhibit the highest quality (Q) factor and were the first optical resonator used in biosensing [5, 15, 40, 41]. Silicon ring resonators are micro-fabricated using top-down photolithography combined with reactive ion etching and are easily integrated into larger sensor arrays [20, 25, 30, 42–47]. Hollow-core glass capillaries that confine light along their perimeter, so called liquid core optical ring resonators, LCORRs, are sensitive to molecules binding to their interior, are fabricated by pulling thin-walled capillaries and are easily combined with microfluidics [36, 38, 39, 48–52]. The availability of the many geometries and materials that can be used for label-free biosensing on chip-scale devices makes the optical resonator microsystem biodetection technology particularly attractive for many research groups around the world that have an interest in engineering these sensitive devices for applications in the life sciences, healthcare, point-of-care, environmental monitoring and emergency response. Table 9.2 gives an overview over the variety of WGM resonator biosensors that are being championed for biosensing applications.

Table 9.2 Examples for optical resonator biosensors that have been utilized or proposed in biosensing applications

Optical resonator	Device example	Potential single molecule resolution	Multiplexing capability	Resonator functionalization
microsphere waveguide coupled [5, 15, 53–58]		proposed	limited	dip-coating, (hang-drop) [41]
microsphere, prism coupled [40]		–	yes	bulk chemistry, micropatterning
microsphere, angle polished fiber coupled [59, 60]		–	no	no biosensor demonstration yet
microtoroid, fiber coupled [19, 31, 32, 61]		proposed	limited	microfluidics, micropatterning
ring resonator, waveguide coupled [25, 28, 62–65]		not proposed	yes	microfluidics or micropattern-ing [66]
fluorescent microsphere [67–72]		no	yes	bulk chemistry
capillary, fiber coupled (LCORR) [36, 50]		not proposed	limited	microfluidics
disk resonator [73, 74], waveguide coupled		not proposed	yes	micropatterning, microfluidics

Table 9.2 (Continued)

Optical resonator	Device example	Potential single molecule resolution	Multiplexing capability	Resonator functionalization
bottleneck resonator [75]		not proposed	limited	no biosensor demonstration yet
microtube ring resonators [76]		not proposed	yes	microfluidics
photonic-plasmonic WGM: microsphere coupled to nanoantenna [77–80]		yes	yes	micropatterning, bulk chemistry

9.2.2 Reactive Sensing Principle

9.2.2.1 A Simple Geometric Model for Monolayer Binding

Light remains confined within the microsphere and recirculates for many tens of thousands of times since the dominant absorption loss in silica glass is typically less than ~7 dB/km. Furthermore, the number of total internal reflections per orbit of a WGM is typically very large (many more than depicted in Fig. 9.4) so that the polygonal optical path resembles more that of a circular optical path circumnavigating close to the microsphere surface as shown in Fig. 9.4. This simple picture leads us to the approximate WGM resonance condition:

$$N \times \frac{\lambda_r}{n} = 2\pi R_0, \qquad (9.3)$$

which states that N number of wavelengths λ_r have to fit on the circular optical path of length $2\pi R_0$, with microsphere radius R_0 and microsphere refractive index n. The resonance frequency of the WGM is then calculated as:

$$\omega = 2\pi f = 2\pi \frac{c}{\lambda_r} = \frac{cN}{nR_0}.$$

Similar to a nanomechanical resonator biosensor, the miniature optical resonator detects the binding of analyte molecules from changes in the WGM resonance frequency. The binding of a biomolecule will shift the resonance frequency by a miniscule amount, see Fig. 9.4. The shift occurs since the bound DNA or protein

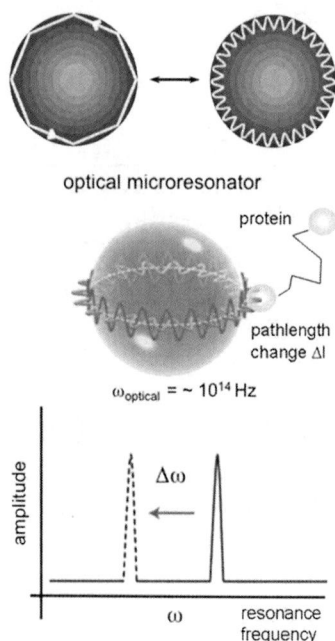

optical microresonator

protein

pathlength
change Δl

$\omega_{optical} = \sim 10^{14}$ Hz

biomolecule will 'pull' part of the optical field to the outside of the microsphere, effectively increasing the path length by $2\pi \Delta l$. This increase in path length produces a shift $\Delta\omega$ in resonance frequency ω according to:

$$\frac{\Delta\omega}{\omega} = -\frac{2\pi \Delta l}{2\pi R_0} = \frac{\Delta l}{R_0} \qquad (9.4)$$

Figure 9.5 shows the first experimental setup that was built to measure the WGM resonant frequency shift upon specific binding of Streptavidin protein molecules from solution [15, 41]. The microsphere resonators used in this demonstration were fabricated by simply melting the tip of an optical fiber either with a hydrogen torch or a CO_2 laser. In both cases the glass is heated and softened so that surface tension forms a spherical object sitting on the fiber stem. The fiber stem allows positioning and alignment of the microsphere with respect to a tapered optical fiber that is used to excite the WGM along the microsphere equator, where the orbiting light does not interfere with the stem. This tapered optical fiber remains in contact with the microsphere for evanescent coupling. A continuous-wave near infrared tunable distributed feedback DFB laser is coupled into one end of this taper and used to probe for the resonant frequency of the microsphere. As the laser frequency is tuned and matches the resonant frequency of the microsphere, the light evanescently couples into the WGM and no longer transmits to the other end of the fiber. A photodetector connected to the fiber end now records a Lorentzian dip in intensity for every WGM. This transmission spectrum (Fig. 9.5b) is recorded every millisecond in 'real-time' for each scan of the laser and is stored and analyzed on a computer. Sev-

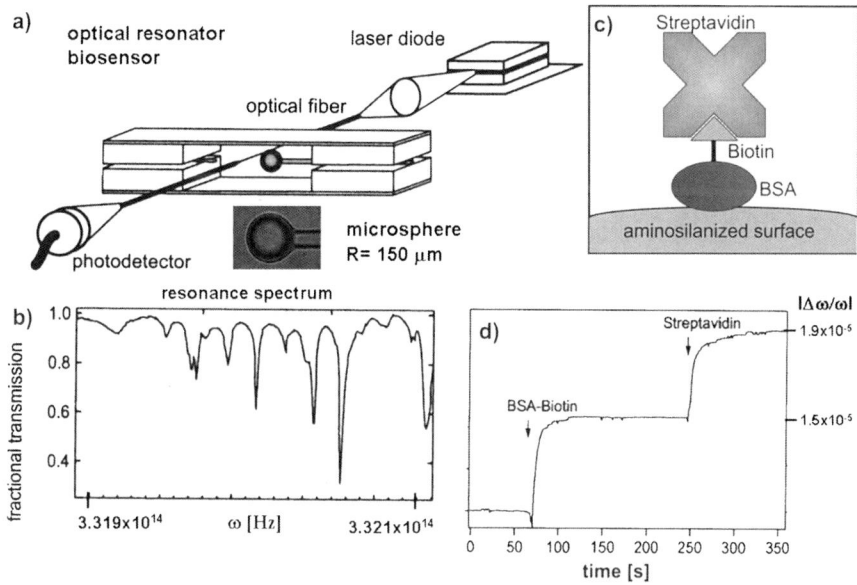

Fig. 9.5 (**a**) First experimental setup to demonstrate optical resonance WGM frequency shifts for specific detection of a biomolecule called Streptavidin, adapted from [15]. (**b**) WGM resonances appear as Lorentzian dips in the fiber-coupled microsphere transmission spectrum. (**c**) Specific detection of Streptavidin is achieved by molecular recognition through the receptor element BSA-Biotin. (**d**) Resonance frequency shifts vs time for Streptavidin binding to BSA-biotin forming monolayer surface coverage at saturation. The *arrow* indicates the time point at which Streptavidin was added to the liquid sample cell

eral Lorentzian dips corresponding to different WGMs are observed in the transmission spectrum for one scan of the laser frequency. For the biosensing experiments discussed in the following it is sufficient to monitor only one of the WGMs.

It was shown that the binding of Streptavidin molecules to the microsphere from solution will induce a red shift of the WGM resonance frequency [15]. The shift in resonance frequency is plotted as $|\frac{\Delta\omega}{\omega}|$ versus time. Such a plot is shown in Fig. 9.4d. The frequency shift signal increases and saturates as Streptavidin molecules occupy all of the binding sites provided by previously surface-immobilized biorecognition elements. In this example, BSA-biotin (bovine serum albumin protein conjugated to biotin) has been used as the specific recognition element for Streptavidin, see Fig. 9.4c. As the frequency shift signal saturates, a complete monolayer of Streptavidin molecules forms on top of the BSA-Biotin layer. At saturation, Streptavidin forms an added optically contiguous layer which exhibits the same refractive index as the glass microsphere ($n \sim 1.45$). The monolayer thus effectively increases the circumference of the glass microsphere by $2\pi \Delta t$ where Δt is the diameter of a Streptavidin molecule with $\sim$2.5 nm radius of gyration. The frequency shift $|\frac{\Delta\omega}{\omega}|$ associated with Streptavidin monolayer formation on the microspheres ($R = 150$ μm) is estimated from Eq. (9.4):

$$\left| \frac{\Delta\omega}{\omega} \right| = \left| -\frac{\Delta l}{R_0} \right| = \left| -\frac{2.5\text{ nm}}{150000\text{ nm}} \right| = 1.45 \times 10^{-5}, \qquad (9.5)$$

which is in good agreement with the measurement presented in Fig. 9.5d.

Detecting the frequency shift associated with monolayer formation is an easy task for our optical resonator biosensor since the WGM line width $\Delta\omega_{line}$ is much smaller than the frequency shift $\Delta\omega$. The narrow line width of a WGM resonator is associated with the high quality Q factor of the optical resonance. The Q factor is determined from the line width of the resonance as $Q = \omega/\Delta\omega_{line}$, about 2×10^6 in this example. In fact, Q factors of more than 10^7 have been obtained for optical resonators immersed in aqueous environments and ultimate Q's only limited by the absorption in the glass material of up to 10^{10} have been reported in $\sim$100 μm glass microspheres in air [81]. A single Streptavidin monolayer shifts the WGM resonance line of a $Q = 10^7$ and $R = 150$ μm microsphere through 100 line widths, and through about 10^5 line widths for a $Q = 10^{10}$, certainly an easy signal to detect! Because of their large Q factors optical resonator biosensors surpass the detection limits of commercial label-free biosensors such as quartz crystal microbalances and surface plasmon resonance detectors. The extremely sensitive method for detecting biomolecules from frequency shift measurements in high Q resonators has been coined the 'reactive sensing principle' [5, 16] since this principle applies to any optical resonator and to any biomolecule. We note that the Q factor in solution is reduced to about 10^{7-8} due to absorption in water; however, such Q factors are still high enough to reach single molecule detection, if the signal can be boosted with plamonic nanoantennas [12, 78, 80, 101], an approach that will be discussed in following sections.

9.2.2.2 Reactive Sensing Principle

To understand the light-matter interaction of a molecule with the microsphere cavity we have to take a closer look at the perturbation induced by the biomolecule bound at the resonator surface. Once bound to the equator of the microsphere, a Streptavidin protein molecule is exposed to the evanescent field generated in the sphere by TIR. Similar to the case of a prism, the evanescent field of a WGM extends about a few hundred nm into the surrounding aqueous medium (Fig. 9.9 (top)). Once bound within this evanescent field, every atom of the Streptavidin molecule (Fig. 9.9 (bottom)) will be polarized (at optical frequency) and the overall induced dipole moment in excess of the displaced water in response to the electric field strength, $E_0(r)e^{i\omega t}$, is calculated as $\delta P = \alpha_{ex} E_0(r_i)$, where α_{ex} is the real part of the excess polarizability of Streptavidin (here $\alpha_{ex} = 4\pi\epsilon_0 \times 3.3 \times 10^{-21}$ cm^3) [16]. Considering the energy of interaction as a first-order perturbation, the shift in the photon energy of the resonant state caused by the induced dipole moment is given by $\hbar\Delta\omega = -\delta P \cdot \frac{E_0^*(r_i)}{2}$. The fractional frequency shift for a Streptavidin protein

Fig. 9.6 The reactive sensing principle. A molecule binding to the microsphere surface is polarized within the evanescent field (*yellow ring*) of a WGM optical resonance. The energy that is needed to polarize the molecule causes the resonance frequency shift (adapted from [100]) (Color figure online)

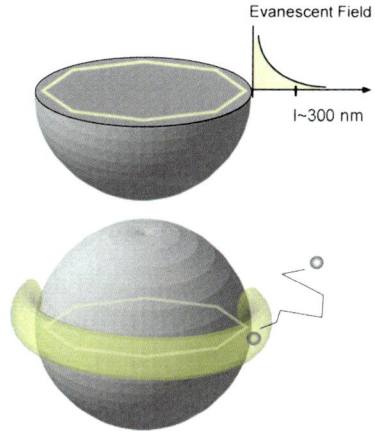

positioned at r_i is given by the result of dividing the perturbation by the energy of the mode (i.e. $\hbar\omega$), as represented by integrating over the energy density in the interior [16]:

$$\left(\frac{\Delta\omega}{\omega}\right)_i \cong \frac{-\alpha_{ex}|E_0(r_i)^2|}{2\int \epsilon_s |E_0(r)|^2 dV}. \tag{9.6}$$

The integral in the denominator is taken over the interior of the mode energy (>94 %). This approximation simplifies the analysis by allowing the homogeneous permittivity, ϵ_s, of the sphere to be pulled through the integral. Note that for biomolecules such as proteins or DNA, α_{ex} is approximately proportional to the mass of the macromolecule.

Equation (9.6) represents the shift due to a single molecule at an arbitrary position on the surface of the microsphere. In a biosensing experiment, there are many molecules that are distributed over random locations on the sphere surface. To account for all these molecules, the singular contribution in Eq. (9.6) is summed over N randomly located molecules and then this discrete sum is turned into an integral over surface differentials, which gives

$$\frac{\Delta\omega}{\omega} \cong \frac{\alpha_{ex}\sigma_p}{2\epsilon_0\epsilon_{rs}} \frac{\int |E_0(r)|^2 dA}{\int |E_0(r)|^2 dV}. \tag{9.7}$$

For a general TE mode, the interior field, $E_0(r)$, at distance r from the sphere center is given as $E_0(r) = A_{in} j_l(k_0 r \sqrt{\epsilon_{rs}})\hat{L}Y_{lm}$, where A_{in} is the amplitude, $j_l(z)$ is a spherical Bessel function, $\hat{L}$ is a dimensionless angular momentum operator ($\hat{L} = -ir \times \nabla$), $k_0 = \frac{\omega}{c}$, and Y_{lm} is a spherical harmonic function [16]. Consequently,

$$\frac{\Delta\omega}{\omega} \cong \frac{\alpha_{ex}\sigma_p}{2\epsilon_0\epsilon_{rs}} \frac{[j_l(k_0 R_0 \sqrt{\epsilon_{rs}})]^2 R^2}{\int_0^R [j_l(k_0 R_0 \sqrt{\epsilon_{rs}})]^2 r^2 dr}. \tag{9.8}$$

After integrating Eq. (9.8) over the complete resonator surface we can obtain a surprisingly simple formula [16]:

$$\frac{\Delta\omega}{\omega} = -\frac{\alpha_{ex}\sigma_p}{\epsilon_0(n_s^2 - n_m^2)R_0} \tag{9.9}$$

where σ_p is the surface density of the bound proteins, ϵ_0 is the vacuum permittivity, n_s and n_m are the refractive indices of the sphere and the surrounding medium, and R is the radius of the sphere. Applied to the microsphere of radius R_0, Eq. (9.9) allows us to calculate the surface density of a bound protein for any given time from the measured change in resonance frequency $\frac{\Delta\omega}{\omega}$. The excess polarizability, α_{ex}, for protein or DNA biopolymer is determined from measurements of the refractive index increment $\frac{dn}{dc}$ of the pure biopolymer solution. The excess polarizability is then related to the refractive index increment by

$$\alpha_{ex} = 2\epsilon_0 n_m \left(\frac{dn}{dc}\right)_{\text{biopolymer}} m_{\text{biopolymer}} \tag{9.10}$$

where $m_{\text{biopolymer}}$ is the molecular weight of the biopolymer (Streptavidin: $\sim 1 \times 10^{-19}$ gram) and $\frac{dn}{dc}$ of protein [14] and DNA are ~ 0.183 cm^3/g and ~ 0.166 cm^3/g, respectively. With this Eq. (9.9) in hand, we can calculate the mass loading associated with the resonance frequency shift and plot the sensor response (sensogram) in units of mass loading [12]:

$$\text{mass.l.} \left[\frac{\text{pg}}{\text{mm}^2}\right] = -\frac{\Delta\omega}{\omega} \frac{(n_s^2 - n_m^2)R \, [\text{mm}]}{2n_m(\frac{dn}{dc})_{\text{biopolymer}} [\frac{\text{mm}^2}{\text{pg}}]}. \tag{9.11}$$

This reactive sensing principle applies to any microcavity and to any biomolecule and allows researchers to quantify the frequency shift of any optical resonator biosensor in response to molecular or nanoparticle binding events.

9.3 Detection of Bio-Samples

9.3.1 Protein Detection

Figure 9.7 shows the example of a WGM biosensor used for Fibronecting (FN) protein adsorption detection. Sensograms plotted in units of mass loading (ng/cm^2) vs time (s) show FN protein adsorption for differently functionalized glass microspheres—one with hydrophobic 13F-coating and another one with hydrophilic polyethyleneglycol PEG-coating—from different solution concentration levels [53]. As expected, the amount of mass loading that is reached in equilibrium depends on the Fibronectin solution concentration level. By fitting of these results to a kinetic binding model such as the Langmuir model one can determine kinetic-on (k_{on}) and

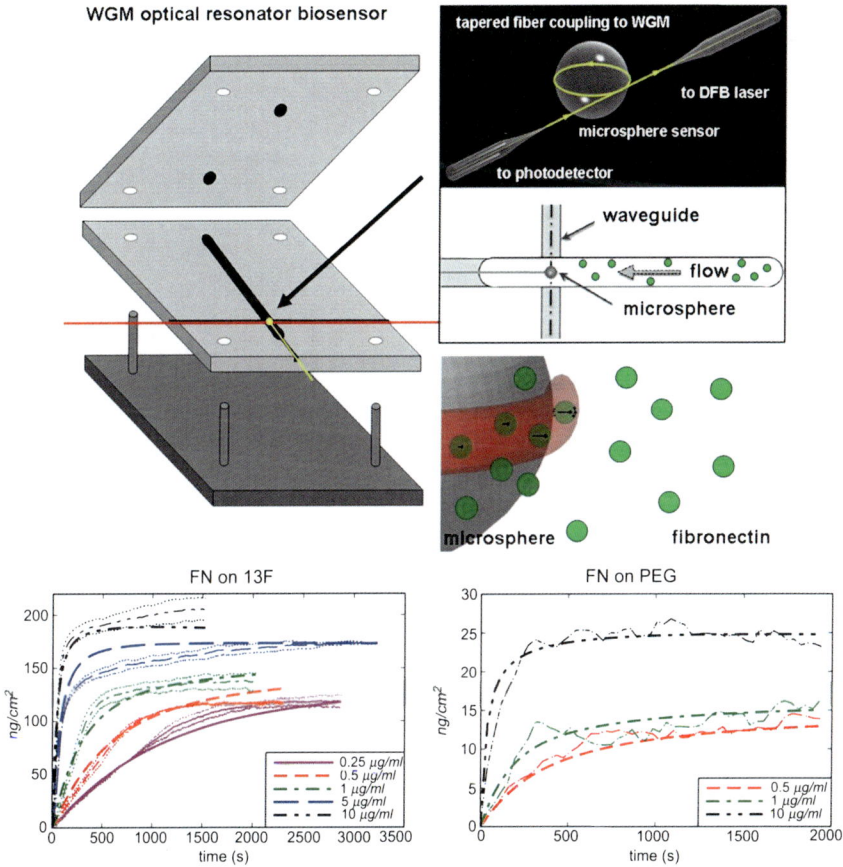

Fig. 9.7 *Top*: WGM optical resonator biosensor integrated in flow cell. Resonance in a glass microsphere is excited via evanescent coupling from tapered fiber. Mass loading associated with binding of protein molecules (*green*) to the microsphere sensor surface is determined from WGM frequency shift, see also Eq. (9.11). *Bottom* adapted from [53]: mass-loading sensograms for Fibronectin (FN) protein binding to differently modified glass microsphere surfaces functionalized with hydrophobic 13F (*left*) and hydrophilic PEG (*right*) coating. The sensogram in mass-loading versus time is determined for different solution concentration levels which here range from 10 µg/ml to 0.25 µg/ml (Color figure online)

-off rates (k_{off}) as well as the dissociation constant $K_d = k_{off}/k_{on}$ which characterizes the affinity of the Fibronectin protein towards the particular microsphere coating 13F or PEG. Care has to be taken at higher solution concentration levels so that measurements are not affected by mass transport limitations [53]. Alternatively, the dissociation constant can be determined from a plot of surface coverage in equilibrium vs solution concentration level. Such a plot is called Langmuir isotherm and the dissociation constant K_d equals just the concentration level at which surface coverage is 50 % [14]. If the sensor surface is modified with specific recognition elements such as antibodies or antigens [19], the above analysis is used to determine

Table 9.3 Overview of biomarkers that have been detected using recognition elements immobilized on the resonator surface. The minimal detectable mass loading and concentration levels are indicated. In some experiments mass loading was not reported separately

Biomarker, approx. MW in kD	Optical resonator type	Recognition element	Approx. mass loading sensitivity	Approx. solution concentration sensitivity
Streptavidin [15, 54] 56 kD	glass microsphere	Biotin	1 pg/mm^2	1 nM
Fibronectin [53] 440 kD	glass microsphere	none	1–10 pg/mm^2	1–10 pM
Glucose oxidase 160 kD (manuscript in prep)	glass microsphere	none	1–10 pg/mm^2	1 nM
Bacteriorhodopsin [82] 26 kD	glass microsphere	none	1 pg/mm^2	n/a
HER2 [50] 138 kD	glass capillary (LCORR)	HER2 antibody	n/a	0.1 nM
DNA [83] 5–10 kD	glass capillary (LCORR)	DNA	6 pg/mm^2	1 nM
Methylated DNA [50] 5–10 kD	glass capillary (LCORR)	methyl binding protein	n/a	1 nM
microRNA [27, 84] 5–10 kD	silicon ring resonator	antibody-DNA duplex, DNA	n/a	<1 nM
PSA [85] 28 kD	silicon ring resonator	antibody	n/a	0.4 nM
Interleukin 2,4,5 [44, 45] 15 kD, 15 kD, 43 kD	silicon ring resonator	antibody	n/a	6–100 pM
TNF [45] 51 kD	silicon ring resonator	antibody	n/a	100 pM
Influenza A virus [10, 32, 61] 300 000 kD	glass microsphere, glass toroid	none	170 attogram	1 fM

the affinity of the recognition element towards the analyte. The above analysis establishes the relationship between the sensor response—the amount or mass loading of specifically bound protein—and the protein solution concentration level. Once such sensor response has been experimentally determined or analytically calculated for different solution concentration levels, the sensor can then be used to determine the unknown concentration of the analyte molecule. It is important to note that the specific detection of an analyte or biomarker can only be achieved if the optical resonator biosensor is functionalized with specific biorecognition elements, such as an antibody, oligonucleotide, aptamer, etc. Table 9.3 gives an overview of some of the biorecognition elements, the detected biomarkers and the minimal solution concentration levels that have been identified with WGM biosensors.

9.3.2 Single Virus Detection

Single Influenza A virus particles have successfully been detected by observing the discrete changes in the frequency of the WGMs due to the binding of the virus particles [10]. We use the reactive sensing principle to predict the maximum frequency shift for the binding of a single nanoparticle to the equator of the microsphere. For this analysis we have to take into account the radius of the nanoparticle a as well as the evanescent field length L:

$$\left(\frac{\Delta \omega}{\omega}\right)_{max} \cong -D \frac{a^3}{R^{5/2}\lambda^{1/2}} e^{-a/L}, \tag{9.12}$$

with

$$D = \frac{2n_m^2 (2n_s)^{\frac{1}{2}} (n_n^2 p - n_m^2)}{(n_s^2 - n_m^2)(n_{np}^2 + 2n_m^2)} \quad \text{and} \quad L \approx \left(\frac{\lambda}{4\pi}\right)(n_s^2 - n_m^2)^{\frac{1}{2}},$$

where n_{np} is the refractive index of the nanoparticle. Equation (9.12) shows that the relative wavelength shift is proportional to $R^{-5/2}$, which is in good agreement with the experimental results of detecting poly styrene beads with radius of 250 nm (Fig. 9.8B). Equation (9.12) in conjunction with Eq. (9.10) has been used to extract the size and mass of a single Influenza A virus particle from the discrete resonance frequency shifts that were observed upon binding of the virion to the equator of a $R = 30$ μm glass microsphere with resonances excited at $\lambda \sim 763$ nm wavelength [10], see Fig. 9.8C. The sizing of virions and nanoparticles with optical resonator biosensors is a fast growing area of research and recent demonstrations of the detection and sizing of nanoparticles down to few 10's of nanometer size range [32, 61, 86] are rapidly approaching the single molecule detection limit where a biomolecule has a radius of ~ 3 nm.

An alternative to the reactive biosensing principle for virus detection is the mode splitting effect, i.e., one resonant mode splitting into two resonances due to interaction of light with nanoscale objects, such as nanoparticles, or virus in the mode volume. When a WGM field encounters a light scattering center such as a particle or virus the light is scattered elastically from the particle. A portion of light is scattered and lost to the environment while some is scattered back into the mode volume and takes the optical path of the counter-propagating mode—coupling between the CW and CCW modes. This lifts the degeneracy of the two modes. Two standing wave modes are formed in the cavity and this is reflected in the splitting of the single resonance into a doublet structure, i.e. two resonances, in the transmission spectra of the resonator (see Fig. 9.9). Because the two split modes reside in the same resonator they share many noises sources, such as temperature variation, and since they are compared to one another, this is a self-referencing sensing technique [86]. A comprehensive study on this topic as well as the first observation of mode splitting in aqueous environment (particle size 50 nm in radius) for the case of microsphere resonators is given in Ref. [87]. The polarizability of a nanoparticle

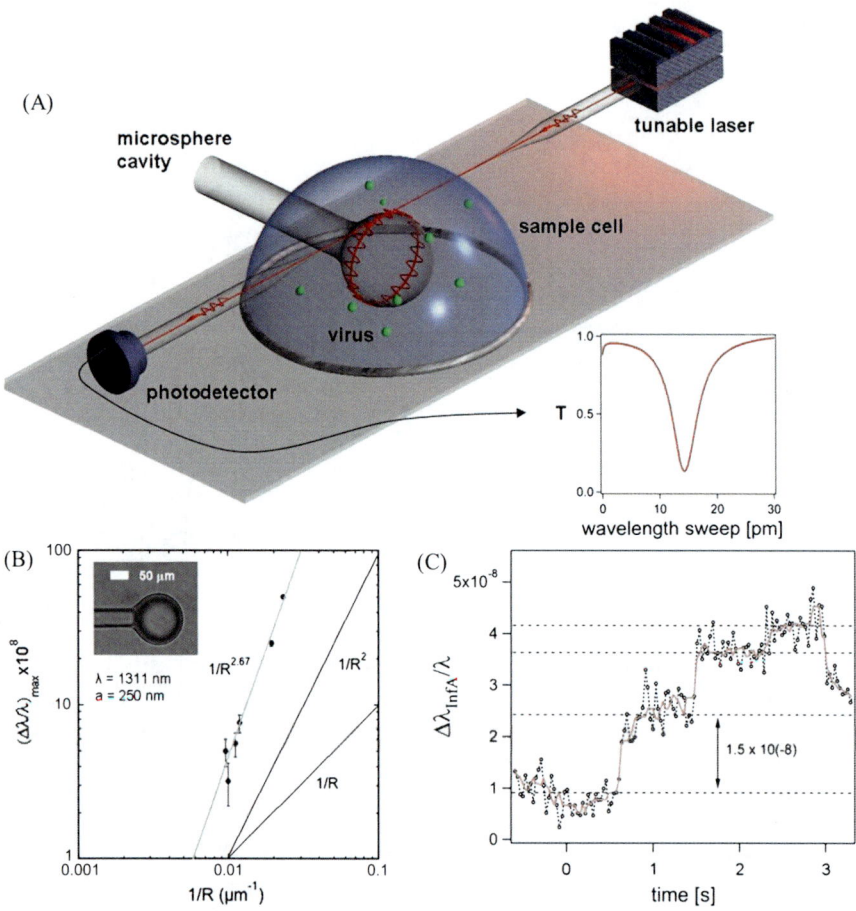

Fig. 9.8 (**A**) WGM resonator biosensor setup for detection of single virus particles in a droplet of fluid, adapted from [10]. An optical resonance (WGM) is excited with a DFB tunable laser diode at ~763 nm wavelength in ~60 μm diameter glass microsphere. (**B**) Maximum step height vs. microsphere curvature for polystyrene particles with radius of 250 nm at wavelength of 1311 nm; inset shows an optical image of one microsphere. (**C**) WGM frequency shifts recorded for the binding of single Influenza A virus particles

can be calculated from the mode-splitting transmission spectrum of WGM resonator after a nanoparticle is adsorbed on its surface, independent of the binding location. The mode-splitting effect in WGM resonators was first reported as an interesting phenomenon in high-Q microspheres [88, 89]. It has since been demonstrated in various WGM resonators with different geometries, including microspheres, micro-toroids and microdisks [86, 90–92].

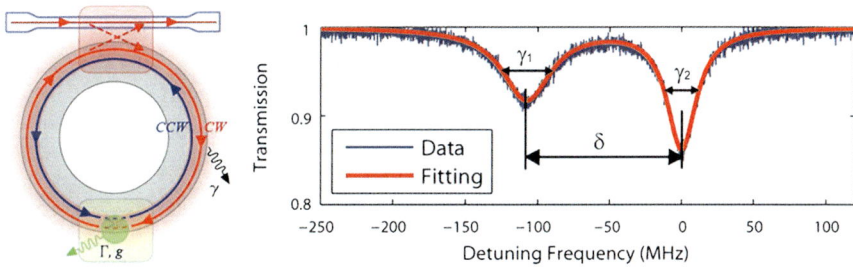

Fig. 9.9 Mode splitting in a WGM resonator. *Left*: WGM resonators supports degenerate coun- ter- clockwise (CW) and counterclockwise (CCW) propagating modes. Light scattering from a nanoparticle couples the initially degenerate CW and CCW modes, lifting the degeneracy. *Right*: Experimentally observed mode-splitting spectrum (*blue*) after the deposition of a single nanopar- ticle [86] (Color figure online)

9.3.3 Multiplexed Sensing Platform

Multiplexing capability is another aspect of the WGM biosensors and one of the demonstrations for this utilized two microsphere resonators coupled to one ta- pered optical fiber waveguide [83]. Many other multiplexing demonstrations can be found, and a highly integrated silicon ring resonator platform is already com- mercialized [25, 30, 93–96]. With the multiplexed approach, label-free detection of bacterial tmRNA [97], cardiac biomarker C-reactive protein [46], and Bean pod mottle virus [98] have been demonstrated.

For multiplexing with WGM biosensors, spheres can be positioned several mi- crons apart so that there is no cross coupling between the WGMs of the two spheres, see Fig. 9.10a. Two Lorentzian dips corresponding to the WGMs from two spheres are shown in Fig. 9.10b. Both spheres S1 and S2 are surface-modified with dif- ferent single stranded DNA probes P1 and P2. In the sample cell, the two spheres were equilibrated at room temperature in PBS until the resonances reach a steady state, Fig. 9.10c. Then, the oligonucleotide that is complementary to the one immo- bilized on sphere S1 was injected into the sample cell. This was followed minutes later by the injection of oligonucleotide complementary to the ssDNA immobilized on sphere S2. A time trace of the two sphere-specific resonance positions allowed to follow the two independent hybridization events, Fig. 9.10c: after injection of the complementary oligonucleotide, the resonance wavelength of the corresponding sphere showed a large increase. There was no detectable unspecific hybridization in this experiment: one injected oligonucleotide increased only the resonance wave- length of the corresponding sphere whereas the resonance wavelength of the other sphere remained unchanged. The oligonucleotides were injected to a final concen- tration of 1 μM. Despite the rather large background concentration, the sequential injection of the second oligonucleotide produced a similar resonance wavelength shift for hybridization to S2 as compared to the first injected oligonucleotide hy- bridizing to S1. The spikes shortly after the injections are due to temperature and refractive index fluctuations caused by turbulences after mixing with a hypodermic needle.

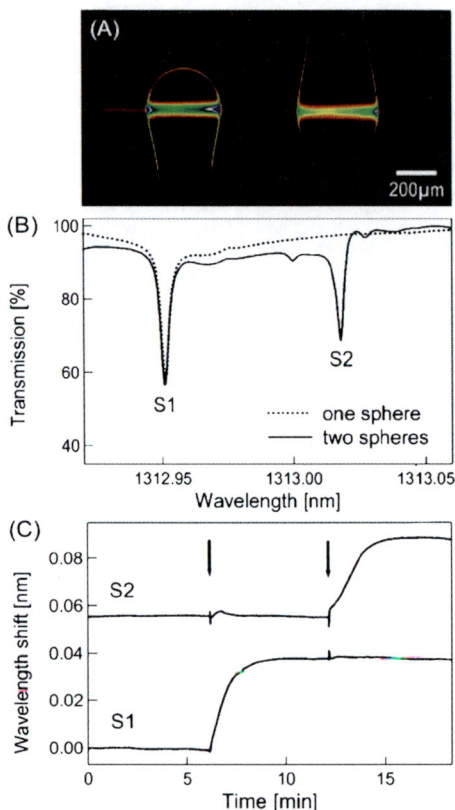

Fig. 9.10 (**A**) Micrograph of two spheres coupled to the optical fiber running horizontally through the center of the image. The image shows two resonances of light orbiting inside each sphere. (**B**) Transmission spectrum for one (*dotted line*) and two (*solid line*) spheres coupled to the same optical fiber, immersed in a PBS solution at room temperature. Both spheres are ∼200 μm in radius. The narrow infrared spectrum ranging from 1312.92 to 1313.06 nm is recorded every 10 ms. The position of the resonant wavelengths from each of the spheres, S1 and S2, is located by a parabolic minimum fit in a resolution of ∼1/50 of the linewidth, allowing detection of a fractional wavelength change $\frac{\delta\lambda}{\lambda}$ as small as ∼3 × 10⁻⁷. Both spheres were modified with unrelated, 27-mer oligonucleotides. (**C**) Shows the time trace of the two resonance positions from S1 and S2. The arrows indicate when the two complementary DNA oligonucleotides are injected into the sample solution at a final concentration of 1 μM each. Hybridization saturates within minutes and the resonance wavelength of the corresponding sphere increased by ∼38 pm each. The noise before adding the complementary DNA was only ∼0.04 pm

9.3.4 WGM Sensors with Plasmonic Enhancement

From Eq. (9.6), we predict that the wavelength shift caused by the binding of a single molecule is proportional to the field strength $|E_0(r_i)^2|$ at the binding site. If $|E_0(r_i)^2|$ can be amplified in a process that does not significantly degrade the Q factor, then sensitivity will be enhanced in proportion to the amplified field strength.

Fig. 9.11 Adapted from Ref. [78]: Field intensity distribution of a whispering gallery mode (WGM) shown in cross section for a microsphere with (**a**) and without (**b**) plasmonic nanoparticle. The microsphere surface is indicated by dashed line. A 55 nm gold nanoparticle is positioned within the evanescent field of one WGM forming a hybrid photonic-plasmonic resonance where large field amplitudes are generated at the nanoparticle site (**a**). The large field intensities of the microsphere+nanoparticle system (**a**) as compared to the plain WGM (**b**) can lead to large sensitivity enhancements for single-particle/molecule detection (**c**), see also Eq. (9.6)

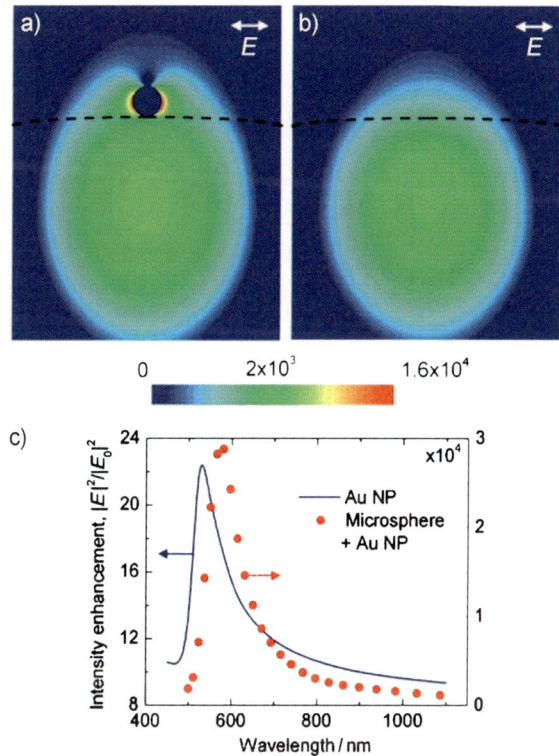

Very recently we suggested to achieve such amplification of WGM field strength by localized surface plasmon resonance (LSPR) generated by plasmonic nanostructures attached to the microcavity [78, 80, 101].

Once a plasmonic nanostructure, i.e. gold nanoparticle, is sitting inside the mode volume of a WGM on the surface of a microsphere and the frequency of the WGM is close to resonant with the nanoparticle, a hybrid photonic-plasmonic resonance will form and create a 'hot spot' at the nanoparticle site, where the electric field is greatly amplified. If particles or biomolecules land on these 'hot spots', large frequency shifts are expected according to Eq. (9.6).

The idea of plasmonic enhancements was first introduced in Ref. [78]. Here it is demonstrated for the first time that a hybrid photonic-plasmonic mode produced by immobilizing gold (Au) nanoparticles within the evanescent field of a microsphere can greatly enhance sensitivity for protein detection (Fig. 9.11). Orders-of-magnitude of frequency shift enhancement has been predicted by simulation and first experimental results indicate indeed such an unprecedented increase in sensitivity [78, 80, 99].

In a recent demonstration using this hybrid photonic-plasmonic approach, the plasmonic enhancement from a gold nanoshell has successfully boosted the sensitivity for detecting small RNA virus MS2 by a factor of 70 [99]. The gold nanoshells are attached on the equator of microspheres by strong WGM gradient forces [58].

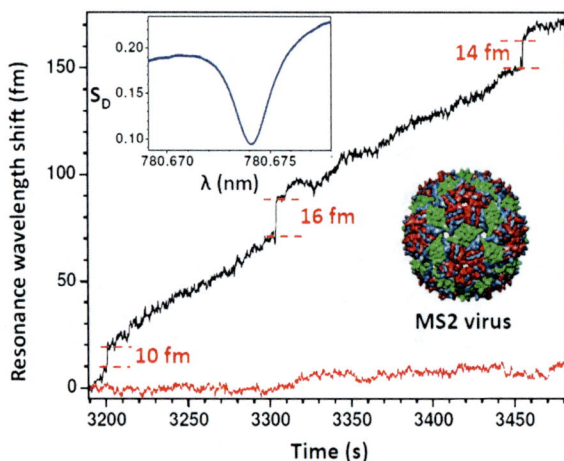

Fig. 9.12 Adapted from Ref. [99]: Resonance shift (at around 780.674 nm) of a WGM micro-sphere ($R \approx 45$ μm) having a gold nanoshell attached on its equator due to the adsorption of MS2 viruses (*upper trace*). Large shifts indicated by the steps are attributed to the landing of the virus on the 'hot spot'. The *lower trace* shows the background without MS2 or the gold nanoshell (r.m.s. noise 2 fm). Insets show the recorded spectrum of one WGM in the sphere ($Q \approx 4 \times 10^5$) and an illustration of MS2 virus (radius ~13.6 nm)

Once the MS2 virions land on the 'hot spots' of the nanoshells, they generate large wavelength shifts of the WGMs as indicated by steps in the resonance wavelength trace (Fig. 9.12). Note that the mass of MS2 is only ~1 % of the Influenza A (6 vs 512 attogram) which was detected by a microsphere sensor without any plasmonic enhancement [10], see Sect. 9.3.2.

9.4 Outlook

Optical resonator biosensors are emerging as one of the most sensitive microsystems biodetection technology that does not require labeling of the biomarker. The label-free optical resonator biosensor platform is highly versatile and future implementations will be tailored to specific detection needs for example in the life sciences, in environmental monitoring and in healthcare. We envision that clinical diagnostic assays—currently performed by trained medical personnel in well-equipped clinical laboratories—will be replaced by inexpensive and portable chip-scale optical resonator biosensor devices that can be operated in low-resource settings and by untrained personnel. To achieve this goal it will be particularly important to overcome challenges associated with integrating multiplexed optical resonator biosensors with other microfluidic, photonic and electronic components. First successfully integrated prototypes are already being tested in the commercial world.

Several physical mechanisms have been proposed to boost sensitivity of optical resonator biosensors towards achieving the ultimate goal of label-free single molecule detection [12] which would revolutionize molecular diagnostics as well as provide for a deeper understanding of the biomolecular world, their self-assembly, functions and interactions. A particularly promising approach towards achieving this goal is based on the use of plasmonic nanoantennas for enhancing the frequency shift signal [78] and we refer to other [12] as well as future reviews on this emerging topic.

9.5   Keywords and Notes

Biosensor   A biosensor is an analytical device that transduces a molecular binding event into a detectable optical or electrical signal.

Biomolecule   A biomolecule is synthesized by cells or bacteria. The synthesis of biomolecules from naturally occurring pre-cursor molecules such as aminoacids or nucleotides is directly or indirectly encoded by a genetic blueprint. Changes in the genetic blueprint for example by random mutations enable the evolution of new and improved biomolecules which manifest themselves by their coding in specific genes.

Virus   A virus is an infectious bioparticle in the $\sim$20–500 nm size range that spreads outside of cells but requires the cell to replicate.

Optical Microcavity   An optical microcavity confines coherent light by total internal reflection which interferes to produce an optical resonance signal. The resonance signal can be identified as a narrow line in the spectral response of the microcavity. Examples for optical microcavities are glass beads, Fabry Perot resonators or ring resonators. Optical microcavities are also referred to as optical resonators.

Whispering Gallery Mode   Multiple total internal reflections guide the light around the inside of a dielectric (glass) microsphere so that the returning light wave starts to interfere with itself, resulting in a traveling wave called 'Whispering Gallery Mode (WGM)'.

Q Factor   Q factor is the figure of merit of a resonant system and is inversely proportional to the relative amount of energy $\Delta E/E_{\text{total}}$ that is lost per oscillation: $Q = 2\pi E_{\text{total}}/\Delta E$. For example, the Q-factor of Foucault's Pendulum in the Pantheon in Paris may reach $Q = 1000$, a typical electronic LC circuit filter has a $Q \sim 20$, a nanomechanical system in vacuum may exhibit Q's up to 10^{5-6} and optical resonator biosensors can reach ultimate Q's of up to 10^{10}.

Nanomechanical Resonator Biosensor   Nanomechanical resonator biosensors such as silicon nanobeams or silicon cantilevers detect the added weight Δm after binding of a biomolecule or virus particle to the miniature sensor element with resting mass m_0. The added weight is detected and quantified from the shift $\Delta\omega$ of the mechanical resonance frequency ω which changes upon binding [11, 22] according to $\frac{\Delta\omega}{\omega} = -\frac{\Delta m}{2m_0}$.

Optical Fiber   A single mode optical fiber is a $\sim$125 µm-diameter glass fiber that transmits light by total internal reflection (TIR). The light is guided by TIR through a higher-refractive index glass core ($\sim$6 µm core diameter) surrounded by the slightly lower refractive index glass cladding ($\sim$125 µm total diameter).

Evanescent Coupling   Light that is total internally reflected at the surface of a prism can couple to a symmetrically opposed prism by evanescent coupling if the two prisms are brought in close proximity so that the distance between the surfaces is on the order of the evanescent field length, typically a fraction of the wavelength or about 100–300 nm for visible light.

DFB Laser   A distributed feedback laser is a laser diode internally modified with a grating structure to produce a narrow laser line down $\sim$100 kHz width. The center wavelength of a DFB laser can be tuned continuously by changing the laser diode current in the milli-Ampere (mA) range. Tuning range is typically achieved within a range of $\sim$0.1–0.3 nm with a typical tuning coefficient $\sim$0.01 nm/mA. DFB lasers are widely used in the telecommunications industry in form of inexpensive chip-scale devices.

Evanescent Field   Total internal reflection of light at a prism surface produces a non-propagating evanescent field that decays exponentially with distance from the reflecting surface. The evanescent field length extends typically a distance given by a fraction of the wavelength, or about 100–300 nm at visible wavelengths.

Polarizability   The polarizability of a molecule describes the tendency of its charge distribution—given by the electron clouds around its atoms—to be distorted from an external electric or electromagnetic field.

Sensogram   Plot of sensor response versus time.

Antibody   An antibody is a protein with a characteristic Y-shaped structure that specifically binds to a target molecule called the antigen. Antibodies are produced by the B-cells of the immune system and are widely used in biosensor applications because of their ability to only recognize the target antigen even against a background of many other biomolecules in solution.

Langmuir Model   The Langmuir model describes the equilibrium between molecules in solution and their independent and fixed number of binding sites at the surface of a substrate such as a biosensor.

Influenza A Virus   Influenza A Virus is also called the flu virus. It causes the seasonal flu and mutations of the virus can trigger pandemics such as the one known as the Spanish Flu which broke out in 1918.

Multiplexing   Multiplexing in WGM biosensing refers to the use of multiple sensor elements for detection of multiple biomarkers, as well as for referenced measurements.

LSPR A localized surface plasmon resonance (LSPR) is a collective oscillation of conduction electrons in a nanoparticle excited by an optical field.

Hybrid Photonic-Plasmonic Resonance A hybrid photonic-plasmonic resonance can be generated by coupling of a WGM to a nanoparticle or nanostructure that exhibits a plasmon resonance. The WGM frequency can be tuned with respect to the plasmon resonance frequency so that large field enhancements appear at the nanoparticle site that greatly enhance sensitivity in WGM biosensing.

References

1. J.L. Arlett et al., Comparative advantages of mechanical biosensors. Nat. Nanotechnol. **6**, 203–215 (2011)
2. P.S. Waggoner, H.G. Craighead, Micro- and nanomechanical sensors for environmental, chemical, and biological detection. Lab Chip **7**, 1238–1255 (2007)
3. F. Patolsky et al., Nanowire sensors for medicine and the life sciences. Nanomedicine **1**, 51–65 (2006)
4. J. Li et al., Carbon nanotube sensors for gas and organic vapor detection. Nano Lett. **3**, 929–933 (2003)
5. F. Vollmer, S. Arnold, Whispering-gallery-mode biosensing: label-free detection down to single molecules. Nat. Methods **5**, 591–596 (2008)
6. P. Alivisatos, The use of nanocrystals in biological detection. Nat. Biotechnol. **22**, 47–52 (2004)
7. J.N. Anker et al., Biosensing with plasmonic nanosensors. Nat. Mater. **7**, 442–453 (2008)
8. K.K. Jain, Nanotechnology in clinical laboratory diagnostics. Clin. Chim. Acta **358**, 37–54 (2005)
9. G.M. Whitesides, The 'right' size in nanobiotechnology. Nat. Biotechnol. **21**, 1161–1165 (2003)
10. F. Vollmer et al., Single virus detection from the reactive shift of a whispering-gallery mode. Proc. Natl. Acad. Sci. USA **105**, 20701–20704 (2008)
11. A.K. Naik et al., Towards single-molecule nanomechanical mass spectrometry. Nat. Nanotechnol. **4**, 445–450 (2009)
12. M. Baaske, F. Vollmer, Optical resonator biosensors: molecular diagnostic and nanoparticle detection on an integrated platform. ChemPhysChem **13**, 427–436 (2012)
13. F. Patolsky et al., Electrical detection of single viruses. Proc. Natl. Acad. Sci. USA **101**, 14017–14022 (2004)
14. K. Wilson, F. Vollmer, *Whispering Gallery Mode Resonator Biosensors*, vol. 4 (Springer, Berlin, 2012)
15. F. Vollmer et al., Protein detection by optical shift of a resonant microcavity. Appl. Phys. Lett. **80**, 4057–4059 (2002)
16. S. Arnold et al., Shift of whispering-gallery modes in microspheres by protein adsorption. Opt. Lett. **28**, 272–274 (2003)
17. T.P. Burg et al., Weighing of biomolecules, single cells and single nanoparticles in fluid. Nature **446**, 1066–1069 (2007)
18. F. Patolsky et al., Fabrication of silicon nanowire devices for ultrasensitive, label-free, real-time detection of biological and chemical species. Nat. Protoc. **1**, 1711–1724 (2006)
19. H.K. Hunt et al., Bioconjugation strategies for microtoroidal optical resonators. Sensors **10**, 9317–9336 (2010)
20. A.L. Washburn, R.C. Bailey, Photonics-on-a-chip: recent advances in integrated waveguides as enabling detection elements for real-world, lab-on-a-chip biosensing applications. Analyst **136**, 227–236 (2011)

21. M. Curreli et al., Real-time, label-free detection of biological entities using nanowire-based FETs. IEEE Trans. Nanotechnol. **7**, 651–667 (2008)

22. B. Ilic et al., Virus detection using nanoelectromechanical devices. Appl. Phys. Lett. **85**, 2604–2606 (2004)

23. X.D. Fan, I.M. White, Optofluidic microsystems for chemical and biological analysis. Nat. Photonics **5**, 591–597 (2011)

24. A.J. Qavi et al., Label-free technologies for quantitative multiparameter biological analysis. Anal. Bioanal. Chem. **394**, 121–135 (2009)

25. M. Iqbal et al., Label-free biosensor arrays based on silicon ring resonators and high-speed optical scanning instrumentation. IEEE J. Sel. Top. Quantum Electron. **16**, 654–661 (2010)

26. K. Vahala (ed.), *Optical Microcavities*. Advanced Series in Applied Physics (World Scientific, Hackensack, 2004)

27. A.J. Qavi, R.C. Bailey, Multiplexed detection and label-free quantitation of MicroRNAs using arrays of silicon photonic microring resonators. Angew. Chem., Int. Ed. Engl. **49**, 4608–4611 (2010)

28. A. Ramachandran et al., A universal biosensing platform based on optical micro-ring resonators. Biosens. Bioelectron. **23**, 939–944 (2008)

29. Y.Z. Sun, X.D. Fan, Optical ring resonators for biochemical and chemical sensing. Anal. Bioanal. Chem. **399**, 205–211 (2011)

30. C.A. Barrios, Integrated microring resonator sensor arrays for labs-on-chips. Anal. Bioanal. Chem. **403**, 1467–1475 (2012)

31. L.N. He et al., Ultrasensitive detection of mode splitting in active optical microcavities. Phys. Rev. A **82** (2010)

32. T. Lu et al., High sensitivity nanoparticle detection using optical microcavities. Proc. Natl. Acad. Sci. USA **108**, 5976–5979 (2011)

33. Y. Lan et al., A self-reference sensing technique for ultra-sensitive chemical and biological detection using whispering gallery microresonators, Proc. SPIE Int. Soc. Opt. Eng. **7913**, 791312 (2011)

34. J.G. Zhu et al., Single virus and nanoparticle size spectrometry by whispering-gallery-mode microcavities. Opt. Express **19**, 16195–16206 (2011)

35. C. Shi et al., Leveraging bimodal kinetics to improve detection specificity. Opt. Lett. **37**, 1643–1645 (2012)

36. I.M. White et al., Liquid-core optical ring-resonator sensors. Opt. Lett. **31**, 1319–1321 (2006)

37. S.M. Harazim et al., Lab-in-a-tube: on-chip integration of glass optofluidic ring resonators for label-free sensing applications. Lab Chip **12**, 2649–2655 (2012)

38. J.D. Suter et al., Label-free quantitative DNA detection using the liquid core optical ring resonator. Biosens. Bioelectron. **23**, 1003–1009 (2008)

39. H.Y. Zhu et al., Analysis of biomolecule detection with optofluidic ring resonator sensors. Opt. Express **15**, 9139–9146 (2007)

40. J. Lutti et al., A monolithic optical sensor based on whispering-gallery modes in polystyrene microspheres. Appl. Phys. Lett. **93** (2008)

41. F. Vollmer, Resonant detection of micro to nanoscopic OBjects using whispering gallery modes. PhD Thesis (2004)

42. K. De Vos et al., Silicon-on-insulator microring resonator for sensitive and label-free biosensing. Opt. Express **15**, 7610–7615 (2007)

43. L. Jin et al., Highly-sensitive silicon-on-insulator sensor based on two cascaded micro-ring resonators with vernier effect. Opt. Commun. **284**, 156–159 (2011)

44. M.S. Luchansky, R.C. Bailey, Silicon photonic microring resonators for quantitative cytokine detection and T-cell secretion analysis. Anal. Chem. **82**, 1975–1981 (2010)

45. M.S. Luchansky, R.C. Bailey, Rapid, multiparameter profiling of cellular secretion using silicon photonic microring resonator arrays. J. Am. Chem. Soc. **133**, 20500–20506 (2011)

46. M.S. Luchansky et al., Sensitive on-chip detection of a protein biomarker in human serum and plasma over an extended dynamic range using silicon photonic microring resonators and

sub-micron beads. Lab Chip **11**, 2042–2044 (2011)

47. B.Q. Su et al., Compact silicon-on-insulator dual-microring resonator optimized for sensing. J. Lightwave Technol. **29**, 1535–1541 (2011)

48. H.Y. Zhu et al., Integrated refractive index optical ring resonator detector for capillary electrophoresis. Anal. Chem. **79**, 930–937 (2007)

49. I.M. White et al., SERS-based detection in an optofluidic ring resonator platform. Opt. Express **15**, 17433–17442 (2007)

50. J.T. Gohring et al., Detection of HER2 breast cancer biomarker using the opto-fluidic ring resonator biosensor. Sens. Actuators B, Chem. **146**, 226–230 (2010)

51. J.T. Gohring, X.D. Fan, Label free detection of CD4+ and CD8 + T cells using the optofluidic ring resonator. Sensors **10**, 5798–5808 (2010)

52. J.D. Suter et al., Label-free DNA methylation analysis using opto-fluidic ring resonators. Biosens. Bioelectron. **26**, 1016–1020 (2010)

53. K.A. Wilson et al., Whispering gallery mode biosensor quantification of fibronectin adsorption kinetics onto alkylsilane monolayers and interpretation of resultant cellular response. Biomaterials **33**, 225–236 (2012)

54. C.E. Soteropulos et al., Determination of binding kinetics using whispering gallery mode microcavities. Appl. Phys. Lett. **99** (2011)

55. N.M. Hanumegowda et al., Refractometric sensors based on microsphere resonators. Appl. Phys. Lett. **87** (2005)

56. I. Teraoka, S. Arnold, Enhancing the sensitivity of a whispering-gallery mode microsphere sensor by a high-refractive-index surface layer. J. Opt. Soc. Am. B, Opt. Phys. **23**, 1434–1441 (2006)

57. H.C. Ren et al., High-Q microsphere biosensor—analysis for adsorption of rodlike bacteria. Opt. Express **15**, 17410–17423 (2007)

58. S. Arnold et al., Whispering gallery mode carousel—a photonic mechanism for enhanced nanoparticle detection in biosensing. Opt. Express **17**, 6230–6238 (2009)

59. V.S. Ilchenko et al., Pigtailing the high-Q microsphere cavity: a simple fiber coupler for optical whispering-gallery modes. Opt. Lett. **24**, 723–725 (1999)

60. V.S. Ilchenko, A.B. Matsko, Optical resonators with whispering-gallery modes, part II: applications. IEEE J. Sel. Top. Quantum Electron. **12**, 15–32 (2006)

61. L. He et al., Detecting single viruses and nanoparticles using whispering gallery microlasers. Nat. Nanotechnol. **6**, 428–432 (2011)

62. A. Yalcin et al., Optical sensing of biomolecules using microring resonators. IEEE J. Sel. Top. Quantum Electron. **12**, 148–155 (2006)

63. C.Y. Chao et al., Polymer microring resonators for biochemical sensing applications. IEEE J. Sel. Top. Quantum Electron. **12**, 134–142 (2006)

64. A. Ksendzov, Y. Lin, Integrated optics ring-resonator sensors for protein detection. Opt. Lett. **30**, 3344–3346 (2005)

65. C. Delezoide et al., Vertically coupled polymer microracetrack resonators for label-free biochemical sensors. IEEE Photonics Technol. Lett. **24**, 270–272 (2012)

66. J.T. Kirk et al., Multiplexed inkjet functionalization of silicon photonic biosensors. Lab Chip **11**, 1372–1377 (2011)

67. A. Francois, M. Himmelhaus, Whispering gallery mode biosensor operated in the stimulated emission regime. Appl. Phys. Lett. **94** (2009)

68. M. Himmelhaus et al., Optical sensors based on whispering gallery modes in fluorescent microbeads: response to specific interactions. Sensors **10**, 6257–6274 (2010)

69. M. Himmelhaus, A. Francois, In-vitro sensing of biomechanical forces in live cells by a whispering gallery mode biosensor. Biosens. Bioelectron. **25**, 418–427 (2009)

70. J. Yang, L.J. Guo, Optical sensors based on active microcavities. IEEE J. Sel. Top. Quantum Electron. **12**, 143–147 (2006)

71. A. Weller et al., Whispering gallery mode biosensors in the low-Q limit. Appl. Phys. B, Lasers Opt. **90**, 561–567 (2008)

72. E. Nuhiji, P. Mulvaney, Detection of unlabeled oligonucleotide targets using whispering gallery modes in single, fluorescent microspheres. Small **3**, 1408–1414 (2007)

73. A. Schweinsberg et al., An environmental sensor based on an integrated optical whispering gallery mode disk resonator. Sens. Actuators B, Chem. **123**, 727–732 (2007)

74. R.W. Boyd, J.E. Heebner, Sensitive disk resonator photonic biosensor. Appl. Opt. **40**, 5742–5747 (2001)

75. M. Pollinger et al., Ultrahigh-Q tunable whispering-gallery-mode microresonator. Phys. Rev. Lett. **103** (2009)

76. E.J. Smith et al., Lab-in-a-tube: detection of individual mouse cells for analysis in flexible split-wall microtube resonator sensors. Nano Lett. **11**, 4037–4042 (2011)

77. J.D. Swaim et al., Detection limits in whispering gallery biosensors with plasmonic enhancement. Appl. Phys. Lett. **99** (2011)

78. M.A. Santiago-Cordoba et al., Nanoparticle-based protein detection by optical shift of a resonant microcavity. Appl. Phys. Lett. **99** (2011)

79. S.I. Shopova et al., Plasmonic enhancement of a whispering-gallery-mode biosensor for single nanoparticle detection. Appl. Phys. Lett. **98** (2011)

80. M.A. Santiago-Cordoba et al., Ultrasensitive detection of a protein by optical trapping in a photonic-plasmonic microcavity. J. Biophotonics **5**, 629–638 (2012)

81. M.L. Gorodetsky et al., Ultimate Q of optical microsphere resonators. Opt. Lett. **21**, 453–455 (1996)

82. J. Topolancik, F. Vollmer, Photoinduced transformations in bacteriorhodopsin membrane monitored with optical microcavities. Biophys. J. **92**, 2223–2229 (2007)

83. F. Vollmer et al., Multiplexed DNA quantification by spectroscopic shift of two microsphere cavities. Biophys. J. **85**, 1974–1979 (2003)

84. A.J. Qavi et al., Anti-DNA:RNA antibodies and silicon photonic microring resonators: increased sensitivity for multiplexed microRNA detection. Anal. Chem. **83**, 5949–5956 (2011)

85. A.L. Washburn et al., DNA-encoding to improve performance and allow parallel evaluation of the binding characteristics of multiple antibodies in a surface-bound immunoassay format. Anal. Chem. **83**, 3572–3580 (2011)

86. J.G. Zhu et al., On-chip single nanoparticle detection and sizing by mode splitting in an ultrahigh-Q microresonator. Nat. Photonics **4**, 46–49 (2010)

87. W. Kim et al., Observation and characterization of mode splitting in microsphere resonators in aquatic environment. Appl. Phys. Lett. **98** (2011)

88. D.S. Weiss et al., Splitting of high-Q mie modes induced by light backscattering in silica microspheres. Opt. Lett. **20**, 1835–1837 (1995)

89. M.L. Gorodetsky et al., Rayleigh scattering in high-Q microspheres. J. Opt. Soc. Am. B, Opt. Phys. **17**, 1051–1057 (2000)

90. A. Mazzei et al., Controlled coupling of counterpropagating whispering-gallery modes by a single Rayleigh scatterer: a classical problem in a quantum optical light. Phys. Rev. Lett. **99** (2007)

91. M. Borselli et al., Beyond the Rayleigh scattering limit in high-Q silicon microdisks: theory and experiment. Opt. Express **13**, 1515–1530 (2005)

92. T.J. Kippenberg et al., Purcell-factor-enhanced scattering from Si nanocrystals in an optical microcavity. Phys. Rev. Lett. **103** (2009)

93. K. De Vos et al., Multiplexed antibody detection with an array of silicon-on-insulator microring resonators. IEEE Photonics J. **1**, 225–235 (2009)

94. H.A. Huckabay, R.C. Dunn, Whispering gallery mode imaging for the multiplexed detection of biomarkers. Sens. Actuators B, Chem. **160**, 1262–1267 (2011)

95. S.Y. Lin, K.B. Crozier, Planar silicon microrings as wavelength-multiplexed optical traps for storing and sensing particles. Lab Chip **11**, 4047–4051 (2011)

96. A.L. Washburn et al., Quantitative, label-free detection of five protein biomarkers using multiplexed arrays of silicon photonic microring resonators. Anal. Chem. **82**, 69–72 (2010)

97. O. Scheler et al., Label-free, multiplexed detection of bacterial tmRNA using silicon photonic microring resonators. Biosens. Bioelectron. **36**, 56–61 (2012)

98. M.S. McClellan et al., Label-free virus detection using silicon photonic microring resonators. Biosens. Bioelectron. **31**, 388–392 (2012)

99. V.R. Dantham et al., Taking whispering gallery-mode single virus detection and sizing to the limit. Appl. Phys. Lett. **101** (2012)

100. F. Vollmer, L. Yang, Label-free detection with high-Q microcavities: a review of biosensing mechanisms for integrated devices, in *Nanophotonics* (de Gruyter, Berlin, 2012). doi:10.1515/nanoph-2012-0021

101. M.R. Foreman, F. Vollmer, Level repulsion in hybrid photonic-plasmonic microresonators for enhanced biodetection. Phys. Rev. A **88** (2013)

Chapter 10
Cavity-Enhanced Spectroscopy on Silica Microsphere Resonators

Jack A. Barnes, Gianluca Gagliardi, and Hans-Peter Loock

Abstract Microcavities are a powerful tool for chemical detection and sensing, but also to study chemical processes at interfaces. In most experiments the cavity mode spectrum is used to infer the chemical composition of the resonator medium and its immediate environment. For example, frequency shifts of cavity modes can be related to either cavity length changes or refractive index changes. Photon lifetime measurements, on the other hand, allow in principle for an independent measurement of the optical loss experienced by a cavity mode, but time-resolved cavity ring-down measurements are difficult due to the small dimensions of the cavity and the consequent short photon lifetimes (ring-down times). This chapter describes how phase-shift cavity ring-down spectroscopy can be adapted to extract the optical loss of a whispering gallery mode in a microresonator. By combining different phase-shift measurements of the total optical loss one can furthermore separate the contributions from intracavity loss due to absorption and scattering from the contributions of optical loss due to coupling to a light delivery waveguide.

The experimental focus is on the use of silica sphere microresonators. The *frequency* of the high-Q whispering gallery modes in these microspheres is strongly dependent on the size of the sphere whereas the *intracavity loss* is influenced by surface absorption and scattering. A sub-monolayer of ethylene diamine on a 300 μm sphere has the effect of, simultaneously, changing the resonance frequencies and the ring-down times of the whispering gallery modes. The absolute surface coverage can be extracted from the resonance frequency, and can be combined with the measurement of intracavity loss to determine the absolute absorption cross section of ethylene diamine at sub monolayer coverage.

J.A. Barnes (✉) · H.-P. Loock
Dept. of Chemistry, Queen's University, Kingston, ON, K7L 3N6 Canada
e-mail: jbarnes@chem.queensu.ca

H.-P. Loock
e-mail: hploock@chem.queensu.ca

G. Gagliardi
Istituto Nazionale di Ottica, Consiglio Nazionale delle Ricerche, Comprensorio "A. Olivetti",
via Campi Flegrei 34, 80078 Pozzuoli (Napoli), Italy
e-mail: gianluca.gagliardi@ino.it

G. Gagliardi, H.-P. Loock (eds.), *Cavity-Enhanced Spectroscopy and Sensing*,   351
Springer Series in Optical Sciences 179, DOI 10.1007/978-3-642-40003-2_10,
© Springer-Verlag Berlin Heidelberg 2014

10.1 Introduction: Microcavities in Chemical Sensing

Optical microresonators, of different geometries, have been used as label-free and ultrasensitive chemical sensors over the past several years [1–3]. The interaction of a chemical species with a microresonator occurs through the modification of the optical environment of the resonator. A change in the real part of the ambient refractive index leads to a wavelength shift in the whispering gallery modes (WGM) supported by the microcavity [4]. This method has been used to detect sub-monolayer adsorption of large molecules on silica microspheres, even enabling single-molecule detection in some cases [3]. Nevertheless, any approach based on refractive index measurements has no intrinsic selectivity to the targeted molecular species and chemical (and biological) speciation requires surface modification of the microresonator [5]. On the other hand, if the imaginary part of the molecule's refractive index leads to absorption lines or bands in the vicinity of the resonance wavelength, the lifetime of the cavity modes is reduced [6] leading also to a reduction in the power transferred to the WGM [7] and a broadening of resonance line due to a corresponding reduction in the quality (Q) factor. Absorption detection of molecules in the microresonator mode volume has therefore some potential as a chemically specific detection scheme. Previously, trace gas detection [7] or trace solute detection [8] have been implemented using this cavity-enhanced absorption spectroscopic approach. In this chapter we demonstrate that optical loss measurements and WGM frequency measurements can be combined to give simultaneously the optical loss and refractive index of molecules absorbed on the surface of the resonator.

Time-domain measurements have been routinely used to measure optical lifetimes in monolithic devices [9], fiber-optic loops [10] and microresonators [11, 12] and other waveguiding cavities. The intensity of a short optical pulse injected into the cavity decays exponentially with time as the circulating power is dissipated. The time-constant for the decay reflects the optical lifetime of the cavity. While this method is practical for macroscopic cavities, the photon lifetime of microcavities is frequently on the order of nanoseconds requiring picosecond laser pulses and fast data acquisition systems. This will not only increase the cost but also result in a low duty cycle.

An alternative approach to time-domain measurements uses light from an intensity-modulated continuous-wave source. The finite optical lifetime of the resonator induces a phase-shift between the light exiting the cavity and the incident light. The variation in the phase-shift with modulation frequency can be related to the optical lifetime [13]. This phase-shift ring-down approach has been extensively applied to fiber-optic loops [14] and mirror cavities [15]. By applying the robust, inexpensive and simple phase-shift cavity ring-down (PS-CRD) method to whispering gallery modes of a microresonator sphere we take a first step to a chemical sensor, which is suited to absorption detection of very small amounts of analyte.

An excellent survey of ultrasensitive detection using microcavities has been provided by Wu and Vollmer in the previous chapter of this book. Their chapter also describes the different types of microcavities such as micro-toroids, microspheres, silicon nanowire resonators, microdisk resonators, silica bottleneck resonators, microtube resonators, etc. These microresonators support low-loss whispering-gallery modes (WGM) when evanescently excited from adjacent waveguiding structures. WGM may be understood as rays of light that are guided along the circumference of the resonator by total internal reflection. As expected for any resonantly excited optical cavity, a transfer of energy to the resonator may be observed as an increase of the intensity of the scattered light, and a decrease in the light transmitted by the waveguide used for coupling. Spheres, rings, disks, toroids, cylinders and capillaries have all been investigated as resonators over the past several years [2, 16–24]. The high quality factor (Q) and small mode volume associated with WGMs also makes these microresonators suitable for numerous optical devices, such as microlasers [25, 26] and add-drop elements [27], as well as chemical sensors [4, 7].

   In this chapter we focus on the oldest form of optical microresonators—the silica microsphere—and investigate its use for direct optical loss measurements through phase-shift cavity ring-down spectroscopy. As will be shown it is possible to combine the advantages of

silica-sphere microcavities: very high finesse and Q-factor, compact size, simple fabrication

active frequency locking: immunity to temperature drifts and ability to extract WGM frequency shifts

cavity-enhanced absorption spectroscopy: long absorption path

and in particular with *phase-shift cavity ring-down spectroscopy*: absolute optical loss measurements, immunity to intensity fluctuations, high duty cycle, ability to measure short ring-down times

   We demonstrate the capabilities of the method using ethylene diamine as an overtone absorber at sub monolayer coverage.

   The next section is reviewing the relevant theoretical background, with an emphasis on the interaction of the WGM with an absorbing overlayer through the evanescent field. The resulting frequency shift is quantified using a simplified model. It is also demonstrated how the increase in optical loss can be quantified using phase-shift CRD. Section 10.3 presents previously published and new experimental observations on phase-shift CRD spectroscopy using a laser that is frequency-locked to a WGM mode of a silica microsphere. In particular, we measure the absolute absorption cross section of aligned ethylene diamine molecules at sub monolayer coverage on the microsphere through an overtone transition at around 1.55 μm.

10.2 Theoretical Background

10.2.1 Modes Inside a Dielectric Sphere

Light circulating within a monolithic resonator of index n_s, surrounded by a uniform medium of index n_m, is trapped by total internal reflection at the resonator surface [28]. In the case of a spherical resonator, the ray of light can be envisioned as following a zig-zag path around the equator of the sphere. Using the more appropriate mode picture of light this ray corresponds to a mode field that peaks just below the surface of the sphere but extends both into the interior of the sphere and into the low-index medium outside the sphere through the evanescent field. Three integers, n, l and m, describe the mode within the cavity. The number of maxima in the polar direction is given by $l - m + 1$ where $|m| \leq l$. The integer n counts the number of radial maxima in the field. The inclination of the zig-zag optical path, with respect to the equator, is represented by m, with $m = l$ having the smallest inclination angle and $m = 0$ being inclined at $90°$ with respect to the equator. A change in sign for m indicates a change in propagation direction about the sphere. For a perfect sphere, the resonant wavelength depends only on n and l, and modes with different of m are degenerate.

The electromagnetic field within a microsphere, of radius R_0, is a vector field which satisfies the Helmholtz equation for a fixed polarization state. The two polarization states, i.e. the transverse electric (TE), and transverse magnetic (TM), represent different solutions and should be distinguished. In the TE mode the electric field is parallel to the surface, while the TM mode has the magnetic field parallel to the surface, with the electric field being predominantly in the radial direction. The TE and TM electric fields can be described by the vectorial spherical harmonic functions

$$
\begin{aligned}
\mathbf{X}_l^m &= \nabla Y_l^m \times \mathbf{r}/\sqrt{l(l+1)}, \\
\mathbf{Y}_l^m &= r\nabla Y_l^m/\sqrt{l(l+1)}, \\
\mathbf{Z}_l^m &= Y_l^m \hat{\mathbf{r}}.
\end{aligned}
\tag{10.1}
$$

Accordingly, the fields can be expressed as

$$
\begin{aligned}
\mathbf{E}_{lm}^{\mathrm{TE}}(\mathbf{r}) &= E_0 \frac{f_l(r)}{k_0 r} \mathbf{X}_l^m(\theta, \phi), \\
\mathbf{E}_{lm}^{\mathrm{TM}}(\mathbf{r}) &= \frac{E_0}{n_s^2} \left(\frac{f_l'(r)}{k_0^2 r} \mathbf{Y}_l^m(\theta, \phi) + \sqrt{l(l+1)} \frac{f_l(r)}{k_0^2 r^2} \mathbf{Z}_l^m(\theta, \phi) \right).
\end{aligned}
\tag{10.2}
$$

The radial function inside the sphere ($r < R_0$) is $f_l(r) = \psi_l(n_s k_0 r)$, whereas the radial decay of the evanescent field outside the sphere ($r > R_0$) is given by $f_l(r) = A\psi_l(k_0 r) + B\chi_l(k_0 r)$. The functions $\psi_l(\rho) = \rho j_l(\rho)$ and $\chi_l(\rho) = \rho n_l(\rho)$ where $j_l(\rho)$ and $n_l(\rho)$ are, respectively, spherical Bessel and Neumann functions [29]. Frequently, it is reasonable to approximate the radial decay of the evanescent field by an exponential function with a decay length of [30]

$$
r_{\mathrm{ev}} = \frac{1}{k_0 \sqrt{n_s^2 - n_m^2}}
\tag{10.3}
$$

where $k_0 = 2\pi/\lambda_0$ is the angular wavenumber, n_s is again the sphere's refractive index and n_m is the refractive index of the surrounding medium. The resonant wavelengths of the microsphere are given by the characteristic Eq. (10.4) which is determined by requiring continuity of the tangential components of the electric and magnet fields at the dielectric boundary [31]

$$\left(\eta_s \sqrt{\frac{l(l+1)}{R_0^2} - k_{l,n}^2 n_m^2} + \frac{l}{R_0} \right) j_l(k_{l,n} n_s R_0) = k_{l,n} n_s j_{l+1}(k_{l,n} n_s R_0) \qquad (10.4)$$

with $\eta_s = 1$ for TE modes and $\eta_s = n_s^2/n_m^2$ for TM modes. For near equatorial modes $l \gg 1$, and the solutions to the characteristic equation are

$$k_{l,n} \simeq \frac{1}{R_0 n_s} \left[(l + 1/2) + 2^{-1/3} a_n (l + 1/2)^{1/3} - \frac{P}{\sqrt{n_s^2 - 1}} + \frac{3 a_n^2}{20} \left(\frac{2}{l+1} \right)^{1/3} \right.$$
$$\left. + O\{(l+1)^{-2/3}\} \right]$$

$$(10.5)$$

where $P = n_s$ for TE modes and $P = 1/n_s$ for TM modes and a_n is the n-th root of the Airy function [32]. For large l and small n values, this expression is approximately equal to

$$k_l = n_s \frac{2\pi}{\lambda} = \frac{l}{R_0}. \qquad (10.6)$$

Equation (10.6) is equivalent to requiring an integer number of wavelengths to fit around the circumference of the microsphere on resonance. For a perfect sphere, each mode is $(2l+1)$-fold degenerate since the solutions of the characteristic equation do not depend on m. Any deviation from the spherical shape lifts this degeneracy, and produces a much more complicated microsphere resonator spectrum.

10.2.2 Estimate of the Volume Fraction of the Evanescent Wave

To estimate the WGM volume that interacts with the adsorbed molecular layer we need to know the fraction of the WGM propagating in the evanescent wave. Assuming, first, that the layer is homogeneous and thick compared to the penetration depth of the evanescent wave (Fig. 10.1b and Eq. (10.3)) we can, for a TE mode, write the ratio of the mode volume of the evanescent field to that inside the sphere as [4]

$$f = \frac{\varepsilon_m \int_{r=R_0}^{\infty} j_l^2(k n_s R_0) \exp(-(r - R_0)/r_{ev}) r^2 dr \int_{\phi=0}^{2\pi} \int_{\theta=0}^{\pi} [Y_l^m(\phi, \theta)]^2 \sin\theta d\theta d\phi}{\varepsilon_s \int_{r=0}^{R_0} j_l^2(k n_s r) r^2 dr \int_{\phi=0}^{2\pi} \int_{\theta=0}^{\pi} [Y_l^m(\phi, \theta)]^2 \sin\theta d\theta d\phi}$$

$$\approx \frac{\varepsilon_0 n_m^2 j_l^2(k n_s R_0) \int_{R=R_0}^{\infty} \exp(-(r - R_0)/r_{ev}) r^2 dr}{\varepsilon_0 n_s^2 j_l^2(k n_s R_0) \frac{R_0^3}{2} \frac{n_s^2 - n_m^2}{n_s^2}}$$

$$\approx \left(R_0^2 r_{ev} + 2 R_0 r_{ev}^2 + 2 r_{ev}^3 \right) \frac{2}{R_0^3} \frac{n_m^2}{n_s^2 - n_m^2}$$

$$\approx \frac{2R_0^2 r_{ev}}{R_0^3} \frac{n_m^2}{n_s^2 - n_m^2}$$

$$\approx \frac{\lambda_0}{\pi R_0} \frac{n_m^2}{(n_s^2 - n_m^2)^2}. \tag{10.7}$$

Here we integrated the spherical Bessel function j_l according to Lam et al. [32], and acknowledged that the evanescent field decay length is much smaller than the sphere's radius, $r_{ev} \ll R_0$. For the general case of a TE or TM mode expression (10.7) changes to

$$f \approx \frac{\lambda_0}{\pi R_0} \frac{n_m^2}{(n_s^2 - n_m^2)N(n_s, n_m, l, k, R_0)} \tag{10.8}$$

where [32]

$$N(n_s, n_m, l, k, R_0) = \begin{cases} n_s^2 - n_m^2 & \text{for TE modes,} \\ (n_s^2 - n_m^2)[(\frac{l+1/2}{kR_0 n_m})^2 + (\frac{l+1/2}{kR_0 n_s})^2 - 1] & \text{for TM modes.} \end{cases} \tag{10.9}$$

For example, the refractive index of the microresonator sphere material at $\lambda = 1.56\,\mu m$ was calculated from the Sellmeier coefficients for silica as $n_s = 1.444$. Assuming that the sphere is suspended in vacuum ($n_m = 1$) only 0.28 % of the WGM field is in the evanescent wave. This number increases to 4.0 % if the surrounding medium is water ($n_m = 1.31$) and 29 % if $n_m = 1.40$.

For an overlayer that is thin compared to the decay length of the evanescent wave the case is different (Fig. 10.1a). In the example given in Sect. 10.3 the WGM interacts only with a thin ethylene diamine adsorbant layer of subwavelength thickness and with a refractive index ($n_{EDA} = 1.4413$) that is close to that of the index of the silica microsphere, n_s. Given that only a small fraction of the WGM power resides in air we can approximate the fraction of the TE mode field in the overlayer as the ratio between the field in the EDA layer and that inside the sphere, i.e. as

$$f \approx \frac{\varepsilon_{EDA} \int_{r=R_0}^{R_0+\Delta R} j_l^2(k n_s R_0) r^2 dr \int_{\phi=0}^{2\pi} \int_{\theta=0}^{\pi} [Y_l^m(\phi, \theta)]^2 \sin\theta \, d\theta \, d\phi}{\varepsilon_s \int_{r=0}^{R_0+\Delta R} j_l^2(k n_s r) r^2 dr \int_{\phi=0}^{2\pi} \int_{\theta=0}^{\pi} [Y_l^m(\phi, \theta)]^2 \sin\theta \, d\theta \, d\phi}$$

$$\approx \frac{\varepsilon_0 n_{EDA}^2 j_l^2(k n_s R_0)(\frac{(R_0+\Delta R)^3}{3} - \frac{R_0^3}{3})}{\varepsilon_0 n_s^2 j_l^2(k n_s(R_0+\Delta R))\frac{(R_0+\Delta R)^3}{2} \frac{n_s^2 - n_m^2}{n_s^2}}$$

$$\approx \frac{2}{3} \frac{(R_0+\Delta R)^3 - R_0^3}{(R_0+\Delta R)^3} \frac{n_{EDA}^2}{n_s^2 - n_m^2}. \tag{10.10}$$

Here, we assumed that the field (value of the squared spherical Bessel function) is constant over the thickness of the EDA layer and that $n_s \approx n_{EDA}$. Accordingly, the general case valid for TE and TM modes is given by

$$f \approx \frac{2}{3} \frac{(R_0+\Delta R)^3 - R_0^3}{(R_0+\Delta R)^3} \frac{n_{EDA}^2}{N(n_s, n_m, l, k, R_0)}. \tag{10.11}$$

Fig. 10.1 Sketch of microsphere with taper. The insert shows the interaction of the evanescent wave (**a**) with a thin and (**b**) with a thick adsorbant overlayer

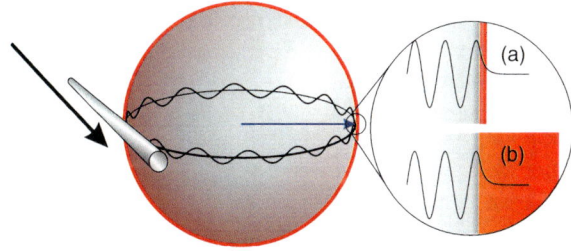

Using $n_s = 1.4439$, $n_{EDA} = 1.4413$ and $n_m = 1.0$ and with $\Delta R = 1.2$ nm one obtains $f = 3.1 \times 10^{-5}$ independent of the wavelength of the light. For a TM polarized WGM the fraction, f, depends on the value assumed for l in (10.9). For example, the fraction of the WGM propagating in the evanescent wave is $f = 7.4 \times 10^{-5}$ if $l = 590$ and 1.6×10^{-5} if $l = 850$. The actual fraction of the total WGM residing the overlayer is slightly less, since we neglected the fraction of the light traveling outside the overlayer (in air) in the denominator of (10.10).

10.2.3 Excitation and Detection of Cavity Modes

Free-space excitation of WGMs is very inefficient for spheres greater in radius than a few wavelengths [33], however, a laser tightly focused tangential to a liquid droplet has been used to excite its whispering-gallery modes (G. Gagliardi, unpublished). Efficient excitation of a microresonator requires phase matching of the evanescent field of an external coupler to the resonator modes.

Prism couplers, utilizing total internal reflection, were among the first coupling devices used. A laser is focused inside a high refractive index prism such that total internal reflection is achieved at the prism face above which the microsphere is situated. Adjustment of the incident angle allows for phase matching between the evanescent field at the prism face and the microcavity resonance. The beam shape can be adjusted to optimize mode overlap, and adjustment of the microsphere-prism distance tunes the coupling efficiency. Coupling efficiencies of $\sim$80 % have been achieved [34]. Hybrid fiber-prisms have also been used to excite microspheres [35].

Side-polished fiber half-block couplers have been successfully used [36] as have various waveguide structures. Silica strip-line pedestal waveguides can couple to microspheres with high efficiency [37]. Waveguide couplers have been constructed in BK-7 glass slides using K^+ ion exchange to excite borosilicate microspheres [38].

Microresonators are excited most efficiently using a tapered optical fiber. Coupling efficiencies of 99.99 % have been observed to fused silica microspheres [39]. A tapered region is produced in a single mode optical fiber through heating and stretching of the fiber [40, 41], or by chemical erosion [42–44]. The tapered region is typically a few microns in diameter, and in the near-infrared a significant fraction of the transmitted power is then carried in the evanescent field of the taper. Coupling is achieved by overlapping the evanescent field of the taper and microsphere mode.

The coupling efficiency decreases exponentially with the taper-microsphere separation distance [45]. The propagation constant, β, of light in the taper is a function of the taper diameter.

$$\beta^2 = k^2 n_f^2 - (2.405)^2/\rho^2. \tag{10.12}$$

Here, k is the free-space wavenumber of the light, n_f is the cladding refractive index, and ρ is the taper diameter [45]. By moving the microsphere along the taper, the propagation constant in the taper can be varied in a continuous fashion, allowing phase matching to the microsphere WGM. Fiber tapers can be used not only to excite WGMs, but also to extract optical energy from these modes.

In a common case the WGM spectrum is recorded by detecting scattered light from the microresonators using, e.g., a microscope objective. The intensity of the detected light is then directly proportional to the light transferred to and scattered from the WGM, i.e. to the Q-factor of the WGM.

Alternatively, the transmission spectrum of a fiber taper used to excite a microresonator shows characteristic attenuation peaks whenever a WGM is excited. The light propagating through a fiber taper consists of a superposition of transmitted incident light and circulating light within the microsphere, which has coupled out of the resonator and back into the fiber taper. There is a 180° phase-shift between these two fields, with the relative amplitudes depending on the coupling conditions [34]. Due to destructive interference a decrease in transmitted intensity is observed only when a WGM is resonantly excited. In the overcoupled regime, the amplitude of the field exiting the microsphere exceeds that of the light transmitted through the taper, whereas in the undercoupled regime, the non-interacting field transmitted through the taper is larger. At critical coupling, both fields are of equal amplitude, so that the transmitted light intensity drops to zero on resonance. The undercoupled and overcoupled cases can be distinguished by rapidly scanning the laser over the WGM resonance [46]. Under these conditions there is a beating of the input and intracavity fields which causes a ringing in the transmitted signal. The ringing pattern is characteristic of the coupling regime, allowing the two coupling regimes to be distinguished. A detailed analysis of the ringing can also give coupling and dispersion properties, the strength of intracavity Rayleigh scattering, and an estimation of the power launched into the mode [47]. Another method relies on using intensity modulated light to excite the cavity resonance and measuring the relative phase-shift between the modulation envelopes of the light exiting the fiber taper and the incident light [48]. In the case of overcoupling, there is a positive relative phase-shift, while a negative relative phase-shift is seen in the undercoupling regime. Phase-shift cavity ring-down spectroscopy will be discussed in more detail below. When recording the spectrum of transmitted light through the taper (or indeed through any of the devices used to couple light into the sphere) one has to remember that the intensity of the attenuation features in the spectrum relate to how close a particular WGM is to critical coupling. The intensity drop in the transmitted light spectrum is therefore not just dependent on the circulating intensity of the light in the WGM.

Beside detection of scattered light through a lens, and detection of attenuation peaks in the light transmitted through a coupling waveguide, a third mode of detection involves monitoring the Rayleigh backscattered light. Rayleigh backscatter can populate WGMs that counter-propagate relative to the initially excited WGM. These counterpropagating WGM can couple light back into the taper but in the opposite direction to the excitation laser light. Rayleigh backscattering is caused by inhomogeneities within the microsphere [49]. An optical circulator can be used to isolate the backscattered light coupled into the fiber taper and direct it to a photodetector. When observing backscattered light an increase in intensity is seen on WGM resonance. While the WGM spectrum of Rayleigh backscattered light is expected to be similar to that of light scattered from the sphere and detected though a lens, the relative intensities are different. In the former case, those modes are selected that can efficiently couple into Rayleigh backscattered WGMs able to couple well with propagating taper modes. In the latter case, no such selection takes place. The WGM spectrum of the Rayleigh backscattered light is therefore more sparse and biased towards low-m and high-Q modes.

10.2.4 Resonance Shifts in WGMs due to External Perturbations

Perturbations in the optical environment of a microsphere lead to a wavelength shift in the whispering-gallery-modes. This was demonstrated by Vollmer *et al.* in the case of protein adsorption on the surface of the microsphere [17]. In the so called reactive mechanism, the wavelength shift can be understood in terms of the perturbation of the resonance due to a change $\delta\varepsilon_r(r)$ in the relative electric permittivity in a volume V_p surrounding the microsphere. The wavelength shift of a microresonator is given by [30]

$$\left(\frac{\delta\lambda}{\lambda_0}\right) = \frac{\int_{V_p} \delta\varepsilon_r \mathbf{E}_0^* \cdot \mathbf{E}_p d\mathbf{r}}{2\int_V \varepsilon_r \mathbf{E}_0^* \cdot \mathbf{E}_0 d\mathbf{r}}, \tag{10.13}$$

$\mathbf{E}_0$ is the unperturbed field in the microsphere and $\mathbf{E}_p$ is the perturbed field within V_p. The relative shift is given by the ratio of the excess electrical energy within V_p to the total mode energy over all space V [30]. Equation (10.13) applies to both TE and TM modes.

For a TE mode and in the limit of a very thin layer on the surface of the sphere with thickness $t \ll \lambda_0$, with uniform refractive index n_p we can approximate the wavelength shift due to adsorption as

$$\left(\frac{\delta k}{k_0}\right)_{TE} = -\frac{1}{R_0(n_1^2 - n_2^2)} \int_{R_0}^{R_0+t} \left(n_p(r)^2 - n_2^2\right) dr. \tag{10.14}$$

If the refractive index of the adsorbed material is assumed constant,

$$\left(\frac{\delta\lambda}{\lambda_0}\right)_{TE} = \frac{(n_p^2 - n_m^2)t}{(n_s^2 - n_m^2)R_0}. \tag{10.15}$$

The shift for TM modes is related to those for TE modes by the relationship

$$\frac{(\delta\lambda/\lambda_0)_{TM}}{(\delta\lambda/\lambda_0)_{TE}} = 1 + \frac{l(l+1)(n_p^{-2} - n_s^{-2})(k_0 R_0)^{-2}}{(\chi_l'/\chi_l)^2 + l(l+1)(n_s k_0 R_0)^{-2}} \tag{10.16}$$

with $\chi_l(z)$ being the spherical Ricatti-Neumann function. When $n_p > n_s$, the TM mode shifts less than the TE mode. If the polarizability of the medium surrounding the microsphere is anisotropic, the TE and TM modes will interact with different components of the polarizability tensor. This is also the case for a thin layer of molecules adsorbed on the surface of a microsphere [50]. In this case, the wavelength shift for the TE mode is given by

$$\left(\frac{\Delta\lambda}{\lambda_0}\right)_{\text{TE}} = \frac{G_{\text{TE}}\alpha_t\theta}{\varepsilon_0 R_0}, \tag{10.17}$$

$$G_{\text{TE}} = \frac{1}{n_s^2 - n_m^2} \tag{10.18}$$

where ε_0 is the vacuum permittivity and G_{TE} represents the response of the WGM to the adsorbed molecules. The surface density of the molecules is given by θ and the tangential component of the polarizability is α_t. This equation was already given in the previous chapter by Wu and Vollmer who also highlighted that α_t describes the excess polarizability resulting from displacement of the surrounding medium by the thin adsorbed layer.

The main component of the electric field for the TM mode is normal to the surface, however, there is a non-negligible tangential component to the field. The fraction of the tangential component in the evanescent field at the sphere surface is

$$\phi_t = \frac{1}{1 + l(l+1)(R_0/r_{\text{ev}})^{-2}} \tag{10.19}$$

where r_{ev} is the decay length of the evanescent field given by (10.3). The wavelength shift for the TM mode is influenced by the normal and tangential component of the polarizability, α_n and α_t, which gives

$$\left(\frac{\Delta\lambda}{\lambda_0}\right)_{\text{TM}} = \frac{G_{\text{TM}}[\phi_t\alpha_t + (1 - \phi_t)\alpha_n]\theta}{\varepsilon_0 R_0} \tag{10.20}$$

and G_{TM} is defined by

$$G_{\text{TM}} = G_{\text{TE}}\frac{(R_0/r_{\text{ev}})^2 + l(l+1)}{(R_0/r_{\text{ev}})^2 + l(l+1)(n_m/n_s)^2}. \tag{10.21}$$

Equations (10.17) and (10.20) can be used to calculate the surface coverage of adsorbed molecules from the wavelength shifts for a TE and a TM mode, if α_t and α_n are known. For small molecules, the polarizability components can be calculated from molecular orbital theory. For larger molecules, where the calculation is unreliable, one can use the readily available isotropic polarizability, α_{iso}, together with the experimentally observed wavelength shift at TE and TM polarization to obtain α_t, α_n and θ from Eqs. (10.17) and (10.20) and (10.22)

$$\alpha_{\text{iso}} = (2\alpha_t + \alpha_n)/3. \tag{10.22}$$

The isotropic polarizability can be determined experimentally from the concentration dependence of the refractive index.

10.2.5 Loss Mechanisms

Different loss mechanisms exist which will limit the photon lifetime of whispering gallery modes [12]. Intrinsic absorption by the resonator material or absorption by material surrounding the microsphere leads to attenuation. Scattering can occur within the resonator or at its surface. Optical tunneling, and coupling to external waveguides used to excite the cavity also dissipate energy. The extent of energy dissipation is usually expressing in terms of a quality-factor (Q-factor) which is defined as the ratio of the energy stored to the power dissipated, normalized to the optical period of oscillation.

$$Q \equiv \omega \frac{E_{\text{stored}}}{P_{\text{diss}}} = \omega\tau = \frac{\omega}{\Delta\omega}. \tag{10.23}$$

Here ω is the resonant angular frequency, $\Delta\omega$ is the resonance width, and τ is the photon lifetime or ring-down time. Each loss mechanism can be considered to have its own Q-factor, with the total Q-factor being given by

$$\frac{1}{Q_{\text{tot}}} = \sum_i \frac{1}{Q_i}. \tag{10.24}$$

Silica is a nearly ideal material for microsphere resonators since it has low optical loss in the near infrared. Its attenuation of only 0.2 dB/km at $\lambda = 1.55$ μm arises from intrinsic absorption and Rayleigh scattering [51]. The intrinsic absorption-limited Q-value can then be calculated as

$$Q_{\text{abs}} = \frac{2\pi n_{\text{eff}}}{\lambda_0 \alpha_{\text{abs}}} \approx 2.92 \times 10^{10}. \tag{10.25}$$

Q-values approaching the material absorption limit have been observed [23]. Estimates have also been made for Rayleigh scattering due to molecular-sized surface clusters under grazing incidence. This leads to the expression

$$Q_{\text{scatt}} = \frac{\lambda_0^2 R_0}{2\pi^2 n_{\text{eff}}^2 L_{\text{scatt}}^2 B_{\text{scatt}}} \tag{10.26}$$

where L_{scatt} and B_{scatt} are the root-mean-square size and correlation length of surface inhomogeneities [6]. Using the values $L_{\text{scatt}} = 0.3$ nm and $B_{\text{scatt}} = 3$ nm reported for glass [52], we find $Q_{\text{scatt}} = 6.8 \times 10^{10}$ for a 300 μm diameter sphere at wavelength $\lambda_0 = 1.55$ μm.

In the ray picture light is confined within a microsphere by total internal reflection, however, total internal reflection is strictly valid only at a flat interface. The curved surface of a microsphere leads to optical tunneling, which results in a loss of optical energy. The Q-factor based on whispering gallery mode leakage decreases exponentially with decreasing sphere radius. For a sphere of radius >12 μm, a value of $Q_{\text{tunnel}} > 10^8$ can be maintained at $\lambda_0 \sim 1.5$ μm [12].

Finally, in the near-infrared, the adsorption of water onto silica can prove a limiting constraint. Apart from contributing to scattering losses, water also possesses vibrational overtone absorption band at around 1.47 μm which dissipates energy through optical absorption of the evanescent wave [23].

10.2.6 Determination of Optical Loss Using Cavity Ring-Down Methods

In most cases the concentration of the molecules within the mode volume of a microresonator is obtained from the wavelength shift of a cavity mode even though the Q-factor (or finesse) of the cavity could also be used to determine the concentration, as long as the sample absorbs or scatters light. Scattering and absorption from even sub-monolayer coverages measurably reduces the finesse of the cavity and can be observed through various means. If the molecules exhibit absorption lines or bands in the vicinity of the resonance wavelength, the lifetime of the cavity modes will be reduced, [6] leading to a reduction in power transferred to the WGM [7] and a broadening of the resonator line due to a corresponding reduction in the Q-factor. Previously, trace gas detection [7] or trace solute detection [8] have been implemented using this cavity-enhanced absorption spectroscopic approach. In time-domain cavity ring-down spectroscopy the lifetime of the cavity modes is measured directly, but the ring-down times can be very short (a few nanoseconds even for $Q > 10^6$). Phase-shift CRD spectroscopy measures the reduction of the phase-difference between intensity modulated light entering and exiting the WGM, and provides the same information while not requiring short laser pulses and fast data acquisition systems.

Time-domain measurements have already been used to measure optical lifetimes in a variety of resonator configurations [53–56] including monolithic silica resonator cavities [9, 57] and microresonators [11, 35] For such cavity ring-down time measurements the time-dependent intensity decay of a short optical pulse injected into the resonator is monitored. Gorodetsky *et al.* monitored the change in lifetime of a WGM in a silica microsphere as water was adsorbed onto its surface [6]. The lifetime was determined by recording the exponential intensity decay after the input laser was shut off using an acousto-optic shutter. The same technique was used by Pöllinger *et al.* to determine the cavity lifetime of a bottle microresonator [58]. Also, Savchenkov *et al.* used time-resolved CRD spectroscopy to measure the lifetime of light that was generated in a CaF_2 microresonator by non-linear Raman scattering effects [59]. Such time-domain measurements require fast optical shutters or picosecond laser pulses together with fast detectors and acquisition electronics.

An alternative approach involves the use of intensity-modulated light from a continuous-wave source and measurement of the modulation frequency dependent phase-shift of light exiting the resonator. This approach is referred to as phase-shift cavity ring-down (PS-CRD) and has been used for mirror cavities [13, 60], fiber-optic loops [61], and, recently, silica microsphere resonators [48, 62]. In phase-shift CRD, the input laser is intensity modulated. The WGM intensity is therefore also modulated, but the modulation envelope is shifted in phase, with respect to that of the input light, due to the finite photon lifetime of the cavity mode. The phase-shift can be determined using an RF-lock-in amplifier, without the need for fast switching of the laser or fast detectors. The phase-shift is directly proportional to the modulation frequency, with the proportionality constant being related to the optical parameters (loss, scattering, or coupling coefficient) of the system. The phase-shift technique was first applied to determine the lifetime of mirror cavities, and later, fiber-optic cavities [13, 63].

To determine the cavity ring-down times, scattered light is commonly collected through a microscope lens. Alternatively, one can monitor the intensity and frequency of the light bypassing the cavity. On resonance this intensity is reduced. A third method records the frequency and the intensity of the Rayleigh backscattered light, i.e. the light that is generated when scattering of the WGM in the cavity inverts the direction of propagation and the light is directed back to the laser light source. In the following paragraphs we discuss these three cases and derive the respective expressions for the shift of the intensity modulation phase. Experimental spectra are shown for all three configurations in Fig. 10.2. It will then become apparent that the combination of two or three measurements allows us to disentangle the intracavity loss term from the optical loss due to coupling to the input waveguide and even provide a means to determine the absolute optical loss of a film at sub-monolayer coverage.

10.2.6.1 Phase-Shift CRD of Isotropically Scattered Light

In our theoretical models we assume that a single whispering gallery mode is excited. To calculate the phase-shift between intensity-modulated light entering and exiting any cavity one can use the Laplace transform of the system's impulse response function [64–66]

$$I(s) = \mathcal{L}\{I(t)\} = \int_0^\infty e^{-st} I(t) dt \qquad (10.27)$$

with $s = \mu + i\Omega$. Here, Ω is the angular modulation frequency of the intensity-modulated light and μ is a damping factor associated with the intensity modulation (here $\mu = 0$). The phase-shift can be obtained from the complex function $I(\Omega) = iN(\Omega) + D(\Omega)$ by noting that $\tan\phi(\Omega) = N(\Omega)/D(\Omega)$.

In a previous article we described the frequency dependence of the phase angle, ϕ, when the resonant light was scattered normal to the sphere [48]. The phase-shift is calculated from the Laplace transform of an exponential decay function, i.e. the impulse response function of a single mode decay of a loaded cavity. On resonance

$$I(t) = A \exp(-t/\tau_L) \qquad (10.28)$$

and with (10.27) we determine the cavity response to an intensity-modulated beam as

$$I(\Omega) = \frac{A}{\tau(\Omega^2 + 1/\tau_L^2)} - i\frac{A\Omega}{(\Omega^2 + 1/\tau_L^2)}. \qquad (10.29)$$

The phase-shift $\Delta\phi$ between the intensity-modulated light entering and exiting the cavity is therefore simply given by

$$\tan\Delta\phi(\Omega) = -\Omega\tau_L. \qquad (10.30)$$

In an experiment one needs to consider an additional phase-shift due to light transmission delays, electronic signal delays, delayed detector response, etc. We

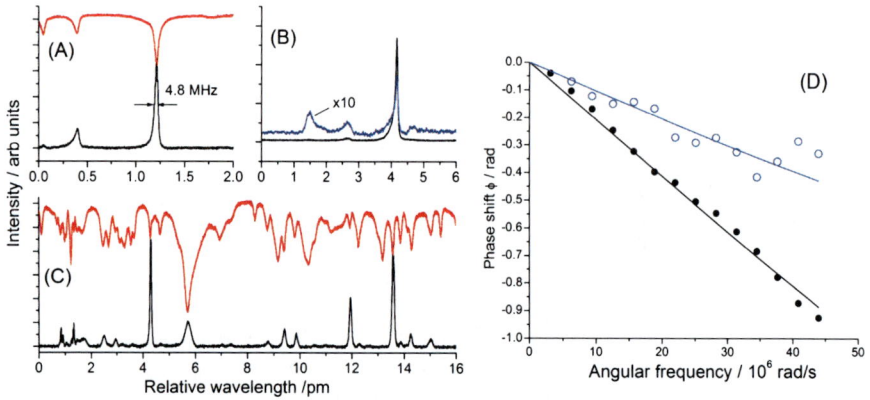

Fig. 10.2 [*Left*] WGM spectra near 1550 nm of a 300 μm silica microsphere (**A**) obtained by monitoring the transmitted intensity of the fiber taper (*red, top*) and the Rayleigh backscattered light (*black, bottom*). The highlighted resonance has a width of 4.8 MHz corresponding to a Q-factor of 42×10^6 and a ring-down time of about 33 ns. (**B**) A different isolated WGM monitored by recording the scattered intensity through a microscope lens above the sphere (*blue*) and again through Rayleigh backscatter (**C**) as in (**A**) but showing a larger section of the WGM spectrum. [*Right*] (**D**) Phase shift observed for a WGM when detecting light scattered radially from the sphere (*open, blue circles*) and through Rayleigh backscatter (*solid, black circles*). The WGM was isolated as in (**B**). The *solid lines* are obtained from Eq. (10.30) (*blue line*) and Eq. (10.40) (*black line*) with $\tau_L = 10.4$ ns in both cases

therefore experimentally determine the offset phase angle, ϕ_0, by e.g. measuring the phase-shift from a Fresnel reflection at the end of the fiber containing the taper.

We need only concern ourselves with the losses experienced by the field circulating within the microresonator. This case is very similar to those described in the cavity ring-down literature for an optical cavity with two or more mirrors (Fig. 10.3). The ring-down time depends on the optical loss in the cavity with effective length/circumference L and can be calculated from [48]

$$\tau = \frac{n_{\mathrm{eff}} L}{c_0(\alpha_{\mathrm{abs}} L - 2 \ln \Gamma)} \tag{10.31}$$

where $\sqrt{1 - \Gamma^2}$ is the coupling coefficient between the microsphere and taper, and α_{abs} is the distributed loss coefficient. For a decay involving more modes, we may have a multiexponential decay function and beating in the time domain. The corresponding function for $\tan \phi$ was given by Bescherer *et al.* [64] who also showed that the depth of the modulation amplitude is dependent on Ω and τ.

Figure 10.2D shows $\Delta \phi$ as a function of the modulation frequency, Ω, when the light was collected through a microscope lens above the sphere. Here the laser was locked to a WGM using the Pound-Drever-Hall method. Despite the high noise level it is apparent that the values can be fit with Eq. (10.30). Since the angles are small the function appears to be a straight line and one may use the small angle approximation to simply fit to $\Delta \phi(\Omega) = -\Omega \tau$. In contrast to what has been stated before [62] this approximation is not needed, of course.

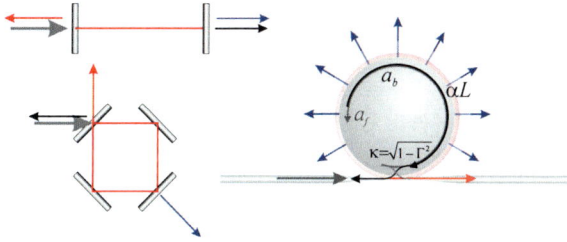

Fig. 10.3 Illustration of light emitted from a two mirror cavity, a four mirror ring-cavity, and a microsphere resonator (from top). In the ring cavities when light (*grey arrow*) enters the resonator one may observe light scattered from the resonator (*blue arrow*), Rayleigh backscattered into counter-propagating modes (*black*) or transmitted through the cavity (*red*). In a two mirror cavity the first two cases correspond to light that is passing through the cavity and the third case corresponds to light reflected from the cavity

10.2.6.2 Phase-Shift CRD of Rayleigh Backscattered Light

The intensity of the light in the WGM can also be detected by collecting light that is Rayleigh backscattered using a circulator before the fiber taper that directs light returning to the laser to a photodetector. Experimental frequency-dependent phase-shift data for Rayleigh backscattered light is shown in Fig. 10.2D. It can be seen that this data is not represented by Eq. (10.30) in that the phase-shift, ϕ (and not $\tan\phi$!) varies approximately linearly with modulation frequency. The phase-shift can even exceed $-\pi/2$ radians. With regards to the phase-shift of the intensity modulated light the microsphere acts similarly to an optical delay line.

Modeling Rayleigh backscattered light is more complicated than for direct scatter. The impulse response is no longer a simple exponential function as was used in deriving (10.30). An obvious approach to the problem would be to assume a simple coupled-mode model [67]. The "forward" circulating light, a_f and the backscattered light a_b interconvert at a slow rate given by rate constant k_S, while both modes decay with the same, faster rate given by k_L to taper or radiative modes denoted by $[q]$:

$$[a_f] \overset{k_S}{\underset{k_S}{\rightleftharpoons}} [a_b]; \qquad [a_f] \overset{k_L}{\rightarrow} [q]; \qquad [a_b] \overset{k_L}{\rightarrow} [q]. \tag{10.32}$$

The corresponding differential equations were given by Kippenberg [12, 67]

$$
\begin{aligned}
\frac{da_f}{dt} &= i\Delta\omega \cdot a_f - \left(\frac{1}{2\tau_0} + \frac{1}{2\tau_{ex}}\right) \cdot a_f + \frac{a_b}{2\gamma_{12}} + s\sqrt{1-\Gamma^2}, \\
\frac{da_b}{dt} &= i\Delta\omega \cdot a_b - \left(\frac{1}{2\tau_0} + \frac{1}{2\tau_{ex}}\right) \cdot a_b + \frac{a_f}{2\gamma_{21}}.
\end{aligned}
\tag{10.33}
$$

Here, a_f and a_b are the amplitudes of the mode field. The excitation frequency, ω, is detuned by $\Delta\omega$ from the whispering gallery mode resonance [67], the input power is given by $P = |s|^2$, and the coupling coefficient between fiber taper and sphere $\sqrt{1-\Gamma^2} = \sqrt{1/\tau_{ex}}$ is related to the associated decay time. The coefficients describing the interconversion between forward and backward propagated

WGM are imaginary and are related to the Rayleigh backscattering time constant as $\gamma_{12}\gamma_{21} = -\tau_S^2$. The decay rate $k_L = 1/\tau_L = 1/\tau_0 + 1/\tau_{ex}$ depends on the losses through the fiber taper coupler (expressed as τ_{ex}) and on absorption and scattering (τ_0). On resonance $\Delta\omega = 0$ and the equations in (10.33) are readily solved using the boundary conditions $a_f(t=0) = 1$, $a_b(t=0) = 0$ and $s = 0$ to give the impulse response function

$$I(t) = |a_b|^2 = \left(\frac{1}{2} - \frac{1}{2}\cos\left[\frac{t}{\tau_S}\right]\right)\exp\left(-\frac{t}{\tau_L}\right). \tag{10.34}$$

For a microresonator the ring-down time due to scattering and absorption $\tau_L = 1/k_L$ can only be shorter than the characteristic scattering time $\tau_S = 1/k_S$ [68]. When $\tau_S \gg \tau_L$ [69, 70] and the impulse response function of Eq. (10.34) passes through a maximum at time $t_{max} = 2\tau_L$. In this respect the microsphere resonator behaves similar to an optical delay line of length $L = 2\tau_L c_0/n$. For reference, the phase-shift of a simple delay line with impulse response function

$$I(t) = \delta(t - \tau_L) \tag{10.35}$$

can be calculated with the Laplace transform of (10.27) as

$$\Delta\phi(\Omega) = -\Omega\tau_L. \tag{10.36}$$

Similarly, the response function given by (10.34) produces a phase angle that is derived as

$$\tan\Delta\phi(\Omega) = -\frac{\Omega\tau_S^{-2} + 3\Omega\tau_L^{-2} - \Omega^3}{\tau_S^{-3} - 3\Omega^2\tau_L^{-1} + \tau_S^{-2}\tau_L^{-1}}. \tag{10.37}$$

One can simplify this equation realizing that $\tau_S > \tau_L$ and therefore

$$\tan\Delta\phi(\Omega) \approx -\frac{\tau_L^3\Omega^3 - 3\Omega\tau_L}{3\tau_L^2\Omega^2 - 1}. \tag{10.38}$$

Note that (10.38) has a discontinuity at $\Omega\tau_L = 3^{-1/2}$ and the respective function for the phase-shift $\phi(\Omega)$ is therefore continuous as it would be for the delay line of Eq. (10.36). In contrast to a delay line, (10.38) predicts, however, that $\tan\Delta\phi(\Omega)$ approaches $-3/2\pi$ for large Ω. Unfortunately, this result does not accurately represent the experimental data (Fig. 10.4c), though.

A third impulse response function may be considered a hybrid between (10.28) and (10.35). In this approximation we consider the time response to be a delayed exponential decay, where the delay time and the decay constant are both given by τ_L

$$I(t) = \exp\left(-k_L(t - \tau_L)\right) \quad \text{if } t > \tau_L,$$
$$I(t) = 0 \quad \text{if } t \leq \tau_L. \tag{10.39}$$

For this impulse response function the phase-shift response is a simple sum of (10.30) and (10.36)

$$\Delta\phi(\omega) = -\Omega\tau_L - \tan^{-1}(\Omega\tau_L). \tag{10.40}$$

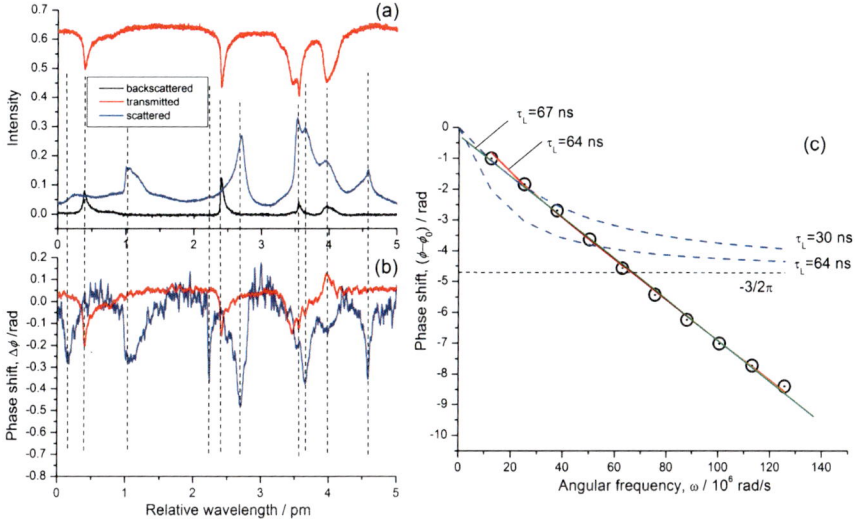

Fig. 10.4 [*Left panel*] Phase-shift spectrum and intensity spectra of a 300 μm diameter microsphere recorded using three photodetectors which were placed to detect the transmitted light through the fiber taper, the scattered light through a microscope objective and the Rayleigh backscattered light counter-propagating to the excitation light after a fiber circulator. (**a**) The intensity of a WGM spectrum when detected using the three photodetectors. (**b**) The associated phase shift of the intensity modulated light ($\Omega = 2$ MHz $\times 2\pi$). The Rayleigh backscattered light has very little background signal and the off-resonance phase is therefore very noisy and not shown. [*Right panel*] (**c**) Measured phase-shift of the Rayleigh backscattered light as a function of the angular modulation frequency (*circles*) analogous to that shown in Fig. 10.2D, but showing the phase shift of a different mode. The *solid green line* is a fit using Eq. (10.36) with $\tau_L = 67$ ns, the *red line* uses Eq. (10.40) with $\tau_L = 64$ ns, and the short-dashed, blue lines are modeled using Eq. (10.38) using $\tau_L = 30$ ns and 64 ns as indicated. The *solid line* indicates the limiting value of -1.5π

This phase-shift expression for backscattered light consists of two terms. The first represents the phase-shift that would result from an optical delay line, while the second corresponds to a resonator. The delayed onset of the peak Rayleigh backscatter intensity accounts for the delay line contribution while the second term results from the resonator behaviour. This function does accurately reproduce the experimental phase-shift data (Fig. 10.4c), but it cannot be derived from the coupled mode model presented in [12, 67].

It is possible to provide alternative experimental justification for the validity of (10.40) by calculating the ring-down time for WGMs observed in backscatter from the linewidth of a whispering gallery mode through the relation $\Delta \nu = 1/2\pi\tau$. For an isolated mode, the ring-down time obtained from modulation frequency-dependent phase-shift data, using (10.40), can be compared to the value calculated from its measured linewidth. These two values are found to agree (Fig. 10.2A). Another check can be made by simultaneously obtaining phase-shift data for a mode observed in direct scatter and backscatter. In the direct scatter case, Eq. (10.30) is expected to apply. Again, identical ring-down times are obtained. Despite a lack of theoretical justification for Eq. (10.40) it does appear to provide correct ring-down times (Fig. 10.2D). At present Eq. (10.40) must be regarded as a provisional result that provides a good approximation to our experimental observations. Its form does reflect the cavity behavior of the microsphere and a timescale for transfer of energy between modes. However, a revised theoretical approach is obviously required to arrive at a more satisfactory result. We currently work on an improved theoretical model that fully describes the phase shift experienced by Rayleigh backscattered light from a microresonator.

10.2.6.3 Phase-Shift CRD of Light Transmitted Through the Taper

When recording the WGM spectrum of the sphere by detecting the light that is transmitted through the taper, we note that on resonance the light is attenuated, where the attenuation depth indicates how close the WGM is to "critical coupling" ($-2 \ln \Gamma = \alpha L$).

Theoretically, the behavior of the microsphere can be modeled using a 4-mirror ring-resonator proposed by Rezac (Fig. 10.3) [71]. This model can also be related to a monolithic ring cavity [57] or to the more common 2-mirror cavity that is used in most CRD setups. Rezac derived equations that describe the frequency dependent intensity of light transmitted through a tapered fiber waveguide in contact with a microsphere resonator, whereas we wish to determine the phase-shift in the envelope of an intensity modulated source coupling to the WGMs of the microsphere. In our case, rather than a single frequency electric field coupling to the resonator, an amplitude-modulated (AM) field is required, which implies the presence of frequency sidebands. Using complex notation for the electric field, we can write the incident AM field as

$$E_{inc} = E_0 e^{i\omega t} \left(\sqrt{1-\beta} + i\sqrt{2\beta} \cos(\Omega t/2) \right)$$

$$= E_0 \sqrt{1-\beta} e^{i\omega t} + E_0 \sqrt{\frac{\beta}{2}} \left(e^{i(\omega+\Omega/2)t+\pi/2} + e^{i(\omega-\Omega/2)t+\pi/2} \right) \quad (10.41)$$

where β is the modulation depth, ω is the carrier frequency and Ω is the amplitude-modulation frequency of the intensity. Note that the electric field is then modulated by $\Omega/2$. The carrier and each frequency sideband will experience their own attenuation and phase-shift (ϕ, ϕ_+ and ϕ_-, respectively) through interaction with the resonator, and the output field will be

$$E_{trans} = E_\omega \sqrt{1-\beta} \exp(i(\omega t + \phi))$$

$$+ E_{\omega+\Omega/2} \sqrt{\frac{\beta}{2}} \exp(i[(\omega+\Omega/2)t + \phi_+ + \pi/2])$$

$$+ E_{\omega-\Omega/2} \sqrt{\frac{\beta}{2}} \exp(i[(\omega-\Omega/2)t + \phi_- + \pi/2]). \quad (10.42)$$

The transmitted intensity is given by

$$
\begin{aligned}
I_{\text{trans}} = \frac{n_{\text{eff}}\varepsilon_0 c}{2} \bigg[& (1-\beta)E_\omega^2 + \frac{\beta}{2}\left(E_{\omega+\Omega/2}^2 + E_{\omega-\Omega/2}^2\right) \\
& + E_\omega E_{\omega+\Omega/2}\sqrt{2\beta(1-\beta)}\sin(-\Omega t/2 + \phi - \phi_+) \\
& + E_\omega E_{\omega-\Omega/2}\sqrt{2\beta(1-\beta)}\sin(\Omega t/2 + \phi - \phi_-) \\
& + \beta E_{\omega+\Omega/2} E_{\omega-\Omega/2}\cos(\Omega t + \phi_+ - \phi_-) \bigg].
\end{aligned}
\tag{10.43}
$$

In our case, we only record the signal modulated at Ω with the lock-in amplifier and the associated phase-shift of the modulation envelope is given by $(\phi_+ - \phi_-)$.

Rezac's formalism can now be used to calculate the phase-shift at each sideband frequency; the difference $(\phi_+ - \phi_-)$ yields the desired modulation envelope phase-shift [71]. The ring-resonator model assumes a single frequency electric field incident upon a multi-mirror ring-cavity (Fig. 10.3). The input mirror has a field amplitude reflectivity, Γ, while the other mirrors are assumed totally reflective in our model. As the field propagates within the resonator, it experiences a fractional loss of $\exp(-\alpha_{\text{abs}}L/2)$ per pass, i.e. the intensity loss is $\exp(-\alpha_{\text{abs}}L)$ per pass. Here α_{abs} is the distributed loss coefficient and L is the round-trip path length. The reflected field from the ring-resonator consists of the superposition of the partially reflected incident field and the partially transmitted field inside the resonator. In our case of a fiber-coupled microsphere resonator the *reflected* field of this model corresponds to the light *transmitted* through the tapered delivery fiber. The normalized amplitude of the respective field for a single frequency incident field, is given by [71]

$$
E = E_{\text{ref}}/E_{\text{inc}} = \frac{\Gamma - e^{-\alpha L/2}e^{iLk}}{1 - \Gamma e^{-\alpha L/2}e^{iLk}}
\tag{10.44}
$$

where the wave vector $k = \omega n_{\text{eff}}/c$. Accordingly, the intensity is calculated from the absolute square of (10.44) as

$$
I = \frac{\Gamma^2 - 2\Gamma e^{-\alpha L/2}\cos(Lk) + e^{-\alpha L}}{1 - 2\Gamma e^{-\alpha L/2}\cos(Lk) + \Gamma^2 e^{-\alpha L}}.
\tag{10.45}
$$

To calculate the phase angle associated with the field of (10.44) the expression is expanded by the complex conjugate of the denominator

$$
E = \frac{\Gamma - (\Gamma^2 + 1)e^{-\alpha L/2}\cos(Lk) + \Gamma e^{-\alpha L} + i(\Gamma^2 - 1)e^{-\alpha L/2}\sin(Lk)}{1 - 2\Gamma e^{-\alpha L/2}\cos(Lk) + \Gamma^2 e^{-\alpha L}}.
\tag{10.46}
$$

With this expression the phase angle associated with the field of (10.44) is described in the complex plane as

$$
\tan\phi = \frac{\text{Im}(E)}{\text{Re}(E)} = \frac{(\Gamma^2 - 1)e^{-\alpha L/2}\sin(Lk)}{\Gamma - (\Gamma^2 + 1)e^{-\alpha L/2}\cos(Lk) + \Gamma e^{-\alpha L}}.
\tag{10.47}
$$

The phase-shifts at the two sideband frequencies ϕ_+ and ϕ_- can be calculated from Eq. (10.47) using $k_+ = (\omega + \Omega/2)n_{\text{eff}}/c$ and $k_- = (\omega - \Omega/2)n_{\text{eff}}/c$ respectively.

The phase-shift for an intensity-modulated input is given by the difference in the phase angles associated with the sideband frequencies of Eq. (10.43). In our experiments the modulation frequency is $<10^{-8}$ of the carrier frequency so we can use a Taylor expansion for the tangent function about the carrier frequency. On resonance, where $Lk = m2\pi$, this results in

$$\tan(\Delta\phi) \approx -\frac{\Omega n_{eff} L}{c} \frac{x(1 - \Gamma^2)}{\Gamma(1 + x^2) - x(1 + \Gamma^2)} \tag{10.48}$$

where $x = e^{-\alpha L/2}$. Insertion and rearrangement leads to

$$\tan(\Delta\phi) \approx -\frac{\Omega n_{eff} L}{c} \frac{2\ln(\Gamma)}{(\ln \Gamma)^2 - (\alpha_{abs} L/2)^2}. \tag{10.49}$$

In Ref. [48] Eq. (10.49) was derived assuming $\Delta\phi$ is small and hence $\tan \Delta\phi \approx \Delta\phi$, however, this approximation need not be made.

It is helpful to distinguish different coupling regimes before examining the implications of the above expressions. When $-2\ln \Gamma > \alpha L$, the cavity is 'overcoupled'. In this regime, coupling to the fiber is the dominant loss mechanism controlling the resonance linewidth [12]. When $-2\ln \Gamma < \alpha L$, the cavity is referred to as 'undercoupled', and 'critically coupled' when $-2\ln \Gamma = \alpha L$. At the point of critical coupling, no light is transmitted through the delivery fiber on resonance. Equation (10.49) shows that the phase-shift is negative in the case of an undercoupled excitation and positive if the resonator is operated in the overcoupled regime. From (10.49) it is also apparent that the phase-shift increases linearly with increasing modulation frequency Ω, and with decreasing dB loss. As can be shown, the resonance line width decreases accordingly. Also, for the undercoupled case the slope $d\tan \Delta\phi/d\Omega$ is negative in the vicinity of the resonance, and the magnitude of the slope increases with decreasing loss α. The same holds true for the phase shift of the scattered or backscattered light, however, the magnitude of the phase shift of the transmitted light is proportional to how close the WGM is to the critical coupling case, whereas in the scattering cases the phase shift of is proportional to the ring-down time, i.e. to the inverse of the optical loss.

10.2.6.4 Summary of Phase-Shift CRD Equations

In summary the phase-shift calculated for the intensity scattered from a single resonantly excited WGM of the sphere is given by

$$\Delta\phi = \tan^{-1}(-\Omega\tau). \tag{10.50}$$

For the backscattered intensity, the phase-shift expression is approximated by

$$\Delta\phi = -\Omega\tau + \tan^{-1}(-\Omega\tau). \tag{10.51}$$

In both cases τ is the photon lifetime, or ring-down time, of the WGM given by

$$\tau = \frac{Ln_{eff}}{c} \cdot \frac{1}{\alpha_{abs} L - 2\ln \Gamma}. \tag{10.52}$$

The phase for light transmitted through the delivery fiber is given by

$$\Delta\phi \approx \tan^{-1}\left(\frac{-\Omega n_{eff}L}{c}\frac{2\ln\Gamma}{(\ln\Gamma)^2 - (\alpha_{abs}L/2)^2}\right) \tag{10.53}$$

where $\Omega = 2\pi\nu$ is the angular intensity modulation frequency, n_{eff} is the effective index of the cavity mode, L is the round-trip path length, $\sqrt{1 - \Gamma^2}$ is the taper-sphere field coupling constant, and α_{abs} is the absorption coefficient [48]. In the undercoupled regime, where $-2\ln\Gamma < \alpha L$, we find that $\Delta\phi < 0$. When the WGM is overcoupled ($-2\ln\Gamma > \alpha L$) it is found that $\Delta\phi > 0$.

It should be noted that in the backscattered and isotropically scattered case, the phase-shift will always be negative, whereas the phase shift for transmitted light can be positive or negative. For example in Fig. 10.4b the WGM at a relative wavelength of 4 pm is overcoupled whereas all other WGMs are undercoupled.

10.3 Experimental Studies

10.3.1 Introduction

To demonstrate the practicality of phase-shift CRD measurements for simultaneous and absolute determinations of absorption coefficients, coupling coefficients and surface coverages we performed a series of experiments using ethylene diamine (EDA) as an adsorbant on a silica microsphere. To the best of our knowledge only one more article describes the use of phase-shift CRD spectroscopy on micro-resonators. In a thorough study Cheema *et al.* used transmitted light combined with phase-shift CRD to determine the optical loss in a microtoroidal cavity. [62] When the cavity is functionalized with biotin and exposed to streptavidin dissolved in phosphate buffer solution a change in the phase-shift was observed as expected from Eq. (10.49). This change was correlated to a change in Q-factor, but since the authors used the *scattering* model of Sect. 10.2.6.1. on light that was *transmitted* through the waveguide (as in Sect. 10.2.6.3) their model may not be appropriate.

The following sections describe the simultaneous measurements of phase-shift and intensity of the light that is, transmitted through the fiber taper and Rayleigh backscattered. We use an approach similar to that by Cheema et al. but detect and quantify weaker absorption features of volatile molecules that formed a monolayer by adsorption on silica.

10.3.2 Properties and Spectra of Ethylene Diamine

In this work we use the PS-CRD technique to measure the optical absorption of near-IR light by ethylene diamine monolayers that are adsorbed on the surface of high-Q silica microspheres. Ethylene diamine (1,2-diamino ethane, H_2N-CH_2CH_2-NH_2) is a compound of considerable industrial and environmental importance and is produced in large quantities (150,000 tonnes annually in 2003 [72]) as e.g. an intermediate for the formulation of polymers, chelating agents, drugs, and pesticides but also as a corrosion inhibitor. Bulk liquid ethylene diamine has a vibrational overtone transition at 6500 cm^{-1} (1540 nm) with a full width half maximum of about 300 cm^{-1} and a peak absorption coefficient (given with respect to base-e) of about $\alpha = 28$ cm^{-1} [73]. At our laser emission wavelength of 1549 nm the absorption coefficient is estimated to be $\alpha = 26$ cm^{-1}.

Using the experimentally determined value for the ethylene diamine refractive index and an estimated absorption coefficient we calculate the surface density, and compare this value to the surface density obtained from the frequency shift of the same WGM resonance.

10.3.3 Experiment

The silica microsphere (diameter: 264 μm) was formed by melting the end of a single-mode optical fiber in an electric arc. The residual optical fiber stem allows for manipulation and positioning of the microsphere. A tapered single-mode optical fiber, with a waist diameter of about 3 μm, was used to couple light into the evanescent-field of the microsphere whispering-gallery modes (WGM). A detailed description of the experimental setup regarding the microsphere and tapered waveguide may be found elsewhere [48].

In this experiment a DFB laser was employed to interrogate the microsphere (Fig. 10.5). The laser output was sinusoidally intensity-modulated between 2 and 20 MHz using a Mach-Zehnder modulator. A polarization controller, positioned after the modulator, allowed the TE or TM resonator modes to be selectively excited. The mode polarization was determined by placing a photodetector, coupled to a ×10 microscope objective, above the microsphere and by viewing the scattered radiation through an infrared polarizer. Rayleigh back-scattering, from imperfections in the silica microsphere, equilibrates degenerate counter-propagating WGMs [49, 67]. A fiber optic circulator directed the backscattered light to a fiber-coupled detector. In addition, WGM resonances are observed by monitoring the light transmitted through the waveguide using a second detector. Two lock-in amplifiers collected the photodiode signals, thus providing intensity and phase angle information referenced to the laser intensity modulation. Spectra of WGM resonances, viewed simultaneously in backscatter and transmission mode, are shown in Figs. 10.2A, 10.2C, 10.4a and a WGM viewed in scatter and backscatter mode is shown in Figs. 10.2B and 10.4a.

While the phase shift can be recorded while tuning the laser frequency across the WGM resonance (Fig. 10.4b), to obtain a more accurate modulation phase measurement, the laser should remain locked to a selected WGM. Frequency locking of the

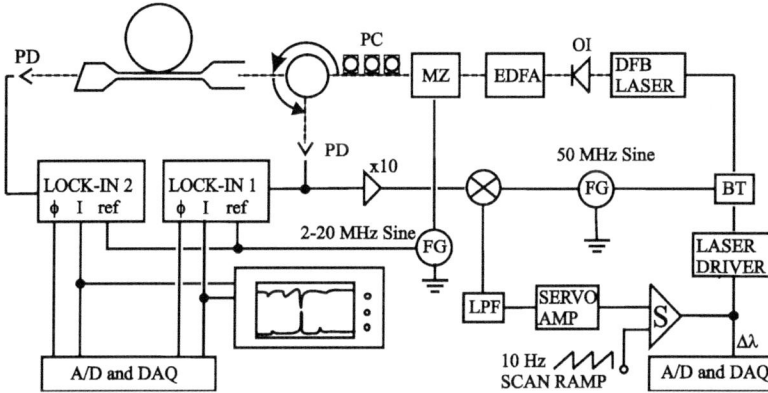

Fig. 10.5 Diagram of the experimental setup. The frequency of the light from a DFB laser is modulated at typically 20–50 MHz using a function generator (FG) and a bias tee (BT). It passes through an optical isolator (OI), is amplified in an erbium doped fiber amplifier (EDFA) and AM modulated at 2–20 MHz with a function generator (FG) and a Mach-Zehnder (MZ) modulator. After polarization control (PC) the modulated light is coupled into the sphere. The three photodetectors (PD) record the light either in transmission, after Rayleigh backscattering behind a circulator, or through a microscope lens (not shown). The phase and amplitude is recorded with two lock-in amplifiers using the function generator output as a reference. The backscattered light is also used to lock the laser to a WGM resonance. The PD signal is then amplified and mixed with the sine-wave providing the FM modulation. Low pass filtering (LPF) producing an error signal (*below*) which is fed into a servo amp that provides the control signal to the laser driver

laser was achieved using the Pound-Drever-Hall (PDH) technique [74]. This technique is commonly employed to lock lasers to high finesse optical cavities. Briefly, the DFB laser is current modulated resulting in a frequency modulation (FM) of the laser beam. When the resulting detector signal is mixed with the FM modulation frequency and low-pass filtered, an error signal is obtained that is proportional to the derivative of the WGM line when the modulation frequency is small compared to the WGM linewidth (Fig. 10.6). The polarity of this signal indicates the relative tuning of the laser with respect to the resonance center frequency. This error signal is applied to a servo amplifier, where it is amplified, filtered and fed back to the laser driver as a correction signal. As a consequence, the laser frequency-locks to the WGM resonance. If the WGM shifts in frequency due to, for example, adsorption of molecules on the microsphere surface, the magnitude of the correction signal applied to the laser changes accordingly. Knowledge of the transfer function of the laser driver (mA/mV) and tuning characteristics of the DFB laser (pm/mA) allows us to calculate the frequency shift from the control voltage change. Amplitude modulation (AM) of the laser intensity is applied through a Mach-Zehnder modulator that is fiber-coupled to the amplified laser output. A function generator drives this modulator with sine modulation at $\Omega = 2$–20 MHz to permit phase shift measurements at different frequencies. The FM and AM modulation frequencies are kept sufficiently different to allow the lock-in amplifiers to accurately ex-

Fig. 10.6 Spectrum of a whispering gallery mode (*red curve*) with corresponding error signal (*black*)

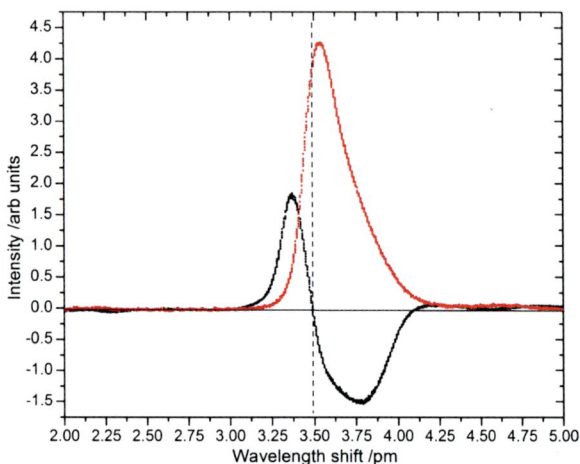

tract intensity and, simultaneously, phase angle values without cross talk. Typically the Rayleigh-backscattered signal is used to lock the laser to the WGM resonance.

It may surprise that simultaneous amplitude and frequency modulation does not give rise to cross talk and erroneous phase-shift measurements and these experiments were even deemed impossible [62]. In fact, we observed that cross talk leads to a considerable increase of the phase noise whenever the frequency differed by less than about 20 %. Since we work in the small frequency regime of the PDH-locking scheme [74], other adverse effects were not observed.

A quartz crystal microbalance (QCM) is a device commonly used to measure sub-monolayer coverages of adsorbed material. To compare the sensitivity of a QCM with that of the microsphere, a quartz crystal microbalance was positioned directly below the fiber taper and microsphere. The surface of the quartz crystal supports a gold electrode which occupies about 25 % of the surface area. While the gold electrode of the QCM presents a different surface compared to the silica microsphere, both are expected to adsorb ethylene diamine and coverages may be compared. Surface coverages of the QCM were calculated from the quartz crystal resonance shifts using the Sauerbrey equation [75]. An A/D converter was used to simultaneously measure the servo control voltage, the QCM frequency, the backscattered signal intensity, and the phase angles from the lock-in amplifiers, while the laser was locked to the WGM.

The fiber taper, microsphere and microbalance sensor element were located under a 5 cm^3 glass dosing chamber into which ethylene diamine vapour was introduced. A low flow of nitrogen was passed over neat ethylene diamine (vapour pressure 11.3 torr at room temperature), carrying the vapour through a Teflon^{TM} valve into the dosing chamber. This entire apparatus was contained within a Plexiglas^{TM} enclosure, purged with dry nitrogen, which served to reduce contamination and the concentration of water vapour.

Fig. 10.7 Thermal response of a microsphere resonator on resonance. Two WGM show different peak shapes when the laser is scanned to longer wavelength (*black curve*) or to shorter wavelength (*red curves*). The WGM remains "locked" to the laser wavelength due to thermal expansion of the sphere as the laser is slowly scanned to longer wavelengths. When the laser current, and therefore the output power is increased, the photothermal effect is more pronounced

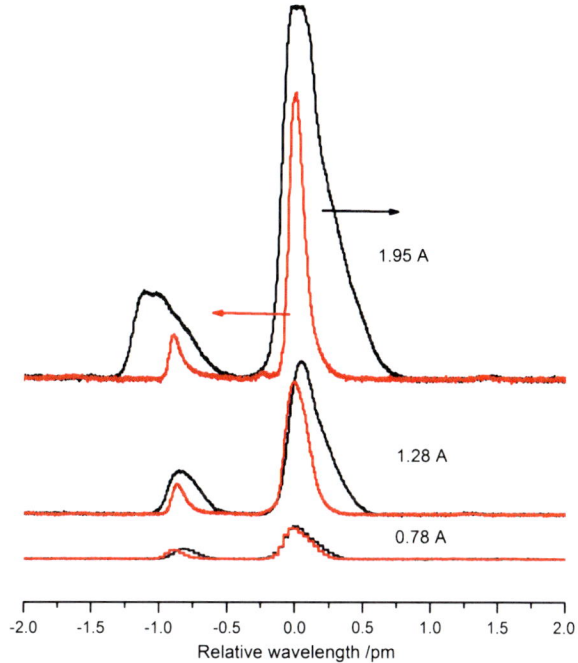

10.3.4 Photothermal Effect

The high Q-values and small mode volumes for WGMs result in high circulating powers within the resonator even with only microwatts of input laser power. The inevitable absorption of a portion of this circulating radiation leads to a temperature increase within the resonator. Through thermal expansion, a temperature increase will result in an increase in the sphere diameter. Additionally, silica has a positive thermo-optic coefficient which causes an increase in refractive index with increasing temperature. Nonlinear processes, due to high electric field values, may also contribute to an increase in refractive index through the Kerr effect. From Eq. (10.6) it can be seen that all of these effects cause the WGM resonance to shift to longer wavelengths, once light starts to couple into the mode. If the laser sweeps towards shorter wavelengths, the laser frequency and WGM resonance frequency will be moving in opposite directions, resulting in a reduced observed linewidth. If the laser is scanned towards longer wavelengths, the WGM is swept along with the laser, increasing the measured linewidth (Fig. 10.7).

These photothermal effects must be considered with regard to the adsorption process. We found that with increasing input power, the resonance shifts to longer wavelengths and attribute that shift to an increase in the microsphere's temperature. Using published values for the thermo-optic coefficient and linear expansion coefficient of silica [76], it is estimated that the temperature of the microsphere is 0.5 K higher than the QCM when the laser is locked to the WGM resonance. It has been shown that ethylene diamine will strongly bind through chemisorption on silica via a proton transfer reaction with the surface silanol groups [77]. Additional physisorbed layers will bond through amine hydrogen bonding interactions which have a lower energy on the order of 13 kJ/mol [78]. Given the relatively high binding energies involved, the temperature increase of the microsphere by 0.5 K is expected to have a negligible effect on the degree of adsorption.

10.3.5 Determination of Ethylene Diamine Coverage and Absorption

It has been shown that ethylene diamine adsorbs strongly to silica through interaction with surface silanol groups [77]. Given the response from the QCM (Fig. 10.8a) we can assume that, in our case, the observed perturbations of the WGM cavity resonance are due to adsorbed ethylene diamine, rather than vapour. In addition, we also have to consider that water adsorbs to silica and the silica surface may originally be covered with a few monolayers of water that are (partially) displaced by ethylene diamine.

As can be seen in Fig. 10.8a the QCM response is discretized due to its 1 Hz frequency resolution whereas the microsphere frequency shift is continuous. The microsphere frequency shift also responds more quickly than the QCM, indicating a greater sensitivity.

Figure 10.8b shows that upon exposing the sphere to ethylene diamine the WGM resonance frequency shifts to longer wavelengths. With the assumptions given in Sect. 10.2.4 it is possible to determine the coverage from the frequency shift as was shown in Eqs. (10.17) or (10.20). These expressions were derived by Teraoka et al. to describe the spectral response due to molecules adsorbed on a microsphere surface [2, 4, 30]. Their calculation is based on the polarization induced in an adsorbed molecule by the evanescent tail of the WGM. This expression holds for a layer that is much thinner than the wavelength of the light. Here $n_s = 1.444$ is the microsphere refractive index, $n_m = 1.0$ is that of the surrounding medium.

If the refractive index of the adsorbed material is assumed constant, (10.17) or (10.20) can be related to the surface coverage, θ, where α_t is the transverse polarizability of the adsorbed molecule. The transverse component of the polarizability depends on the components of the polarizability tensor and on the orientation of the adsorbed molecule on the surface [50]. Using calculated values for the diagonal tensor components for gas phase ethylene diamine, it can be shown that α_t varies by, at most, 6 % from the isotropic polarizability value of EDA ($\alpha_{iso} = 6.7 \times 10^{-24}$ cm^3) with molecular orientation [79], hence, orientation effects can be ignored. Figures 10.8a and 10.8b give the surface coverage values obtained from the QCM and those calculated from the WGM resonance shift using (10.17). The maximum dosage level indicated by the QCM is a factor of four to five greater than that based on the WGM resonance shift. The difference is not surprising since the microsphere surface is atomically smooth silica, whereas the sensitive region of the QCM is principally gold surface created by evaporative deposition. It has also been shown

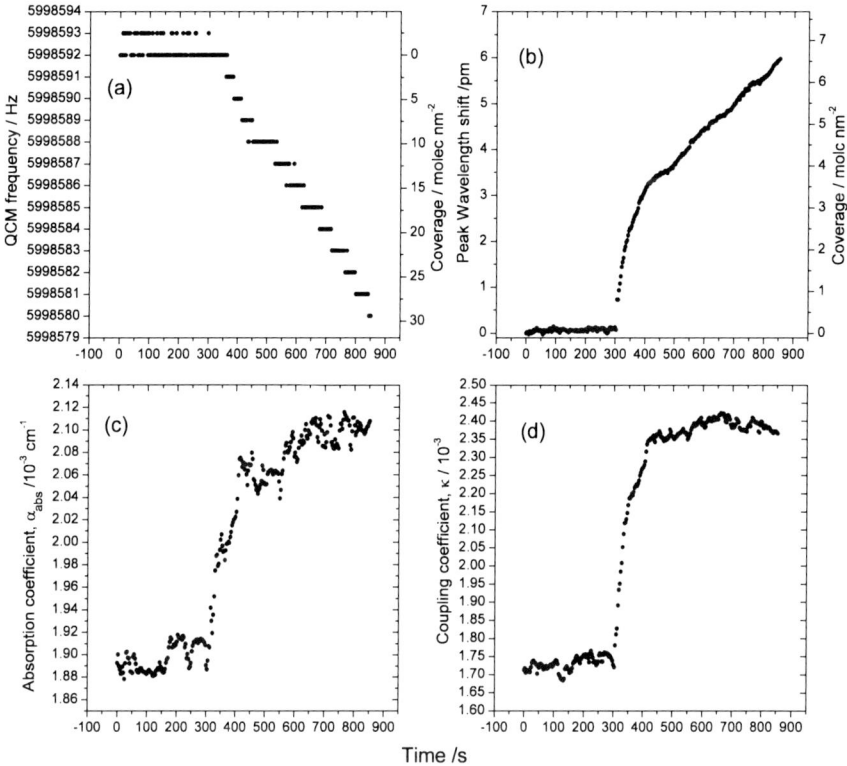

Fig. 10.8 Response of WGM to dosing of ethylene diamine onto the silica microsphere surface. Very low concentrations of ethylene diamine has been mixed into a stream of dry nitrogen starting at $t = 300$ s. (**a**) Frequency response of the quartz crystal micro balance. (**b**) Wavelength shift $\Delta\lambda$ of the WGM resonance as measured by the correction voltage supplied to the PDH-locked DFB laser. (**c**) Absorption coefficients, α_{abs}, and (**d**) coupling coefficients, κ calculated from the observed phase shifts and Eqs. (10.54), (10.55), (10.56)

that gold reversibly adsorbs amines in a manner highly dependent on surface roughness [80]. Photothermal effects, such as desorption of ethylene diamine upon heating of the sphere on resonance, were also considered, but given the high binding energy to silica and the moderate (about 0.5 K) temperature increase of the silica sphere on resonance, these effects are negligible (see Sect. 10.3.4).

From now on we use the more reliable coverages obtained from (10.17) instead of the coverage calculated from the response of the quartz crystal microbalance. Ethylene diamine adsorbs onto silica through interactions with surface silanol groups which are the major surface species, even in the presence of water vapour [77]. The average surface density of silanol groups on silica is estimated at 4.9 nm^{-2} [81] which suggests that, based on the resonance shift data, less than two monolayers of ethylene diamine are deposited on the microsphere over the course of this experiment. By comparison, the "footprint" of an ethylene diamine molecule in the bulk liquid phase can be estimated from the density (0.9 g/mL) and the molar mass

(60.1 g/mol) as 0.23 nm^2/molecule which is somewhat larger than expected from the *ab initio* diameter of the isolated molecule (0.13 nm^2/molecule). The surface density of each ethylene diamine layer in bulk liquid should therefore be quite similar to the surface density of silenol groups on silica, e.g. about 4–7 molecules/nm^2.

In our experiment, optical absorption of ethylene diamine through its N–H stretch overtone band at 1549 nm is measured with the phase-shift cavity ring-down technique. As described in Sect. 10.2.6.3., the ring-cavity model of Rezac [71] was used to calculate the phase-shift observed in the light transmitted through the delivery fiber [48]. This transmitted light represents the superposition of a portion of the incident field and the forward scattered field of the microsphere. The phase-shift observed in the transmitted light is given by

$$\Delta\phi \approx \tan^{-1}\left(\frac{-\Omega n_{eff}L}{c} \frac{2\ln\Gamma}{(\ln\Gamma)^2 - (\alpha_{abs}L/2)^2}\right) \tag{10.54}$$

where Ω is the laser intensity modulation frequency, n_{eff} is the effective refractive index of the WGM, L is the circumference of the microsphere, $\kappa = \sqrt{1 - \Gamma^2}$ is the coupling constant between the taper and microsphere, and α_{abs} is the attenuation coefficient [48]. In this experiment the phase-shift of the backscattered field is also recorded. We showed above that the phase-shift for the backscattered light is given by [68]

$$\Delta\Phi = -\Omega\tau - \tan^{-1}(\Omega\tau) \tag{10.55}$$

where τ is the ring-down time of the microsphere given by

$$\tau = \frac{L n_{eff}}{c_0(\alpha_{abs}L - 2\ln\Gamma)}. \tag{10.56}$$

Based on the ring-down time at zero coverage, the Q-factor for the particular TE mode used is $\sim 1.6 \times 10^7$. Calculating Q-factors based on ring-down time eliminates complications due to thermal effects which are known to influence Q-factor determinations based on linewidth measurements [82].

By recording the phase-shift in transmission mode and backscatter mode simultaneously, at a fixed Ω, one can calculate the ring-down time (τ), and from it, the absorption coefficient (α_{abs}) and the coupling coefficient (κ), continually and independently, during the dosing process.

Finally, we can relate the adsorbant coverage obtained from the frequency shift (Fig. 10.8b) to the absorption coefficient and the coupling constant. In Fig. 10.9B the coupling coefficient, κ, is seen to increase linearly until a coverage of about 5 molec/nm^2, i.e. roughly one monolayer, is reached, whereupon κ remains largely unchanged. Since κ depends on the overlap of the WGM with the modefield in the fiber taper, one can rationalize this observation through the modification of the overlap integral by one monolayer of ethylene diamine. Even more interesting is the behavior of the overtone absorption cross section as a function of coverage (Fig. 10.9A). The absorption cross section is calculated from

$$\sigma = \frac{(\alpha_0 - \alpha_{abs})l_{EDA}}{f\theta} \tag{10.57}$$

Fig. 10.9 (**A**) Ethylene
diamine absorption cross
section, σ, and (**B**) coupling
coefficient, κ, as a function of
the mass loading, which was
calculated from the frequency
shift of the WGM
(Fig. 10.8b) and Eq. (10.17)

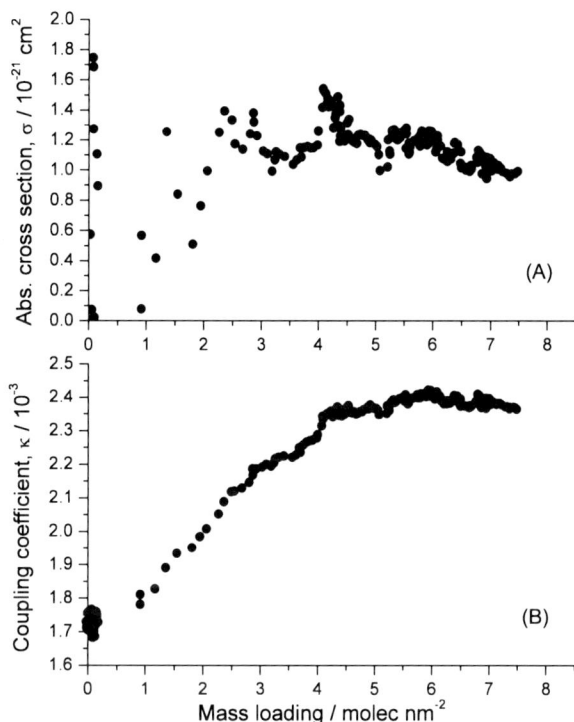

where $\alpha_0 = 0.0019$ cm^{-1} is the optical loss of the WGM before dosing, $l_{EDA} = 360$ pm is the estimated length of an EDA molecule, and our estimate for the thickness of one monolayer, and $f = 1 \times 10^{-5}$ is the fraction of the WGM travelling in the evanescent wave as in Eq. (10.10). While the value is found to be close to $\sigma = 1.2 \times 10^{-21}$ cm^2 we also see, as yet unexplained, oscillations in the cross section—especially at sub monolayer coverages. By comparison, the bulk absorption cross section of ethylene diamine liquid at 1550 nm is $\sigma_{bulk} = 2.9 \times 10^{-21}$ cm^2 which was calculated from the liquid absorption coefficient to base-e of $(\alpha_0 - \alpha_{abs}) = 26$ cm^{-1} [73]. The difference is readily explained by a combination of several effects which all tend to lower the apparent cross section. First, it is likely that a monolayer of weakly absorbing water is displaced upon adsorption of the first layer of EDA. Water binds very strongly to silica and is expected to be present even under a constant flow of dry nitrogen. The absorption cross section associated with a submonolayer EDA is therefore the difference between the cross section of EDA and of water. In addition, the WGM is TE polarized, i.e. by dipole selection rules only the component of the transition dipole that is aligned parallel to the silica surface is contributing to the transition probability. Since N–H bonds are predicted to be largely normal to the silica surface, we expect this contribution to be lower than the isotropic value in bulk liquid. Finally, we emphasize that many of the values used in this estimate of the absorption cross section, such as l_{EDA}, f, θ

and even $\alpha_{abs} L$ have errors estimated to be in the 10–20 % range and even the bulk value for the liquid σ_{bulk} may be in error.

When benchmarking this phase-shift cavity ring-down experiment to those reported before, we calculate from Fig. 10.8 that the minimal detectable absorption loss is about $\alpha_{min} = 10^{-5} \, cm^{-1}$, or $\alpha_{min} L = 4.2 \times 10^{-6}$ for our $R = 132 \, \mu m$ sphere. We therefore propose that *absolute* measurements of absorption cross sections of single molecules are possible by this method as long as the optical loss per molecule exceeds $\alpha_{min} L = 10^{-5}$. The minimal detectable cross section is calculated from Eq. (10.57) as $\sigma_{min} = 2.7 \times 10^{-12} \, cm^2$ assuming a coverage $\theta = 1/1500 \, \mu m^{-2}$ (i.e. one molecule in the evanescent mode volume). This absorption cross section and the molecular dimensions are comparable to those of e.g. gold nanoparticles and aerosol particles.

In conclusion, we demonstrated that the absolute absorption cross sections and coupling coefficients can be measured by supplementing the Pound-Drever-Hall frequency locking method with amplitude modulation and phase-shift measurements. Considering that even microcavities with comparably short ring-down times and monolayer coverages of a weak (overtone) absorber are susceptible to this method, we propose that the technique may also find applications in other optical cavities that require simultaneous measurements of optical loss and of the coupling coefficient.

Acknowledgements   We thank *Photonics Research Ontario* (Ontario Centres of Excellence) for financial support of this work. Contributions from the *Canadian Institute for Photonic Innovations* (CIPI), from the *Natural Science and Engineering Research Council* (NSERC) and *Queen's University* are acknowledged by the Canadian researchers. GG also acknowledges financial support from the Italian Ministry for Education, University and Research (PON-SIMONA) and assistance from the Consiglio Nazionale delle Ricerche by the RSTL-project (cod. 3007) and the CNR Short-Term Mobility Program 2008.

Finally, we thank Zhaobing Tian for providing some of the tapered fiber waveguides, Saverio Avino for his help in setting up the PDH-coupling scheme, and Scott Yam, James Fraser, and Mark Wilson for their help in deriving Eq. (10.49) in Sect. 10.2.6.3.

References

1. A.M. Armani et al., Label-free, single-molecule detection with optical microcavities. Science **317**, 783–787 (2007)
2. I. Teraoka et al., Perturbation approach to resonance shifts of whispering-gallery modes in a dielectric microsphere as a probe of a surrounding medium. J. Opt. Soc. Am. B, Opt. Phys. **20**, 1937–1946 (2003)
3. F. Vollmer, S. Arnold, Whispering-gallery-mode biosensing: label-free detection down to single molecules. Nat. Methods **5**, 591–596 (2008)
4. S. Arnold et al., Shift of whispering-gallery modes in microspheres by protein adsorption. Opt. Lett. **28**, 272–274 (2003)
5. S. Arnold et al., MicroParticle photophysics illuminates viral bio-sensing. Faraday Discuss. **137**, 65–83 (2008)
6. M.L. Gorodetsky et al., Ultimate Q of optical microsphere resonators. Opt. Lett. **21**, 453–455 (1996)
7. G. Farca et al., Cavity-enhanced laser absorption spectroscopy using microresonator whispering-gallery modes. Opt. Express **15**, 17443–17448 (2007)

8. A.M. Armani, K.J. Vahala, Heavy water detection using ultra-high-Q microcavities. Opt. Lett. **31**, 1896–1898 (2006)

9. A.C.R. Pipino, Ultrasensitive surface spectroscopy with a miniature optical resonator. Phys. Rev. Lett. **83**, 3093–3096 (1999)

10. R.S. Brown et al., Fiber-loop ring-down spectroscopy. J. Chem. Phys. **117**, 10444–10447 (2002)

11. D.K. Armani et al., Ultra-high-Q toroid microcavity on a chip. Nature **421**, 925–928 (2003)

12. T.J. Kippenberg, Nonlinear optics in ultra-high-Q whispering-gallery optical microcavities. Ph.D. thesis, California Institute of Technology, 2004

13. R. Engeln et al., Phase shift cavity ring down absorption spectroscopy. Chem. Phys. Lett. **262**, 105–109 (1996)

14. Z.G. Tong et al., Phase-shift fiber-loop ring-down spectroscopy. Anal. Chem. **76**, 6594–6599 (2004)

15. M.C. Chan, S.H. Yeung, High-resolution cavity enhanced absorption spectroscopy using phase-sensitive detection. Chem. Phys. Lett. **373**, 100–108 (2003)

16. A.M. Armani, Label-free, single-molecule detection with optical microcavities (August, pg 783, 2007). Science **334**, 1496 (2011)

17. F. Vollmer et al., Protein detection by optical shift of a resonant microcavity. Appl. Phys. Lett. **80**, 4057–4059 (2002)

18. A.M. Armani et al., Ultra-high-Q microcavity operation in H_2O and D_2O. Appl. Phys. Lett. **87**, 151118 (2005)

19. R.W. Boyd, J.E. Heebner, Sensitive disk resonator photonic biosensor. Appl. Opt. **40**, 5742–5747 (2001)

20. C. Chao, L.J. Guo, Polymer microring resonators fabricated by nanoimprint technique. J. Vac. Sci. Technol. B **20**, 2862–2866 (2002)

21. E. Krioukov et al., Integrated optical microcavities for enhanced evanescent-wave spectroscopy. Opt. Lett. **27**, 1504–1506 (2002)

22. T. Ling, L.J. Guo, A unique resonance mode observed in a prism-coupled micro-tube resonator sensor with superior index sensitivity. Opt. Express **15**, 17424–17432 (2007)

23. D.W. Vernooy et al., High-Q measurements of fused-silica microspheres in the near infrared. Opt. Lett. **23**, 247–249 (1998)

24. I.M. White et al., Liquid-core optical ring-resonator sensors. Opt. Lett. **31**, 1319–1321 (2008)

25. T.J. Kippenberg et al., Ultralow-threshold microcavity Raman laser on a microelectronic chip. Opt. Lett. **29**, 1224–1226 (2004)

26. S.I. Shopova et al., Microsphere whispering-gallery-mode laser using HgTe quantum dots. Appl. Phys. Lett. **85**, 6101–6103 (2004)

27. M. Cai et al., Fiber-optic add-drop device based on a silica microsphere whispering gallery mode system. IEEE Photonics Technol. Lett. **11**, 686–687 (1999)

28. B.E. Little et al., Analytic theory of coupling from tapered fibers and half-blocks into microsphere resonators. J. Lightwave Technol. **17**, 704–715 (1999)

29. G.C. Righini et al., Whispering gallery mode microresonators: fundamentals and applications. Riv. Nuovo Cimento **34**, 435–488 (2011)

30. I. Teraoka, S. Arnold, Theory of resonance shifts in TE and TM whispering gallery modes by nonradial perturbations for sensing applications. J. Opt. Soc. Am. B, Opt. Phys. **23**, 1381–1389 (2006)

31. B.E. Little et al., Analytical theory of coupling from tapered fibers and half-blocks into microsphere resonators. J. Lightwave Technol. **17**, 704–715 (1999)

32. C.C. Lam et al., Explicit asymptotic formulas for the positions, widths, and strengths of resonances in Mie scattering. J. Opt. Soc. Am. B, Opt. Phys. **9**, 1585–1592 (1992)

33. A.B. Matsko, V.S. Ilchenko, Optical resonators with whispering-gallery modes, part I: basics. IEEE J. Sel. Top. Quantum Electron. **12**, 3–14 (2006)

34. M.L. Gorodetsky, V.S. Ilchenko, Optical microsphere resonators: optimal coupling to high-Q whispering-gallery modes. J. Opt. Soc. Am. B, Opt. Phys. **16**, 147–154 (1999)

35. V.S. Ilchenko et al., Pigtailing the high-Q microsphere cavity: a simple fiber coupler for optical whispering-gallery modes. Opt. Lett. **24**, 723–725 (1999)
36. N. Dubreuil et al., Eroded monomode optical fiber for whispering-gallery mode excitation in fused silica microspheres. Opt. Lett. **20**, 813–815 (1995)
37. B.E. Little et al., Pedestal antiresonant reflecting waveguides for robust coupling to microsphere resonators and for microphotonic circuits. Opt. Lett. **25**, 73–75 (2000)
38. K. Grujic et al., Whispering gallery modes excitation in borosilicate glass microspheres by K+ ion-exchanged channel waveguide coupler. Proc. Soc. Photo-Opt. Instrum. Eng. **6101**, 6101L-1–6101L-5 (2006)
39. S.M. Spillane et al., Ideality in a fiber-taper-coupled microresonator system for application to cavity quantum electrodynamics. Phys. Rev. Lett. **91**, 043902 (2003)
40. A. Serpenguzel et al., Excitation of resonances of microspheres on an optical-fiber. Opt. Lett. **20**, 654–656 (1995)
41. G. Griffel et al., Morphology-dependent resonances of a microsphere-optical fiber system. Opt. Lett. **21**, 695–697 (1996)
42. J.P. Laine et al., Etch-eroded fiber coupler for whispering-gallery-mode excitation in high-Q silica microspheres. IEEE Photonics Technol. Lett. **11**, 1429–1430 (1999)
43. H.S. Haddock et al., Fabrication of biconical tapered optical fibers using hydrofluoric acid. Mater. Sci. Eng. B **97**, 87–93 (2003)
44. E.J. Zhang et al., Hydrofluoric acid flow etching of low-loss subwavelength-diameter biconical fiber tapers. Opt. Express **18**, 22593–22598 (2010)
45. Y. Lu et al., Optimal conditions of coupling between the propagating mode in a tapered fiber and the given WG mode in a high-Q microsphere. Optik **112**, 109–113 (2003)
46. Y. Dumeige et al., Determination of coupling regime of high-Q resonators and optical gain of highly selective amplifiers. J. Opt. Soc. Am. B, Opt. Phys. **25**, 2073–2080 (2008)
47. S. Trebaol et al., Transient effects in high-Q whispering gallery mode resonators: modelling and applications, in *13th International Conference on Transparent Optical Networks* (2011)
48. J.A. Barnes et al., Loss determination in microsphere resonators by phase-shift cavity ring-down measurements. Opt. Express **16**, 13158–13167 (2008)
49. M.L. Gorodetsky et al., Rayleigh scattering in high-Q microspheres. J. Opt. Soc. Am. B, Opt. Phys. **17**, 1051–1057 (2000)
50. I. Teraoka, S. Arnold, Estimation of surface density of molecules adsorbed on a whispering gallery mode resonator: utility of isotropic polarizability. J. Appl. Phys. **102**, 076109 (2007)
51. Corning SMF-28 Optical Fiber Product Information (2002)
52. K.H. Guenther, P.G. Wierer, Surface-roughness assessment of ultrasmooth laser mirrors and substrates. Proc. Soc. Photo-Opt. Instrum. Eng. **401**, 266–279 (1983)
53. G. Berden et al., Cavity ring-down spectroscopy: Experimental schemes and applications. Int. Rev. Phys. Chem. **19**, 565–607 (2000)
54. M.D. Wheeler et al., Cavity ring-down spectroscopy. J. Chem. Soc. Faraday Trans. **94**, 337–351 (1998)
55. M. Mazurenka et al., Cavity ring-down and cavity enhanced spectroscopy using diode lasers. Annu. Rep. Prog. Chem., Sect. C, Phys. Chem. **101**, 100–142 (2005)
56. A. O'Keefe, D. Deacon, Cavity ring-down optical spectrometer for absorption measurements using pulsed laser sources. Rev. Sci. Instrum. **59**, 2544–2551 (1988)
57. A.C.R. Pipino et al., Evanescent wave cavity ring-down spectroscopy with a total-internal-reflection minicavity. Rev. Sci. Instrum. **68**, 2978–2989 (1997)
58. M. Pollinger et al., Ultrahigh-Q tunable whispering-gallery-mode microresonator. Phys. Rev. Lett. **103**, 053901 (1988)
59. A.A. Savchenkov et al., Ringdown spectroscopy of stimulated Raman scattering in a whispering gallery mode resonator. Opt. Lett. **32**, 497–499 (2007)
60. J.M. Herbelin et al., Sensitive measurement of photon lifetime and true reflectances in an optical cavity by a phase-shift method. Appl. Opt. **19**, 144–147 (1980)
61. Z. Tong et al., Phase-shift fiber-loop ring-down spectroscopy. Anal. Chem. **76**, 6594–6599 (2004)

62. M.I. Cheema et al., Simultaneous measurement of quality factor and wavelength shift by phase shift microcavity ring down spectroscopy. Opt. Express **20**, 9090–9098 (2012)
63. M. Jakubinek et al., Configuration of ring-down spectrometers for maximum sensitivity. Can. J. Chem. **82**, 873–879 (2004)
64. K. Bescherer et al., Measurement of multi-exponential optical decay processes by phase-shift cavity ring-down. Appl. Phys. B, Lasers Opt. **96**, 193–200 (2009)
65. J.R. Lakowicz et al., Analysis of fluorescence decay kinetics from variable-frequency phase-shift and modulation data. Biophys. J. **46**, 463–477 (1984)
66. J.R. Lakowicz, Frequency-domain lifetime measurements, in *Principles of Fluorescence Spectroscopy* (Springer, New York, 2006)
67. T.J. Kippenberg et al., Modal coupling in traveling-wave resonators. Opt. Lett. **19**, 1669–1671 (2002)
68. J.A. Barnes et al., Phase-shift cavity ring-down spectroscopy on a microresonator by Rayleigh backscattering. Phys. Rev. A, At. Mol. Opt. Phys. **87**, 053843 (2013)
69. K. Iwatsuki et al., Effect of Rayleigh backscattering in an optical passive ring-resonator gyro. Appl. Opt. **23**, 3916–3924 (1984)
70. M. Nakazawa, Rayleigh backscattering theory for single-mode optical fibers. J. Opt. Soc. Am. **73**, 1175–1180 (1983)
71. J. Rezac, Properties and applications of whispering-gallery mode resonances in fused silica microspheres. Ph.D. thesis, Oklahoma State University, 2002
72. Rapid market demand increase of ethylenediamine in China, in *Focus on Surfactants* (2005), p. 2
73. M. Buback, H.P. Vogele, *FT-NIR Atlas* (VCH, Weinheim, 1993)
74. E.D. Black, An introduction to Pound-Drever-Hall laser frequency stabilization. Am. J. Phys. **69**, 79–87 (2001)
75. V. Tsionsky, E. Gileadi, Use of the quartz crystal microbalance for the study of adsorption from the gas phase. Langmuir **10**, 2830–2835 (1994)
76. G. Adamovsky et al., Peculiarities of thermo-optic coefficient under different temperature regimes in optical fibers containing fiber Bragg gratings. Opt. Commun. **285**, 766–773 (2012)
77. M. Xu et al., Ethylenediamine at air/liquid and air/silica interfaces: protonation versus hydrogen bonding investigated by sum frequency generation spectroscopy. Environ. Sci. Technol. **40**, 1566–1572 (2006)
78. J.D. Lambert, E.D.T. Strong, The dimerization of ammonia and amines. Proc. R. Soc. Lond. Ser. A **200**, 566–572 (1950)
79. L.V. Lanshina et al., Structure of liquid ethylenediamine according to data on molecular-scattering of light. J. Struct. Chem. **30**, 684–687 (1989)
80. M.T.S.R. Gomes et al., Detection of volatile amines using a quartz crystal with gold electrodes. Sens. Actuators B, Chem. **57**, 261–267 (1999)
81. V. Dong et al., Detection of local density distributions of isolated silanol groups on planar silica surfaces using nonlinear optical molecular probes. Anal. Chem. **70**, 4730–4735 (1998)
82. C. Schmidt et al., Nonlinear thermal effects in optical microspheres at different wavelength sweeping speeds. Opt. Express **16**, 6285–6301 (2008)

Chapter 11
Cavity Ringdown Spectroscopy for the Analysis of Small Liquid Volumes

Claire Vallance and Cathy M. Rushworth

Abstract Cavity-enhanced absorption techniques show considerable promise for applications in which a high detection sensitivity is required while maintaining a small probed volume of liquid sample. Conventional two-mirror cavity ringdown or cavity-enhanced absorption spectroscopy appears to provide a simple general 'benchtop' platform for measurements on flow cells and microfluidic chips with suitable optical properties, and recent progress in developing integrated cavities on-chip hints at the potential for future miniaturised self-contained microfluidic devices.

11.1 Introduction

When developing chemical sensors, minimising the sample volume required for a successful test is often a key goal. In some situations this may be because only very limited amounts of sample are available, for example in forensics and some medical tests, while in others there may be a desire to increase the portability of the sensor, to reduce the costs associated with expensive reagents, or to minimise the amount of chemical waste generated. In the medical arena, reduced sample volume generally reduces the invasiveness of the test, for example replacing a venous blood draw with a fingerprick blood test, or allows an increased number of tests to be carried out on a given sample volume.

One of the limiting factors in reducing the sample volume is the availability of detection techniques with sufficient sensitivity to detect extremely small quantities of the chemical species of interest. A variety of such techniques are available, each with their own strengths and weaknesses. In the field of microfluidics, for example, where sample volumes tend to lie in the microlitre to picolitre range, detection

C. Vallance (✉) · C.M. Rushworth
Department of Chemistry, Chemistry Research Laboratory, University of Oxford,
12 Mansfield Rd., Oxford OX1 3TA, UK
e-mail: claire.vallance@chem.ox.ac.uk

C.M. Rushworth
e-mail: catherine.rushworth@chem.ox.ac.uk

G. Gagliardi, H.-P. Loock (eds.), *Cavity-Enhanced Spectroscopy and Sensing*,
Springer Series in Optical Sciences 179, DOI 10.1007/978-3-642-40003-2_11,
© Springer-Verlag Berlin Heidelberg 2014

methods include electrochemical techniques [1], fluorescence [2] and Raman spectroscopy [3, 4], nuclear magnetic resonance spectroscopy [5], and mass spectrometry [6], amongst others [7–15]. To date, absorption spectrosocopy has been fairly conspicuous by its absence in microfluidic applications. While having considerable appeal as a detection method for chemical sensing, since every chemical species absorbs light in some region of the electromagnetic spectrum and may therefore be detected assuming suitable light sources and detectors are available, difficulties arise when dealing with very small sample volumes. This is easily illustrated by considering the well-known Beer-Lambert law law, which describes the fraction of light transmitted through a sample as a function of the analyte concentration, C, the path length through the sample, l, and the absorption coefficient α or extinction coefficient ε of the sample molecule, which quantify the efficiency with which the molecule absorbs light.

$$\frac{I}{I_0} = e^{-\alpha C l} = 10^{-\varepsilon C l}. \tag{11.1}$$

Here, I_0 and I denote the light intensity incident on and transmitted through the sample. In general laboratory applications of absorption spectroscopy, the liquid sample is contained within a cuvette of path length ~ 1 cm. Reducing the sample volume to microfluidic dimensions or similar often yields a path length so short that the absorption becomes undetectable. Inspection of Eq. (11.1) reveals that the detection sensitivity can be improved by increasing the absorption/extinction coefficient, the sample concentration, or the path length through the sample. Since the absorption coefficient is a constant of the molecule under study,[1] and the concentration is generally a constant of the sample, increasing the path length through the sample is often the only option. As we shall see, cavity-enhanced spectroscopic techniques have the potential to achieve significant increases in absorption path length, and therefore detection sensitivity, without increasing the sample volume, and show considerable promise as a means of amplifying the absorption signal in small-volume liquid samples.

When dealing with *gas-phase* samples, we have seen from other chapters that cavity-enhanced spectroscopic methods provide a huge improvement in detection sensitivity over single-pass absorption methods. The minimum detectable absorption per unit path length, $\kappa = \alpha C$, is typically in the range from 10^{-6} to 10^{-9} cm^{-1} [16], with detection limits as low as 8×10^{-11} cm^{-1} having been achieved using some of the more sophisticated approaches [17]. As a result, cavity-enhanced approaches have become the spectroscopic detection methods of choice for applications ranging from fundamental studies of molecular spectroscopy, kinetics and dynamics, through to atmospheric and environmental sensing, and air and process quality monitoring [16, 18–21]. There has also been considerable recent interest in the potential for developing medical diagnostic and monitoring devices

---

[1] See Sect. 11.5 for a discussion of colourimetric cavity ringdown spectroscopy, which uses colourimetric reactions to convert the molecule of interest into a strongly absorbing derivative compounds, thereby effectively increasing α.

based on the absorption signatures of biomarkers present in trace amounts within exhaled breath [22–26]. In the liquid phase, the much higher density of molecules means that there are fewer situations in which the level of signal amplification offered by a cavity-enhanced absorption technique is required. Also, for a variety of reasons (explored in Sect. 11.3), the sensitivity enhancements achievable for liquid-phase samples are generally less than for gaseous samples. These two facts taken together have meant that applications of cavity-enhanced spectroscopies to measurements on liquid-phase samples are much less common than in the gas phase, with most reports to date comprising proof-of-concept studies on possible cavity configurations [27–36]. However, as noted above, there is considerable scope for using such techniques when probing small-volume liquid samples.

We begin this chapter with a brief introduction to the two most commonly-used cavity-enhanced absorption techniques, namely cavity ringdown spectroscopy (CRDS) and cavity-enhanced absorption spectroscopy (CEAS), focusing on the simplest experimental configuration, in which the cavity is formed from a pair of carefully-aligned high-reflectivity plano-concave mirrors. We then outline the challenges associated with incorporating liquid samples into an optical cavity and summarise the various approaches that have been adopted. Finally, we explore a number of potential application areas in microfluidics and chemical sensing.

11.2 Fundamentals of Cavity Ringdown and Cavity-Enhanced Absorption Spectroscopy

11.2.1 General Principles of Cavity Ringdown Spectroscopy

A typical experimental setup for a cavity ringdown or cavity-enhanced absorption spectroscopy measurement is shown schematically in Fig. 11.1. The basis of the cavity ringdown technique is a measurement of the decay of light intensity within a high-finesse optical cavity. In a typical measurement, a pulse of light is directed along the cavity axis towards the back face of one of the mirrors, and the small amount of light coupled through the mirror into the cavity undergoes repeated reflections between the two mirrors. A small fraction of light is lost on each round-trip of the cavity, primarily due to transmission or absorption by the mirrors or absorption or scattering within the cavity, with the result that the light intensity within the cavity decays exponentially with time. The exponential decay may be recorded by positioning a sensitive photodetector behind the second mirror to detect the light leaking out through the mirror on each pass. The time constant, τ, of the decay, also known as the 'ringdown time' is given by

$$\tau = \frac{nd}{c(L + \alpha Cl)} \tag{11.2}$$

where d is the cavity length, c/n the speed of light within the cavity (n is the refractive index of the cavity medium), L the cavity loss per pass in the absence of a

Fig. 11.1 Schematic of the experimental setup for a cavity ringdown or cavity-enhanced absorption spectroscopy measurement

sample, and α the absorption coefficient of an absorbing species present at concentration C over a path length l within the cavity. Because the measured decay constant is independent of the absolute intensity of light injected into the cavity, CRDS is largely insensitive to noise caused by shot-to-shot fluctuations in the intensity of the pulsed laser source.

For a gas-phase measurement, the cavity loss L for a simple two-mirror cavity is simply given by $1 - R$, with R the mirror reflectivity. For liquid-phase experiments, L also includes refraction, reflection and scattering losses from the (ideally non-absorbing) solvent and any container surfaces introduced into the cavity. Equation (11.2) is valid both for two-mirror cavities and for the fibre loop cavities considered in Sect. 11.3.1 and Chap. 12. Note that apart from constants of the experimental setup, such as the length of the cavity, the intrinsic cavity losses, and the path length through the sample, the only parameters that determine the ringdown time are the absorption coefficient and concentration of the sample. Consequently, by recording the ringdown time in the presence and absence of the sample (denoted τ and τ_0, respectively), the absorption per unit path length, $\kappa = \alpha C$, can be determined, yielding the concentration C if the absorption coefficient is known at the wavelength of interest

$$\kappa = \alpha C = \frac{nd}{cl}\left(\frac{1}{\tau} - \frac{1}{\tau_0}\right). \tag{11.3}$$

11.2.2 General Principles of Cavity-Enhanced Absorption Spectroscopy

Cavity ringdown measurements require the use of a pulsed laser or rapidly switched continuous wave (CW) laser in combination with a fast detector and associated electronics in order to record the decay transient of the cavity. Cavity-enhanced absorption spectroscopy [37–39] (CEAS), also known as integrated cavity output spectroscopy (ICOS), is an experimentally simpler variant of cavity ringdown spectroscopy in which a CW (or pseudo-CW) light beam is coupled into the cavity, and the intensity coupled out from the cavity is monitored. It can be shown that the resulting signal is proportional to the ringdown time of the cavity [38, 40], and therefore provides the necessary information on absorption without the need to record the decay transient. In many cases this experimental simplification is enough to make CEAS the preferred approach over CRDS. In addition, CEAS often has a larger dynamic range than CRDS. While in CRDS measurement, intense absorptions often result in a ringdown time that is too short to measure with sufficient accuracy [32], such absorptions do not pose a problem for CEAS measurements, in which only an intensity must be measured. As we shall see in Sect. 11.3.2, CEAS has a further advantage over CRDS when broadband measurements are desired. However, while CEAS has a number of advantages over CRDS, it also has two disadvantages. Firstly, CRDS yields an absolute absorption measurement, while CEAS provides only a relative absorption measurement unless a calibration measurement is carried out in order to determine the mirror reflectivity (often this is achieved in a separate ringdown measurement). Secondly, since the method involves a measurement of light intensity rather than rate of decay of intensity, the achievable signal-to-noise ratio is now dependent on the stability of the light source.

Assuming a low loss per pass, a valid approximation in any cavity-enhanced spectroscopy measurement, Fiedler $et\ al.$ [41] showed that the absorption per unit path length for a CEAS measurement is given by

$$\kappa = \frac{1}{d}\left(\frac{I_0}{I} - 1\right)(1 - R) \tag{11.4}$$

where d is the cavity length and R the mirror reflectivity, as before, and I_0 and I are the transmitted light intensity in the absence and presence of an absorbing sample.

Equations (11.4) and (11.8) indicate that a CEAS measurement is more sensitive than a single-pass absorption measurement by a factor of $(1 - R)^{-1}$. This factor is equal to the number of passes through the sample during the measurement, and is usually referred to as the cavity enhancement factor [34] (CEF). The CEF may be determined experimentally in any one of a number of ways; for example, by measuring the mirror reflectivity, as noted above, by making an absorption measurement on a sample with a known κ, or by measuring the ratio of the cavity-enhanced to single-pass absorption under equivalent experimental conditions.

11.2.3 Detection Limits in Cavity Ringdown and Cavity-Enhanced Absorption Measurements

The detection limit in a cavity ringdown measurement is determined by the minimum measureable change in ringdown time, $\Delta\tau_{min}$, on introduction of a sample to the cavity. Based on IUPAC recommendations [42], this is often defined quantitatively as three times the standard deviation in the baseline ringdown time, τ_0. Once $\Delta\tau_{min}$ has been determined, by making the (very good) approximation that at low sample concentrations $\tau\tau_0 \approx \tau_0^2$, we can rearrange Eq. (11.3) to determine the minimum detectable absorption per unit path length, κ_{min} [43].

$$\kappa_{min} = \frac{nd\,\Delta\tau_{min}}{cl\tau_0^2}. \tag{11.5}$$

Inspection of Eq. (11.5) reveals that to improve the sensitivity of a ringdown measurement, we can either minimise the baseline cavity losses in order to maximise

τ_0, or increase the path length travelled through the sample on each pass. When dealing with very small sample volumes, the scope for increasing the path length is limited, and improving the detection sensitivity relies primarily on minimising the cavity losses. Since the cavity length, and therefore the ringdown time, varies widely between different experiments, a more useful parameter for comparing the cavity losses associated with different experimental setups is the number of passes N through the cavity per unit ringdown time

$$N = \frac{\tau_0 c}{nd}. \tag{11.6}$$

Since signal is usually acquired over more than one ringdown time, the cavity enhancement factor (CEF) relative to a single-pass measurement is generally larger than N.

A similar approach can be used to determine the detection limit in a cavity-enhanced absorption measurement. Making the (again, very good) approximation that

$$\frac{I_0 - I}{I} \approx \frac{I_0 - I}{I_0}, \tag{11.7}$$

the minimum detectable absorption per unit path length is given by

$$\kappa_{min} = \frac{1}{d} \frac{\Delta I_{min}}{I_0} (1 - R) \tag{11.8}$$

where ΔI_{min} is the minimum detectable change in light intensity transmitted through the cavity, again taken to be three times the standard deviation in the baseline noise of the measurement.

11.3 Applying Cavity-Enhanced Spectroscopies to the Liquid Phase

Achieving high detection sensitivity in cavity-enhanced spectroscopic measurements relies on keeping cavity losses attributable to anything but absorption by the species of interest to a minimum. Measurements on condensed-phase samples raise considerable challenges in this context. The most straightforward way to make measurements on liquid samples is to place the solution inside the cavity in direct contact with the mirrors. However, this requires a fairly large volume of liquid sample, which may not always be available, and also introduces high scattering losses. The much higher refractive index of liquids relative to air can alter the reflectivity properties of dielectric cavity mirrors, and in extreme cases, corrosive samples can cause permanent damage to the mirror surfaces. To avoid placing liquid in contact with mirrors, the sample is generally enclosed within a cuvette or other container. This introduces four additional surfaces into the cavity, and minimising scattering, refraction, and reflection losses at these surfaces is critical to maintaining the low overall cavity losses and correspondingly long ringdown times needed for high-sensitivity absorption measurements. A number of innovative cavity configurations have been employed for liquid-phase absorption measurements, and we will explore a variety of these in Sect. 11.3.1. A point worth noting is that because the round-trip cavity losses tend to be considerably higher in liquid-phase CRDS and CEAS measurements than in their gas-phase counterparts, the reflectivity of the mirrors is usually no longer the dominant factor determining the overall round-trip losses. As a result, lower-cost mirrors with somewhat lower reflectivity may often be used without significantly affecting the performance of the cavity.

A second consideration when making absorption measurements on liquids is the structure of the spectral absorption features for the molecule of interest. This is particularly true for sensing applications. While gas-phase species generally exhibit sharp absorption lines and detailed rovibrational structure in their absorption spectra, at least when probed at sufficiently high spectral resolution, intermolecular interactions in the liquid phase blur this structure into broad, relatively featureless absorption bands. For sensing applications in the gas phase, it is often sufficient to identify a single rovibrational absorption line that is distinct to the molecule of interest, and to record the absorption signal at the corresponding single wavelength. This is less likely to be possible in liquid-phase sensing applications, as it tends to be much more difficult to identify a single wavelength within an absorption band of the analyte of interest that does not overlap with absorption signals from solvent or other chemical species present in the sample. An alternative is to record the absorption spectrum over a range of wavelengths and employ a fitting procedure based on the known absorption features of the various chemical species present in the sample to extract the contribution from the species of interest. As in all absorption measurements, solvents must be chosen carefully, as common solvents such as water, methanol and ethanol often have significant absorption features in spectroscopic regions of interest. Historically, most CRDS and CEAS spectrometers have employed a tuneable monochromatic laser source to record absorption spectra over a range of wavelengths. The ringdown time or transmitted intensity is measured one wavelength at a time as the laser is scanned, and data acquisition rates are correspondingly rather slow. In more recent years, a considerable amount of effort has been invested into developing 'broadband' CRDS and CEAS methods. These employ a broadband light source and spectrally resolving detector, allowing the complete absorption spectrum over the wavelength region of interest to be recorded in a single measurement. Broadband cavity-enhanced spectroscopy methods are described in Sect. 11.3.2.

11.3.1 Cavity Configurations for Liquid-Phase Analysis

A number of different cavity configurations that have been employed successfully for liquid-phase CRDS and CEAS measurements are illustrated in Fig. 11.2. Configuration (a), in which the entire cavity is filled with the liquid sample, has already been discussed in Sect. 11.3. The high scattering losses occasioned by the liquid

Fig. 11.2 Possible configurations for liquid phase CRDS and CEAS: (**a**) liquid-filled cavity; (**b**) normal-incidence absorption cell; (**c**) Brewster-angle absorption cell; (**d**) liquid sheet jet; (**e**) evanescent wave absorption. Adapted from Ref. [44]

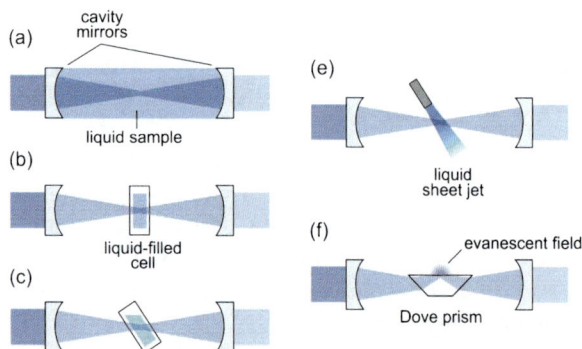

usually limit the cavity length in this configuration to a few millimetres, rather than the more usual dimensions of tens of centimetres. This reduces the ringdown time to such a degree that sub-nanosecond time resolution is required in order to make an accurate measurement in a CRDS measurement, though CEAS measurements are still feasible, and there have been some notable proof-of-concept studies employing this approach. Indeed, McGarvey et al. [33] achieved the highest cavity enhancement factor of any liquid phase CEAS measurement reported to date in their measurements on Bacteriochlorophyll a at 783 nm, employing a cavity of length 1.7 mm formed from 99.998 % mirrors. The authors even speculated that single-molecule detection may be possible within a very short cavity of length $\sim$10 µm. A more modest CEF of 429 was obtained by Seetohul et al. [36], who employed a 20 cm cavity formed from 99.9 % mirrors in combination with an LED light source. This was still sufficient to detect methylene blue at the picomolar level. The CEF was lower than the value of around 1000 that was expected based on the known mirror reflectivity and cavity length, which the authors took to indicate a reduction in the mirror reflectivity on contact with the liquid sample.

Either one of two strategies may be adopted when introducing an absorption cell into a two-mirror cavity to contain a liquid sample. In the first (illustrated in Fig. 11.2(b)), the cell is mounted at normal incidence to the laser beam, such that all reflections are retained along the cavity axis and reflection losses are therefore minimised. In the second (illustrated in Fig. 11.2(c)), the cell is mounted at Brewster's angle, in which case the reflection losses at the surfaces are theoretically zero for light polarised parallel to the plane of incidence [27]. While either approach sounds appealing in theory, in practice there are problems with both. In the first geometry, all surfaces within the absorption cell must be parallel to each other to a high degree of precision in order to ensure that the light circulating within the cavity strikes all surfaces at exactly normal incidence. In practice, this is rarely the case, and some light will almost always be reflected into unstable cavity modes and be lost from the cavity on each encounter with the absorption cell. This problem can be mitigated by reducing the cavity length, which relaxes the alignment requirements to some extent, though at the cost of a reduced ringdown time. As an example, van der Sneppen et al. [29] investigated CRDS as a detection technique for liquid chromatography, inserting a cuvette into a short (4 cm) cavity at normal incidence to the beam axis, and demonstrated detection limits (in terms of absorption per unit path length) of around 10^{-4} cm^{-1} for a range of nitro-polyaromatic hydrocarbons. In the second geometry, the different refractive indices present at the four interfaces within the absorption cell mean that, while there is an 'optimum angle' at which surface reflection losses are minimised, there is in fact no single Brewster angle condition that is satisfied at all four interfaces. In addition, the optimum angle for the absorption cell depends strongly on the refractive index of the liquid sample, and the cell positioning within the cavity must therefore be reoptimised for each sample.

Reported CEAS measurements on liquids have relied almost exclusively on either the liquid-filled cavity or normal-incidence absorption cell configurations. However, for CRDS measurements, the Brewster-angle geometry has proved considerably more popular than the normal-incidence geometry. Xu *et al.* [28] have published one of the lowest detection limits, employing both the cavity geometry shown in Fig. 11.2(c) and a second geometry in which two Brewster cells were incorporated into the cavity, and recording a minimum detectable absorption per unit path length of 10^{-7} cm^{-1} for the fifth vibrational overtones of the C–H stretch in pure benzene and benzene in solution with hexane. Snyder *et al.* [27] went to considerable lengths to optimise the Brewster-cell arrangement, designing a custom liquid flow cell which ensured that the Brewster angle condition was satisfied at every interface. Using a pulsed laser to excite the cavity, the resulting cavity enhancement factor was similar to that achieved by Xu *et al.*, although the detection limit was significantly poorer due to a shorter path length through the sample on each pass, at 6.2×10^{-4} cm^{-1}. By replacing the pulsed laser with a narrowband CW diode laser [30], which was carefully mode matched into the cavity to excite a single cavity mode, the CEF was improved by more than a factor of two. This, together with the greatly improved signal-to-noise ratio achieved with the more stable light source, reduced the detection limit to 7.8×10^{-6} cm^{-1}.

An innovative approach which solves the problem of having liquid sample in contact with the mirrors, while at the same time eliminating the need to introduce a liquid container into the cavity, is the flowing liquid-sheet jet configuration developed by Alexander [31] and illustrated in Fig. 11.2(d). The resulting cavity enhancement factor is similar to that of Xu *et al.* and the pulsed-laser work of Sneppen *et al.*, but the very short path length per pass through the liquid jet of 23 μm yields a relatively poorer detection sensitivity of 1.6×10^{-2} cm^{-1}. Also, though the probed volume using this approach is very small, the flowing jet requires a relatively large reservoir volume of sample.

All of the cavity configurations we have considered so far involve what might be termed 'direct' absorption measurements, in which the absorption is measured via direct transmission of the laser beam through the liquid sample. An alternative approach exploits the evanescent wave generated at a surface within the cavity where the light undergoes total internal reflection. A liquid or solid sample placed on the surface absorbs the evanescent field, leading to a reduction in the ringdown time of the cavity. An example of such an approach, in which the evanescent field is created at the surface of a Dove prism placed within the cavity, is shown in Fig. 11.2(e), though numerous other configurations are possible [45–47], including

Fig. 11.3 (a) Linear fibre
cavity, employing either
mirror-coated fibre ends or
fibre Bragg gratings to
confine the light within the
cavity; (b) Fibre-loop cavity,
with input and output
coupling from the side of the
optical fibre

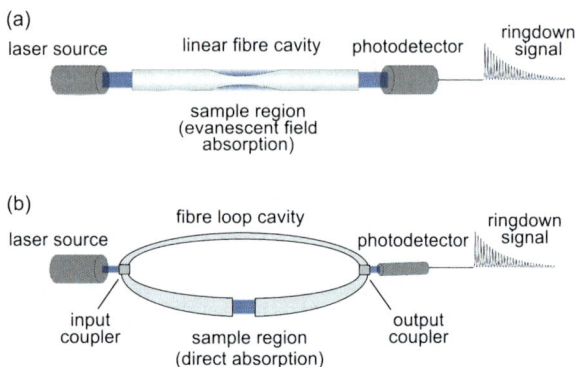

(a)

laser source   linear fibre cavity   photodetector   ringdown
                                                                     signal

                              sample region
                             (evanescent field
                               absorption)

(b)
                              fibre loop cavity   ringdown
laser source   photodetector   signal

              input   output
              coupler   sample region   coupler
                              (direct absorption)

some in which a custom-designed prism acts as both the optical cavity and the sample interface [48–50]. Since only a relatively small fraction of the propagating light interacts with the sample in such an arrangement, one might expect a considerable reduction in sensitivity relative to direct absorption methods. However, since cavity configurations for evanescent absorption measurements usually introduce fewer additional surfaces into the cavity than a liquid container, this is often compensated for by a reduction in the overall cavity losses relative to a direct absorption measurement. Because the evanescent wave only propagates a very short distance from the surface, on the order of a wavelength, cavity configurations of this type are well suited to the study of thin films and processes at interfaces [46, 47, 49, 50].

An increasingly popular type of cavity exploits total internal reflection within an optical fibre. Since optical-fibre cavities are the subject of Chap. 12 of this book, we will only provide a brief description of such cavities here. Fibre cavities can be grouped into two classes, namely linear fibre cavities and fibre-loop cavities, both of which are illustrated schematically in Fig. 11.3. Linear fibre cavities consist of a length of fibre with either a mirror-coating [51–53] or a fibre Bragg grating [54] at either end to confine the light within the fibre. Light is injected into one end of the fibre and undergoes repeated reflections along the length of the fibre, with the waveguiding properties of the fibre ensuring that the cavity is permanently and robustly aligned. Linear fibre cavities almost exclusively employ evanescent field absorption: a sensing region is created by stripping the plastic coating from a small region of fibre and tapering the fibre either by etching in hydrofluoric or flurosilicic acid, or by heating and pulling, to expose the evanescent field of the waveguided light. Fibre-loop cavities [25, 55, 56] consist of a loop of optical fibre in which the excitation light circulates, with the sample probed by direct absorption, having been introduced into a small gap between the fibre ends where they join to form the loop. There are considerable challenges associated with optimising the side-coupling of light into and out of the cavity and minimising losses associated with the coupling and sample regions when using this type of cavity. There have been numerous demonstrations of fibre cavities for sensing of a range of properties, including temperature [57], pressure [58], stress [59] and refractive index [60]. In applications to chemical sensing, optical fibre cavities are well suited to the interrogation of extremely small sample volumes, being inherently size-matched to microfluidic channel widths, for example. They are also inherently broadband in nature, facilitating measurements at multiple wavelengths or across a continuum of wavelengths. However, fibre-based cavities are much more lossy than their mirror-based counterparts, such that their robust and compact nature is usually offset by significantly poorer detection limits.

11.3.2 Broadband Cavity Ringdown Techniques

As noted earlier, the ability to perform absorption measurements across a range of wavelengths is extremely useful in the context of the broad absorption features associated with a liquid phase analyte. Under these conditions, measurements at a single wavelength provide limited information, and absorption by other species present in the sample is often problematic. By recording across a complete absorption band, broadband measurements can provide specificity, and in the case where several absorbing species are present, they can provide the necessary spectral coverage to monitor multiple species simultaneously.

There have been a number of approaches to combining a broadband light source with cavity ringdown measurements [21]. Most early methods lacked a broadband detector and used a monochromator to scan the wavelength either at the entrance to or exit from the cavity, making them essentially equivalent to experiments carried out with a tuneable (rather than broadband) light source. A variation on this approach was the Fourier transform CRDS method developed by Engeln and Meijer [61], which built up the wavelength-resolved ringdown signals from a series of interferograms. Engeln and coworkers later developed a phase-shift version of the technique [62] with significantly higher sensitivity. All of these approaches essentially involve a wavelength scan. An alternative approach is to record all wavelengths simultaneously and perform a scan in time to build up the ringdown transients at each wavelength. This was the approach adopted by Czyzewski [63], who dispersed the broadband light through a spectrograph at the cavity exit and imaged the resulting spectrum using a time-gated CCD camera. By sweeping the CCD time gate through the ringdown decay over consecutive laser pulses, the broadband ringdown trace is obtained. This method neatly makes use of the broadband character of the light source, but because the ringdown trace is built up over multiple laser shots, any shot-to-shot fluctuations in the intensity of the light source manifest as additional noise on the ringdown traces.

There are only very few examples of 'true' broadband cavity ringdown spectroscopy, in which the ringdown decay transient is recorded for each wavelength on every laser shot. The first of these is a technique known as 'ringdown spectral photography', developed by Scherer [64, 65] and illustrated schematically in Fig. 11.4(a). Broadband radiation leaving the cavity is incident first on a rotating mirror, and then on a diffraction grating, before being imaged on a two-dimensional CCD array. The rotating mirror resolves the time-dependence of the signal, with light from different time segments of the ringdown decay striking different points

Fig. 11.4 (a) Ringdown spectral photography; and (b) clocked CCD camera approaches to broadband cavity ringdown spectroscopy. Adapted from Ref. [68]

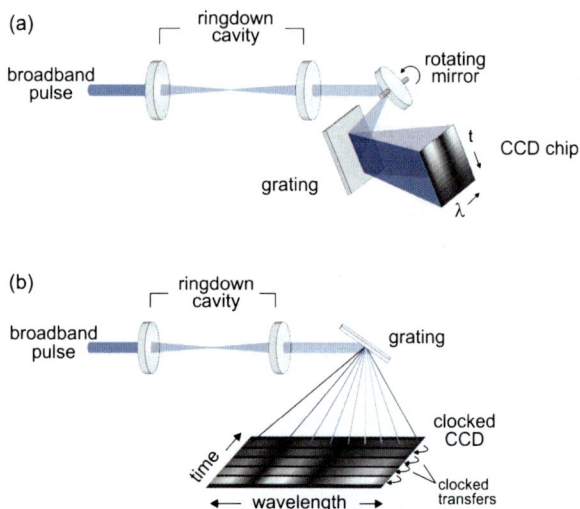

on the diffraction grating. The diffraction grating then disperses the signal in wavelength along a perpendicular axis. The resulting two-dimensional image is resolved in time along one axis and wavelength along the second axis. The resolution of the grating and CCD array used in these measurements yielded a spectral resolution of around 1.5 cm^{-1}, sufficient to resolve rotational structure, while the time resolution, determined by the rotation speed of the mirror and the number of pixels in the CCD sensor, was typically around 40 ns per pixel. The second broadband CRDS technique, developed by Ball *et al.* [21, 66], also employs a grating to disperse the broadband light in wavelength as it emerges from the cavity. However, the time resolution is achieved using a two-dimensional clocked 512×1024 CCD array (see Fig. 11.4(b)). The CCD array is masked so that the dispersed cavity output falls only on the first ten rows of pixels. After acquiring signal for 5 μs, the pixel voltages are clocked such that the charge collected on these rows is transferred to the next ten rows, allowing a new image to be recorded on the (now empty) first ten rows. The process is repeated, shunting the acquired charge further down the sensor every 5 μs. This procedure permits around ten images to be recorded during the ~50 μs ringdown decay, with the final image read out from the CCD sensor comprising a two-dimensional 'wavelength vs time' image similar to that recorded in the ringdown spectral photography method. Following initial proof of concept work [21, 66], the instrument was used in field measurements during the 2002 North Atlantic Marine Boundary Layer Experiment [67] to measure ambient levels of NO_3, N_2O_4, I_2 and OIO.

The need to achieve both wavelength-resolved and time-resolved detection makes broadband cavity ringdown measurements experimentally complex. In contrast, broadband CEAS requires only a broadband light source and a means of wavelength dispersion of the cavity output, often a commercial spectrograph or spectrometer. It is therefore experimentally much more straightforward than the corresponding CRDS technique, and has been adopted enthusiastically by the cavity-enhanced spectroscopy community for high-sensitivity broadband absorption measurements. A variety of light sources have been used for broadband CEAS measurements, including Xe arc lamps [41, 69–74], light emitting diodes [75–81], supercontinuum sources [82–88], and femtosecond frequency combs [89–92]. Extremely high detection sensitivities have been achieved, down to the pptv level in some cases [71, 74, 77, 80–82], with signal averaging times of seconds to minutes depending on the intensity of the light source.

11.4 Example Application: Microfluidic Sensors

There is currently a great deal of research aimed at developing microfluidic devices to carry out chemical processes. Known by a variety of names, including 'lab on a chip' and 'μTAS' (micro-total-analysis systems), such devices integrate a raft of laboratory functions, such as sample preparation, mixing, reaction, and separation, onto a glass or plastic chip a few centimetres in size. There are many situations in which miniaturising chemical analysis in this way holds a number of advantages over bulk processing of chemical samples. A microfluidic platform is ideal when attempting to develop portable chemical tests, for example, and such chips have considerable potential for applications such as medical diagnostics, forensics, and environmental monitoring. Fabrication costs are generally low, and only very small volumes of sample and reagents are required for analysis, which is often particularly important when dealing with biological samples. A secondary advantage of the small reagent volumes employed is improved safety relative to lab-based tests. There is also the potential for very high-throughput chemical processing, achievable by running many experiments, reactions, or analyses in parallel on an array of chips. However, there are considerable challenges associated with developing such devices, not least the difficulty associated with the quantitative detection of chemical species in volumes on the order of picolitres. Cavity-enhanced spectroscopic techniques provide one potential approach to meeting this challenge [44].

As noted in Sect. 11.1, once path lengths are reduced to microfluidic dimensions of typically 50–100 μm, the absorption signal often becomes too small to measure reliably using conventional single-pass absorption spectroscopy. Many researchers have attempted to address this issue with a range of extended-path length single-pass methods [93–104], and a variety of multi-pass microfluidic absorption cell designs [103, 105–113], many have which have recently been reviewed [44]. However, both of these approaches come at the expense of a larger sample volume, and to implement them in a particular application requires fairly major modifications to the microfluidic chip design. In contrast, embedding a microfluidic chip within an optical cavity (or, alternatively, embedding an optical cavity within a microfluidic chip) provides a large increase in path length, and a corresponding enhancement in detection sensitivity, with no accompanying increase in the probed volume of sample.

Incorporating an optical cavity into a microfluidic chip is equivalent to cavity configuration (a) of Fig. 11.2, and avoids many of the problems outlined in

Sect. 11.3.1 associated with incorporating a liquid container into the cavity. In this approach, the very short cavity length precludes ringdown measurements, but CEAS measurements are still possible. The simplest approach to creating a cavity within a chip is simply to mirror-coat the internal surfaces of a microchannel. This was the method adopted by Billot *et al.* [114], who coated 150 nm thick gold mirrors onto the inner surfaces of two glass substrates and then bonded them with PDMS (polydimethylsiloxane) to form a microfluidic chip containing a 50 µm channel. Using the focused output from a halogen lamp as the light source and coupling the cavity output to a spectrometer, the spectrum of Rhodamine B was acquired over the wavelength range from 500–600 nm, with a measured cavity enhancement of 50 relative to an uncoated channel. Further development in integrated-cavity microfluidic chip designs may well be inspired by advances in microfluidic laser technology [115–117], with both applications sharing the requirement for a high-finesse on-chip optical cavity. Cavity designs for microfluidic lasers include ring dye lasers [118], and mirror-coated optical fibres positioned either side of one or more microfluidic channels [119]. By incorporating two microfluidic channels within the cavity, it has already been demonstrated by Galas *et al.* [119] that the latter design may be used for chemical sensing. The first channel was filled with Rhodamine 6G to provide laser gain, and the second with the analyte of interest. The sensitivity of the laser output to the loss introduced to the cavity by the absorbing analyte allowed the detection of micromolar solutions of methylene blue within an optical path length of only 140 µm.

An alternative to incorporating a cavity into a chip is to insert the microfluidic chip into a conventional two-mirror cavity. While the presence of a chip within the cavity increases the cavity losses, this approach has the advantage that a single cavity can be used to make measurements on a variety of different microfluidic chips and provides a fairly general platform for absorption measurements on microfluidic samples [120]. Any microfluidic chip to be used in such measurements must introduce the lowest possible optical loss to the cavity, and must therefore have a high degree of optical transparency and low surface roughness, with all surfaces within the chip fabricated to be as parallel as possible. It is also important that the beam waist within the cavity is carefully matched to the size of the microfluidic channel to be probed, in order to minimise scattering losses. Using a chip fabricated from glass slides sandwiched together with thiolene optical adhesive, inserted at Brewster's angle within a two-mirror optical cavity, James *et al.* carried out a series of proof-of-concept experiments to demonstrate the capabilities of CRDS in monitoring both static and time-varying analyte concentrations, with a cavity enhancement factor of around 150. The minimum detectable absorption per unit path length was determined from a series of measurements on $KMnO_4$ solutions of varying concentrations to be 1.0×10^{-3} cm^{-1}. Measurement of pH to a precision of ± 0.04 pH points was demonstrated on a 20 nL sample by monitoring the absorption of phenolphthalein indicator present at millimolar concentrations in a series of buffered pH standards. A typical titration curve from these measurements is shown in Fig. 11.5(a). Finally, the Belousov-Zhabotinsky oscillating reaction was followed on-chip to demonstrate the potential of CRDS for chemical kinetics studies. The reaction involves the catalytic oxidation and bromination of malonic acid, and in the

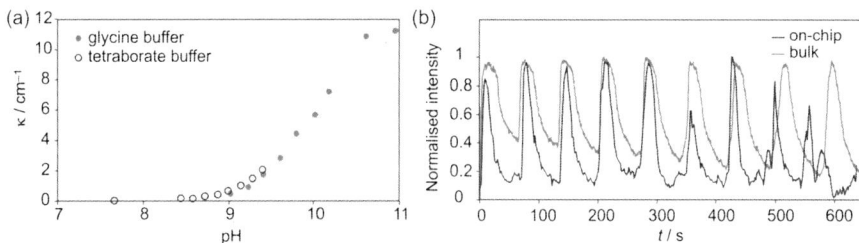

Fig. 11.5 (a) Optical measurement of pH on a microfluidic chip via CRDS. The measured absorption at 532 nm of an indicator dye for a series of buffered pH standards tracks the titration curve of the indicator; (b) Reaction monitoring on a microfluidic chip. Oscillatory behaviour of the Belousov-Zhabotinsky reaction monitored via single-pass absorption in a 1 cm cuvette (*grey*) and CRDS on a microfluidic chip (*black*). Adapted from [120]

presence of a ferroin indicator, an oscillating red to blue colour change is observed with a period of several tens of seconds. The time-dependent absorption of the reaction mixture at 532 nm allows the oscillation to be followed on-chip in real time. Figure 11.5(b) shows a data set comparing the microfluidic measurements with a set of measurements made on a bulk reaction mixture using a conventional UV/vis spectrometer. While the oscillation period is similar in both cases, the shapes of the time-resolved absorption peaks are quite different, implying that the detailed reaction kinetics depend on the size and shape of the reaction vessel. At later times the microfluidic measurements were degraded somewhat by the formation of bubbles of CO_2 (one of the reaction products) within the microfluidic channel. In principle, CRDS offers the possibility of tracking chemical kinetics down to the microsecond timescale of the ringdown decay and, if a time-varying concentration is allowed when fitting the ringdown traces to determine the absorption, to even shorter timescales.

The discussion of microfluidic samples so far has focused on continuously flowing liquids. However, many microfluidic applications rely on the generation and manipulation of a sample consisting of uniform droplets, which essentially constitute miniature 'test tubes' with volumes of femtolitres to picolitres. Droplet microfluidics has been used for applications ranging from synthesis of nanoparticles [121] and optimisation of protein crystallisation conditions [122] through to reaction kinetics studies [123–126]. Both CRDS and CEAS have sufficient sensitivity for single-droplet analysis [127, 128], and by making measurements as both the droplets and the carrier phase pass through the beam, the carrier phase measurement can be used as the 'blank' for the absorption measurement, assuming any effects of refractive index mismatch between the two phases can be accounted for. The results of proof-of-principle investigations on dye droplets using both CRDS at 532 nm [120] and broadband CEAS [85] are shown in Fig. 11.6. While in both studies the data acquisition systems employed were relatively slow, limiting the studies to large droplets (or 'slugs') and low droplet production rates, with suitably fast data acquisition hardware it should be entirely feasible to study considerably smaller droplets at high droplet production rates. In both cases the signals are char-

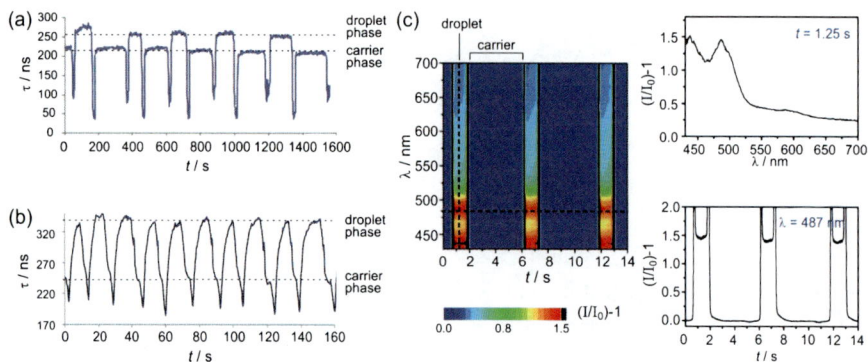

Fig. 11.6 Droplet flow analysis via CRDS of slugs of 1 µM aqueous Rhodamine 6G in a toluene carrier phase at a flow rate of (**a**) 0.008 mL min^{-1}, and (**b**) 0.075 mL min^{-1} on a glass-thiolene chip placed at Brewster's angle within the cavity; (**c**) Droplet flow analysis via broadband CEAS of slugs of 40 µM aqueous K_2IrCl_6 dye solution in a toluene carrier phase. The cuts through the full data set show the absorption spectrum of a single droplet and the time dependence at the peak of the absorption curve of the dye. Adapted from [120] and [85]

acterised by a large reduction in measured ringdown time or transmitted intensity as the boundaries of each droplet pass through the laser beam. This can be attributed to the lensing effect of the droplet meniscus scattering light out of the cavity. In the centre of the droplet, the absorption signal of the droplet contents is recovered. In the case of the broadband CEAS measurements, the entire absorption spectrum of the aqueous K_2IrCl_6 dye used for the measurements is measured. For the CRDS measurements, the chip was placed at Brewster's angle within the cavity, and the observed ringdown times depend both on absorption by the droplets and carrier solution, and also on the refractive indices of the two solutions, since these determine the optimum 'Brewster' angle of the chip within the cavity.

Finally, there has been considerable interest in using optical-fibre-based cavities for microfluidic measurements. The serendipitous size matching between the diameter of an optical fibre and a microfluidic channel makes the use of optical-fibre cavities in microfluidic sensing an attractive prospect. Fibre-based cavities fall into two categories: linear fibre cavities [51–54]; and fibre-loop cavities [25, 56]. These are shown schematically in Fig. 11.3. Fibre cavities are covered in detail in Chaps. 12 and 13 of this book, so only a very brief overview will be given here, with the focus on applications in microfluidics.

As noted previously in Sect. 11.3.1, fibre cavities intrinsically probe extremely small sample volumes, and are inherently broadband in nature, allowing measurements to be made at multiple wavelengths or across a continuum of wavelengths simultaneously. Unfortunately, the compact and robust nature of such cavities tends to be offset by the fact that the optical losses are much higher than for a two-mirror cavity, typically of the order of a few percent, such that the minimum detectable absorption per unit path length is often significantly lower. While the fibre itself occasions small losses due to imperfect transmission, the majority of the optical losses within a fibre-loop cavity are associated with the sample region and the couplers used to couple light into and out of the loop. As a consequence, it is possible to achieve relatively long ringdown times even with relatively high losses on each pass simply by increasing the length of fibre used to create the loop cavity, noting that it takes the circulating light around 5 ns to traverse 1 m of fibre. The ability to achieve long ringdown times in this way makes fibre-loop cavities well suited to phase-shift [129], rather than direct, measurements of the ringdown time, and this approach has been implemented with some success by Loock and coworkers [130–133]. Phase-shift measurements employ a CW laser source modulated at a frequency ω, and use a lock-in amplifier to record the phase shift introduced by the cavity to the modulated light coupled out from the cavity. The phase shift is related to the ringdown time τ and the modulation frequency ω by

$$\phi = -\arctan(\omega\tau). \tag{11.9}$$

By injecting light into the cavity of multiple wavelengths, each with a different modulation frequency, the phase-shift approach allows absorption measurements to be made at two or more wavelengths simultaneously [133].

Whether making a direct measurement of the ringdown time or employing phase-shift techniques, the overriding aim when employing fibre cavities in microfluidics (or any other application) is to minimise the losses associated both with the sample region and with the light coupling scheme used to couple light into and out of the fibre. In linear fibre cavities, light may be coupled through a mirrored fibre end in much the same way as for a two-mirror cavity, while in a fibre-loop cavity the excitation light must be coupled into the loop through the side of the fibre, a much more challenging prospect. A variety of side-coupling schemes have been developed, including bend coupling [55], use of commercially available 99:1 coupler/splitters [134], and methods exploiting microfabricated mirrors machined into the fibre core [135, 136].

The simplest way to incorporate a microfluidic sample into a fibre cavity is to introduce a break into the optical fibre and align the two fibre ends on either side of a microfluidic channel. In order to maintain a low optical loss, in the absence of an absorbing species it is important that virtually all of the light emitted from the first fibre end is coupled back into the second fibre end. In practice this is very difficult to achieve for gaps larger than a few tens of microns [55], and divergence of the light emitted from the first fibre end means that end separations are limited to distances on the order of the fibre core diameter. A modest improvement in the achievable path length per pass can be obtained by lensing the ends of one or both fibres, such that light emitted from the first fibre is focussed through the sample into the second fibre. Achieving the highest detection sensitivity relies on finding the best compromise between minimising the loop losses by optimising the coupling scheme and keeping the width of the sample gap as small as possible, and maximising the path length through the sample on each pass by increasing the gap width. Depending on the relative balance between these two factors, cavity enhancement factors ranging from 50 or so [55, 136] down to less than 5 [137, 138] have been obtained. The highest cavity enhancement factors allow the detection of a variety of analytes down to sub-micromolar concentrations in a detection volume of tens of nanolitres [133, 136].

An alternative approach to improving the sensitivity of fibre-loop cavity ring-down measurements is to accept that the loop losses will always be relatively large in comparison with two-mirror cavities, and to attempt to compensate for the losses by introducing a variable gain element into the loop. While this has not yet been implemented for microfluidic measurements, the approach has been investigated by Stewart *et al.* [139] in fibre-loop ringdown measurements on gas-phase samples. Incorporating a 5 cm open path micro-optical gas cell into a 58 m fibre loop occasioned an insertion loss of around 20 % per pass, which was offset by including an erbium-doped fibre amplifier (EDFA) into the loop. The optical gain element allows the ringdown time to be increased from around 100 ns to several tens of microseconds, but unfortunately the less than optimal stability of the amplifier (and therefore the ringdown time) becomes the limiting factor in performing sensitive absorption measurements [140]. Improving the amplifier stability is not trivial, but if the stability issues can be addressed then amplified fibre-loop cavities offer considerable potential for making highly sensitive absorption measurements on microfluidic samples.

11.5 Example Application: Sensing of Trace Compounds in Water

In the previous section, we outlined the general approaches employed for cavity-enhanced spectroscopic measurements on microfluidic samples, and some of the important experimental considerations associated with such measurements. We now focus on a specific application, namely the quantitative determination of trace compounds in water. Such determinations are important in a range of contexts, including water quality testing and environmental monitoring, and there is considerable interest in the development of compact, robust sensors capable of operating remotely for long periods of time without human intervention.

One of the key applications for such sensors is in chemical oceanography. Understanding the rich chemistry of the world's oceans requires measurements of the distribution and dynamics of a broad range of elemental and molecular chemical species. In addition to probing the fundamental physical, thermodynamic, and kinetic properties of chemical species in marine environments, understanding the interactions between ocean chemistry and other biological, geological, and physical processes is of vital importance in understanding the past evolution of the earth, thereby allowing us to model its future. Several quantities of interest in chemical oceanography, such as temperature and salinity, are relatively straightforward to measure, and the technology for measuring them is highly advanced. As an example, a network of over 3000 research buoys making up the Argo float project continuously monitors temperature, pressure and salinity within the upper 2000 m of the ocean, transmitting data back to shore by satellite [141]. For other quantities, the development of suitable sensing systems is still very much a work in progress. As an example, photosynthetic phytoplankton require a range of nutrients (e.g. NO_3^-,

PO_4^{3-}, and SiO_2) and micronutrients (mostly trace metals) in order to synthesise their cellular machinery. Phytoplankton play an important role in the synthesis of the majority of marine organic compounds from aquatic carbon dioxide, and represent a considerable CO_2 sink; so much so that seeding oceans with the micronutrient Fe^{3+} in order to encourage the formation of phytoplankton ('algal') blooms is under serious consideration as a route to CO_2 sequestration from the atmosphere [142]. To further our understanding of such processes there is an urgent need for robust chemical sensors suitable for field studies.

Sensors based on microfluidic platforms represent an attractive approach due to their small size and low power and reagent consumption. However, as noted previously, a sensitive detection technique is required in order to detect low concentrations of analytes in volumes of nanolitres or less. The problem is compounded by the fact that many of the analytes of interest do not have any characteristic absorption features in spectroscopic regions for which suitable light sources are available. In the following, we will explore one possible solution to this problem, in which cavity-enhanced spectroscopies are combined with colourimetry in order to achieve selective and highly sensitive detection of a range of analytes.

Colourimetric assays employ a chemical reaction to convert the (non-absorbing) analyte of interest selectively into an intensely coloured derivative, which can then be quantified via an absorption measurement. Knowledge of the stoichiometry of the colourimetric reaction then allows the concentration of the initial analyte to be determined. The large absorption coefficient of the derivative compound provides a large signal amplification, which can be amplified further by employing cavity ringdown or cavity-enhanced absorption spectroscopy to make the absorption measurement. Colourimetric assays have already been developed on a microfluidic platform for analytes such as haemoglobin [104], botulinum toxin [143], and glucose [144, 145], amongst others, using a single-pass absorption measurement in the detection step.

To demonstrate the potential applications of cavity-enhanced spectroscopic detection for water sensing, we review the results of recent proof-of-concept experiments aimed at determining the detection limits for nitrite and iron using colourimetric CRDS. Nitrite (NO_2^-) plays an important part in the global carbon and nitrogen cycles, as well as being a key marine nutrient required for phytoplankton growth [146, 147]. At high levels, it is toxic to humans and animals, and its levels are carefully monitored in waterways and drinking water. The conentration of nitrite in sea water ranges from 0.1 to 50 µM, and the World Health Organisation guidelines set limits for short-term exposure to nitrites at 62.5 µM [148]. The second analyte investigated, iron, is essential for all known living organisms [149]. The concentration of dissolved iron (Fe^{2+} and Fe^{3+}) varies widely in marine and aquatic environments, depending primarily on the dissolved oxygen concentration. In rivers, the concentration ranges from around 3 to 25 µM [148], while in oceans the concentration is much lower, at around 0.02–2 nM [149].

Two different colourimetric assays were used for detection of nitrite and Fe(II), respectively. The coloured products of both assays absorb strongly at 532 nm, the wavelength of the frequency-doubled microchip Nd:YAG laser used for the CRDS measurements. Nitrite was detected via the Griess assay [150], which involves acidifying the nitrite sample, diazotising the nitrite with a primary amine (e.g. sulfanilamide), and coupling the product to an aromatic molecule to produce an intensely coloured azo dye with an absorption maximum [151, 152] in water of around $80,000 \ M^{-1} cm^{-1}$ at 520 nm. The dye is formed highly specifically in a 1:1 ratio from nitrite in the sample, so a quantitative absorption measurement on the dye yields the concentration of nitrite in the original sample. Iron concentrations can be determined colourimetrically using any one of a number of complexation reactions [153–155], and for these measurements, complexation of Fe^{2+} with 4,7-diphenyl-1,10-phenanthroline [156] (bathophenanthroline), which reacts with Fe(II) in a 3:1 ratio, was chosen due to the large absorption coefficient ($51,600 \ M^{-1} cm^{-1}$) of the coloured product at 532 nm.

For the absorption measurements, the products of the colourimetric reaction are contained within a commercial short (1 mm) path length fused silica flow cell positioned at Brewster's angle within a conventional two-mirror cavity. Recall from Sect. 11.2.1 that determination of an absolute absorption per unit path length, κ, from a cavity ringdown measurement requires the measurement of two ringdown times, τ and τ_0, corresponding to the sample of interest and a 'blank'. An important consideration in colourimetric CRDS measurements is the correct choice of blank for the τ_0 measurement. An obvious choice for the blank would appear to be the deionised water used to make up the standard solutions of nitrite and iron. However, even the highest quality deionised water is likely to contain small but significant concentrations of the analyte of interest, and in fact a more appropriate blank is the reaction mixture obtained when the appropriate colourimetric reaction is carried out on a sample of deionised water. In practice, when dealing with extremely low concentrations of analyte, the quality of the blank solution is often one of the key limiting factors in determining the detection sensitivity, and the choice of blank should be given very careful consideration.

The minimum detectable absorbance per unit path length, κ_{min} ($= \alpha C_{min}$), determined from Eq. (11.5), with $\Delta\tau_{min}$ taken to be three times the standard deviation in repeated measurements of the baseline τ_0, was found to be around $1 \times 10^{-4} \ cm^{-1}$. Using the known absorption coefficients, α for the absorbing species, this yielded limits of detection of 1.4 nM for nitrite and 3.8 nM for Fe(II). A somewhat higher detection limit of 16 nM for Fe(II) was obtained when the measurements were repeated in a 0.1 mm path length flow cell, a path length commensurate with a microfluidic channel. The detection limits are comparable with or better than those achieved previously in experiments employing much longer path lengths, and therefore requiring much larger sample volumes [152, 157–163]. To put the results into context, when the illuminated volume and path length per pass within the flow cell is taken into account, we find that the measured limits correspond to the detection of around one billion molecules.

This case study demonstrates that colourimetric CRDS may be used to detect both nitrite and iron in path lengths similar to those encountered in microfluidic systems, and that with further development this technique could potentially form the basis for ship-based or remote sensors for monitoring trace species in both marine and aquatic environments. The sensitivity to nitrite is already sufficient for monitoring ambient levels in all such environments, and while further improvements are required in order to satisfy the requirements for monitoring Fe(II) in the open ocean, the sensitivity is already sufficient to monitor concentrations in river or drinking water. Incorporating a pre-concentration step into the analysis, while complicating the analytical procedure, would in principle allow ambient marine concentrations to be detected. Such steps have been demonstrated previously by a number of researchers [157, 164, 165], yielding sub-nanomolar limits of detection.

11.6 The Future

We hope to have shown in this chapter that, while they do not achieve the often staggeringly high cavity enhancement factors seen in gas-phase measurements, cavity-enhanced absorption measurements on liquid samples provide a valuable means of improving the detection sensitivity, particularly in situations where the path length or sample volume are restricted. There is considerable scope for further improvement in sensitivity, and we conclude this chapter with a brief consideration of the various approaches that may be taken to achieve this.

Detection sensitivity in a cavity-enhanced spectroscopy measurement can be improved either by reducing the cavity losses, thereby increasing the ringdown time; by increasing the path length through the sample on each pass; or by reducing the shot-to-shot variation in the measured ringdown time. We will consider each of these factors in turn.

The cavity losses L (in the absence of a sample) may be reduced either by increasing the mirror reflectivity or by reducing the losses associated with the microfluidic chip or other liquid container within the cavity. Inspection of Eqs. (11.5) and (11.2) reveals that the minimum detectable absorption per unit path length is proportional to L^2. As an example, the 1 mm flow cell used in the trace detection experiments described in Sect. 11.5 introduced a loss per pass of around 0.22 %, similar to the 0.2 % loss introduced by the the 99.8 % reflecting mirrors. If the mirrors were replaced with the 'best available' 99.999 % reflectivity mirrors, and the flow cell was replaced by the custom-designed Brewster-angle flow cell developed by Snyder and Zare [27] (see Sect. 11.3.1), which has a reported loss of 0.06 %, then the cavity transmission would be increased from 99.58 % to 99.94 %, the loss per pass would be reduced from 0.42 % to 0.06 %, and the minimum detectable absorption per unit path length, κ_{min}, would be reduced by a factor of nearly 50.

Considering the path length per pass, Eq. (11.5) implies that increasing the path length should improve the detection sensitivity (i.e. reduce the detection limit). This is certainly true for gas-phase measurements, but unfortunately is not always the case for liquid-phase measurements. Increasing the path length per pass often leads to a signficant increase in scattering and absorption losses associated with the solvent, and also increases any uncertainties associated with blank measurements where these are critical to the measurement. The effect of path length on detection sensitivity is best investigated on a case by case basis depending on the particular application at hand.

Improving the reliability of the ringdown measurement itself is perhaps best achieved by moving from a pulsed laser source to a continuous wave (CW) source. This was demonstrated convincingly by Zare and coworkers in their experiments employing the custom Brewster cell described above. Replacing the pulsed light source employed in their earlier experiments [27] with a single-mode, CW source allowed excitation of a single cavity mode and reduced the shot-to-shot variation in the measured value of the ringdown time from 1 % to 0.04 %, reducing the associated experimental uncertainty accordingly.

In summary, we have seen that cavity-enhanced spectroscopies employing a standard two-mirror optical cavity provide a fairly general detection method for microfluidics and other small-volume liquid-phase sensing applications, and there is considerable scope for developing a commercial instrument with these capabilities. The requirements for developing compact stand-alone sensors employing cavity-enhanced detection techniques are more demanding, and will require the development of high-finesse miniature cavities suitable for integration into microfluidic channels. There are a number of candidate technologies for fabricating microcavities, with perhaps one of the most interesting being the recently-reported femtolitre tunable optical cavity arrays developed by Dolan *et al.* [166]. Such sensors could potentially be developed for a broad range of applications spanning the chemical, biological, environmental, forensic and medical sciences.

References

1. M. Trojanowicz, Anal. Chim. Acta **653**, 36 (2009)
2. P.S. Dittrich, A. Manz, Anal. Bioanal. Chem. **382**, 1771 (2005)
3. S.A. Leung, R.F. Winkle, R.C.R. Wootton, A.J. deMello, Analyst **130**, 46 (2005)
4. G. Cristobal, L. Arbouet, F. Sarrazin, D. Talaga, J.L. Bruneel, M. Joanicot, L. Servant, Lab Chip **6**, 1140 (2006)
5. C. Massin, F. Vincent, A. Homsy, K. Ehrmann, G. Boero, P.A. Besse, A. Daridon, E. Verpoorte, N.F. de Rooij, R.S. Popovic, J. Magn. Reson., Ser. A **164**(2), 242 (2003)
6. S. Koster, E. Verpoorte, Lab Chip **7**, 1394 (2007)
7. C. Yi, Q. Zhang, C.W. Li, J. Yang, J. Zhao, M. Yang, Anal. Bioanal. Chem. **384**(6), 1259 (2006)
8. B. Kuswandi, Nuriman, J. Huskens, W. Verboom, Anal. Chim. Acta **601**, 141 (2007)
9. K.B. Mogensen, J.P. Kutter, Electrophoresis **30**, S92 (2009)
10. M.A. Schwarz, P.C. Hauser, Lab Chip **1**, 1 (2001)
11. K.B. Mogensen, H. Klank, J.P. Kutter, Electrophoresis **25**, 3498 (2004)
12. A.J. deMello, Nature **442**, 394 (2006)
13. F.B. Myers, L.P. Lee, Lab Chip **8**, 2015 (2008)
14. M. O'Toole, D. Diamond, Sensors **8**, 2453 (2008)
15. J.C. Jokerst, J.M. Emory, C.S. Henry, Analyst **137**(1), 24 (2012)
16. G. Berden, R. Peeters, G. Meijer, Int. Rev. Phys. Chem. **19**(4), 565 (2000)
17. F.M. Schmidt, A. Foltynowicz, W. Ma, T. Lock, O. Axner, Opt. Express **15**(7), 10822 (2007)
18. A. O'Keefe, D.A.G. Deacon, Rev. Sci. Instrum. **59**, 2544 (1988)
19. M.D. Wheeler, S.M. Newman, A.J. Orr-Ewing, M.N.R. Ashfold, J. Chem. Soc. Faraday Trans. **94**(3), 337 (1998)

20. J.J. Scherer, J.B. Paul, A. O'Keefe, R.J. Saykally, Chem. Rev. **97**, 25 (1997)
21. S.M. Ball, R.L. Jones, Chem. Rev. **103**(12), 5239 (2003)
22. B. Cummings, M.L. Hamilton, L. Ciaffoni, T.R. Pragnell, R. Peverall, G.A.D. Ritchie, G. Hancock, P.A. Robbins, J. Appl. Physiol. **111**(1), 303 (2011)
23. M.R. McCurdy, Y. Bakhirkin, G. Wysocki, R. Lewicki, F.K. Tittel, J. Breath Res. **1**, 014001 (2007)
24. G. Neri, A. Lacquaniti, G. Rizzo, N. Donato, M. Latino, M. Buemi, Nephrol. Dial. Transplant. **27**(7), 2945 (2012)
25. C. Wang, Sensors **9**, 7595 (2009)
26. J. Wojtas, Z. Bielecki, T. Stacewicz, J. Mikolajczyk, M. Nowakowski, Optoelectron. Rev. **20**(1), 26 (2012)
27. K.L. Snyder, R.N. Zare, Anal. Chem. **75**(13), 3086 (2003)
28. S.C. Xu, G.H. Sha, J.C. Xie, Rev. Sci. Instrum. **73**(2), 255 (2002)
29. L. van der Sneppen, F. Ariese, C. Gooijer, W. Ubachs, J. Chromatogr. A **1148**(2), 184 (2007)
30. K.L. Bechtel, R.N. Zare, A.A. Kachanov, S.S. Sanders, B.A. Paldus, Anal. Chem. **77**(4), 1177 (2005)
31. A.J. Alexander, Anal. Chem. **78**(15), 5597 (2006)
32. S.E. Fiedler, A. Hese, A.A. Ruth, Rev. Sci. Instrum. **76**, 023107 (2005)
33. T. McGarvey, A. Conjusteau, H. Mabuchi, Opt. Express **14**(22), 10441 (2006)
34. M. Islam, N. Seetohul, Z. Ali, Appl. Spectrosc. **61**(6), 649 (2007)
35. N. Seetohul, Z. Ali, M. Islam, Anal. Chem. **81**, 4106 (2009)
36. N. Seetohul, Z. Ali, M. Islam, Analyst **134**, 1887 (2009)
37. A. O'Keefe, Chem. Phys. Lett. **293**, 331 (1998)
38. R. Engeln, G. Berden, R. Peeters, G. Meijer, Rev. Sci. Instrum. **69**(11), 3763 (1998)
39. P.K. Dasgupta, J.S. Rhee, Anal. Chem. **59**(5), 783 (1987)
40. B. Bakowski, L. Corner, G. Hancock, R. Kotchie, R. Peverall, G.A.D. Ritchie, Appl. Opt. **75**(6–7), 745 (2002)
41. S.E. Fiedler, A. Hese, A.A. Ruth, Chem. Phys. Lett. **371**(3–4), 284 (2003)
42. IUPAC compendium of chemical terminology. http://goldbook.iupac.org
43. M. Mazurenka, A.J. Orr-Ewing, R. Peverall, G.A.D. Ritchie, Annu. Rep. Prog. Chem., Sect. C, Phys. Chem. **101**, 100 (2005)
44. C.M. Rushworth, J. Davies, J.T. Cabral, P.R. Dolan, J.M. Smith, C. Vallance, Chem. Phys. Lett. **554**, 1 (2012)
45. A.C.R. Pipino, J.W. Hudgens, R.E. Huie, Chem. Phys. Lett. **280**(1–2), 104 (1997)
46. A.M. Shaw, T.E. Hannon, F.P. Li, R.N. Zare, J. Phys. Chem. B **107**(29), 7070 (2003)
47. J.R. O'Reilly, C.P. Butts, I.A. l'Anson, A.M. Shaw, J. Am. Chem. Soc. **127**(6), 1632 (2005)
48. A.C.R. Pipino, J.W. Hudgens, R.E. Huie, Rev. Sci. Instrum. **68**(8), 2978 (1997)
49. A.C.R. Pipino, Phys. Rev. Lett. **83**(15), 3093 (1999)
50. A.C.R. Pipino, Appl. Opt. **39**(9), 1449 (2000)
51. T. Von Lerber, M.W. Sigrist, Appl. Opt. **41**(18), 3567 (2002)
52. D.E. Vogler, M.G. Mülller, M.W. Sigrist, Appl. Opt. **42**(27), 5413 (2003)
53. D.E. Vogler, A. Lorencak, J.M. Rey, M.W. Sigrist, Opt. Lasers Eng. **43**(3–5), 527 (2005)
54. M. Gupta, H. Jiao, A. O'Keefe, Opt. Lett. **27**(21), 1878 (2002)
55. R.S. Brown, I. Kozin, Z. Tong, R.D. Oleschuk, H.P. Loock, J. Chem. Phys. **117**(23), 10444 (2002)
56. H. Waechter, J. Litman, A.H. Cheung, J.A. Barnes, H.P. Loock, Sensors **10**(3), 1716 (2010)
57. C. Wang, Opt. Eng. **44**, 030503 (2005)
58. C. Wang, S.T. Scherrer, Appl. Opt. **43**(35), 6458 (2004)
59. P.B. Tarsa, D.M. Brzozowski, P. Rabinowitz, K.K. Lehmann, Opt. Lett. **29**, 1339 (2004)
60. N. Ni, C.C. Chan, L. Xia, P. Shum, IEEE Photonics Technol. Lett. **20**(16), 1351 (2009)
61. R. Engeln, G. Meijer, Rev. Sci. Instrum. **67**(8), 2708 (1996)
62. E. Hamers, D. Schram, R. Engeln, Chem. Phys. Lett. **365**, 237 (2002)
63. A. Czyżewski, S. Chudzyński, K. Ernst, G. Karasizński, L. Kilianek, A. Pietruczuk, W. Skubiszak, K. Stacewicz, T. Stelmaszczyk, B. Koch, P. Rairoux, Opt. Commun. **191**(3–6), 271

(2001)
64. J.J. Scherer, Chem. Phys. Lett. **292**, 143 (1998)
65. J.J. Scherer, J.B. Paul, H. Jiao, A. O'Keefe, Appl. Opt. **40**(36), 6725 (2001)
66. S.N. Ball, I.M. Povey, E.G. Norton, R.L. Jones, Chem. Phys. Lett. **342**, 113 (2001)
67. M. Bitter, S.M. Ball, I.M. Povey, R.L. Jones, Atmos. Chem. Phys. **5**, 2547 (2005)
68. C. Vallance, New J. Chem. **29**, 867 (2005)
69. S.E. Fiedler, G. Hoeheisel, A.A. Ruth, A. Hese, Chem. Phys. Lett. **382**(3–4), 447 (2003)
70. A.A. Ruth, J. Orphal, S.E. Fiedler, Appl. Opt. **46**, 3611 (2007)
71. D. Venables, T. Gherman, J. Orphal, Environ. Sci. Technol. **40**, 6758 (2006)
72. J. Orphal, A.A. Ruth, Opt. Express **16**, 1595 (2008)
73. S. Vaughan, T. Gherman, A.A. Ruth, J. Orphal, Phys. Chem. Chem. Phys. **10**, 4471 (2008)
74. A. Ruth, R. Varma, D. Venables, EGU Gen. Assem. **11**, 12954 (2009)
75. S.M. Ball, J.M. Langridge, R.L. Jones, Chem. Phys. Lett. **398**(1–3), 68 (2004)
76. J.M. Langridge, S.M. Ball, R.L. Jones, Analyst **131**, 916 (2006)
77. J.M. Langridge, S.M. Ball, A.J.L. Shillings, R.L. Jones, Rev. Sci. Instrum. **79**, 123110 (2008)
78. T. Wu, W. Zhao, W. Chen, W. Zhang, X. Gao, Appl. Phys. B **94**, 85 (2008)
79. T. Wu, W. Chen, E. Fertein, F. Cazier, D. Dewaele, X. Gao, Appl. Phys. B **106**, 501 (2011)
80. M. Triki, P. Cermak, G. Mejean, D. Romanini, Appl. Phys. B **91**, 195 (2010)
81. I. Ventrillard-Courtillot, E. Sciamma O'Brien, S. Kassi, G. Mejean, D. Romanini, Appl. Phys. B **101**, 661 (2010)
82. J.M. Langridge, T. Laurila, R.S. Watt, R.L. Jones, C.F. Kaminski, J. Hult, Opt. Express **16**, 10178 (2008)
83. T. Watt, R.S. ad Laurila, C.F. Kaminski, J. Hult, Appl. Spectrosc. **63**, 1389 (2009)
84. P.S. Johnston, K.K. Lehmann, Opt. Express **16**, 15013 (2008)
85. S.R.T. Neil, C.M. Rushworth, C. Vallance, S.R. Mackenzie, Lab Chip **11**, 3953 (2011)
86. S.S. Kiwanuka, T. Laurila, C.F. Kaminski, Anal. Chem. **82**, 7498 (2010)
87. M. Schnippering, P.R. Unwin, J. Hult, T. Laurila, C.F. Kaminski, J.M. Langridge, R.L. Jones, M. Mazurenka, S.R. Mackenzie, Electrochem. Commun. **10**, 1827 (2008)
88. L. van der Sneppen, C. Hancock, G. Kaminski, T. Laurila, S.R. Mackenzie, S.R.T. Neil, R. Peverall, G.A.D. Ritchie, M. Schnippering, P.R. Unwin, Analyst **135**, 133 (2010)
89. T. Gherman, S. Kassi, A. Campargue, D. Romanini, Chem. Phys. Lett. **383**, 353 (2004)
90. C. Gohle, B. Stein, A. Schliesser, T. Udem, T. Hänsch, Phys. Rev. Lett. **99**, 1 (2007)
91. M. Thorpe, J. Ye, Appl. Phys. B **91**, 397 (2008)
92. B. Bernhardt, A. Ozawa, P. Jacquet, M. Jacquey, Nat. Photonics **4**, 2009 (2009)
93. K.B. Mogensen, J. El-Ali, A. Wolff, J.P. Kutter, Appl. Opt. **42**(19), 4072 (2003)
94. K.B. Mogensen, F. Eriksson, O. Gustafsson, R.P.H. Nikolajsen, J.P. Kutter, Electrophoresis **25**, 3788 (2004)
95. M.H. Wu, H. Cai, X. Xu, J.P.G. Urban, Z.F. Cui, Z. Cui, Biomed. Microdevices **7**(4), 323 (2005)
96. R. Jindal, S.M. Cramer, J. Chromatogr. A **1044**, 277 (2004)
97. J.P. Landers, *Handbook of Capillary and Microchip Electrophoresis and Associated Microtechniques* (CRC Press, Boca Raton, 2007)
98. Y. Xue, E.S. Yeung, Anal. Chem. **66**, 3575 (1994)
99. R. Alves-Segundo, N. Ibañez Garcia, M. Baeza, M. Puyol, J. Alonso-Chamarro, Mikrochim. Acta **172**(1), 225 (2011)
100. N.J. Petersen, K.B. Mogensen, J.P. Kutter, Electrophoresis **23**, 3528 (2002)
101. S.E. Moring, R.T. Reel, R.E.J. van Soest, Anal. Chem. **65**(23), 3454 (1993)
102. K.W. Ro, K. Lim, B.C. Shim, J.H. Hahn, Anal. Chem. **77**(16), 5160 (2005)
103. T.M. Tiggelaar, T.T. Veenstra, R.G.P. Sanders, J.G.E. Gardeniers, M.C. Elwenspoek, A. van der Berg, Talanta **56**, 331 (2002)
104. J. Steigert, M. Grumann, M. Dube, W. Streule, L. Riegger, T. Brenner, P. Koltay, K. Mittmann, R. Zengerle, J. Ducrée, Sens. Actuators A **130–131**, 228 (2006)
105. T. Wang, J.H. Aiken, C.W. Huie, R.A. Hartwick, Anal. Chem. **63**(14), 1372 (1991)

106. H. Salimi-Moosavi, Y. Jiang, L. Lester, G. McKinnon, D.J. Harrison, Electrophoresis **21**, 1291 (2000)
107. P.S. Ellis, A.J. Lyddy-Meaney, P.J. Worsfold, I.D. McKelvie, Anal. Chim. Acta **499**(1–2), 81 (2003)
108. E. Verpoorte, A. Manz, H. Lüdi, A.E. Bruno, F. Maystre, B. Krattiger, H.M. Widmer, B.H. van der Schoot, N.F. De Rooij, Sens. Actuators B **6**, 66 (1992)
109. A. Llobera, R. Wilke, S. Büttgenbach, Lab Chip **4**, 24 (2004)
110. A. Llobera, R. Wilke, S. Büttgenbach, Lab Chip **5**, 506 (2005)
111. A. Llobera, S. Demming, R. Wilke, S. Büttgenbach, Lab Chip **7**, 1560 (2007)
112. S. Demming, A. Llobera, R. Wilke, S. Büttgenbach, Sens. Actuators B **139**, 166 (2009)
113. J.Z. Pan, B. Yao, Q. Fang, Anal. Chem. **82**(8), 3394 (2010)
114. L. Billot, A. Plecis, Y. Chen, Microelectron. Eng. **85**, 1269 (2008)
115. D. Psaltis, S.R. Quake, C. Yang, Nature **442**, 381 (2006)
116. B. Helbo, A. Kristensen, A. Menon, J. Micromech. Microeng. **13**, 307 (2003)
117. H. Shao, D. Kumar, S.A. Feld, K.L. Lear, J. MEMS **14**(4), 756 (2005)
118. M. Gersborg-Hansen, S. Balslev, N.A. Mortensen, A. Kristensen, Microelectron. Eng. **78**, 185 (2005)
119. J.C. Galas, C. Peroz, Q. Kou, Y. Chen, Appl. Phys. Lett. **89**, 224101 (2006)
120. D. James, B. Oag, C.M. Rushworth, J.W.L. Lee, J. Davies, J.T. Cabral, C. Vallance, RSC Adv. **2**, 5376 (2012)
121. I. Shestopalov, J.D. Tice, R.F. Ismagilov, Lab Chip **4**, 316 (2004)
122. B. Zheng, L.S. Roach, R.F. Ismagilov, J. Am. Chem. Soc. **125**(37), 11170 (2003)
123. H. Song, J.D. Tice, R.F. Ismagilov, Angew. Chem., Int. Ed. Engl. **42**(7), 767 (2003)
124. H. Song, R.F. Ismagilov, J. Am. Chem. Soc. **125**(47), 14613 (2003)
125. Z.T. Cygan, J.T. Cabral, K.L. Beers, E.J. Amis, Langmuir **21**(8), 3629 (2005)
126. H. Song, D.L. Chen, R.F. Ismagilov, Angew. Chem., Int. Ed. Engl. **45**(44), 7336 (2006)
127. S. Rudic, R.E.H. Miles, A.J. Orr-Ewing, J.P. Reid, Appl. Opt. **46**, 6142 (2007)
128. R.E.H. Miles, S. Rudic, A.J. Orr-Ewing, J.P. Reid, Phys. Chem. Chem. Phys. **12**(15), 3914 (2010)
129. R. Engeln, G. von Helden, G. Berden, G. Meijer, Chem. Phys. Lett. **262**, 105 (1996)
130. Z. Tong, A. Wright, T. McCormick, R. Li, R.D. Oleschuk, H.P. Loock, Anal. Chem. **76**(22), 6594 (2004)
131. R. Li, H.P. Loock, R.D. Oleschuk, Anal. Chem. **78**(16), 5685 (2006)
132. H. Waechter, K. Bescherer, C.J. Dürr, R.D. Oleschuk, H.P. Loock, Anal. Chem. **81**, 9048 (2009)
133. H. Waechter, D. Munzke, A. Jang, H.P. Loock, Anal. Chem. **83**, 2719 (2011)
134. P.B. Tarsa, P. Rabinowitz, K.K. Lehmann, Chem. Phys. Lett. **383**(3–4), 297 (2004)
135. C.M. Rushworth, D. James, C.J.V. Jones, C. Vallance, Opt. Lett. **36**(15), 2952 (2011)
136. C.M. Rushworth, D. James, J.W.L. Lee, C. Vallance, Anal. Chem. **83**, 8492 (2011)
137. Z. Tong, M. Jakubinek, A. Wright, A. Gillies, H.P. Loock, Rev. Sci. Instrum. **74**(11), 4818 (2003)
138. M. Andachi, T. Nakayama, M. Kawasaki, S. Kurokawa, H.P. Loock, Appl. Phys. B **88**, 131 (2007)
139. G. Stewart, K. Atherton, H. Yu, B. Culshaw, Meas. Sci. Technol. **12**, 843 (2001)
140. G. Stewart, K. Atherton, B. Culshaw, Opt. Lett. **29**(5), 442 (2004)
141. http://www.argo.ucsd.edu/
142. W.P. Boyd, T. Jickells, C.S. Law, S. Blain, E.A. Boyle, K.O. Buesseler, K.H. Coale, J.J. Cullen, H.J.W. de Baar, M. Follows, M. Harvey, C. Lancelot, M. Levasseur, N.P.J. Owens, R. Pollard, R.B. Rivkin, J. Sarmiento, V. Schoemann, V. Smetacek, S. Takeda, A. Tsuda, S. Turner, A.J. Watson, Science **315**, 612 (2007)
143. J. Moorthy, G.A. Mensing, D. Kim, S. Mohanty, D.T. Eddington, W.H. Tepp, E.A. Johnson, D.J. Beebe, Electrophoresis **25**, 1705 (2004)
144. V. Srinivasan, V.K. Pamula, R.B. Fair, Anal. Chim. Acta **507**, 145 (2004)

145. M. Grumann, J. Steigert, L. Riegger, I. Moser, B. Enderle, K. Riebeseel, G. Urban, R. Zengerle, J. Ducrée, Biomed. Microdevices **8**, 209 (2006)
146. K. Grasshoff, K. Kremling, M. Ehrhardt, *Methods of Seawater Analysis*, 3rd edn. (Wiley-VCH, New York, 2007)
147. F.J. Millero, *Chemical Oceanography*, 3rd edn. (CRC Press, Boca Raton, 2005)
148. WHO, *Guidelines for Drinking-Water Quality*, 4th edn. (World Health Organization, Geneva, 2011)
149. D.R. Turner, K.A. Hunter, *The Biogeochemistry of Iron in Seawater* (Wiley, New York, 2001)
150. P. Griess, Ber. Dtsch. Chem. Ges. **12**, 426 (1879)
151. D.F. Boltz, J.A. Howell, *Colorimetric Determination of Nonmetals* (Wiley, New York, 1978)
152. J.V. Sieben, C.F.A. Floquet, I.R.G. Ogilvie, M.C. Mowlem, H. Morgan, Anal. Methods **2**, 484 (2010)
153. E.B. Sandell, *Colorimetric Determination of Traces of Metals* (Wiley, New York, 1959)
154. E.P. Achterberg, T.W. Holland, A.R. Bowie, R.F.C. Mantoura, P.J. Worsfold, Anal. Chim. Acta **442**, 1 (2001)
155. S. Pehkonen, Analyst **120**, 2655 (1995)
156. G.F. Smith, W.H. McCurdy, H. Diehl, Analyst **77**, 418 (1952)
157. D.W. King, J. Lin, D.R. Kester, Anal. Chim. Acta **247**, 125 (1991)
158. P. Raimbault, G. Slawyk, B. Coste, J. Fry, Mar. Biol. **104**(2), 347 (1990)
159. C.F.A. Floquet, V.J. Sieben, A. Milani, E.P. Joly, I.R.G. Ogilvie, H. Morgan, M. Mowlem, Talanta **84**, 235 (2011)
160. E.T. Steimle, E.A. Kaltenbacher, R.H. Byrne, Mar. Chem. **77**, 255 (2002)
161. R.D. Waterbury, W. Yao, R.H. Byrne, Anal. Chim. Acta **357**(1–2), 99 (1997)
162. J.Z. Zhang, Deep-Sea Res., Part 1, Oceanogr. Res. Pap. **47**(6), 1157 (2000)
163. J.Z. Zhang, C. Kelbe, F.J. Millero, Anal. Chim. Acta **438**, 49 (2001)
164. S. Blain, P. Tréguer, Anal. Chim. Acta **308**, 425 (1995)
165. J.L. Manzoori, S. Soflaee, Anal. Lett. **34**(2), 231 (2001)
166. P.R. Dolan, G.M. Hughes, F. Grazioso, B.R. Patton, J.M. Smith, Opt. Lett. **35**(21), 3556 (2010)

Chapter 12
Fiber Loop Ringdown Sensors and Sensing

Chuji Wang

Abstract Fiber loop ringdown (FLRD) spectroscopy evolves directly from cavity ringdown spectroscopy, using a section of optical waveguide to replace the mirror-based cavity to achieve the multi-pass approach. Over the last several years, FLRD has gone far beyond the original applications of cavity ringdown spectroscopy to trace gas measurements to a broad range of applications in chemical, physical, and biological sensing. Using a uniform sensing scheme—measuring *time* to sense a quantity, FLRD is not only able to adopt a wide variety of sensing mechanisms for individual sensor fabrication, but is also uniquely suitable for large-scale, multi-function sensor network development.

12.1 Introduction

The concept of fiber loop ringdown (FLRD) spectroscopy was introduced about a decade ago. FLRD utilizes a section of optical fiber to replace a mirror-based ringdown cavity while still taking advantage of the ringdown approach—the increased path-length. FLRD has now become a universal technique that combines the attractively generic features of optical fibers with the enhanced effect of cavity ringdown spectroscopy (CRDS) for the development of sensors and sensing systems. Significant strides have been made in FLRD-based sensing during the last several years. Chemical, physical, and biological quantities can all be detected by the FLRD technique. In trace chemical (gas or liquid) sensing, if based on spectral fingerprints (i.e. absorption wavelengths), FLRD still holds its original taste of "spectroscopy". In physical (i.e. pressure, strain, etc.) sensing, however, FLRD has grown more and more distant from "spectroscopy" and become only a universal means to enhance detectivity via the multi-round trip (path) approach. Sensing different quantities using the FLRD technique is all achieved through measuring a time constant (*ringdown time*). Therefore, this uniform sensing signal is favorable for multiplexing individual FLRD sensors regardless of their measurands (parameters, quantities, or

C. Wang (✉)
Department of Physics and Astronomy, Mississippi State University, P.O. Drawer 5167, Mississippi State, MS 39762, USA
e-mail: cw175@msstate.edu

G. Gagliardi, H.-P. Loock (eds.), *Cavity-Enhanced Spectroscopy and Sensing*, Springer Series in Optical Sciences 179, DOI 10.1007/978-3-642-40003-2_12, © Springer-Verlag Berlin Heidelberg 2014

functions) to form a sensor network that shares a single light source and the same detector. FLRD spectroscopy and its optical waveguides have been discussed in several excellent reviews as well as in Chap. 9 [1–4], but this chapter pays more attention to recent developments in FLRD-based sensors and sensing. The content is put together in such a way so that new players who wish to set up their own FLRD systems may find this chapter useful for their jump-start and to get a fresh update with FLRD.

12.2 Fiber Loop Ringdown (FLRD) Basics

12.2.1 Principle of FLRD

The FLRD technique is fundamentally evolved from the well-known cavity ring-down spectroscopy (CRDS) technique [5–12]. CRDS obtains high sensitivities because of the multi-pass nature of the optical absorption path, as illustrated in Fig. 12.1. Since its introduction, the CRD technique has rapidly developed and matured, from initial applications focusing on weak absorption spectroscopic measurements to the fully commercialized process for trace gas analysis and sensing now. Although new ideas and the latest laser technologies have prompted the evolution of the CRDS technique with various forms of ringdown cavities [13–22], different CRD systems are all based on the same measuring principle: measuring time decay rates (ringdown times) of the light intensity to determine absolute absorbance of a gas that is confined in an optical cavity, which is often formed by a pair of high reflectivity mirrors. This measuring principle has also been achieved by implementing a "conceptual cavity"—a fiber loop formed by a section of optical fiber (not necessarily single mode optical fiber), thus named fiber loop ringdown (FLRD) [23–26].

A light pulse is coupled into a fiber loop and travels (*rings*) inside the loop for many round trips. In each round trip, a small fraction of the light couples out of the loop into a photodetector; and the rest of the light travels in the fiber, experiencing internal fiber transmission losses. The output signal observed by the detector follows an exponential decay. This behavior can be modeled by [27, 28]

$$\frac{dI}{dt} = -\frac{IAc}{nL} \tag{12.1}$$

where I is the light intensity at time t when the light source is shut off and a light pulse is injected into the loop, L is the total length of the fiber loop, c is the speed of light in a vacuum, n is the averaged-refractive index of the fiber loop, and A is the total fiber transmission loss (by percentage) of the light in each round trip. The total fiber transmission loss includes the fiber absorption loss, the coupling losses, and the fiber scattering loss; and $A = \alpha L + E + \gamma$, where α is the wavelength dependent absorption coefficient for the fiber core material with units of, i.e. cm^{-1}, E is the total coupling loss at the two coupling points (in and out), and γ is the total fiber

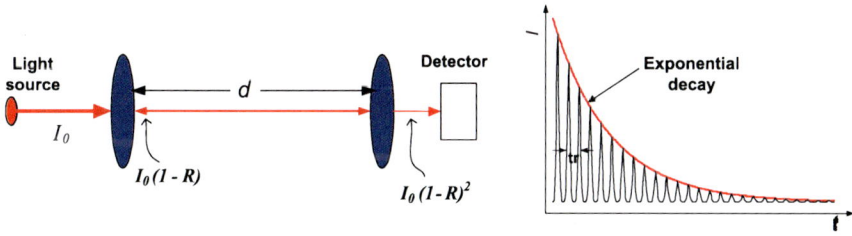

Fig. 12.1 The concept of cavity ringdown spectroscopy (CRDS)

scattering loss. The solution of Eq. (12.1) describes the temporal behavior of the light intensity observed by the detector:

$$I = I_0 e^{-\frac{c}{nL}At} \tag{12.2}$$

Equation (12.2) shows that FLRD measures the light intensity decay rate, not the absolute intensity change, ΔI. Therefore, the measurement of A is insensitive to fluctuations of the incident light intensity I_0.

The time required for the light intensity I to decrease to $1/e$ of the initial light intensity I_0, as observed by the detector, is referred to as the ringdown time τ_0 and is given by Eq. (12.3a)

$$\tau_0 = \frac{nL}{cA}, \tag{12.3a}$$

$$\tau = \frac{nL}{c(A+B)}. \tag{12.3b}$$

For a given FLRD sensor unit (a ringdown loop), the total transmission loss, A, is a constant, which is determined by the physical parameters of the loop, such as the fiber absorption loss, the coupling losses, the refractive index, and the fiber length. Apparently, the lower the losses of the light in the fiber are, the longer the decay time constants will be. When an external sensing activity, such as absorption, or a change of any measurands (pressure, temperature, strain, etc.) occurs at one section (sensor head) of the fiber loop, the result is an additional optical loss, B, of the light pulse in the fiber loop, which causes a change in the ringdown time from τ_0 to τ that is given by Eq. (12.3b).

From Eqs. (12.3a) and (12.3b), we have

$$\left(\frac{1}{\tau} - \frac{1}{\tau_0}\right) = \frac{c}{nL}B. \tag{12.4}$$

Equation (12.4) expresses the sensing principle of the FLRD technique. For a given fiber loop ringdown sensor, a change in a sensing activity (i.e. absorption, fiber mechanical deformation, etc.) is determined by measuring the ringdown time without the sensing activity and the ringdown time with the activity, and that the term $(\frac{1}{\tau} - \frac{1}{\tau_0})$ has a linear relationship with the activity-induced optical loss, B.

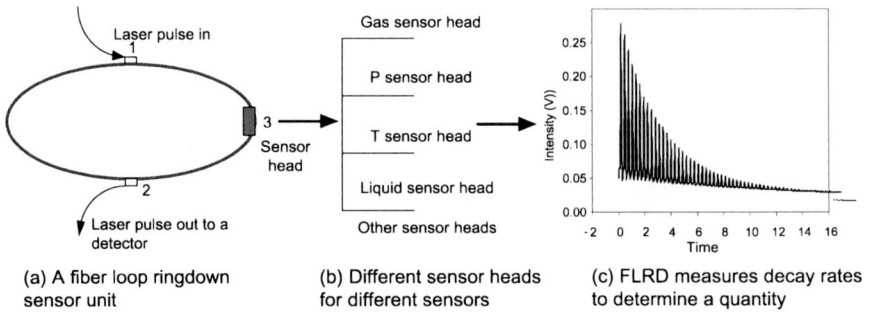

(a) A fiber loop ringdown
sensor unit

(b) Different sensor heads
for different sensors

(c) FLRD measures decay rates
to determine a quantity

Fig. 12.2 Fiber loop ringdown (FLRD)—a uniform sensing scheme

12.2.2 FLRD—A Uniform Sensing Scheme

Figure 12.2(a) shows a schematic of a typical FLRD sensor unit—a fiber loop consisting of one section of optical fiber, two couplers for light coupling in and out, and a sensor head. Figure 12.2(b) shows that multiple FLRD-based sensor units can be fabricated by just replacing different sensor heads for detection of different measurands. Figure 12.2(c) shows the typical light intensity decay behavior observed by the detector. Each of the separated spikes represents the intensity of the light coming out of the loop after each round trip. The time between two adjacent spikes is the round trip time of the light inside the loop. The envelope follows approximately a single exponential decay. Therefore, the decay rate is immune to pulse-to-pulse light intensity fluctuations. A slower decay rate (longer ringdown time) means a lower optical loss of the light in the loop, and vice versa. Figures 12.2(a), (b), and (c) illustrate the concept of the FLRD sensing scheme that FLRD measures time to determine a quantity. Different from the frequency-domain or light intensity-based sensing methods, FLRD is a *uniform* time-domain sensing scheme.

12.2.3 Configuration of the FLRD Sensing Scheme

A FLRD sensing scheme may consist of two major components: a fiber loop and an interrogator that includes a laser source, a detector, and electronics for system control and data processing. Figure 12.3 illustrates the FLRD sensing scheme. The two parts, the FLRD sensor unit and the interrogator, may be connected through two standard FC/APC fiber connectors. This configuration allows one FLRD interrogator to be able to power multiple FLRD sensor units regardless of their sensing parameters.

Coupling a laser pulse into a fiber loop can be achieved in three ways, the direct radiation of a laser pulse onto a bent bare fiber [24, 26], using a fiber coupler [27, 28], or using a notch coupler [29, 30]. Each of the coupling methods is illustrated in Figs. 12.4, 12.5, and 12.6, respectively. The direct radiation coupling is

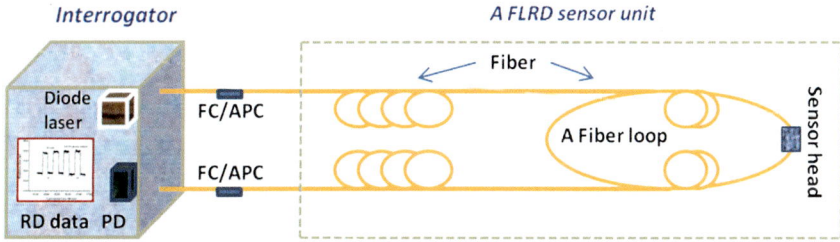

Fig. 12.3 Configuration of a fiber loop ringdown system has two major components, a FLRD interrogator and a FLRD sensor unit

Fig. 12.4 A ringdown fiber loop using the bend coupling method

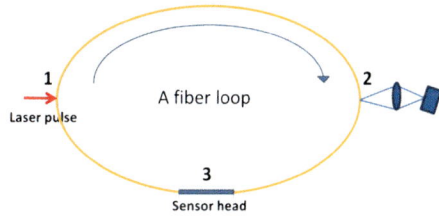

Fig. 12.5 A ringdown fiber loop using two off-the-shelf fiber couplers

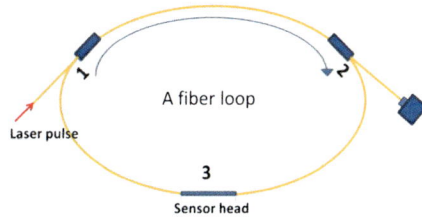

Fig. 12.6 Different configurations of the notch coupler used in FLRD (Reproduced with permission from Ref. [29]. Copyright 2011 Optical Society of America)

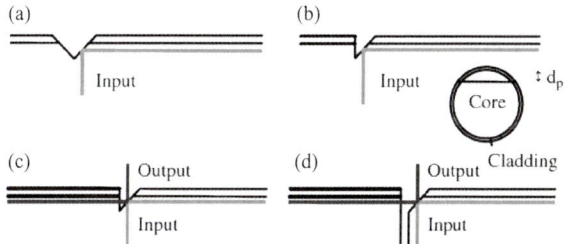

easy to use and it has no need of additional fabrication or using optical components. This method is more suitable for coupling a laser pulse with relatively high pulse energy. Taking advantage of inexpensive by-products in the telecommunications industry, using an off-the-shelf fiber coupler to couple a laser beam into and out of a fiber loop is low cost. Current fiber splicing equipment can achieve a minimum coupling ratio of 0.1/99.9 in a single mode fiber coupler. This coupling method is rugged and stable (the coupling ratio is fixed). It is especially suitable for using a low power pig-tailed DFB diode laser as the laser source. The new notch coupling method has been introduced only very recently [29, 30]. In this method, a notch is carefully cut on the surface of fiber core and the cut depth is of a few percent of the core diameter. An incident beam is focused onto the notch reflective surface and up to 100 % of the incident light can be coupled into the fiber. Like the bend coupling method, the notch method can also have a tunable coupling ratio. But alignments and preparation in the notch method is more complicated. Different configurations of a notch coupler are shown in Fig. 12.6.

Similar to the control of the laser source in CRDS [5, 6], if a continuous wave (CW) laser source is used in FLRD, an additional electronic component is needed to create a short laser pulse from the CW source [11, 13]. Modulation of diode laser current is an effective way, in which a rapid shutdown of the laser driver current to zero or drop below the lasing current threshold can be achieved by an in-house electric board [31]. One can also use a multi-channel pulse generator combined with a frequency mixer to modulate the driver current if no designated electrical board is available [32]. Probably the most convenient light source for FLRD is pulsed fiber lasers that can have a pulse duration of nanoseconds and a repetition rate of up to kHz [33]. Using a pulsed fiber laser as the light source in FLRD would significantly simplify a FLRD experimental system in terms of the system's geometry, weight, and cost. Both a photomultiplier tube (PMT) and a photodiode (PD) can be used as a FLRD detector. The key specification of the detector is its responding time. For instance, the round trip time of a laser pulse in a 100-m strand single mode fiber (SMF) loop is approximately 480 ns. The detector's response must be in nanoseconds in order to resolve each individual laser spike coming out of each round trip in the loop. The data processing software used in the mirror-based CRDS can be directly adopted for data process in FLRD. The interrogator shown in the FLRD sensing scheme may have two different configurations, a free-space, lab-based system or a fully-packaged, standalone instrument box. The former can be seen in many CRDS and FLRD studies reported in the literature [1, 6–10] and in Chap. 9; the latter will be described, for example, in Sect. 12.5.1.

12.2.4 FLRD Sensing Signal

FLRD has a uniform sensing signal, *time*, regardless of the sensing quantity, gas concentration, pressure, temperature, or microfluids, etc. Generation of ringdown time comes from fitting of the ringdown decay curve as processed in the conventional mirror-based ringdown. In many cases, a ringdown decay curve may not follow a single exponential decay [22, 34–36]. Special care should be taken when the FLRD is used for quantification, such as measurements of absolute concentration. Once a FLRD unit is constructed, the baseline stability of the ringdown signal is not affected by any mechanical instability resulting from connections, alignments, or mechanical vibrations, as normally encountered in a mirror-based CRDS system. Depending on the length of the fiber loop and the total optical loss of the laser pulse inside the loop, the ringdown time in FLRD can be up to hundreds of μs or longer.

In many cases, the sensing signal, ringdown time, can be related to the sensing quantity of interest. For example, B in Eq. (12.4) is related to absolute number density of a gas molecule by

$$\left(\frac{1}{\tau} - \frac{1}{\tau_0}\right) = \frac{c}{nL}B = \frac{c}{nL}\sigma(v)nl \tag{12.5}$$

where n is the number density, σ is absorption cross-section at laser frequency v, and l is the absorption path-length. If an air gap is introduced in one section of the fiber loop, l is the spacing of the gap. In pressure sensing [27, 28], B can be related to a pressure by

$$\left(\frac{1}{\tau} - \frac{1}{\tau_0}\right) = \frac{c}{nL}B = \frac{c\beta lS}{nL}P = kP, \tag{12.6a}$$

$$\left(k = \frac{c\beta lS}{nL}\right) \tag{12.6b}$$

where β is the pressure-induced loss coefficient with units of, i.e. $Pa^{-1} m^{-3}$ and l is the length of the fiber (sensor head) that has direct contact with the pressure being applied.

In the evanescent field (EF)-FLRD glucose sensor [37], B is expressed as

$$\left(\frac{1}{\tau^{glu}} - \frac{1}{\tau_0^{air}}\right) = B\frac{c}{nL} = \frac{c}{nL}\left(S_0 + SC + S_1C^2\right) \tag{12.7}$$

where τ_0^{air} is the ringdown time when the sensor head is exposed to air, τ^{glu} is the ringdown time when the sensor head is exposed to the external medium to be detected, such as a glucose solution, and B is related to an absolute difference (a.u.) between the EF attenuation in air and in the glucose solution of the sensor head. S's are constants for a given EF-FLRD glucose sensor exposed in a given medium and C is the glucose concentration (% or mg/dl) in the medium.

Many more examples can be seen in the literature [38, 39]. For instance, the constant B can be related to FLRD sensing parameters, μM or nM in liquid sensing [40], degree (°C) in temperature sensing [41], Δn in refractive index (RI) sensing [42], number of molecular cells in biomedical sensing [43], etc. The examples above show that the FLRD sensing scheme has a uniform sensing signal, ringdown time, from which different quantities can be detected. This sensing uniformity offers unique advantages in sensor networking [38], outcompeting other sensing schemes used in the current fiber optic sensors (see details in Sect. 12.6).

12.2.5 Optical Losses in a Fiber Loop and Detection Sensitivity of FLRD

The total optical loss of a laser pulse in a fiber loop includes low fiber absorption and scattering losses (optical fiber transmission loss), coupling losses of the light into

Table 12.1 Losses of a fiber loop using fused silica single mode fiber (SMF-28, O.D. = 125 μm in cladding). The SMF attenuation rate is approximately 0.3 dB/km at 1550 nm

Typical optical loss	Loss (dB)	Loss (%)
Coupling ratio, 0.1 % (2)	0.04 ∗ 2	0.08
Coupling ratio, 0.1 % (2)	0.0043 ∗ 2	0.0086
Fiber splicing loss (2)	0.02 ∗ 2	0.9
Loss in 100-meter fiber	0.03	0.7
Total fiber scattering loss	0	0
Total loss per round trip	0.158	3.54

and out of the fiber loop at the coupling points, and loss in the sensor head due to a sensing event. For example, a fiber loop sensor unit is constructed by a 100-m single mode silica fiber with a refractive index of 1.464, as illustrated in Fig. 12.2(a), the two identical fiber couplers at points 1 and 2 have a coupling ratio of 0.1/99.9. The absorption loss rate of the single mode fiber (SMF-28) is approximately 0.3 dB/km at 1550 nm. Table 12.1, as an example, lists some typical optical losses related to the off-the-shelf fiber couplers and SMF attenuation at 1550 nm. The total optical loss in a fiber loop using two off-the-shelf fiber couplers can be estimated. For instance, the total optical loss per round trip of the laser pulse traveling in a 100-m long SMF loop is 0.158 dB, corresponding to 3.54 % optical loss.

Theoretically, the loss in a bend coupler can be as small as the loss in a commercial fiber coupler (i.e. at a split ratio of 0.1/99.9). Practically, the loss to be adjusted at each coupling point in the bend coupling method is typically of 0.09–0.15 dB (2–3.5 %). This would result in a total loss per round trip in a 100-m loop of up to 0.21–0.33 dB (4.7–7.3 %), including the 0.03 dB fiber attenuation in the 100-m loop.

Although the fabrication of a notch coupler does not require additional equipment, making a perfect notch (the lowest loss) is a fine work. Nevertheless, Rushworth et al. has demonstrated the new coupling technique [29, 30]. A fiber loop using the notch couplers can have a loss per round trip as low as 4 %.

For a given coupling method, we may estimate the maximum length of a fiber loop that can be realistically constructed for FLRD. For example, the optical transmission loss of a laser beam in SMF at the telecommunications C band (1530–1565 nm) has a rate of 0.3 dB/km. The lowest total insertion loss (sum of the excessive loss and splicing loss) in the two fiber couplers with a split ratio of 0.1/99.9 is 0.128 dB. Table 12.2 lists the length of the fiber loop, total optical loss per round trip, ringdown time, and number of the round trip in one ringdown time.

The detection sensitivity is often characterized by the minimum detectable optical loss. Rearranging Eq. (12.4), we have

$$B = \frac{t_r}{\tau_0}\frac{\Delta\tau}{\tau} = \frac{1}{m}\frac{\Delta\tau}{\tau} \qquad (12.8a)$$

$$(\Delta\tau = \tau_0 - \tau) \qquad (12.8b)$$

Table 12.2 A set of fiber loops with typical parameters for each loop. The RI is assumed to be 1.464 at 1515 nm for SMF. Commercially available fiber couplers are used in the estimates

Length of the fiber loop (m)	Loss in two fiber couplers (dB)	Loss in fiber (dB)	Total loss per round trip (dB)	Loss (%)	Ringdown time (μs)	M number
1	0.128	0.0003	0.1283	2.91	0.17	34
10	0.128	0.003	0.131	3.031	1.61	33
100	0.128	0.03	0.158	3.058	15.96	33
1000	0.128	0.3	0.428	3.328	146.63	30
10000	0.128	3	3.128	6.028	809.56	17

where t_r is the round trip time of the laser pulse in the fiber loop, and m is the number of round trips (m-number). Therefore, the minimum detectable optical loss B_{min}, which is defined as the one-σ detection limit, is given by

$$B_{min} = \frac{1}{m} \frac{\Delta \sigma_\tau}{\bar{\tau}} \qquad (12.9)$$

where $\Delta \sigma_\tau$ is the one-σ standard deviation of the averaged ringdown time.

Several points with regard to B_{min} defined in Eq. (12.9) need to be noted. (1) The B_{min} defined in Eq. (12.9) represents the theoretical detection limit (or limit of detection) based on the one-σ standard. When the optical loss B in Eq. (12.8a) approaches to the minimum detectable optical loss B_{min}, $\Delta \tau$ can be approximated by the one-σ standard deviation of the averaged ringdown time $\bar{\tau}$ and $\bar{\tau} \approx \tau_0$. (2) Although the one-$\sigma$ standard detection limit is often used as a reference for discussion on the detection sensitivity of a ringdown system, the International Union of Pure and Applied Chemistry (IUPAC) has the specific guidelines about detection limits and the 3-σ standard criterion is required for the definition of the detection limit. (3) The experimental detection limit based on the 3-σ standard, i.e. the low detectable concentration or the minimum detectable absorbance, needs to be extrapolated from an experimental calibration curve. Care needs to be taken in the extrapolation of the experimental calibration curve since the calibration curve does not always pass through the origin of the coordinates. A thorough discussion about the determination of detection limits and their methods can be seen in a very recent publication [44].

In Eq. (12.9), $\Delta \sigma_\tau / \bar{\tau}$ can be experimentally achieved at 1×10^{-3}, which is a typical level of the minimum detectable fraction intensity $\Delta I / I_0$ in a conventional intensity-based sensing scheme. Therefore, if a conventional intensity-based fiber optic sensor has a detection limit B, a FLRD optic sensor will have a detection limit B/m, thus improving the detection sensitivity by a factor of m. Therefore, sensitivity of the FLRD is determined by the number of the round trip (m-number) during one ringdown time; and the m-number is fundamentally determined by the optical loss per round trip. If an air gap is introduced in the fiber loop, on the one hand, the gap space is as small as possible to minimize the optical transmission loss at the gap, i.e. a few mm. On the other hand, a longer absorption path-length (l in

Fig. 12.7 A fiber loop
ringdown gas sensor using an
ultra-low loss U-bench as the
gas cell

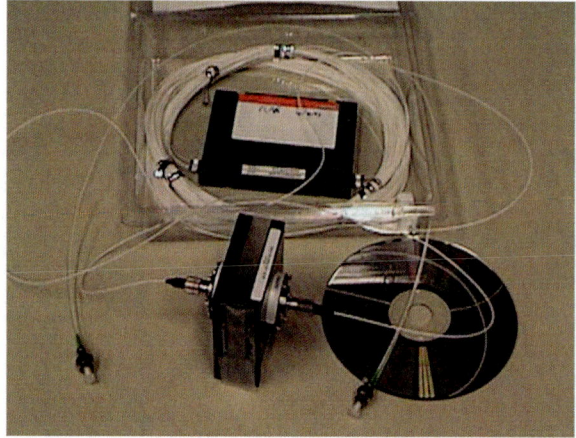

Eq. (12.5)) would offer a higher sensitivity. In practice, these two points compromise each other. Figure 12.7 shows a FLRD gas sensor unit, which utilizes a high precision U-bench coupler to create a 1.5-cm air gap for gas sensing [45]. Although the U-bench has the highest free space coupling efficiency, there is still a $\sim$0.1 dB loss. Due to the large optical loss (a small m number) and the limited gap space, even 100 round trips can only create an effective path-length of 1.5 m (assumed that the air gap is 1.5 cm)!

In a different configuration, in which, instead of a fiber loop, two fiber Bragg gratings (FBGs) separated by a section of SMF are used to form a fiber cavity [17, 20]; this approach also has the ringdown-enhanced multiple interaction effect. However, the enhancement effect in this linear fiber cavity should be counted by 2 m; the factor of 2 comes from the twice interactions in the sensor head in each round trip. For instance, in a recent study of chemical sensing, the sensor used a polymer-coated long period grating (LPG) as the sensor head that was located in a dual FBGs fiber cavity [46]. The FBG separation was 6 m, RI of the fiber cladding at 1550 nm for the SMF was 1.4453. When the total cavity round trip loss was 21 % ($\sim$1 dB), the ringdown time was approximately 140 ns. This gave the round trip number $m = 5$. However, due to the fact that the sensor head (the coated LPG) is located between the two FBGs, the effective enhancement is m $= 10$ (two interactions in each round trip, as compared to one interaction in each round trip in the fiber loop!).

12.3 Fabrication of a FLRD Sensor Head

Fabrication of a FLRD sensor head depends on what kind of sensing mechanism is used and its sensing environment. The former governs the configuration, components, size, and rigidness of the sensor head; and the latter determines whether an additional engineering effort is needed in the construction of the sensor head [38, 39, 47]. In the following, typical sensing mechanisms used in FLRD are described first, followed by a brief review on the current FLRD sensor heads.

12.3.1 Sensing Mechanisms

To date, sensing mechanisms (or transduction principles) of fiber optic sensors (FOS) have been well established [48–50]. A wide variety of transduction principles have been reported. For example, fluorescence- [51–55], direct absorption- [56, 57], capillary electrophoresis absorption-, and EF absorption-based transduction principles have been used in fiber optic chemical sensors [40, 58, 59]. Mechanical deformation-, thermal expansion-, and interferometric-based sensing mechanisms have been utilized in fiber optic physical sensors (i.e. Fabry-Perot interferometric (FPI) pressure and temperature sensors, FBG strain and temperature sensors, etc.) [60–72]. EF absorption/scattering and or in combination with using chemical coatings or optical components such as FBGs or LPGs can be seen in biological and biomedical sensing [40, 46, 73–77]. Theoretically, all of the sensing mechanisms currently used in fiber optic sensors can be directly adopted in the FLRD sensing scheme [38, 39].

In the earliest report of fiber loop ringdown spectroscopy, direction absorption was investigated for potential gas sensing [23]. EF in combination with chemical modifications of the tapered single mode fiber surface was also employed to sense biological cells [43]. Combination of using a LPG with polymer coatings was reported to achieve detection of different volatile organic compounds [46]. Capillary electrophoresis absorption was used for detection of human serum albumin [40]. Coating chemicals layer-by-layer on the surface of one section of partially-etched SMF was used for label-free DNA sensing [76]. FLRD glucose sensor utilized immobilization of glucose oxidase (GOD) on the etched fiber surface to achieve index-difference-based sensing [37]. In general, chemical coatings (immobilization) employed in FLRD serve as modification of RI at the interface between the fiber and the coating layer or as a substrate to bind chemical agents to be detected.

Interference phenomena are another major sensing mechanism being used in the FLRD technique. Light travels along two different paths (or along the same path-length in media with different RI), and an optical phase shift is created. This sensing mechanism can be achieved in several different approaches, for instance, one light beam travels in a fiber core (the first mode), the other beam travels in the fiber cladding (the second mode). To date, the second mode has also been generated using LPGs, tilted FBGs, coupling from a single-core fiber to a multi-core fiber, or core-offset [78]. More recently, a SMF taper or photonic crystal fiber (PCF) has also been used to create an interferometer to measure RI [79, 80].

The FBG is one of the most used sensing elements in FOS and its sensing mechanism is based on detection of a wavelength (or frequency) shift resulting from a physical modification of the period or change in RI. Most FBG- or LPG-based fiber optic sensors are based on frequency-domain sensing using an optical spectral analyzer (OSA) or a designated FBG-interrogator to measure the frequency shift resulting from a sensing event, i.e. a change in temperature, strain, or both. However, in the time-domain FLRD sensing scheme, the sensing mechanism of using a FBG is based on detection of a transmission loss of the laser beam passing through the FBG, rather than evaluating the reflection of the light at the central wavelength of

Fig. 12.8 Illustration of a
FBG's spectral response
curve, which is divided into
three zones, Zones I to III, of
which Zones I and III may be
used for FBG-FLRD sensor.
Points *a* and *b* mark two
different locations in the
bandwidth curve for
temperature sensing. No
specific wavelength is labeled
in the *x-axis*

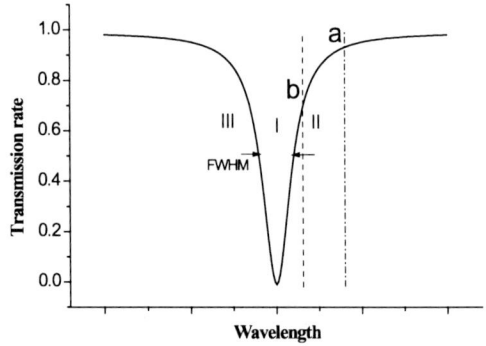

the FBG [47, 71]. Since the use of a FBG in the FLRD sensing scheme is different from those in the conventional FOS, using a FBG-FLRD temperature sensor as
an example [81, 82], a little more detailed description of the sensing mechanism is
given here.

The Bragg wavelength of a FBG is given as,

$$\lambda_B = 2n\Lambda \tag{12.10}$$

where n is the RI of the fiber core and Λ is the grating period. For a given temperature variation, both n and Λ change, and the resultant Bragg wavelength as well as
the whole bandwidth curve shift. The spectral response of the FBG to a change in
temperature is described by [83, 84]

$$\Delta\lambda = \lambda_B\left(\alpha + \frac{1}{n}\frac{dn}{dT}\right)\Delta T, \tag{12.11}$$

where $\Delta\lambda$ is the wavelength shift of the Bragg wavelength λ_B resulting from the
FBG's temperature increase ΔT; α is the thermal expansion coefficient; and dn/dT
is the change rate of the fiber RI. Typically, a temperature increase shifts the Bragg
wavelength and the entire bandwidth curve to the longer wavelength side. In FBG-
FLRD sensors, a laser is tuned to a particular wavelength that corresponds to a preset
location, i.e. point *a*, in the FBG bandwidth curve, as shown in Fig. 12.8, the shift
of the spectral response curve resulting from the temperature variation changes the
relative position of the laser wavelength in the bandwidth curve, i.e. from point *a*
at room temperature to point *b* at a higher temperature; thus, the light transmission
loss incurred in the FBG (the parameter B in Eq. (12.4)) changes, which, therefore,
changes the observed ringdown time from τ_0 to τ ($\tau < \tau_0$ in this case). This is
the basic concept of how a FBG-FLRD or LPG-FLRD temperature sensor works
by measuring the ringdown time. Since temperature, pressure, strain all can affect
FBG's characteristics; this FBG-based sensing mechanism can be used in FLRD for
other physical sensing. Note that the linear range of response of a FBG-FLRD sensor
is limited only to the narrow parts of the spectral range in which FBG's spectral
response is linear; on the other hand, the frequency response of a conventional FBG
sensor has an unlimited linear range.

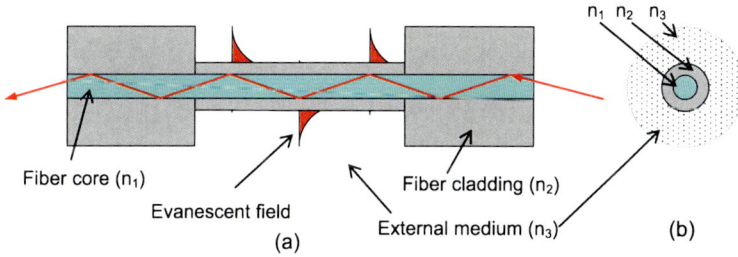

Fig. 12.9 Illustration of the concept of the EF-FLRD sensing using a section of partially-etched single mode fiber as the sensor head. (**a**) Illustration of the EF around the etched fiber surface, (**b**) different indices in the three regions as shown in the cross-section of the fiber

The incorporation of the EF sensing mechanism into the fiber ringdown technique to create a FOS was first reported by von Lerber and Sigrist using a section of straightened, end-coated SMF to form a linear fiber ringdown cavity (not a loop cavity) to explore EF absorption [17] and later by Tarsa et al. using a fiber ring resonator (a loop cavity) to demonstrate the detection of the EF absorption of water and EF scattering of biomolecular cells [74]. Recently, the EF absorption sensing mechanism has been used for detection of dimethyl sulphoxide solutions in a fiber ringdown loop [85]. More recently, EF-scattering due to RI difference at the interface has been adopted for sensing biomedical agents [37, 76]. Detailed discussion on the EF fundamentals [86–92] and its application in CRDS can be seen elsewhere [17, 39, 73, 92].

Briefly, the sensing mechanism that is responsible for the EF induced optical losses at the sensor heads can be described by [75]

$$B = \frac{nL}{c}\left(\frac{1}{\tau} - \frac{1}{\tau_0}\right) = l_e(\gamma_\alpha + \gamma_\beta), \tag{12.12}$$

where l_e is the length of the sensor head where the EF is created and γ_α and γ_β are the light attenuation coefficients due to EF absorption and EF scattering, respectively. Based on the near-field optics theory [86], the attenuation coefficient γ_α of the EF at the medium interface is a function of the refractive indices of the fiber, n_1 (the fiber core) and n_2 (the fiber cladding), shown in Fig. 12.9, the wavelength of the light, λ, the incident angle, θ (where $\theta >$ the critical angle θ_c), the etched fiber diameter, D, and the EF absorption coefficient of the absorbing cladding. The attenuation coefficient due to the EF absorption in the absorbing cladding γ_α is expressed by [86]

$$\gamma_\alpha = \zeta \frac{n_1 n_2 \cos^2 \theta}{D \sin \theta (n_1^2 - n_2^2)(n_1^2 \sin^2 \theta - n_2^2)^{1/2}}, \tag{12.13}$$

where ζ is a constant that is a function of λ and n_2^α, the EF absorption coefficient of the absorbing cladding. n_2^α is often expressed as the imaginary part of the RI of the absorbing cladding, namely, $n_2^* = n_2 + i n_2^\alpha$. When the cladding is non-absorbing, $n_2^* = n_2$. The ratio γ_α/ζ is often used as a factor to characterize the sensitivity of

EF absorption. Simple simulations using Eq. (12.13) show that γ_α/ζ is proportional to n_2. This means that under the same experimental parameters the EF absorption induced optical loss is proportional to n_2. However, this EF absorption mechanism is not applicable to the cases in which the observed optical losses are lower when n_2 is larger. In such cases, EF scattering is responsible for the optical loss with the light attenuation coefficient, γ_β. The EF scattering loss is dependent on several parameters of the sensor head, including the refractive indices of the fiber, diameter of the fiber core (or the effective fiber core if the cladding is partially etched), and EF penetration depth [89–92]. For a given EF sensor head with the same etched fiber diameter and the same etched fiber length, the external medium with a higher refractive index (n_3), equivalently, a smaller index difference Δn ($\Delta n = n_2 - n_3$ and $n_2 > n_3$), results in a lower EF scattering loss [91].

12.3.2 Sensor Heads

A sensor head in FLRD is only a section of waveguide in the fiber loop. Figure 12.10 illustrates four different configurations of the sensor heads, as an example, for sensing pressure (P), temperature (T), and gas concentration. For sensing pressure, a section of bare SMF (without plastic jacket) is typically placed in a micro-bending platform to form the sensor head. The sensitivity of FLRD pressure sensors depicted in Fig. 12.10(a) can be characterized by the minimum detectable pressure change per ringdown time change (s), Pa/s. Since force is related to pressure by $F = PS$, with a known surface area S of the platform, the pressure sensor also senses the force F applied. For FLRD-based temperature sensing, for example, a piece of FBG written on a section of fiber can be directly spliced with a long section of fiber to form a fiber loop; and the FBG serves the sensor head, as shown in Fig. 12.10(b). Wavelength selections of the FBG-FLRD temperature sensors are determined by the Bragg wavelength of the FBG. A narrow bandwidth FBG (with a narrow wing in each side of the bandwidth curve) can provide high sensitivity because a small shift drastically changes the transmission at the mid-reflection point, i.e. the wavelength of highest sensitivity. FBGs, LPGs, and linear fiber gratings (LFGs) can all be employed to form temperature sensor heads, and have different properties, such as sensitivity, temperature measuring range, response linearity, etc. The sensitivity of FLRD temperature sensors can be characterized by the minimum detectable temperature change per ringdown time change, K/s.

In gas sensing, an air gap in a fiber loop may be created by simply separating two fiber ends facing each other. High transmission coatings on the surface of the fiber ends can be used to reduce optical loss in the coupling. A U-bench air-gap with an ultra-low insertion loss (i.e. <0.1 dB) and high thermal and mechanical stability can be directly connected through two FC/APC fiber couplers to form a fiber loop, as shown in Fig. 12.7, for direct absorption gas sensing. Gas samples can be directly flowing through or statically confined in the air-gap. The separation of the gap is governed by a combination of the best sensitivity and the maximum optical loss per

Fig. 12.10 Different types of sensor heads used in FLRD. (**a**) Micro-bending for pressure and force sensing, (**b**) a FBG for temperature sensing, (**c**) an air gap for gas sensing, and (**d**) a section of hollow core photonic crystal fiber for gas sensing

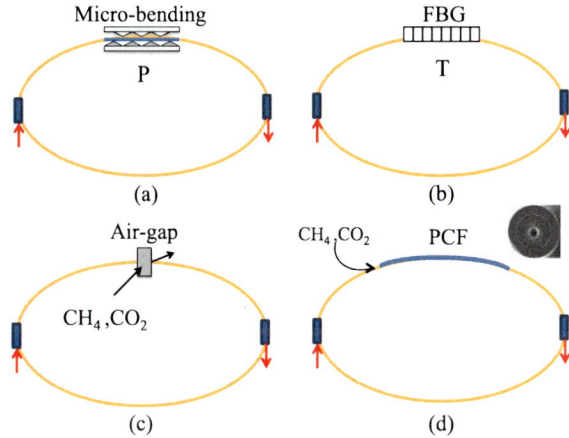

pass. For example, for CH_4 and CO_2 sensors, two NIR DFB laser diodes can be used for detections of CH_4 and CO_2 at 1651 nm and 1572 nm, respectively [93]. Wavelength selections should be based on the combined consideration of sensitivity and possible spectral interferences, and they can be characterized based on spectral simulations using HITRAN 96 [94]. Due to the limited length of the air gap, i.e. 1.5 cm (longer air gaps, higher sensitivities, but also larger optical losses), the detection limit of this configuration is typically at the levels of a few percent [45]. Note that with the same path-length (1.5 cm), CH_4 is not detectable by the single-pass absorption scheme at 1651 nm. This air-gap configuration is advantageous for chemical identification, but not for trace gas sensing. In order to further enhance the detection sensitivity, a photonic crystal fiber (PCF) with air holes in the fiber core (Fig. 12.10(d)) can be fabricated into a fiber loop. Current technology allows for the lowest insertion loss in splicing SMF with PCF to be <0.2 dB. Using a section of PCF as a gas cell in the fiber loop is desirable. In this way, a long portion of the fiber loop can be filled with a gas sample. The detection sensitivity will be doubly enhanced by both the nature of the multiple round trips of the laser pulse in the loop and the long path-length of the laser pulse in the sample in each round trip (i.e. from a 1-cm path-length in the air-gap to a 1-m path-length in the air-hole PCF per round trip), this directly results in an additional 100-fold sensitivity enhancement. However, no study has reported on such a sensor head in the FLRD-based gas sensing to date.

Overall the fabrication of the FLRD sensor heads, as shown in Fig. 12.10, and integration of them into a fiber loop are a trivial effort. Figure 12.11 illustrates several more different sensor heads. As compared to the fabrication of the sensor heads shown in Fig. 12.10, more efforts are needed in the fabrication of the sensor heads shown in Fig. 12.11 [39].

More complicated procedures are required to fabricate a chemical coating-based sensor head. The fabrication typically involves several simple yet needed steps: (1) the fiber preparation that may include etching or tapering one section of SMF or integrating a FBG or a LPG into a fiber loop, (2) the surface modifications that often

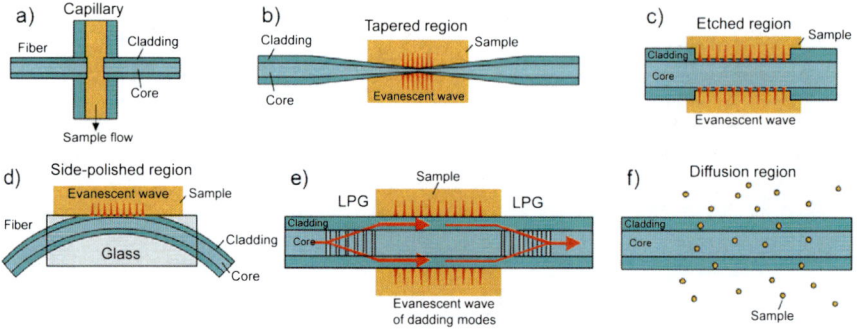

Fig. 12.11 Different types of sensor heads that require additional efforts in the sensor head construction as compared to the direct integration of an off-the-shelf fiber optic component into the fiber loop, as shown in Fig. 12.10. (**a**) A capillary gap created in one section of the fiber loop for direct absorption of all propagated light through the sample. (**b**) A fiber taper that allows EF excited at the interface between the fiber core and the sensing sample (the medium). (**c**) A section of etched fiber as the sensor head that allows EF to be extended into the sensing medium. (**d**) A section of side-polished fiber and EF is launched at the interface between the polished surface and the surrounding medium. (**e**) A pair of identical LPGs that couple the light from the core mode into a single cladding mode and then back into the core. Between the two LPGs the light travels in the cladding mode and EF interacts with the sensing medium. (**f**) A section of porous fiber that allows gaseous molecules to be diffused into the fiber for direction absorption or scattering sensing (Reproduced with permission from Ref. [39]. Copyright 2010 Multidisciplinary Digital Publishing Institute)

involve cleaning, coating of one specific chemical or several layers of coatings of different functional materials, and (3) coatings of a probing layer.

The four most widely exploited means to construct micron-sized tapers are using flame, CO_2 laser heating, microfurnace, and a fusion splicer [95, 96]. The flame method can pull a fiber diameter down to 50 nm with lengths of tens of mm for single mode laser and beam propagation loss rate through the tapered fiber is typically of 0.1 dB mm^{-1}. The laser heating method can readily achieve a taper diameter of as small as 3–4 μm. A smaller taper waist and a longer fiber length have a larger optical loss. The taper profile may be described by an exponential function of the tapered fiber waist radius r, the initial fiber radius r_0, the pull-length z, and the length of the fiber thermally treated [97],

$$r(z) = r_0 e^{\frac{-z}{L}}. \tag{12.14}$$

The key parameters for controlling the tapering processes include the drawing temperature, precise heated length, drawing speed, chemical contamination in the processing environment, etc. The taper fabrication starts with removal of a section of the plastic jacket followed by a thorough cleaning of the exposed fiber surface with alcohol. Single mode silica fiber has a core diameter of 8–9 μm and a cladding diameter of 125 μm. In FLRD, a sensor head using a SMF taper typically has a tapered fiber waist of 9–50 μm with a taper fiber length of a few mm to 20 mm, for example. The taper can also be coated with a protection layer using a low RI silicone gel for some sensing purposes in physically harsh environments [98]. Coatings of other chemicals on the surface of the taper are selected either for micropartitioning of the analyte of interest or for selective chemical binding. Assumed that the fabrication processes are reproducible, taper diameters can be estimated using Scanning Electron Microscopy (SEM). Overall fabrication of a fiber taper with waist diameters of tens of μm for FLRD sensor heads is straightforward. Using a fiber taper as a sensor head has one limitation that the tapered fiber length is short, typically, up to a 1–2 cm. In some applications, a sensor head with a long fiber length, i.e. up to tens of cm, is desired.

Another popular method to physically modify fiber structure for construction of a FLRD sensor head is chemical etching. Theoretically, the etching method can allow the length of the etched fiber to be unlimited, regardless of total optical loss of a laser beam passing through this section of the etched fiber. Furthermore, the etched fiber section can be uniform in terms of the etched fiber diameter or an optical loss rate (loss per cm). An etching process is simple. A section of fiber in a fiber loop is selected and the plastic outer jacket is removed. The bare section of fiber is immersed in a 48 % hydrofluoric acid (HF) solution and the section of the fiber in contact with the acid is gradually etched away (The process needs be to handled under an ventilation hood! Also HF is extremely toxic and a contact poison). Etching time determines the etching depth (diameter of the etched fiber). For single mode silica fiber, the etched fiber diameter is approximately 10 μm after 64–65 minutes etching process in the ambient air [99]. The etched fiber diameter is typically estimated using SEM, which has to be done by cutting the etched fiber off the fiber loop. Mostly, this section of etched fiber cannot be re-spliced back to the fiber loop. Accurate estimation of the etched fiber diameter can be done by immersing several identical fiber segments into the etching solution, going through the same etching and cleaning processes. Therefore, the diameter of the etched fiber segments is assumed to be the same as the diameter of the etched fiber in the fiber loop. In order to control and determine the etched fiber diameter in a fiber loop, a highly reproducible process is required, which is often a challenge based on monitoring the etching time.

A recent work shows that the etching process can be monitored in real-time, on-line by the FLRD technique [75, 99]. A bare section of fiber is immersed in a 48 % HF solution and the section of the fiber in contact with the acid is gradually etched away. When the EF propagating through the cladding starts to leak out to the external medium, the observed ringdown time begins to decrease due to the increase of the optical loss. Figure 12.12 shows the results of the on-line monitoring of etching SMF in the fiber loop. In the absence of optical absorption, the loss is attributed to the EF scattering loss due to the RI difference between the "fiber core" and the external medium. It should be noted that the fiber core here is actually the effective fiber core composed of the original fiber core and the unetched cladding. The images of the etched fibers are shown in Fig. 12.13. Diameters of the etched fiber at different etching stages were measured, for example, at the etching points A, C, and G, the etched fiber diameters were 90, 45, and 10 μm, respectively.

Chemical coatings on a fiber taper or etched fiber or a FBG/LPG can have several functions: modification of RI, increase of chemical selectivity, formation of a substrate for further functional coatings, or all of them. Polydimethylsiloxane (PDMS)

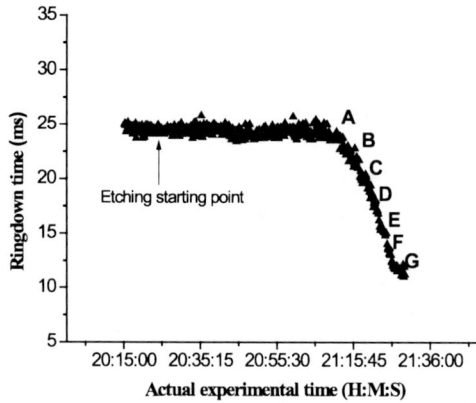

Fig. 12.12 Real-time, on-line monitoring of the fiber etching process using the fiber loop ring-down technique. The arrow marks the time when the fiber was immersed in the HF solution and the etching process began. Points A–G indicate the sequential stages of the etching process. The measured ringdown times at the indicated points (A, B, C, D, E, F, and G) are 23.90, 23.70, 22.90, 20.40, 17.10, 14.00, and 12.10 μs, respectively (Reproduced with permission from Ref. [75]. Copyright 2010 The Institute of Physics)

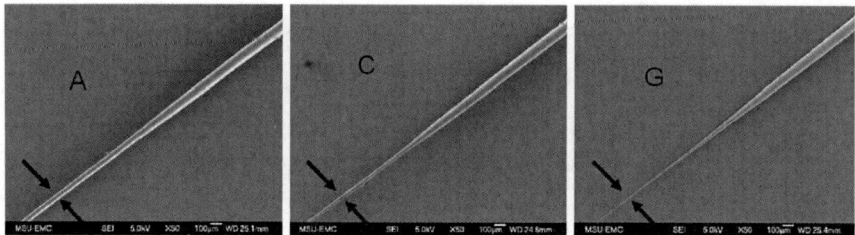

Fig. 12.13 Scanning electron microscopy characterization of the etched SMF fiber. At the etching stages A, C, and G, as labeled in Fig. 12.12, the etched fiber diameters were measured to be 90, 45, and 10 μm, respectively. The *arrows* mark where the etched fiber diameter is determined

is a popular material to be used to coat LPGs for chemical sensing. Loock et al. [100] introduced this approach in the development of FLRD chemical sensors, as shown in Fig. 12.14. They developed a specific polymer (PDMS) that has RI matched to the cladding material of the LPG used. The sensor head was exposed to two groups of volatile organic compounds (VOCs), the polymer had differential responses, demonstrating the selectivity of the sensor. Sensitivity of the polymer coated LPG sensor head was optimized by tuning the RI of the polymer coating to be slightly below that of the cladding, i.e. between 1.423 and 1.438 at 1550 nm. The index formulation and analyte uptake in the chemical matrix were evaluated by using a commercial refractometer to measure refractive indices at the specific wavelength. For example, the composition of the polymer with 8.5 % diphenylsiloxane, 0.5 % titanium cross-linker, and 91 % PDMS resulted in a measured RI of 1.4237. In order to stabilize the RI of the formula, the coated LPG needed to be treated in an elevated temperature

Fig. 12.14
A chemically-coated long period grating was used as the sensor head in the FLRD gas vapor sensor based on the phase shift detection approach (Reproduced with permission from Ref. [100]. Copyright 2008 Royal Society of Chemistry)

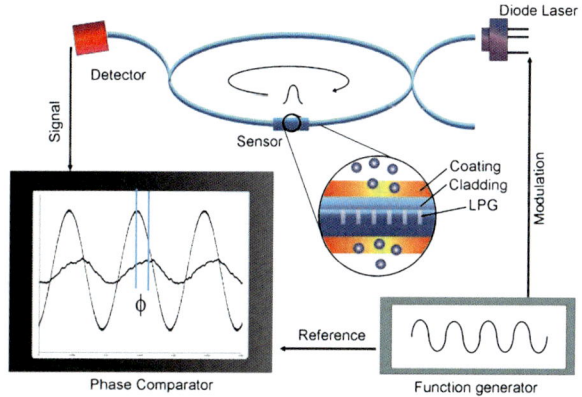

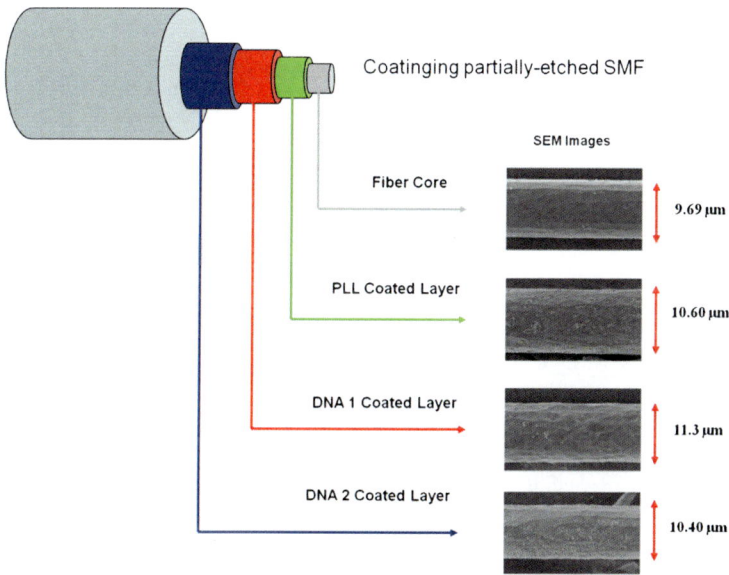

Fig. 12.15   A label-free FLRD DNA sensor using multi-layer coatings on the etched single mode fiber

for up to several days. In addition to PDMS, syndiotactic polystyrene was also used for coating a LPG for chemical sensing in FLRD [100].

Direct coatings of chemicals on one section of etched SMF have also been reported in FLRD sensors very recently [76]. The multi-layer coatings serve for different functions. Figure 12.15 is an illustration of the multi-layer coatings on a section of partially-etched SMF for label-free DNA sensing using the FLRD scheme. The first layer of coating was used as a substrate, on which further coatings were conducted to chemically bind a sensing probe (probe DNA) that will selectively sense a specific DNA (target DNA). Prior to the coatings, the partially-etched fiber was

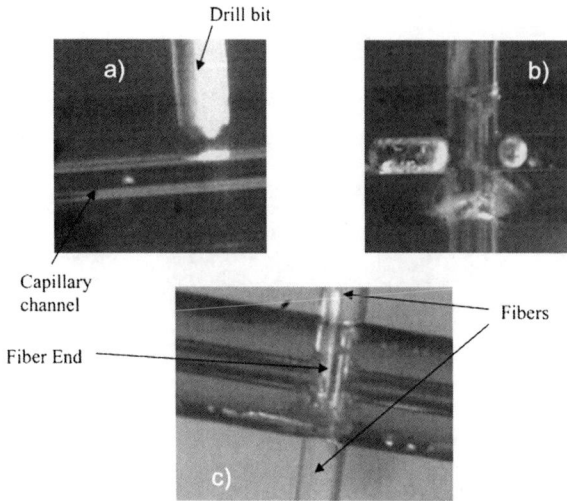

Fig. 12.16 The art of the capillary-fiber interface sensor head used in FLRD fluid sensors. (a) Drilling through the capillary. The drill bit (150-µm in diameter) can be seen just above the channel (100 µm) of the 360-µm-O.D. capillary, (b) the channel drilled completely through, and (c) fiber ends introduced into the channel (lens fiber on the top and square-cut receiving fiber on the bottom). The fibers were then sealed with epoxy glue (Reproduced with permission from Ref. [40]. Copyright 2006 American Chemical Society)

first cleaned with phosphate-buffered saline solution (NaH$_2$PO$_4$/Na$_2$HPO$_4$ pH 7.4, 150 mM) (PBS). In order to be effectively coated with poly-L-lysine (PLL) solution (0.1 % W/V in water, the molecular weight $=150,000$–300,000 g/mol), the sensor head was immersed in PLL for 160 minutes. The sensor head was then cleaned with PBS to remove excess PLL and immersed for 130 minutes in 20 µM 16 ssDNA S1 (probe DNA). Again cleaned with PBS, the sensor head was then ready for sensing the matched ssDNA S2 (the target DNA). Since the positively charged NH$_3^+$ in the PLL chain is binding to the negatively charged DNA, and the probe DNA selectively binds with the target DNA, the label-free DNA sensing is achievable by the FLRD DNA sensor. The entire stepwise coating process, illustrated in Fig. 12.15, was performed at room temperature and coating effects were monitored by ringdown times that were continuously collected throughout the coating process.

An even more complicated procedure has been used in the fabrication of the FLRD sensor head for microfluid sensing [40]. In this device, a section of fused silica capillary (I.D. 100 µm/O.D. 360 µm) was embedded in polymethylmethacrylate (PMMA) polymer that was used to hold the capillary. Then a 150-µm hole was drilled perpendicular to the axis of the capillary through the PMMA and the capillary; and two ends of the fiber in the open loop were inserted through the hole and aligned to allow the fiber ends to meet with the inner capillary wall ($\sim$60 µm). Finally, the capillary-fiber interface was sealed by epoxy glue. Figure 12.16 shows the image of the capillary-fiber interface sensor head developed in Loock's group.

Numerous sensing mechanisms will continue to be adopted in the FLRD sensing scheme. Correspondingly, more different types of sensor heads will be constructed using different fabrication procedures. Certainly, the aforementioned fabrications of different FLRD sensor heads are not inclusive.

12.3.3 Sensing Parameters (Functions)

Since its introduction of the concept of FLRD in 2001, many different quantities, in all of the three classes (chemical, physical, and biomedical or biological) have been measured, sensed, or monitored using the uniform FLRD sensing scheme combined with different sensing mechanisms. Table 12.3 lists the measurands that have been detected by FLRD sensors to date. Due to the rapid update of the field, some reported measurands may not be included in this table.

12.4 Individual FLRD Sensors

12.4.1 Chemical Sensors

The first report of fiber loop ringdown was in 2001 [23]. Stewart et al. constructed an optical fiber loop of several tens of meters long; and a 5 cm open path gas cell was fabricated in the loop to explore direct gas absorption [23]. Due to the large optical loss in the open gap, i.e. 1 dB, the observed ringdown times were as short as 100 ns. By adding a 15–20 m long erbium-doped fiber into the fiber loop (still a single-loop), the large optical loss was compensated by the erbium-doped fiber amplifier (EDFA) that was pumped by a pumping laser. In the original setup with a fiber loop of 58 m, the ringdown time was extended up to 2.6 µs, giving the total round trip loss of 0.48 dB (10.5 % loss), corresponding to $m = 9.5$. They successfully demonstrated the concept of the fiber cavity ringdown spectroscopy technique. The main limitation of this direct-pass gas phase absorption approach is its low sensitivity—the effect of the ringdown enhanced multiple pass is compromised by the short pathlength (the gap length). The large optical loss in the open gap resulted in a low m number. Based on the same detection scheme, the same group later demonstrated an improved performance by introducing automatic EDFA gain control through the use of lasing action in the loop and observed much longer ringdown times, i.e. 200 µs in a 38-m long fiber loop. That ringdown time gave a net round trip loss of 0.004 dB (or 0.09 % loss), corresponding to $m = 1111!$ [101]. Driven by achieving high detection sensitivity in the direct-pass absorption FLRD technique, the amplified FLRD approach was revisited later by Loock and his coworkers. They adopted the concept of the two-loop approach [23] and demonstrated, for the first time, high-sensitivity gas sensing using fiber loop ringdown spectroscopy. In the two-loop approach, one loop was used to stabilize the gain and the other loop (ringdown measuring loop) with a

Table 12.3 Parameters detected by FLRD-based sensors

Quantity	Sensing function	Specific species	Sensing mechanism	Sensor head configuration	Report year and after
Chemical quantity	Gas concentration	H_2, CH_4, H_2O, 1-octyne, xylene, trichloroethylene, cyclohexane, gasoline	Direct absorption, EF absorption, modification of reflective index	Air-gap, tapered bare fiber, coated LPG	2001
	Microfluids	DDCI dye solution, human serum albumin, methylene blue, tartrazine, myoglobin	Direct absorption, EF absorption	Air-gap, capillary	2002
	P, Fe, Mg, Cd, Co	Trace elements in water	EF scattering (index difference)	Etched bare SMF	2011
	Jet fuel and oil	Oil concentration in fuel	Direct absorption (a volume of 25 nL)	Capillary-fiber interface (200 μm gap)	2012
	Water	Presence of water in concrete	EF scattering (index difference)	Etched bare SMF	2012
	Liquid explosives	Streptavidin-Cy3	Direct absorption and EF absorption	End-gap (0–15 μm) and coated tapers	2012
Physical quantity	Pressure, force	Pressure and force applied, positive gas pressure in a chamber	Mechanical deformation	Bare SMF	2004
	Strain	Micro strain, and large strain	Mechanical deformation	FBG, bare SMF	2004
	Temperature	High precision in a small range, low precision in a large range	Thermal expansion	FBG, LPG	2005
	Refractive index	Index oils, salinity in water	EF scattering	LPG, bare SMF	2008
	Cracking	Cracking in concrete	Micro-bending	Bare SMF	2012
	Current	100–300 A	Faraday rotation	Optical solenoid	2012
Biomedical or biological quantity	Single molecular cell	Mammalian cancer cells, single cell	Light scattering	Tapered fiber	2004
	Human serum albumin	Human serum albumin labeled in dye	Absorption	Capillary-fiber interface	2006
	Bacteria, DNA	E-Coli, 16 ssDNA	Index difference EF	Multi-layer chemical coatings	2011
	Glucose	Glucose solutions, glucose in urines	EF-scattering	Chemical coatings	2012

Fig. 12.17 FLRD
measurements of acetylene.
Absorption spectral scans of
the P(13) rotational line of
acetylene near 1532 nm
(Reproduced with permission
from Ref. [39]. Copyright
2010 Multidisciplinary
Digital Publishing Institute)

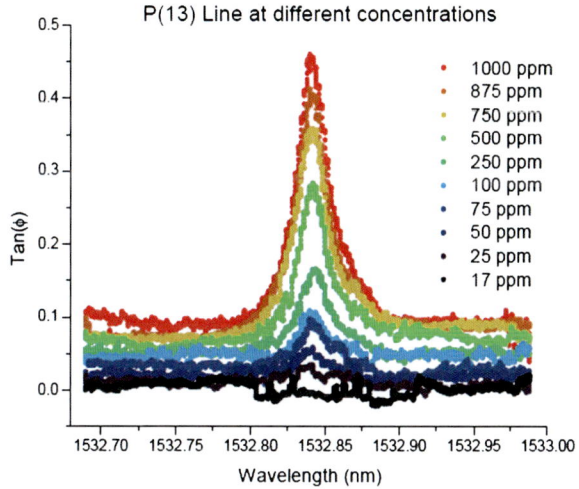

6-cm absorption gap was operated right below the gain threshold [39]. The relative
gain in these two loops was balanced via adjusting a variable optical attenuator. Unlike the original (single-loop) amplified FLRD system, this setup (two-loop FLRD)
does not require the measuring laser wavelength to be close to the lasing wavelength
to achieve a long ringdown time.[1] Absorption spectral scans of the P(13) rotational
line near 1532 nm of acetylene in atmospheric helium in different concentrations
ranging from 17 to 1000 parts per million by volume (ppm) were obtained, as shown
in Fig. 12.17. A detection limit for acetylene of 25 ppm was achieved. Measurement
accuracy of the amplified FLRD is often affected the ASE noise generated by the
EDFA.

Ni et al. introduced a digital least-mean-square filter in their amplified FLRD
system to improve the measurement accuracy by reducing measurement errors
[102–104]. At the same wavelength 1532 nm, they reported a detection limit of
70.1 ppm, eight times lower than that without using the noise filter. A lower detection limit can be further pushed in the direct-pass gas absorption using further
improved approaches in the loop structure. For example, in order to minimize the
optical coupling loss in the air gap, a commercially available ultra-low coupling
loss U-bench can be introduced into a fiber loop for direct gas absorption measurement. Combination of the two-loop structure with the digital least-mean-square
noise reduction algorithm or using a long section of photonic crystal fiber to replace the air gap in the fiber loop would potentially improve the detection sensitivity.

Combing polymer coatings on a LPG with micro extraction of an organic vapor into the functionalized polymer matrix, Loock's group reported a structurally
simplified (without using an air gap and the structure of the two-loop) fiber loop

---

[1]In the text below FLRD means the fiber loop ringdown with a single loop only.

ringdown gas sensor and demonstrated detections of xylene and cyclohexane vapors with a detection limit of 300 ppm for xylene [100]. More recently, the same group used the same sensing mechanism to achieve detections of five other VOCs, m-xylene, cycloheane, trichloroethylene, and gasoline with reasonably low detection limits of hundreds ppm [46]. But this more recent work, like the earlier study of hydrogen diffusion [18], does not fall into the category of fiber *loop* ringdown, since the fiber cavity consists of a pair of high reflectivity FBGs separated by a section of optical fiber. Nevertheless, the similar gas sensing approach based on a coated LPG can certainly be employed in a loop cavity for detection of other VOCs. Tarsa et al. [74] reported their study on spectroscopic measurements using the FLRD scheme, in which the sensor head was made of a section of tapered fiber. The sensor head was immersed in 1-octyne and EF absorption of 1-octyne at 1532.5 nm was detected. The external cavity diode laser was scanned from 1525 to 1555 nm and the broad absorption peak was compared with the one from FTIR.

Compared to the limited fiber loop ringdown gas sensors, more FLRD sensors for detection of a small volume of liquid have been reported. Loock's group first reported on a FLRD liquid sensor by introducing a micro-gap into a section of fiber in the fiber loop for detection of a small volume of solutions [24, 26]. They demonstrated the measurement of 7×10^{-15} mole dye solution using both a CW laser and a pulsed laser [26]. The same group further advanced the FLRD technique by introducing a phase-shift measurement, which greatly improved the data acquisition rate to close to real-time (10–100 ms) [19, Chaps. 9 and 14]. This technique has been demonstrated to be suitable for low cost, real-time, and online detection of capillary electrophoresis absorption with a detection limit at micromole concentration levels. Using flow injections, the device can detect a series of solution samples at different concentrations. The demonstrated detection limit is 5.3×10^{-12} mole samples in a 530 pL (10^{-12} liter) volume. Later, a minimum fractional absorption of 6 cm^{-1} has been demonstrated by using the FLRD technique with a fast gain switch diode laser [2, 40, 100]. More recently, the same group has used the similar flow injection device in the FLRD liquid sensor and reported a detection limit of 6 nM based on absorption of tartrazine, myoglobin, a pharmaceutical ingredient, and 5 μm polystyrene microbeads at 405 nm in a 200 μm path-length [105]. Some results are shown in Fig. 12.18.

In the aforementioned FLRD measurements of liquids, the light was coupled into and out of the loop using either the bend coupling method or commercial fiber couplers. Very recently, Vallance's group has explored the new notch coupling method in FLRD [30]. They conducted a side-by-side comparison of the light coupling effects on the detection limit using both the bend coupling method and the notch coupling method in the same FLRD system (the same light source and the same detector). A detection limit of 0.11 cm^{-1} at 532 nm or 0.93 μM Rh6G or a volume of 19 nL was achieved by the notch coupling method, which was one order of magnitude better than the result from the bend coupling method employed in the same experimental system. Detection of small volumes of liquids using the mirror-based CRDS can be seen

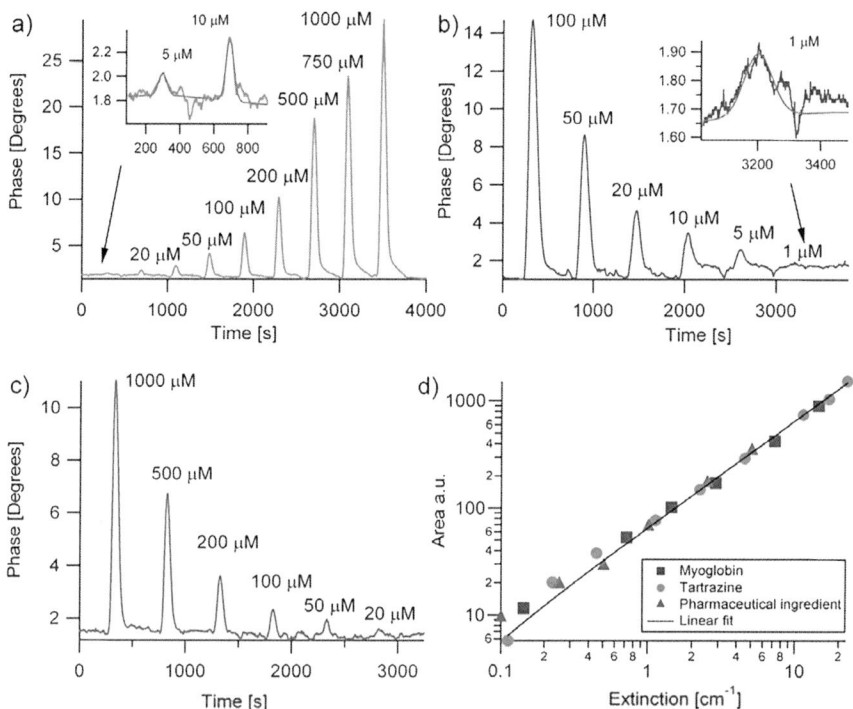

Fig. 12.18 (a) Phase-shift measurements of tartrazine samples in phosphate buffer (pH 7.2) were measured at concentrations from 5 to 1000 μM (buffer flow, 10 μL/min; sample volume, 5 μL). (b) Phase-shift measurements of myoglobin in phosphate buffer (pH 7.2) with concentrations from 100 to 1 μM. (c) Phase-shift measurements of heterocyclic pharmaceutical ingredient provided by Eli Lilly Canada Inc. in HCl buffer (pH 2.0) with concentrations from 1000 to 20 μM. (d) To compare the results from tartrazine, the pharmaceutical ingredient, and myoglobin, the integrated peak areas of the concentration time profiles are correlated to the calculated extinction. As expected, the data of all three substances fall on the same curve (Reproduced with permission from Ref. [105]. Copyright 2009 American Chemical Society)

elsewhere [Chap. 9]. Different from the pursuit of detection of an extremely small volume in the FLRD liquid sensors, sensing the presence or absence of water inside a concrete structure using the FLRD technique has been reported very recently by Kaya et al. [106]. The sensing principle is based on EF scattering (see details in Sect. 12.3.1). This work may extend the high sensitivity FLRD liquid sensing to another application domain, structure health monitoring (SHM).

Different from chemical sensing in gas concentration or liquid volume, trace elements in water solutions have also been explored very recently in our group [107]. The RI difference-based EF sensing mechanism is adopted in the FLRD technique. The high index detection sensitivity can discriminate Mg, Fe, P, Cd, and Co elements in water solutions at a concentration of 500 ppm. Figure 12.19 shows the results of the detections of the trace elements using the EF-FLRD technique.

Fig. 12.19 Detections of
trace elements (Mg and P) in
water solutions using an
EF-FLRD sensor
Concentrations of the
elements are 500 ppm by
weight

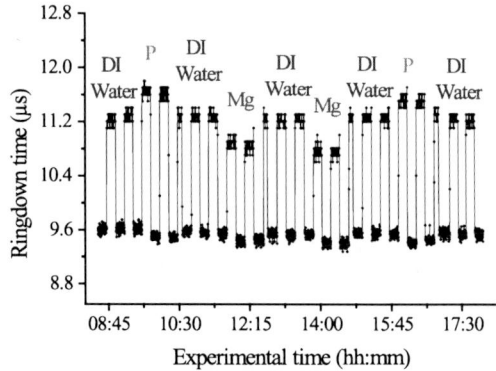

12.4.2 Physical Sensors

Due to the ringdown enhanced detection sensitivity, high speed of measurement,
and low cost for instrumentation, the FLRD sensing scheme was rapidly recognized for physical sensing. The first FLRD physical sensor was reported in 2004
by demonstrating pressure and force sensing [27, 28, 108]. A section of bare SMF
with a length of 1 cm was used as the sensor head. The sensing principle is based
on the fact that micro mechanical deformation of the fiber increases optical loss
in the fiber loop. The sensor showed repeatable response and good reversibility to
pressure changes, as shown in Fig. 12.20(a). Each step in Fig. 12.20(a) contained
many data points which were collected repeatedly at one pressure. The sensor's response to changes in pressure was less than one second. By converting the changes
in ringdown time to optical losses, the sensor's response to pressure change had
good linearity, as shown in Figs. 12.20(b) and (c). By using different configurations
of the sensor heads, later on, different sensing ranges and different sensitivities of
pressure were achieved. For example, Qiu et al. used a section of multimode fiber
for pressure sensing in the large dynamic range of 3.1–62 MPa [109]. Instead of
using the bare SMF, Jiang et al. used a section of partially-etched SMF as the sensor
head for pressure sensing in the range of 0 to 32.5 MPa [110]. The sensitivity of the
sensor was 38 times better than the pressure sensor using the bare SMF. Recently,
Jiang et al. has used a mechanically induced LPG as the sensing element spliced into
a fiber loop for pressure sensing. The pressure sensor demonstrates a detection limit
for pressure of 0.0068 MPa based on the one-σ standard [109]. More recently, positive gas pressure sensing using FLRD has also been reported by Tang et al [112].
A FBG was spliced into a fiber loop and the sensing element (the FBG) was housed
in a chamber and positive pressure generated a micro strain to the FBG, resulting
a shift of the FBG bandwidth curve. The sensor displayed a linear response to the
pressure ranging from 0.10 to 4.90 MPa with a minimum detectable pressure of
0.20 MPa [112].

Based on the micro-bending mechanism, a fiber loop ringdown strain sensor was
first reported by Tarsa et al. [113] in 2004. A 10 mm single mode fiber taper with a
waist diameter of 30 μm was fabricated in one section of the fiber loop and the taper

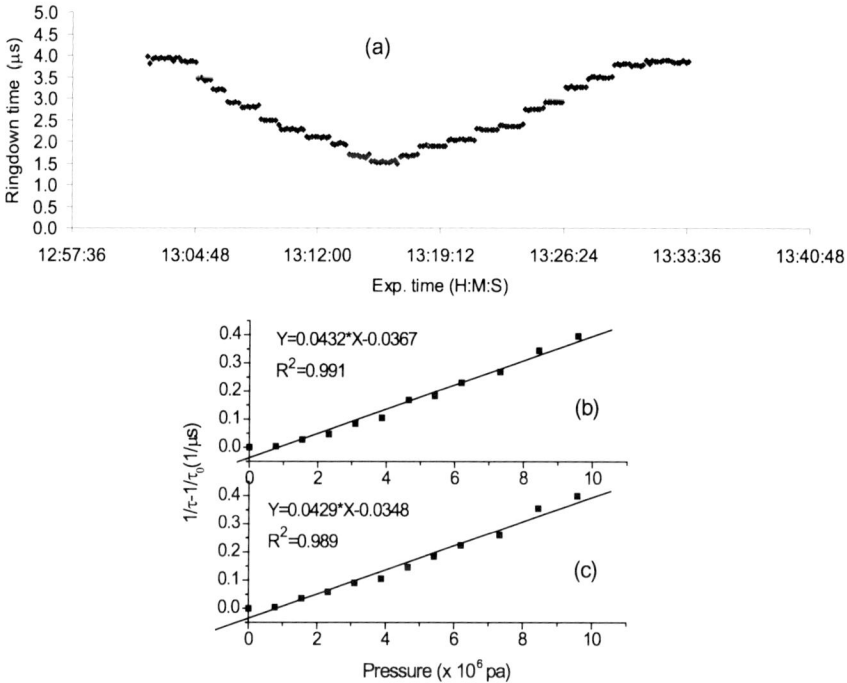

Fig. 12.20 Performance of the FLRD pressure sensor. (**a**) A pressure response curve of the sensor device. From the left to the right, the applied pressure increases from zero to the maximum and decreases from the maximum to zero. The *y-axis* indicates the change in the ringdown times and the *x-axis* represents the actual experimental time. The experimental time at each pressure point was not intentionally controlled to be the same in the testing. (**b**) and (**c**) The sensor device shows a linear response to pressure. (**b**) and (**c**) Were obtained by using the data presented in the left part and the right part of the curve in (**a**), respectively; the *dot* and the *line* denote the experimental data and the fitted curve, respectively (Reproduced with permission from Ref. [28]. Copyright 2004 Optical Society of America)

was suspended between two micro translation mounts, on which each end of the taper was fixed. The stain sensor could detect a minimum displacement of 75 nm over the 10 mm long taper, yielding a detection limit of 7.4 $\mu\varepsilon$, derived from the linear section in the sensor's response curve. The nonlinear behavior stems from the excitation of higher-order modes that have different coupling efficiency from the lowest-order modes. This nonlinear response feature poses a limitation of the linear range of the SMF taper strain sensors. Later, Ni et al. reported a FLRD strain sensor using a LPG as the sensor head spliced in a fiber loop [114]. When the LPG is strained, the grating bandwidth curve shifts due to the change of the period under influence of the axial strain; the relative location of the laser wavelength in the spectral bandwidth curve changes. The different optical transmission loss is reflected by a change in ringdown time. Advantageous over the taper strain sensor, this LPG strain sensor shows a linear response over a larger range of 0–300 $\mu\varepsilon$. A detection limit of 9 $\mu\varepsilon$ was derived from the sensor's baseline noise of 0.04 %. An even larger strain response range was reported using a fiber mode converter as the sensing element integrated in a multiple-mode fiber loop [115]. The strain sensors showed responses in the range of 0–1100 με, with a linear range of 0–800 με. Using a fast digital processing card with a time resolution of 0.1 ns, the sensors can have a strain resolution of 0.33 με. More recently, Gan et al. has used a linear chirped FBG as the sensor head to achieve distributed strain sensing with a spatial resolution of 2 mm [116]. Zhou et al. has used a section of PCF to construct a Mach-Zehnder interferometer (MZI) in the fiber loop as a strain sensing element. The MZI was fabricated by splicing the PCF between two sections of SMF and a detection limit of $\sim$3.6 με was obtained [117]. Still, in terms of pursuing low detection sensitivity limits, these FLRD-based strain sensors have a long way to go, since other types of fiber optic strain sensors can readily reach a low detection limit of 1 με, or even $\sim$100 fε ($1 \text{ fε} = 10^{-15} \text{ ε}$) [118, 119].

Moving toward another end of strain sensing, an extremely large dynamic range yet a relatively lower strain sensitivity, very recently, our group has explored strain-based FLRD crack sensor embedded in concrete structures for SHM (see Sect. 12.5.2) [120].

The first temperature sensing based on the FLRD scheme was reported in 2005 by using a single mode FBG as a sensing element spliced into a fiber loop [41, 81, 121]. Since the Bragg wavelength of a FBG is temperature dependent, changes in the temperature in the sensor head (FBG) are related to corresponding optical losses of the laser beam through the FBG. Different optical losses due to the shift of the FBG curve, resulting from the temperature change in the FBG, are detected by measuring the ringdown time. One of the advantages of the fiber FBG-FLRD temperature sensors is high temperature accuracy, which is not limited by the bandwidth of a FBG and the spectral resolution of an OSA. In that work [81], a precision of 0.02 °C was demonstrated in the temperature range of 92–114 °C. Another advantage of the FBG-FLRD temperature sensor is the low cost as compared with the current FBG-OSA temperature sensors that use an expensive OSA or a FBG interrogator. Limitation of the FBG-based FLRD temperature sensors are the small temperature response range that is limited by the narrow bandwidth curve of a FBG, i.e. 0.1–1 nm. The thermal sensitivity of a standard bare FBG is approximately 0.01 nm/°C [122]. With different fiber materials and/or using different fabrication approaches to construct a sensor head, i.e. using a metal embedded FBG or a Teflon substrate, the typical sensitivity can be up to 0.039 nm/°C [123]. A FBG of a 0.5 nm bandwidth can have a response range of 50 °C. To overcome this drawback, several FBGs with bandwidth curves adjacent to each other (partially overlapped) can be connected in serial powered by a broad band laser source to achieve a larger range of temperature sensing [81]. Figure 12.21(a) shows the sensor's response in the range of 40–110 °C by using three FBGs and each FBG covers a small temperature range, i.e. 10–15 °C.

Our group also used a broadband LPG as the sensor head and demonstrated the LPG-FLRD temperature sensor for a larger temperature measuring range, from −169 to 950 °C [41, 81]. Figure 12.21(b) shows the responses of the LPG-FLRD temperature sensor in the temperature range of 21–450 °C. Figures 12.21(b-1) and (b-2) show good reproducibility of the sensor's performance. Using high temperature composed fiber materials to fabricate a sensor head, such as a FBG or a LPG,

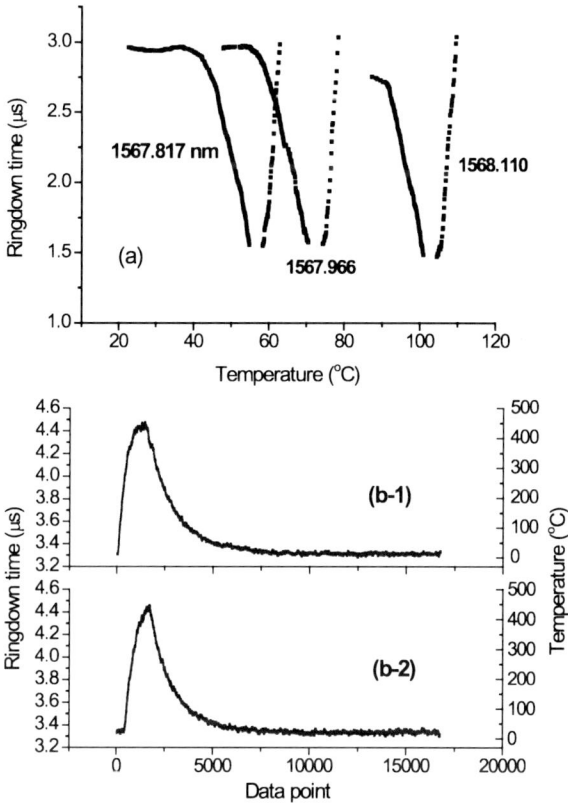

Fig. 12.21 FLRD temperature sensors based on FBG and LPG. (**a**) Demonstration of the adjustable temperature operation regions of the Type I sensor, FGB-FLRD sensor, (**b**) demonstration of the Type II sensor, LPG-FLRD sensor, with a large dynamic range. Typical temperature response curves of the Type II sensor, in which a LPG is used as the sensing element and the temperature is measured by the ringdown time. The left-hand portion of the curves shows the sensor's response to the temperature increase from 21 to 450 °C through the heating process; the right-hand portion shows the sensor's response to the temperature decrease from 450 to 21 °C through the natural cooling process. (**b-1**) and (**b-2**) demonstrate good reproducibility of the sensor (Reproduced with permission from Ref. [81]. Copyright 2006 The Institute of Physics)

high temperature, fast-response, low-cost, FLRD temperature sensors can be developed for a variety of applications in high temperature environments.

Refractive index is another important physical parameter. The first RI sensing was reported by Ni et al. using a fiber loop with a EDFA and a LPG as the sensing element [42]. The RI sensor was demonstrated in the index range of 1.35–1.43 at room temperature. This type of FLRD index sensor is based on a shift of the Bragg wavelength resulting in different optical losses at the preset wavelength. Recently, Zhou et al. reported a FLRD index sensor using a section of optical fiber with micro-holes fabricated by chemically assisted femtosecond laser drilling. When gels were infused into the micro-holes, RI of the section of the modified fiber changed and

Fig. 12.22 Image of
mammalian cancer cells that
are adsorbed on the surface of
the fiber taper and detected by
FLRD (Reproduced with
permission from Ref. [43].
Copyright 2004 American
Institute of Physics)

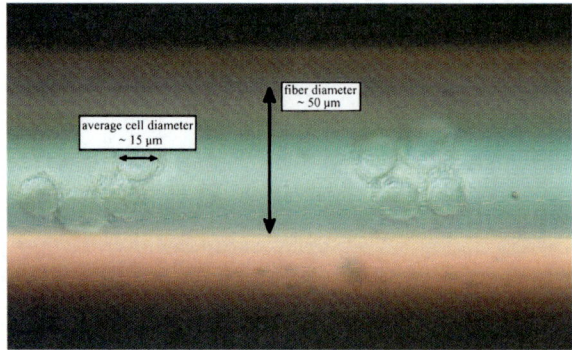

different ringdown times were recorded. A detection sensitivity limit in terms of
change of RI of 1.4×10^{-4} was achieved [124]. Very recently, Wong et al. reported
a FLRD index sensor using a section of PCF to form a Mach–Zehnder interferometer
as the sensor head and achieved a detection limit of 7.8×10^{-5} RIU [80].

Another way to realize RI sensing using the FLRD scheme is to use the EF
sensing mechanism. A conventional EF-based fiber index sensor uses a section of
partially-etched SMF as the sensing element, when the etched SMF is exposed to a
medium that has a RI index different from the index of the original fiber cladding
an additional optical loss occurs due to the EF scattering at the interface. This type
of index sensor is based on the single-pass detection scheme and has relatively low
detection sensitivity [91]. The same sensing mechanism has been adopted into the
FLRD scheme, and a high sensitivity FLRD RI sensor has been demonstrated very
recently in our group. The sensor head is just a short section of partially-etched
SMF and a detection limit for an optical index change of 3.2×10^{-5} RIU has been
demonstrated in the initial study using certified refractive index oils and lab-made
sodium chloride solutions [77]. The potential detection limit can be as low as 10^{-6},
close to a theoretically derived detection limit of 10^{-7} [125].

One interesting work has been reported more recently for current sensing using
the FLRD scheme. Ironically, we often acclaim that one of the advantages of FOS
is its immunity to EM signals or interferences. This work, however, uses a SMF to
construct a fiber solenoid forming a close loop that is coupled into an additional
fiber loop (the ringdown loop) to sense currents in the range of 0–500 A [126].

12.4.3 Biomedical and Biological Sensors

Biomedical and biological sensing using FLRD is a late addition to the FLRD sensing. To date, mammalian cancer cells [43], human serum albumin (HSA) [40], bacteria, DNA, and glucose [76] have been detected using the FLRD scheme combined
with different sensing approaches in the sensor head. Figure 12.22 shows the image of the mammalian cancer cells attached to the surface of a poly-D-lysine (PDL)
coated SMF taper. The fiber taper was 10 mm long with a waist diameter of 25 μm.

The two ends of the taper were spliced with SMF to form a fiber loop. The coating of PDL was to form an attaching substrate on the bare fiber surface and the cells could then be effectively bonded to the PDL chemically. The binding layer formed by the target agents (cells) changed the RI of the interface between the taper surface and the coating matrix, resulting in different EF scattering losses. In addition to the modification of RI (surface index modification), the large size of the cancer cells, (~ 10 μm), allowed a single cell to be detected. The sensor had the standard error of 0.044 μs resulting from average over 200 ringdown events and the change in ring-down time due to the EF scattering of a single cell was 0.23 μs, corresponding to a S/N ratio of 5. More cells are attached to the coating surface; more EF scattering loss is induced. Assumed the cells are uniformly attached to the surface to form a single layer, change in ringdown time versus number of cells shows a linear behavior.

Detections of biomolecules using FLRD scheme were also demonstrated by Loock and his coworkers using the capillary electrophoresis absorption approach [40]. They combined the FLRD scheme with the phase-shift detection approach to achieve on-line detection of HSA in μM quantity. The HSA labeled with NIR dye (ADS805WS) by mixing the dye with protein solutions was continuously flowing through a 40 cm long capillary with a flow injection speed of 10 μL/min. The flow interacted with the ringdown laser beam at the capillary-fiber interface (sensor head) that had an absorption path-length of 25 μm. The sensor was interrogated with flow concentration of 10 μM–2.0 mM with a detection limit of 10 μM, corresponding to absorption coefficient of 1.6 cm^{-1}. The relationship of the single peak area time $\times \tan(\Delta\phi)$ with HSA concentration shows excellent linearity, yielding a theoretical detection limit of 300 nM.

Bacteria in water solutions have also been detected using the FLRD technique [76]. A section of etched SMF was used as the sensor head. EF scattering loss depends on RI of the medium, to which the etched sensor head is exposed. The etched fiber (sensor head) length was 240 mm with an etched fiber diameter of 10 μm. As compared to the 10 mm fiber taper, this long sensor head greatly enhanced the detection sensitivity of the EF scattering. In this work, bacteria (*Escherichia coli* strain DH5α) were suspended in DI-water. The size of the bacteria was ~ 2 μm long and 0.5 μm in diameter. The concentration of the bacteria was 1.4×10^9 cell/mL. Adding chemical coatings on the etched fiber surface, label-free DNA sensing has also been demonstrated in the same sensor system [76].

Very recently, sensing glucose in different situations, such as glucose in water, glucose in synthetic urine, glucose in human urine, and human diabetic urines has been explored in the same group [37]. The sensor head consists of a section of partially-etched SMF with glucose oxidase (GOD) immobilized on the surface of the etched fiber. When GOD is exposed to a glucose solution, GOD reacts with glucose and generates gluconic acid, resulting in a change in the refractive index around the surface of the sensor head. The sensors showed linear responses to glucose concentrations in the lower concentration range (0.1–1.0 %) but non-linear responses for glucose concentrations in the higher range (1.0–10 %). The demonstrated detection sensitivities of the sensors in glucose solutions and artificial urines were 75 mg/dl and 50 mg/dl respectively. Potential detection sensitivity of such sensors

Fig. 12.23 Response
behavior of a EF-FLRD
glucose sensor, which had
GOD coatings at the sensor
head. The sensor was tested
with the synthetic urine
samples in different glucose
concentrations ranging from
0.1 % to 1 % (Reproduced
with permission from
Ref. [37]. Copyright 2012
SPIE)

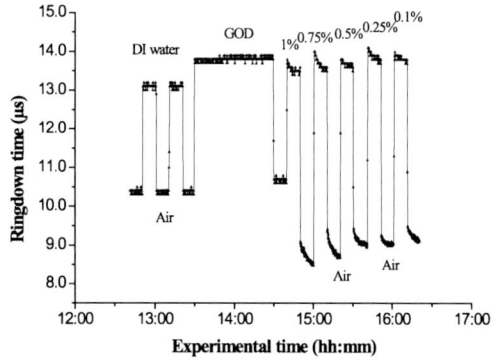

can reach up to 0.01 % or 10 mg/dl. Figure 12.23 shows the sensor's response to glucose solutions in different concentrations.

FLRD sensors are still in their infancy. With its universally applicable sensing scheme and attractive application features, many new FLRD sensors are expected to come [1, 38, 39, 127, 128]. In gas phase chemical sensing, using a section of hollow core photonic crystal fiber as a long-path gas cell would resume the original attempt to achieve high sensitivity gas sensing and analysis. Current challenge to this end is the high splicing loss between SMF and PCF. Another possible concern is the long gas sample flushing time. For biomedical and biological sensing, capillary electrophoresis absorption- and EF-index difference-based sensing would have high potential to be realized. Probably, in physical sensing, more attention will be paid to large-scale, multi-function sensing.

12.5 Applications of FLRD Sensors

12.5.1 FLRD Instrumentation

Like CRDS, FLRD will also experience the entire process cycle, from lab-based research to instrument prototype to full commercialization. It took almost seven years to convince the analytical instrumentation community that CRDS could be commercialized, which was ultimately benchmarked by the first CRDS trace water analyzer [129]. Nowadays, a suite of CRDS trace gas analyzers are being used in diverse applications, including the semiconductor industry, environmental monitoring, specialty gases characterization, breath analysis, etc. Trace gas molecules, elements, isotopes, all can be measured by a standalone CRDS spectrometer in different scenarios, i.e. indoor, outdoor, land, ocean, and atmosphere. FLRD, however, has not been commercialized to date although FLRD instrumentation is presumably less challenging because instrument components of FLRD are more mechanically rigid and stable. Also the FLRD sensor heads themselves are commercially available or readily fabricated. The main FLRD instrumentation effort lies therefore in

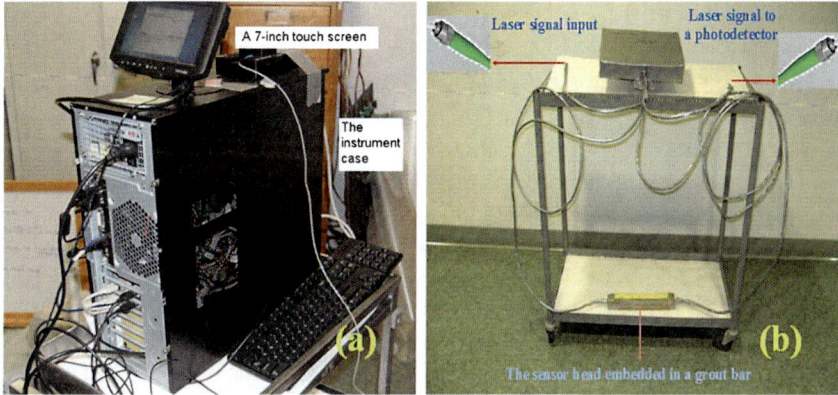

Fig. 12.24   A standalone FLRD sensor. (**a**) The FLRD interrogator; (**b**) a FLRD sensor unit. The two parts are connected by two standard FC/APC connectors

the light source control and data processing. Using a CW-fiber loop ringdown system as an example, we have taken several key steps toward a FLRD instrument. First, the current driver and the temperature controller of a diode laser will be replaced by an integrated electric board. The pulse generator used to generate pulsed driving currents for the laser diode will be replaced by a piece of electrical board. The oscilloscope and the data processing computer will be replaced by a CPU with a small display screen. All of the electronic parts, including an A/D card, CPU, a temperature control board, a current driver board, a hard drive, a power supply, a display screen, and a wireless card, will be housed in a standalone instrument case, i.e. a CPU tower. Figure 12.24(a) shows the picture of the FLRD instrument (or FLRD interrogator) developed in our group recently. A standalone sensor unit is shown in Fig. 12.24(b). The FLRD interrogator can be powered by AC or a battery. Starting power is approximately 90 W. A car battery can run the instrument for up to 8 h continuously. Ringdown data is displayed by a 7-inch touch screen. The instrument can be operated locally via the touch-screen or remotely through a wireless network or an Ethernet connection. In addition to one electrical cable for the AC power supply if powered by AC, only two SMF lines link the instrument. Each fiber line has an AC/APC fiber coupler for a laser pulsed to be coupled to the loop or a ringdown signal coupled to the detector. Furthermore, the fiber loop (Fig. 12.24(b)) can be non-directional if two identical fiber couplers are used for the light coupling into and out of the loop. Nowadays pulsed fiber lasers are readily available, the laser operation is simply modulated by an external trigger with a desired repetition rate of up to 5 kHz. This will make the instrument shown in Fig. 12.24(a) likely ten times smaller and lighter.

Due to the uniform sensing signal, ringdown time, the same FLRD interrogator can drive multiple FLRD sensor units (loops) individually or simultaneously. When the interrogator operates multiple FLRD sensor units, two $1 \times n$ fiber splitters and a fast MEMS switch may be used, depending on the network configurations (see Sect. 12.6).

12.5.2 FLRD in Remote Sensing

The features of high flexibility, light weight, small footprint, and low light transmission loss of optical fiber makes FLRD attractive in remote sensing. Here the meaning of "remote" is two-fold. A FLRD with a long fiber loop, i.e. 200 m, can allow the interrogator and the sensor head to be separated in two "remote" locations. The second layer of the "remote" is to utilize a wireless network to achieve remote operation of a sensor deployed remotely. Following the footprints of successful applications of FOS in large-scale structure heath monitoring (SHM), FLRD sensors has just begun to demonstrate their potential in this field, yet with more advantageous features in potential sensor networking. In our recent exploratory study, FLRD has been moved out of the research labs for the first time to a real-world application—embedment of FLRD sensors in concrete structures for real-time, long-term monitoring of water, cracks, and temperature in a 10 ft × 10 ft × 8 ft test bed.

Detection of water is critically important for mitigation of water penetration and early maintenance of concrete structures. The challenge faced by current FOS used for water monitoring in concrete structures is the irreversibility [106]. Secondly, current FOS, based on different sensing schemes [38], have a limited capability of being multiplexed and integrated into a network for large-scale SHM. Therefore, there has been no fiber optic water sensor network available for water monitoring in concrete to date. Most FOS for water monitoring uses a coated FBG as a sensing element. When water is present at the polymer coated FBG sensor head, the polymer absorbs water and the swollen polymer induces a stress that results in a FBG's wavelength shift [106]. This type of water sensor when embedded in concrete structures is not reversible; only a mono trend of dry–wet transition can be monitored. The recently developed EF-FLRD water sensor, however, is reproducibly reversible and can be used to monitor the successive increasing and decreasing levels of water in concrete [106]. Figure 12.24(b) shows a typical FLRD water sensor unit. The EF sensor head was a 22 cm long etched SMF that was embedded in a concrete bar to form a rigid water sensor head module with approximate dimensions of 30 cm × 5 cm × 5 cm. When 10 ml water was poured onto the surface of the concrete bar, the sensor had a S/N ratio of 86, corresponding to a detection sensitivity limit of 0.12 ml. The EF-FLRD water sensors embedded in concrete demonstrated to be highly reversible when tested in water-dry-water-dry duty cycles for as long as 43 hours. Figure 12.25 shows some testing results of the three EF-FLRD water sensors embedded in concrete (Units A and D) and grout (Unit H). Presence or absence of water at the interface between the etched fiber surface and the surrounding medium (concrete) changes the RI; and different EF scattering losses yield different ringdown times [75, 77]. This sensing principle allows for the embedded sensor head to have a reproducibly reversible response to the change of water, as shown in Fig. 12.25(d). One limitation of this type of water sensor is that the sensor has a steady signal (a constant ringdown time) when the sensor head is completely saturated with water.

The FLRD crack sensors are based on the micro-bending sensing principle. A section of bare SMF in a fiber loop was directly embedded in a concrete bar

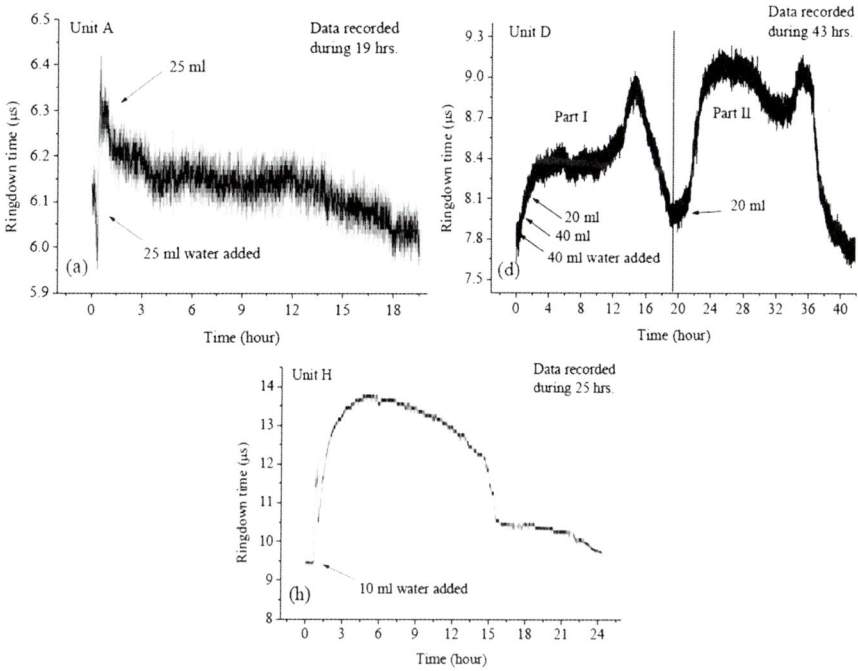

Fig. 12.25 Reproducibly reversible FLRD water sensors embedded in concrete and grout. (**a**) Reversibility of the water sensor (Unit A) embedded in concrete. (**d**) Reproducible reversibility of the water sensor (Unit D) embedded in concrete. (**h**) High sensitivity FLRD water sensor (Unit H) embedded in grout

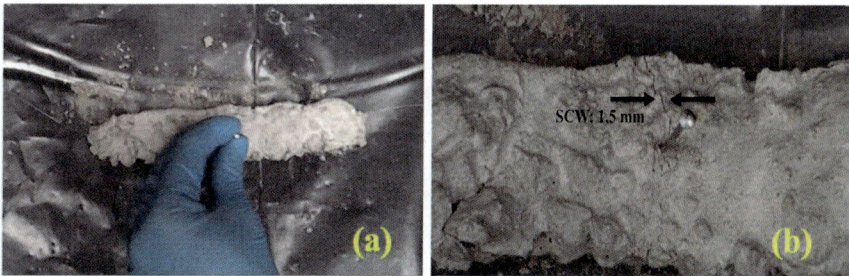

Fig. 12.26 A FLRD crack sensor is embedded in a concrete bar for testing. (**a**) Manual generation of a surface crack by hammering a nail into the bar. (**b**) A crack is generated with a surface crack width (SCW) of 1.5 mm

for initial testing. As shown in Fig. 12.26(a), cracks were manually generated on the surface of the bar. When the crack propagates into the location where the sensor head is embedded, strain resulting from the deformation of the concrete induces an additional optical loss and a change in ringdown time is observed [120]. Increase in the surface crack width (SCW) results in decrease in ringdown time, as shown

Fig. 12.27 Response of the FLRD crack sensor embedded in a concrete bar to the surface crack widths

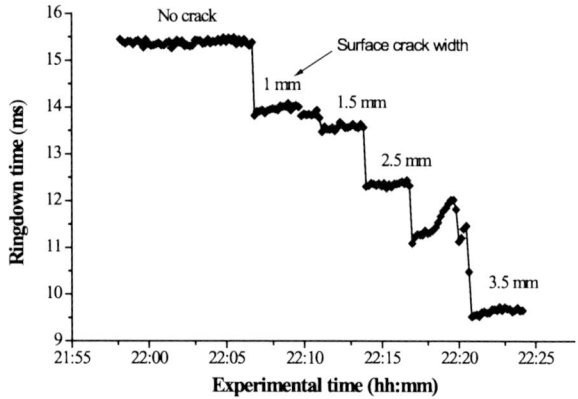

in Fig. 12.27. Significant drops in ringdown time were recorded for each cracking event. Cracks produced in steps with SCW of 1, 1.5, 2.5, and 3.5 mm, resulted in ringdown times of 13.8, 13.5, 12.3, and 9.5 μs, respectively. Averaged detection sensitivity for crack in terms of SCW was 0.5 mm. Using the one-σ standard deviation, $\Delta\tau_{min}$ of 0.0511 μs and the ringdown baseline, 15.50 μs, the baseline stability was determined to be 0.33 %, yielding a minimum measurable SCW, $\Delta d_{min} = 31$ μm. On the average over 100 ringdown event, the sensor's response time was 1.5 s, corresponding to a response time of 15 ms based on a single ringdown event. Advantages of the crack sensor include temperature-independence, fast response, and high sensitivity (in term of needed sensitivity in SHM). One limitation of this crack sensor is that an exact crack width in the sensor head location cannot be determined. The relationship between the SCW and the actual crack width in the sensor head location needs to be calibrated under specifically controlled testing conditions. Nevertheless, this FLRD crack sensor has demonstrated potential for long-term monitoring of cracking events in SHM.

Two standalone FLRD-FBG temperature sensors were also packaged to be used for the later deployment in concrete structures. Different from the FLRD water sensors and crack sensors, which used a diode laser operating at 1550 nm, the FLRD-FBG temperature sensors used a laser diode at the central wavelength of 1672.5 nm, which was located in the left wing area (Zone III in Fig. 12.8) of the FBG curve that had a bandwidth of 0.5 nm and reflectivity of 35 % at the central wavelength. In order to be embedded in concrete, the FBG was sealed in a copper tube with an inner diameter of 4 mm. In this way, the FBG was not stressed when the sensor head was immersed in wet grout later on and copper had excellent thermal conductivity. A small volume of air inside the copper tube can quickly reach thermal equilibrium, so that the temperature monitored by the FBG can indicate the surrounding temperature of the sensor head.

After rigorous testing of each sensor module, i.e. as shown in Fig. 12.24(b), in the lab-simulated environment, six FLRD sensor units were installed in a 10 ft × 10 ft installation panel (Panel #5) that was entombed in the test cube by grout later on. Figure 12.28(a) shows the locations of the six FLRD sensors for water, crack, and

Fig. 12.28 A multi-function
FLRD system installed for
testing. (**a**) Six FLRD sensors
were installed on Panel #5
that would be entombed in
the 10 ft × 10 ft × 8 ft test
cube. *Blue*: FLRD water
sensors; *Red*: FLRD crack
sensors; *Orange*: FLRD
temperature sensors. (**b**) The
standalone FLRD interrogator
is linked to the six FLRD
sensors installed on the panel
by two 120-m long SMF lines
(*yellow*)

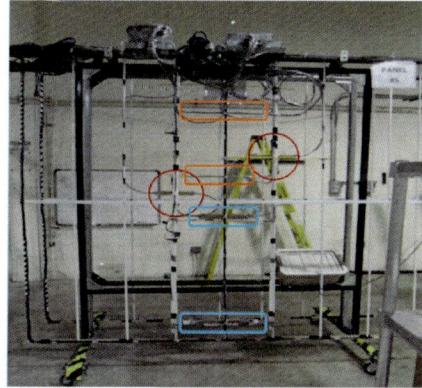

(a)

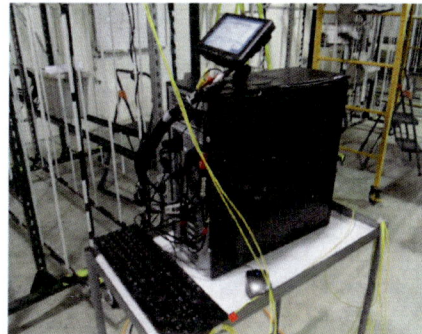

(b)

Fig. 12.28 A multi-function FLRD system installed for testing. (**a**) Six FLRD sensors were installed on Panel #5 that would be entombed in the 10 ft × 10 ft × 8 ft test cube. *Blue*: FLRD water sensors; *Red*: FLRD crack sensors; *Orange*: FLRD temperature sensors. (**b**) The standalone FLRD interrogator is linked to the six FLRD sensors installed on the panel by two 120-m long SMF lines (*yellow*)

temperature sensing. Each sensor unit has two fiber cables coming out of the sensor, one for the laser in and the other for ringdown signal out. The 12 fiber cables were multiplexed through two 1 × 6 fiber splitters for laser in and ringdown signal out. The fiber cables and the fiber splitters were all packaged and mounted on the top of the panel since the rest of the panel would be completely embedded in the test cube by the grout. Out of the installation panel, there were only two 120-m long SMF lines linked to the FLRD interrogator, as shown in Fig. 12.28(b). Figure 12.29(a) shows a top view the installed panels including Panel #5 inside the test cube and the grout was poured into the cube and moving from the bottom slowly to the top, entombing the panels. Figure 12.29(b) shows the part of the test cube (the other half was in the subsurface). The FLRD interrogator is located in the control office that is 30 feet away from the test cube. Along with other 244 sensors using different sensing techniques, the once considered delicate FLRD sensors survived the 3-h long harsh grouting process. Note that the other 244 sensors were embedded in the concrete for different sensing purposes as well as for a side-by-side comparison when the same parameters were monitored by different types of sensors.

Since the grouting began, the FLRD sensors were operated. The FLRD water sensor, which was installed in the bottom of the panel (the water sensor Bottom), was first turned on and change of water in the sensor head was recorded in real-time.

Fig. 12.29 Test of the FLRD
water, temperature, and crack
sensors in the test bed.
(a) Top view of the test cube.
Six FLRD sensors were
installed on the panel (#5)
that was being entombed by
the grout moving from the
bottom of the test cube to the
top. (b) Image of the test cube
(the other half is 4 ft below
the surface) that houses eight
installation panels, on which
250 sensors are installed. The
six FLRD sensors embedded
in the cube are connected to
the FLRD interrogator housed
in the control office by two
120-m SMF lines

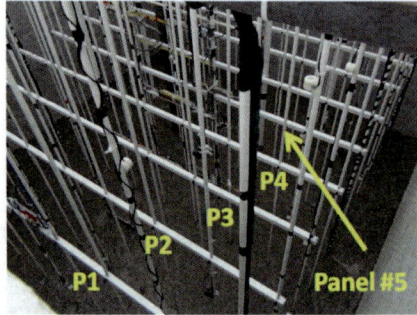

(a)

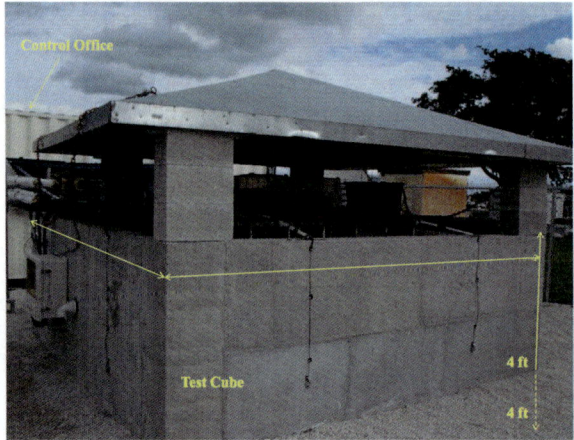

(b)

When the grout was moving up, the second water sensor (the water sensor Top) was immersed into the wet grout; and the presence of water in the second sensor head location was monitored. Figure 12.30 shows the changes of water in the location of the water sensor Bottom recorded by the FLRD water sensor. The gap was due to the stop when the FLRD interrogator was switched to run the water sensor Top. The water temporal profile during the 3-h grouting period was recorded in real-time. The sensor survived the harsh environment and the results were as expected and agreeable with the data from other conventional water sensors installed. Since then the FLRD sensors have been continuously running through a remote access at MSU (Starkville, MS) while the test cube is located in Miami, FL.

Figure 12.31 shows the testing data from one of the two FBG-FLRD temperature sensors (the temperature sensor Bottom and the temperature sensor Top, as show in Fig. 12.28(a)). The data shows the real-time change of the grout temperature at the location of the temperature sensor Bottom during the first 14 days. The results are in a good agreement with the temperature data from the conventional thermocouple sensors. The data gaps in the figure were due to the pausing of temperature sensor readout for other operational activities. As can be seen in the figure, a temperature

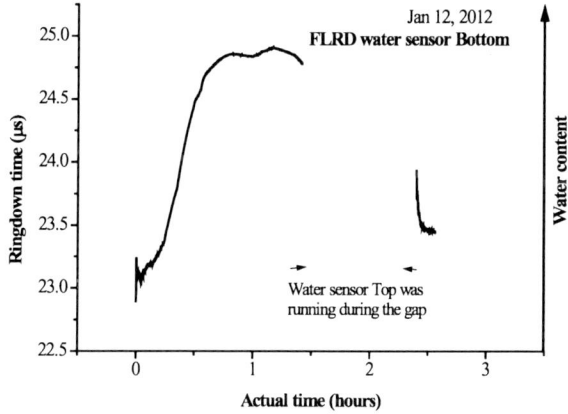

Fig. 12.30 Response of the FLRD water sensor Bottom embedded in the test cube

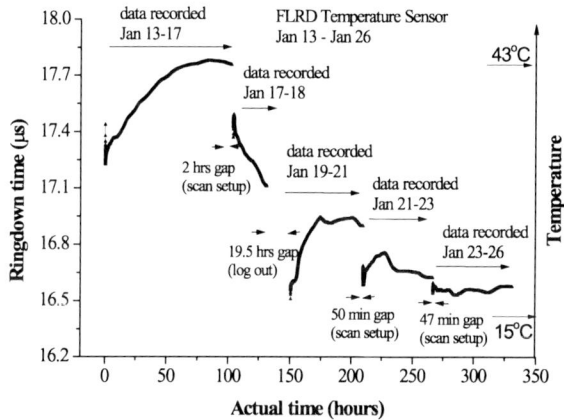

Fig. 12.31 Long-term monitoring of temperature by the FBG-FLRD temperature sensor embedded in the test cube

surge during the setting of the grout was recorded and after a long time, the grout began to cool down and reached the environmental temperature at one foot above the surface inside the cube. It is different from the high resolution temperature sensing conducted under the lab-controlled condition (Fig. 12.21(a)), this embedded FBG-FLRD temperature sensor only showed a temperature contour over the monitoring period. The temperature features were determined by the local climate as well as the thermodynamics of the grout structures. The designed measuring range of the temperature sensors was from 10 to 45 °C, covering the seasonal temperature variations in the site as well as the highest temperature in the grout setting.

Figure 12.32 shows one piece of data collected by the crack sensor Bottom (see Fig. 12.28(a)) from the 21st day after the grouting. The significant drop of the ringdown time was due to the physical strain exerted on the crack sensor. The signal amplitude suggests that a severe cracking event happened around the sensor head; however, the location of the cracking event inside the cube could not be pinpointed. It could also be due to the extended effect of a crack that happened in a distant location from the crack sensor head.

Fig. 12.32 A cracking event
was detected by the FLRD
crack sensor embedded in the
test cube

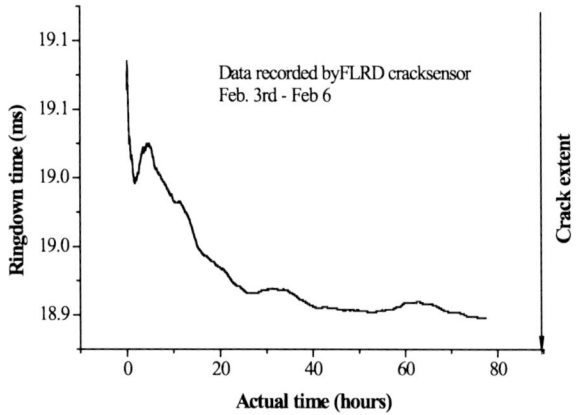

All of the six FLRD sensors embedded in the grout in the test bed are still operational. The objective of this project was to test the feasibility of FLRD sensors
in the real-world (functionality, deployment capability, and ruggedness) but not focused on scientific data collection. The results show that (1) FLRD sensors with
different functionalities can be operated by a single FLRD interrogator, (2) FLRD
sensors are rugged and can survive harsh environments, (3) FLRD sensors have
good durability, fast response, high data collection efficiency, and are suitable for
long-term remote monitoring, and (4) FLRD sensors can be embedded in concrete
structures for SHM. This work is just a starting point of the application of the FLRD
technique in SHM.

One limitation of the FLRD sensor system deployed is that switching from one
sensor to another has to be done manually in the site. This initial design limits the
system operation to one sensor at a time.

12.6 Future FLRD Sensor Network

As shown in Sect. 12.4, various types of FLRD-based sensors have been reported
during the last ten years. Each individual sensor has demonstrated high sensitivity
and high speed of detection gained from the nature of the FLRD technique [23–26].
This new technique allows a large-scale fiber optic sensor network to be developed
even for multi-function sensing. A complicated FLRD sensor network may include
two basic and the simplest configurations, parallel and serial configurations for connections of individual FLRD sensor units, from which a sensor network platform
can be constructed. The following sections discuss briefly some perfectives on future FLRD sensor networks—design, operation, and functionality.

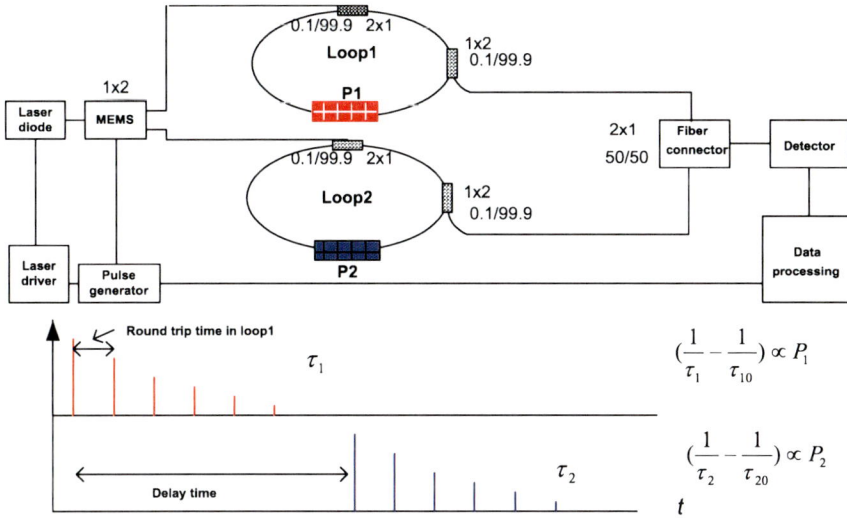

Fig. 12.33 The parallel configuration of the FLRD sensing platform. Only two sensor units are used to illustrate the concept (Reproduced with permission from Ref. [38]. Copyright 2009 Multi-disciplinary Digital Publishing Institute)

12.6.1 Parallel Configuration

Figure 12.33 illustrates a speculative design of a parallel configuration of the simplest FLRD sensor network [38]. In this simplest parallel configuration that consists of two loops only, signals from the two sensors are decoupled. The sensor system measures P1 and P2 parameters alternatively. The multiplexing and control can be achieved by using the MEMS optical switching technique. Figure 12.34 shows the layout of the 1×2 MEMS switch that controls two EF-FLRD sensors connected in the parallel configuration as illustrated in Fig. 12.33. The laser beam was guided by a section SMF with a FC/APC connector that was connected to the 1-channel end of the switch. Two SMF lines from the 2-channel end were connected to the loops (two sensor units) via two FC/APC connectors. The switch was connected to the laptop via a USB cable. Switching from one loop to another was operated readily by clicking the selected channel bottom (1 or 2) on the computer screen. Responses from the individual sensors are shown in Fig. 12.35.

In Fig. 12.34, the two sensor units have different physical parameters, thus they have two different ringdown baselines and two different sensitivities. For instance, $\tau_{01} = 15.0$ μs and $\tau_{02} = 10.4$ μs. The sensor heads were fabricated based on the EF scattering principle and initially tested when the sensor heads were exposed to water and air alternatively. Due to the different RI in air and water, the sensors read different ringdown times. Operation of these two sensors was electronically controlled by a 1×2 channel MEMS switch that was interfaced to a laptop.

Current MEMS can readily have 1×64 channels or more with switching frequencies up to MHz. Therefore, a multiple-channel sensor network using the par-

Fig. 12.34 A 1 × 2 channel
MEMS switch to
electronically operate the two
FLRD sensor units connected
in the parallel configuration.
The "2" displayed on the
switch board indicates that
Loop 2 is operating now

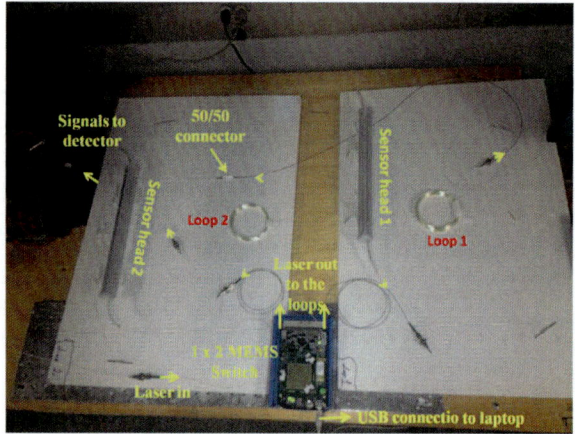

Fig. 12.35 Results of two
EF-FLRD sensors connected
in the parallel configuration
and electrically controlled by
a 1 × 2 MEMS switch

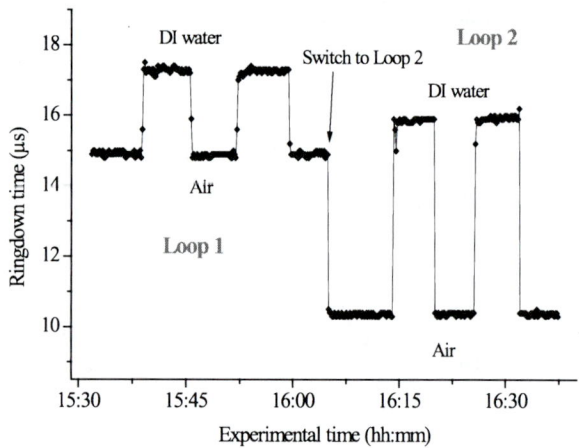

allel configuration can be constructed. The maximum number of channels is only
limited by the combination of switching frequency and the duration of each ring-
down measurement event. For instance, if a ringdown event is on the order of μs,
then the highest switching frequency can only be ∼100 kHz. This assumes use of a
fast A/D converter and data transmission processing.

12.6.2 Serial Configuration

Intuitively, FLRD sensor units can also be connected in a serial configuration [38].
Figure 12.36 illustrates a design perspective of a FLRD sensor platform in the serial
configuration. In this simplest configuration, the laser power is split into the two

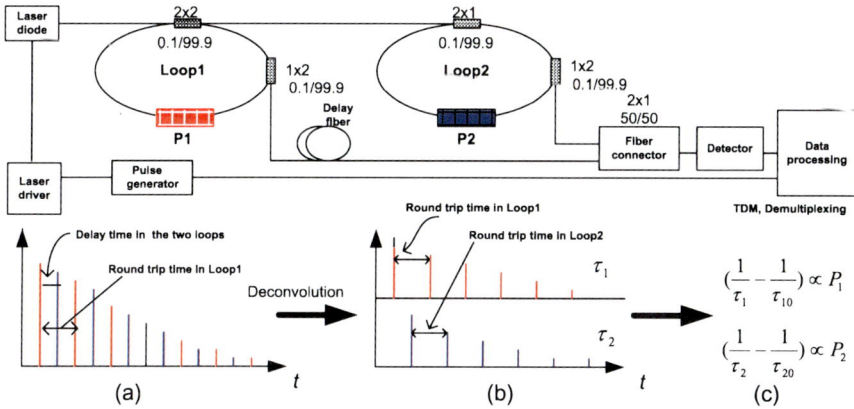

Fig. 12.36 The serial configuration of the FLRD sensing platform. Only two sensor units are used to illustrate the concept. (**a**) Signals from the two sensor units are coupled; *Red*: from Loop 1; *Blue*: from Loop 2. (**b**) Decoupled signals for determination of P_1 and P_2 (Reproduced with permission from Ref. [38]. Copyright 2009 Multidisciplinary Digital Publishing Institute)

sensor units consecutively. Ringdown signals from the two loops are coupled and the detector observes a coupled signal. In order to obtain ringdown times (consequently, the sensing parameters) from individual fiber loops, the signal needs to be decoupled. The time division multiplexing (TDM) technique used in digital signal processing can be adopted directly to demultiplex the coupled ringdown signal into two individual ringdown decays, each of which yields a separate ringdown time. [130–133]. In this configuration, the time sequence specifies each sensor unit. Different from the parallel configuration, the maximum number of sensor units in this serial configuration is determined by the laser power budget. Theoretically, a 30 mW DBF diode laser can power more than 4000 sensor units connected in the serial configuration [38].

12.6.3 Large-Scale FLRD Sensing Platform

The combination of the two basic network configurations can create a large-scale FLRD sensing platform, as illustrated in Fig. 12.37 [38]. Physical and chemical FLRD sensors all can be integrated in the same sensing platform. Due to the uniform sensing signal, *time*, all of the sensing signals from each individual sensor unit can be fused and transmitted uniformly through a single fiber linked to the detector. Laser diodes operating at different wavelengths are multiplexed by using WDM [134–136], and the MEMS selectively controls the laser beam with the needed wavelengths to be injected into the fiber loops for detection of different quantities. Location of each sensor will be determined by the time sequence. In principle, such

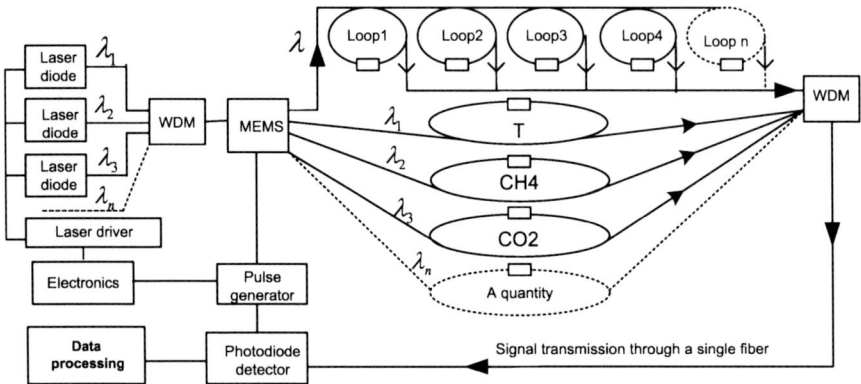

Fig. 12.37 A multi-function FLRD sensor network formed by adding FLRD sensor units to the sensing platform that consists of a light source, a detector, and FLRD sensor units. The light source may be from multiplexing of several laser diodes (Reproduced with permission from Ref. [38]. Copyright 2009 Multidisciplinary Digital Publishing Institute)

a sensor platform can form a sensor network that has multiple sensor units for detecting multiple parameters at different locations. The most challenging part in the development of such a sensor network lies in the data processing.

12.7 Conclusion

Indeed, FLRD is a neat, simple, and universal sensing scheme, suitable for chemical, physical, and biomedical or biological sensors and sensing applications. Although FLRD is only ten years old, the technology has been ready for instrumentation and commercialization. If we say that the rapid advancement of the field of FOS took a full ride on the telecommunications industry boom, then the growth of FLRD sensors and sensing would have been advantageously fortressed by the well-established field of FOS. More simply, many innovative sensing mechanisms currently used in FOS can be directly adopted by the uniform FLRD sensing scheme to create new FLRD sensors with multi-pass enhanced detectivity. Over the last several years, various individual FLRD sensors have emerged, but new FLRD sensors will continue to grow. With regard to the future of the FLRD technique, two points that need to be stressed here are FLRD instrumentation and FLRD sensor network.

FLRD instrumentation: Due to its simplicity and low cost, especially, the low cost of the terminal detection device (i.e. using a photodiode as a detector), FLRD has high commercialization potential. A side-by-side comparison of a FBG-FLRD temperature sensor with a FBG-OSA temperature sensor, for example, would justify this argument. One uses an inexpensive photodiode as the detector; the other utilizes an expensive OSA or a designated FBG interrogator as the spectral analyzer. Furthermore, as described previously, multi-function FLRD sensors can be powered by a single FLRD interrogator, due to the uniform sensing signal *time*; this feature would profitably yield an array of FLRD sensors (a product line) with low instrument cost.

FLRD sensor network: Future research effort in FLRD may lean toward FLRD sensor networks, since the networking potential has not been much explored yet. The mature technologies, tools, and components in the telecommunications industry, such as WDM, TDM, MEMS switch, laser diode array, pulsed fiber lasers, etc. can be directly adopted for development and assembly of a FLRD sensor network for large-scale, multi-function sensing in diverse applications. The smaller demand in power budget and the insensitivity to laser power fluctuations in FLRD provide unique advantages that allow one to avoid the essential issues in the current networking architecture. A simple estimate shows that a 30 mW DFB laser diode can power more than 4000 FLRD sensor units configured in series without needing an amplifier in the network. One likely promising application of such a network lies in structure health monitoring or sensing and control in a sophisticated operation complex.

FLRD originates from the cavity ringdown spectroscopy. With its versatility in spectroscopic measurements and fiber-guided sensing, the research related to FLRD has become truly interdisciplinary, involving not only spectroscopy but also fiber optics, electro-mechanics, and others. Due to the limited knowledge of the author in the field, this chapter only touches one part of FLRD—sensors and sensing with an emphasis on its engineering efforts. Other aspects of FLRD can be seen in previous reviews as well as in Chap 9. Hopefully, readers would catch a glimpse of FLRD through reading this chapter.

Acknowledgements The work and highlights covered in this chapter bridge a period of 10 years of the emerging technique of FLRD. Many results from Mississippi State University included in this chapter came from the collaborative effort of former and current graduate students working in my research group. Their names and substantial scientific contributions are partially reflected in the references below. In particular, I want to thank Susan Scherrer, Armstrong Mbi, Chamini Herath, Malik (Burak) Kaya, Peeyush Sahay, and Haifa Alali.

Our FLRD research is currently supported by the National Science Foundation (#CMMI-0927539) and the US Department of Energy (#AC84132N through Savannah River Nuclear Solutions LLC).

References

1. G. Berden, R. Engeln (eds.), *Cavity Ring-Down Spectroscopy: Techniques and Applications* (Wiley-Blackwell, West Sussex, 2009)
2. H.-P. Loock, Ring-down absorption spectroscopy for analytical microdevices. TrAC, Trends Anal. Chem. **25**, 655–664 (2006)
3. H.-P. Loock, J.A. Barnes, G. Gagliardi, R. Li, R.D. Oleschuk, H. Wächter, Absorption detection using optical waveguide cavities. Can. J. Chem. **88**, 401–410 (2010)
4. C. Vallance, Innovations in cavity ringdown spectroscopy. New J. Chem. **97**, 867–874 (2005)
5. A. O'Keefe, D.A.G. Deacon, Cavity ring-down optical spectrometer for absorption measurements using pulsed laser sources. Rev. Sci. Instrum. **59**, 2544–2551 (1988)
6. K.W. Busch, M.A. Busch (eds.), *Cavity-Ringdown Spectroscopy: An Ultratrace-Absorption Measurement Technique.* ACS Symposium Series, vol. 720 (American Chemical Society, Washington, 1999)

7. G. Berden, R. Peeters, G. Meijer, Cavity ring-down spectroscopy: experimental schemes and applications. Int. Rev. Phys. Chem. **19**, 565–607 (2000)
8. B.A. Paldus, A.A. Kachanov, An historical overview of cavity-enhanced methods. Can. J. Phys. **83**, 975–999 (2005)
9. M.I. Mazurenka, A.J. Orr-Ewing, R. Peverall, G.A.D. Ritchie, Cavity ring-down and cavity enhanced spectroscopy using diode lasers. Annu. Rep. Prog. Chem., Sect. C, Phys. Chem. **101**, 100–142 (2005)
10. C. Wang, G.P. Miller, C.B. Winstead, Cavity ringdown laser absorption spectroscopy, in *Encyclopedia of Analytical Chemistry: Instrumentation and Applications* (Wiley, Chichester, 2008)
11. K.K. Lehmann, Ring-down cavity spectroscopy cell using continuous wave excitation for trace species detection. U.S. Patent No. 5,528,040, 1996
12. D. Romanini, A.A. Kachanov, N. Sadeghi, F. Stoeckel, CW cavity ringdown spectroscopy. Chem. Phys. Lett. **264**, 316–322 (1997)
13. B.A. Paldus, J.S. Harris Jr., J. Martin, J. Xie, R.N. Zare, Laser diode cavity ring-down spectroscopy using acousto-optic modulator stabilization. J. Appl. Phys. **82**, 3199–3204 (1997)
14. A.C.R. Pipino, J.W. Hudgens, R.E. Huie, Evanescent wave cavity ring-down spectroscopy with a total-internal-reflection minicavity. Rev. Sci. Instrum. **68**, 2978–2989 (1997)
15. K.K. Lehmann, P. Rabinowitz, High-finesse optical resonator for cavity ring-down spectroscopy based upon Brewster's angle prism retrorefrectors. U.S. Patent No, 5,973,864, 1999
16. T. Von Lerber, M.W. Sigrist, Time constant extraction from noisy cavity ring-down signals. Chem. Phys. Lett. **353**, 131–137 (2002)
17. T. Von Lerber, M.W. Sigrist, Cavity ring-down principle for fiber-optic resonators: experimental realization of bending loss and evanescent-field sensing. Appl. Opt. **41**, 3567–3575 (2002)
18. D.E. Vogler, M.G. Muller, M.W. Sigrist, Fiber-optical cavity sensing of hydrogen diffusion. Appl. Opt. **42**, 5413–5417 (2004)
19. Z. Tong, A. Wright, T. McCormick, R. Li, R.D. Oleschuk, H.-P. Loock, Phase-shift fiber-loop ring-down spectroscopy. Anal. Chem. **76**, 6594–6599 (2004)
20. M. Gupta, H. Jiao, A. O'Keefe, Cavity-enhanced spectroscopy in optical fibers. Opt. Lett. **27**, 1878–1880 (2002)
21. C. Wang, Plasma-cavity ringdown spectroscopy (P-CRDS) for elemental and isotopic measurements. J. Anal. At. Spectrom. **22**, 1347–1363 (2007)
22. M. Andachi, T. Nakayama, M. Kawasaki, S. Kurokawa, H.-P. Loock, Fiber-optic ring-down spectroscopy using a tunable picosecond gain-switched diode laser. Appl. Phys. B **88**, 131–135 (2007)
23. G. Stewart, K. Atherton, H. Yu, B. Culshaw, An investigation of an optical fibre amplifier loop for intra-cavity and ring-down cavity loss measurements. Meas. Sci. Technol. **12**, 843–849 (2001)
24. R.S. Brown, I. Kozin, Z. Tong, R.D. Oleschuk, H.-P. Loock, Fiber-loop ring-down spectroscopy. J. Chem. Phys. **117**, 10444–10447 (2002)
25. P.B. Tarsa, K.K. Lehmann, P. Rabinowitz, A passive optical fiber resonator for cavity ring-down spectroscopy, in *Abstracts of Papers, 224th ACS National Meeting*, Boston, MA, United States, August 18–22 (2002)
26. Z. Tong, M. Jakubinek, A. Wright, A. Gillies, H.-P. Loock, Fiber-loop ring-down spectroscopy: a sensitive absorption technique for small liquid samples. Rev. Sci. Instrum. **74**, 4818–4826 (2003)
27. C. Wang, S.T. Scherrer, Fiber ringdown pressure sensors. Opt. Lett. **29**, 352–354 (2004)
28. C. Wang, S.T. Scherrer, Fiber loop ringdown for physical sensor development: pressure sensor. Appl. Opt. **43**, 6458–6464 (2004)
29. C.M. Rushworth, D. James, C.J.V. Jones, C. Vallance, Fabrication of an optical fiber reflective notch coupler. Opt. Lett. **36**, 2952–2954 (2011)
30. C.M. Rushworth, D. James, J.W.L. Lee, C. Vallance, Top notch design for fiber-loop cavity ring-down spectroscopy. Anal. Chem. **83**, 8492–8500 (2011)

31. R. Augustine, C. Krusen, C. Wang, W. Yan, System and method for controlling a light source for cavity ring-down spectroscopy. U.S. Patent No. 7,277,177 B2, 2007

32. C. Wang, S.P. Koirala, S.T. Scherrer, Y. Duan, C.B. Winstead, Diode laser microwave induced plasma cavity ringdown spectrometer: performance and perspective. Rev. Sci. Instrum. **75**, 1305–1313 (2004)

33. M. Kakui, Fiber lasers: pulsed fiber lasers reach 50 kW peak power at <100 ps pulse duration (2012). Available online: http://www.laserfocusworld.com/articles/2011/05/pulsed-fiber-lasers-reach-50-kw-peak-power-at-100-ps-pulse-duration.html

34. P. Zalicki, R.N. Zare, Cavity ring-down spectroscopy for quantitative absorption measurements. J. Chem. Phys. **102**, 2708–2717 (1995)

35. J.T. Hodges, J.P. Looney, R.D. van Zee, Laser bandwidth effects in quantitative cavity ring-down spectroscopy. Appl. Opt. **35**, 4112–4116 (1996)

36. K.K. Lehmann, H. Huang (eds.), *Optimal Signal Processing in Cavity Ring-Down Spectroscopy in Frontiers of Molecular Spectroscopy* (Elsevier, Oxford, 2008), pp. 623–658

37. C. Wang, M. Kaya, C. Wang, Evanescent field-fiber loop ringdown glucose sensor. J. Biomed. Opt. **17**, 037004 (2012)

38. C. Wang, Fiber loop ringdown—a time-domain sensing technique for multi-function fiber optic sensor platforms: current status and design perspectives. Sensors **9**, 7595–7621 (2009)

39. H. Waechter, J. Litman, A.H. Cheung, J.A. Barnes, H.-P. Loock, Chemical sensing using fiber cavity ring-down spectroscopy. Sensors **10**, 1716–1742 (2010)

40. R. Li, H.-P. Loock, R.D. Oleschuk, Capillary electrophoresis absorption detection using fiber-loop ring-down spectroscopy. Anal. Chem. **78**, 5685–5692 (2006)

41. C. Wang, Fiber ringdown temperature sensors. Opt. Eng. **44**, 030503 (2005)

42. N. Ni, C.C. Chan, L. Xia, P. Shum, Fiber cavity ring-down refractive index sensor. IEEE Photonics Technol. Lett. **20**, 1351–1353 (2008)

43. P.B. Tarsa, A.D. Wist, P. Rabinowitz, K.K. Lehmann, Single-cell detection by cavity ring-down spectroscopy. Appl. Phys. Lett. **85**, 4523–4525 (2004)

44. H.-P. Loock, P. Wentzell, Detection limits of chemical sensors: applications and misapplications. Sens. Actuators B **173**, 157–163 (2012)

45. C. Wang, Unpublished data (2004)

46. J.A. Barnes, R.S. Brown, A.H. Cheung, M.A. Dreher, G. Mackey, H.-P. Loock, Chemical sensing using a polymer coated long-period fiber grating interrogated by ring-down spectroscopy. Sens. Actuators B **148**, 221–226 (2010)

47. H. Li, D. Li, G. Song, Recent applications of fiber optic sensors to health monitoring in civil engineering. Eng. Struct. **26**, 1647–1657 (2004)

48. K.T.V. Grattan, T. Sun, Fiber optic sensor technology: an overview. Sens. Actuators A **82**, 40–61 (2000)

49. O.S. Wolfbeis, Fiber-optic chemical sensors and biosensors. Anal. Chem. **76**, 3269–3283 (2004)

50. C. McDonagh, C.S. Burke, B.D. MacCraith, Optical chemical sensors. Chem. Rev. **108**, 400–422 (2008)

51. C.S. Chu, Y.L. Lo, High-performance fiber-optic oxygen sensors based on fluorinated xerogels doped with Pt(II) complexes. Sens. Actuators B **124**, 376–382 (2007)

52. T.S. Yeh, C.S. Chu, Y.L. Lo, Highly sensitive optical fiber oxygen sensor using Pt(II) complex embedded in sol-gel matrices. Sens. Actuators B **119**, 701–707 (2006)

53. J.R. Epstein, D.R. Walt, Fluorescence-based fibre optic arrays: a universal platform for sensing. Chem. Soc. Rev. **32**, 203–214 (2003)

54. C.D. Geddes, J.R. Lakowicz (eds.), *Glucose Sensing in Topics in Fluorescence Spectroscopy* (Springer, New York, 2006), pp. 351–375

55. Z. Zhang, K.T.V. Grattan, A.W. Palmer, Fiber-optic high temperature sensor based on the fluorescence lifetime of alexandrite. Rev. Sci. Instrum. **63**, 3869–3873 (1992)

56. J. Mulrooney, J. Clifford, C. Fitzpatrick, E. Lewis, Detection of carbon dioxide emissions from a diesel engine using a mid-infrared optical fibre based sensor. Sens. Actuators A **136**, 104–110 (2007)

57. B. Alfeeli, G. Pickrell, A. Wang, Sub-nanoliter spectroscopic gas sensor. Sensors **6**, 1308–1320 (2006)
58. S. Tao, S. Gong, J.C. Fanguy, X. Hu, The application of a light guiding flexible tubular waveguide in evanescent wave absorption optical sensing. Sens. Actuators B **120**, 724–731 (2007)
59. W. Peng, G.R. Pickrell, F. Shen, A. Wang, Experimental investigation of optical waveguide-based multigas sensing. IEEE Photonics Technol. Lett. **16**, 2317–2319 (2004)
60. W. Yuan, H.P. Ho, C.L. Wong, S.K. Kong, C. Lin, Surface plasmon resonance biosensor incorporated in a Michelson interferometer with enhanced sensitivity. IEEE Sens. J. **7**, 70–73 (2007)
61. Ch. Stamm, R. Dangel, W. Lukosz, Biosensing with the integrated-optical difference interferometer: dual-wavelength operation. Opt. Commun. **153**, 347–359 (1998)
62. F. Shen, W. Peng, K.L. Cooper, G. Pickrell, A. Wang, UV-induced intrinsic Fabry-Perot interferometric fiber sensors. Proc. SPIE **5590**, 47–56 (2004)
63. K.A. Chang, H.J. Lim, C.B. Su, A fibre optic Fresnel ratio meter for measurements of solute concentration and refractive index change in fluid. Meas. Sci. Technol. **13**, 1962–1965 (2002)
64. R. Kashyap (ed.), *Fiber Bragg Gratings* (Academic Press, San Diego, 1999)
65. R.O. Claus, K.A. Murphy, A. Wang, R.G. May (eds.), *High-Temperature Optical Fiber Sensors in Optical Fiber Smart Materials and Structures* (Wiley, New York, 1995), pp. 537–562
66. Y. Zhu, K.L. Cooper, G.R. Pickrell, A. Wang, High-temperature fiber-tip pressure sensor. J. Lightwave Technol. **24**, 861–869 (2006)
67. W. Peng, G.R. Pickrell, A. Wang, High temperature fiber optic cubic-zirconia pressure sensor. Opt. Eng. **44**, 124402 (2005)
68. Z. Huang, W. Peng, J. Xu, G.R. Pickrell, A. Wang, Fiber temperature sensor for high-pressure environment. Opt. Eng. **44**, 104401 (2005)
69. Z. Huang, X. Chen, Y. Zhu, A. Wang, Wavefront splitting intrinsic Fabry-Perot fiber optic sensor. Opt. Eng. Lett. **44**, 070501 (2005)
70. F. Shen, A. Wang, Frequency estimation-based signal processing algorithm for white-light optical fiber Fabry-Perot interferometers. Appl. Opt. **44**, 5206–5214 (2005)
71. Y. Zhao, C. Yu, Y. Liao, Differential FBG sensor for temperature-compensated high-pressure (or displacement) measurement. Opt. Laser Technol. **36**, 39–42 (2004)
72. S. Pal, T. Sun, K.T.V. Grattan, S.A. Wade, S.F. Collins, G.W. Baxter, B. Dussardier, G. Monnom, Stain-independent temperature measurement using a type-I and type-IIA optical fiber Bragg grating combination. Rev. Sci. Instrum. **75**, 1327–1331 (2004)
73. L. Van der Sneppen, F. Ariese, C. Gooijer, W. Ubachs, Liquid-phase and evanescent-wave ring-down spectroscopy in analytical chemistry. Annu. Rev. Anal. Chem. **2**, 13–35 (2009)
74. P.B. Tarsa, P. Rabinowitz, K.K. Lehmann, Evanescent field absorption in a passive optical fiber resonator using continuous-wave cavity ring-down spectroscopy. Chem. Phys. Lett. **383**, 297–303 (2004)
75. C. Wang, C. Herath, Fabrication and characterization of fiber loop ringdown evanescent field sensors. Meas. Sci. Technol. **21**, 085205 (2010)
76. C. Herath, C. Wang, M. Kaya, D. Chevalier, Fiber loop ringdown DNA and bacteria sensors. J. Biomed. Opt. **16**, 050501 (2011)
77. C. Wang, C. Herath, High-sensitivity fiber-loop ringdown evanescent-field index sensors using single-mode fiber. Opt. Lett. **35**, 1629–1631 (2010)
78. Z. Tian, S.S.-H. Yam, H.-P. Loock, Single-mode fiber refractive index sensor based on core-offset attenuators. IEEE Photonics Technol. Lett. **20**, 1387–1389 (2008)
79. Z. Tian, S.S.-H. Yam, J. Barnes, W. Bock, P. Greit, J.M. Fraser, H.-P. Loock, R.D. Oleschuk, Refractive index sensing with Mache-Zehnder interferometer based on concatenating two single-mode fiber tapers. IEEE Photonics Technol. Lett. **20**, 626–628 (2008)
80. W. Wong, W. Zhou, C.C. Chan, X. Dong, K.C. Leong, Cavity ringdown refractive index sensor using photonic crystal fiber interferometer. Sens. Actuators B **161**, 108–113 (2011)

81. C. Wang, A. Mbi, An alternative method to develop fiber grating temperature sensors using the fiber loop ringdown scheme. Meas. Sci. Technol. **17**, 1741–1745 (2006)
82. A. Mbi, Novel fiber optic temperature sensors: fiber grating loop ringdown. M.S. thesis, Mississippi State University, May 2006
83. M.B. Reid, M. Özcan, Temperature dependence of fiber optical Bragg grating at low temperature. Opt. Eng. **37**, 237–240 (1998)
84. K.T.V. Grattan, B.T. Meggitt (eds.), *Optical Fiber Sensor Technology, vol. 2: Devices and Technology* (Springer, Berlin, 1998)
85. M. Jiang, W. Zhang, Q. Zhang, Y. Liu, B. Liu, Investigation on an evanescent wave fiber-optic absorption sensor based on fiber loop cavity ring-down spectroscopy. Opt. Commun. **283**, 249–253 (2010)
86. A.W. Synder, J.D. Love (eds.), *Optical Waveguide Theory* (Kluwer Academic, Norwell, 2000)
87. N.J. Harrick (ed.), *Internal Reflection Spectroscopy* (Wiley-Interscience, New York, 1967)
88. F. De Fornel (ed.), *Evanescent Wave from Newtonian Optics to Atomic Optics* (Springer, Berlin, 2001)
89. H. Matsuoka, Evanescent wave light scattering: a fusion of the evanescent wave and light scattering techniques to the study of colloids and polymers near the interface. Macromol. Rapid Commun. **22**, 51–67 (2001)
90. L. Xu, J.C. Fanguy, K. Soni, S. Tao, Optical fiber humidity sensor based on evanescent-wave scattering. Opt. Lett. **29**, 1191–1193 (2004)
91. P. Polynkin, A. Polynkin, N. Peyghambarian, M. Mansuripur, Evanescent field-based optical fiber sensing device for measuring the refractive index of liquids in microfluidic channels. Opt. Lett. **30**, 1273–1275 (2005)
92. M. Schnippering, S.R.T. Neil, S.R. Mackenzie, P.R. Unwin, Evanescent wave cavity-based spectroscopic techniques as probes of interfacial processes. Chem. Soc. Rev. **40**, 207–220 (2011)
93. C. Wang, N. Srivastava, B.A. Jones, R.B. Reese, A novel multiple species ringdown spectrometer for in situ measurements of methane, carbon dioxide, and carbon isotope. Appl. Phys. B **92**, 259–270 (2008)
94. HITRAN 96 database. www.hitran.com
95. G.A. Valaskovic, M. Hoton, G.H. Morrison, Parameter control, characterization, and optimization in the fabrication of optical fiber near-field probes. Appl. Opt. **34**, 1215–1228 (1995)
96. L. Tong, R.R. Gattass, J.B. Ashcom, S. He, J. Lou, M. Shen, I. Maxwell, E. Mazur, Subwavelength-diameter silica wires for low-loss optical wave guiding. Nature **426**, 816–819 (2003)
97. T.A. Birks, Y.W. Li, The shape of fiber tapers. J. Lightwave Technol. **10**, 432–438 (1992)
98. B.S. Kawasaki, R.G. Lamont, Biconical-taper single-mode fiber coupler. Opt. Lett. **6**, 327–329 (1981)
99. C. Herath, C. Wang, High precision fiber loop ringdown chemical corrosion sensors. Earth Space 1609–1614 (2010)
100. J. Barnes, M. Dreher, K. Plett, R.S. Brown, C.M. Crudden, H.-P. Loock, Chemical sensor based on a long-period fibre grating modified by a functionalized polydimethylsiloxane coating. Analyst **133**, 1541–1549 (2008)
101. G. Stewart, K. Atherton, B. Culshaw, Cavity-enhanced spectroscopy in fiber cavities. Opt. Lett. **29**, 442–444 (2004)
102. N. Ni, C.C. Chan, Improving the measurement accuracy of CRD fibre amplified loop gas sensing system by using a digital LMS adaptive filter. Meas. Sci. Technol. **17**, 2349–2354 (2006)
103. N. Ni, C.C. Chan, T.K. Chuah, L. Xia, P. Shum, Enhancing the measurement accuracy of a cavity-enhanced fiber chemical sensor by an adaptive filter. Meas. Sci. Technol. **19**, 115203 (2008)

104. C.C. Chan, N. Ni, J. Sun, Improving the detection accuracy in fiber Bragg grating-sensors by using a wavelet filter. J. Optoelectron. Adv. Mater. **9**, 2376–2379 (2007)
105. H. Waechter, K. Bescherer, C.J. Dürr, R.D. Oleschuk, H.-P. Loock, 405 nm absorption detection in nanoliter volumes. Anal. Chem. **81**, 9048–9054 (2009)
106. M. Kaya, P. Sahay, C. Wang, Reproducibly reversible fiber loop ringdown water sensor embedded in concrete and grout for water monitoring. Sens. Actuators B **176**, 803–810 (2012)
107. C. Wang, Site testing of six fiber loop ringdown sensors for ISD structures monitoring. Report submitted to the U.S. Department of Energy, March 2012
108. C. Wang, Fiber ringdown pressure/force sensors. U.S. Patent No. 7,241,986, 2007
109. H. Qiu, Y. Qiu, Z. Chen, B. Fu, X. Chen, G. Li, Multimode fiber ring-down pressure sensor. Microw. Opt. Technol. Lett. **49**, 1698–1700 (2007)
110. Y. Jiang, D. Yang, D. Tang, J. Zhao, Sensitivity enhancement of fiber loop cavity ring-down pressure sensor. Appl. Opt. **48**, 6082–6087 (2009)
111. Y. Jiang, J. Zhao, D. Yang, D. Tang, High-sensitivity pressure sensors based on mechanically induced long-period fiber gratings and fiber loop ring-down. Opt. Commun. **283**, 3945–3948 (2010)
112. D. Tang, D. Yang, Y. Jiang, J. Zhao, H. Wang, S. Jiang, Fiber loop ring-down optical fiber grating gas pressure sensor. Opt. Lasers Eng. **48**, 1262–1265 (2010)
113. P.B. Tarsa, D.M. Brzozowski, P. Rabinowitz, K.K. Lehmann, Cavity ringdown stain gauge. Opt. Lett. **29**, 1339–1341 (2004)
114. N. Ni, C.C. Chan, X.Y. Dong, J. Sun, P. Shum, Cavity ring-down long-period fibre grating strain sensor. Meas. Sci. Technol. **18**, 3135–3138 (2007)
115. H. Qiu, Y. Qiu, Z. Chen, B. Fu, G. Li, Strain measurement by fiber-loop ring-down spectroscopy and fiber mode converter. IEEE Sens. J. **8**, 1180–1183 (2008)
116. J. Gan, Y. Hao, Q. Ye, Z. Pan, H. Cai, R. Qu, Z. Fang, High spatial resolution distributed strain sensor based on linear chirped fiber Bragg grating and fiber loop ringdown spectroscopy. Opt. Lett. **36**, 879–881 (2011)
117. W. Zhou, W. Wong, C.C. Chan, L. Shao, X. Dong, Highly sensitive fiber loop ringdown strain sensor using photonic crystal fiber interferometer. Appl. Opt. **50**, 3087–3092 (2011)
118. S. Avino, J.A. Barnes, G. Gagliardi, X. Gu, D. Gutstein, J.R. Mester, C. Nicholaou, H.-P. Loock, Musical instrument pickup based on a laser locked to an optical fiber resonator. Opt. Express **25**, 25057–25065 (2011)
119. G. Gagliardi, M. Salza, S. Avino, P. Ferraro, P. De Natale, Probing the ultimate limit of fiber-optic strain sensing. Science **330**, 1081–1084 (2010)
120. P. Sahay, M. Kaya, C. Wang, Fiber loop ringdown sensors for potential real-time monitoring of cracks in concrete structures: an exploratory study. Sensors **13**, 39–57 (2013)
121. Wang, C. Fiber, Bragg grating loop ringdown method and apparatus. U.S. Patent No. 7,323,677, 2008
122. X.C. Li, F. Prinz, J. Seim, Thermal behavior of a metal embedded fiber Bragg grating sensor. Smart Mater. Struct. **10**, 575–579 (2001)
123. T. Mizunami, H. Tatehata, H. Kawashima, High-sensitivity cryogenic fiber-Bragg-grating temperature sensors using teflon substrates. Meas. Sci. Technol. **12**, 914–917 (2001)
124. K. Zhou, D.J. Webb, C. Mou, M. Farries, N. Hayes, Optical fiber cavity ring down measurement of refractive index with a microchannel drilled by femtosecond laser. IEEE Photonics Technol. Lett. **21**, 1653–1655 (2009)
125. D.K.C. Wu, B.T. Kuhlmey, B.J. Eggleton, Ultrasensitive photonic crystal fiber refractive index sensor. Opt. Lett. **34**, 322–324 (2009)
126. H. Zhang, Y. Qiu, H. Li, A. Huang, H. Chen, G. Li, High-current-sensitivity all-fiber current sensor based on fiber loop architecture. Opt. Express **20**, 18591–18599 (2012)
127. H. Omrani, A. Dudelzak, H.-P. Loock, Fiber-coupled fluorescence and absorption spectroscopy for oil and fuel characterization. Appl. Ind. Opt. JW2A.1 (2012)
128. C. Rushworth, C. Vallance, Fibre loop cavity ring-down spectroscopy for the sensitive and selective detection of minute sample volumes of liquid explosives. Proc. SPIE **7838**, 78380Y (2010)

129. TigerOptics LLC, CRDS trace water analyzer, MTO-1000-H_2O

130. G. Li, Y. Qiu, S. Chen, S. Liu, Z. Huang, Multichannel-fiber ringdown sensor based on time-division multiplexing. Opt. Lett. **33**, 3022–3024 (2008)

131. C. Wang, A. Mbi, Optical superposition in double fiber loop ringdown. Proc. SPIE **6377**, 637702 (2006)

132. Y. Gao, Y. Qiu, H. Chen, Y. Huang, G. Li, Four-channel fiber loop ring-down pressure sensor with temperature compensation based on neural networks. Microw. Opt. Technol. Lett. **52**, 1796–1799 (2010)

133. J. Shang, W. Zhang, S. Wei, H. Zhang, Two-channel fiber microcavity strain sensor based on fiber loop ring-down spectroscopy technology. Microw. Opt. Technol. Lett. **54**, 1305–1309 (2012)

134. S. Liu, Y. Yu, J. Zhang, S. Fei, A novel interrogation technique for time-division multiplexing fiber Bragg grating sensor arrays. Proc. SPIE **6781**, 67812M (2007)

135. D.J.F. Cooper, T. Coroy, P.W.E. Smith, Time-division multiplexing of large serial fiber-optic Bragg grating sensor arrays. Appl. Opt. **16**, 2643–2654 (2001)

136. G.A. Cranch, P.J. Nash, Large-scale multiplexing of interferometric fiber-optic sensors using TDM and DWDM. J. Lightwave Technol. **19**, 687–699 (2001)

Chapter 13
Fiber-Optic Resonators for Strain-Acoustic Sensing and Chemical Spectroscopy

Saverio Avino, Antonio Giorgini, Paolo De Natale, Hans-Peter Loock, and Gianluca Gagliardi

Abstract Over the past several years, fiber-optic resonators have been used as mechanical probes by virtue of their intrinsic sensitivity to length changes, and chemical sensors based on sensitivity to molecular absorption or refractive index changes. The capabilities of high-finesse fiber Bragg-grating cavities for quasi-static and dynamic strain sensing are discussed. Pound-Drever-Hall (PDH) frequency locking techniques are considered for low-noise, fast and wide dynamic range active interrogation. Such methods can ultimately be used in combination with highly-stabilized laser sources or optical frequency combs (OFCs) which would provide both an exotic coherent radiation source and an ultra-stable optical reference at the same time. The implementation of similar systems for applications to seismic monitoring and acoustic pickup for musical instruments are described. Also, we describe detection schemes for chemical sensing and evanescent-wave spectroscopy using fiber-ring resonators, for which we propose lasers as well as supercontinuum comb generators as light sources.

S. Avino (✉) · A. Giorgini · G. Gagliardi
Istituto Nazionale di Ottica, Consiglio Nazionale delle Ricerche, Comprensorio "A. Olivetti",
via Campi Flegrei 34, 80078 Pozzuoli (Napoli), Italy
e-mail: saverio.avino@ino.it

A. Giorgini
e-mail: antonio.giorgini@ino.it

G. Gagliardi
e-mail: gianluca.gagliardi@ino.it

P. De Natale
Istituto Nazionale di Ottica, Consiglio Nazionale delle Ricerche, Via N. Carrara 1,
50019 Sesto Fiorentino, Italy
e-mail: paolo.denatale@ino.it

H.-P. Loock
Dept. of Chemistry, Queen's University, Kingston, ON, K7L 3N6, Canada
e-mail: hploock@chem.queensu.ca

G. Gagliardi, H.-P. Loock (eds.), *Cavity-Enhanced Spectroscopy and Sensing*,
Springer Series in Optical Sciences 179, DOI 10.1007/978-3-642-40003-2_13,
© Springer-Verlag Berlin Heidelberg 2014

13.1 Introduction

Optical fiber systems have had a great impact in the field of optical sensing, thanks to the growth of the optoelectronics and fiber-optic communication industries [1]. The inherent advantages of fiber-optic sensors include their light weight, low cost, small size and ruggedness, making it possible to create sensor networks and fully integrate them with gas, solid, or liquid environments, even those with access difficulties. The immunity to electromagnetic interference, the small weight and fast response are crucial for high sensitivity, wide bandwidth and high precision sensing. The past 20 years have witnessed an intensive research effort on the use of optical fiber sensors to measure physical and chemical quantities [2, 3].

Among the optical sensors, fiber Bragg gratings (FBGs) have proved very promising as mechanical probes for a number of applications [4]. A significant contribution came from FBG-based resonant structures, such as fiber Fabry-Pèrot cavities and π-phase shifted FBGs, whose highly dispersive power near resonance can be exploited to measure sub-pm length perturbations over a wide range of Fourier frequencies. The best performances have been achieved thanks to sophisticated systems based on narrow-band laser sources and laser frequency stabilization methods, making sensitivity limits better than 10^{-10} feasible both for quasi-static and dynamic strain monitoring [5]. Similarly, acceleration and acoustic vibration sensing is possible thanks to the fast response of fiber-resonator strain sensors.

In the arena of chemical sensors, optical resonators based on fiber Bragg structures or ring cavities have been successfully employed for refractive index, gas and liquid sensing. Cavity-enhanced and ring-down techniques allow light-matter interaction to be probed in either direct or evanescent-wave spectroscopy schemes [6], as already demonstrated using lasers or broad-band light sources. This class of sensors has the advantage of being inexpensive and easy to use, and, in addition, their small size make them minimally invasive and ideal for analysis of small-volume liquid samples.

Very recently, optical frequency comb (OFC) synthesizers have become of age as new light sources for spectroscopy possessing both high coherence and broad emission at the same time [7, 8]. Direct absorption and cavity-enhanced spectroscopy with combs have been demonstrated with combs in different fashions for molecular fingerprinting of gas species [9–11]. Here we will give a review of fiber-cavity strain sensors and their applications, and also show how it is possible to extend such sensors to the realm of evanescent-wave spectroscopy in the liquid phase.

13.2 Strain Sensing

13.2.1 Fiber Bragg Grating Fabry-Pérot Resonators

A fiber Fabry-Pérot resonator can easily be fabricated using gratings directly written in single-mode (SM) optical fibers. A *Bragg* grating written in an optical fiber

(FBG) consists of a periodic modulation of the fiber core refractive index, created using UV interferometric inscription techniques, that generates an intrinsic narrow-band reflector whose reflectivity spectrum is centered around the Bragg wavelength $\lambda_B = 2n_{eff}P$, where P is the grating pitch and n_{eff} the effective refractive index of the propagating mode. This center wavelength can be tuned by altering the fiber length or refractive index (e.g. by changing the temperature), and the peak reflectivity can be >99.9 %. A high-finesse resonator is readily built by writing two high-reflectivity single-mode FBGs separated by a fixed distance L along the same fiber. The so-formed optical cavity behaves in the same way as a standard mirror cavity except that in a SM fiber only the axial modes are allowed to oscillate. The transmitted (as well as the reflected) field can be described with the same mathematical expressions as any Fabry-Pérot resonator [12], thus showing the typical periodic resonance spectrum. Finesse values exceeding 1000 are feasible. Upon stretching or compression of the fiber, optical pathlength variations ΔL of the intra-cavity fiber are turned into frequency shifts of the resonance frequency position Δv. In fact, it is easy to see [2] that

$$\Delta v \cong -0.78 \cdot v \cdot \frac{\Delta L}{L} \qquad (13.1)$$

where the 'correction factor' 0.78 is necessary to account for the elasto-optic change of refractive index due to mechanical strain of the fiber $\varepsilon = \frac{\Delta L}{L}$. The cavity thereby transforms a strain into a very large shift or modulation (depending whether it is static or dynamic) of its transmission and reflection peak position. In principle, also the cavity photon lifetime, or analogously the free-spectral-range, is affected by intra-cavity length although this is a much weaker effect. The strain-to-frequency response of the resonator in Eq. (13.1) appears as a sort of *leverage* process via the optical frequency v and it is amplified as a consequence of multiple-passes of light through the internal cavity fiber which is affected by length changes. The quality of the resonator plays the relevant role in this regard: the larger is the finesse, the longer is the effective pathlength travelled by the light in the fiber. On the other hand, since mechanical strain is detected from resonance shift, the cavity enhancement leads to higher spectral resolution because the cavity mode linewidth is proportional to the ratio between the mode spacing (free spectral range) and the finesse. Measuring the resonance frequency position is thus very convenient to fully exploit the FBG cavity as a mechanical sensor. Although the working principle of fiber resonators is quite straightforward, in order to achieve very low noise interrogation, sophisticated read-out methods need to be developed. Conventional interrogation approaches are based on broad-band, incoherent light sources that illuminate the sensor and provide a signal that is modulated with the strain on the fiber. Readout of amplitude changes from that signal provides information on the strain experienced by the fiber. Using narrow-band lasers instead of incoherent light, not only increases the power density upon detection above the photodetector's noise-equivalent power density (NEP), but also increases the optical resolution, thanks to the intrinsically narrower emission spectrum. In addition, when using tunable lasers, active tracking of the sensor is possible thereby improving bandwidth and dynamic range. The resulting frequency

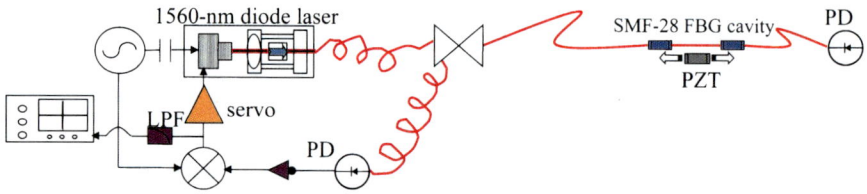

Fig. 13.1 Typical optical set-up for strain sensing with PDH-locking of a laser diode to a FBG resonator [15] (© IOP Publishing. Reproduced by permission of IOP Publishing. All rights reserved)

response can be exceedingly high, as we will show in the following. A first possible set-up is represented in Fig. 13.1. A diode laser is actively locked to the resonator by an opto-electronic feedback loop that is based on the Pound-Drever-Hall (PDH) scheme [13, 14, see also Chap. 3, §3]. Basically, the laser beam that is reflected by an optical cavity is the coherent sum of two contributions: a promptly reflected beam that bounces back from the input mirror and a leakage beam that comes from the cavity due to imperfect reflectivity of the mirror. When the laser is perfectly resonant, the promptly reflected and the cavity back-transmitted fields are equal in amplitude but out of phase (180°) and thus they interfere destructively. In PDH method, the laser is phase modulated at frequency f. Two new fields (sidebands) are thus symmetrically generated on the laser beam at a frequency distance f from the carrier. These sidebands are used as a phase reference with which one can measure the phase of the reflected beam. In general, they are totally reflected by the input mirror and may interfere with the cavity back-reflected field. The beats between the sidebands and the reflected carrier provide the phase difference between the reflected beam and the cavity resonance, i.e. an error signal that quantifies the deviation of the laser from the resonance. On perfect resonance, the sideband beats with the carrier are identical and yield an overall zero signal. Any shift from this condition leads to a signal whose amplitude depends on the relative position between the laser and the cavity mode while its sign indicates the direction in which the laser is moving. In Fig. 13.1, the PDH error signal is generated first modulating the diode-laser phase via a high-frequency bias-tee current input and then collecting the reflected power through a fiber circulator[1] and heterodyning it with the modulation signal in a double-balanced mixer.[2] The latter is sent to a proportional-integrative amplifier (servo). With the loop gain at its highest value (before self-oscillation), the laser remains perfectly stable on the cavity mode. In this way, the laser is kept always on the resonance peak of the sensor while the feedback signal contains the desired strain information [15]. This action is effective from DC to the loop unity-gain frequency. With this in mind, any mechanical or thermal stress applied to the cavity

---

[1] A fiber circulator is a two-ways device that collects the reflected beam and separates it from the incident beam.

[2] A double balance mixer is a nonlinear electronic circuit that produces new signals at the sum $f_1 + f_2$ and difference $f_1 - f_2$ of the original frequencies f_1 and f_2 applied to a mixer.

Fig. 13.2 Servo locking signal for static elongation of the fiber of an FBG resonator (length 10 cm, finesse 300). The *upper curve* corresponds to the sensor readout and refers to the *left vertical axis*. Servo voltage may be expressed in terms of the fiber strain. The *lower curve* instead shows the voltage steps applied to the a piezo transducer used to test and calibrate the sensor

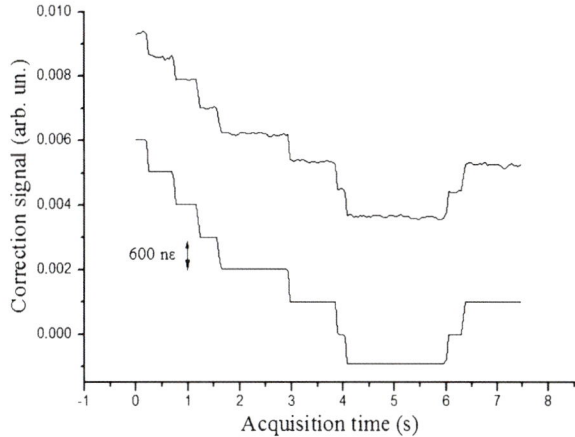

Fig. 13.3 Spectral (FFT) response of the locking signal for different excitation frequencies in the SM-fiber cavity. Spurios peaks are present due to self-oscillations of the locking loop and ac line noise (50 Hz)

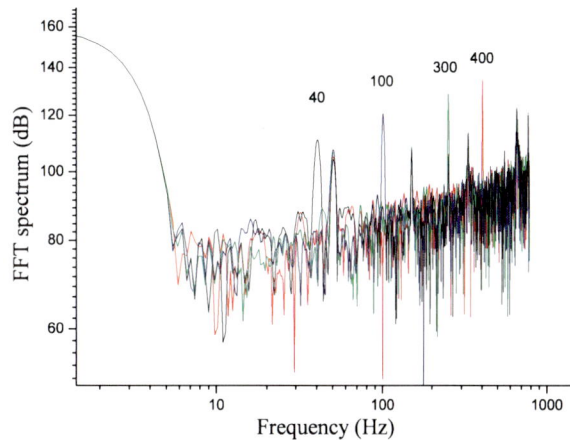

immediately translates into amplitude changes at the servo output, thus providing a fast and sensitive strain monitor at frequencies within the servo bandwidth.

In Fig. 13.2, the response of the locking servo signal to a 600-nε peak-to-peak quasi-static strain on the intracavity fiber is shown. The noise level along single steps is on the order of 2×10^{-4} that would correspond to a strain resolution of about 100 nε (1 n$\varepsilon = 10^{-9}$ fractional length change. The lower plot enables conversion of the arbitrary units in strain) for quasi-static deformations.

In Fig. 13.3, an example of analysis of the spectral distribution of noise on dynamic strain signals is shown. The traces correspond to the fast Fourier transform (FFT) of the servo signals recorded with the laser tightly-locked to the fiber resonator. An acoustic wave is excited in the sensing fiber by a piezo-electric transducer (PZT). The latter is driven by an ac voltage with a strain-equivalent rms amplitude of about 10 nε. Strain calibration can be carried out by sweeping the intra-cavity length over about 10–20 % of the cavity FSR and observing its transmission spectrum along with the laser sidebands (for a given radiofrequency (RF) value). In this way, the depth of the laser sweep can be accurately calibrated using the RF side-bands as a precise ruler and the strain can be related to PZT bias voltage at different modulation frequencies through Eq. (13.1). In addition to the applied perturbations, satellite oscillations are also visible from 50 to 350 Hz, regardless of the servo action. These may be consequence of cross-talk from the ac power supply as well as environmental noise that couples to the fiber. Given the signal-to-noise ratio of the FFT spectrum of Fig. 13.3, i.e. the peak values on the noisefloor, the minimum detectable strain can be extrapolated for different excitation frequencies. Resolution levels ranging from $0.1 \, n\varepsilon/\sqrt{Hz}$ to $10 \, n\varepsilon/\sqrt{Hz}$ are achievable between 1000 and 10 Hz using commercial free-running DFB diode lasers for interrogation. The limit to the minimum detectable strain is dictated by the laser frequency noise, as will be shown in detail in the following sections.

A different way to interrogate a FBG Fabry-Pérot sensor consists in locking a diode laser to a fiber resonator by an extension of the polarization-spectroscopy scheme originally developed by Hansch et al. [16]. It was theoretically investigated and experimentally demonstrated that the anisotropy induced into the FBG during the inscription process [17], combined with the natural birefringence of the pristine fiber, gives rise to the same ellipticity effect as would a polarizer internal to the cavity. Examples of the FBG Fabry-Pérot operating as a static and as a dynamic strain sensor using the intracavity fiber as a sensitive element in this fashion have been demonstrated by our group [18] with minimum detection level comparable with that obtained with PDH using an unstabilized laser diode.

13.2.2 Fiber-Loop Cavities

A different though substantially analogous sensor for strain and pressure sensing can be developed using a ring optical cavity built around a fiber bi-conical taper [19]. The interrogation system is based on cavity ring-down measurement of the intra-cavity loss and its variation induced by mechanical perturbation of the taper. The method is attractive because it doesn't use any complicated set-up or sophisticated sensor head. The detection limit seems appropriate for many applications ($\sim 80 \, n\varepsilon/\sqrt{Hz}$) although no characterization has been performed on its response to dynamic perturbations (e.g. vibration, acceleration). However, the performance is not comparable to that achievable in a FBG cavity by tracking of the cavity-mode shift, which benefits from the huge leverage effect seen from Eq. (13.1). Using a biconical fiber taper, the ability to extract the strain information depends strongly on the exponential time-decay fitting procedure employed and appears not suitable for real-time monitoring/tracking of fast strain signals unless very sophisticated acquisition electronics is adopted. Moreover, the sensing element is extremely fragile and shows a non-linear response due to excitation of higher-order fiber modes when a deformation is applied.

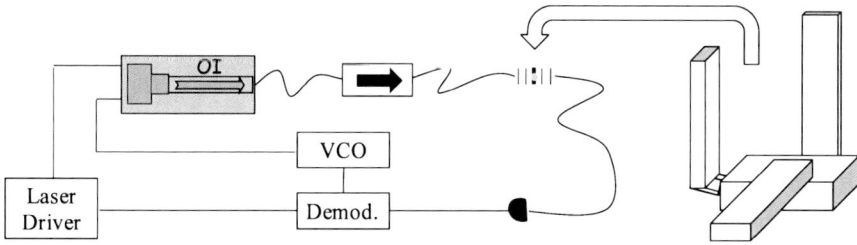

Fig. 13.4 Schematic diagram of an accelerometer head. Stainless steel cantilevers are clamped together using aluminum plates. All cantilevers have the same dimensions and nominal resonant frequencies of about 1.5 kHz (see Ref. [22])

13.2.3 π-Phase Shifted Fiber Bragg Gratings

Similarly to in-fiber Bragg resonators, π-phase shifted FBGs (PSFBGs) have been proposed for strain sensing. PSFBGs are very attractive for strain sensing as they have a short gauge length (~few mm), similar to standard FBGs, but exhibit the unique spectral features of optical resonators, i.e. a very high dispersive power around the resonance [20]. On the other hand, PSFBGs are easier to fabricate than standard FBGs since they comprise just one grating, and no wavelength matching is required as is the case for FBG-cavities. Furthermore, PSFBGs can be written with a reflectivity >99 % and are thus well suited for developing small-size (large bandwidth) opto-mechanical sensors of high quality. Very recently, high-sensitivity interrogation schemes based on a frequency-stabilized laser were proposed for PSF-BGs [21]. A strain resolution in the order of 10 p$\varepsilon/\sqrt{Hz}$ was demonstrated in the audio and ultra-sonic range ($\geq$1 kHz) with an extended-cavity diode laser stabilized by a PDH system. The achievable resolution is more limited at lower frequencies, most likely due to laser excess (flicker) noise as well as mechanical instabilities of the laser cavity. Subsequently, using PSFBGs, a miniature fiber-based 3-axis accelerometer suitable for seismic and vibration sensing was built [22]. Indeed, velocities and accelerations can be efficiently measured using a fiber-optic probe integrated with an inertial mass or beam, provided the mechanical response of the solid element is known. Three separate PSFBGs were attached to three cantilever beams aligned along orthogonal planes, as shown in Fig. 13.4. The sensors were interrogated by three telecom distributed-feedback (DFB) diode lasers actively locked to their central resonances by the PDH technique. Near-infrared DFB lasers have the advantage of being cheap, easily available from the telecom market and rather insensitive to environmental acoustic noise. An acceleration noisefloor of 10 $\mu g/\sqrt{Hz}$ between 10 Hz and 1 kHz was achieved. This sensitivity level compares well with typical performance of seismic accelerometers, with the advantage of a much wider frequency response. At very low frequencies, the sensitivity is reduced due to the characteristic noise roll-up when approaching DC. Basically, the main limitation to the acceleration resolution comes from the free-running laser phase and frequency noise. A significant improvement can be obtained if the interrogating lasers are pre-stabilized onto a stable reference. As an example, in Fig. 13.5, we show the effect

Fig. 13.5 Noise reduction of the accelerometer's readout by locking the laser to an external fiber-loop cavity. The plot is obtained from the FFT of the accelerometer signal

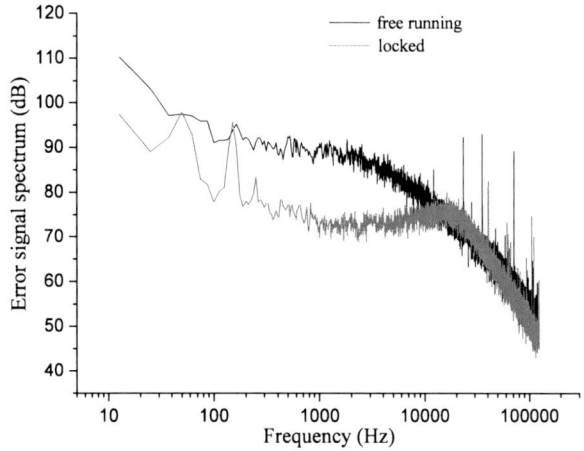

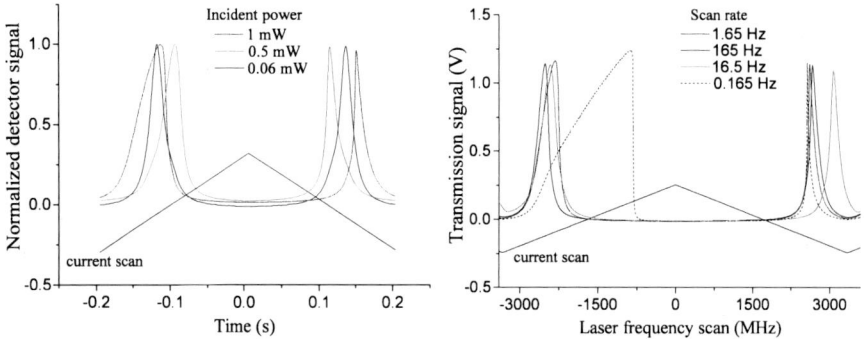

Fig. 13.6 *On the left*: asymmetry of the grating's resonance for increasing incident power. *On the right*: marked push (negative slope) and pull (positive slope) effect for increasing scan rate

of locking one of these lasers to an external 2-m fiber-loop cavity (finesse 300), obtaining a noise reduction of more than 10 dB between 100 Hz and 10 kHz.

As mentioned above, the PSFBGs present a characteristic response which is quite similar to other optical resonators. The phase defect in the periodic structure indeed modifies its photonic bandgap and creates a sharp resonance exactly at the Bragg wavelength. The narrow resonance improves the capability of detecting small wavelength shifts caused by mechanical action on the fiber but also points out to other interesting phenomena. Light trapping into these tiny resonators can be very efficient and gives rise, for instance, to a strong thermo-optic effect.

A typical push-pull effect on the central resonance can be observed when scanning the laser's injection current by a triangular wave with laser power maintained at <1 mW as already observed in fiber Fabry-Pérot cavities [23]. As illustrated in Fig. 13.6, the resonant mode lineshape becomes markedly asymmetric due to the heating and expansion of the fiber grating/PSFBG, with different symmetry along opposite slopes of the scan (i.e. whether the laser is tuned towards or away from the resonance).

The effect becomes more evident when either the input power or the scan rate increases. The latter is likely a consequence of the larger power build-up during slower (or narrower) sweeps around the resonance. Since each sensor is glued on a steel substrate which also partially acts as a heat sink, one expects that the thermo-optic effect would be hugely enhanced if the fiber had no thermal contact with the surroundings. A full explanation of this behavior however requires a detailed physical model of the PSFBG thermal response. Development of such a model is presently underway.

In principle, a rigorous assessment of fiber-resonator sensitivity limits in mechanical sensing should start from the shot-noise-limited frequency noise spectral density of the PDH locking method [24]

$$S_{PDH}(\text{Hz}/\sqrt{\text{Hz}}) = \frac{\Delta \nu_c}{J_0(\beta)} \sqrt{\frac{h\nu}{8\eta P_i}}. \tag{13.2}$$

Assuming a cavity mode linewidth $\Delta \nu_c = 2$ MHz, detector efficiency $\eta \sim 0.9$, incident power $P_i = 0.1$ mW and the modulation depth $\beta = 0.3$, which are quite common values, a minimum strain noise $\varepsilon_{PDH} = S_{PDH}/0.78\nu \cong 0.14$ f$\varepsilon/\sqrt{\text{Hz}}$ can be calculated from Eq. (13.2). However, a more realistic approach includes consideration of the unavoidable free-running phase and frequency fluctuations associated with the laser. Assuming that the laser is dominated by spontaneous-emission frequency noise, its emission spectrum will be Lorentzian. For DFB diode lasers, for instance, a linewidth in the order of at least 10–20 MHz is expected, which can be converted from Eq. (13.1) to cavity length fluctuations, i.e. in strain noise, yielding an ultimate strain resolution of about 10–20 p$\varepsilon/\sqrt{\text{Hz}}$. This result agrees well with experimental findings, as shown above.

Pre-stabilized lasers may provide benefits in this regard. Chow et al. locked a low-power extended-cavity diode laser emitting at 1550 nm to a high-finesse free-space reference cavity to suppress its free-running frequency noise and then used the resulting variation for interrogation of a FBGFP sensor [24]. Adopting the PDH frequency-locking technique for active tracking of the sensor, this scheme was capable of detecting subpicostrain signals, at frequencies from 100 Hz to beyond 100 kHz.

Most of these approaches are rewarding when strain signals in the acoustic and ultrasonic range have to be monitored. Nonetheless, a sensor capable of tracking very slow deformations with high resolution and accuracy would find important applications in several fields, including telescope control, inertial sensing, seismic monitoring and nanotechnology process analysis, among others. Also, low-frequency fibre interrogation techniques may have an impact in wavelength-encoded chemical sensing, e.g. with opto-acoustic resonators and optical microcavities. Recently, fibre hydrophones have been even proposed for detection of ultra-high energy neutrinos in large deep-sea detectors. In all these applications, the involved physical quantities change very slowly. On the other hand, quasi-static sensing has always been challenging, as it must contend with unwanted laser frequency fluctuations and drifts that degrade the signal-to-noise ratio and set a high detection limit at infrasonic frequencies. The use of optical frequency references improves the long-term stability, leading to a greatly enhanced sensor readout.

Notable attempts towards laser stabilization for fibre sensing have been made in the past few years, relying on atomic and molecular transitions as well as long interferometers. Nevertheless, using external devices for noise suppression in the infrasonic range makes the (thermal and mechanical) stability prerequisites extremely demanding even for well-controlled laboratory systems. Very recently, Chow et al. used an easily available near-IR ro-vibrational transition of HCN as a stable reference for strain measurements by a FBG cavity sensor [25]. A fiber laser emitting at a wavelength of 1550 nm is locked to the HCN absorption line. The experiment demonstrates a noticeable reduction of low-frequency laser instabilities with subsequent gain in strain resolution, thereby leading to a limit in the order of $10 \, p\varepsilon/\sqrt{Hz}$ in the 1-Hz frequency range. However, though the optical scheme looks very practical and effective, it is also clear that any change in the gas cell pressure and/or temperature could manifest as a small drift that adds to the background strain noise.

On the other hand, it now appears as if the ultimate performance of fibre lasers and passive interferometric fibre systems is hampered by thermally induced phase noise. The physical problem is similar to Johnson's noise in electrical circuits, and can be treated in a way analogous to Nyquist's theorem starting from thermal equilibrium energy fluctuations [26]. This is the case for long fibre-optic interferometers and Sagnac gyroscopes [27, 28]. The same kind of noise appears in the low-frequency spectrum of fibre lasers [29]. This aspect has been investigated in depth over the last decades both theoretically and experimentally, and different models have been proposed [30, 31], none of which fully agreeing with experimental findings. An extremely stable laser interrogation may give the opportunity to disclose the contribution of thermodynamic phase fluctuations over other effects that usually dominate the noise budget, such as laser instabilities and drifts, electronic noise etc.

Nowadays, optical frequency comb synthesizers (OFCSs) provide an exceptionally stable, absolute frequency grid from the XUV to the mid-infrared [32]. Combs originate from short, mode-locked laser pulses that are equally time-spaced by a radio-frequency (RF) clock. In 2010, our group [33] reported an unprecedented resolution level in strain sensing when exploying a passive FBG Fabry-Pèrot (FBGFP) resonator interrogated by a diode laser that is stabilised against an OFCS. The comb is phase-locked to a 10-MHz oven-controlled quartz oscillator (OCXO) that is linked to a Rb-clock to suppress long-term frequency jitter and drifts. The free-running quartz exhibits a single-sided phase noise $\mathcal{L}(f) \approx -122$ dBc/Hz at 1 Hz. For time scales longer than 400 s, the OCXO is phase linked to a local Rb-clock, which is ultimately disciplined by the primary Cs-standard via a GPS receiver for compensation of very-long term instabilities. The comb beam spans about 40 nm around 1560 nm and is directly available from an optical fibre output with a 10 mW power.

The diode laser beam is phase locked to the nearest comb tooth. Thus, the laser frequency f can be related to f_r and f_O through the expression

$$f = \pm f_{beat} \pm f_O + m \cdot f_r, \qquad (13.3)$$

where f_{beat} is the laser-comb beat-note frequency (given by an external synthesizer). Treating the OFCS as a 'rigid' multiplication gear of the OCXO with a scale factor N ($N \sim 192\,\text{THz}/10\,\text{MHz} = 1.9 \times 10^7$), the laser noise can be calculated from $S_\nu(f) = N \cdot f \sqrt{2L(f)} \approx 20\,\text{Hz}/\sqrt{\text{Hz}}$ at 1 Hz. Any laser frequency fluctuation $\delta\nu$ is converted into strain ε using Eq. (13.1).

Two 'secondary carriers' are created by deep phase modulation using another tuneable synthesizer which is locked to a cavity mode by a PDH scheme. In this way, active interrogation of the FBGFP is possible by the SC, only adding a small frequency jitter from the synthesizer, while the main laser carrier remains phase locked to the OFCS. In this way, the response of the sensor is brought very close to the fundamental limit imposed by the fiber thermodynamic noise, in a previously inaccessible frequency range. Strain readout is taken from the locking loop correction signal. A lower bound of $550\,\text{f}\varepsilon_{rms}/\sqrt{\text{Hz}}$ around 2 Hz can be measured, with a minimum of $350\,\text{f}\varepsilon_{rms}/\sqrt{\text{Hz}}$ at 5 Hz. The measurement performance achieved by comb-referenced interrogation of the FBGFP sensor is the best ever reported so far at infrasonic frequencies. The sensor might no longer be limited by the residual locked laser stability but rather affected by thermodynamic phase noise of the fiber. However, a complete theoretical model for finite-cladding fibres is not yet available [34]. Moreover, a reliable estimate of the noise level is significantly affected by the uncertainty of the involved parameters while the correctness of the model has never been tested for very low frequencies. New experiments are still necessary to confirm or modify general theories of thermodynamic phase noise in optical fibers, particularly for very low frequencies, and to contribute to better understanding of the underlying physical mechanisms.

13.3 Fiber Optic Sensors for Musical Recordings

As was shown above, single FBGs and especially cavities made of two identical FBGs can be configured into very sensitive sensors for strain and vibration. This would also make them broadband and low-noise pick-ups for musical instruments. All "acoustic" string instruments such as guitars, violins, and even cembalos or harps have a soundboard that amplifies the sound generated by the string movement. The exact shape of the soundboard determines the coloration of the instrument through, e.g. the presence of harmonic excitations and vibrational nodes. Most serious musicians prefer to record the sound of their instrument using a high quality microphone, but this may be impractical in an environment with large background "noise" such as a performance stage. Many acoustic guitars and, to a much lesser extent, other string instruments are therefore equipped with piezoelectric transducers (PZT "pick-ups") that are placed on the soundboard and convert its vibrations into an electrical signal. While pick-ups have a fairly flat frequency response between about 100 Hz–20 kHz and are inexpensive, they also have comparably high inertia, and are difficult to amplify due to their high impedance. When many of these PZTs are mounted onto a single instrument, the vibrations of the soundboard may be affected and the instrument coloration may be altered.

Fiber optic transducers are preferred when size and inertial mass is a concern. Recently, the feasibility of acoustic transduction has been demonstrated by affixing a single FBG or a FBG Fabry-Pèrot resonator on a guitar body and comparing the recordings with those made by an attached PZT [35]. An in-fiber Fabry-Perot (FFP) cavity made from two identical FBGs can be used for vibration and strain measurements, since the resonances in the cavity spectrum depend on the cavity length and thereby, again, on the strain applied to the waveguide as shown by Eq. (13.1). By detection and audio sampling of the reflected light from a FFP cavity, compact and very lightweight transducers ("pick-ups") for guitars, violins, harmonicas and other musical instruments, in which the sound is generated to a large extent by the vibration of a soundboard, can be realized. The subjective quality of the recorded music was comparable to that of commercial piezoelectric transducers, and the frequency range was found to encompass the entire audible region of the audio spectrum i.e. from about 20 Hz to 20 kHz. These earlier measurements were performed by recording the intensity of the light reflected either from the mid-reflection point of a FBG or the mid-reflection point of a cavity fringe of a low-finesse FFP [36].

Using an interrogation method that locks the frequency of the probe laser tightly to the audio-modulated cavity-reflection feature has some advantage over this approach: it makes the measurement almost immune to the laser amplitude noise (and some external disturbances), reduces the influence of electronic noise, has a much larger vibration amplitude (=sound intensity) range, and widens the dynamic range of the intrinsic sensor with a bandwidth that covers the entire audio spectrum. The acoustic strain information is contained in the correction signal that is generated to keep the laser locked. Also, infrasound variations of the FFP cavity reflection spectrum due to, e.g., thermal expansion or material fatigue will be compensated by the feedback mechanism.

The acoustic transducer designed for this work consists of a fiber Fabry-Perot cavity containing two 23 dB FBGs positioned 2 centimeters apart (see Fig. 13.7). The cavity was attached to the guitar body (*Dagmar Guitar*) close to the bridge, where a conventional piezoelectric pick-up was mounted and used for comparison. The piezo pick-up is a passive device and designed for use with classical guitars. A DFB diode laser at 1549 nm was locked to the peak of a cavity fringe using the PDH technique. The correction signal contains frequency components corresponding to the vibrations of the guitar body, and hence, may be amplified and recorded, or it may be fed to a speaker to reproduce the sound generated by the guitar.

The PDH error signal is fed to a proportional-integral servo amplifier whose output is added to the laser current. The servo amplifier was designed to have a bandwidth of at least 30 kHz. A portion of the feedback correction signal was applied to a ×20 gain acoustic amplifier and then sampled by Adobe Audition 3.0 software for recording and analysis.

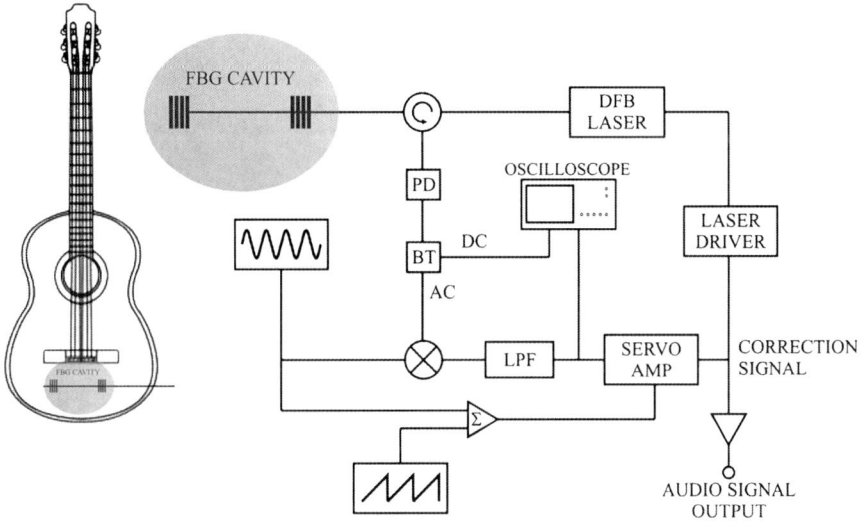

Fig. 13.7 Diagram of the experimental scheme for music recording by a FBG cavity sensor. DFB: distributed feedback; PD: Photodiode; BT: Bias Tee; LPF: 50 kHz low pass filter

Our analysis attempts to address the parameters that are most important for assessing musical instrument transducer quality, i.e. response time and frequency response curve, the signal-to-noise level, and dynamic range of the amplitude measurement. An ideal transducer exhibits a very low noise floor, a sensor bandwidth spanning the range from a few Hertz to 44 kHz (the rate at which Compact Disks are sampled) and a flat frequency response curve from about 50 Hz to 20 kHz, i.e. over the entire audible region. Figure 13.8 and experiments described in [36] give these parameters but they are not able to represent the quality of the sound recording. which is best assessed using audio recordings such as those published as supplementary material in [36]. The amplitude of the soundboard vibration could not be measured directly while the acoustic guitar was played, but it can be estimated from the measured modulation of the cavity fringe wavelength. A modulation by about 5.7 pm (2.2 pm) was observed when the E_2 (E_4) string was played. Using a gauge factor of 0.78, this modulation amplitude corresponds to a strain of 4.7 $\mu\varepsilon$ (1.8 $\mu\varepsilon$). Using a calibration obtained for a different guitar in an earlier publication, we can estimate the amplitude of vibration as approximately 90 μm (35 μm), which is consistent with earlier measurements (30–70 μm) for a slightly less "loud" instrument [37].

As was mentioned in earlier work [36], the pickup is lightweight and may be multiplexed into a sensor array. It is therefore conceivable that multiple FFP cavity sensors may be embedded in the guitar body and—depending on their position—provide different timbres. Some high-end musical instruments already incorporate two or even more piezo pickups, giving a more balanced sound. Of course, one then requires a separate laser, laser driver, detector and locking electronics for each

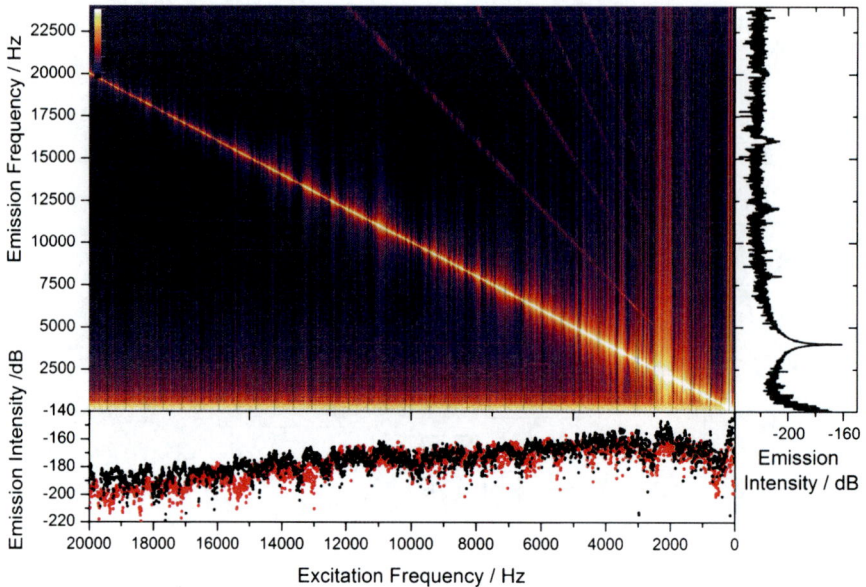

Fig. 13.8 Excitation-emission matrix spectrum showing the response of the guitar's soundboard to sound emitted from a speaker at different excitation frequencies. The FFP transducer reproduces the fundamental frequencies from about 30 Hz to 20000 Hz, as well as its overtones. Horizontal lines at multiples of 1 kHz are due to a weak clock signal from the computer's USB port. At frequencies above 13 kHz the FFP transducer response near the fundamental of the excitation frequency (*black curve*) shows a 5–10 dB stronger signal compared to the PZT pickup (*red*). The colour scale bar ranges from −220 to −180 dB. The *curve in the right panel* is the Fourier transform of the FFP cavity response at an excitation frequency 4 kHz (Ref. [36])

channel, whereas the power supplies, high-frequency generator and preamplifer may be shared. A prototype of such an instrument was recently built (Fig. 13.9).

Finally, we want to emphasize that audio recordings are extremely revealing with regards to the performance of *any* audio-range vibration sensor, since very slight distortions are easily audible even by untrained ears. A sensor that performs to high-fidelity audio standards should be more than adequate for almost all mechanical sensing applications.

13.4 Evanescent-Wave Chemical Sensing

Chemical sensors using fiber-optic methodology are the subject of extensive research and development activity with potential applications in industrial, environmental and biomedical monitoring. In this context, a miniature chemical sensor combining laser spectroscopy and state-of-the-art optical fiber technology may be suitable for in-situ, non-invasive gas or liquid analysis with high selectivity and sensitivity. This can be based on either direct or indirect (indicator-based) detection

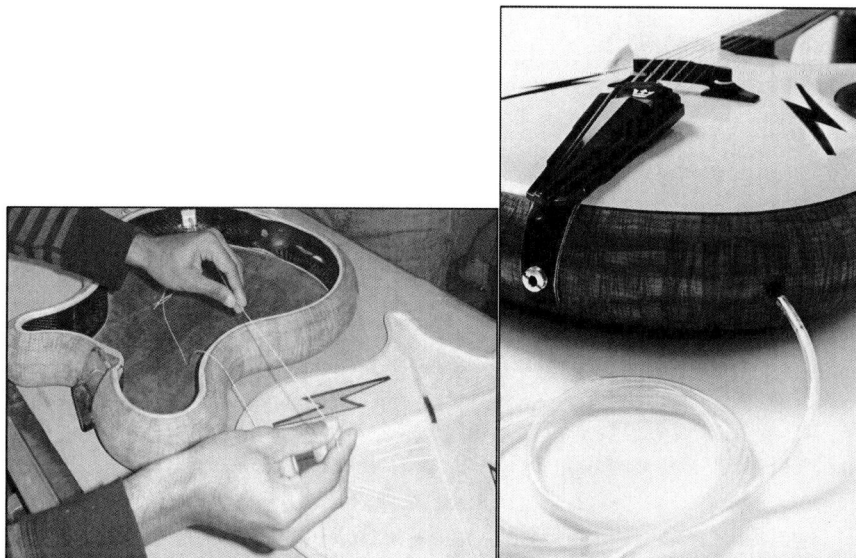

Fig. 13.9 Acoustic Guitar prototype designed and manufactured by *Dagmar Guitars* (Niagara-on-the-Lake, ON). *Left*: the guitar body and top plate of the guitar are seen before assembly showing 7 FBG FP cavities that can be used as vibration sensors. *Right*: the finished guitar

techniques. In the direct scheme, the optical properties of the analyte, such as refractive index (RI), absorption or emission, are measured directly. In the indirect scheme, the color or fluorescence of an immobilized label compound, or any other optically-detectable bioprocess, are monitored [38, 39]. In recent years, interrogation techniques have further advanced with the use of evanescent-wave spectroscopy as well as surface-plasmon resonance [40, 41]. Sensors have also been incorporated into passive optical cavities consisting of fiber loops or linear fiber cavities defined e.g., by two identical FBGs [42, 43]. These cavities have been shown to be effective means of amplifying the sensors' response. Their application to mechanical sensing has been reviewed in the previous sections of this chapter. Optical microresonators, of different geometries, have been also used as label-free and ultrasensitive chemical sensors over the past several years [44, 45]. In all cases above, a change in ambient refractive index may lead to a wavelength shift of the cavity modes, if part of the evanescent wave of the mode is exposed to the environment. On the other hand, if the molecules exhibit absorption lines or bands in the vicinity of the resonance wavelength, the cavity lifetime, namely the ring-down time (RDT), will be reduced, leading also to a reduction in power transmitted through the resonator and in the quality (Q-) factor.

Since their invention, optical frequency combs (OFCs) have been used for a large number of studies and applications, such as frequency metrology, characterization of materials dispersion, length measurements, strain sensing, and molecular spectroscopy. At the same time, optical resonator-based detection methods, such as cavity-enhanced absorption spectroscopy (CEAS) and cavity ring-down spectroscopy (CRDS), either with conventional mirror cavities or microcavities, have demonstrated impressive performance in chemical spectroscopy and sensing experiments thanks to their inherent immunity to laser amplitude noise and the associated quality factor [46]. A recent breakthrough was represented by coherent coupling of OFCs to high finesse optical cavities used as sample compartments [10, 11]. In such cases, spectral analysis of cavity transmitted light can be performed by dispersive elements and Fourier-transform methods [47] to extract the absorption features over several tens of nm. However, so far comb-based absorption spectrometers have rested on conventional linear cavities and used only for gas spectroscopy.

Optical spectroscopy in the liquid phase also has an unrealized potential in the analysis of molecular species relevant to biomedical processes [48, 49]. In recent years, there have been only a few works on cavity-enhanced absorption spectroscopy of liquids using lasers or broadband light [50]. An effective and minimally-invasive method relies on using total internal reflection in optical fibers. In this case, the interaction with liquid chemicals in the surrounding environment may occur if the evanescent field is exposed along the external interface. Using fiber-optic resonators, multiple round-trips provide a longer effective absorption path-length and thus stronger signals. At the same time, the possibility of devising analyzers whose sensitive element acts as a separate probe, eventually in remote operation, is highly desirable for liquid sensing. Optical fibers are particularly suitable for in-situ, non-invasive sensing, even in environments with access difficulty, and lend themselves to the realization of multiplexed chemical probes. Furthermore, all-fiber resonators are cheap, compact, easy to build and they usually do not require special care in terms of alignment, cleaning and isolation.

On the other hand, the choice of the radiation source is crucial. When liquid spectroscopy is the main target, the required tunability can be demanding for a laser, and therefore broadband light beams are necessary for this purpose. If incoherent sources are used in conjunction with high-finesse cavities, efficient coupling to the resonator modes is prevented, thereby dramatically reducing the available optical power. Furthermore, such a scheme is not immune to light-source intensity fluctuations, making it less sensitive than other CEAS and CRDS techniques.

13.4.1 Optical Fiber Loop Spectroscopy

The easiest way to build a resonator with an optical fiber is to create a closed fiber loop within which light remains confined. To inject light into the ring cavity one can splice commercial fiber couplers into the ring. Couplers can also be used to leak a small portion of the circulating light out of the cavity with each round trip. Alternative fiber loop designs are presented in Chaps. 9 and 10 of this book. It is well known that a fiber loop ring cavity made from single-mode fiber behaves in a similar way to a conventional Fabry-Pérot resonator [12], exhibiting resonance modes spaced by the free-spectral-range, namely c/Λ ($\Lambda = nL$ where n is the fiber-core refractive index and L the total ring length). Light interaction with an absorbing sample, whether it be gaseous or liquid, can be probed either by the internal cavity field or the evanescent wave around the fiber, which can be exposed along the fiber by partial removal of the cladding or by tapering processes. The evanescent-wave scheme is ideal for minimally-invasive analysis of small-volume liquid samples even in environments presenting access difficulties [51].

Tarsa et al. [52] combined cavity ring-down spectroscopy (CRDS) with single-mode fiber loops to benefit from the versatility of fiber-optic sensing and the improved sensitivity of CRDS. They used a km-long fiber-loop cavity and interrogated it by means of a high-quality extended-cavity diode laser for evanescent-wave detection of a liquid absorption band. For this purpose, a tapered sensing region is fabricated along the fiber by a simple modification of the fiber shape that exposes the evanescent field. The resulting detection limit is in the order of $1.7 \times 10^{-4}/\sqrt{\text{Hz}}$. Despite the unique tunability range of their laser, which is well beyond the typical performance of other diode lasers, the spectral coverage of this system is just sufficient to recover a liquid band of only 25-nm width. Also, the bi-conical fiber taper represents a further limitation since it is hard to reproduce, quite fragile and rather impractical for real applications. Evanescent-wave access blocks that can be obtained by side-polishing of single-mode fibers [53] are more robust and reliable although special machines are necessary for fabrication. This is one of the reasons why other groups have chosen a different approach, which relies on leaving a small gap within the fiber loop [54] or devising special liquid interfaces to enable light-matter interaction [55, 56]. Chapters 9 and 10 provide reviews of different fiber loop ring-down methods and their applications.

In the following, we present a novel method that extends the capabilities of a near-infrared optical frequency comb (OFC) to cavity ring-down evanescent-wave spectroscopy with a fiber loop [57]. The OFC teeth are kept resonant with the cavity modes using a Pound-Drever-Hall (PDH) locking scheme. Thanks to the strong group-velocity dispersion in the fiber, the loop cavity behaves as a highly dispersive element allowing only a narrow interval of the OFC wavelengths to be resonant with the cavity. A fast step scan of the repetition rate permits covering the whole comb emission range and thus performing fiber-optic cavity comb spectroscopy (FOCCS) on the broad absorption features of liquid samples. Using a spectrally-broadened output of the comb laser, obtained by extension of the original Er-fiber fs-pulsed laser using a nonlinear optical fiber, one octave of the near-IR can be spanned, from 1.05 to 2.1 micron.

The experimental setup is shown in Fig. 13.10. The sensing element, an evanescent-field access block (EAB), consists of a side-polished single-mode optical fiber that allows interaction of the cavity evanescent field with a liquid sample. A 20-m fiber-loop cavity is built around the EAB. Light from a mode-locked erbium-fiber laser comb is injected into the cavity and collected through 0.6-% evanescent fiber couplers. The system is all-fiber made without any free-space gap.

To make the OFC resonant with the fiber loop, the repetition rate is tuned in order to match a *magic condition* [58] for our cavity, which is achieved when the comb teeth spacing is an integer multiple of the cavity FSR for a given wavelength. However, the fiber cavity acts as a dispersive element, which selects a narrow group

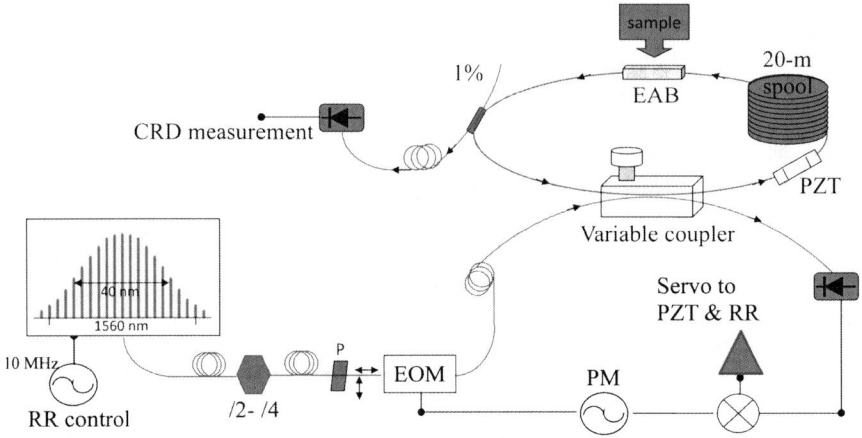

Fig. 13.10 Experimental set-up for evanescent-wave spectroscopy of liquids by means of optical combs injected in fibre-optic ring cavities. RR: repetition rate; PM: phase modulation; PZT: piezoelectric transducer; EOM: electrooptic modulator; EAB: Evanescent field access block

of teeth that obey the above-mentioned magic condition. At other wavelengths, the magic condition is satisfied by a slightly different repetition rate. The central resonant wavelength is then swept by tuning the repetition rate synthesizer, thereby operating the cavity as a tunable spectrometer. In Fig. 13.2, a typical spectrometer response around 1550 nm is shown, as measured with an optical spectrum analyzer (OSA). Besides the fundamental magic condition (FMC) at 1550 nm, having a full width at half maximum (FWHM) of 0.7 nm (~250 teeth), there is a large population of narrower secondary magic conditions (SMCs) at different wavelengths. In principle, these extra resonances can be attributed to group velocity dispersion (GVD) of the fiber acting on the OFC pulse. Actually, only the teeth included in the spectral interval selected by the FMC resonate within the cavity and can be used for ring-down spectroscopy. Due to the presence of SMCs, the OFC cannot be completely switched off when modulating the total incident light power as in previous systems [11]. This problem is overcome by switching off only the FMC by amplitude modulation of PDH sidebands via the phase modulator.

From 100 repeated ring-down measurements in the same experimental conditions, a mean ring-down time $\tau = 3.04$ μs with a standard error of 0.004 μs (standard deviation divided by the square root of the number of samples) was obtained. Using this value, it is possible to estimate a minimum detectable single-pass absorbance of 3×10^{-5} for each spectral point, i.e. 1×10^{-6} Hz$^{-1/2}$ considering an effective detection bandwidth of $\frac{1}{2\pi\tau N} \approx 524$ Hz (N number of averaged acquisitions). This figure of merit points to an impressive sensitivity level if compared to other laser-based fiber-cavity systems [52, 55] and translates into a noise-equivalent absorption coefficient of 3×10^{-4} cm$^{-1}/\sqrt{\text{Hz}}$ per spectral point, considering the effective evanescent-wave interaction length of the sensor head (~30 μm). In addition, thanks to the spectrally-extended OFC emission, the sensor is capable of providing also information on the absorption spectrum over a wide wavelength range.

Fig. 13.11 Evanescent-wave absorption spectrum obtained with a sample containing tetraethylenepentamine (TEPA) diluted in D_2O at different concentrations (from 50 to 62 %)

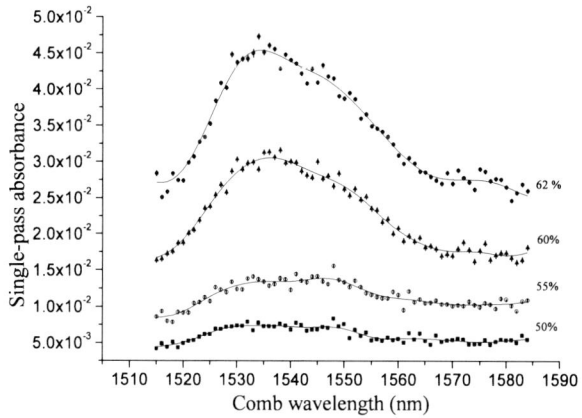

The spectroscopic performance is assessed by a test experiment with liquid samples containing polyamines, which exhibit strong absorption bands in the near infrared due to overtone vibrations. All the samples are diluted with aqueous solutions. During repetition rate steps, the fiber cavity remains locked to the OFC teeth while a LabView™ code controls both the radio frequency synthesizer that rules the repetition rate frequency for fine tuning of the FMC and the PDH signal to ensure a correct frequency-lock operation. Simultaneously, it acquires intracavity power data and performs ring-down time retrieval for each point of the FDS wavelength scan. Each value results from the average of the fitting parameters of 100 decay events. A full spectrum can be recorded in about 120 s. In Fig. 13.11, we show CRDS measurements along a fast scan of the FDS around an absorption band of tetra-ethylenepentamine (TEPA) using the direct Er-laser comb emission in the telecom range. The result is in good qualitative agreement with an FTIR spectroscopic database [59] although these recordings were obtained with a drop of solution prepared from TEPA in D_2O at volume concentrations varying around 90 %.

As opposed to other comb-based systems [10, 11], where dispersion causes strong bandwidth limitations, here strong GVD effects due to the fiber cavity can be exploited for direct analysis of cavity spectra, without using any additional optical compensation element. Different sensing configurations can be considered using the same experimental scheme. Special engineered fibers may be adopted to replace the evanescent-field access element for amplifying light-matter interaction and to optimize the spectroscopic performance also by tailoring cavity dispersion.

13.5 Conclusions

In this chapter we presented an overview of the most advanced optical-fiber cavity-enhanced systems developed for mechanical, inertial and chemical sensing. Using state-of-the-art techniques for laser stabilization, noise reduction and fast readout, we show that it is possible to extend the knowhow of precision laser spectroscopy and optical metrology to the realm of fiber optic sensors. High finesse fiber cavities are nowadays technologically feasible with relatively-simple optical fiber devices while working in the telecommunication wavelength range provides huge benefits in terms of device quality and costs. Mode-locked laser comb synthesizers are also considered both, for their amazing potential as frequency references and their unique properties as broadband, coherent light sources. In particular, we show how optical combs can be used in this context for strain sensing as well as broadband liquid evanescent-wave spectroscopy.

References

1. D.J. Richardson, Filling the light pipe. Science **330**, 327–328 (2010)
2. A.D. Kersey, A review of recent developments in fiber optic sensor technology. Opt. Fiber Technol. **2**, 291–317 (1996)
3. F.T.S. Yu, S. Yin (eds.), *Fiber Optic Sensors* (Marcel Dekker, New York, 2002), p. 449
4. Y.J. Rao, In-fibre Bragg grating sensors. Meas. Sci. Technol. **8**, 355–375 (1997)
5. G. Gagliardi et al., Optical-fiber sensing based on reflection laser spectroscopy. Sensors **10**, 1823–1845 (2010)
6. H.-P. Loock et al., Absorption detection using optical waveguide cavities. Can. J. Chem. **88**, 401–410 (2010)
7. J. Ye, H. Schnatz, L.W. Hollberg, Optical frequency combs: from precision frequency metrology to optical phase control. IEEE J. Sel. Top. Quantum Electron. **9**, 1041–1058 (2003)
8. S.T. Cundiff, Y.J. Colloquium, Femtosecond optical frequency combs. Rev. Mod. Phys. **75**, 325–342 (2003)
9. S.A. Diddams, L. Hollberg, V. Mbele, Molecular fingerprinting with the resolved modes of a femtosecond laser frequency comb. Nature **445**, 627–630 (2007)
10. M.J. Thorpe, K.D. Moll, R.J. Jones, B. Safdi, J. Ye, Science **311**, 1595 (2006)
11. R. Grilli, G. Mèjean, C. AbdAlrahman, I. Ventrillard, S. Kassi, D. Romanini, Phys. Rev. A **85**, 051804(R) (2012)
12. L.F. Stokes, M. Chodorow, H.J. Shaw, All-single-mode fiber resonator. Opt. Lett. **7**, 288–290 (1982)
13. R.W. Drever et al., Laser phase and frequency stabilization using an optical resonator. Appl. Phys. B **31**, 97–105 (1983)
14. E.D. Black, An introduction to Pound–Drever–Hall laser frequency stabilization. Am. J. Phys. **69**, 79–87 (2001)
15. G. Gagliardi, M. Salza, P. Ferraro, P. De Natale, Interrogation of FBG-based strain sensors by means of laser radio-frequency modulation techniques. J. Opt. A **8**, S507–S513 (2006)
16. T.W. Hansch, B. Couillaud, Laser frequency stabilization by polarization spectroscopy of a reflecting reference cavity. Opt. Commun. **35**, 441–444 (1980)
17. T. Erdogan, V. Mizrahi, Characterization of UV-induced birefringence in photosensitive Ge-doped silica optical fibers. J. Opt. Soc. Am. **11**, 2100–2105 (1994)
18. G. Gagliardi, P. Ferraro, S. De Nicola, P. De Natale, Interrogation of fiber Bragg-grating resonators by polarization-spectroscopy laser-frequency locking. Opt. Express **15**, 3715–3728 (2007)
19. P.B. Tarsa, D.M. Brzozowski, P. Rabinowitz, K.K. Lehmann, Cavity ringdown strain gauge. Opt. Lett. **29**, 1339–1341 (2004)
20. J. Canning, M.G. Sceats, π-phase-shifted periodic distributed structures in optical fibres by UV post-processing. Electron. Lett. **30**, 1344–1345 (1994)
21. D. Gatti, G. Galzerano, D. Janner, S. Longhi, P. Laporta, Fiber strain sensor based on a π-phase-shifted Bragg grating and the Pound-Drever-Hall technique. Opt. Express **16**, 1945–1950 (2008)

22. T.T.-Y. Lam, M. Salza, G. Gagliardi, J.H. Chow, P. De Natale, Optical fiber 3-axis accelerometer based on lasers locked to π-phase-shifted Bragg gratings. Meas. Sci. Technol. **21**, 094010 (2010)
23. J.H. Chow, B.S. Sheard, D.E. McClelland, M.B. Gray, I.C.M. Littler, Photothermal effects in passive fiber Bragg grating resonators. Opt. Lett. **30**, 708–710 (2005)
24. J.H. Chow, D.E. McClelland, M.B. Gray, I.C.M. Littler, Opt. Lett. **30**, 1923–1925 (2005)
25. T.T.-Y. Lam, J.H. Chow, D.A. Shaddock, M.B. Gray, G. Gagliardi, D.E. McClelland, High resolution absolute frequency referenced fiber optic sensor for quasi-static strain sensing. Appl. Opt. **49**, 4029–4033 (2010)
26. W.H. Glenn, IEEE J. Quantum Electron. **25**, 1218–1224 (1989)
27. S. Knudsen, A.B. Tveten, A. Dandridge, IEEE Photonics Technol. Lett. **7**, 90–92 (1995)
28. R.P. Moeller, W.K. Burns, Opt. Lett. **21**, 171–173 (1996)
29. S. Foster, A. Tikhomirov, M. Milnes, IEEE J. Quantum Electron. **43**, 378–384 (2007)
30. K.H. Wanser, Electron. Lett. **28**, 53–54 (1992)
31. L.Z. Duan, Electron. Lett. **46**, 1515 (2010)
32. P. Maddaloni, P. Cancio, P. De Natale, Meas. Sci. Technol. **20**, 052001 (2009)
33. G. Gagliardi, M. Salza, S. Avino, P. Ferraro, P. De Natale, Probing the ultimate limit of fiber-optic strain sensing. Science **330**, 1081–1084 (2010)
34. L.Z. Duan, General treatment of the thermal noises in optical fibers. Phys. Rev. A **86**, 023817 (2012)
35. H.P. Loock, W.S. Hopkins, C. Morris-Blair, R. Resendes, J. Saari, N.R. Trefiak, Recording the sound of musical instruments with FBGs: the photonic pickup. Appl. Opt. **48**, 2735–2741 (2009)
36. S. Avino, J.A. Barnes, G. Gagliardi, X. Gu, D. Gutstein, J.R. Mester, C. Nicholaou, H.-P. Loock, Musical instrument pickup based on a laser locked to an optical fiber resonator. Opt. Express **19**, 25057–25065 (2011)
37. N. Ballard, D. Paz-Soldan, P. Kung, H.P. Loock, Musical instrument recordings made with a fiber Fabry-Perot cavity: photonic guitar pickup. Appl. Opt. **49**, 2198–2203 (2010)
38. W.R. Seitz, Chemical sensors based on fiber optics. Anal. Chem. **56**, A16 (1984)
39. M.M. Lopez, A.A. Atherton, W.G. Tong, Ultrasensitive detection of proteins and antibodies by absorption-based laser wave-mixing detection using a chromophore label. Anal. Biochem. **399**, 147–151 (2010)
40. L. van der Sneppen, F. Ariese, C. Gooijer, W. Ubachs, Liquid-phase and evanescent-wave cavity ring-down spectroscopy in analytical chemistry. Annu. Rev. Anal. Chem. **2**, 13–35 (2009)
41. J. Homola, Surface plasmon resonance sensors for detection of chemical and biological species. Chem. Rev. **108**, 462–493 (2008)
42. M. Gupta, H. Jiao, A. O'Keefe, Cavity-enhanced spectroscopy in optical fibers. Opt. Lett. **27**, 1878–1880 (2002)
43. T. von Lerber, M.W. Sigrist, Cavity-ring-down principle for fiber-optic resonators: experimental realization of bending loss and evanescent-field sensing. Appl. Opt. **41**, 3567–3575 (2002)
44. A.M. Armani, K.J. Vahala, Heavy water detection using ultra-high-Q microcavities. Opt. Lett. **31**, 1896–1898 (2006)
45. G. Farca, S.I. Shopova, A.T. Rosenberger, Cavity-enhanced laser absorption spectroscopy using microresonator whispering-gallery modes. Opt. Express **15**, 17443–17448 (2007)
46. I. Galli, S. Bartalini, S. Borri, P. Cancio, D. Mazzotti, P. De Natale, G. Giusfredi, Molecular gas sensing below parts per trillion: radiocarbon-dioxide optical detection. Phys. Rev. Lett. **107**, 270802 (2011)
47. B. Bernhardt, A. Ozawa, P. Jacquet, M. Jacquey, Y. Kobayashi, T. Udem, R. Holzwarth, G. Guelachvili, T.W. Hänsch, N. Picqué, Cavity-enhanced dual-comb spectroscopy. Nat. Photonics **4**, 55 (2010)
48. J.J. Burmeister, M.A. Arnold, Evaluation of measurement sites for non-invasive blood glucose sensing with near-infrared transmission spectroscopy. Clin. Chem. **45**, 1621–1627 (1999)

49. N. Ghosh, S.K. Majumder, P.K. Gupta, Polarized fluorescence spectroscopy of human tissues. Opt. Lett. **27**, 2007–2009 (2002)
50. L. Nitin Seetohul, Z. Ali, M. Islam, Liquid-phase broadband cavity enhanced absorption spectroscopy (BBCEAS) studies in a 20 cm cell. Analyst **134**, 1887–1895 (2009)
51. M. Schnippering et al., Evanescent wave broadband cavity enhanced absorption spectroscopy using supercontinuum radiation: a new probe of electrochemical. Electrochem. Commun. **10**, 1827–1830 (2008)
52. P.B. Tarsa, P. Rabinowitz, K.K. Lehmann, Evanescent field absorption in a passive optical fiber using continuous wave cavity ring-down spectroscopy. Chem. Phys. Lett. **383**, 297–303 (2004)
53. A. Sharma, J. Kompella, P.K. Mishra, Analysis of fiber directional couplers and coupler half-blocks using a new simple model for single-mode fibers. J. Lightwave Technol. **8** (1990)
54. H. Waechter, J. Litman, A.H. Cheung, J.A. Barnes, H.-P. Loock, Chemical sensing using fiber cavity ring-down spectroscopy. Sensors **10**, 1716–1742 (2010)
55. H. Waechter, D. Munzke, A. Jang, H.P. Loock, Simultaneous and continuous multiple wavelength absorption spectroscopy on nanoliter volumes based on frequency-division multiplexing fiber-loop cavity ring-down spectroscopy. Anal. Chem. **83**, 2719–2725 (2011)
56. C.M. Rushworth, D. James, C.J.V. Jones, C. Vallance, Fabrication of an optical fiber reflective notch coupler. Opt. Lett. **36**, 2952–2954 (2011)
57. S. Avino, A. Giorgini, M. Salza, M. Fabian, G. Gagliardi, P. De Natale, Evanescent-wave comb spectroscopy of liquids with strongly-dispersive optical fiber cavities. Appl. Phys. Lett. **102**, 201116 (2013)
58. T. Gherman, D. Romanini, Modelocked cavity-enhanced absorption spectroscopy. Opt. Express **10**, 1033 (2002)
59. M. Buback, H.-P. Vogele, *FT-NIR Atlas* (VCH, Weinheim, 1993)

Chapter 14
Broadband Cavity-Enhanced Absorption Spectroscopy with Incoherent Light

A.A. Ruth, S. Dixneuf, and R. Raghunandan

Abstract Although broadband incoherent light does not efficiently couple into a high-finesse optical cavity, its transmission is readily detectable and enables applications in cavity-enhanced absorption spectroscopy in the gas phase, liquid phase and on surfaces. This chapter gives an overview of measurement principles and experimental approaches implementing incoherent light sources in cavity-enhanced spectroscopic applications. The general principles of broadband CEAS are outlined and general "pros and cons" discussed, detailing aspects like cavity mirror reflectivity calibration or the establishment of detection limits. Different approaches concerning light sources, cavity design and detection schemes are discussed and a comprehensive overview of the current literature based on a methodological classification scheme is also presented.

14.1 Introduction

Since the initial development of cavity ring-down spectroscopy (CRDS) in the late 1980s, the application of temporally and spatially incoherent light sources in cavity-enhanced spectroscopy has long been considered rather inappropriate, if not impossible. The reason for this perception was due to the generally significantly lower brightness of incoherent light sources (in comparison to lasers), and their low coupling efficiency to high finesse optical cavities [1]. At first glance, these apparently adverse features make cavity-enhanced absorption spectroscopy (CEAS) with incoherent light appear unattractive. However, there is also a multitude of advantages

A.A. Ruth (✉) · S. Dixneuf · R. Raghunandan
Physics Department & Environmental Research Institute, University College Cork, Cork, Ireland
e-mail: a.ruth@ucc.ie

S. Dixneuf
e-mail: s.dixneuf@ucc.ie

R. Raghunandan
e-mail: r.raghunandan@ucc.ie

G. Gagliardi, H.-P. Loock (eds.), *Cavity-Enhanced Spectroscopy and Sensing*,
Springer Series in Optical Sciences 179, DOI 10.1007/978-3-642-40003-2_14,
© Springer-Verlag Berlin Heidelberg 2014

of using incoherent broadband light for CEAS that have motivated a significant number of experimental implementations in analytical and environmental applications.

Experimentally all broadband CEAS methods[1] have in common that wavelength selection takes place after the cavity. This removes the need for scanning the wavelength and introduces the possibility of multiplexed light detection. Even though the coupling efficiency for truly white light[2] to high finesse cavities is rather low, at any given time, incoherent light (or spectrally broad light of limited temporal coherence) contains frequencies that correspond to the eigen-modes of a cavity for a given geometry, i.e. for a certain cavity length, mirror radius of curvature and diameter. Therefore a certain fraction of the light will always couple to the cavity. In a typical experiment white light entering the cavity, both on- and off-axis, will excite all accessible axial and transverse modes of the cavity for the given excitation conditions. As the mirrors are subject to small fluctuations (e.g. vibrations or minor mechanical instabilities) and slow thermal drift, the cavity modes are moreover continuously modified. Since the superposition principle holds inside the cavity [3], the transmitted light as detected with moderate resolution will not exhibit a mode structure (or mode effects), as observed in case of coherent excitation of cavities with narrow band lasers. The mode structure of the cavity plays no role in the measurement principle, which is solely based on the increase of the absorption path length by the cavity. The spectral resolution of broadband CEAS thus depends only on the approach chosen to select a wavelength of the light transmitted through the cavity.

The various experimental implementations of broadband cavity-enhanced absorption methods have been the subject of several reviews [4, 5]. Generally one can distinguish between methods that determine the characteristic photon storage time in the cavity (ring-down approach [6] or phase shift measurements [7]), and methods that measure the light intensity transmitted through the cavity [8]. This book chapter predominantly aims at the discussion of the latter type of approach. In Sect. 14.2 the general principles of broadband CEAS are outlined and general "pros and cons" discussed. Section 14.3 outlines different aspects concerning light sources, cavity considerations and detection schemes. In Sect. 14.4, a comprehensive overview of the current literature based on a methodological classification scheme is presented.

---

[1]The principle of measuring the cavity transmission directly to determine the loss inside a cavity was first implemented by O'Keefe [2] who used a **pulsed** cavity ring-down setup and named the approach "integrated cavity output spectroscopy" (ICOS). Besides CEAS as an acronym for **cw** applications, the abbreviation ICOS is also found in the literature.

[2]The term "white light" will be used synonymously for "incoherent broadband over a certain spectral region".

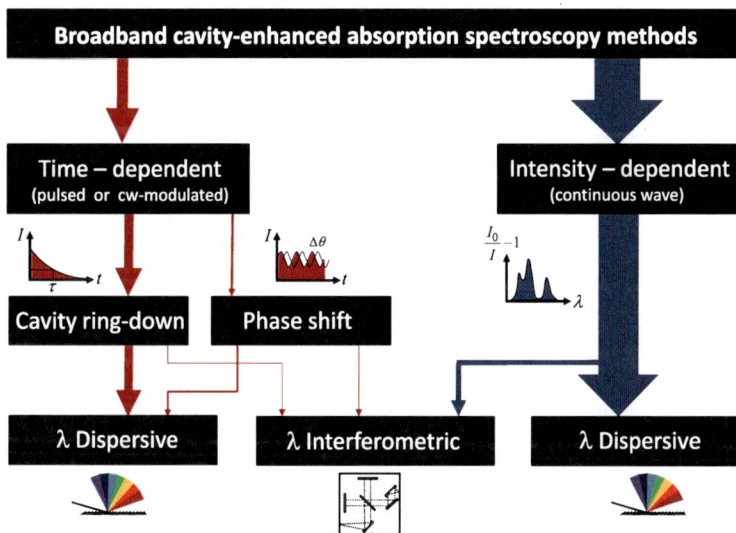

Fig. 14.1 Classification of broadband cavity-enhanced absorption approaches and detection schemes. The *widths of the arrows* indicate the relative number of occurrences of the type of approach in the literature

14.2 Broadband Cavity-Based Absorption Spectroscopy

14.2.1 Classification of Experimental Broadband Approaches

Figure 14.1 gives a general overview of broadband cavity-enhanced absorption methods, which can be broadly divided into two categories.

(A) *Time-dependent methods* are based on the measurement of the time that light can be stored inside the cavity (referred to as cavity ring-down spectroscopy, CRDS—Sect. 14.2.2). The excitation light source is generally pulsed (or cw-modulated).

(B) *Intensity-dependent methods* are based on the measurement of the broadband light transmitted through the cavity (referred to as cavity-enhanced absorption spectroscopy, CEAS—Sect. 14.2.3). The excitation light source is commonly continuous wave.

All approaches use experimental multiplexing features to establish a wavelength-dependent optical loss spectrum "in one go". The time resolution in most applications can be very high, but depends ultimately on the detection scheme and spectral resolution. The basic measurement principles are briefly outlined in this section followed by a discussion of several advantages and drawbacks of specific approaches/properties for certain applications.

14.2.2 Time-Dependent Broadband Methods

14.2.2.1 Cavity Ring-Down Spectroscopy (CRDS)

The intensity, I, of broadband light in a cavity after pulsed injection decays exponentially according to the wavelength-dependent mirror reflectivity, $R(\lambda)$,[3] and losses, $L(\lambda)$, associated with the optical extinction (absorption and scattering) of the sample inside the cavity [9]:

$$I(t, \lambda) = I_{t=0} \exp\left(-\left[(1 - R(\lambda)) + L\right]\frac{tc}{d}\right), \tag{14.1}$$

where d is the length of the cavity and c is the speed of light. The time it takes the light intensity to reduce to e^{-1} of its initial value, $I_{t=0}$, is known as the ring-down time (see Chap. 1):

$$\tau(\lambda) = \frac{d}{c[(1 - R(\lambda)) + L(\lambda)]}. \tag{14.2}$$

For an empty cavity, the ring-down time is given by $\tau_0 = d/[c(1 - R)]$ and hence the optical loss can be calculated as

$$L(\lambda) = \frac{d}{c}\left(\frac{1}{\tau(\lambda)} - \frac{1}{\tau_0(\lambda)}\right). \tag{14.3}$$

The loss L depends on the sample in the cavity and/or the way it has been prepared. In most cases L is in good approximation a Lambert-Beer loss and can be described as $L = \varepsilon d'$, where ε is the extinction coefficient of the sample containing contributions from absorption and scattering, and $d' \leq d$ is the physical length of the sample inside the cavity.

The difference between broadband and conventional CRD spectroscopy is that broadband approaches measure multiple ring-down events as a function of wavelength simultaneously and hence possess multiplexing character. Wavelength selection and detector electronics typically require a time resolution of less than one µs to capture ring-down events [4]. As an example, a fast-rotating mirror in conjunction with a diffraction grating was used by Scherer et al. [10] to record the time and wavelength response of an optical cavity along orthogonal axes (time along the vertical axis, wavelength along the horizontal axis) of a CCD array detector. Wavelength-dependent ring-down times are acquired for each independent decay event. In another experimental implementation by Ball et al. [6] a two dimensional clocked CCD array was used at the exit of a spectrograph dispersing the light exiting the cavity. By applying suitably phased voltages to the electrodes along one of the axes of the pixel array, charge can be transferred efficiently between rows of CCD pixels in either direction. This enables a clocking sequence for images to be transferred between vertical pixels, permitting the continuous two dimensional acquisition of ring-down events [5, 11]. The technique has been applied to atmospheric

---

[3] $R(\lambda) = \sqrt{R_1 R_2}$, where R_1, R_2 are the reflectivities of the individual cavity mirrors (Fig. 14.2).

field studies for ambient concentration measurements of I_2, OIO, NO_2, NO_3, N_2O_5 [12–14]. More recent variants of broadband CRDS have seen the combination of the technique with white light supercontinuum sources [15–18].

14.2.2.2 Phase-Shift CRD Spectroscopy

In phase shift CRDS [19, 20], intensity modulated narrow band light (from a cw laser) is injected into a high finesse cavity. Due to the delay introduced by the optical cavity, the modulation on the light exiting the cavity is shifted in phase by $\phi(\lambda)$, which is related to the ring-down time, $\tau(\lambda)$, and the angular modulation frequency, Ω, by:

$$\tau(\lambda) = -\Omega^{-1} \tan[\phi(\lambda)]. \tag{14.4}$$

Thus absorption information is deduced from the wavelength-dependent phase shift that the modulated beam undergoes upon passing through the cavity. In the broadband version by Hamers et al. [7] the intensity-modulated light of a high pressure cw xenon arc was directed into the cavity after passing through a Fourier Transform (FT) spectrometer, and the light transmitted through the cavity was detected by a photomultiplier tube. The ratio of the in-phase and out-of-phase components of the intensity modulated light exiting the cavity (determined with a lock-in amplifier) gave the absorption information of the sample. This allowed sensitive broadband absorption measurements (over 250 nm) of the spin forbidden $b(0,0) \leftarrow X$ transition (A-band) of molecular oxygen in air, at a resolution determined by the interferometer.

14.2.3 Intensity-Dependent Broadband Methods

In CEAS with incoherent broadband (IBB) light sources, the time-integrated steady-state light intensity transmitted through the cavity is measured, which is inversely proportional to the extinction coefficient, ε. The use of incoherent light sources was first experimentally demonstrated with a short-arc Xe lamp in 2003 [8] by measuring the forbidden $b(2,0) \leftarrow X$ transition (γ-band) of O_2 and the $S_1 \leftarrow S_0$ absorption of gaseous azulene. Figure 14.2 shows a basic schematic of the implementation of an IBBCEAS setup.

The spectral resolution of IBBCEAS depends solely on the way the wavelength of the transmitted light is selected. Dispersive methods (typically grating spectrographs with multichannel detectors) enable high time resolution but limited selectivity, while interferometric approaches permit a high spectral resolution but a limited time resolution (see Sect. 14.3.3).

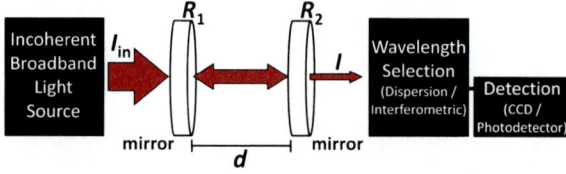

Fig. 14.2 Schematic of the experimental implementation of incoherent broadband cavity-enhanced absorption spectroscopy (IBBCEAS). The cavity is formed by two highly reflective spherical mirrors of reflectivity R_1 and R_2 which are separated by a distance d. Intensities of light incident on, and transmitted through the cavity are denoted I_{in} and I, respectively. Note: $R = (R_1 R_2)^{0.5}$

14.2.3.1 Measurement Principles of IBBCEAS

Consider a cavity of length d formed by two mirrors of reflectivity $R(\lambda)$ which is continuously excited with incoherent light of intensity I_{in} (see Fig. 14.2). Irrespective of the cavity's mode structure (see introduction) the intensity of light transmitted through the cavity, I, corresponds to the sum of photons leaking out of the cavity after an odd number of passes. For a cavity containing a sample with losses, $L(\lambda)$, this superposition of light transmitted through the cavity can be represented by the geometric series [8]:

$$I = I_{in}(1 - R)^2(1 - L) \sum_{n=0}^{\infty} (1 - L)^{2n}. \tag{14.5}$$

For Eq. (14.5) to hold, the transmission integration time t is assumed to be substantially larger than the ring-down time, τ. Since $R < 1$ and $L < 1$ this series converges to:

$$I = I_{in} \frac{(1 - R)^2(1 - L)}{1 - R^2(1 - L)^2}. \tag{14.6}$$

For a resonator without sample losses ($L = 0$) Eq. (14.6) simplifies to:

$$I_0 = I_{in} \frac{1 - R}{1 + R}. \tag{14.7}$$

Provided that the attenuation of light per pass is caused by Lambert-Beer losses, i.e. $(1 - L(\lambda)) = \exp(-\varepsilon(\lambda)d)$, the extinction coefficient, ε can be written as:

$$\varepsilon(\lambda) = \frac{1}{d} \left| \ln \left[\frac{1}{2R^2(\lambda)} \left(\sqrt{4R^2(\lambda) + \left(\frac{I_0(\lambda)}{I(\lambda)}(R^2(\lambda) - 1) \right)^2} \right. \right. \right.$$
$$\left. \left. \left. + \frac{I_0(\lambda)}{I(\lambda)}(R^2(\lambda) - 1) \right) \right] \right|, \tag{14.8}$$

irrespective of the magnitude of the loss per pass, εd. Note that the mirror separation d must be replaced by the physical length of the sample inside the cavity, $d' \leq d$, if the sample does not fill the entire cavity (see Sect. 14.2.2). In case of small losses per pass ($L \to 0$) and high reflectivities of the mirrors ($R \to 1$), ε can be approximated by:

$$\varepsilon(\lambda) \approx \frac{1}{d}\left(\frac{I_0(\lambda)}{I(\lambda)} - 1\right)(1 - R(\lambda)). \tag{14.9}$$

Equation (14.9) demonstrates that the effective path length in IBBCEAS is increased by a factor of $(1 - R)^{-1}$ in comparison to a single pass. The signal-to-noise ratio, however, is only improved by a factor of $[2(1 - R)]^{-1/2}$ assuming a photon-noise limited experiment and negligible absorption losses in the mirrors [21]. Fiedler et al. have studied in detail the influence of cavity parameters like the cavity length, mirror curvature and reflectivity, different light injection geometries on the IBBCEAS signal [22].

As opposed to time-dependent approaches where the mirror reflectivity can be determined through measurements with an empty cavity, R needs to be calibrated in case of intensity-dependent methods for the measurement of absolute extinction coefficients. A number of ways to determine $R(\lambda)$ will be given in Sect. 14.3.2.3.

14.2.3.2 Advantages and Drawbacks of Broadband CEAS

In order to get an overview of the benefits and weaknesses involved, the general properties of broadband cavity-enhanced absorption methods with incoherent light sources will be outlined in the context of the commonly desirable attributes of spectroscopic absorption methods:

(1) **Sensitivity:** A key criterion for the quality of an absorption technique is the smallest possible detection limit for one or more target species. The detection limit is essentially based on the effective absorption path length over which the light can interact with a sample for a given signal-to-noise ratio, which depends on a given integration time and long-term stability of the experimental setup (see Sect. 14.3.3.3). As a result of the low photon flux and coupling efficiency of incoherent light, ultimate sensitivities, as they can be achieved by mode matching a narrow band laser to the modes of a high finesse cavity, cannot be reached. However, 3σ minimum detectable extinction coefficients of 2.4×10^{-9} cm^{-1} Hz$^{-1/2}$ have been reported [23].

The sensitive detection of species can be done on an absolute (quantitative) or on a relative scale. Absolute extinction measurements require the rigorous calibration of the mirror reflectivity (Sect. 14.3.2.3), which is especially critical in broadband techniques where the high reflectivity of low loss mirrors can exhibit significant wavelength-dependent variations depending on the spectral bandwidth of the mirrors.

In contrast to ring-down approaches, there is in principle no constraint on the upper limit of the dynamical range over which incoherent broadband CEAS methods are applicable [8]. A practical limit is reached when optical losses can be measured in a single pass. That makes broadband methods attractive to systems that exhibit inherent losses such as transparent liquids or solids (see item (5) below).

(2) **Selectivity and multi-component detection:** The selectivity of an instrument is its ability to distinguish between different species absorbing at similar wavelengths; it depends on the thermodynamic state of the sample and the way it has been prepared. Although many different species may absorb light at one or more wavelengths, the total spectral profile of any particular species is unique. In practice, the high reflectivity range of the cavity mirrors and the spectral characteristics of light sources always limit the wavelength range available to probe a given sample. As a result, selecting an appropriate wavelength range that is specific to the target species can pose a serious challenge, especially if sample mixtures (i.e. "interfering" species) are present or several loss processes dominate the experiment. If an absorption method is not combined with another non-spectroscopic analytical approach to separate sample species (e.g. chromatographic techniques, filtering, titration, mass selection by ionization, to name a few), the spectral resolution is generally a measure for the method's capability to distinguish between different species.

Broadband methods allow a significant spectral range to be covered simultaneously. This multiplex advantage enables several species to be monitored at once, while gathering information over the entire spectral region. While dispersive methods (typically grating spectrograph with multichannel detectors) enable high time resolution (short integration times) but limited selectivity, interferometric approaches permit a high spectral resolution but a limited time resolution.

(3) **Time resolution:** An ideal spectroscopic absorption approach should achieve a good sensitivity with a high time resolution (i.e. short integration times) to enable the study of dynamics, e.g. reaction kinetics, molecular processes or flow dynamics. Obviously sensitivity and time resolution are connected via the achievable signal-to-noise ratio and hence the detection limit. In case of an interferometric detection scheme the required spectral resolution also affects the time resolution. Integration times from the ms range [24, 25] (condensed phase) to many hours [26] (high resolution FT experiment) have been reported in the literature. A natural restriction of the time resolution for cavity-enhanced methods is the ring-down time. For integration times that are shorter than the ring-down time, Eq. (14.5) does not hold anymore. If the system's dynamics proceeds on a time scale on the order of or smaller than the ring-down time, the analysis procedure outlined in Sect. 14.2.3.1 will become time-dependent and the geometric series in Eq. (14.5) becomes a time-dependent integral.

(4) **Versatility/Adaptability:** Since absorption spectroscopy is a universal tool in atomic and molecular sciences, ideally an absorption method should be applicable to gaseous, liquid, or solid systems and the corresponding interfaces. The principle applicability of a method over a wide spectral range is a criterion for its versatility. Depending on the light source and the high reflectivity range of the mirrors, broadband CEAS methods offer a high degree of adaptability. Even though cavity-enhanced methods have been primarily developed for gas phase absorption measurements, the absorption enhancement principle by increasing

the interaction path length with the sample is, of course, also applicable for liquid phase systems, transparent solids, films and surface layers on substrates and internal reflection optics (Chaps. 10 and 11, [21, 24, 27–31]).

Generally, the higher the mirrors' reflectivity (i.e. the higher the method's sensitivity), the smaller is the usable wavelength range and hence the versatility of the approach. An experimental approach to overcome these restrictions is outlined in Sect. 14.3.2.1.

(5) **Practical parameters:** Other aspects that constitute practical advantages of sensitive detection methods relate to the experimental approach being (a) compact or even potentially portable, (b) uncomplicated to implement and "user-friendly", (c) robust and stable/durable, (d) inexpensive with low maintenance requirements. Some of the broadband approaches summarized in Sect. 14.4 possess several of these advantages. The sensitivity to optical alignment, complexity, size or cost of some cavity-enhanced absorption methods and also that of outright different techniques, such as degenerate four wave mixing, photoacoustic spectroscopy, resonance enhanced multi-photon ionization spectroscopy, or intracavity laser absorption spectroscopy, deter their widespread adoption for field observations or common application in commercial analytical laboratories.

14.3 Experimental Aspects

14.3.1 Light Source Considerations

The choice of light source (pulsed, modulated or continuous wave) essentially determines the type of experimental broadband scheme (cf. Fig. 14.1) and also governs the signal-to-noise ratio for a given detection system. The favourable attributes of light sources listed below under (a)…(d) should be considered for a successful implementation of the benefits of broadband cavity-enhanced setups, as discussed in the previous section:

(a) *Broad unstructured emission spectrum:* For light sources whose emission bandwidth is broader than the high reflectivity range of the cavity mirrors, rigorous filtering of wavelengths outside the high reflectivity range is essential in order to suppress the detection of unwanted stray light. Very broad emission spectra provide a high degree of flexibility in terms of switching between different excitation ranges, without major alterations on the excitation side of the experiment. If the emission spectrum is smooth and unstructured, then spectral intensity fluctuations that are inherent to the source have less negative effects on cavity absorption measurements (see (d)).

(b) *High spectral brightness:* The number of photons incident on the cavity per unit area and time ought to be as large as possible because of the unfavourable coupling efficiency of broadband (temporally incoherent) light to high finesse cavities. The smaller the emitting area for a specific output power (i.e. the higher the spatial coherence at the source's exit aperture) the better are the imaging properties of the source. However, increasing spatial coherence through spatial filtering is accompanied by a reduction in intensity [$W\,m^{-2}$]. The key quantity that should therefore be maximized for broadband cavity-enhanced applications is the spectral radiance [$W\,m^{-2}\,sr^{-1}\,nm^{-1}$].

(c) *High emission stability and low spectral noise/drift:* For (intensity-dependent) absolute extinction measurements the spatial and spectral stability of the light source is of great importance. (Narrow) line features in the emission spectrum or significant spectral intensity gradients can be difficult to account for by reference measurements (without sample) even if the light source exhibits only small intensity fluctuations. Attention should be paid to not commence measurements during warm-up of the light source, when potential spectral drifts and output intensity changes are very common.

(d) *Robustness, compactness, low maintenance, low cost, long lifetime:* Some of these very practical attributes of light sources constitute key advantages of broadband CEAS and may well determine the choice of light source for a specific application as well as outweigh some of the experimentally important features (a) to (c). Preference for one of the attributes at the expense of another is often given based on the aspects listed under (d).

The following section summarizes the most important properties of light sources commonly used in broadband CEAS (the list is exclusive of frequency comb sources, which are discussed in Chap. 8.

Arc Lamps Corresponding to the colour temperature (typically several thousand K) arc lamps produce a white light continuum that usually covers the spectral region from the onset of the vacuum UV to the near IR [32]. For Xe arcs atomic lines contribute mainly in the region between 750 and 1000 nm to the total optical power and to a lesser extent around 475 nm, leaving the remainder of the visible output rather insensitive to line fluctuations. Since generally the size of the plasma arc scales with the wattage rating of the bulb, higher power lamps are not necessarily more appropriate for broadband cavity applications. On the contrary, light emerging from an extended area cannot be imaged as well as the light from a small (quasi point) source. While most Xe-arc lamps have a *diffuse* arc where the area of light emission is distributed across the entire plasma, Varma et al. [33] used a short-arc Xe lamp with a *hot spot*, where ca. 80 % of the emission come from a plasma spot of the dimensions of 150 μm yielding a spectral radiance of 18 $W\,cm^{-2}\,sr^{-1}\,nm^{-1}$ at 400 nm [34]. However, the arc is prone to erratic jumps and wandering of the hot spot, which requires an active spot stabilization.

For high resolution applications requiring Fourier transform detection (see Sect. 14.3.3.2), *superquiet* current stabilized Xe lamps (type Hamamatsu) with typical intensity fluctuations of 0.2 % and drifts of ±0.5 %/h are useful, because longer integration times are generally required and unwanted high frequency noise introduced by arc movement is to be avoided to maximize the signal-to-noise ratio [35].

For many years Xe-arc lamps have been the "work horse" for spectroscopic techniques such as long-path differential optical absorption spectroscopy (LP-

DOAS) [36]. In the context of cavity-enhanced absorption spectroscopy they have been first applied in 2003 [8]. There have been numerous applications in incoherent broadband cavity-enhanced absorption spectroscopy (IBBCEAS) [8, 21, 26, 35, 37–44], but also broadband phase-shift CRD spectroscopy (PSCRDS), where the intensity of the light needs to be modulated before entering the cavity [7]. Furthermore, Walsh and Linnartz showed recently that the plasma emission in a supersonic jet can also be used to enhance the observability of self-absorption in the plasma [45].

Halogen Lamps   Halogen lamps emit a smooth continuous spectrum from the near UV to the near IR. Their advantage over Xe lamps is the absence of emission lines, and price. However, their inferior brightness and the large size of the emitting filament make them less suitable for broadband CEAS. Their main application is in absorption systems with inherently large losses occurring for instance in condensed phase studies [21, 25, 46].

Light Emitting Diodes (LED)   High power *LEDs* are power efficient, compact, generally inexpensive and long lived (up to > 10000 h).

Coloured LEDs which are sufficiently bright for CEAS are now available in spectral windows from the near infrared to the near UV (>350 nm). Being sensitive to temperature fluctuations they are prone to (thermal) spectral drifts and hence require adequate temperature and current stabilization for cavity applications. LED spectra typically shift towards longer wavelengths when the temperature increases [47]; slight tuning of the wavelength range can be achieved in that way to adapt the output spectrum to the reflectivity range of the mirrors [48]. E.g. based on a temperature stabilization of ± 1 mK, Ventrillard et al. [49] reported that the spectral drift of the emission spectrum of their red LED (625 nm) was smaller than 0.015 nm, while the total intensity fluctuation was better than 0.02 % per hour. When stabilized appropriately, LEDs emit a rather smooth quasi-Gaussian [47] or Lorentzian [48] spectrum over typically a few tens of nm (FWHM), although emission bandwidth can vary from manufacturer to manufacturer (up to 50 nm, [48]). Depending on the FWHM of the LED spectral filtering is less stringent but still recommended.

White LEDs (e.g. based on the excitation of an emitting phosphor in the blue [31]) cover large parts of the visible spectrum simultaneously. Often LED arrays are used to achieve high optical output powers. However, due to the larger overall surface area (typically a few mm^2) the imaging properties of arrays are generally inferior.

The application of LEDs in IBBCEAS was first demonstrated by Ball et al. [48] in the green (535 nm) and red (661 and 665 nm) parts of the visible spectrum. LED based IBBCEAS has since been widely applied for gas phase spectroscopy in the visible [49–58] and in the near-UV [59–61], but also for liquid phase spectroscopy [27, 28, 62]. LEDs have also been used in conjunction with cavity attenuated phase shift (CAPS) spectroscopy [63, 64].

Superluminescent light emitting diodes (*SLEDs*) are based on the generation and amplification of spontaneous emission in a semiconductor waveguide over a broad wavelength range. Their emission bandwidth can vary from 5 to 100 nm.

SLEDs therefore combine beam divergence and power density comparable to that of a single-mode laser diode with the low temporal coherence of a conventional LED [65]. SLEDs have not been extensively used as light sources for CEAS measurements yet [66–68]. Some studies compare the performance of SLEDs to that of supercontinuum sources (see below).

Supercontinuum Sources: "White Fibre Lasers"   In supercontinuum (SC) sources a broad spectral emission is generated by pumping a certain length of a highly non-linear micro-structured photonic crystal fibre (PCF) with short (fs up to ns) pulses from a seed laser, generally using a high repetition rate (kHz to MHz) [69]. The well collimated fibre output has a high power density (up to several mW per nm) and is spectrally very broad; it typically extends from the blue region of the spectrum (>400 nm) to the near-IR (<2.5 μm) for a pump wavelength around 1060 nm. Significant performance instabilities can be caused by optical feedback into the PCF, making operation with an optical cavity critical due to the strong back reflection from the entrance cavity mirror (especially when working around the seed wavelength). SC sources are also prone to exhibit considerable power as well as spectral fluctuations, which have adverse effects in cavity applications.

In 2008 the first applications of SC sources to broadband CEAS were reported. In the wavelength range 630–700 nm Langridge et al. [23] used an SC source for NO_3 and NO_2 detection and reported the stability of the transmission spectra to vary between 0.2 and 0.5 % over 100 consecutive (2 s) acquisitions. Other examples of application of SC sources are briefly outlined in Sect. 14.3.2.1 and listed in Sect. 14.4. A somewhat different approach to create continuum radiation for broadband CRD spectrography was taken by Stelmaszczyk et al. [16], who generated plasma filaments in quartz instead of using a PCF.

Comparative Studies   Van der Sneppen et al. [31] compared the performance of a 1 W white LED with an emission spectrum covering 420–650 nm to that of an SC source in the context of evanescent-wave BBCEAS. In this specific comparison the measurements taken with the SC source were reported to have much lower baseline noise (by an order of magnitude) and higher precision (by a factor of $\sim$1.7) [31]. Denzer et al. [67] compared the performance of two near-IR SLEDs, whose outputs were combined using a polarizing beam-splitter, to that of an SC source in the context of Fourier transform CEAS. The optical power from the SC source (within the range 1600–1700 nm) was four times higher than the total optical power incident on the cavity from the combined SLEDs. Using a FT spectrometer with a maximal spectral resolution of 4 cm^{-1} a significant sensitivity improvement from a minimal detectable absorption of 2×10^{-8} cm^{-1} to 5×10^{-9} cm^{-1} in a 4 min acquisition time was observed upon swapping the SLEDs for the SC source.

Generally, when SC sources are used in conjunction with high resolution Fourier-transform detection, instabilities in the SC output or due to the presence of intrinsic etalon effects can be observed, negatively affecting performance [66, 70].

Non Mode-Locked Nanosecond Broadband Lasers Among the sources with high directionality, nanosecond dye lasers without dispersive optical element have been commonly applied in pulsed broadband cavity applications [4, 6, 71, 72]. They produce lasing and/or strong amplified spontaneous emission (ASE) over a bandwidth which is determined by the laser dye (or dye mixture). Broadband dye lasers typically emit light over tens of nanometer in the visible (or near UV or IR) and do not require rigorous spectral filtering of wavelengths outside the high reflectivity range of the mirrors. Due to the good spatial coherence and high power density of the emitted light the injection into a high finesse cavity can be effective even without mode-matching. Since broadband dye lasers can exhibit significant shot-to-shot fluctuations and spectral structure in the emission, they are not particularly appropriate for applications depending on the cavity transmission [1]. While the first application of a "broadband" nanosecond dye laser was in Fourier transform CRDS enabling simultaneous monitoring of a 400 cm^{-1} broad spectral range in the near-IR [73], most applications have been in broadband CRDS [4, 6, 12, 48], broadband ring-down spectral photography [10], and cavity-ring-down spectrography [71]. In the latter the tuning element and the output mirror of the dye laser were removed, enabling suppression of commonly excited high order transverse modes. A drawback of (expensive) dye lasers is that they are not particularly compact and only moderately flexible in term of quick and straightforward wavelength range selection, which requires changing the dye.

A *short pulsed free-electron laser* running at high repetition rate (ca. 12 MHz), characterized by very good beam quality and stability over a 25 nm FWHM, was used for broadband (pulse-stacked) CRDS in the infrared [74].

14.3.2 Cavity Considerations

14.3.2.1 Types of Cavity Used for BBCEAS with Incoherent Light Sources

Linear Cavities By far the most common geometry is the conventional linear (two-) mirror cavity (Fig. 14.3(a)). Fiedler et al. [22] studied the influence of various cavity parameters on the signal-to-noise ratio (SNR) for IBBCEAS. According to this study, a symmetric confocal linear cavity with converging light injection produces the optimal SNR. The maximum SNR with a symmetric near-confocal cavity configuration is easier to achieve for large mirror separations than for short cavities. The optimum use of linear cavities requires the f number of the cavity to be adapted to that of the spectrometer.

The large majority of studies carried out with linear (two-) mirror cavities (Fig. 14.3(a)(i)) focus on the detection of **gas phase** species (see Sect. 14.4), mostly employing setups where the cavity is part of a flow [38–40] or static cell [8, 41, 48] at high or modest pressures, or forming a vacuum-tight chamber for low pressure measurements [37, 45, 75]. In some atmospheric applications open-path configurations were also used, e.g. in conjunction with atmospheric simulation chambers.

Fig. 14.3 Typical cavity
geometries used for
broadband cavity-enhanced
absorption spectroscopy.
(**a**) linear cavity: (**i**) directly
filled with gas [5] or
liquid [62], (**ii**) containing a
coated or doped
substrate [33], or an
absorption cell [21] or
microfluidic chip [76],
(**b**) prism cavity [77],
(**c**) folded cavity (different
internal reflection elements
can be used) [25, 79],
(**d**) end-coated fiber
cavity [65], (**e**) ring (bow-tie)
cavity [68]

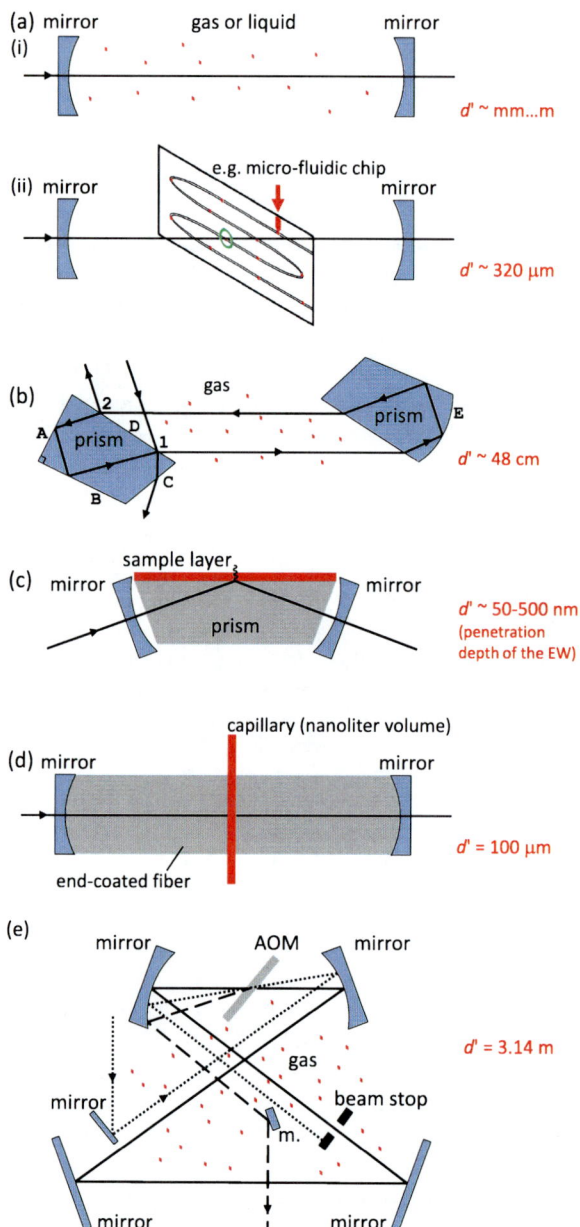

(a) mirror   gas or liquid   mirror
(i)

$d' \sim$ mm...m

(ii) mirror   e.g. micro-fluidic chip   mirror

$d' \sim 320$ μm

(b)   gas   prism   E
A   prism   D 1

$d' \sim 48$ cm

B   C

(c)   mirror   sample layer   mirror

prism

$d' \sim 50\text{-}500$ nm
(penetration
depth of the EW)

(d) mirror   capillary (nanoliter volume)   mirror

end-coated fiber

$d' = 100$ μm

(e)
mirror   AOM   mirror

$d' = 3.14$ m

mirror   gas   beam stop

m.

mirror   mirror

Closed-path and open-path configurations present advantages and drawbacks which will be discussed in detail in Sect. 14.3.2.2.

Linear (two-mirrors) cavities have also been used for **liquid phase** applications, where the liquid is either contained in a cuvette or HPLC cell [21, 27–29], or enclosed in the liquid-tight cavity itself (the cavity mirrors acting as windows of the setup) [62]. Islam and co-workers, for instance, studied liquid analytes with cavities where the high reflectivity concave mirrors were separated by up to 20 cm and sealed with liquid-tight perfluoro polymer O-rings [62]. With a reported sensitivity of 2.8×10^{-7} cm^{-1} in a 2.5 s integration time the 20 cm cavity study demonstrated the potential of IBBCEAS for spectral separation of multiple liquid-phase analytes with overlapping absorption profiles.

Neil et al. [76] reported the application of IBBCEAS to *in situ* analyte detection within microfluidic droplets. In [76] a microfluidic chip was inserted into the optical cavity normal to the optical axis such that the light within the cavity passes through a single microfluidic channel (Fig. 14.3(a)(ii)). The number of light passes through the cavity in this approach is primarily limited by the optical quality of the microfluidic chip itself.

Prism Cavity   If the spectral coverage in a broadband cavity experiment is important, the high reflectivity range of the mirrors can become the limiting experimental factor.[4] Using a prism cavity the high reflectivity bandwidth is practically only limited by the spectral regions where the prism material exhibits high internal transmission losses. Depending on the prism material (e.g. calcium fluoride—UV/Vis, fused silica—Vis/NIR, barium fluoride NIR) large regions can be covered with high effective reflectivities. Figure 14.3(b) shows the optical cavity designed by Johnston and Lehmann [77], i.e. two retroreflectors facing each other such that corresponding faces are parallel. Light enters the cavity at point **1** and exits at **2**. The dimensions, refractive index and angles of the first prism are chosen such that a ray entering surface **D** at the Brewster angle will be totally internally reflected at surfaces **A** and **B** at 45°. The second prism is identical except that surface **E** is convex (with a radius of curvature of 6 m in [77]). The input beam passes through surface **C** also at Brewster's angle. The effective reflectivity depends on the detuning angle of the input prism around Brewster's angle, $\delta\theta(\lambda)$ [rad] [77]:

$$R(\lambda) = \frac{(n^4(\lambda) - 1)^2}{4n^6(\lambda)}\delta\theta^2(\lambda), \tag{14.10}$$

where $R(\lambda)$ is the fractional reflected intensity, $n(\lambda)$ is the prism material index of refraction. Scattering losses at the prism interfaces are minimized by super polishing the optical surfaces (<1 Å rms roughness, scratch-dig 0/0). The calculated Fresnel losses per prism are minimal at the design wavelength of the cavity (1064 nm) and monotonically increases from "negligible" to longer and shorter wavelengths (to a

---

[4]High resolution spectra of the weak $b - X (1 \leftarrow 0)$ transition of O_2 at 14529 cm^{-1} and of the fifth overtone of the acetylene C–H stretch vibration at 18430 cm^{-1} are reported in [77] with a FWHM of 0.18 and 0.44 cm^{-1}, respectively.

value of ca. 4 ppm at ~650 and ~1750 nm in [77]). The round trip time in the cavity was 3.6 ns for a cavity length of 0.5 m, including the propagation time inside the prisms. More details on the choice of prism material and fabrication errors, optical beam propagation calculations to determine the specification of prism angles and relative dimensions, sensitivity of the optic axis to changes in prism alignment and other experimental features can be found in [78].

Folded Cavities for Evanescent Wave Applications   Cavities involving a total internal reflection element (e.g. monolithic or Pellin-Broca cavities and other folded- or ring-cavities) have been applied in numerous evanescent wave (EW) CRDS applications (not including whispering gallery modes in solid microspheres or droplets). A comprehensive review of EW-cavity based spectroscopy (including cavity design and light source considerations as well as applications) has been given recently by Schnippering et al. [79]. For **broadband applications** only **folded cavities** have been used comprising two high reflectivity mirrors and a trapezoidal or right angle prism (Fig. 14.3(c)). The entrance and exit of the beam are normal to the prism faces, ensuring the reflective losses to be trapped within the cavity [21]. The first broadband (390–625 nm) application demonstrated by Ruth and Lynch [30] used a right angle prism cavity with xenon lamp illumination to investigate thin dry films of metallo-porphyrins deposited from solution at the silica–air interface. In this study, a detailed evaluation of the prism losses within the folded cavity is given together with the description of the loss behaviour of an evaporating layer of sample solution. The 3σ minimum loss per pass (εd) was found to be 2×10^{-5} for an integration time of 100 ms.

An SC source was also applied in conjunction with a folded evanescent-wave cavity for thin layer diagnostic at the interface of a silica substrate [24] to detect electrogenerated species at the silica-water interface in the range 400–750 nm. Schnippering et al. report an effective minimum detectable loss per pass of 3.9×10^{-5} in an integration time of 0.5 s. Real time fast interfacial kinetics of electrogenerated Ir(IV) complexes in a thin-layer electrochemical cell arrangement was shown feasible with a dove prism folded cavity and SC source excitation by van der Sneppen et al. [31].

Other Cavities   An implementation of broadband cavity-enhanced absorption spectroscopy with a "bow-tie" **ring cavity** (Fig. 14.3(e)) has been demonstrated by Petermann and Fischer for a gas phase application [68]. An intracavity acousto-optic modulator (AOM) is used to directly couple light into (and out-of) the cavity at specified times, however, at the expense of additional losses caused by the AOM. The AOM by-passes the cavity mirrors and makes it possible to load the cavity efficiently and to actively "dump" the total energy in the cavity at any point onto a photo-detector, enabling a time-dependent CRD approach. The "bow-tie" geometry helps minimizing the beam waist in the AOM, necessary for fast switching.

Other cavity types for broadband absorption enhancement are used for systems on the micro- to nano-scale. Gomez et al. [65] used an **end-coated fibre** for IB-BCEAS measurements of aqueous samples in nanoliter volumes; the liquid sample is contained within a vertical 100 μm wide capillary that is integrated at the center of the horizontal optical fiber (Fig. 14.3(d)). Recently the use of low quality factor whispering gallery modes in spherical silicon nanoshells has been demonstrated by Yao et al. [80] (see also Chap. 9). The low quality factor enables good coupling conditions, broadens the resonant absorption peak, and makes the absorption enhancement region wider.

14.3.2.2 Linear Cavity: Open-Path Versus Closed-Path for In Situ Trace Gas Monitoring

The largest signal-to-noise enhancements in absorption spectroscopy with cavities can be achieved for gas phase systems. The good time resolution of broadband cavity approaches, in conjunction with a wide spectral coverage at sufficient spectral resolution and their general robustness, are attractive features for *in situ* monitoring of many typical atmospheric trace constituents, either in the natural environment or in atmospheric simulation chambers. Under atmospheric conditions Mie scattering and absorption of aerosol particles can reduce the intensity of light in a cavity, apart from molecular absorption of the target species and the elastic scattering by air. Aerosol extinction can be accounted for through the analysis of the light attenuation (caused by aerosol) with differential fitting techniques applied to the acquired broadband spectra [33, 50]. The retrieval of species concentration from the spectra is described in detail elsewhere and the reader is referred to Refs. [5, 33, 40, 55].

Broadband cavity-based instruments with standard linear mirror cavities have been used for atmospherically relevant research either in **open-path** configurations outdoors [12–14], indoors [50, 56, 61], or in simulation chambers [33, 38, 44, 57, 81]), and in **closed-path** configuration [43, 49, 52, 54, 82, 83] by continuously extracting gas from a sample volume and flowing it through an enclosed cavity for detection. Both, open-path and closed-path setups present advantages and drawbacks for trace gas monitoring which will be outlined in the following.

Closed-Path   Extractive sampling enables the design of compact broadband cavity systems (typically ca. 1 m cavity length) providing good mechanical stability and compactness. Instruments can be made portable and can be sheltered inside a small room or container (during measurement campaigns), where thermal stability of the setup components, notably the spectrometer and optics, can be more easily ensured. A major advantage of enclosed cavities is that any type of calibration procedure, most importantly the *in situ* determination of the mirror reflectivity, can be performed in a straightforward manner, by simply filling the cavity with e.g. zero air (for CRD applications) or a calibration gas (for CEAS applications)—see Sect. 14.3.2.3. Moreover, aerosol optical losses can be eliminated (if required) by filtering the extracted air sample. The main drawback of extractive systems is related to the walls of the setup which are associated with different losses for different target species. Wall losses occur in the inlet line and on the walls of the enclosed cavity over time. The length of the extraction line and the positioning of the instrument with respect to the sampling point is an important consideration; in general short inlet lines are to be preferred. Wall losses of target species in the inlet line, including effects of potential filters or scrubbers and cavity wall losses, which depend on the sample residence time and hence the flow speed settings, need to be thoroughly considered and require extensive calibration or correction procedures [83, 84]. Unfortunately for real outdoor gas samples the wall losses may change depending on the composition of the gas mixture and its chemistry, so that some uncertainty of wall losses always remains. Mirrors are (not necessarily but) typically protected by a small purge flow of zero air or N_2.

Open-Path   Instruments operating an open-path cavity have the advantage that they are free from wall losses of target species. Since the cavity mirrors are not enclosed, the length of the cavity can be extended arbitrarily. Extending the cavity will increase the sensitivity, provided a light source is used that can be well collimated/imaged (cf. Sect. 14.3.1), the mechanical stability is guaranteed, and the gaseous sample is homogenously distributed between the mirrors. For instance, in Refs. [33, 85] an IBBCEAS setup installed at the SAPHIR atmospheric simulation chamber (Jülich, Germany) is described employing a ∼20 m long open-path cavity for *in situ* detection of the NO_3 radical between 630 and 690 nm. Using a longer cavity offers the possibility of using mirrors with a lower reflectivity without affecting the sensitivity. Lower mirror reflectivities increase the overall cavity transmission (provided the same amount of light can be coupled through the cavity, e.g. with a SC source) and at the same time enable wider spectral ranges to be covered, which generally improves the retrieval accuracy of target species. Both aerosol extinction and absorption of gaseous molecular and radical species can be retrieved in open-path setups [33, 86]. However, in case of a highly variable atmosphere, especially with regard to aerosol load, the concentration retrieval of target species can be challenging. If strong aerosol (or target species) fluctuations occur on a time scale smaller than the experimental integration time, the accuracy of the data analysis is seriously affected. Moreover, for very high aerosol concentrations the molecular absorption signal can be swamped, lowering the effective sensitivity of the open-path design owing to a significant reduction of the effective path length. There is a genuine drawback of open-path configurations which is inherent to all (time- and intensity-dependent) cavity methods using open-path, i.e. a "clean air" (target species free) reference point (e.g. transmission I_0 in Eq. (14.9) or τ_0 in Eq. (14.3)) is not necessarily available. All concentration measurements including the mirror reflectivity calibration (see Sect. 14.3.2.3) refer to a reference measurement, which cannot be accurately defined in real outdoor/field environments. In simulation chambers [33, 38, 44] and other controlled environments where open-path configurations are applied, this issue can be adequately addressed by filling the chamber with zero-air or a calibration gas. For short enough cavities and depending on the experimental design, open-path cavities can be easily turned into closed-path configurations and vice versa, by including and removing a tube between the mirrors. This straightforward feature of having a removable tube incorporated in the cavity, gives the instrument maximum flexibility in terms of reference measurements and calibration. Open-path cavity mirrors are generally also purged with zero-air or N_2. On the detector side of the cavity means for stray light reduction (such as short tubes) are usually implemented for daytime measurements.

14.3.2.3 Determination of Mirror Reflectivities

For the measurement of absolute extinction coefficients the (effective) mirror reflectivity, R, needs to be calibrated in case of intensity-dependent methods. This is different from time-dependent approaches where R can be determined through measurements without a sample in the cavity. The calibration is especially critical in broadband techniques where the high reflectivity of low loss mirrors can exhibit significant variations depending on the broadness of the wavelength range. A number of approaches have been established to determine $R(\lambda)$:

(a) Measuring the transmission of a cavity without and with a sample of known concentration and optical loss (absorption cross-section) leaves only $R(\lambda)$ in Eq. (14.9) as unknown parameter. The error in the reflectivity depends on the uncertainty of the (resolution corrected) absorption cross-section of the calibration sample apart from the experimental uncertainties associated with determining concentrations and transmission intensities. For atmospheric gas phase measurements typical calibration gases have been NO_2 [38, 59] or the dimer O_2-O_2 [50, 53], and for surface applications organic dyes have been applied [31].

(b) Rather than using a molecular absorption loss for calibration the elastic (Rayleigh) scattering losses of a calibration gas (ideally at different pressures) can be used to determine the mirror reflectivity. For example, Washenfelder et al. determined the reflectivity, $R(\lambda)$, by measuring the ratio of transmitted intensities when the cavity was filled with He and zero air, (I_{air}/I_{He}) [40]. Since Rayleigh scattering cross sections for He and air are well known, the reflectivity of the mirrors can be directly determined as

$$R(\lambda) = 1 - \left(\frac{(\frac{I_{air}}{I_{He}})\alpha_r^{air} - \alpha_r^{He}}{1 - (\frac{I_{air}}{I_{He}})} \right) d, \qquad (14.11)$$

where α_r^{air} and α_r^{He} denote the extinction coefficients for Rayleigh scattering of air and He respectively.[5] The same principle was used by Chen and Venables [43] who used N_2 and CO_2. The advantage of this approach is that elastic

---

[5]Note that the gases used (He and air) have large differences in their Rayleigh scattering cross sections.

scattering losses vary uniformly with wavelength over the entire spectrum, enabling reflectivity measurements over a wide spectral range covering the near UV and visible.

(c) Mirror reflectivities can also be determined by using an antireflection-coated optical substrate of known loss, $L(\lambda)$, instead of a calibration gas [30, 33], which is mostly used in open-path configurations for calibration and referencing purposes. The reflectivity is then obtained as

$$R(\lambda) = 1 - \left(\frac{I_1(\lambda)}{I_0(\lambda) - I_1(\lambda)} L(\lambda) \right), \tag{14.12}$$

where $I_1(\lambda)$ and $I_0(\lambda)$ are the respective transmission spectra of the cavity with and without the low loss substrate in a clean cavity. $L(\lambda)$ must be determined independently either through a CRD experiment or through a calibration approach such as (a) or (b).

(d) Mirror reflectivity can also be measured if the experimental setup provides the possibility of a cavity ring-down measurement [15, 16]. A few studies have used a type of "hybrid setup" [29, 48, 77, 87] where CEAS and CRD are combined. Ideally the mirrors do not need to be changed between the CRD measurement and the broadband CEAS measurement.

(e) An alternative "hybrid" calibration approach without altering the cavity alignment is using PSCRDS, where the mirror reflectivity $R(\lambda)$ is determined directly from the measurement of the phase delay [19], $\phi(\lambda)$, introduced in the cavity when purging it with a dry gas [29, 54, 82, 83, 88]:

$$R(\lambda) = 1 + d \left(\frac{\Omega}{c \tan \phi(\lambda)} + \alpha(\lambda) \right), \tag{14.13}$$

where Ω is the angular modulation frequency of the excitation source and α the wavelength-dependent extinction coefficient due to elastic Rayleigh losses. Mirror reflectivity calibration using PSCRDS was demonstrated using acousto-optic tunable filters for modulating a single supercontinuum source, used for both the calibration and the main absorption experiment [88]. The filters permit narrowband portions of the broadband radiation to be scanned over the full bandwidth of the cavity mirrors. After the calibration the system is switched over into a IBBCEAS configuration without any loss in optical alignment.

Note that if the full surface area of the mirror is illuminated in a cavity experiment (typical for IBBCEAS), then the corresponding reflectivity is generally somewhat smaller than in applications using only on-axis excitation of the cavity. Depending on the cavity geometry the distribution of modes excited in the cavity determines the reflectivity obtained in the calibration. Since the contribution of transverse modes is larger if the full mirror is illuminated, one essentially measures a (spatial) mean of reflectivities across the mirror surface for given excitation conditions [51]. This should be kept in mind when different beam diameters are used for the calibration and the main absorption measurement. The problem of finding a quantitative relationship between the spatial and spectral excitation conditions for a given cavity geometry and the resulting effective reflectivities of the cavity mirrors has not been adequately addressed in the literature at the time of publication.

14.3.3 Detection Schemes

All broadband CEA approaches have in common that the wavelength selection takes place after the cavity[6] by either dispersive or interferometric means, as sketched in Fig. 14.4. While the spectral resolution of the setup is solely determined by the type of wavelength selection, the sensitivity is affected by both the wavelength selection scheme and the subsequent photo detection efficiency.

14.3.3.1 Dispersive Wavelength Selection Approaches (Predominantly Near UV to Near IR)

The light transmitted through the cavity is either directly focused onto the entrance slit of a monochromator or coupled into a multimode fibre (or fibre bundle) connected to the entrance aperture of a monochromator. The light in the entrance slit is typically imaged onto a (potentially cooled) charged coupled device (CCD) array via a dispersive optical element; typically a reflection grating selecting a single order, see Fig. 14.4. If the diffraction pattern exhibits overlapping orders, then sometimes additional optical filtering is applied to limit the width of the diffraction orders and to avoid the spectral/spatial overlap. Depending on the dispersion of the monochromator the entire broadband spectrum can be acquired simultaneously with a time resolution that is determined by the intensity $[W\,cm^{-2}]$ on the detector. In case of CCD detector arrays the signal-to-noise ratio is greatly improved by vertical binning of the CCD pixel columns. The monochromator's slit width, the groove density of the grating and its distance from the multi-channel CCD detector determine the spectral resolution in combination with the corresponding pixel size. For IBBCEAS compact monochromators with a few cm focal lengths and modest stray light suppression (typically $10^{-3}\ldots10^{-4}$) are commonly used, providing typical spectral resolutions between ~0.1 and ~1 nm. This resolution range is sufficient to resolve homogeneously or inhomogeneously broadened spectral features in the UV–VIS as they occur in many molecular environments and systems, e.g. through collisional and Doppler broadened spectra in the gas phase, or through various inhomogeneous broadening mechanisms for molecules in the condensed phase and on surfaces. In these cases spectral interpretation, species monitoring, and diagnostic applications all benefit more from fast detection over a broad wavelength range rather than from a high spectral resolution.

---

[6]An exception is the setup in Ref. [7] where interferometric processing of the signal was done before the cavity and the transmitted signal was measured immediately after the cavity by a photomultiplier tube.

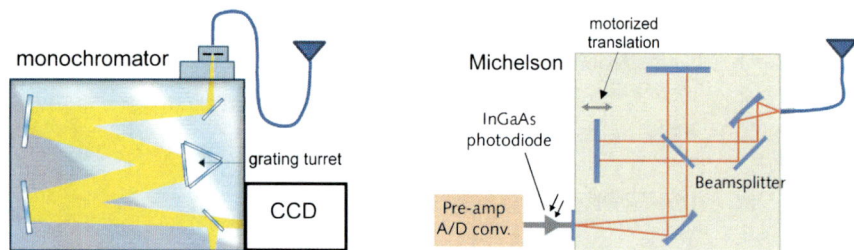

Fig. 14.4 Sketches of typical detection schemes. *Left*: Dispersive approach—grating monochromator and multi-channel detector (CCD). *Right*: Interferometric approach—FT spectrometer (Michelson) with single-channel detector (photodiode or PMT)

Dispersive wavelength selection methods have been successfully implemented in a wide variety of IBBCEAS applications, including gas phase [8, 23, 37, 61, 82, 89], liquid phase [21, 27–29, 62], thin film and surface [24, 30, 31] studies.

14.3.3.2 Interferometric Wavelength Selection Approaches (Typically in the VIS to Near IR)

The light transmitted through the cavity is imaged onto the entrance aperture of a interferometer, which generates an amplitude modulated signal (interferogram) by constructive and destructive interferences of the broadband light travelling over different path lengths inside the interferometer. While a continuously moving (or stepwise scanned) mirror generates optical path differences (δ) between two light beams emerging from the interferometer's beamsplitter, the intensity (I') of the re-combined beams is measured by a single detector, typically a (potentially cooled) photodiode or PMT in the focal plane of a positive lens (for interferences local-ized at infinity). The spectrum $I(\lambda)$ is reconstructed after Fourier transforming the recorded interferogram $I'(\delta)$. The spectral resolution is determined by the com-bination of both the size of the circular aperture and the maximum optical path difference δ_{max} (i.e. retardation) induced between the two interfering beams. The longer the maximum distance between the moving mirror and the beamsplitter, the higher is the achievable resolution. Typical resolutions reported in the context of broadband cavity experiments are between 0.02 cm^{-1} and 4 cm^{-1} [35, 67]. With the option of a much higher spectral resolution than dispersive methods, FTS in conjunction with broadband CEAS has been mainly applied in the (near) IR region, where narrow band ro-vibrational absorption features of molecular species can be exploited to achieve good selectivity. However, high resolution and sensitivity come at the expense of numerous δ scans, which generally require rather long acquisi-tion times. Moreover, high resolution Fourier transform spectrometers are neither particularly compact, nor very robust (for field applications) and can furthermore be costly. These aspects illustrate that FT-IBBCEAS is less appropriate for kinetic studies and for environmental trace gas monitoring and that its main benefits are in

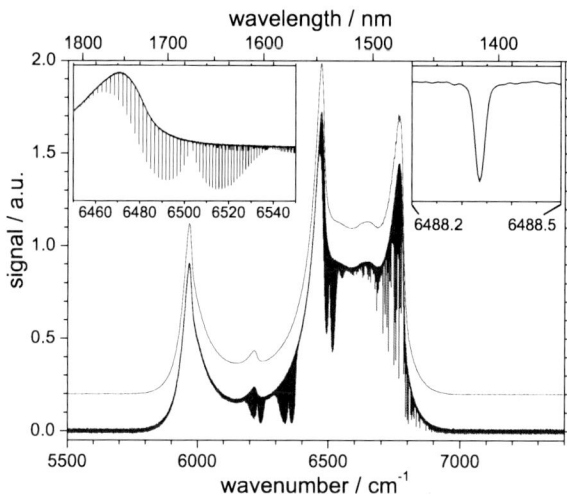

Fig. 14.5 High-resolution spectrum of CO_2 in the near IR obtained using an optical cavity in conjunction with a Fourier-transform spectrometer (a.u. = arbitrary units of intensity) [35]. The upper trace (shifted upwards for clarity) shows the spectrum of the empty cavity, the lower trace shows the spectrum obtained with 26.7 mbar of CO_2 in the cavity. The acquisition time is 90 min, the spectral resolution is 0.02 cm^{-1}. The *inset in the left upper corner* shows a weak overtone band $(30^0 1 \leftarrow 00^0 0)$ of CO_2 centered at 6503.08 cm^{-1}, and the *inset in the right upper corner* shows the P(18) line in this band to illustrate the signal-to-noise ratio, the spectral resolution, and the symmetry of the instrumental line shape.

spectroscopic investigations of steady state gas phase systems. It is especially useful for applications where small sample volumes are required (e.g. in discharges, combustion plasmas, flames or chemical flow reactors) or where small quantities of rare and/or expensive compounds (e.g. isotopologues) are of interest.

FT-IBBCEAS was first demonstrated by Ruth et al. [26] who used a current-stabilized short-arc Xe lamp as light source. The spin- and symmetry-forbidden B-band of gaseous oxygen at 688 nm as well as weak absorption transitions of water vapour in ambient air in the same region were measured in this proof-of-principle study. Further FT-based IBBCEAS studies include the detection of overtone bands of CO_2, OCS [35], CH_3CN [70] and $HD^{18}O$ in the near IR [90]. Sensitivities of 1.5×10^{-8} cm^{-1} [66] using a fiber coupled SLED and $\sim 4 \times 10^{-9}$ cm^{-1} [67] using a supercontinuum source have been reported for an acquisition time of 240 s. It should be noted that fluctuations of the light source as a function of wavelength are more critical when using an interferometric (FT) detection scheme, because the noise of the light source is fully transferred to the detector as opposed to a dispersive approach where an inherent "spectral averaging" applies.

Figure 14.5 shows moderately strong bands of the $14^0 1 \leftarrow 00^0 0$, $22^0 1 \leftarrow 00^0 0$ and $30^0 1 \leftarrow 00^0 0$ transitions of the Fermi tetrad of CO_2 centred at ~ 6227 cm^{-1}, 6347 cm^{-1} and 6503 cm^{-1}, respectively [91], measured using FT-IBBCEAS [35]. There are also some associated hot bands as well as water absorption lines in the 1.6 µm region. This example demonstrates another advantage of combining broadband coverage (ca. 1200 cm^{-1} in Fig. 14.5) with high spectral resolution (and high sensitivity). The zero absorption baseline (I_0) and the transmitted intensity with target species (I) can be performed simultaneously as long as the observed line density is low enough for I_0 to be unambiguously observable. The simultaneous measurement of I_0 and I strongly reduces the systematic error on the resultant extinction coefficient. Moreover, an effective mirror reflectivity measurement becomes possible if a calibration gas (such as CO_2) is present in the cavity or can be added to the experimental gas mixture (the calibration gas spectrum can obviously be used for wavelength calibration at the same time).

If the partial pressure of the calibration gas is known (P_{cal} [Pa]), R can be determined from the known integrated line intensity of the calibration gas (S_{cal} [cm molecule^{-1}]). Let d [cm] be the cavity length, T [K] the temperature, and k the Boltzmann constant, then the (effective) reflectivity as a function of maximal absorption wavenumbers ($\tilde{v}_{max}$ [cm^{-1}]) of the calibration gas is given by [35]:

$$R(\tilde{v}_{max}) = 1 - \frac{d \times \frac{P_{cal}}{kT} \times S_{cal}(\tilde{v}_{max})}{\int_{\tilde{v}_0}^{\tilde{v}_0 + \Delta\tilde{v}} \left(\frac{\bar{I}_0(\Delta\tilde{v})}{I_{cal}(\tilde{v})} - 1\right) d\tilde{v}}, \tag{14.14}$$

where $I_{cal}(\tilde{v})$ is the intensity transmitted through the cavity on an absorption line of the calibration gas at wavenumber $\tilde{v}$, and $\bar{I}_0(\Delta\tilde{v})$ is the zero-absorption baseline averaged over a small wavenumber interval $\Delta\tilde{v}$ in the vicinity of corresponding absorption line of the calibration gas. $\tilde{v}_0$ is a spectral position off the absorption line ($\tilde{v}_0 < \tilde{v}_{max}$) that defines the integral boundaries in Eq. (14.14). The choice of the integration limits depends on the line profile and line width. For example, in the case of a Lorentzian line shape (with broad wings) integration over 6 half widths at half maximum (HWHM) yields only 91 % of the line area. In such cases, fitting of the measured data with an analytical expression for the line profile might be helpful, but requires knowledge of the line shape and, if possible, a negligible (or well known) contribution of the instrument function.

14.3.3.3 Detection Limit

The integration time corrected noise-equivalent extinction (nee) coefficient, ε_{nee}, at a given wavelength (typically taken at the maximal mirror reflectivity) is defined as the minimal extinction coefficient, ε_{min} [cm^{-1}], that can be measured reliably by the instrument, multiplied by the square root of the integration time, $t_{int}^{1/2}$ [Hz$^{-1/2}$]. The resulting noise-equivalent extinction coefficient at λ is independent of the integration time and can thus be compared for instruments using different detection methods:

$$\varepsilon_{nee}(\lambda) = \varepsilon_{min}(\lambda) \times t_{int}^{1/2}. \tag{14.15}$$

For IBBCEAS the minimum extinction coefficient, ε_{min}, depends on the minimal detectable intensity difference ($\Delta I_{min} = (I_0 - I)_{min}$) and can be written as (cf. Sect. 14.2.3.1) [21, 22]:

$$\varepsilon_{min}(\lambda) \geq \frac{1}{d}\left(\frac{\Delta I_{min}(\lambda)}{I(\lambda)}\right)(1 - R(\lambda)). \qquad (14.16)$$

In Eq. (14.16) the "=" sign corresponds to a signal-to-noise ratio of 1, which does not serve as a particularly meaningful limit of detection. In practice, the minimal detectable intensity variation $\Delta I_{min}(\lambda)$ is established considering the 3σ standard deviation of a statistically significant data set of transmissions of a target-free cavity.

Increasing the integration time, t_{int}, generally increases the signal-to-noise ratio (SNR) and therefore minimizes the noise-equivalent extinction coefficient up to a certain limit that depends on the way different forms of noise affect the measurement. The noise in cavity-enhanced absorption experiments is generally not dominated by detector noise (i.e. readout noise and dark noise), because the application of a cavity (which inherently lowers light intensities substantially) requires the detector noise to be smaller than the photon noise in order to achieve an enhancement. Many IBBCEAS experiments are photon noise-limited (also termed "quantum noise" or "shot noise") for reasonably short integration times. In this case the SNR is proportional to $t_{int}^{1/2}$ up to a time where other forms of noise (e.g. light source fluctuations, mechanical stabilities, thermal drifts and others) start dominating the noise. Beyond that point, the detection is said to be "instrumental" or "environmental" noise-limited [92].[7] The two regimes are illustrated in Fig. 14.6(a) where the species-specific limit of detection (see below) is shown as a function of integration time.

The species-specific limit of detection (LOD) depends additionally on the retrieval procedure applied to the measured extinction spectra. The extinction coefficient $\varepsilon(\lambda)$ can be expressed in a general form as:

$$\varepsilon(\lambda) = \sum_i \sigma_i(\lambda) \int_0^{d'} n_i(x)dx + \varepsilon_{backgr}(\lambda), \qquad (14.17)$$

where $\sigma_i(\lambda)$ and n_i are the cross-section and number density of species i at λ, while the second term, $\varepsilon_{backgr}(\lambda)$, accounts for any unspecified ("extinction") background. Knowledge of $\sigma_i(\lambda)$ from reference databases (e.g. HITRAN, PNNL, GEISA and others) and appropriate modelling of $\varepsilon_{backgr}(\lambda)$ (e.g. by an adequate polynomial), enables the fitting of Eq. (14.17) to measured extinction data for the retrieval of number densities (commonly also expressed as mixing ratios) of sample species i. The retrieval procedure depends on the method used for least square minimization, on the choice of the final form of Eq. (14.17), on the reliability of the reference cross-sections and their convolution to instrumental resolution, and finally on the wavelength range used for the retrieval. Retrieval methods will not be discussed

---

[7]The "instrumental/environmental" noise does not include systematic error sources and needs to be considered from experiment to experiment.

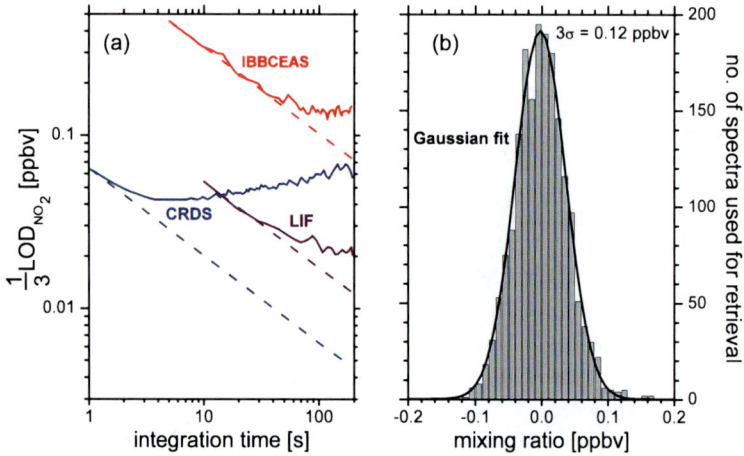

Fig. 14.6 (**a**) Log-log plot of the 1σ signal-to-noise ratio ($=1/3$ LOD) of the target-free (zero-air) SAPHIR chamber in Jülich versus integration time for three different instruments (IBBCEAS, CRDS and laser induced fluorescence (LIF)). Data were determined during the NO3comp campaign in 2007 [33, 85] and the figure was adapted from Ref. [84]. The main IBBCEAS target species were NO_3, NO_2 and H_2O. The spectral evaluation region for NO_2 was 630–645 nm. The optimum integration time for the IBBCEAS setup is ca. 100 s. (**b**) The viewgraph shows the number of zero-air spectra taken with an IBBCEAS instrument at SAPHIR in 2011 (1800 in total over 30 discontinued hours) as a function of the retrieved NO_2 mixing ratios. The 3σ value of the Gaussian distribution represents the LOD for NO_2 of 0.12 ppbv for a 1 min integration time. The LOD depends on the spectral range covered (352–386 nm) and on the retrieval method (which in the present case also included the target species HONO)

here—more details can be found in Refs. [5, 33, 55]. The species-specific LOD can be defined as the number density of a given species i that can be reliably retrieved for a certain integration time, t_{int}:

$$LOD_i(t_{int}) \geq \Delta n_{i,min} \qquad (14.18)$$

where $\Delta n_{i,min}$ is the 3σ deviation of the (typically Gaussian) distribution of number densities retrieved for a target-free cavity. This distribution can be obtained from a statistically significant number of measured spectra in the target-free cavity (for atmospheric applications typically flushed with zero-air). From each of these spectra, number densities of all target species accounted for in Eq. (14.17) are retrieved simultaneously. An example is shown Fig. 14.6(b) for an open-path IBBCEAS instrument that has been implemented at the SAPHIR atmospheric chamber for the near-UV detection of NO_2 and HONO.

14.4 Summary of Literature

Table 14.1 summarizes the main literature on **broadband cavity-enhanced absorption methods with incoherent light sources**. The literature is ordered according to

Table 14.1 Summary of broadband cavity-enhanced absorption methods with broadband light sources until 2012 (inclusive). This lookup table does not contain frequency comb applications and whispering gallery mode sensors

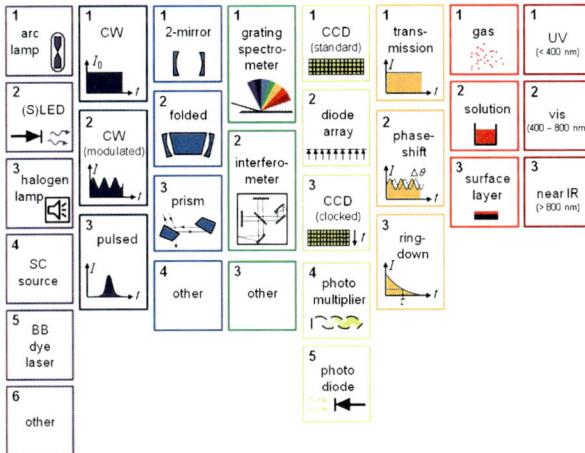

Legend (icon key):

Light source: 1 arc lamp; 2 (S)LED; 3 halogen lamp; 4 SC source; 5 BB dye laser; 6 other

Mode: 1 CW (I_0); 2 CW (modulated); 3 pulsed

Cavity: 1 2-mirror; 2 folded; 3 prism; 4 other

Wavelength selection: 1 grating spectrometer; 2 interferometer; 3 prism; 4 other

Detector: 1 CCD (standard); 2 diode array; 3 CCD (clocked); 4 photomultiplier; 5 photo diode

Exp. method: 1 transmission; 2 phase-shift ($\Delta\theta$); 3 ring-down

Sample phase: 1 gas; 2 solution; 3 surface layer

Spectral range: 1 UV (<400 nm); 2 vis (400–800 nm); 3 near IR (>800 nm)

Light source	Mode	Cavity	Wave-length selection	Detector	Exp. method	Sample phase	Spectral range	References
1	1	1	1	1	1	1	1	[43,44]
1	1	1	1	1	1	1	2	[33,38-42,81,84-86]
1	1	1	1	1	1	2	2	[21, 102]
1	1	1	1	2	1	1	2	[8,37]
1	1	1	2	5	1	1	2	[26]
1	1	1	2	5	1	1	3	[35,90]
1	1	2	1	2	1	3	2	[30]
1	2	1	2	4	2	1	2	[7]
1[a]	3	1	1	1	1	1	2	[45]
1	3	1	1	1	1	1	2	[75]
2	1	1	1	1	1	1	1	[59-61,89,103]
2	1	1	1	1	1	1	2	[48[H],49-54,56,57,82[H],83[H],85,89,104,105]
2	1	1	1	1	1	2	2	[27,28,62]
2	1	1	2	5	1	1	3	[66,67]
2	1	2	1	1	1	3	2	[31]
2	1	4[b]	3	5	1	2	3	[65]
2	2[c]	1	1	5[c]	1	1	3	[66]
2	2	1	3	4	2	1	2	[63]
2	2	1	3	5	2	1	2	[64,106]
2	3	1	1	1	1	1	1	[58]
2	3[d]	4	1	1	3	1	2	[68]
3	1	1	1	4	1	2	2	[21]
3	1	1	3	4	1	1	2	[46]
4	3	1	1	1	1	1	2	[23,88[H]]
4	3	1	1	1	1	2	2	[29[H],76]
4	3	1	1	1	3	1	2	[15,17]
4	3	1	1	2	3	2	2	[18]
4	3	1	1	4	3	1	2	[16]
4	3	1	1[e]	?[e]	1	1	3	[87[H]]
4	3	1	2	5	1	1	3	[67,70]
4	3	1	3	4	3	2	2	[18]
4	3	2	1	1	1	3	2	[24,31]
4	3	3	1	1	1	1	2	[77[H]]
5	3	1	1	1	1	1	2	[107]
5	3	1	1	1	3	1	2	[10,72,108]
5	3	1	1	3	3	1	2	[6,12-14,84,85,109]
5	3	1	2	4	3	1	2	[73]
6[f]	3	1	1	5	3	1	3	[74]

[a] Plasma self-absorption in a pulsed jet

[b] End-coated fiber cavity

[c] Mechanical chopper modulation and lock-in (phase sensitive) detection

[d] An intracavity accousto-optic modulator (AOM) was used to directly couple light into (and out-of) the cavity

[e] Wavelength selection and light detection were integrated into a commercialized optical spectrum analyser (ASO, Yokogawa AQ6317Q)

[f] High repetition rate free electron laser

[H] Hybrids where different experimental approaches are used for calibration and main absorption measurements

an 8-digit code which can be looked up in the diagram above the table. The code represents the combination of the following experimental parameters: light source, (time-)mode, cavity type, wavelength selection method, detector type, measurement principle, phase of the target sample, and spectral range. The listed literature does **not** comprise spectroscopic approaches involving frequency combs (see Chap. 8), for instance in mode-locked cavity-enhanced absorption spectroscopy [1, 93–95], in cavity-enhanced femtosecond frequency comb spectroscopy [96–99] or in vernier spectroscopy [100, 101]. It neither covers cavity approaches involving whispering gallery mode resonators which will be dealt with in Chap. 8 of this book.

Acknowledgements The authors gratefully acknowledge support by Science Foundation Ireland (11/RFP.1/PHY/3233), by the European Marie Curie Programme (FP7 IEF-302109, Alma Mater), and the IRCSET INSPIRE post-doc fellowship scheme cofounded by the FP7 Marie Curie programme (COFUND).

References

1. T. Gherman, D. Romanini, Mode-locked cavity-enhanced absorption spectroscopy. Opt. Express **10**, 1033–1042 (2002)
2. A. O'Keefe, Integrated cavity output analysis of ultra-weak absorption. Chem. Phys. Lett. **293**, 331–336 (1998)
3. K.K. Lehmann, D. Romanini, The superposition principle and cavity ring-down spectroscopy. J. Chem. Phys. **105**, 10263–10277 (1996)
4. S.M. Ball, R.L. Jones, Broad-band cavity ring-down spectroscopy. Chem. Rev. **103**, 5239–5262 (2003)
5. S. Ball, R. Jones, Broadband cavity ring-down spectroscopy, in *Cavity Ring Down Spectroscopy: Techniques and Applications*, ed. by G. Berden, R. Engeln (Wiley, New York, 2009). ISBN: 978-1-4051-7688-0
6. S.M. Ball, I.M. Povey, E.G. Norton, R.L. Jones, Broadband cavity ring-down spectroscopy of the NO_3 radical. Chem. Phys. Lett. **342**, 113–120 (2001)
7. E. Hamers, D. Schram, R. Engeln, Fourier transform phase shift cavity ring down spectroscopy. Chem. Phys. Lett. **365**, 237–243 (2002)
8. S.E. Fiedler, A. Hese, A.A. Ruth, Incoherent broad-band cavity-enhanced absorption spectroscopy. Chem. Phys. Lett. **371**, 284–294 (2003)
9. A. O'Keefe, D.A.G. Deacon, Cavity ring-down optical spectrometer for absorption measurements using pulsed laser sources. Rev. Sci. Instrum. **59**, 2544–2554 (1988)
10. J.J. Scherer, J.B. Paul, H. Jiao, A. O'Keefe, Broadband ringdown spectral photography. Appl. Opt. **40**, 6725–6732 (2001)
11. I.M. Povey, A.M. South, A. t'Kint de Roodenbeke, C. Hill, R.A. Freshwater, R.L. Jones, A broadband lidar for the measurement of tropospheric constituent profiles from the ground. J. Geophys. Res. **103**, 3369–3380 (1998)
12. M. Bitter, S.M. Ball, I.M. Povey, R.L. Jones, A broadband cavity ringdown spectrometer for in-situ measurements of atmospheric trace gases. Atmos. Chem. Phys. **5**, 2547–2560 (2005)
13. A. Saiz-Lopez et al., Modelling molecular iodine emissions in a coastal marine environment: the link to new particle formation. Atmos. Chem. Phys. **6**, 883–895 (2006)
14. R.J. Leigh et al., Measurements and modelling of molecular iodine emissions, transport and photodestruction in the coastal region around Roscoff. Atmos. Chem. Phys. **10**, 11823–11838 (2010)
15. G. Schmidl, W. Paa, W. Triebel, S. Schippel, H. Heyer, Spectrally resolved cavity ring down measurement of high reflectivity mirrors using a supercontinuum laser source. Appl. Opt. **48**, 6754–6759 (2009)

16. K. Stelmaszczyk, M. Fechner, P. Rohwetter, M. Queißer, A. Czyżewski, T. Stacewicz, L. Wöste, Towards supercontinuum cavity ring down spectroscopy. Appl. Phys. B **94**, 369–373 (2009)

17. K. Stelmaszczyk, P. Rohwetter, M. Fechner, M. Queißer, A. Czyżewski, T. Stacewicz, L. Wöste, Cavity ring-down absorption spectrography based on filament-generated supercontinuum light. Opt. Express **17**, 3673–3678 (2009)

18. S.S. Kiwanuka, T.K. Laurila, J.H. Frank, A. Esposito, K. Blomberg von der Geest, L. Pancheri, D. Stoppa, C.F. Kaminski, Development of broadband cavity ring-down spectroscopy for biomedical diagnostics of liquid analytes. Anal. Chem. **84**, 5489–5493 (2012)

19. J.M. Herbelin, J.A. McKay, M.A. Kwok, R.H. Ueunten, D.S. Urevig, D.J. Spencer, D.J. Benard, Sensitive measurement of photon lifetime and true reflectances in an optical cavity by a phase-shift method. Appl. Opt. **19**, 144–147 (1980)

20. R. Engeln, G. von Helden, G. Berden, G. Meijer, Phase shift cavity ring down absorption spectroscopy. Chem. Phys. Lett. **262**, 105–109 (1996)

21. S.E. Fiedler, A. Hese, A.A. Ruth, Incoherent broad-band cavity-enhanced absorption spectroscopy of liquids. Rev. Sci. Instrum. **76**, 023107 (2005) [Erratum: Rev. Sci. Instrum. **76**, 089901 (2005)]

22. S.E. Fiedler, A. Hese, U. Heitmann, Influence of the cavity parameters on the output intensity in incoherent broadband cavity-enhanced absorption spectroscopy. Rev. Sci. Instrum. **78**, 073104 (2007)

23. J.M. Langridge, T. Laurila, R.S. Watt, R.L. Jones, C.F. Kaminski, J. Hult, Cavity enhanced absorption spectroscopy of multiple trace gas species using a supercontinuum radiation source. Opt. Express **16**, 10178–10188 (2008)

24. M. Schnippering, P.R. Unwin, J. Hult, T. Laurila, C.F. Kaminski, J.M. Langridge, R.L. Jones, M. Mazurenka, S.R. Mackenzie, Evanescent wave broadband cavity enhanced absorption spectroscopy using supercontinuum radiation: a new probe of electrochemical processes. Electrochem. Commun. **10**, 1827–1830 (2008)

25. K. Lynch, Incoherent broad-band cavity-enhanced total internal reflection spectroscopy of surface-adsorbed metallo-porphyrins, Ph.D. thesis, Physics Department, University College, Cork, Ireland, 2008

26. A.A. Ruth, J. Orphal, S.E. Fiedler, Fourier-transform cavity-enhanced absorption spectroscopy using an incoherent broadband light source. Appl. Opt. **46**, 3611–3616 (2007)

27. M. Islam, L.N. Seetohul, Z. Ali, Liquid-phase broadband cavity-enhanced absorption spectroscopy measurements in a 2 mm cuvette. Appl. Spectrosc. **61**, 649–658 (2007)

28. L.N. Seetohul, Z. Ali, M. Islam, Broadband cavity enhanced absorption spectroscopy as a detector for HPLC. Anal. Chem. **81**, 4106–4112 (2009)

29. S.S. Kiwanuka, T. Laurila, C.F. Kaminski, Sensitive method for the kinetic measurement of trace species in liquids using cavity enhanced absorption spectroscopy with broad bandwidth supercontinuum radiation. Anal. Chem. **82**, 7498–7501 (2010)

30. A.A. Ruth, K.T. Lynch, Incoherent broadband cavity-enhanced total internal reflection spectroscopy of surface adsorbed metalloporphyrins. Phys. Chem. Chem. Phys. **10**, 7098–7108 (2008)

31. L. Van der Sneppen, G. Hancock, C. Kaminski, T. Laurila, S.R. Mackenzie, S.R.T. Neil, R. Peverall, G.A.D. Ritchie, M. Schnippering, P.R. Unwin, Following interfacial kinetics in real time using broadband evanescent wave cavity-enhanced absorption spectroscopy: a comparison of light-emitting diodes and supercontinuum sources. Analyst **135**, 133–139 (2010)

32. A.T.M. Wilbers, G.M.W. Kroesen, C.J. Timmermans, D.C. Schram, The continuum emission of an arc plasma. J. Quant. Spectrosc. Radiat. Transf. **45**, 1–10 (1991)

33. R.M. Varma, D.S. Venables, A.A. Ruth, U. Heitmann, E. Schlosser, S. Dixneuf, Long optical cavities for open-path monitoring of atmospheric trace gases and aerosol extinction. Appl. Opt. **48**, B159–171 (2009)

34. B. Welz, H. Becker-Ross, S. Florek, U. Heitmann, *High-Resolution Continuum Source AAS: The Better Way to do Atomic Absorption Spectrometry* (Wiley-VCH, New York, 2005)

35. J. Orphal, A.A. Ruth, High-resolution Fourier-transform cavity-enhanced absorption spectroscopy in the near-infrared using an incoherent broad-band light source. Opt. Express **16**, 19232–19243 (2008)
36. U. Platt, J. Stutz, *Differential Optical Absorption Spectroscopy: Principles and Applications* (Springer, Berlin, 2008)
37. S.E. Fiedler, G. Hoheisel, A.A. Ruth, A. Hese, Incoherent broad-band cavity-enhanced absorption spectroscopy of azulene in a supersonic jet. Chem. Phys. Lett. **382**, 447–453 (2003)
38. D.S. Venables, T. Gherman, J. Orphal, J.C. Wenger, A.A. Ruth, High sensitivity in situ monitoring of NO_3 in an atmospheric simulation chamber using incoherent broadband cavity-enhanced absorption spectroscopy. Environ. Sci. Technol. **40**, 6758–6763 (2006)
39. S. Vaughan, T. Gherman, A.A. Ruth, J. Orphal, Incoherent broad-band cavity-enhanced absorption spectroscopy of the marine boundary layer species I_2, IO and OIO. Phys. Chem. Chem. Phys. **10**, 4471–4477 (2008)
40. R.A. Washenfelder, A.O. Langford, H. Fuchs, S.S. Brown, Measurement of glyoxal using an incoherent broadband cavity enhanced absorption spectrometer. Atmos. Chem. Phys. **8**, 7779–7793 (2008)
41. S. Dixneuf, A.A. Ruth, S. Vaughan, R.M. Varma, J. Orphal, The time dependence of molecular iodine emission from *Laminaria digitata*. Atmos. Chem. Phys. **9**, 823–829 (2009)
42. U. Nitschke, A.A. Ruth, S. Dixneuf, D.B. Stengel, Molecular iodine emission rates and photosynthetic performance of different thallus parts of *Laminaria digitata* (Phaeophyceae) during emersion. Planta **233**, 737–748 (2011)
43. J. Chen, D.S. Venables, A broadband optical cavity spectrometer for measuring weak near-ultraviolet absorption spectra of gases. Atmos. Meas. Tech. **4**, 425–436 (2011)
44. J. Chen, J.C. Wenger, D.S. Venables, Near-ultraviolet absorption cross sections of nitrophenols and their potential influence on tropospheric oxidation capacity. J. Phys. Chem. A **115**, 12235–12242 (2011)
45. A. Walsh, D. Zhao, H. Linnartz, Cavity enhanced plasma self-absorption spectroscopy. Appl. Phys. Lett. **101**, 091111 (2012)
46. J.E. Thompson, H.D. Spangler, Tungsten source integrated cavity output spectroscopy for the determination of ambient atmospheric extinction coefficient. Appl. Opt. **45**, 2465–2473 (2006)
47. C. Kern, S. Trick, B. Rippel, U. Platt, Applicability of light-emitting diodes as light sources for active differential optical absorption spectroscopy measurements. Appl. Opt. **45**, 2077–2088 (2006)
48. S.M. Ball, J.M. Langridge, R.L. Jones, Broadband cavity enhanced absorption spectroscopy using light emitting diodes. Chem. Phys. Lett. **398**, 68–74 (2004)
49. I. Ventrillard-Courtillot, E. Sciamma O'Brien, S. Kassi, G. Méjean, D. Romanini, Incoherent broad-band cavity-enhanced absorption spectroscopy for simultaneous trace measurements of NO_2 and NO_3 with a LED source. Appl. Phys. B **101**, 661–669 (2010)
50. J.M. Langridge, S.M. Ball, R.L. Jones, A compact broadband cavity enhanced absorption spectrometer for detection of atmospheric NO_2 using light emitting diodes. Analyst **131**, 916–922 (2006)
51. M. Triki, P. Cermak, G. Méjean, D. Romanini, Cavity-enhanced absorption spectroscopy with a red LED source for NO_x trace analysis. Appl. Phys. B **91**, 195–201 (2008)
52. T. Wu, W. Zhao, W. Chen, W. Zhang, X. Gao, Incoherent broadband cavity enhanced absorption spectroscopy for in situ measurements of NO_2 with a blue light emitting diode. Appl. Phys. B **94**, 85–94 (2009)
53. S.M. Ball, A.M. Hollingsworth, J. Humbles, C. Leblanc, P. Potin, G. McFiggans, Spectroscopic studies of molecular iodine emitted into the gas phase by seaweed. Atmos. Chem. Phys. **10**, 6237–6254 (2010)
54. A.K. Benton, J.M. Langridge, S.M. Ball, W.J. Bloss, M. Dall'Osto, E. Nemitz, R.M. Harrison, R.L. Jones, Night-time chemistry above London: measurements of NO_3 and N_2O_5 from the BT tower. Atmos. Chem. Phys. **10**, 9781–9795 (2010)

55. U. Platt, J. Meinen, D. Pöhler, T. Leisner, Broadband cavity enhanced differential optical absorption spectroscopy (CE-DOAS)—applicability and corrections. Atmos. Meas. Tech. **2**, 713–723 (2009)
56. R. Thalman, R. Volkamer, Inherent calibration of a blue LED-CE-DOAS instrument to measure iodine oxide, glyoxal, methyl glyoxal, nitrogen dioxide, water vapour and aerosol extinction in open cavity mode. Atmos. Meas. Tech. **3**, 1797–1814 (2010)
57. J. Meinen, J. Thieser, U. Platt, T. Leisner, Technical note: using a high finesse optical resonator to provide a long light path for differential optical absorption spectroscopy: CE-DOAS. Atmos. Chem. Phys. **10**, 3901–3914 (2010)
58. D.J. Hoch, J. Buxmann, H. Sihler, D. Pöhler, C. Zetzsch, U. Platt, A cavity-enhanced differential optical absorption spectroscopy instrument for measurement of BrO, HCHO, HONO and O_3. Atmos. Meas. Tech. Discuss. **5**, 3079–3115 (2012)
59. T. Gherman, D.S. Venables, S. Vaughan, J. Orphal, A.A. Ruth, Incoherent broadband cavity-enhanced absorption spectroscopy in the near-ultraviolet: application to HONO and NO_2. Environ. Sci. Technol. **42**, 890–895 (2008)
60. J.M. Roberts, P. Veres, C. Warneke, J.A. Neuman, R.A. Washenfelder, S.S. Brown, M. Baasandorj, J.B. Burkholder, I.R. Burling, T.J. Johnson, R.J. Yokelson, J. de Gouw, Measurement of HONO, HNCO, and other inorganic acids by negative-ion proton-transfer chemical-ionization mass spectrometry (NI-PT-CIMS): application to biomass burning emissions. Atmos. Meas. Tech. Discuss. **3**, 301–331 (2010)
61. T. Wu, W. Chen, E. Fertein, F. Cazier, D. Dewaele, X. Gao, Development of an open-path incoherent broadband cavity-enhanced spectroscopy based instrument for simultaneous measurement of HONO and NO_2 in ambient air. Appl. Phys. B **106**, 501–509 (2012)
62. L.N. Seetohul, Z. Ali, M. Islam, Liquid-phase broadband cavity enhanced absorption spectroscopy (BBCEAS) studies in a 20 cm cell. Analyst **134**, 1887–1895 (2009)
63. P.L. Kebabian, S.C. Herndon, A. Freedman, Detection of nitrogen dioxide by cavity attenuated phase shift spectroscopy. Anal. Chem. **77**, 724–728 (2005)
64. P.L. Kebabian, E.C. Wood, S.C. Herndon, A. Freedman, A practical alternative to chemiluminescence-based detection of nitrogen dioxide: cavity attenuated phase shift spectroscopy. Environ. Sci. Technol. **42**, 6040–6045 (2008)
65. A.L. Gomez, R.F. Renzi, J.A. Fruetel, R.P. Bambha, Integrated fiber optic incoherent broadband cavity enhanced absorption spectroscopy detector for near-IR absorption measurements of nanoliter samples. Appl. Opt. **51**, 2532–2540 (2012)
66. W. Denzer, M.L. Hamilton, G. Hancock, M. Islam, C.E. Langley, R. Peverall, G.A.D. Ritchie, Near-infrared broad-band cavity enhanced absorption spectroscopy using a superluminescent light emitting diode. Analyst **134**, 2220–2223 (2009)
67. W. Denzer, G. Hancock, M. Islam, C.E. Langley, R. Peverall, G.A.D. Ritchie, D. Taylor, Trace species detection in the near infrared using Fourier transform broadband cavity enhanced absorption spectroscopy: initial studies on potential breath analytes. Analyst **136**, 801–806 (2011) [Erratum: Analyst **136**, 5308 (2011)]
68. C. Petermann, P. Fischer, Actively coupled cavity ringdown spectroscopy with low-power broadband sources. Opt. Express **19**, 10164–10173 (2011)
69. J.M. Dudley, G. Genty, S. Coen, Supercontinuum generation in photonic crystal fiber. Rev. Mod. Phys. **78**, 1135–1184 (2006)
70. D.M. O'Leary, A.A. Ruth, S. Dixneuf, J. Orphal, R. Varma, The near infrared cavity-enhanced absorption spectrum of methylcyanide. J. Quant. Spectrosc. Radiat. Transf. **113**, 1138–1147 (2012)
71. A. Czyżewski, S. Chudzyński, K. Ernst, G. Karasiński, Ł. Kilianek, A. Pietruczuk, W. Skubiszak, T. Stacewicz, K. Stelmaszczyk, B. Koch, P. Rairoux, Cavity ring-down spectrography. Opt. Commun. **191**, 271–275 (2001)
72. J.J. Scherer, J.B. Paul, H. Jiao, A. O'Keefe, Broadband ringdown spectral photography. Appl. Opt. **40**, 6725–6732 (2001)
73. R. Engeln, G. Meijer, A Fourier transform cavity ring down spectrometer. Rev. Sci. Instrum. **67**, 2708–2714 (1996)

74. E.R. Crosson, P. Haar, G.A. Marcus, H.A. Schwettman, B.A. Paldus, T.G. Spence, R.N. Zare, Pulse-stacked cavity ring-down spectroscopy. Rev. Sci. Instrum. **70**, 4–10 (1999)

75. A. Walsh, D. Zhao, W. Ubachs, H. Linnartz, Optomechanical shutter modulated broad-band cavity-enhanced absorption spectroscopy of molecular transients of astrophysical interest. J. Phys. Chem. A (2012). doi:10.1021/jp310392n

76. S.R.T. Neil, C.M. Rushworth, C. Vallance, S.R. Mackenzie, Broadband cavity-enhanced absorption spectroscopy for real time, in situ spectral analysis of microfluidic droplets. Lab Chip **11**, 3953–3955 (2011)

77. P.S. Johnston, K.K. Lehmann, Cavity enhanced absorption spectroscopy using a broadband prism cavity and a supercontinuum source. Opt. Express **16**, 15013–15023 (2008)

78. K.K. Lehmann, P.S. Johnston, P. Rabinowitz, Brewster angle prism retroreflectors for cavity enhanced spectroscopy. Appl. Opt. **48**, 2966–2978 (2009)

79. M. Schnippering, S.R.T. Neil, S.R. Mackenzie, P.R. Unwin, Evanescent wave cavity-based spectroscopic techniques as probes of interfacial processes. Chem. Soc. Rev. **40**, 207–220 (2011)

80. Y. Yao, J. Yao, V.K. Narasimhan, Z. Ruan, C. Xie, S. Fan, Y. Cui, Broadband light management using low-Q whispering gallery modes in spherical nanoshells. Nat. Commun. **664** (2012). doi:10.1038/ncomms1664

81. E.R. Ashu-Ayem, U. Nitschke, C. Monahan, J. Chen, S.B. Darby, P.D. Smith, C.D. O'Dowd, D.B. Stengel, D.S. Venables, Coastal iodine emissions. 1. Release of I_2 by *Laminaria digitata* in chamber experiments. Environ. Sci. Technol. **46**, 10413–10421 (2012)

82. J.M. Langridge, S.M. Ball, A.J.L. Shillings, R.L. Jones, A broadband absorption spectrometer using light emitting diodes for ultrasensitive, *in situ* trace gas detection. Rev. Sci. Instrum. **79**, 123110 (2008)

83. O.J. Kennedy et al., An aircraft based three channel broadband cavity enhanced absorption spectrometer for simultaneous measurements of NO_3, N_2O_5 and NO_2. Atmos. Meas. Tech. **4**, 1759–1776 (2011)

84. H. Fuchs et al., Intercomparison of measurements of NO_2 concentrations in the atmosphere simulation chamber SAPHIR during the NO3Comp campaign. Atmos. Meas. Tech. **3**, 21–37 (2010)

85. H.P. Dorn et al., Intercomparison of NO_3 radical detection instruments in the atmosphere simulation chamber SAPHIR. Atmos. Meas. Tech. Discuss. **6**, 303–379 (2013)

86. C. Monahan, E.R. Ashu-Ayem, U. Nitschke, S.B. Darby, P.D. Smith, D.B. Stengel, D.S. Venables, C.D. O'Dowd, Coastal iodine emissions, part 2: chamber experiments of particle formation from *Laminaria digitata*-derived and laboratory-generated I_2. Environ. Sci. Technol. **46**, 10422–10428 (2012)

87. R.S. Watt, T. Laurila, C.F. Kaminski, J. Hult, Cavity enhanced spectroscopy of high-temperature H_2O in the near-infrared using a supercontinuum light source. Appl. Spectrosc. **63**, 1389–1395 (2009)

88. T. Laurila, I.S. Burns, J. Hult, J.H. Miller, C.F. Kaminski, A calibration method for broad-bandwidth cavity enhanced absorption spectroscopy performed with supercontinuum radiation. Appl. Phys. B **102**, 271–278 (2011)

89. J.L. Axson, R.A. Washenfelder, T.F. Kahan, C.J. Young, V. Vaida, S.S. Brown, Absolute ozone cross section in the Huggins Chappuis minimum (350–470 nm) at 296 K. Atmos. Chem. Phys. **11**, 11581–11590 (2011)

90. M.J. Down, J. Tennyson, J. Orphal, P. Chelin, A.A. Ruth, Analysis of an ^{18}O and D enhanced water spectrum and new assignments for $HD^{18}O$ and $D_2^{18}O$ in the near-infrared region (6000–7000 cm^{-1}) using newly calculated variational line lists. J. Mol. Spectrosc. **282**, 1–8 (2012)

91. C.E. Miller, L.R. Brown, Near infrared spectroscopy of carbon dioxide I. $^{16}O^{12}C^{16}O$ line positions. J. Mol. Spectrosc. **228**, 329–354 (2004)

92. B. Ouyang, R.L. Jones, Understanding the sensitivity of cavity-enhanced absorption spectroscopy: pathlength enhancement versus noise suppression. Appl. Phys. B **109**, 581–591 (2012)

93. T. Gherman, S. Kassi, A. Campargue, D. Romanini, Overtone spectroscopy in the blue region by cavity enhanced absorption spectroscopy with a mode-locked femtosecond laser: application to acetylene. Chem. Phys. Lett. **383**, 353–358 (2004)

94. T. Gherman, D. Romanini, I. Sagnes, A. Garnache, Z. Zhang, Cavity-enhanced absorption spectroscopy with a mode-locked diode-pumped vertical external-cavity surface-emitting laser. Chem. Phys. Lett. **390**, 290–295 (2004)

95. J. Morville, S. Kassi, M. Chenevier, D. Romanini, Fast, low-noise, mode-by-mode, cavity-enhanced absorption spectroscopy by diode-laser self-locking. Appl. Phys. B **80**, 1027–1038 (2005)

96. M.J. Thorpe, K.D. Moll, R.J. Jones, B. Safdi, J. Ye, Broadband cavity ringdown spectroscopy for sensitive and rapid molecular detection. Science **311**, 1595–1599 (2006)

97. M.J. Thorpe, F. Adler, K.C. Cossel, M.H.G. de Miranda, J. Ye, Tomography of a supersonically cooled molecular jet using cavity-enhanced direct frequency comb spectroscopy. Chem. Phys. Lett. **468**, 1–8 (2009)

98. K.C. Cossel, F. Adler, K.A. Bertness, M.J. Thorpe, J. Feng, M.W. Raynor, J. Ye, Analysis of trace impurities in semiconductor gas via cavity-enhanced direct frequency comb spectroscopy. Appl. Phys. B, Lasers Opt. **100**, 917–924 (2010)

99. A. Foltynowicz, P. Maslowski, T. Ban, F. Adler, K.C. Cossel, T.C. Briles, J. Ye, Optical frequency comb spectroscopy. Faraday Discuss. **150**, 23–31 (2011)

100. C. Gohle, B. Stein, A. Schliesser, T. Udem, T.W. Hänsch, Frequency comb Vernier spectroscopy for broadband, high-resolution, high-sensitivity absorption and dispersion spectra. Phys. Rev. Lett. **99**, 263902 (2007)

101. B. Hardy, M. Raybaut, J.B. Dherbecourt, J.M. Melkonian, A. Godard, A.K. Mohamed, M. Lefebvre, Vernier frequency sampling: a new tuning approach in spectroscopy—application to multi-wavelength integrated path DIAL. Appl. Phys. B **107**, 643–647 (2012)

102. P.K. Dasgupta, J.-S. Rhee, Optical cells with partially reflecting windows as nonlinear absorbance amplifiers. Anal. Chem. **59**, 783–786 (1987)

103. J.M. Langridge, R.J. Gustafsson, P.T. Griffiths, R.A. Cox, R.M. Lambert, R.L. Jones, Solar driven nitrous acid formation on building material surfaces containing titanium dioxide: a concern for air quality in urban areas? Atmos. Environ. **43**, 5128–5131 (2009)

104. S. Nakao, Y. Liu, P. Tang, C.-L. Chen, J. Zhang, D.R. Cocker, Chamber studies of SOA formation from aromatic hydrocarbons: observation of limited glyoxal uptake. Atmos. Chem. Phys. **12**, 3927–3937 (2012)

105. W. Zhao, M. Dong, W. Chen, X. Gu, C. Hu, X. Gao, W. Huang, W. Zhang, Wavelength-resolved optical extinction measurements of aerosols using broad-band cavity-enhanced absorption spectroscopy over the spectral range of 445–480 nm. Anal. Chem. **85**, 2260–2268 (2013)

106. P.L. Kebabian, W.A. Robinson, A. Freedman, Optical extinction monitor using CW cavity enhanced detection. Rev. Sci. Instrum. **78**, 063102 (2007)

107. L. Biennier, F. Salama, M. Gupta, A. O'Keefe, Multiplex integrated cavity output spectroscopy of cold PAH cations. Chem. Phys. Lett. **387**, 287–294 (2004)

108. A. Czyżewski, K. Ernst, G. Karasinski, H. Lange, P. Rairoux, W. Skubiszak, T. Stacewicz, Cavity ring-down spectroscopy for trace gas analysis. Acta Phys. Pol. B **33**, 2255–2265 (2002)

109. A.J.L. Shillings, S.M. Ball, M.J. Barber, J. Tennyson, R.L. Jones, An upper limit for water dimer absorption in the 750 nm spectral region and a revised water line list. Atmos. Chem. Phys. **11**, 4273–4287 (2011)

Index

Symbols

$1/f$ noise, 179
2λ spectroscopy, 76
$3\text{-}\sigma$ standard, 419
3-body reaction, 63
3-pentanone, 68, 82
4-mirror ring-resonator, 368
$\angle COO$ bend, 70
β-HEP, 86, 87
β-hydroxyethylperoxy, 86
T-butyl peroxy, 89
π-phase shifted FBG, 464
σ_p, 81
$\tilde{A}-\tilde{X}$, 64
$\tilde{B}-\tilde{X}$, 64

A

Absorbance, 412
Absorption cell, 392
Absorption coefficient-α, 149, 151, 155, 157, 378
Absorption cross section, 77, 417
Absorption loss, 412
Absorption path-length, 417
Accelerator mass spectrometry-AMS, 159
Accelerometer, 469
Acetylene, 219, 222, 236–238, 246, 247, 262
Acoustic transducer, 474
Acousto-optic modulator, 232, 256
Acousto-optic modulator: fiber-coupled, 233
Active mode-locking, 275
Adsorbant coverage, 378
Aerosol extinction, 502
AgGaSe$_2$, 282
Airy function, 189
Allan deviation, 145
Allan variance, 28, 119, 120, 132, 244

Amine hydrogen bonding, 376
Amplified fibre-loop cavities, 402
Amplified spontaneous emission, 41
Amplitude attenuation, 214, 215
AOM, 83, 232–234
Arc lamp, 494
Arsine (AsH$_3$), 302
Atmospheric monitoring, 129
Atmospheric sensing, 127
Atmospheric windows, 99
Attenuation, 215, 230
Audio recordings, 475
Audio spectrum, 474
Averaged-refractive index, 412

B

Bacteria, 432
Ballistic assumption, 106
Bandwidth curve, 422
Bathophenanthroline, 404
Beam diameter, 504
Beer-Lambert, 95, 386
Beer-Lambert absorption, 191
Beer-Lambert law, 107
Belousov-Zhabotinsky oscillating reaction, 398
Bennet hole, 219
Biconical fiber taper, 468
Biomarkers, 387
Biomedical sensing, 421
Biomolecules, 323
Biosensor, 323
Biotin, 371
Birefringent components, 241
Blank, 404
Bottle microresonator, 362
Bragg wavelength, 422, 465

G. Gagliardi, H.-P. Loock (eds.), *Cavity-Enhanced Spectroscopy and Sensing*, 519
Springer Series in Optical Sciences 179, DOI 10.1007/978-3-642-40003-2,
© Springer-Verlag Berlin Heidelberg 2014

Printed by Publishers' Graphics LLC
DBT140129.15.20.265